DAIRY MICROBIOLOGY HANDBOOK

THIRD EDITION

DAIRY MICROBIOLOGY HANDBOOK

THIRD EDITION

DAIRY MICROBIOLOGY HANDBOOK

THIRD EDITION

Edited by

RICHARD K. ROBINSON

WILEY-INTERSCIENCE

A JOHN WILEY & SONS, INC., PUBLICATION

This book is printed on acid-free paper. ♾

Published simultaneously in Canada.

For ordering and customer service, call 1-800-CALL-WILEY.

Library of Congress Cataloging-in-Publication Data is available.

ISBN 0-471-38596-4

Printed in the United States of America.

10 9 8 7 6 5 4 3 2 1

CONTENTS

PREFACE

In many countries, milk and milk products are indispensable components of the food supply chain. Individual consumers use liquid milk in beverages, families use milk for cooking, and the food manufacturing industry utilizes vast quantities of milk powder(s), concentrated milks, butter, and cream as raw materials for further processing. When fermented dairy products like cheese and yogurt are added to the list, it is easy to appreciate the importance of the dairy industry in less developed and industrialized countries alike.

Equally important is the fact that milk is an excellent source of nutrients for humans, and yet in a different context these same nutrients provide a most suitable medium for microbial growth and metabolism. Many important pathogens like *Salmonella* spp. and *Listeria monocytogenes* will grow in liquid milk or high-moisture milk products; and even if these vegetative forms can be eliminated by pasteurization, the spore-forming *Bacillus cereus* may cause problems. It is not surprising, therefore, that the microbiology of milk and milks products remains a priority interest for everyone associated with the dairy industry.

The fact that John Wiley & Sons has agreed to publish this Third Edition of *Dairy Microbiology* reflects this concern because, since the Second Edition appeared some 10 years ago, the need for effective quality assurance has, if anything, increased. Pathogenic strains of *Escherichia coli* are now a major concern, milk-borne strains of *Mycobacterium avium* sub-sp. *paratuberculosis* have been identified as a possible cause of Crohn's disease, and even little-known parasites like *Cryptosporidium* have caused disease outbreaks. To combat this ever-expanding list of microbial hazards, microbiologists have been forced to devise new strategies to protect the consumer. Consequently, a hazard analysis of (selected) control/critical control points (HACCP) in a food manufacturing process has become central to any program geared toward preventing the contamination of food, but verification by end-product testing still remains essential. In some situations, standard methods of microbiological analysis are widely used, but, in others, new techniques are available which allow a pathogen to be detected in a retail sample in a matter of hours rather than days.

A critical evaluation of these changes and, in particular, of their impact on the diary industry is vital if the excellent safety record of milk and milk products is to be maintained, and it is to be hoped that this book will contribute to this aim. If it does, then the credit lies with the authors who have so generously given of their time and expertise because, in keeping with most editors, my interference with the manuscripts has been minimal. This reluctance to modify an approach selected by a given author(s) of a chapter has led to minor degrees of repetition, but if a particular pathogen, for example, is important in a number of disparate products, then the relevant behavior of the pathogen may well merit additional emphasis.

Finally, a word of appreciation for Janet Bailey, Michael Penn, and Danielle Lacourciere from John Wiley & Sons. Working with authors from across the world can never be easy, and their patience in handling this venture has been a major factor in ensuring its completion.

Richard K. Robinson

CONTRIBUTORS

RODNEY J. BENNETT, Institute of Food, Nutrition and Human Health, Massey University, Palmerston North, New Zealand

THOMAS P. BERESFORD, Dairy Products Research Centre, Teagasc, Fermoy, County Cork, Ireland

THOMAS BINTSIS, Laboratory of Food Microbiology and Hygiene, Department of Food Science and Technology, Faculty of Agriculture, Aristotelian University of Thessaloniki, Thessaloniki, Greece

KATHRYN J. BOOR, Milk Quality Improvement Program, Department of Food Science, Cornell University, Ithaca, New York

JAMES V. CHAMBERS, Department of Food Science, Purdue University, West Lafayette, Indiana

TIMOTHY M. COGAN, Dairy Products Research Centre, Teagasc, Fermoy, County Cork, Ireland

J. KEVIN COLLINS, Department of Microbiology, University College, Cork, Ireland

NANA Y. FARKYE, Dairy Products Technology Center, California Polytechnic State University, San Luis Obispo, California

GERALD FITZGERALD, Department of Microbiology, University College, Cork, Ireland

GILLIAN E. GARDINER, Teagasc, Dairy Products Research Centre, Moorepark, Fermoy, County Cork, Ireland. *Present address*: Lawson Research Institute, London, Ontario, Canada

PARIYAPORN ITSARANUWAT, School of Food Biosciences, The University of Reading, Reading, England

DAVID JERVIS (DECEASED), Unigate Group Technical Centre, Wootton Bassett, Wiltshire, England

PETER J. JOOSTE, ARC—Animal Nutrition and Animal Products Institute, Irene, South Africa

PHIL M. KELLY, Teagasc, Dairy Products Research Centre, Moorepark, Fermoy, County Cork, Ireland

J. FERDIE MOSTERT, ARC—Animal Nutrition and Animal Products Institute, Irene, South Africa

STEVEN C. MURPHY, Milk Quality Improvement Program, Department of Food Science, Cornell University, Ithaca, New York

PHOTIS PAPADEMAS, P. Roussounides Enterprises Ltd., Pralina, Nicosia, Cyprus

RICHARD K. ROBINSON, School of Food Biosciences, The University of Reading, Reading, England

R. PAUL ROSS, Teagasc, Dairy Products Research Centre, Moorepark, Fermoy, County Cork, Ireland

HARJINDER SINGH, Institute of Food, Nutrition and Human Health, Massey University, Palmerston North, New Zealand

CATHERINE STANTON, Teagasc, Dairy Products Research Centre, Moorepark, Fermoy, County Cork, Ireland

ADNAN Y. TAMIME, Scottish Agricultural College, Ayr, Scotland. *Present address*: 24 Queens Terrace, Ayr, Scotland

EBENEZER R. VEDAMUTHU, 994 NW Hayes, Corvallis, Oregon

R. ANDREW WILBEY, School of Food Biosciences, The University of Reading, Reading, England

MONIKA WSZOLEK, University of Agriculture, Animal Products Technology Department, Kraków, Poland

CHAPTER 1

MILK AND MILK PROCESSING

HARJINDER SINGH and RODNEY J. BENNETT
Institute of Food, Nutrition and Human Health, Massey University,
Palmerston North, New Zealand

Milk is the secretion of the mammary gland of female mammals (over 4000 species), and it is often the sole source of food for the very young mammal. The role of milk is to nourish and provide immunological protection. The milks produced by cows, buffaloes, sheep, goats, and camels are used in various parts of the world for human consumption. For much of the world's population, cow's milk accounts for the large majority of the milk processed for human consumption.

Milk is a complex biological fluid probably containing about 100,000 different molecular species in several states of dispersion, but most have not been identified. However, most of the major components—proteins, lactose, fat, and minerals—can be separated and isolated from milk relatively easily. Consequently, the main milk components have been thoroughly studied and the principal characteristics of various constituents are well known.

Milk in its natural state is a highly perishable material because it is susceptible to rapid spoilage by the action of naturally occurring enzymes and contaminating microorganisms. Many processes have been developed over the years—in particular, during the last century—for preserving milk for long periods and to enhance its utilization and safety. Milk is converted into a wide variety of milk products using a range of advanced processing technologies. These include the traditional products, such as the variety of cheeses, yogurts, butters and spreads, ice cream, and dairy desserts, but also new dairy products containing reduced fat content and health-promoting components. Milk is

Dairy Microbiology Handbook, Third Edition, Edited by Richard K. Robinson
ISBN 0-471-38596-4

also an excellent material for producing multifunctional ingredients that can be used in many food products.

This chapter provides an overview of the composition of milk and the properties and structures of the main milk components. Some general aspects of dairy processes and their applications in the manufacture of dairy products are briefly described. The changes that occur in milk during various processing operations are discussed in some detail. These topics have been covered in greater depth in several text and reference books, mentioned throughout this chapter.

1.1 MILK COMPOSITION

The major component of milk is water; the remainder consists of fat, lactose, and protein (casein and whey proteins) (Table 1.1). Milk also contains smaller quantities of minerals, specific blood proteins, enzymes, and small intermediates of mammary synthesis. The structures and properties of these components profoundly influence the characteristics of milk and have important consequences for milk processing.

The composition of milk varies with the dairy breeds. The most commonly found breeds—Friesian, Jersey, Guernsey, Ayrshire, Brown Swiss, and Holstein—have fairly similar lactose levels, but milk fat and protein vary considerably (Table 1.1). These differences are partly genetic in origin and partly the results of environmental and physiological factors. Within a herd of cows of a single breed, there are considerable variations in milk composition between individual cows. For example, the milk fat content in Jerseys can range from 4% to 7%, with an average of about 5.0%.

The composition of milk changes considerably with the progress of lactation. The first secretion collected from the udder at the beginning of lactation, known as colostrum, has a high concentration of fat and

TABLE 1.1. Typical Composition of Milks of Some Breeds of Cow (g/100 g)

Breed	Protein	Fat	Lactose	Ash
Jersey	4.0	5.2	4.9	0.77
Friesian	3.4	4.2	4.7	0.75
Brown Swiss	3.5	4.0	4.9	0.74
Guernsey	3.7	3.7	4.7	0.76
Holstein	3.3	3.5	4.7	0.72
Ayrshire	3.5	3.9	4.6	0.72

protein, particularly immunoglobulins, and a low content of lactose. The composition of the secretion gradually changes to that of mature milk within 2–4 weeks.

The percentage of lactose and protein in milks from the same cow varies very little from one milking to another. However, milk fat content is much more variable; the more frequent the milking, the greater is the variation. Generally, during milking, the fat content increases. Morning milk is usually richer in fat than evening milk.

The kind and the quantity of feed affect milk composition, especially fat content and fat composition. Other factors, such as mastitis, extreme weather conditions, stress, and exhaustion, can also exert an influence on milk composition.

The percentages of the main constituents of milk vary to a considerable extent among different species. The milks of the sheep and the buffalo have much higher fat and casein contents than those of the other species. The most obvious characteristic of human milk, compared with other milks, is its low casein and ash contents.

1.2 MILK COMPONENTS

1.2.1 Lipids

The major lipid component of cow's milk is triglyceride, which makes up about 98% of milk fat. The other 2% of milk lipids consists of diglycerides, monoglycerides, cholesterol, phospholipids, free fatty acids, cerebrosides, and gangliosides. The lipid composition of bovine milk is given in Table 1.2 (see Christie, 1995 for more details).

Only 13 fatty acids are present in milk at reasonable concentrations (Table 1.3), and these can be arranged in many different ways to give

TABLE 1.2. Lipid Composition of Bovine Milk

Lipid	Percentage in Milk Fat (w/w)
Triacylglycerols (triglycerides)	98.3
1,2-Diacylglycerols (diglycerides)	0.30
Monoacylglycerols (monoglycerides)	0.03
Free fatty acids	0.10
Phospholipids	0.80
Cholesterol	0.30
Cholesterol ester	0.02
Cerebrosides	0.10
Gangliosides	0.01

TABLE 1.3. Major Fatty Acid Constituents of Bovine Milk Fat

Fatty Acid	Weight Percent	Fatty Acid	Weight Percent
4:0	3.8	15:0	1.1
6:0	2.4	16:0	43.7
8:0	1.4	18:0	11.3
10:0	3.5	14:1	1.6
12:0	4.6	16:1	2.6
14:0	12.8	18:1	11.3
		18:2	1.5

hundreds of different triglycerides, although the distribution of fatty acids on the triglyceride chain is not random. Because various triglycerides have different melting points, milk fat has a large melting point range. Milk fat has a relatively high content of short-chain saturated fatty acids, such as butyric (C_4) and capric (C_{10}) acids (Table 1.3). These fatty acids are important to the flavor of milk products and in off-flavors that may develop in milk. The distribution of fatty acids between various positions of the glycerol moiety of triglycerides varies with the fatty acids. In general, the longer-chain fatty acids ($C_{16:0}$ and $C_{18:0}$) are found in position *sn*-1, whereas the shorter-chain fatty acids ($C_{4:0}$, $C_{6:0}$) and unsaturated fatty acids are mostly present in position *sn*-3.

In milk, nearly all of the fat (>95%) exists in the form of globules ranging in size from 0.1 to 15 μm in diameter. Approximately 90% of the fat is in globules with diameters of 1.0–6.0 μm. There are a large number of small fat globules (<1.0 μm) present in milk, but these contain only 2–3% of the total fat. Each globule is surrounded by a thin membrane, 8–10 nm in thickness, usually called a milk fat globule membrane (MFGM). Its composition and properties are completely different from those of either milk fat or plasma. MFGM is derived from the apical cell membrane of Golgi vacuoles and other materials of the lactating cell, although there may be some rearrangement of the membrane after release into the lumen, as amphiphilic substances from the plasma adsorb onto the fat globule and parts of the membrane dissolve into either the globule core or the serum. The membrane acts as a natural emulsifying agent, enabling the fat to remain dispersed throughout the aqueous phase of milk, preventing to some extent flocculation and coalescence.

The total mass of fat globules that is accounted for by membrane material has not been determined with certainty. An estimated mass of the membrane is 2–6% of that of the total fat globules. Proteins and phospholipids together account for over 90% of the membrane dry weight, but the relative proportions of lipids and proteins may vary widely. The lipid component of the MFGM is composed of ~62% high-melting triacylglycerols, ~22% phospholipids, ~9% diacylglycerols, ~7% free fatty acids, and small quantities of unsaponifiable lipids and monoacylglycerols.

Protein accounts for 25–60% of the mass of the membrane material, depending upon the isolation method chosen. Most of the proteins of the membrane are highly specific, and their composition and structures are not well known. There are at least 10 different protein species with molecular weights ranging from 50 to 155 kDa. They are predominantly glycoproteins, among which are sialoglycoproteins. In general, membrane glycoproteins have, besides highly glycosylated regions, strongly hydrophobic regions, which are needed for binding to the lipids of the membrane. Butyrophilin, with an estimated molecular weight of 64–66 kDa, is the major glycoprotein in MFGM, accounting for about 40% of the mass of membrane proteins. Xanthine oxidase, with a molecular weight of 155 kDa, accounts for 20% of the membrane proteins. The MFGM contains at least 25 different enzymes, more than half of which are members of the hydrolase class followed by oxidoreductases and transferases. The most abundant enzymes are alkaline phosphatase and xanthine oxidase (Kennan and Dylewski, 1994).

1.2.2 Proteins

Normal bovine milk contains about 3.5% protein, which can be fractionated into two main groups. On acidification of milk to pH 4.6 at 20°C, about 80% of the total protein precipitates out of solution; these proteins are called casein. The proteins that remain soluble under these conditions are referred to as whey proteins or serum proteins. Both the casein and whey protein groups are heterogeneous. The concentrations of different proteins in milk are shown in Table 1.4.

1.2.2.1 Caseins and Casein Micelles. Several reviews and monographs on the structures and properties of caseins and casein micelles have been published (e.g., Walstra, 1990; Holt, 1992; Rollema, 1992; Swaisgood, 1992). Caseins can be fractionated into four distinct proteins: α_{s1}-, α_{s2}-, β-, and κ-caseins (Table 1.4). There are also several derived caseins, resulting from the action of indigenous milk proteinases, especially plasmin, on the main caseins. These are usually referred to as γ-caseins. The caseins are all phosphoproteins with the phosphate groups being esterified to the serine residues in the protein

TABLE 1.4. Typical Concentration of Proteins in Bovine Milk

	Grams/Liter	% of Total Protein
Total protein	33	100
Total caseins	26	79.5
α_{s1}-Casein	10	30.6
α_{s2}-Casein	2.6	8.0
β-Casein	9.3	28.4
κ-Casein	3.3	10.1
Total whey proteins	6.3	19.3
α-Lactalbumin	1.2	3.7
β-Lactoglobulin	3.2	9.8
BSA	0.4	1.2
Immunoglobulins	0.7	2.1
Proteose peptone	0.8	2.4

chains. The phosphate groups bind large amounts of calcium, and they are important to the structure of casein micelles.

Calcium binding by individual caseins is proportional to their phosphate content. Both α_{s1}- and α_{s2}-caseins are most sensitive to calcium, precipitating at Ca^{2+} concentrations in the range 3–8 mM; β-casein precipitates in the range 8–15 mM Ca^{2+}, but remains in solution at concentrations up to 400 mM Ca^{2+} at 1°C; κ-casein remains soluble at all levels of calcium. κ-Casein is not only soluble in Ca^{2+} but also capable of stabilizing α_s- and β-caseins against precipitation by Ca^{2+}.

The primary structures of the four principal caseins are now well established (Swainsgood, 1992). In comparison with typical globular proteins, the structures of caseins are quite unique. The most unusual feature is the amphiphilicity of their primary structure. The hydrophobic residues in caseins are not uniformly distributed along the polypeptide chain; for example, α_{s1}-casein has three hydrophobic regions, residues 1–44, 90–113, and 132–199, and α_{s2}-casein has two hydrophobic segments, 90–120 and 160–207. The C-terminal two-thirds of β-casein, the most hydrophobic of the caseins, is strongly hydrophobic, whereas segments 5–65 and 105–115 of κ-casein are strongly hydrophobic. Many of the charged residues, particularly the phosphoserine residues, in the caseins are also clustered. For example, the sequence 41–80 in α_{s1}-casein contains all but one of the eight phosphate residues and has a charge of −21 at pH 6.6, whereas the rest of the molecule has no net charge. The C-terminal 47-residue sequence of α_{s2}-casein has a net charge of +9.5, whereas the N-terminal 68-residue sequence has a net charge of −21. β-Casein has a strong negatively charged sequence between residues 13 and 21, with four of the five phosphoserine residues and three Glu residues; segment 42–48 is also strongly charged with three Glu, one Asp, and one Lys residues. The N-terminal 21-residue sequence of β-casein has a net charge of −12. the amino half of κ-casein is hydrophobic with no net charge, whereas the C-terminal 50 residues, which contain no cationic residue, contain 10 or 11 anionic amino acid residues as well as the negatively charged sugar residues, giving a negative charge of −15 to −16.

The clustering of similar residues and high levels of proline residues and their uniform distribution throughout the polypeptide chain influence the secondary and tertiary structure of the caseins which are relatively open and not very ordered. Therefore, all four caseins have a distinctly amphipathic character with separate hydrophobic and hydrophilic domains. Without much tertiary structure, there is considerable exposure of hydrophobic residues to the aqueous environment. Consequently, their hydrophobic domains interact to form polymers. For example, α_{s1}-casein polymerizes to give tetramers of molecular weight ~110,000 Da, and the degree of polymerization increases with increasing protein concentration. At 40°C, β-casein exists in solution as monomers of molecular weight ~25,000 Da. As the temperature is raised, the monomers polymerize to long thread-like chains of 20 units at 85°C, with the degree of association being dependent on protein concentration. In unreduced form, κ-casein is present largely as a polymer of molecular weight ~600,000 Da with some larger polymers also being present.

The α_{s1}-, α_{s2}-, and β-caseins can bind considerable concentrations of metal ions, mainly Ca^{2+}, leading to strong aggregation. Under normal circumstances, α_{s1}-casein can bind up to 10 mol Ca^{2+}/mol protein. κ-Casein, which has only one phosphoserine residue, does not bind Ca^{2+} strongly and is soluble in Ca^{2+}. Furthermore, κ-casein associates with α_{s1}- or β-caseins and, in the presence of Ca^{2+}, stabilizes α_{s1}- and β-caseins against precipitation. These associations lead to the formation of stable colloidal particles that are generally similar to the native casein micelles that exist in milk.

In normal milk, 95% of the casein exists as coarse colloidal particles, called micelles, with diameters ranging from 80 to 300 nm (average ~150 nm). On a dry weight basis, the micelles consist of ~94% protein and ~6% small ions, principally calcium, phosphate, magnesium and citrate, referred to collectively as colloidal calcium phosphate (CCP). These particles are formed within the secretory cells of the mammary gland and undergo relatively little change after secretion.

Some physicochemical characteristics of casein micelles are presented in Table 1.5. The precise structure of the casein micelle is a

TABLE 1.5. Physicochemical Characteristics of Casein Micelles

Diameter	80–300 nm
Surface area	8×10^{-10} cm^2
Volume	2×10^{-15} cm^3
Density	1.063 g/cm^3
Molecular weight (hydrated)	1.3×10^9 Da
Hydration	2 g H_2O/g protein
Water content (hydrated)	63%

matter of considerable debate at the present time. A number of models have been proposed over the past 40 years, but none can describe completely all aspects of casein micelle behavior. The models include (a) coat–core models, which postulate that the interior of the micelle is composed of proteins that are different from those on the exterior (Waugh and Noble, 1965; Hansen et al., 1996), and (b) subunit structure models to which the term submicelle is attached (Slattery, 1976; Schmidt, 1982; Walstra, 1990).

In the subunit models (e.g., Schmidt, 1982), caseins are aggregated to form submicelles (10–15 nm in diameter). It has been suggested that submicelles have a hydrophobic core that is covered by a hydrophilic coat. The polar moieties of κ-casein molecules are concentrated in one area. The remaining part of the coat consists of the polar parts of other caseins, notably segments containing their phosphoserine residues. The submicelles are assumed to aggregate into micelles by CCP, which binds to α_{s1}-, α_{s2}-, and β-caseins via their phosphoserine residues. Submicelles with no or low κ-casein are located in the interior of the micelle, whereas κ-casein-rich submicelles are concentrated on the surface, making the overall surface κ-casein rich.

Other models consider the micelle as a porous network of proteins (of no fixed conformation); the calcium phosphate nanoclusters are responsible for crosslinking the protein and holding the network together (Holt, 1992). A recent model proposed by Horne (1998) assumes that assembly of the casein micelle is governed by a balance of electrostatic and hydrophobic interactions between casein molecules. As stated earlier, α_{s1}-, α_{s2}-, and β-caseins consist of distinct hydrophobic and hydrophilic regions. Two or more hydrophobic regions from different molecules form a bonded cluster. Growth of these polymers is inhibited by the protein charged residues, the repulsion of which pushes up the interaction free energy. Neutralization of the phosphoserine clusters by incorporation into the CCP diminishes that free energy and produces a second type of crosslinking bridge. κ-Casein acts as a terminator for both types of growth, because it contains no phosphoserine cluster and no other hydrophobic anchor point.

A common factor in all models is that most of the κ-casein appears to be present on the surface of casein micelles. The hydrophilic C-terminal part of κ-casein is assumed to protrude 5–10 nm from the micelle surface into the surrounding solvent, giving it a "hairy" appearance. The highly charged flexible "hairs" physically prevent the approach and interactions of hydrophobic regions of the casein molecules. Removal of the hairs by cleavage with rennet or their collapse in ethanol destroys the stabilization effect of κ-casein, allowing micelles to interact and aggregate.

1.2.2.2 Whey Proteins. The principal whey protein fractions are β-lactoglobulin, bovine serum albumin (BSA), α-lactalbumin and immunoglobulins. Major reviews covering the structures and properties of the whey proteins have been published (e.g., Swaisgood, 1982; Whitney, 1988; Kinsella and Whitehead, 1989; Hambling et al., 1992). β-Lactoglobulin is the most abundant whey protein and represents about 50% of the total whey protein in bovine milk. There are eight known genetic variants of β-lactoglobulin: A, B, C, D, E, F, G, and Dr. The A and B genetic variants are the most common and exist at almost the same frequency. β-Lactoglobulin has a molecular weight of 18,000 Da and contains two internal disulfide bonds and a single free thiol group, which is of great importance for changes occurring in milk during heating.

α-Lactalbumin accounts for about 20% of the whey proteins and has three known genetic variants. It has a molecular weight of 14,000 Da and contains four interchain disulfide bonds. α-Lactalbumin binds two atoms of calcium very strongly, and it is rendered susceptible to denaturation when these atoms are removed.

Serum albumin is identical to the serum albumin found in the blood and represents about 5% of the total whey proteins. The protein is synthesized in the liver and gains entrance to milk through the secretory cells. It has one free thiol and 17 disulfide linkages, which hold the protein in a multiloop structure. Serum albumin appears to function as a carrier of small molecules, such as fatty acids, but any specific role that it may play is unknown.

Immunoglobulins are antibodies synthesized in response to stimulation by macromolecular antigens foreign to the animal. They account for up to 10% of the whey proteins and are polymers of two kinds of polypeptide chain: light (L) of molecular weight 22,400 Da and heavy (H) of molecular weight 50,000–60,000 Da. Four types of immunoglobulins have been found in bovine milk: IgM, IgA, IgE, and IgG.

Several other proteins are found in small quantities in whey; these include β-microglobulin, lactoferrin, and transferrin, both of which are iron-binding proteins, proteose peptones, and a group of acyl glycoproteins.

1.2.3 Milk Salts

Milk salts consist mainly of chlorides, phosphates, citrates, sulfates, and bicarbonates of sodium, potassium, calcium, and magnesium. Some of the milk salts (i.e., the chlorides, sulfates, and compounds of sodium and potassium) are soluble and are present almost entirely as ions dissolved in milk whey. Others—calcium and phosphate in particular—are much less soluble and at the normal pH of milk exist partly in dissolved and partly in insoluble (i.e., colloidal) form, in close association with the casein micelles (Walstra and Jenness, 1984). The partition of calcium phosphate between the dissolved and colloidal states significantly influences the properties of milk.

A large number of mineral elements, such as zinc, iron, and manganese, are present in normal milk in trace amounts.

1.2.4 Lactose

Lactose, the major carbohydrate in milk, is found in cow's milk at levels of ~4.8%. This level of sugar does not make milk unduly sweet because lactose is less sweet than sucrose as well as less sweet than an equimolar mixture of its components, galactose and glucose. Lactose makes a major contribution to the colligative properties of milk (osmotic pressure, freezing point depression, boiling point elevation). For example, it accounts for ~50% of the osmotic pressure of milk. Lactose exists in both α- and β-lactose forms, although in solution an equilibrium mixture of the two forms is attained, the composition of which is dependent on the temperature. Compared with many other sugars, lactose is relatively less soluble in water; its solubility at 25°C is only 17.8 g/100 g solution. This relatively low solubility can cause some manufacturing problems because lactose crystals are gritty in texture. Crystallization of lactose is also responsible for caking and lumping of dried milk during storage, particularly if moisture is absorbed from the air. Lactose, like other reducing sugars, can react with free amino groups of proteins to give products that are brown in color.

Milk also contains many vitamins (e.g., vitamins A and C), enzymes (e.g., lactoperoxidase and acid phosphatase) and somatic cells. Some of the minor constituents may perform an important function, and others may be accidental contaminants (antibodies and disinfectants).

1.3 MILK PROCESSING

Milk, in its natural state, is a highly perishable material, subject to microbial and chemical degradation. Many processes have been developed over the years to enhance its utilization and safety. These processes can be grouped and analyzed in a variety of ways. A useful classification is provided under the following headings:

- Fractionation
- Concentration
- Preservation

Most of the processes used in dairy product manufacture belong in one or more of these groups. The processes in each group are briefly examined, and then their application to the manufacture of a wide variety of products is discussed. Detailed analysis of the processes, operations, and technology can be found in standard food and chemical engineering textbooks such as Earle (1983), Perry et al. (1984), Rosenthal (1991), Robinson (1993), Bylund (1995) and Walstra et al. (1999).

1.3.1 Fractionation

This term is used to describe the fractionation or disassembly of the components of milk, utilizing their various properties as earlier discussed. Processes in this category include the following:

- Centrifugal separation, utilizing density difference of the components. The most common equipment used is the disc bowl separator (Figure 1.1), which allows separation of light and heavy phases and also allows removal of any sediment.
- Membrane separation, utilizing size or charge difference. This is normally a pressure-motivated, flow-dependent process, involving the use of a selective membrane, with a wide range of fractionations possible, from simple water removal to separation of different proteins. The operating principle is illustrated in Figure 1.2.

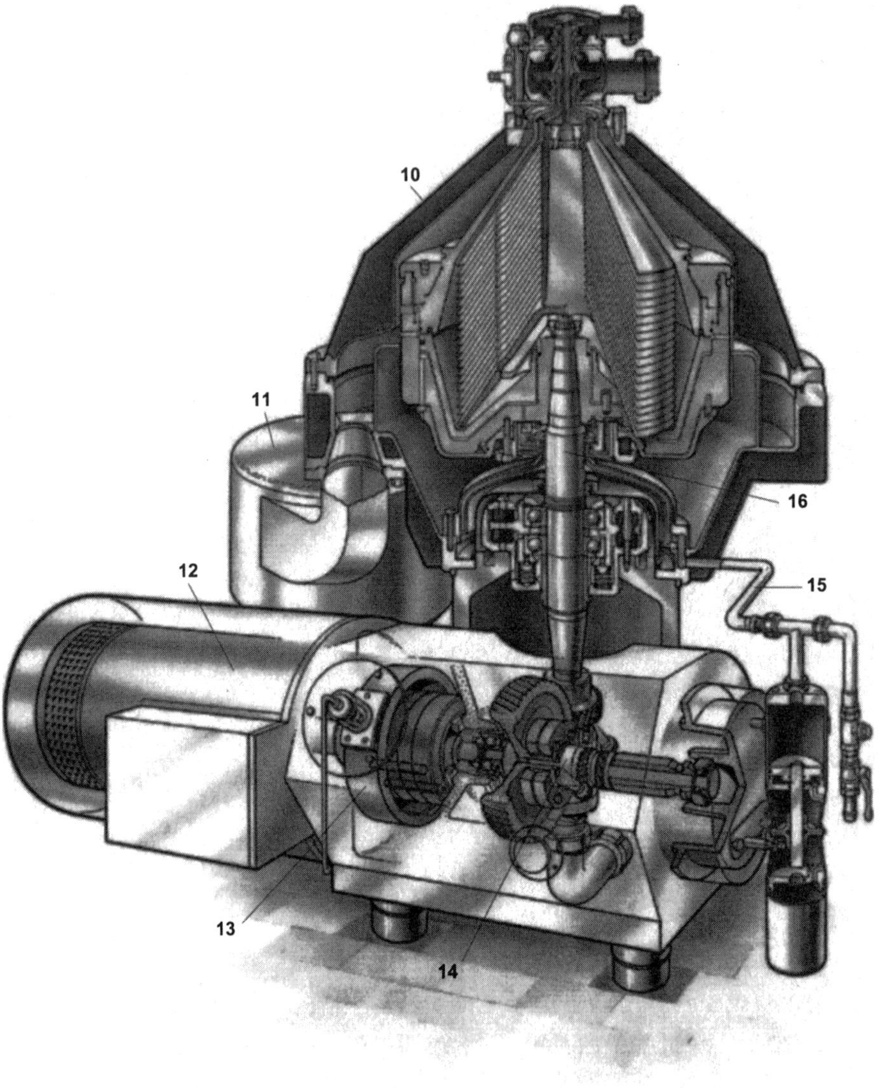

Figure 1.1. Sectional view of a modern hermetic separator. 10, Frame hood; 11, sediment cyclone; 12, motor; 13, brake; 14, gear; 15, operating water system; 16, hollow bowl spindle. (Courtesy of Tetra Pak.)

- Ion exchange, utilising charge difference. In this process, tiny resin beads exchange charged ions on their surface with charged ions or larger charged molecules in solution, removing them for subsequent recovery.
- Precipitation and crystallization, utilizing differences in solubility and suspension stability. An example of equipment used for this

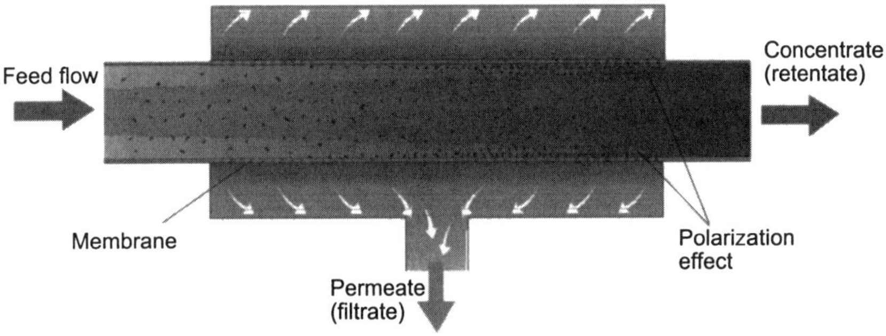

Figure 1.2. Principles of membrane filtration. (Courtesy of Tetra Pak.)

is the cheese vat illustrated in Figure 1.3, in which liquid milk is converted to a gel by destabilization of the casein micelle. Subsequent cutting of the gel and syneresis or whey loss results in *fractionation* of the fat and casein from the remaining milk components.

- Filtration, utilizing size difference. The principle is similar to that already discussed with membrane separation but involves the separation of larger components. An example of the equipment used is a dewheying screen used in separating curd and whey in cheesemaking, shown in Figure 1.4.
- Homogenization is a process of size reduction of the fat globules to *prevent fractionation* of the cream and skim milk by density difference. A combination of a high-pressure pump and special valves provides high shear.

1.3.2 Concentration

Processes in this grouping involve removal of one or more components, resulting in a concentration of the remaining components. Many of these processes also involve fractionation. The processes include the following:

- Evaporation, utilizing phase change of the aqueous component. An evaporator is a specialized heat exchanger operating under vacuum, facilitating efficient water vapor generation and removal from a liquid with minimal thermal damage to the remaining liquid. An example is shown in Figure 1.5.
- Freeze concentration, also utilizing phase change. This involves freezing and crystallization of the aqueous component of a liquid

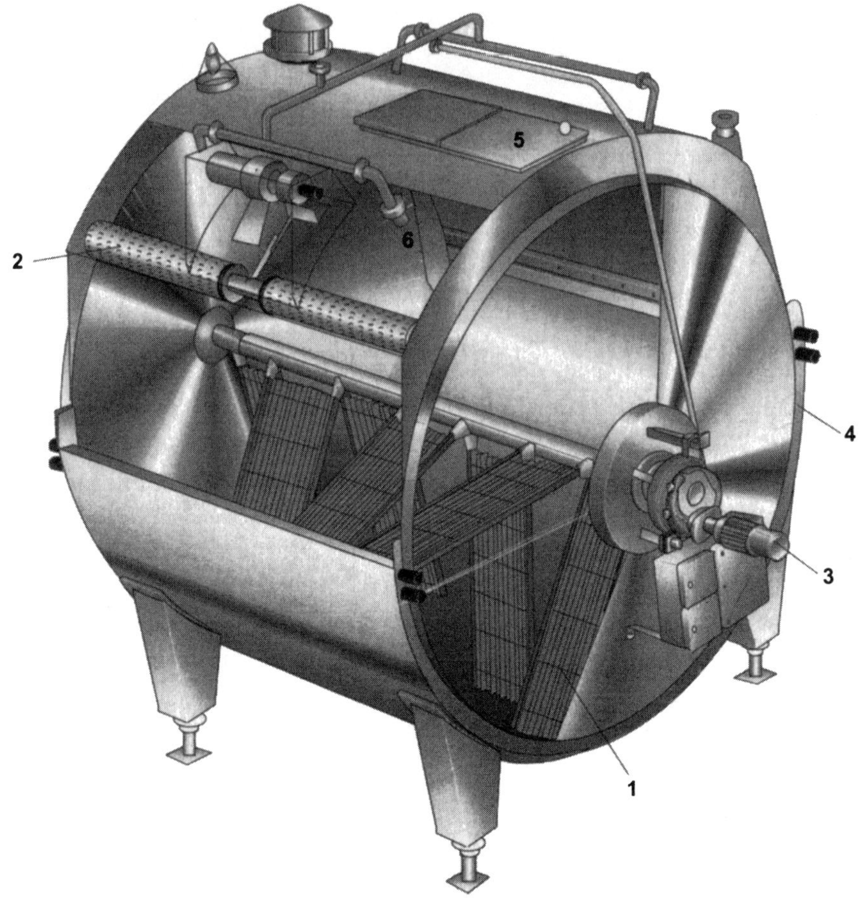

Figure 1.3. Horizontal enclosed cheese tank with combined stirring and cutting tools and hoisted whey drainage system. 1, Combined cutting and stirring tools; 2, strainer for whey drainage; 3, frequency-controlled motor drive; 4, jacket for heat; 5, manhole; 6, CIP nozzle. (Courtesy of Tetra Pak.)

by refrigeration followed by crystal removal. It is not widely used in dairy processing.

- Membrane separation, utilizing size for both concentration and fractionation. This process has already been described above, where it can be seen that the permeate, or material passing through the membrane, includes water, enabling concentration of the retentate or material retained.
- Drying, utilizing phase change. This is a very important process, particularly in the production of milk powder, casein and whey products. It involves water removal from a liquid concentrate or

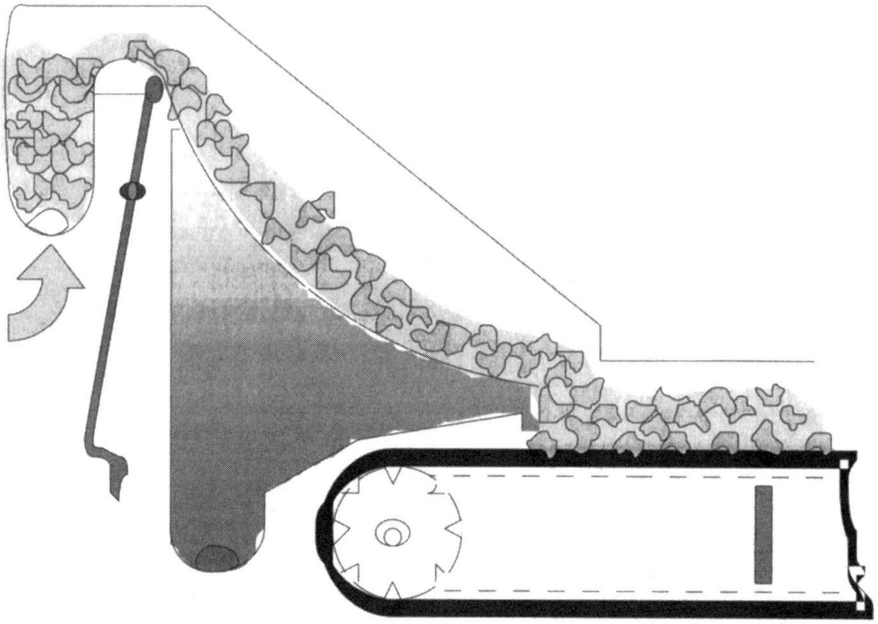

Figure 1.4. Dewheying screen for separating curd and whey. (Courtesy of Tetra Pak.)

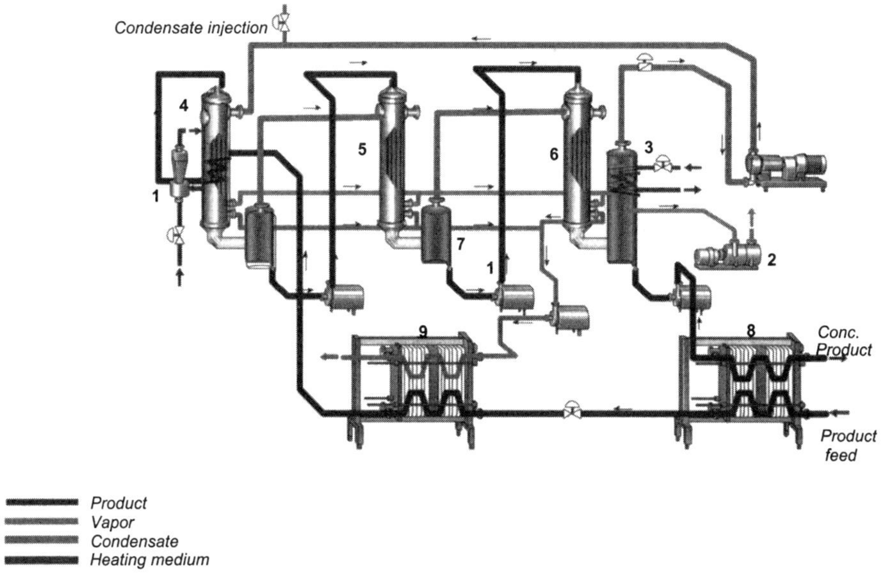

Figure 1.5. Three-effect evaporator with mechanical vapor compression. 1, Thermocompressor; 2, vacuum pump; 3, mechanical vapor compressor; 4, first effect; 5, 2nd effect; 6, third effect; 7, vapor separator; 8, product heater; 9, plate condenser. (Courtesy of Tetra Pak.)

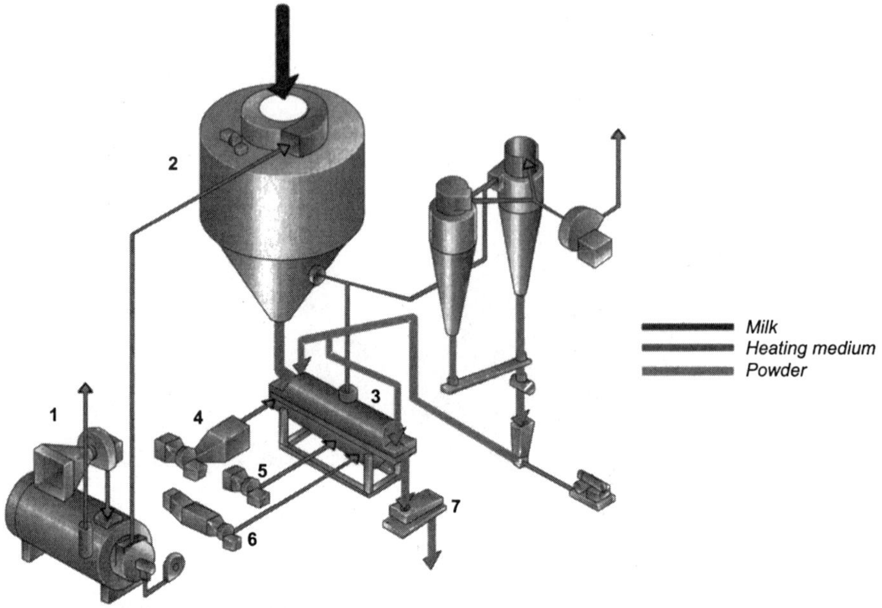

Figure 1.6. Spray drier with fluid bed attachment (two-stage drying). 1, Indirect heater; 2, drying chamber; 3, vibrating fluid bed; 4, heater for fluid bed air; 5, ambient cooling air for fluid bed; 6, dehumidified cooling air for fluid bed; 7, sieve. (Courtesy of Tetra Pak.)

solid by heating with hot air. An example of the equipment used is a spray drier, shown in Figure 1.6.

- Centrifugal separation, utilizing density difference. The principles of this have already been described under fractionation. An example of equipment used for water removal and consequent concentration is the decanter centrifuge.

1.3.3. Preservation

The processes in this category are primarily concerned with reducing microbiological and chemical change. They include the following:

- Pasteurization, thermalization, and sterilization, utilizing heat to kill microorganisms. These processes all involve the transfer of heat into the product in order to raise the temperature to achieve a closely controlled time–temperature process (e.g., 72°C, 15 s) for pasteurization. An example of the equipment used is a plate heat exchanger, shown in Figure 1.7.

Figure 1.7. Principles of flow and heat transfer in a plate heat exchanger. (Courtesy of Tetra Pak.)

- Chilling and freezing, to slow microbial growth and chemical change. This is widely used both during or prior to processing or for final product storage. Heat exchangers of the type shown in Figure 1.7 can be used for liquid products, with cool stores and freezing chambers for finished goods.
- Reduction of pH, to inhibit microbial growth. This may be achieved by addition of acids or by bacterial fermentation of the lactose. An example of the equipment used is the cheese vat shown in Figure 1.3.
- Dehydration (drying), to inhibit microbial growth and chemical change. The equipment used for spray drying has already been described (Figure 1.6).
- Salting, to reduce water activity and inhibit microbial growth. Salt may be added as dry granular salt or by means of a brine solution, with the product being immersed for a period in a tank of concentrated brine.

- Packaging, to contain the product, protect it, and reduce microbiological and chemical change. Examples of commonly used packaging are the cartons and plastic bottles for liquid products, the bulk 25-kg gas-flushed bags for whole milk powder, and the form/fill/seal packages for cheese.

1.4 UTILIZATION OF PROCESSES TO MANUFACTURE PRODUCTS FROM MILK

Figure 1.8 illustrates the processes that are involved in the manufacture of the large range of products that can be produced from the very versatile raw material, whole milk. The products fall into two

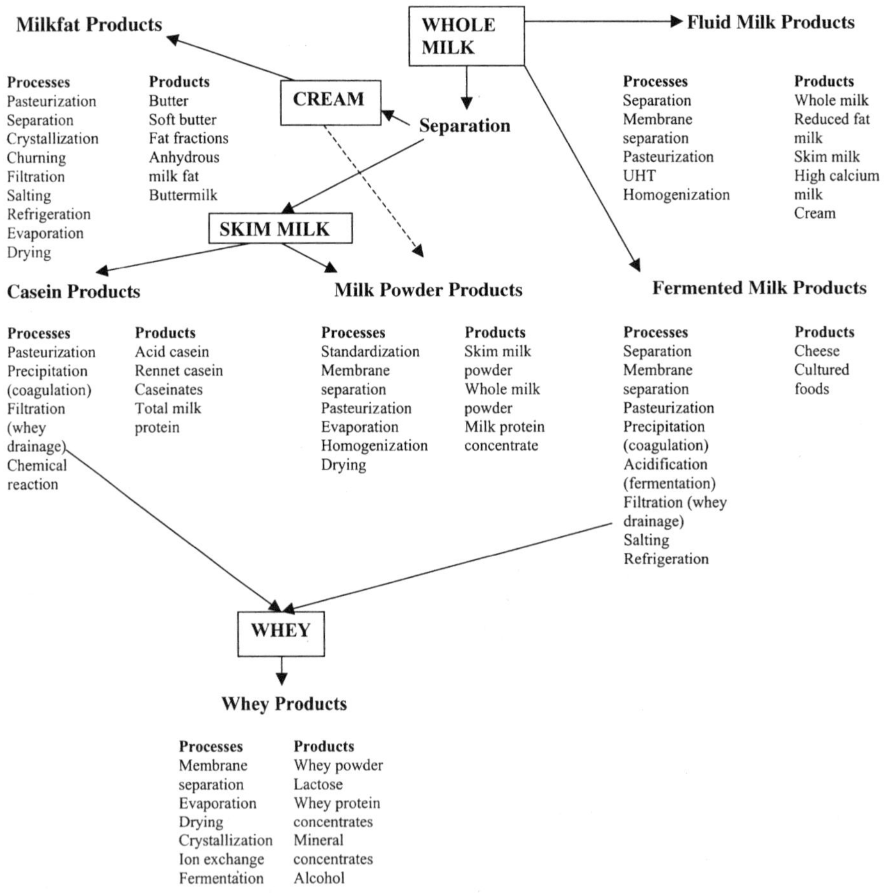

Figure 1.8. Products from milk.

broad categories: those for immediate availability to the consumer (consumer products) and those that are ingredients that will subsequently be utilized to produce consumer dairy products or other foods. The products and processes that are involved in each group are briefly discussed.

1.4.1 Fluid Milk Products

This group of products falls into the consumer products family, competing in the beverage sector of the grocery business. Their manufacture is relatively simple, involving *fractionation* processes such as centrifugal separation to produce cream, skim milk, or reduced fat milk, *concentration* processes such as membrane separation (ultrafiltration) to produce high calcium milks, and *preservation* processes such as pasteurization, ultra-heat treatment (UHT), and refrigeration to extend the safety and shelf life of the product range. Homogenization is used to prevent separation of the fat in the liquid product.

1.4.2 Fermented Milk Products

There are two groups in this family: (a) cheese products in which part of the original liquid is removed during manufacture as whey and (b) products in which there is no whey drainage, such as yogurts. Both groups have a very long history and were probably developed by accident as a means of preserving milk. The basic principles of manufacture are shown in Figure 1.9. All or some of the standard food preservation tools of *moisture* removal, *acid* development, *salt* addition, and *temperature* adjustment may be used. The first letters of the italic words spell the conveniently remembered acronym MAST.

Cheese manufacture is a highly complex process. The composition of the initial milk is adjusted or standardized (*fractionation*) by centrifugal separation and possibly also ultrafiltration. For most cheese types, the milk will then be pasteurized (72°C/15s) to reduce the risk from pathogenic organisms, adjusted to the desired fermentation temperature, and then pumped into a cheese vat. Starter culture consisting of a carefully selected species of lactic acid bacteria and a coagulant (e.g., calf rennet) are then added and the milk is allowed to coagulate. This is by destabilization of the casein micelle. This permits the beginning of the *fractionation* and selective *concentration* processes that form the basis of cheese-making. Once the coagulum is of sufficient

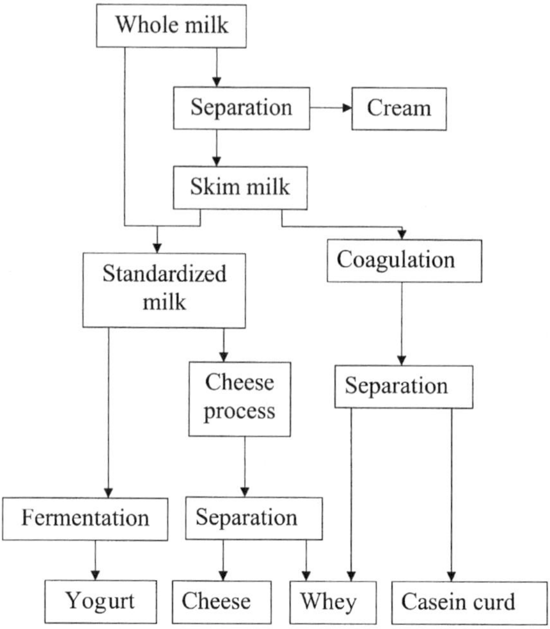

Figure 1.9. Manufacture of fermented dairy products.

strength, it is cut into small particles and, by a process of controlled heating and fermentation, syneresis or expulsion of moisture and minerals (whey) occurs. Separation of the curd from the whey over a screen (filtration) follows. Depending on the cheese type, the curd may be allowed to fuse together (e.g., Cheddar) or may be kept in granules (e.g., Colby). Salt may then be incorporated into the curd for *preservation*, as dry granules or by immersion in brine. The curd is pressed into blocks by either gravity or mechanical compression, and the cheese then goes into controlled storage conditions for final fermentation and maturation.

For fermented milk products, the process of manufacture is somewhat simpler. For example, in yogurt manufacture, the milk to be used is fortified with additional protein (skim milk powder or concentrated milk), severely heated (e.g., 95°C/5 min) to reduce the microbial load and to encourage whey protein/casein interaction, cooled to fermentation temperature (e.g., 36°C), and transferred to a fermentation vessel. Selected cultures are then added, and fermentation is continued until the desired pH of around 4.5 is reached. This causes the coagulation of the vessel contents, and the plain yogurt is then cooled prior to possible incorporation of fruit and flavoring, followed by packaging.

1.4.3 Milk Powder Products

The first steps of the milk powder production process involve the *fractionation* of the raw milk into the desired components for the final product specification. For example, for skim milk powder, almost all the fat component of the milk is removed as cream by centrifugal separation. For whole milk powder, only a proportion of the cream may be removed. The lactose content of the milk may be adjusted, by addition of crystalline lactose or permeate, to achieve the desired protein/carbohydrate ratio in the final product. Alternatively, the skim milk may be partially concentrated by ultrafiltration to achieve not only water removal but also protein concentration in the manufacture of speciality milk powders known as milk protein concentrates. The standardised liquid is pasteurized to help with *preservation*. It undergoes a prescribed heat treatment for the desired final product characteristics and is then *concentrated* by multieffect evaporators to a solids level of about 50%, and the concentrate is fed to a spray drier for final *concentration* and *preservation* by water removal to a moisture content of 3–4%. If whole milk powder is being produced, a homogenization step is included prior to drying. Packaging is a critical component of the *preservation* process, and flushing with an inert gas such as nitrogen is often used to reduce fat oxidation.

Milk powder products have a very wide range of uses from reconstitution into liquid products to ingredients for a wide range of food products. Liquid concentrated milk products are also produced from evaporated milk, or reconstituted from powder to a high solids liquid. These are shelf-stable products commonly presented in cans or cartons.

1.4.4 Casein Products

The manufacture of this family of products involves the *fractionation* and *concentration* of the casein protein fraction of the milk. The first stages involve centrifugal separation of the skim milk from the cream, followed by pasteurization. The casein micelles are then destabilized either by the action of a coagulant such as rennet or by a reduction in pH to the isoelectric point (4.6) by fermentation or addition of mineral acid. The coagulated protein is then heated to firm the curd and encourage syneresis or loss of whey. The curd is then *separated* from the whey by filtration or centrifugation, in combination with countercurrent washing with water. The curd may then be dried as an insoluble casein for *preservation*, or first be reacted with alkali (e.g., sodium hydroxide), followed by drying to produce a water-soluble caseinate. A protein product including both the casein and the whey proteins, known as total milk proteinate, may be manufactured by a variation of the process described.

The protein products described have a wide range of food ingredient and industrial applications, many utilizing the emulsifying, water- and fat-binding, and nutritional properties of the proteins.

1.4.5 Milkfat (Cream) Products

These are products that are derived from the cream or fat-containing portion of the milk, and as such the first step in their manufacture involves the *centrifugal separation* of the incoming milk to produce cream and skim milk. This is normally followed by a pasteurization step, which may include a vacuum/steam treatment step for feed taint removal.

The most familiar product made from cream is butter and two different processes are currently in use. The first is the traditional *Fritz* process, which involves chilling the cream overnight to about 8°C to aid crystallization of part of the fat, followed by churning to invert the oil-in-water emulsion in milk to the water-in-oil emulsion in butter. A filtration or screening step follows to *separate* the butter granules from the surrounding buttermilk, and the granules are further worked or blended under vacuum into a homogeneous mass and extruded into packaging, as tubs or blocks. Salt may be incorporated during the final working stages, for flavor and *preservation*. The product is then refrigerated for *preservation*. The buttermilk may be evaporated and spray dried for subsequent use as a food ingredient.

The more recently developed process for butter manufacture essentially follows a margarine system and is known as the *Ammix* process. It involves initially the production of a highly purified anhydrous milk fat by *selective concentration* of the cream using *centrifugal separation* and vacuum drying. The milk fat is then emulsified and blended with an aqueous phase containing some milk proteins and possibly salt, followed by rapid refrigeration in a scraped-surface heat exchanger and packaging. Again, storage under refrigeration follows for *preservation*.

Soft butter and a variety of milk fat fractions may be prepared by *fractionating* the anhydrous milk fat, utilizing fractional crystallization and filtration. The variation in hardness of the fractions permits a variety of uses, such as pastry manufacture. Anhydrous milk fat is itself an important product, being widely used with skim milk powder during recombining to produce recombined liquid milk products.

1.4.6 Whey Products

Whey is produced as a by-product of cheese and casein manufacture and for many years was regarded as a nuisance, low-value material requiring disposal at least cost. The whey was of a similar volume to the incoming milk; and the components—whey proteins, lactose, and minerals—were present in solution, making recovery difficult. However, the development of new technologies—in particular, membrane separation—has revolutionized the processing of whey into many highly valued products. There are many possible products and manufacturing processes. Figure 1.10 outlines a number of possible processes for cheese whey. The first step involves *separation* and selective *concentration* of residual fat and casein by centrifugation. *Selective concentration* and *fractionation* of the whey proteins follows by the use of membrane separation (ultrafiltration and diafiltration). The protein stream is then further *concentrated* and *preserved* by evaporation and drying to produce whey protein concentrate, with a variety of protein levels and functional properties. Alternatively, further *fractionation* and *concentration* of the whey proteins may be performed using ion exchange to produce whey protein isolates.

The lactose- and mineral-containing side stream from ultrafiltration is known as permeate. Mineral recovery may occur by precipitation, filtration, and drying of the mineral component. The remaining lactose stream may then be *concentrated* by membrane separation (reverse osmosis) and evaporation. Lactose can then be recovered by crystallization, centrifugal separation, and drying, to produce crystalline lactose. Another option is to convert the lactose stream to alcohol by fermentation and distillation.

The whey products produced have a very wide range of uses, such as food ingredients with unique functional (gelling, emulsifying) and nutritional properties.

1.5 CHANGES TO MILK COMPONENTS DURING PROCESSING

Various methods of processing milk, leading to defined products, are shown in Figure 1.8. Processing alters the nature and behavior of milk components, which consequently influences the properties of the dairy products. Some of the process-induced changes that occur in milk are summarized in Table 1.6.

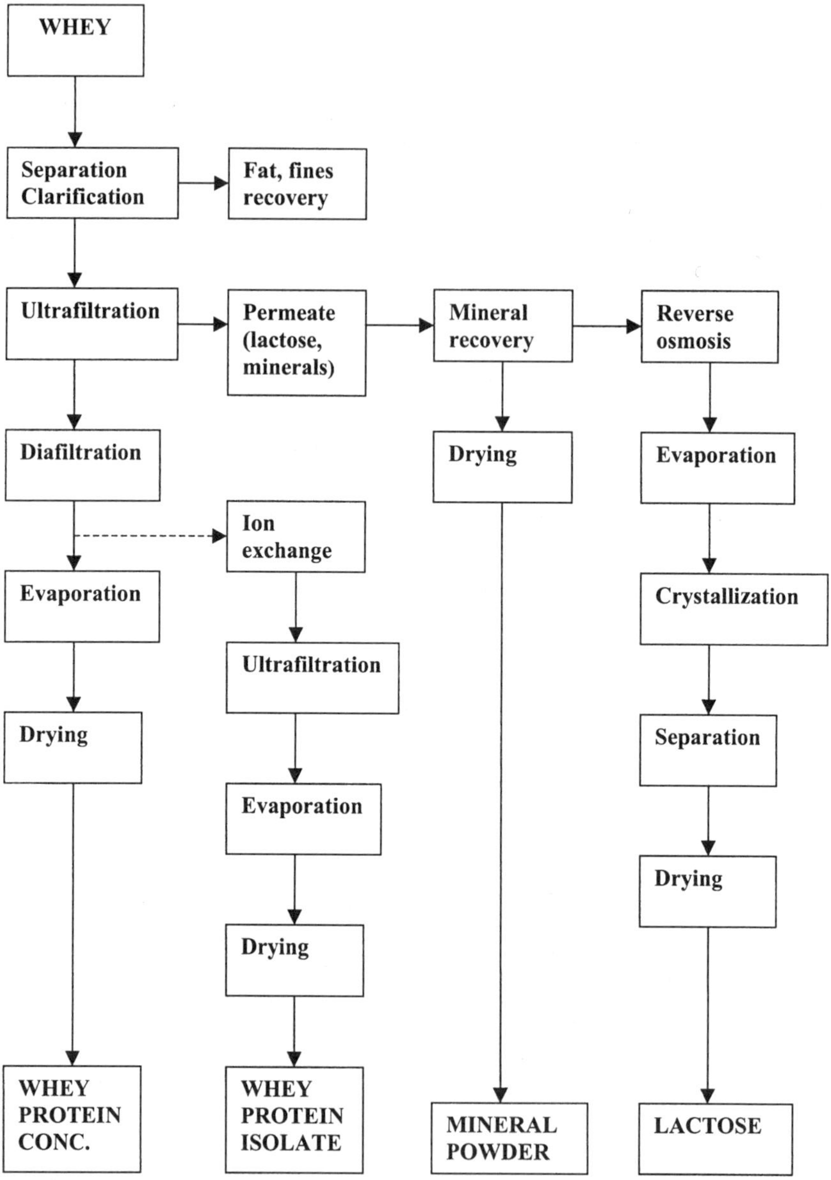

Figure 1.10. Manufacture of products from whey.

1.5.1 Homogenization

Intense turbulence during homogenization markedly reduces fat globule size, with a consequent four- to sixfold increase in surface area. Because the amount of available original membrane material is insuf-

TABLE 1.6. Process-Induced Changes in Milk

Process	Effect
Heating	Destruction of bacteria; inactivation of enzymes; destruction of some vitamins; denaturation of whey proteins; formation of aggregates of whey proteins; formation of a complex between κ-casein and β-lactoglobulin; shift of the soluble salts to the colloidal phase; changes in micelle structure; dephosphorylation of caseins; peptide bond cleavage of proteins; lysinoalanine formation; Maillard reaction; isomerization and degradation of lactose; changes to fat globule membrane; pH decrease.
Evaporation	Concentration of milk solids; increase in colloidal salts; increase in micelle size; decrease in pH; limited denaturation of whey proteins.
Homogenization	Increase in number of fat globules; adsorption of casein on to fat globules; decrease in fat globule size; decrease in protein stability.
Spray drying	Rapid removal of water; relatively minor changes in protein.

ficient to cover this area, plasma proteins adsorb onto the fat globule surface; the new membranes in homogenized milk consist of MFGM material, caseins, casein micelles, and whey proteins (Anderson et al., 1977; Darling and Butcher, 1978).

Casein micelles adsorb preferentially over the whey proteins, and the adsorbed casein micelles partly spread on the fat surface, possibly as submicelles (Walstra and Oortwijn, 1982). Although the whey proteins make up a much smaller proportion (about 5%), they cover a disproportionately greater area (about 25%) of the fat globule surface (Walstra and Oortwijn, 1982). The relative content of original membrane material in the new membrane depends on the increase in surface area, and thus on the intensity of homogenisation. The homogenized fat globules act as large casein micelles and will participate in any reaction of caseins, such as renneting, acid precipitation, and heat coagulation (van Boekel and Walstra, 1989).

After homogenization, clusters of fat globules in which casein micelles are shared by globules may be found (Ogden et al., 1976). Two-stage homogenization may (partly) prevent the occurrence of homogenization clusters (Walstra, 1983). Homogenization retards creaming because the fat globules are smaller. Lipase is activated by homogenization because of transfer of casein-associated lipase to the surface of fat globules, increase in fat surface area, and dissociation of casein micelles (Harper and Hall, 1976).

When skim milk is homogenized under normal milk homogenization conditions (e.g., 15 MPa), there is no change in the casein micelle size, but the micelles can be disrupted at high homogenization pressures.

1.5.2 Heat Treatments

Heating temperatures vary widely, ranging from pasteurization (72°C for 15 s) to low-temperature, long-time heating (e.g., 85°C for up to 30 min) to high-temperature, short-time heating (120°C for 2 min) using direct (steam injection) or indirect (plate heat exchanger) heating. These heat treatments cause a number of important chemical and physical reactions in milk proteins, and many reactions involve nonprotein constituents of milk as well. Some of these changes are listed in Table 1.6. Knowledge of the heat-induced changes in milk has expanded considerably during the last 30 years, and major reviews have been published (Fox, 1981; Singh, 1988; Singh and Creamer, 1992; Singh and Newstead, 1992). A brief overview of the main changes is provided in the following sections.

1.5.2.1 Changes in Proteins. Upon heating milk above 65°C, whey proteins are denatured by the unfolding of their polypeptides, thus exposing the side-chain groups originally buried within the native structure. The unfolded proteins then interact with casein micelles or simply aggregate with themselves, involving thiol–disulfide interchange reactions, hydrophobic interactions, and ionic linkages. The order of sensitivity of the various whey proteins to heat in milk has been reported to be immunoglobulins > BSA > β-lactoglobulin > α-lactalbumin (Dannenberg and Kessler, 1988; Oldfield et al., 1998a).

The kinetics of denaturation of whey proteins are quite complex, with the reaction characteristics showing marked changes above 80–100°C. The denaturation of α-lactalbumin appears to follow a first-order reaction, but agreements on the order of reaction for β-lactoglobulin have not been achieved. Dannenberg and Kessler (1988) suggest that a reaction order of 1.5 is best to fit the rate of denaturation, but other workers favor either a first-order (de Wit and Swinkels, 1980) or a second-order (Hillier and Lyster, 1979) reaction. Despite these differences, there is general agreement that the denaturation reaction can be represented by a nonlinear Arrhenius plot.

The apparent activation energies reported by various workers (Hillier and Lyster, 1979; Dannenberg and Kessler, 1988; Oldfield et al., 1998a) are in the range 260–280 kJ/mol for β-lactoglobulin and 270–280 kJ/mol for α-lactalbumin at temperatures below 90°C. At higher temperatures, the activation energy is lower, ranging from 54 to 60 kJ/mol for β-lactoglobulin and from 55 to 70 kJ/mol for α-lactalbumin. Different techniques have been used by various workers to assess denaturation; these include measurement of loss of solubility at pH 4.6, reactivity of thiol groups, electrophoretic analysis, loss of antigenic activity, and differential scanning calorimetry. Because these methods are based on different physical and chemical properties of the protein, it is not surprising that the data obtained by these different techniques are not totally in agreement. In addition, various heating methods have been used, including heating samples in glass tubes, capillary tubes immersed in water/oil baths, and laboratory-scale heat exchangers. It is likely that whey proteins respond differently depending on how the milk is heated—that is, on the time required to reach the desired temperature, flow conditions, and cooling times and rates.

Ionic strength, pH, and the concentrations of calcium and protein markedly influence the extent of denaturation of the whey proteins (Dannenberg and Kessler, 1988; Oldfield et al., 1998a). Heat denaturation of whey proteins is also influenced by lactose and other sugars, polyhydric alcohols, and protein-modifying agents (Donovan and Mulvihill, 1987).

Denatured whey proteins have been shown to associate with κ-casein on the surface of the casein micelles, giving the appearance under an electron microscope of thread-like appendages, protruding from the micelles (Mohammad and Fox, 1987). The principal interaction is considered to be between β-lactoglobulin and κ-casein, and it involves both disulfide and hydrophobic interactions (Smits and van Brouwershaven, 1980; Singh and Fox, 1987). Not all the denatured whey proteins complex with the casein micelles. Some remain in the serum where they may form aggregates with other whey proteins or with serum κ-casein. The extent of association of denatured whey protein with casein micelles is markedly dependent on the pH of the milk prior to heating, levels of soluble calcium and phosphate, milk solids concentration, and type of heating system (water bath, indirect or direct heating system) (Singh, 1995). Heating at pH values less than 6.7 results in a greater quantity of denatured whey proteins associating with casein micelles, whereas, at higher pH values, whey protein–κ-casein complexes dissociate from the micelle surface, apparently due to dissociation of κ-casein (Singh and Fox, 1985). An indirect heating system tends to result in greater proportions of β-lactoglobulin and α-lactalbumin associating with the micelles than does a direct heating system (e.g., direct steam injection) (Corredig and Dalgleish, 1996a; Oldfield et al., 1998b).

A mechanism of β-lactoglobulin denaturation and its association with casein micelles in milk systems has been proposed recently by Oldfield et al. (1998b), who suggested that there are at least three possible species of denatured β-lactoglobulin that could associate with the micelles: (i) unfolded monomeric β-lactoglobulin, (ii) self-aggregated β-lactoglobulin, and (iii) β-lactoglobulin–α-lactalbumin aggregates. The relative rates of association of these species with the casein micelles depend on temperature and heating rate, which in turn affect the relative rates of unfolding and the formation of the various aggregated species. The β-lactoglobulin aggregates, which have been shown to be stiff and rod-like (Griffen et al., 1993), would protrude from the micelle surface, providing steric effects for further β-lactoglobulin association. In addition, these aggregates may have their reactive sulfhydryl group buried within the interior of the aggregate, and therefore unavailable for sulfhydryl–disulfide interchange reactions with micellar κ-casein. In contrast, unfolded monomeric β-lactoglobulin molecules would be expected to penetrate the κ-casein hairy layer with greater ease and have a readily accessible sulphydryl group. The formation of unfolded β-lactoglobulin may be promoted by long heating times at low temperatures, or by heating at a slow rate to the required temperature. However, at high temperatures and fast heating rates, all whey proteins begin to unfold in a short period of time, thus presenting more opportunity for unfolded monomeric β-lactoglobulin to self-aggregate, which consequently is likely to associate with the casein micelles less efficiently.

Casein micelles are very stable at high temperatures, although changes in zeta potential, size, and hydration of micelles, as well as some association–dissociation reactions, do occur at severe heating temperatures (Fox, 1981; Singh, 1988; Singh and Creamer, 1992). Heating milk up to 90°C causes only minor changes in the size of casein micelles but severe heat treatments increase the micelle size. The increase in micelle size during heating is accompanied by a large increase in the number of small particles, possibly formed by disaggregation of casein micelles. For example, Aoki et al. (1974, 1975) reported that heating whey-protein-free milk at temperatures >110°C caused a considerable increase in the level of "soluble" casein, of which ~40% was κ-casein. In milk, the extent of dissociation of κ-casein-rich protein is dependent on the pH of heating; below pH 6.7, virtually no dissociation occurred on heating at 140°C for 1 min, but, at higher pH values (above 6.9), dissociation increased with an increase in pH (Singh and Fox, 1985).

In concentrated milk, the aggregation of casein micelles increases gradually with increasing intensity of heating (Singh and Creamer,

1991). In these systems, the dissociation of κ-casein-rich protein occurs on heating at normal pH (6.6), and the extent of dissociation increases with the pH at heating.

1.5.2.2 Changes in Fat Globules. When whole milk is heated above 60°C, fat globule membrane proteins denature, resulting in exposure of reactive thiol groups. As a consequence, thiol–disulfide interchange reactions may occur between different membrane proteins, and whey proteins could also participate in these reactions. Both α-lactalbumin and β-lactoglobulin have been shown to bind via intermolecular disulfide bridges to the surface of the milk fat globule on heating milk at 65–85°C (Dalgleish and Banks, 1991; Houlihan et al., 1992; Corredig and Dalgleish, 1996b). Whey proteins in whole milk have more affinity for the MFGM than for the casein micelle surface (Corredig and Dalgleish, 1996b). Phospholipids migrate from the fat globules to the aqueous phase during heating, to some extent (Houlihan et al., 1992).

When whole milk is homogenized before heating, the fat globules behave differently during and after heating from natural milk fat globules. The membranes of the homogenized globules have a different composition, and the size of the globules is smaller. The subsequent heating of homogenized milk causes complex formation between whey proteins and the casein adsorbed on the fat globule surface. Homogenized fat globules act as large casein micelles because of the casein in their new membranes and thus participate in any reaction of the caseins, as stated earlier (van Boekel and Walstra, 1989). When whole milk is heated before homogenization, the whey proteins are denatured (if the temperature is >70°C) and interact with both the κ-casein of casein micelles (at the natural pH) and the native fat globule membrane (Dalgleish and Banks, 1991). During subsequent homogenization, the micellar complex of casein and whey protein will adsorb on to the newly created fat surfaces (Sharma and Dalgleish, 1994). Competition between the caseins and whey proteins for the fat surface will be small or even absent. These two treatments, homogenization and heat treatment, cause major variations in the quality of processed dairy products such as evaporated and sweetened condensed milk (van Boekel and Walstra, 1989).

1.5.2.3 Changes in Milk Salts. Milk salts exist in an equilibrium between the soluble and colloidal phases of milk, and this distribution is affected by heat treatment of the milk (Walstra and Jenness, 1984). Heating milk causes transfer of calcium and phosphate from the soluble to the colloidal state. On subsequent cooling, some of the indigenous or heat-precipitated calcium phosphate may redissolve, especially if the heating temperatures are less than 85°C. However, after heating at higher temperatures, the precipitated material does not resolubilize. The decrease in soluble calcium, phosphate, and citrate during heating in the temperature range 70–90°C involves two steps (Pouliot et al., 1989). The majority of the decrease takes place during the first minute of heating, after which a small decrease occurs over an extended period of time (up to 120 min). Heat treatments have little effect on the monovalent ions Na^+, K^+, and Cl^-. It is not known what subsequent effects these changes in the equilibria of salts and ions have on the processing properties of milk.

1.5.2.4 Changes in Lactose. Lactose can interact with proteins during heat treatment of the milk, involving the Maillard reaction. This reaction involves condensation of lactose with the free amino groups of protein, with subsequent rearrangements and degradations leading to a variety of brown-colored compounds. Manifestations of the Maillard reactions, which are important in milk products, include brown color, off-flavor, loss of available lysine, lowered nutritional value, reduced digestibility, and reduced solubility. The rate of reaction is strongly dependent on pH, time, and temperature of heating, water activity, and temperature during storage. Besides this, lactose may isomerize into other sugars (e.g., lactulose). If the temperature of heating is greater than 100°C, lactose can be converted to acids, especially formic acid, thereby decreasing the pH of the milk.

1.5.3 Concentration

Milk is normally concentrated by evaporation or ultrafiltration. Evaporation is normally carried out at temperatures between 50°C and 70°C. In addition to concentration, evaporation causes numerous other changes in the milk system (Singh and Newstead, 1992). The pH of milk decreases during concentration from an average initial value of 6.7 to approximately 6.3 at 45% total solids. This is partly due to changes in salt equilibria as more calcium phosphate is transferred from the soluble to the colloidal state, with a concomitant release of hydrogen ions. Le Graet and Brule (1982) showed that when milk is concentrated about fivefold by evaporation, soluble calcium and soluble phosphate increase by a factor of about two, the remainder of the soluble calcium and phosphate being transformed into the colloidal state. The activity of calcium ions increases only slightly, but the ratio of monovalent to divalent cations increases markedly (Nieuwenhuijse et al., 1988). There is no significant denaturation of β-lactoglobulin or immunoglobulin G during evaporation of skim milk to 45% total solids, whereas some denaturation of α-lactalbumin may occur (Oldfield, 1997). During evaporation, there is an increase in the amounts of β-lactoglobulin and α-lactalbumin associated with the micelles. Casein micelles may increase in size due to the increase in colloidal calcium phosphate or due to coalescence of the micelles (Walstra and Jenness, 1984). The sensitivity of casein micelles to heat increases with increasing concentration. Maillard browning and the insolubilization of protein begin to occur if the concentration factor is too high, particularly at high temperatures.

Under normal conditions with the evaporation temperatures below 70°C, the alterations to the proteins are likely to be minor compared with those incurred during heating.

Concentration of skim milk by ultrafiltration results in an increase in fat, protein, and colloidal salts. The ultrafiltration temperature has an impact on the composition of the concentrate, with higher temperatures resulting in higher permeation rates and lower retention of lower-molecular-weight components.

Examination of casein micelles in ultrafiltration concentrates by electron microscopy shows that significant differences, compared with the micelles in unconcentrated milk, occur. In skim milk at a volume concentration factor of 5, the average diameter of the casein particles appears to be smaller (Srilaorkul et al., 1991). Milk protein concentrates are more susceptible to heat denaturation of whey proteins. For instance, when heat treatment at 75°C for 5 min is applied, the denaturation degree increases from 31% in skim milk to 64% in ultrafiltration retentate at a concentration factor of 4.4 : 1 (McMahon et al., 1993). However, Waungana et al. (1996) found no noticeable difference in β-lactoglobulin denaturation between normal and twofold ultrafiltration-concentrated skim milk.

Concentrate produced by ultrafiltration is more heat stable than conventional skim milk concentrate prepared by evaporation (Sweetsur and Muir, 1980). Milk proteins and colloidal salts of calcium and phosphates are concentrated by ultrafiltration and therefore exert an increased buffering effect (Brule et al., 1979; Covacevich and Kosikowski, 1979). This increased buffering capacity of the milk may slow the rate of pH reduction in cheese made using ultrafiltered milk. The buffering effect of skim milk ultrafiltrate can be reduced significantly, with the removal of more of the minerals, by carrying out the ultrafiltration in the cold (4°C) and at pH 5.3 (St. Gelias et al., 1992).

1.5.4 Spray Drying

Little work has been reported on the changes induced in milk systems by drying. Atomization of the concentrate gives a large surface area over which drying can take place. The droplets are sprayed into the main drying chamber and are intimately mixed with dry heated air (180–220°C). Drying is usually very rapid, and the temperature of the milk droplets does not exceed 70°C until they have lost almost all their water. The temperature of the droplets approaches that of the outlet air as the drying process nears completion. For this reason, the outlet air temperature is a critical parameter controlling heat damage to dry milk products. The heat exposure of milk during spray drying may vary considerably, depending on the design of the drier, the operating conditions, and the length of time the powder is held before cooling. The native properties of the milk components are essentially unmodified by moderate drying conditions. The normal size distribution of the casein micelles and their heat stabilities and renneting characteristics are substantially recovered on reconstitution of spray-dried milk.

Under normal spray-drying conditions, whey protein denaturation is negligible and most enzymes remain active (Walstra and Jenness, 1984). Oldfield (1997) found no apparent denaturation of immunoglobulin G and only a small loss of BSA (3–7%) during the spray drying of skim milk.

Changes may occur in the salts equilibria during spray drying. The process of drying would be expected to produce the same types of changes in salts equilibria as evaporation—that is, an increase in CCP and a decrease in pH. It has been shown that the concentrations of soluble calcium and soluble phosphate and the calcium ion activity in reconstituted skim milk are about 20% lower than those in the original milk (Le Graet and Brule, 1982).

When whole milk is spray dried, disruption of fat globules may occur during nozzle atomization, because the applied pressures are comparable with that used in homogenizers.

1.5.5 Acidification

Milk can be acidified by bacterial cultures (which ferment lactose to lactic acid), by the addition of chemical acids such as HCI, or by the use of glucono-δ-lactone (GDL) where the hydrolysis of GDL to gluconic acid results in a reduction in pH. Acid-induced coagulation and gel formation in milk have been reviewed recently by Lucey and Singh (1998).

During acidification of milk, many of the physicochemical properties of casein micelles undergo considerable change, especially in the pH range from 5.5 to 5.0. As the pH of milk is reduced, CCP is dissolved and the caseins are dissociated into the milk serum phase (Roefs et al., 1985; Dalgleish and Law, 1988). The extent of dissociation of caseins is dependent on the temperature of acidification: At 30°C, a decrease in pH causes virtually no dissociation; at 4°C, about 40% of the caseins are dissociated in the serum at pH ~5.5 (Dalgleish and Law, 1988). Aggregation of casein occurs as the isoelectric point (pH 4.6) is approached. Apparently little change in the average hydrodynamic diameter of casein micelles occurs during acidification of milk to pH ~5.0 (Roefs et al., 1985). The lack of change in the size of the micelles on reducing the pH of milk to 5.5 may be due to concomitant swelling of the particles as CCP is solubilized.

Heat treatment of milk prior to acidification has little effect on the extent of solubilization of CCP (Singh et al., 1996). When acidification is carried out at 5°C, more caseins dissociate from the micelles in heated milks than in unheated milks; the reverse occurred when milks are acidified at 22°C (Singh et al., 1996).

The exact nature of the casein particles that exist in acidified milk is uncertain. At pH around 5.3, most of the CCP in the micelles is solubilized, the native charges on individual caseins are reduced, and the ionic strength of the solution increases. It is expected that the forces responsible for the integrity of these "micelle-like" CCP-depleted casein particles would be considerably different from those in native micelles even if the average hydrodynamic diameter appears to be largely unchanged. Hydrophobic and electrostatic interactions are also important for the stability of these casein particles as evidenced by the temperature and pH dependence of the dissociation of caseins (Lucey et al., 1997). Thus, the casein particles that aggregate to form an acid-induced gel would appear to be very different from native casein micelles.

1.5.6 Rennet Coagulation

The coagulation of milk by the specific action of selected proteolytic enzymes, particularly rennets, forms the basis for the manufacture of most cheese varieties. These rennets consist of the enzymes chymosin and pepsin, and they are traditionally prepared from the stomachs of calves, kids, lambs, or other mammals in which rennins are the principal proteinases.

The coagulation of milk occurs in phases: a primary enzymatic phase, a secondary nonenzymatic phase, and a less clearly defined tertiary phase (Dalgleish, 1992). During the primary phase, rennet causes specific hydrolysis of κ-casein in the region of the phenylalanine$_{105}$–methionine$_{106}$ bond, with Phe$_{105}$ supplying the carboxyl group and Met$_{106}$ supplying the amino group, resulting in the formation of two peptides of contrasting physical and chemical properties. The glycomacropeptide (GMP) moiety, which comprises amino acid residues 106–169, is hydrophilic, and it diffuses away from the casein micelle after splitting from κ-casein and into the serum. The second peptide, para-κ-casein, which consists of amino acid residues 1–105, is strongly hydrophobic and remains attached to the micelles.

κ-Casein hydrolysis during the primary phase alters the properties of the casein micelles, rendering them susceptible to aggregation, and this marks the onset of the secondary phase. Loss of GMP during the primary phase of renneting results in loss of about half the negative charge as well as surface steric repulsion of κ-casein (Darling and Dickson, 1979; Dalgleish, 1984). Consequently, the micellar surface becomes more hydrophobic, due to the accumulation of para-κ-casein, and micelle aggregation becomes possible.

There is no clear distinction between the end of the primary phase and the beginning of the secondary phase because the two reactions overlap to some extent with some aggregation commencing before complete hydrolysis of κ-casein (Green et al., 1978; Dalgleish, 1979; Chaplin and Green, 1980). However, there is a critical value of κ-casein hydrolysis (86–88%) below which micelles cannot aggregate. The action of rennet can be seen as providing "hot spots" (areas on casein micelle surfaces from which the protective GMP moiety has been "shaved off") via which the micelles can aggregate, with these reactive areas being produced by removal of κ-casein from a sufficiently large area (Green and Morant, 1981; Payens, 1982). As the last of the stabilizing surface is removed (i.e., during the destruction of the final 20% of the κ-casein), the concentration of micelles capable of aggregation and the rate at which they aggregate increase rapidly. Van der Waals and hydrophobic and specific ion-pair interactions are thought to control coagulation.

The process of gel assembly during the secondary phase of rennet coagulation is not well understood. The initial stages of gel formation appear to involve the formation of small aggregates in which the micelles tend to be linked in chains that link together randomly to form a network (Green et al., 1978). Eventually, the network chains group together to form strands. The gel is thought to be assembled by linkage of smaller aggregates rather than by addition of single particles to preformed chains. Initially, linkage of aggregated micelles is through bridges, which slowly contract with time, forcing the micelles into contact and eventually causing them to partly fuse. This process probably progressively strengthens the links between micelles, giving an increase in curd firmness after coagulation.

The tertiary stage involves the rearrangement of the network and processes such as syneresis and the nonspecific proteolysis of the caseins in the rennet gel (Dalgleish, 1992).

1.6 CONCLUSIONS

Milk is a highly versatile, perishable raw material that can be manufactured into a very wide range of products. The processes that are used can be grouped as *fractionation*, *concentration*, and *preservation*. This classification emphasizes the versatility of milk as a multicomponent raw material, from which components can be selectively concentrated and stabilized. The processes and products described are only a proportion of those in commercial use today or in development. There are many more, including those minor components with specialized health benefits, that will become products of the future. Continued developments in sophisticated methods for product characterization will allow greater understanding of the interactions of milk components at a molecular level during processing.

REFERENCES

Anderson, M., Brooker, B. E., Canston, T. E., and Cheeseman, G. C. (1977) *J. Dairy Res.*, **44**, 111–123.

Aoki, T., Suzuki, H., and Imamura, T. (1974) *Milchwissenchaft*, **29**, 589–594.

Aoki, T., Suzuki, H., and Imamura, T. (1975) *Milchwissenchaft*, **30**, 30–35.

Brule, G., Fauquant, J., and Maubois, J. L. (1979) *J. Dairy Sci.*, **62**, 869–875.

Bylund, G. (1995) *Dairy Processing Handbook*. Tetra Pak Processing Systems AB, Lund, Sweden.

Chaplin, B., and Green, M. L. (1980) *J. Dairy Res.*, **47**, 351–358.

Christie, W. W. (1995) Composition and structure of milk lipids. In *Advanced Dairy Chemistry—2. Lipids*, P. F. Fox, ed., Chapman & Hall, London, pp. 1–28.

Corredig, M., and Dalgleish, D. G. (1996a) *Milchwissenchaft*, **51**, 123–127.

Corredig, M., and Dalgleish, D. G. (1996b) *J. Dairy Res.*, **63**, 441–449.

Covacevich, H. R., and Kosikowski, F. V. (1979) *J. Dairy Sci.*, **62**, 204–207.

Dalgleish, D. G. (1979) *J. Dairy Res.*, **46**, 643–661.

Dalgleish, D. G. (1984) *J. Dairy Res.*, **51**, 425–438.

Dalgleish, D. G. (1992) The enzymatic coagulation of milk. In *Advanced Dairy Chemistry—1. Proteins*, P. F. Fox, ed., Elsevier Applied Science Publishers, London, pp. 579–620.

Dalgleish, D. G., and Banks, J. M. (1991) *Milchwissenchaft*, **46**, 70–75.

Dalgleish, D. G., and Law, A. J. R. (1988) *J. Dairy Res.*, **55**, 529–538.

Dannenberg, F., and Kessler, H. G. (1988) *J. Food Sci.*, **53**, 258–263.

Darling, D. F., and Butcher, D. W. (1978) *J. Dairy Res.*, **45**, 197–208.

Darling, D. F., and Dickson, J. (1979) *J. Dairy Res.*, **46**, 441–451.

de Wit, J. N., and Swinkels, G. A. M. (1980) *Biochim. Biophys. Acta*, **624**, 40–50.

Donovan, M., and Mulvihill, D. M. (1987) *Irish J. Food Sci. Technol.*, **11**, 87–100.

Earle, R. L. (1983) *Unit Operations in Food Processing*, 2nd ed., Pergamon Press, Oxford.

Fox, P. F. (1981) *J. Dairy Sci.*, **64**, 2127–2137.

Green, M. L., and Morant, S. V. (1981) *J. Dairy Res.*, **48**, 57–63.

Green, M. L., Hobbs, D. G., Morant, S. V., and Hill, V. A. (1978) *J. Dairy Res.*, **45**, 413–422.

Griffen, W. G., Grifen, M. C. A., Martin, S. R., and Price, J. (1993) *J. Chem. Soc. Faraday Trans.*, **89**, 3395–3406.

Hambling, S. G., McAlpine, A. S., and Sawyer, L. (1992) β-Lactoglobulin. In *Advanced Dairy Chemistry—1. Proteins*, P. F. Fox, ed., Elsevier Applied Science Publishers, London, pp. 141–190.

Hansen, S., Bauer, R., Lomholt, S. B., Qvist, K. B., Pedersen, J. S., and Mortensen, K. (1996) *Eur. Biophys.*, **24**, 143–147.

Harper, W. J., and Hall, C. W. (1976) Processing-induced changes. Chapter 13 in *Dairy Technology and Engineering*, AVI Press, Westport, CT.

Hillier, R. M., and Lyster, R. L. J. (1979) *J. Dairy Res.*, **46**, 95–102.

Holt, C. (1992) Structure and stability of bovine casein micelles. In *Advances in Protein Chemistry*, C. B. Afinsen, J. D. E. D. Sall, F. K. Richards, and D. S. Eisenberg, eds., Academic Press, New York, pp. 63–151.

Horne, D. S. (1998) *Int. Dairy J.*, **8**, 171–177.

Houlihan, A. V., Goddard, P. A., Kitchen, B. J., and Masters, C. J. (1992) *J. Dairy Res.*, **59**, 321–329.

Kennan, T. W., and Dylewski, D. P. (1994) Interacellular origin of milk lipid globules and the nature and structure of milk lipid globule membrane. In *Advanced Dairy Chemistry—2. Lipids*, P. F. Fox, ed., Chapman & Hall, London, pp. 89–130.

Kinsella, J. E., and Whitehead, D. M. (1989) *Adv. Food Nutr. Res.*, **33**, 343–438.

Le Graet, Y., and Brule, G. (1982) *Le Lait*, **62**, 113–125.

Lucey, J. A., and Singh, H. (1998) *Food Res. Int.*, **30**, 529–542.

Lucey J. A., Dick, C., Singh, H., and Munro, P. A. (1997) *Milchwissenchaft*, **52**, 603–606.

McMahon, D. J., Youshif, B. H., and Kalab, M. (1993) *Int. Dairy J.*, **3**, 239–256.

Mohammad, K. S., and Fox, P. F. (1987) *NZ J. Dairy Sci. Technol.*, **22**, 191–203.

Nieuwenhuijse, J. A., Timmermans, W., and Walstra, P. (1988) *Neth. Milk Dairy J.*, **42**, 387–421.

Ogden, L. V., Walstra, P., and Morris, H. A. (1976) *J. Dairy Sci.*, **59**, 1727–1737.

Oldfield, D. J. (1997) Heat-induced whey protein reactions in milk. Kinetics of denaturation and aggregation as related to milk powder manufacture. PhD thesis, Massey University, Palmerston North, New Zealand.

Oldfield, D. J., Singh, H., Taylor, M. W., and Pearce, K. N. (1998a) *Int. Dairy J.*, **8**, 311–318.

Oldfield, D. J., Singh, H., and Taylor, M. W. (1998b) *Int. Dairy J.*, **8**, 765–770.

Payens, T. A. J. (1982) *J. Dairy Sci.*, **65**, 1863–1873.

Perry, R. H., Green, D. W., and Maloney, J. W. (1984) *Perry's Chemical Engineers' Handbook*, 6th ed., McGraw-Hill, New York.

Pouliot, Y., Boulet, M., and Paquin, P. (1989) *J. Dairy Res.*, **56**, 185–192.

Robinson, R. K. (ed.) (1993) *Modern Dairy Technology*, 2nd ed., Elsevier Applied Science, London.

Roefs, S. P. F. M., Walstra, P., Dalgleish, D. G., and Horne, D. S. (1985) *Neth. Milk Dairy J.*, **39**, 119–122.

Rollema, H. S. (1992) Casein association and micelle formation. In *Advanced Dairy Chemistry—1. Proteins*, P. F. Fox, ed., Elsevier Applied Science Publishers, London, pp. 111–140.

Rosenthal, I. (1991) *Milk and Dairy Products: Properties and Processing*. Balaban Publishers, VCH, New York.

Schmidt, D. G. (1982) Association of caseins and casein micelle structure. In *Developments in Dairy Chemistry—1. Proteins*, P. F. Fox, ed., Elsevier Applied Science Publishers, London, pp. 61–86.

Sharma, S. K., and Dalgleish, D. G. (1994) *J. Dairy Res.*, **61**, 375–384.

Singh, H. (1988) *NZ J. Dairy Sci. Technol.*, **23**, 257–273.

Singh, H. (1995) Heat induced changes in casein, including interactions with whey proteins. In *Heat-Induced Changes in Milk*, 2nd ed., P. F. Fox, ed., International Dairy Federation, Brussels, pp. 86–104.

Singh, H., and Creamer, L. K. (1991) *J. Food Sci.*, **56**, 238–246.

Singh, H., and Creamer, L. K. (1992) Heat stability of milk. In *Advanced Dairy Chemistry—1. Proteins*, P. F. Fox, ed., Elsevier Applied Science Publishers, London, pp. 621–656.

Singh, H., and Fox, P. F. (1985) *J. Dairy Res.*, **52**, 529–538.

Singh, H., and Fox, P. F. (1987) *J. Dairy Res.*, **54**, 509–521.

Singh, H., and Newstead, D. F. (1992) Aspects of proteins in milk powder manufacture. In *Advanced Dairy Chemistry—1. Proteins*, P. F. Fox, ed., Elsevier Applied Science Publishers, London, pp. 735–765.

Singh, H., Roberts, M. S., Munro, P. A., and Teo, C. T. (1996) *J. Dairy Sci.*, **79**, 1340–1346.

Slattery, C. W. (1976) *J. Dairy Sci.*, **59**, 1547–1556.

Smits, P., and van Brouwershaven, J. H. (1980) *J. Dairy Res.*, **47**, 313–325.

Srilaorkul, S., Ozimek, L., Ooraikul, B., Hadziyers, D., and Wolfe, W. (1991) *J. Dairy Sci.*, **74**, 50–57.

St. Gelias, D., Hache, S., and Gros-Louis, M. (1992) *J. Dairy Sci.*, **75**, 1167–1172.

Swaisgood, H. E. (1982) Chemistry of milk protein. In *Developments in Dairy Chemistry—1. Proteinss*, P. F. Fox, ed., Elsevier Applied Science Publishers, London, pp. 1–59.

Swaisgood, H. E. (1992) Chemistry of the casein. In *Advanced Dairy Chemistry—1. Proteins*, P. F. Fox, ed., Elsevier Applied Science Publishers, London, pp. 63–110.

Sweetsur, A. W. M., and Muir, D. D. (1980) *J. Dairy Res.*, **47**, 327–335.

van Boekel, M. A. J. S., and Walstra, P. (1989) Physical changes in the fat globules in unhomogenised and homogenised milk. *International Dairy Federation Bulletin*, **238**, 13–16.

Walstra, P. (1983) Physical chemistry of fat globules. In *Developments in Dairy Chemistry—2. Lipids*, P. F. Fox, ed., Elsevier Applied Science Publishers, London, pp. 119–158.

Walstra, P. (1990) *J. Dairy Sci.*, **73**, 1965–1979.

Walstra, P., and Jenness, R. (1984) *Dairy Chemistry and Physics*, Wiley-Interscience, New York.

Walstra, P., and Oortwijn, H. (1982) *Neth. Milk Dairy J.*, **36**, 103–113.

Walstra, P., Geurts, T. J., Noomen, A., Jellema, A., and van Boekel M. A. J. S. (1999) *Dairy Technology: Principles of Milk Properties and Processes*. Marcel Dekker, New York.

Waugh, D. F., and Noble, R. W. (1965) *J. Am. Chem. Soc.*, **87**, 2246–2257.

Waungana, A., Singh, H., and Bennett, R. (1996) *Food Res. Int.*, **29**, 715–721.

Whitney, R. McL. (1988) Proteins of milk. In *Fundamentals of Dairy Chemistry*, N. P. Wong, R. Jenness, M. Keeney, and E. H. Marth, eds., van Nostrand Reinhold Company, New York, pp. 81–169.

CHAPTER 2

THE MICROBIOLOGY OF RAW MILK

JAMES V. CHAMBERS
Department of Food Science, Purdue University, West Lafayette, Indiana

2.1 INTRODUCTION

Milk, by its very nature, is a natural growth medium for microorganisms. Normally, milk is collected from a lactating animal (most commonly a dairy cow) at least twice a day and is recognized as a highly perishable foodstuff easily subjected to microbial contamination. This contamination can vary widely due to milk-handling practices ranging from milking a few cows by hand in the out of doors to milking 3000 cows by a complex, automated system in a well-equipped parlor. In countries with developing economies, it is not uncommon to find small quantities of nonrefrigerated milk being hauled by individual producers to collection centers for entry into the market. In contrast, where dairy farming is more advanced technologically, milk is refrigerated immediately after collection and stored in farm tanks until picked up by bulk tank drivers who deliver the milk to processing plants. Thus, the initial microbiological quality of milk can vary enormously. Nevertheless, there are three basic sources of microbial contamination of milk: (1) from within the udder, (2) from the exterior of the teats and udder, and (3) from the milk handling and storage equipment.

Milk is produced at ambient temperatures ranging from subzero centigrade, where it is necessary to protect milk from freezing, to above 25°C, where refrigeration is needed. Furthermore, the duration of milk storage time on the farm can vary widely. Therefore, the numbers and types of microorganisms present when milk leaves the farm can differ, often unpredictably, even under apparently similar conditions.

Dairy Microbiology Handbook, Third Edition, Edited by Richard K. Robinson
ISBN 0-471-38596-4

In most dairying areas, milk production methods, equipment, and on-farm storage have improved over time. However, udder disease remains widespread because of the presence of mastitis-associated microorganisms. Refrigeration on the farm all too often masks the effects of unsanitary practices, including the use of inadequately cleaned and sanitized milking equipment. As a result, the microbiological quality of raw milk supplies produced under apparently good sanitary conditions and stored under adequate refrigeration may produce off-flavors, yield poor product, and present a risk of food-borne infections to the consumer. Factors influencing the microbiological quality of raw milk were reviewed in a bulletin issued by the International Dairy Federation (1980).

2.2 THE INITIAL MICROFLORA OF RAW MILK

The numbers and types of microorganisms in milk immediately after production (i.e., the initial microflora) directly reflect microbial contamination during production, collection, and handling. The microflora in the milk when it leaves the farm is influenced significantly by the storage temperature and the elapsed time after collection. Where milk is stored at ≤4°C, this low temperature normally will delay bacterial multiplication for at least 24 h. The microflora, therefore, is similar to that present initially. However, if unsanitary conditions exist with the milking equipment or storage tank, the low temperature could mask these conditions.

A useful indicator for monitoring the sanitary conditions present during the production, collection, and handling of raw milk is the "total" bacterial count or standard plate count (SPC). The SPC is determined by plating (or using equivalent procedures) on a standardized plate count agar followed by aerobic incubation for 2 or 3 days at 32°C or 30°C, respectively. Microorganisms failing to form colonies, of course, will not be counted. The SPC does not indicate the source(s) of bacterial contamination or the identity of production deficiencies leading to high counts. Its sole value is to indicate changes in the production, collection, handling, and storage environment. Follow-up microbial assessments for psychrotrophs or thermoduric bacteria, spore-forming bacteria, streptococci, and coliforms can assist in determining of sanitary deficiencies.

Certain groups can be enumerated selectively. For example, psychrotrophs can be counted either by incubating SPC plates for 10 days at 5–7°C or by using a preliminary incubation of the raw milk at 13°C for 16h followed by performing the SPC procedure. Thermoduric bacteria can be determined by laboratory pasteurization of milk before plating. Selective or diagnostic media can be used for coliforms, lactic acid bacteria, mastitis pathogens, Gram-negative rods (GNRs), lipolytic, proteolytic and caseinolytic microbial types, and so on. An increased number of automated methods now are being employed for plating and enumerating bacteria. Also, rapid quantifying techniques are being used, such as the direct epifluorescent filtration technique (Pettipher et al., 1980), adenosine triphosphate method, and impedance measurements (Phillips and Griffiths, 1985).

2.2.1 Total Raw Milk Bacterial Content via the SPC Method

SPC values for raw milk can range from $<1000\,ml^{-1}$, where contamination during production is minimal, to $>1 \times 10^6\,ml^{-1}$. The microorganisms present may be derived from one or any combination of the three main sources of contamination previously identified. Consequently, high initial SPC values (e.g., $>100{,}000\,ml^{-1}$) are evidence of serious deficiencies in production hygiene, whereas SPC values of $<20{,}000\,ml^{-1}$ reflect good sanitary practices (International Dairy Federation, 1974).

In many countries, a standard for Grade A (or Grade 1) raw milk is an SPC of $<1 \times 10^5\,ml^{-1}$ for milk intended for heat treatment before consumption. For milk that is to be consumed raw, a more stringent standard generally is required because consumers of raw milk are at a greater risk for contracting a milk-borne illness such as salmonellosis. In some countries, standards adopted may depend on whether milk is refrigerated or merely water-cooled. For example, in North America, SPC values of $<1 \times 10^6\,ml^{-1}$ or equivalent are acceptable for manufacturing grade milk. In contrast, the United Kingdom makes no distinction between raw milk marketed for manufacture and that marketed for fluid consumption.

In the United Kingdom, the payment system for the producer includes bonuses relating to the SPC of the milk supplied. This is determined as an average (with some exceptions) of four measurements made in each 4-week period. Milk with an SPC $<2 \times 10^4\,ml^{-1}$ is classified as Band A, and the producer receives a $0.23\,pl^{-1}$ bonus. Milk with an SPC between 2×10^4 and $1 \times 10^5\,ml^{-1}$ is classified as Band B, and the producer receives no bonus. A producer of Band C milk, which has an SPC over $1 \times 10^5\,ml^{-1}$, is penalized $1.5\,pl^{-1}$ for exceeding the maximum allowed limit; $6.0\,pl^{-1}$ for a sequential 6-month violation; and $10\,pl^{-1}$ for a continuation of the violation after 6 months from the previously reported SPC (J. Marshall, 1991). The producer and milk volume dis-

TABLE 2.1. Average Percentage of Producers in England and Wales and Total Volume of Milk in Each Quality Band from April 1987 to March 1988

	Band A: <20,000	Band B: >20,000–100,000	Band C: >100,000
	Percentage of Producers		
Average:	75.5	22.6	1.9
Monthly range:	70.4–83.6	14.7–27.9	0.9–3.0
	Total Milk		
Volume %:	83.0	16.2	0.8
Monthly range:	8.10–89.7	9.6–20.8	0.4–1.4

tribution for England and Wales and the respective band payment categories are shown in Table 2.1. Table 2.1 shows the average weighted SPC for all milk to be 16,000 (range 12,000–18,000 ml^{-1}). Also shown is an improvement in hygienic quality over those reported by Panes et al. (1979), where 333 farms were sampled over a year, with 86.4% of the producers having SPCs of <100,000 ml^{-1}, compared to the above 98.1% of producers.

In contrast to the UK system, milk handlers in the United States pay premiums based on seven basic criteria. Within the "quality incentive" programs, payments are made over the market's class of utilization price per 100 lb of milk. Like the UK system, the US program debits the dairy producer for underperformance. Table 2.2 presents an incentive program used by a US midwest regional milk cooperative for the distribution of quality incentive payments/debits to its members. In terms of Grade A milk utilization in the upper midwest region, fluid milk represents 14.5%; creamed products, fermented products, cottage cheese, and frozen desserts represent 3.4%; cheese, butter and milk powder represent 81%; and butterfat and nonfat milk solids represent 1.1% (AMS, USDA, 2000).

2.2.2 Types of Microorganisms Present in Raw Milk

Table 2.3 summarizes the main groups of mesophilic (30–32°C), aerobic microorganisms and their respective genera and species that comprise the microflora commonly found in raw milk taken from individual farms located in the United Kingdom. The microorganisms are identified from isolates from SPC agar following a relatively simple scheme used to characterize these isolates based on morphology, reaction to Gram stain, catalase production, and formation of acid and gas in

TABLE 2.2. Quality Incentive Program Used by a Milk Cooperative in the United States

Quality Attribute	+50¢ Premium Paid per cwt	+35¢ Premium Paid per cwt	No Premium	−35¢ Premium Debit per cwt	−50¢ Premium Debit per cwt
Standard plate count (ml^{-1})	10,000 and below	15,000 and below	16,000–50,000	51,000–100,000	Over 100,000
Preincubation count (ml^{-1})	30,000 and below	50,000 and below	51,000–150,000	Above 150,000	Not applicable
Inhibitor	No	No	No	Not applicable	Not applicable
Sediment pad score	1 or 2	1 or 2	1 or 2	3	4
Added water	No	No	No	Yes	Yes
Maintains Grade A status	No degrade	No degrade	No degrade	Degraded	Degraded
Somatic cell count (ml^{-1})	200,000 and below	201,000–400,000	401,000–550,000	551,000–650,000	Over 650,000

Note: Cwt indicates "100 pounds of milk by weight." "Preincubation count" involves incubating the milk sample at 13°C for 16 h, then performing the aerobic SPC method. While not required by Grade A Standards, this method indicates the sanitary conditions at the farm. If there is a problem, the count usually will be three times the initial aerobic SPC.

TABLE 2.3. Types of Aerobic Mesophilic Microorganisms in Fresh Milk Which Form Colonies on SPC Agar (Carreira et al., 1955)

Micrococci	Streptococci	Asporogenous Gram-Positive Rods	Spore-Formers Group	GNR Group	Miscellaneous Groups
Micrococcus	*Enterococcus*	*Microbacterium*	*Bacillus*	*Pseudomonas*	Streptomycetes
Staphylococcus	(“fecal”)	*Corynebacterium*	(spores or	*Acinetobacter*	Yeasts
		Arthrobacter	vegetative	*Flavobacterium*	Molds
	Group N	*Kurthia*	cells)	*Enterobacter*	
				Klebsiella	
	Mastitis			*Aerobacter*	
	streptococci			*Escherichia*	
	Streptococcus agalactiae			*Serratia*	
	Streptococcus			*Alcaligenes*	
	dysgalactiae				
	Streptococcus uberis				

Note: Special media and/or incubation conditions are needed for the isolation and detection of species that belong to *Clostridium, Lactobacillus* and lactic acid bacteria, *Corynebacterium*, and certain pathogens.

TABLE 2.4. Incidence of Main Groups of Microorganisms in Low Count Raw Milk

Group	Incidence (%)
Micrococcus	30–99
Streptococci	0–50
Asporogenous Gram-positive rods	<10
GNRs (includes coliforms)	<10
Bacillus spores	<10
Miscellaneous (includes streptomycetes)	<10

McConkey's broth as first described by Carreira et al. (1955). This scheme is widely used and provides useful information such as the survey data shown in Table 2.4. These data taken from low count raw milk of <5000 colony-forming units (cfu) ml^{-1} indicate that a micrococci and streptococci population dominate the microflora present. In such milk, minimal bacterial contamination from the exterior of the udder and from milking equipment is reflected in the presence of those bacteria commonly associated with the normal udder and teat skin flora. Similar findings are reported in the United States by Dorner (1930). The information shown in Table 2.4 not only applies to low count milk samples from individual farms, but also is representative of most UK farms (Thomas et al., 1962; Jackson and Clegg, 1966; Thomas, 1974b).

It has been observed that as SPCs increase, the dynamics of the microfloral population shifts. That is, usually an increase in the GNRs begin to dominate the microflora at the expense of the micrococci. In most cases, an SPC increase correlates well with unsanitary conditions existing within the milk collection and handling system in the milk house. Considerable variation in the incidence of thermoduric organisms and psychrotrophs in fresh, raw milk has been reported. Some differences may be regional or seasonal. Other differences are associated with methods of cleaning and disinfecting on farms. Some variation may be accounted for by differences in methods of performing laboratory pasteurization for the estimation of thermoduric counts. Time and temperature of incubation of media may account for other variations in both thermoduric and psychrotrophic counts (Thomas and Thomas, 1975).

2.2.2.1 The Thermoduric Microflora. The genera surviving laboratory pasteurization are shown in Table 2.5. *Microbacterium lacticum* and bacterial spores normally show 100% survival; some *Micrococcus*

TABLE 2.5. Thermoduric and Psychrotrophic Microorganisms in Raw Milk

Thermoduric Genera[a]	Psychrotrophic Genera[b]
Microbacterium	*Acinetobacter-Moraxella*
Micrococcus	*Flavobacterium*
Bacillus spores	*Enterobacter*
Clostridium spores	*Alcaligenes*
Alcaligenes	*Bacillus*
	Arthrobacter

[a]Survive heating at 63°C for 30 min.
[b]Visible growth at 5–7°C in 7–10 days.

spp. are slightly less heat resistant; and only 1–10% of strains of *Alcaligenes tolerans* can survive. Species of streptococci (e.g., *Enterobacter faecalis*), lactobacilli, and some corynebacteria are heat resistant, surviving 60°C for as long as 20 min, but only a small percentage, an estimated <1%, normally survives 63°C for 30 min. Most incidences of coliforms and *Escherichia coli* found in milk after pasteurization usually can be attributed to postpasteurization contamination in the storage and/or filling steps. However, one must go back to the early work of Olson, Macy, and Halvorson (1952) to appreciate the thermal resistance of *E. coli*. These researchers reported that the growth conditions existing in the collecting and cooling (specifically the storage temperature) of raw milk could influence the thermal resistance of this bacterium (e.g., higher growth temperatures promoted greater thermal resistance properties). Thus, the *z*-value term was introduced, and adjustments to the pasteurization temperatures above 63°C were made to assure adequate destruction of *E. coli*. In contrast, the *Bacillus* spore content of raw milk rarely exceeds 5000 ml^{-1}, and the *Bacillus* spore generally is present in higher numbers during winter than in summer. This is because these bacteria dominate the milk's microflora during winter and are derived largely from surfaces of teats that have been in contact with bedding materials and soil associated with the housing of the dairy herd (see Section 2.3.3 on the microflora of the exterior teats and udder).

Ridgeway (1955) found that 12% of farm milk supplies from winter-housed dairy herds demonstrated spore counts of >100 ml^{-1}, and the counts were much lower in the summer months. *Bacillus licheniformis* is the most common specie present while *Bacillus cereus* is found only sporadically in bulk tank milk. Milk cans are known to be a primary source of *B. cereus* spores in milk.

In contrast to spores, micrococci and *Microbacterium* spp. are derived almost exclusively from milking equipment that is sometimes so heavily contaminated with these bacteria (most likely from the presence of biofilm) that thermoduric counts in the milk often exceed $5 \times 10^4 ml^{-1}$. Most thermoduric organisms do not multiply appreciably in raw milk even at ambient temperatures, thus a high thermoduric count in milk held up to 24h could be reliable evidence of gross contamination from milking equipment. For this reason, the thermoduric or laboratory pasteurization count has been proposed in the United Kingdom. However, it is reported that there is a weak correlation between the SPC and the thermoduric count when used as an indicator of poorly cleaned milking equipment (Luck, 1972).

Panes et al. (1979) found a correlation of 0.65 between thermoduric and SPCs when the geometric means were compared for 12 monthly samples from approximately 350 individual farms. When analyzing the individual milk samples, 16.4%, 10.9%, and 2.7% demonstrated thermoduric counts of 5×10^3, 1×10^4, and $1 \times 10^5 ml^{-1}$, respectively, with an overall geometric mean thermoduric count of a low $750 ml^{-1}$. This represents about 4% of the geometric mean of the initial SPCs. From these data, there is no evidence of thermoduric counts being higher in summer than in winter. Milk delivered in cans had higher thermoduric counts than milk collected from refrigerated bulk tanks (Thomas et al., 1967), but it is suggested that this may be due to lower levels of thermoduric organisms in the milking machines of bulk tank producers, rather than due to contamination from the milk cans.

In contrast, the US dairy industry seldom uses the thermoduric count, but rather employs the preincubation of raw milk at 13°C for 16h followed by the SPC procedure to estimate psychrotrophic numbers.

Spores of *Clostridium* spp. demonstrate heat resistance and normally are detected in raw milk that has been heated to destroy bacteria in the vegetative state. As previously mentioned, isolation and detection of the *Clostridium* spp. require the use of suitable media and an anaerobic incubation environment. Clostridial spore counts are found to be highest in winter because they originate mainly from silage used for winter feeding and bedding materials (see Section 2.3.3). When numbers of spores of *Clostridium tyrobutyricum*, associated with bad silage, are present in milk in excess of 1 spore ml^{-1}, industrial experience suggests that the milk may be unsuitable for making Gruyére and Emmenthal cheese. After cows are on pasture, clostridial spore counts decline and usually are $<1 ml^{-1}$. It is reported that *Clostridium* does not multiply in raw milk (Goudkov and Sharpe, 1965).

2.2.2.2 The Psychrotrophic Microflora. Psychrotrophic microflora are those microorganisms that can thrive under refrigerated temperatures (3–7°C). An excellent review about the psychrotrophs found in milk and dairy products is given by Cousin (1982). The most commonly occurring psychrotrophs in raw milk are the GNRs (Table 2.5). *Pseudomonas* spp. accounts for at least 50% of the GNRs, with *Pseudomonas fluorescens* predominating; other species include *Pseudomonas putida*, *Pseudomonas fragi*, and *Pseudomonas Aeruginosa. Flavobacterium*, *Acinetobacter-Moraxella*, *Achromobacter*, *Alcaligenes*, *Chromobacterium*, *Aeromonas*, *Klebsiella*, and the coliform group comprise most of the remaining psychrotrophic GNRs (Witter, 1961; Juffs, 1973; Cousin, 1982). Some of these psychrotrophs, when growing in refrigerated milk, produce extracellular, heat-stable lipases that contribute to the development of a rancid flavor, as well as proteinases that degrade casein. Even among strains of one specie there can be considerable variation in activity such as *Pseudomonas fluorescens*, which in most cases produces both lipolytic and caseinolytic enzymes. The Gram-negative psychrotrophs are destroyed by pasteurization, but their enzymes remain active. As a consequence, these bacteria are of considerable importance in manufactured milk products and are most commonly associated with postpasteurization contamination (Phillips et al., 1981; Cousin, 1982).

The incidence of psychrotrophic strains of the *Bacillus* spp. containing spores in individual producer milk is low, seldom exceeding $10 ml^{-1}$. Those species found in raw milk include *B. coagulans*, *B. circulars*, *B. cereus*, *B. pumilus*, and *B. subtilis* (Witter, 1961; Juffs, 1973; Mikolajcik, 1979; Cousin, 1982; McKinnon and Pettipher, 1983). *Arthrobacter* and other Gram-positive species (e.g., streptococci) also have been reported as associated with raw milk's psychrotrophic microflora. The psychrotrophic microflora derived from teat surfaces are a poorly defined, relatively inactive group that includes microbial spores, corynebacteria, and GNRs. Inadequate cleaning and disinfecting of milking equipment continue to be the main reasons for the psychrotrophic GNRs in raw milk. On average, they comprise between 10 and 50% of the initial SPC, but with individual milk samples, the GNRs may be found to be an even higher percentage of the milk's microflora. Panes et al. (1979) found a correlation coefficient of 0.66 between the psychrotrophic counts and SPCs of bulk tank milk samples when the numerical means for individual farms were compared. In this survey, the geometric mean psychrotrophic count of $1305 ml^{-1}$ was only 7% of the geometric mean of the SPC of ~$20,000 ml^{-1}$. But, 25% of the 5000 samples examined yielded psychrotroph counts of >$5000 ml^{-1}$, and 25%

of farms had at least one psychrotrophic count of >50,000 ml^{-1} during the course of a year. Survey results for similar studies are reported by Luck (1972) and Mikolajcik (1979). Later, Griffiths et al. (1984) described a preincubation test permitting the more rapid enumeration of psychrophiles within 26 h.

2.2.2.3 Coliform Bacteria. Incidences of coliforms belonging to the genera *Enterbacter* and *E. coli* in raw milk have received considerable attention. This attention is partly because of their association with contamination from a fecal origin and the consequent risk of other pathogenic, fecal-associated microorganisms being present and partly because of the spoilage their growth in milk can produce at ambient temperatures.

It is now well recognized that the presence of coliforms in raw milk is not evidence of direct fecal contamination and should not be relied upon to detect inadequate udder cleaning prior to milking. Coliforms can rapidly build up in moist, milky residues (biofilms) on milking equipment and then become a major source for contamination of the milk being collected. However, relatively low coliform counts in milk do not necessarily indicate effective cleaning and disinfecting of equipment. Coliform counts regularly in excess of 100 ml^{-1} are considered by some sanitarians to be evidence of unsatisfactory production practices leading to environmental contamination of milk. However, a sporadic high coliform count could be related to an unrecognized coliform mastitis condition in the dairy herd. In the United States, a sediment pad test for milk samples taken from the farm bulk tank is used in lieu of the coliform count for assessing the dairy herd's teat and udder preparation prior to milking (USDHHS, 1999).

Some species of the genera making up the coliform group of bacteria are psychrotrophic (Table 2.5) and constitute 10–30% of the raw milk microflora isolated at 5–7°C. The majority of these coliforms are the *Aerobacter* spp. (Thomas and Druce, 1972).

Although coliforms are relatively ubiquitous in the environment, their presence can be useful in assessing water supplies for dairy production and milk collection activities. In the United States, it is a Public Health requirement that water supplies be tested every 3 years for the presence of coliforms (USDHHS, 1999). By using the most probable number (MPN) or membrane filter method for quantifying coliforms, an analyzed sample of 100 ml of water may not have more than 1.1 MPN or 1 cfu on a membrane filter to be considered an acceptable potable water supply.

2.3 BIOSECURITY, UDDER DISEASE, AND BACTERIAL CONTENT OF RAW MILK

Since humankind began collecting milk from domesticated lactating animals, biosecurity and udder disease/mastitis have been constant animal health concerns. For many years, researchers have studied dairy herds and individual cows to gain insights into the causative factors contributing to herd health problems. Their findings reveal three common factors: (1) the health and lactation stage of the animal, (2) the production environmental conditions, and (3) the collection practices.

With the current trend in the United States toward expanding the number of cows in dairy herds, biosecurity has become a major concern as new cows are introduced into the milking herd. To address this concern, most farm managers are using a quarantined milking unit to screen incoming cows. Issues of antibiotics, vaccinations, reproduction, udder health, and abomasal disorders are assessed in this unit before cows are moved into the main milking unit(s). Animals failing to meet the defined health requirements are removed from the herd. While there is an initial expense during the screening process, major herd health problems can be averted and protocols can be established to assure the expected productivity of the herd. But, even with healthy cows, dairy producers continue to encounter the challenges of mastitis in the herd.

It has been observed that the bacterial content of raw milk can be influenced by the presence of mastitis among dairy producing animals (Bramley et al., 1984). Usually, mastitis, or udder inflammation, is a consequence of bacterial infection and is responsible for considerable economic loss due to lower milk yields (Bramley and Dodd, 1984). Mastitis may be present in a clinical form, in which macroscopic changes to the milk or udder are readily detectable at the time of milking (Figure 2.1), but is more commonly present as a subclinical condition in which both milk and udder appear normal. Subclinical mastitis can be diagnosed only by testing milk samples, the results of which reveal the presence of pathogenic bacteria (Figure 2.2), an increased somatic cell count (SCC), or a variety of biochemical changes which are signs of udder disease (Wheelock et al., 1966; Schalm et al., 1971). These compositional changes largely reflect an increased movement of blood components into the milk during inflammation. The milk from infected quarters contains increased concentrations of anti-trypsin, bovine serum albumin, immunoglobulin, and sodium and chloride ions, whereas concentrations of lactose and potassium are reduced.

Figure 2.1. Withdrawal of foremilk from an udder quarter infected with clinical mastitis, demonstrating severe clotting of the secretion.

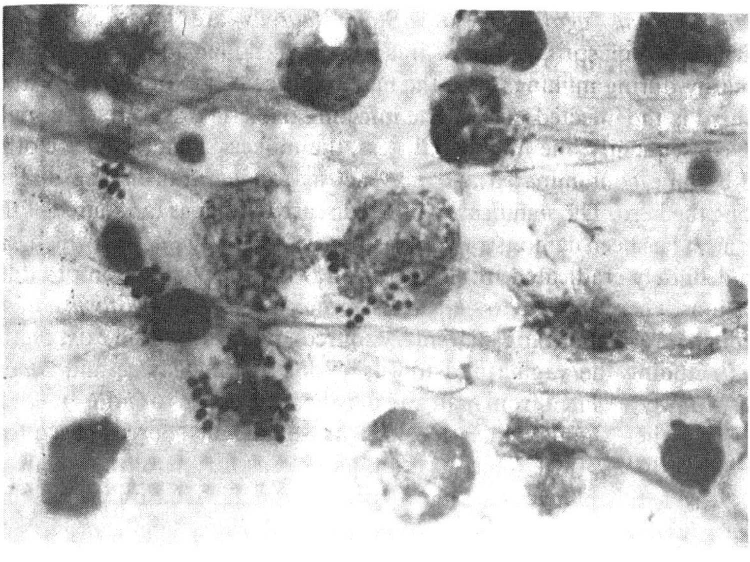

Figure 2.2. Microscopic appearance of milk from an udder quarter subclinically infected with *Staphylococcus aureus*, showing infecting cocci, fibrin, and increased leucocyte count.

The mastitis infection process involves a pathogen carrier source, means of transfer, the opportunity to invade, and a susceptible host. This process begins with teat end penetration, usually during milking, followed by microbial attachment to the tissue of the teat cistern. Microorganisms enter the udder through the duct at the teat tip. The duct varies in length from 5 mm to 14 mm, and its surface is heavily keratinised. This keratin layer retains milk residues and exhibits antibacterial activity (Collins et al., 1988). Teat end damage due to injury is known to defeat the protective barrier the teat duct provides. Penetration of the teat duct by microorganisms continues to be a subject for much research, but it has been shown that under certain conditions the milking machine can be responsible for propelling bacteria through the teat duct (Thiel et al., 1973; Bramley, 1987a). However, because mastitis can occur among beef cattle, other mechanisms such as growth of the bacteria through the duct may exist. Certain species of bacteria, most notably *Staphylococcus aureus*, readily colonize the teat duct, particularly in the region of the teat orifice. Teat orifice colonization may persist for many weeks without the bacteria penetrating to the teat sinus. Once penetration has occurred, an adaptation to the new environment takes place followed by growth and the release of toxins. This results in the conditon known as mastitis (NMC, 1978; Peterson and Quis, 1981; Schalm et al., 1971; Smith, 1977).

A survey of approximately 500 British herds in 1977 suggests that at least one-third of dairy cattle were infected (Anonymous, 1979). Clinical mastitis incidences relating directly to increased SPCs are summarized by Booth (1988) and by Jeffery and Wilson (1987). Booth's data indicate a mean value of 35 cases/100 cows/year. However, values can differ considerably between herds. Of 754 farm bulk milk samples with SPCs above 45,000/ml, 43.8% contained a microflora dominated by mastitis-associated bacteria (Jeffery and Wilson, 1987).

Similar surveys have been conducted in the United States. Findings summarized by the NMC (1978) indicate that 50% of dairy cows are "infected with pathogenic organisms in an average of 2 quarters per cow." Of these infections, 75% are nonclinical cases, 24% are mild cases, and 1% are severe cases.

Infections with *Staphylococcus aureus*, *Streptococcus agalactiae*, *Streptococcus uberis*, *Streptococcus dysgalactiae*, or *Escherichia coli* result in the greatest economic loss. In contrast, *Corynebacterium bovis* and coagulase-negative micrococci commonly are present in milk samples collected aseptically, but rarely reduce milk yield (Bramley, 1975).

Streptococcus agalactiae and *Staphylococcus aureus* are most commonly implicated in mastitis and are transferred between udder quarters and cows primarily during milking. As bacteria are secreted into the milk, the milking clusters, milkers' hands, udder cloths, and so on, become contaminated and serve as vehicles for transferring disease among the herd. The significance of the diseased udder as the source of the organism has been demonstrated very clearly for *Streptococcus agalactiae*, which can be eradicated from a herd following its elimination from infected udders. There are data to suggest that the same may be true for *Staphylococcus aureus*, although a variety of extramammary sources of this bacterium exist on parts of the cow, including the vagina and tonsils. Both *Streptococcus agalactiae* and *Staphylococcus aureus* may be transferred from humans to cattle, but only rarely do these human infections act as significant sources for intramammary disease (see Section 2.3.1 on pathogens for humans in raw milk).

Other mastitis pathogens are less dependent on the milking process for their dissemination within the herd. Most significant among these are *Streptococcus uberis*, *Escherichia coli*, and *Mycoplasma*. *Streptococcus uberis* has been isolated from the lips, teats, rumen, belly, vulva, and rectum of the cow and also may be found in large numbers in bedding (Bramley et al., 1979). Bedding and feces serve as primary sources of *E. coli* contamination (Bramley and Neave, 1975). *Mycoplasma* has become a mastitis concern, particularly with expanding dairy herds (Morin, 1998). The infected mammary gland is the primary source for this pathogen.

Many countries employ control systems to reduce the incidence of bovine mastitis. The systems can be divided into two types. In the United Kingdom, United States, Australia, and other countries, control measures are based upon the application of a disinfecting teat dip after milking and the treatment of all udder quarters with antibiotics at drying-off. Oliver et al. (1992) suggest that an intramammary infusion of cloxacillin or cephapirin 7 days before calving can reduce mastitis incidences in the early stages of lactation. Treatment with antibiotics at drying-off and 7 days before calving acts both prophylactically, to prevent new udder disease arising when the cow is not milked, and therapeutically, to eliminate existing subclinical mastitis when dry and during early lactation. Levels of udder disease have been markedly reduced with this management approach (Figure 2.3), and *Streptococcus agalactiae* has been eradicated from many herds (Wilson and Kingwill, 1975).

Proper sanitary preparation of the cow's teats prior to milking and postmilking teat disinfecting help to reduce the transfer of bacteria during milking and to destroy pathogens left on the teat skin at the end of milking. It is during this time that the teat orifice is open, and the

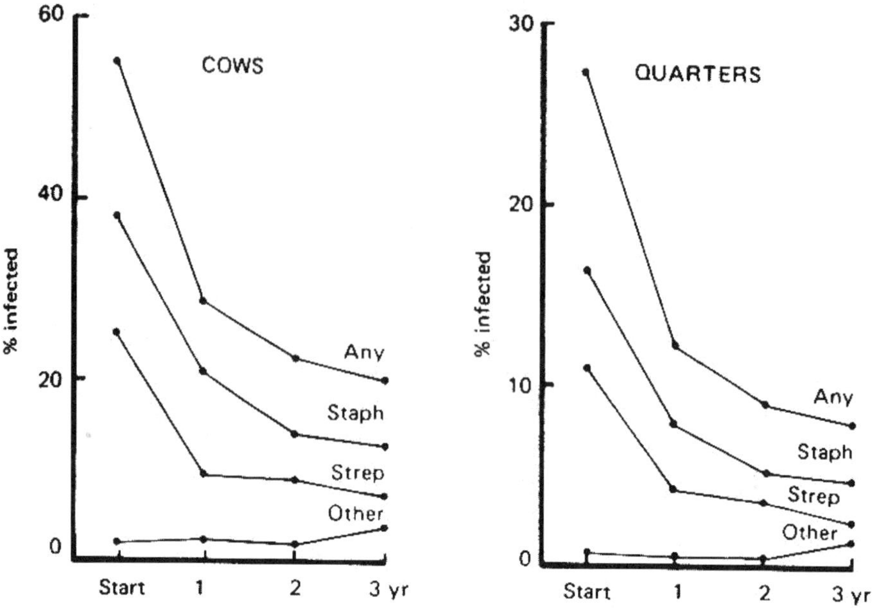

Figure 2.3. Reduction in the proportion of cows and udder quarters with subclinical mastitis in 30 herds over 3 years using a control system including postmilking teat dipping and using antibiotic therapy for cases of clinical mastitis and for all cows at the end of each lactation. [From Kingwill et al. (1979).]

flow of milk provides the link between the environment and the intramammary gland. Preparing the teats for milking includes washing the teats with a combination disinfectant–detergent, such as an iodophore, followed by drying the teats with single-use towels. Once the milk has been collected, a postmilking teat dip is applied to each teat prior to the cow leaving the milking parlor.

Which teat dip is most effective continues to be debated among dairy professionals. The NMC, at its 1994 annual meeting, formed a research subcommittee to examine peer-reviewed scientific literature concerning teat disinfectants. Its findings can be found on the internet website: www.nmconline.org/teatbibl99.htm. Essentially, seven disinfectants were evaluated, namely, chlorhexidine, iodine, linear dodecyl benzene, lauryl sulfate, quaternary ammonium compounds, sodium hypochlorite and iodophors. These disinfectants appear to be most effective against *Staphylococcus aureus* and *Streptococcus agalactiae*.

An alternative approach for reducing mastitis in the United States, Denmark, Sweden, Switzerland, The Federal Republic of Germany, and other countries employs extensive laboratory facilities to detect heavily infected herds by bacteriological or cytological examination of bulk

TABLE 2.6. The Effect of Including Milk from a Cow Infected with (a) Streptococcal and (b) Coliform Mastitis on the Bacterial Count of Bulk Herd Milk[a]

	Count ml^{-1} of Milk $\times 10^{-3}$					
	(a) Streptococcal Mastitis			(b) Coliform Mastitis		
Week Sample Taken	Total Bacteria	Streptococci	Somatic Cells	Total Bacteria	Coliforms	Somatic Cells
1	6.3	0.6	174	29	0.01	213
2	92.0	90.0	134	200	170.0	208
3	4.1	0.5	318	35	0.02	192

[a]Data from Cousins (1978).

tank milk. This approach is similar to the one employed by managers of expanding dairy herds. Problem cows are examined in detail, and infected quarters are treated with antibiotic (Olsen, 1975). Both approaches, with consistent management, effectively reduce the rates of streptococcal and staphylococcal udder disease. These approaches, however, are relatively ineffective in preventing *E. coli* mastitis, a serious problem among housed cows exposed to excessive wet conditions.

New udder infections arise most frequently in early lactation and in the period immediately following drying-off at the end of lactation. Certain pathogens, notably *E. coli* and *Corynebacterium pyogenes*, show marked seasonal trends. For example, *E. coli* mastitis is associated closely with winter housing (Jackson and Bramley, 1983). In Northern Europe, *C. pyogenes* is one of the causative organisms of "summer mastitis," a serious form of the disease prevalent among unmilked cows and heifers between July and September and possibly spread by the head fly, *Iydroteae irritans* (Bramley, 1987b). Infected quarters excrete fluctuating numbers of organisms in the milk, and generally the highest numbers of bacteria are found in foremilk (Murphy, 1943). While subclinical mastitis usually contributes <10,000 bacterial ml^{-1} of herd raw milk, the inclusion of milk from clinical cases can alter the raw milk's bacterial content significantly. Quarters infected with clinical mastitis may be excreting $>1 \times 10^7$ bacteria ml^{-1}; and this excretion, under certain circumstances, can increase the bulk milk count by 1×10^5 organisms ml^{-1} (Table 2.6). This has been found to occur most commonly with cases of streptococcal and coliform mastitis and infrequently with staphylococcal udder disease (Bramley et al., 1984).

Mastitis pathogens survive well in raw milk cooled to 10°C or less, although, with the exception of *Streptococcus uberis*, they will not grow at this temperature. Most will grow at 15°C in raw milk, but less rapidly than saprophytic contaminants. Bacteriological media commonly used for the examination of raw milk can support the growth of mastitis pathogens—with the exception of *C. bovis*, which requires the presence of Tween 80, oleic acid, or milk fat in the media, and *C. pyogenes*, which needs blood or serum and is stimulated by an atmosphere contain 10% carbon dioxide.

2.3.1 Pathogens for Man in Raw Milk

Dr. Michael Taylor is reported to be the first scientist to observe that contaminated raw milk could spread disease to humans. He observed this through epidemiological evidence relative to the spread of typhoid and scarlet fevers within families living in Penrith, England (Swithinbank and Newton, 1903). In the United States, there is similar epidemiological evidence that raw milk can be a vehicle for spreading disease (Armstrong and Parran, 1927).

Raw milk may contain microorganisms that are pathogenic to humans, and their source may lie either within or outside the udder. Historically, the most serious human diseases disseminated by the consumption of contaminated raw milk are tuberculosis and brucellosis. In both diseases, the causative organisms that may be excreted in milk from infected animals are *Mycobacterium bovis* or *M. tuberculosis* (Weir and Barbour, 1950) and *Brucella abortus*, *B. melitensis*, or *B. suis* (Smith, 1934). Often with *Brucella* infections, there is little change in the milk or udder (i.e., mastitis is not present), but, in the case of tuberculosis mastitis, a pronounced and characteristic change in the milk and udder is observed. In many parts of the world, these diseases have been eradicated from cattle and no longer pose a hazard to human health.

Pathogenic bacteria also may be present in raw milk as a direct consequence of udder disease. Among the organisms commonly producing mastitis, *Streptococcus agalactiae*, *Staphylococcus aureus*, and *Escherichia coli* are pathogens known to humans. *Streptococcus agalactiae* can initiate a variety of clinical conditions, the most serious of which are bacteremia and meningitis of newborns, which are potentially fatal to infected infants.

However, for humans the pathogenicity of bovine strains of *Streptococcus agalactiae* is uncertain and is carried by a large proportion of the human population (Sinell, 1973). While it seems likely that the consumption of contaminated raw milk may play a part in infection of the population at large, some researchers have reported higher rates of *S. agalactiae* among consumers of raw milk who do not experience symptoms of a milk-borne illness (Hahn et al., 1970).

Staphylococcal mastitis of the cow poses a more direct threat to public health because some bovine strains produce enterotoxin (Olsen et al., 1970). Consumption of food containing enterotoxin leads to a symptomatic illness, usually of approximately 24-h duration, characterized by nausea, diarrhea, and abdominal pain. The production of enterotoxin usually is associated with the multiplication of staphylococci under favorable growth conditions during storage of the milk. Because enterotoxin is relatively heat stable (Read and Bradshaw, 1966), subsequent pasteurization of the contaminated milk will not make it safe for consumption.

High numbers of *E. coli* may be present in milk as a consequence of mastitis, and this bacterium is responsible for several diseases of man of varying severity. While direct links between *E. coli* udder infection and human disease have not been reported, a wide range of *E. coli* serotypes have been isolated from bovine milk, and it is probable that some of these are pathogenic for humans. Interestingly, microorganisms that have a greater pathogenicity for humans seldom produce bovine mastitis and yet may be present in raw milk. They include *Leptospira* spp., *Listeria monocytogenes*, *Bacillus cereus*, *Pasteurella multocida*, *Clostridium perfringens*, *Nocardia* spp., *Cryptococcus neoformans*, and *Actinomyces* spp. There has been recent concern about the presence of *L. monocytogenes* in milk products, particularly in soft cheeses. While listeria mastitis undoubtedly can occur, the incidence is reportedly low. Cases can be acute, chronic, or subclinical (Gitter et al., 1980). The veterinary aspects of listeriosis have been reviewed by Gitter (1989). Additionally, *Coxiella burnetii*, the causative agent of Q fever, may infect the udder, probably by the hematogenous route and consumption of, or contact with, infected milk leading to human infection.

Further biohazards stem from the adventitious contamination of raw milk by pathogenic bacteria from sources external to the udder. Salmonellae and thermoduric *Campylobacter* strains fall into this category and have produced many outbreaks of enteritis (Robinson et al., 1979; Taylor et al., 1979; Werner et al., 1979). The majority of these strains originate either directly or indirectly from fecal contamination of the milk. However, it has been demonstrated experimentally that the udder may be infected with *Salmonella typhi* (Scott and Minett, 1947) or *Campylobacter coli/jejuni* (Lander and Gill, 1979). Human carriers also may be sources of infection in milk-borne outbreaks. This has been reported for *Salmonella* infections and for cases of scarlet fever or septic sore throat attributed to *Streptococcus pyogenes* (Bryan, 1969).

All of these pathogens are destroyed by pasteurization, except *Clostridium perfringens* and *Bacillus cereus*, which can survive the pasteurization process because of their ability to sporulate. It is improbable, however, that *C. perfringens* will germinate and multiply under the modern-day conditions of milk storage. Most *B. cereus* food-borne illnesses are not related to the drinking of contaminated pasteurized milk because the growth of *B. cereus* will lead to the development of "off-flavors" and an undesirable appearance (Davies and Wilkinson, 1973). According to Davies and Wilkinson, the reported outbreaks of *B. cereus* food-borne illnesses in the United Kingdom were found to be associated with the consumption of cooked rice, rather than milk.

2.3.2 Antimicrobial Systems in Milk

Several antimicrobial systems are detectable in milk operating either to protect the mammary gland from infection or to confer disease resistance to the suckling young. For a more detailed review, readers are referred to Craven and Williams (1985).

2.3.2.1 Immunoglobulin (Ig). Antibodies to potentially pathogenic bacteria often are present in milk. They may be produced locally within the udder or transferred to milk (IgG) from blood circulation. The primary function of these antibodies is to protect the newborn by passive immune transfer, but the complement/antibody system operates within the udder to protect it from infection by strains of coliform bacteria that are susceptible to complement/antibody killing (Carroll et al., 1973). Antibodies also may serve to reduce the severity of udder disease—for example by neutralizing toxins released during the disease process or by acting as opsonins to facilitate the phagocytosis of bacteria by polymorphonuclear (PMN) leucocytes. Antibiotics also may serve to prevent adhesion of bacteria to mucosal surfaces.

2.3.2.2 Phagocytosis. It is generally accepted that protection of the udder from mastitis depends primarily on phagocytosis and the killing of invading bacteria by the PMN cells. The SCC of milk from uninfected udders ranges from approximately 10 to 50×10^5 cells ml^{-1}, of which approximately 10% are PMN. Infected quarters may excrete milk containing 10×10^7 cells ml^{-1}, of which 90% are PMN. The increase in the PMN cell content of milk associated with mastitis allows the assessment of herd levels of udder disease by measurement of the PMN cell expressed as SCC of the herd bulk milk (Westgarth, 1975).

Phagocytosis and killing by the PMN cell is less effective in milk than in blood, largely because the PMN cells ingest large quantities of fat and casein (Russell et al., 1977) and because of the presence of *Staphylococcus aureus*-derived protein A (Pankey et al., 1985). Protein A apparently blocks the attachment of the opsonins to the cell wall of the bacterium that is essential to the engulfment process. As a consequence of this reduced phagocytic activity, the udder is relatively easy to infect even by small numbers of invading bacteria. Increasing the PMN cell content of milk has been shown to increase resistance of the udder to infection (Schalm et al., 1964). Depleting the cow of PMN leucocytes by the use of equine antibovine PMN serum results in chronic subclinical staphylococcal mastitis being converted to a peracute gangrenous form of the disease (Schalm et al., 1976).

One strategy being employed to improve phagocytosis of the PMN cell against *Staphylococcus aureus* is to vaccinate the dairy cow with protein A isolated from this bacterium. The protein A stimulates antibody production against this protein with the phagocytosis process enhanced through bacterial cell surface exposure to the opsonins and ultimate PMN cell engulfment of the bacterium (Pankey et al., 1985; Nickerson, 1999).

2.3.2.3 Nonspecific Defense Mechanisms. Several other antibacterial systems that have a broad spectrum of activity also occur in milk and are discussed below.

2.3.2.3.1 Lactoferrin (LF). LF is an iron-binding protein similar to serum transferring. Its presence and concentration is markedly increased in the milk secretion obtained from unmilked or infected animals. LF inhibits the multiplication of bacteria by depriving them of iron and may protect the dry udder from infection with *E. coli* (Reiter and Bramley, 1975). Although LF is present in bovine milk, the high citrate and low bicarbonate concentrations have been shown to reduce markedly the iron-binding and, therefore, the inhibitory properties of LF.

2.3.2.3.2 Lactoperoxidase/Thiocyanate/Hydrogen Peroxide System (LP System). Lactoperoxidase is synthesized within the mammary gland and is present in high concentrations in bovine milk. Thiocyanate is present in varying concentrations related primarily to the nutrition of the cow. The third component of the system, hydrogen peroxide, may be supplied by hydrogen peroxide-producing organisms within the udder (e.g., streptococci) or by the PMN cell. The LP system will inhibit only temporarily the growth of some organisms (e.g., group B and N streptococci) while exhibiting bacteriocidal activity to others (e.g., group A streptococci, *E. coli*, and *Salmonella typhimurium*). Recent evidence suggests the LP system may play a role in the defense of the mammary gland, in addition to the protection of the calf from enteritis (Marshall et al., 1986). It has been suggested that the LP system might be utilized as a "cold sterilization" process to render milk safe for consumption without damaging, by heat treatment, the various antimicrobial systems present (Reiter, 1978).

Other antimicrobial systems have been demonstrated in milk including lysozyme (Vakil et al., 1969) and vitamin binders for B_{12} and folate (Ford, 1974). The significance of these systems for the protection of the mammary gland or the neonate remains unknown.

2.3.3 The Microflora of the Exterior Udder and Teats

Between milkings, teats often become soiled with dung, mud, and bedding materials such as straw, sawdust, wood shavings, or sand. If not removed before milking, this dirt, together with the large number of microorganisms associated with it, is washed into the milk. Numbers and types of microorganisms vary according to the type and amount of soil on the teats. Milk from cows with teats that are heavily soiled with dung may have an SPC approaching 10×10^5 cfu ml^{-1}.

2.3.3.1 Numbers of Microorganisms from Teats. Bedding materials on which cows are housed during the winter may have very high bacterial counts (10^8–10^{10} cfu g^{-1}) even though the bedding may appear to be relatively clean and dry. The incidence of the main groups of microorganisms in bedding is shown in Table 2.7 (Cousins, 1978). Particles of grossly contaminated bedding materials can adhere, sometimes unobtrusively, to teat surfaces. Teats of cows kept in strewed yards

TABLE 2.7. Incidence of Different Groups of Bacteria in Wood Shavings, Straw, and Sand Bedding

	Geometric Means[a] (cfu g^{-1})			
Bedding	Total	Psychrotrophs	Coliforms	*Bacillus* Spores
Shavings	1.2×10^{10}	1.1×10^9	8.3×10^5	5.0×10^6
Straw	7.4×10^8	9.8×10^7	1.8×10^5	1.5×10^5
Sand	5.4×10^9	1.4×10^9	3.9×10^5	5.0×10^6

[a]Six samples of each type of bedding.

can become heavily and visibly soiled if the straw is not adequately and frequently replenished. Bacterial contamination of the teats is correspondingly high, unless teats are washed thoroughly.

Table 2.8 presents comparative microbial counts and types observed on the teat apex, unwashed and washed, after exposure to sand bedding and pasture conditions. During the summer when cows normally are turned out to pasture, a marked decline in the level of contamination on teats occurs. This seasonal effect is reflected especially on farms where milking equipment is effectively cleaned and disinfected, thus reducing bacterial contamination. The effectiveness of hose washing teats to reduce microbial numbers using a solution of hypochlorite (~600 ppm available chlorine), followed by drying off the teats with paper towels, is presented. Note that even after washing, total microbial counts of 10^6 cfu per teat are recovered. Reductions for other groups of microorganisms are not very effective, and a reduction of coliform numbers is not affected by washing. These numbers are all low, averaging only about 10 cfu per teat (Cousins, 1978). Teat washing using an iodophor solution also is found to be ineffective in reducing the bacterial population of the teat apex, unless that procedure is done carefully and is followed by thorough drying (Zarkower and Scheuchenzuber, 1977). Differences in effectiveness between various types of disinfectant solutions used for teat washing and teat dips have been reviewed by the NMC's Research subcommittee (refer to Section 2.3 on premilking and postmilking teat disinfectants).

Bacterial counts of bulk tank milk from a herd at the National Institute for Research in Dairying (NIRD) demonstrated that even where teats were regularly washed in both winter and summer, winter

TABLE 2.8. Effect of Washing and Drying of Teats on Bacterial Counts of Teats' Apex Via Swab Samples Taken from Cows Bedded on Sand or on Pasture

			Geometric mean[a] (cfu per teat apex)			
Conditions	Herd	Teats	Total	Psychrotrophs	Coliforms	Spores
Bedded on sand	A	Unwashed	8.4×10^6	1.2×10^6	10	5.0×10^4
		Washed	7.3×10^5*	8.3×10^4	12	1.2×10^4
	B	Unwashed	3.3×10^7	1.3×10^6	15	1.0×10^5
		Washed	8.5×10^6*	4.0×10^5	11	4.8×10^4
On pasture	A	Unwashed	7.5×10^4	1.2×10^4	1	1.3×10^2
		Washed	3.1×10^4	2.5×10^3	9	1.1×10^2
	B	Unwashed	1.2×10^5	4.3×10^3	14	4.9×10^2
		Washed	1.4×10^5	3.3×10^3	11	5.8×10^2

[a]Mean of six samples; *$P < 0.05$.

housing increased the numbers of cfu per ml of the milk (Table 2.9). The numbers of bacteria remaining on teat surfaces after washing can be high, as demonstrated by swabs or rinses of teats (Thomas and Druce, 1971; McKinnon and Pettipher, 1983).

The direct effects of teat washing on SPCs should be studied only in milk using cows free from udder infections (i.e., giving milk containing <10 cfu ml^{-1} in aseptically drawn samples) and milked with properly sanitized milking equipment or in-line milk samplers (McKinnon et al., 1973). Where these precautions are taken, average SPCs ranged from 7000 cfu ml^{-1} in milk drawn from teats unwashed or washed with water and left wet to 1500 cfu ml^{-1} in milk from teats hose-washed rapidly with hypochlorite solution and dried with paper towels (McKinnon et al., 1973, 1988). These average differences conceal wide variations for bacterial numbers in milk from individual cows (Table 2.10). This is because some teats, before washing, occasionally are clean even when

TABLE 2.9. Influence of Housing and Teat Washing on Bacterial Content of Bulk Tank Milk from a Single Herd, Sampled Once Weekly

		Geometric Mean[a] (cfu ml^{-1} of Milk)				
Conditions	Teats	Total	Psychrotrophs	Coliforms	Thermoduric organisms	*Bacillus* spores
Bedded on sand	Unwashed	31,700	1,500	43	120	18
	Washed	15,500	990	61	110	14
On pasture	Unwashed	4,250	280	19	990	7
	Washed	3,530	270	26	750	5

[a]Each result is the mean of 8–9 milk samples.

TABLE 2.10. Contamination of Milk with Bacteria from the Surfaces of Cows' Teats, Either Unwashed or After Hose-Washing

	Geometric Mean Counts of 30 Samples (cfu ml^{-1} of Milk)		
Treatment of Teats	Total Count ($\times 10^3$)	Spores	Coliforms
Unwashed	7.5 (0.5–75.6)	34 (4–555)	2.0 (0–20)
Washed with water, left wet	7.9 (0.6–111.0)	31 (3–590)	1.3 (0–10)
Washed with water, dried	4.2 (0.1–54.0)	16 (1–137)	0.5 (0–4)
Washed with NaOCl, left wet	4.1 (0.4–64.2)	38 (6–180)	0.7 (0–4)
Washed with NaOCl, dried	1.5[b] (0.1–22.0)	14 (2–112)	0.03 (0–1)

[a]Ranges of counts are shown in parentheses.
[b]Significantly different, $0.01 < P < 0.05$.

housed on very dirty bedding. Conversely, under clean housing conditions, individual teats are, on occasion, heavily soiled with barnyard debris. Milk collected from washed teats may show variable counts, possibly because it is difficult to ensure that teat ends are adequately cleaned in the limited time (15 s) available for washing. The difficulty of controlling both environmental conditions and effectiveness of teat washing could explain why, in various surveys, ratings of "efficiency" obtained by inspection of teat washing and udder and teat cleanliness generally fail to show a clear, direct correlation with bacterial counts in the milk (Luck, 1972; Panes et al., 1979).

One survey taken on eight commercial farms over 1 year examined the microbial content of milk using in-line sampling techniques and documented the effect of hose-washing of teats with water containing disinfectant, followed by drying. This survey demonstrated the benefit of efficient teat preparation by reducing the total count per ml of the milk by ~50% during winter housing. During the summer when cows normally were on pasture, there was no difference in the SPC of the milk from washed and unwashed cows. The use of a disinfectant-impregnated cloth for drying teats has been shown to be an effective method of udder preparation (McKinnon et al., 1985), reducing the contamination from the teat surfaces and preventing the spread of mastitis pathogens between teats. However, this practice is discouraged by the dairy-related regulatory enforcement agencies in the United States.

2.3.3.2 Types of Microorganisms from Teat Surfaces. Generally, cows housed during the winter months demonstrate SPCs ranging from 10^5 to 10^7 cfu per teat. Micrococci, including coagulase-negative staphylococci, are among the predominant groups present (~10^4 cfu per teat). Streptococci, mainly fecal types, also are numerous, but GNRs, including coliforms, are much less numerous. Coliform counts rarely exceed 10^2 cfu per teat (Thomas et al., 1971; Cousins, 1978). It appears that these organisms, unlike micrococci, for example, do not survive well on teat skin, although they do form a large proportion of the microflora of bedding materials. Psychrotrophs (detected by incubation of plates at 5°C for 10 days) range from 10^3 cfu per teat for washed teats of cows at pasture to 10^6 cfu per teat for unwashed teats of cows bedded on sand (Table 2.8).

The psychrotrophic microflora of teat surfaces is a poorly defined group of organisms, consisting of coryneforms and GNRs, most of which are inactive in litmus milk at 22°C and do not appear to multiply readily in raw milk (Johns, 1962, 1971).

The aerobic thermoduric organisms on teat surfaces are almost entirely *Bacillus* spores. Spore counts can range from 10^2 to 10^5 per teat, depending on the environmental soil conditions. The predominant species derived from soil are *B. Licheniformis*, *B. subtilis*, and *B. pumilis*; *B. cereus* and *B. firmus*; with *B. circulans* occurring less frequently (Underwood et al., 1974). The data in Table 2.10 show that aerobic spore counts in milk obtained from individual cows can range from 1 to 590 ml^{-1}. However, spore counts can be as high as 3000 ml^{-1} in milk from commercial farms as reported by Underwood et al. (1974). Although the total spore count of milk is markedly lower during the summer months than during winter, the psychrotrophic spore count essentially remains the same with the thermoduric spore-forming bacteria decreasing in numbers within the total spore population. The psychrotrophic spore count in summer months is derived primarily from soil contaminating the teat surface (McKinnon and Pettipher, 1983).

It can be reasoned that teat surfaces also are a source of clostridial spores in milk. These spores have been detected in fodder, bedding, and feces and decline markedly in numbers when cows are on pasture. Spores of lactate-fermenting clostridia (*Clostridium tyrobutyricum*) may cause a textural defect known as "late blowholes" in Dutch Emmenthal and Gruyére cheese. This bacterium is associated with "bad" silage and may be transmitted via feces of silage-fed cows to teats and ultimately to the milk, unless the fecal material is washed from the teats (Bergere, 1979).

A wide variety of genera and species of microorganisms in the cow's environment may be present on teat surfaces, but those representing a small proportion of the microflora will not be detected. They will be transmitted, however, to the milk and finally to the milking equipment. Some may become established if conditions are suitable; others may be only transient contaminants. Normally, only genera and species capable of forming colonies on plate count agar incubated aerobically (e.g., SPC) are detected. Specific cultural conditions are required to encourage outgrowth, for example, of lactobacilli, clostridia, and other fastidious microorganisms from dung, soil, herbage, and water. Fresh, raw milk may be inhibitory or may even kill some of the microorganisms existing on the teat surfaces, so that even if present in appreciable numbers, they may not be detected. A close correlation cannot be expected between the population dynamics found on teat surfaces and those present in milk.

A common practice in dairy production is to filter milk through a cotton cloth medium. This filtering process removes only large debris but does not remove fine particles such as bacteria and somatic cells. Therefore, bacteria introduced into the milk can pass through the filter. Hence, bacterial contamination of filtered milk does not correlate well with lower sediment pad grading scores. The sediment pad scores will monitor only the conditions of cow preparation at the time of milking. In order to minimize bacterial contamination from teat surfaces, it is essential to prevent teats from becoming heavily soiled between milkings, to wash the teats with water containing disinfectant, and to dry thoroughly the teats prior to attaching the milking units.

2.4 ENVIRONMENTAL SOURCES

While the lactating animal, the production environment, and the milk handling equipment remain the principal sources for microbial contamination of raw milk, there are other environmental sources that should be considered. These are air, the milk handler, and the water supply.

2.4.1 Air

Air is not considered a significant source for microbial contamination in raw milk. Through its movement, air transfers soil and dust particles from a microbial-laden source into any exposed milk surface, such as (a) small quantities of soil falling into a milking bucket during hand milking or (b) microbes entering into milk via air entering the milking machine during use. Bacterial counts of air in cowsheds or parlors seldom exceed 200 cfu liter^{-1} and usually are much less. Micrococci account for >50% of aerial microflora. Coryneforms, *Bacillus* spores, and small quantities of streptococci and GNRs also may be present and account for the remaining microflora. Calculations indicate that airborne bacteria account for <5 cfu ml^{-1} of milk produced and that *Bacillus* spores constitute <1 cfu ml^{-1} (Benham and Egdell, 1970; Underwood et al., 1974).

2.4.2 The Milk Handler

When cows are hand milked, it is possible that the milk handler can contribute to an increased microbial load in the raw milk by dislodging dirt particles from the udder, increasing aerial contamination through accelerated air movement and contacting the milk with infected hands. Risks of contamination from the milk handler are much less with machine milking, but, because of the possibility of introducing pathogens into the milk by the handler, many countries prohibit certain known human pathogen carriers from taking part in the collection of milk.

2.4.3 The Water Supply

Water used in the milk production process should be of potable quality. This means that the water supply must be from an approved source, free of pathogens and fecal contamination. In many cases, farms obtain water from untreated water supplies (boreholes, wells, lakes, springs, and rivers) that may be contaminated with microorganisms from a fecal origin. Examples of common fecal bacteria found in water are coliforms, fecal streptococci, and clostridia. In addition, a wide variety of saprophytic microorganisms originating from soil or vegetation may be present, including the *Pseudomonas* spp., coliforms and other GNRs, *Bacillus* spores, coryneform bacteria, and lactic acid bacteria. The amount of these contaminants found in water will vary widely. In the United States, a potable water source must be demonstrated through microbial testing to be free of fecal associated coliforms. Also, the elimination of potential cross connections between the potable water supply and a contaminated water source, such as a watering trough, is required by the PMO guidelines (USDHHS, 1999).

Additionally, one must recognize that a potable water supply can become contaminated within the dairy production environment, such as in a farm storage tank that is not properly protected from rodents, birds, insects, and dust. Bacteria also may be introduced into the water supply through dirty wash troughs, buckets, and hoses. If untreated water gains access into milk or is used for rinsing equipment and containers, the microbes present in the water eventually will contaminate the milk.

2.5 THE MICROFLORA OF MILKING EQUIPMENT AND ITS EFFECTS ON RAW MILK

When properly cleaned, disinfected, and drained, milking equipment can be eliminated as a source for microorganisms in raw milk. However, inadequately cleaned and disinfected (sanitized) milk contact surfaces, including milk cans and bulk tanks, are considered to be the major sources of bacteria found in milk after its collection from the udder. Other contributing factors are failure to replace worn or damaged equipment parts and inattention to the cleanliness of valve assemblies.

The simplest form of milking equipment is the bucket used during hand milking, with both the bucket and the milk handler becoming potential microbial sources. Machine milking requires one or more teat cup clusters through which the milk flows into a receptacle or a pipeline, possibly passing through a recorder jar (Figure 2.4). In the case of milk being collected in a bucket, the milk is pooled into either milk cans or a bulk tank. When a pipeline is used, the pipeline transports the milk to a receiver tank from which it is released or pumped into milk cans or a bulk tank. Ancillary equipment may include (a) a strainer or in-line filter, (b) a cooler that may be open surface, in-can, or a plate heat exchanger, and (c) milk flow indicators and meters (Akam, 1979).

Cows normally are milked twice daily, and the milking machine must be cleaned after each milking. Because of the complexity of the milking machine and some of its components, milk residues and associated bacteria may not be completely removed from the equipment and tend to accumulate over time. Except in very cold weather, bacteria remaining in the equipment can multiply between each milking, and their numbers can increase faster than can be visually observed (e.g., milk residues). Thus, visual inspection of the equipment provides little assurance that the equipment has been adequately cleaned and sanitized.

In practice, the contribution of contaminated milking equipment to the microflora of the milk cannot be assessed definitively by bacterial counts on the milk produced because of the variability in numbers and types derived from cows' udders. The most widely used method for determining the extent of bacterial contamination on milk contact surfaces is by using a sterilized liquid rinse and then subjecting this rinse to an SPC in the same way that milk is tested. The types of bacteria in the rinse then can be determined through assessment of the colonies isolated and their ultimate identification. The exact relationship between numbers of bacteria recovered from rinses and available to milk during milking is not known. However, from results of repetitive rinsing or from other assessment methods (Cousins, 1963, 1972), the proportion recovered by rinsing is known to be at least 10% of the number available to the milk. The proportion will be higher where bacteria have multiplied on moist surfaces as biofilms or in milk residual water that has collected in poorly drained milking machines. This is because the bacteria present in the biofilm/residual are released readily from the surfaces.

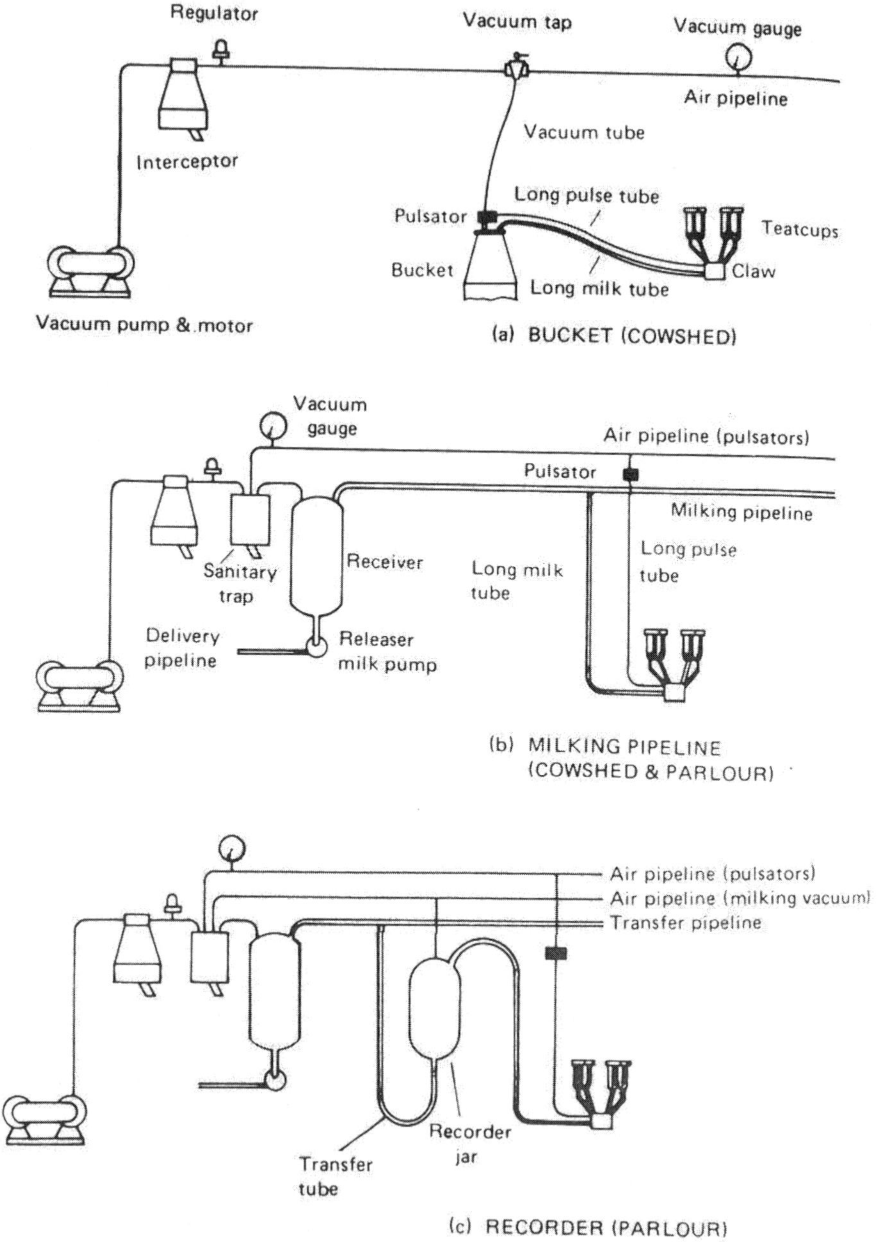

Figure 2.4. Diagrams of principal types of milking machines used in the United Kingdom. [From Akam (1979).]

A method that appears to eliminate the problem of the relative efficiency between microbial removed via rinses versus microbial removal via milk is the in-line sampling technique (McKinnon et al., 1988). Using this method, a direct assessment of bacterial contamination of the milk from the milking equipment and bulk tank can be made.

It has been pointed out that milking equipment must be contaminated heavily to markedly increase the bacterial count per milliliter of the milk passing through it. For example, to increase the bacterial count of 1000 liters (~264 gal) of milk by 1 bacterium ml^{-1}, it requires one million bacteria; thus, to increase the count by 10,000 ml^{-1}, it requires 10 billion bacteria. A milking installation consisting of four or five milking stalls and jars, together with pipelines and a bulk tank, has a milk contact surface area of approximately 10 m^2 (~107 ft^2) and, therefore, would need to contribute 1 billion bacteria m^{-2} (or 100 million ft^{-2}), on average, to its surface area. Clearly, where milking equipment is solely responsible for high raw milk counts (e.g., >50,000 cfu ml^{-1}), the cleaning and sanitizing protocol must be seriously defective. However, milking equipment seldom is contaminated uniformly. Usually the bacteria and milk residues accumulate as biofilms in difficult-to-clean areas and in parts of badly designed components. Figure 2.5 illustrates examples of these problematic areas: crevices, joints, dead ends, and fittings that cannot be cleaned-in-place effectively and need to be dismantled at regular intervals. In contrast, the smooth surfaces of jars and pipelines are cleaned readily by circulation of chemical solutions through the machines and pipelines.

2.5.1 Numbers of Microorganisms on Milking Equipment Surfaces

In the United Kingdom, bacteriological rinses and sample swabs of milking equipment have been used for many years for advisory, investigational, and survey purposes. The methods used for assessment are well documented (Cousins, 1963; McKinnon and Cousins, 1969; BSI, 1975). In the United States, the Standard Methods for the Examination of Milk and Dairy Products (APHA, 1992) and the Pasteurized Milk Ordinance (PMO) (USDHHS, 1999) serve as documented sources.

Results of equipment rinses taken during the course of advisory work in Wales (Thomas et al., 1966) show that relatively high counts are present in some milking equipment. About 20% of the milking machine unit clusters had rinse counts of $>1 \times 10^9$ cfu per cluster, and 33% of the deposits of residues scraped from milk tubes contained

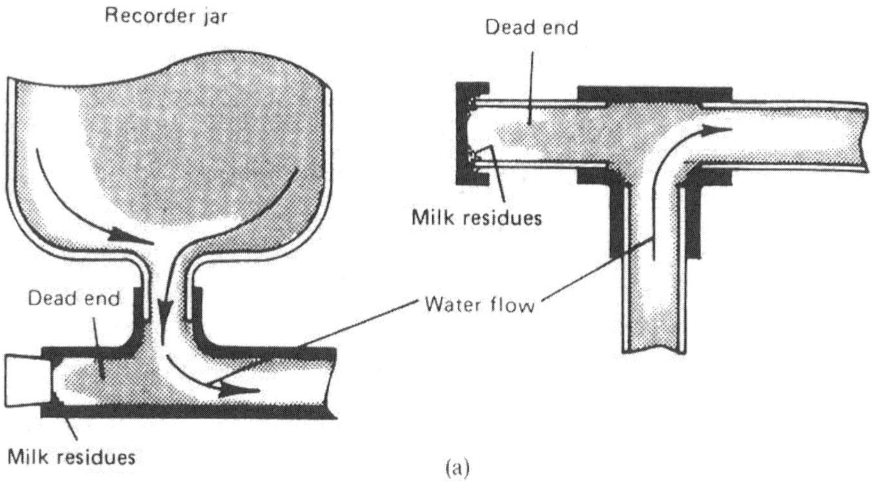
Recorder jar
Dead end
Milk residues
Dead end
Water flow
Milk residues

(a)

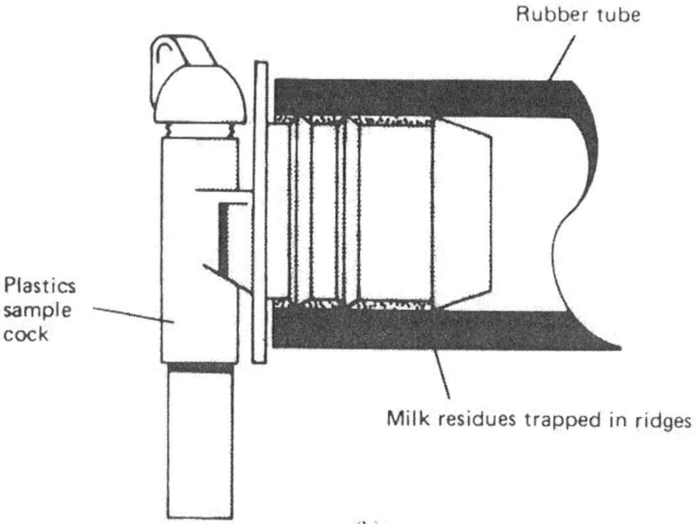
Rubber tube
Plastics
sample
cock
Milk residues trapped in ridges

(b)

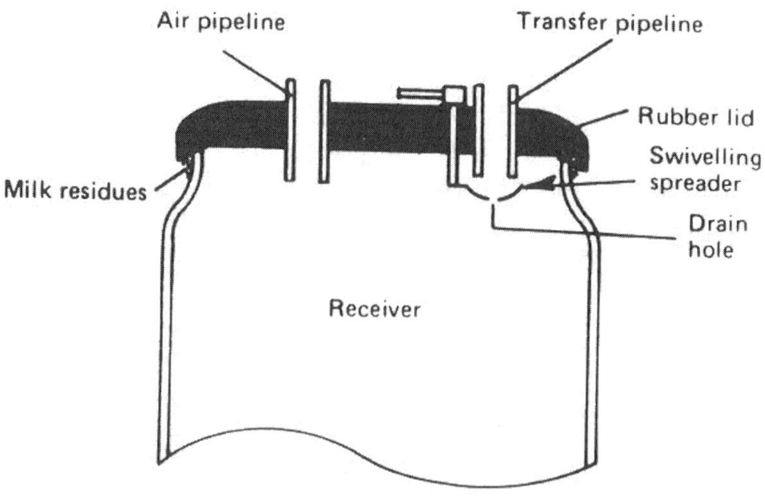
Air pipeline
Transfer pipeline
Rubber lid
Milk residues
Swivelling
spreader
Drain
hole
Receiver

(c)

$>1 \times 10^9$ cfu g^{-1}. Many of these high counts were obtained from equipment on farms that had reported high SPCs.

Other surveys show that good results are attainable where recommended protocols for cleaning and sanitizing are followed; they also show that high rinse counts are, in the main, associated with protocol deviations, namely, using defective rubber parts, using equipment in poor physical condition, ignoring the recommended temperature and chemical concentrations for the circulating solutions, and allowing the buildup of milk residues on milk contact surfaces.

Milking with bucket machines still is used for small herds (~10% of producers in the United Kingdom), but bulk milk collection in farm refrigerated storage tanks is universal. Milking by means of pipeline machines has increased rapidly since the 1950s, and this method accounts for most of the milk produced.

2.5.1.1 Pipeline Milking Machines. The use of steam sterilization, while effective in reducing the bacterial content in milking equipment, now is obsolete in the United Kingdom because it is costly and time-consuming.

Steam sterilization has been replaced by the cleaning-in-place (CIP) method. The CIP protocol offers a reasonably effective system via the circulation of caustic/acidic and sanitizing solutions to clean the milking equipment. When compared to the steam sterilization method, the CIP method is found to be less effective in reducing the bacterial numbers. Compared to rinse counts of $<5 \times 10^4/ft^{-2}$ ($\sim 5 \times 10^5 m^{-2}$) from steam-sterilized equipment, CIP-cleaned equipment has SPCs that are consistently higher. Perhaps an explanation is that optimal CIP process parameters are not met. Parker et al. (1953) and Kaufmann et al. (1960) defined the engineering aspects of cleaning pipelines in-place and examined the milk soil removal properties of various stainless steel surfaces. Four interfacing factors (temperature, time, chemical concentration, and physical force applied) have a direct influence on the efficacy of the CIP process.

Surveys during the 1960s show that a beneficial effect was observed through lower bacteria counts in the milking equipment when hot

Figure 2.5. (a) Milk residues in components forming dead ends at the base of a recorder jar (*left*) and at the end of a transfer pipeline (*right*). (b) Section of a plastic sample cock at the base of a recorder jar showing site of milk residues. [From Cousins and McKinnon (1979).] (c) Section of a receiver and lid showing sites of milk residues. [From Cousins and McKinnon (1979).]

detergent solutions (initial temperature range of 70–80°C) were applied to the CIP process. This temperature range facilitates higher soil removal and imparts some lethality to the reduction of the microfloral population. Today, public health authorities recommend cleaning temperatures above 70°C for the CIP protocol. With the detergent solution temperature dropping usually by 25–35°C during the CIP process, there is an expected corresponding decrease in cleaning efficiency and lethality. Normally, a sanitizer is circulated through the milking equipment following the cleaning and rinse cycles.

Another in-place cleaning system used in the United Kingdom is the acidified boiling water (ABW) process. This process relies primarily on acidic conditions and heat for sanitizing the milking equipment (Cousins and McKinnon, 1979). Both the CIP and ABW methods, when correctly applied, are capable of effectively cleaning and disinfecting milking machines while contributing relatively few bacteria to the milk. However, survey data indicate that some milking machines become contaminated heavily with bacteria, most likely because of faulty equipment design, incorrect installation of components, improper pump timing adjustments leading to an inadequate flow of cleaning and sanitizer solutions, or the use of solutions that lack the proper temperature and/or chemical concentration. Thus, numbers of microbes recovered by rinsing these machines can range from $<5 \times 10^5$ to $>1 \times 10^9\,cfu^{-2}m^{-2}$.

During the survey reported by Panes et al. (1979), rinses were taken of the milking equipment (pipeline, milking machines, and refrigerated bulk tanks) at approximately 350 farms, once in winter and once in summer. Table 2.11 presents the compiled data derived from this survey and shows the extent of bacterial contamination in the milking equipment as affected by the two seasons represented. For milking machines, ABW demonstrated better results than CIP, even though 4.5% of equipment-rinse SPCs for ABW-cleaned machines were $>1 \times 10^9\,cfu\,m^{-2}$. Twenty-five percent of the machines demonstrated considerably higher rinse SPCs of $>1 \times 10^8\,cfu\,m^{-2}$, with another 32.7% of corresponding milk samples showing SPCs of $<1 \times 10^4\,ml^{-1}$ and an additional 42.5% having counts between 1×10^4 and $5 \times 10^4\,ml^{-1}$.

In an investigation initiated by Runnels (1988) to assess the influence of a pre-rinse step cycle on the efficiency of CIP cleaning at five farms, it was found that acceptable standards for milking equipment cleanliness could be achieved consistently with the geometric mean rinse SPCs ranging from 8.9×10^5 to $12 \times 10^6\,cfu\,m^{-2}$.

In Panes et al.'s survey (1979), about half of the milking machines were cleaned adequately when hot cleaning solutions were applied

TABLE 2.11. Bacterial Contamination in Pipeline Milking Machines, in Bulk Milk Tanks, and on Bulk Tank Outlet Plugs[a]

Equipment	Number of Rinses	Percentage of Frequency Distribution[b] ($cfu\,m^{-2}$)				
		$>1 \times 10^5$	$>1 \times 10^6$	$>1 \times 10^7$	$>1 \times 10^8$	$>1 \times 10^9$
Milking machines	702	98.3	91.0	66.7	26.1	7.4
Tanks, cleaned by						
Hand (brush)	284	77.7	56.2	26.5	4.9	1.1
Hand spray	277	80.5	50.2	23.1	4.7	0.7
Automatic spray	194	56.0	33.7	14.0	2.6	0.0
Tank outlet plugs[c]	755	82.5	67.9	46.7	23.8	8.9

[a]Data from Panes et al. (1979).
[b]$1 \times 10^6\,cfu\,m^{-1} = 100\,cfu\,cm^{-2} \simeq 1 \times 10^5\,cfu\,ft^{-2}$.
[c]Results expressed as cfu per plug.

twice daily in the CIP process. The remaining equipment used the hot cleaning step only once daily followed by a cold cleaning solution step during the evening milking. There was no clear evidence that omitting one of the hot treatments had any detrimental effect on the equipment-rinse SPCs. Panes et al. concluded that it is possible that one daily heat treatment could provide an effective prevention of any buildup of detectable milk residues and bacterial numbers, when compared to an exclusive use of cold cleaning.

In Ireland, circulation cleaning of large pipeline machines using a cold caustic-based detergent without a conventional disinfectant is reported to maintain the milk contact surfaces in a clean condition for at least a month. In this case, the bacteriological results compared favorably with those of conventional circulation cleaning, while demonstrating a low incidence of proteolytic GNRs in the microflora (via the rinse SPC) and offering a substantial cost savings (Palmer, 1977). Palmer suggests that the effectiveness of the cold process protocol may be due to the caustic remaining in the system after the cleaning solution has been drained from the machine. No water rinse is used immediately after the cleaning step in the protocol. With the extended contact time, the caustic solution penetrates into joints and crevices, and microbes do not grow. Perhaps the influence of an alkaline environment with a $pH > 10$ contribute to this observed effect. Rinsing the machine with cold water is delayed until just before the next milking. A monthly heat treatment is recommended to remove residual biofilm or milkstone deposits.

2.5.1.2 Farm Bulk Milk Tanks. Most refrigerated farm tanks have smooth, stainless steel surfaces that are cleaned easily. Accessories such as agitators, dipsticks, plugs, or outlet valves and manhole gaskets can be problematic. Thus, sanitary conditions of these accessories must be maintained and monitored for cleanliness. In the United Kingdom, most farm tanks of <4000-liter (1000-gal) capacity are "cold wall" or ice bank tanks. Hot solutions cannot be used for cleaning because the ice bank would become depleted and the refrigeration capacity would be delayed while replacing the lost ice. Also, the ice bank would have a cooling effect on the hot cleaning solution. These ice bank tanks normally are cleaned by means of mechanical or hand sprays using cold iodophor solutions or by manual brushing. In North America and some other dairying areas, direct expansion refrigeration systems are used to cool the raw milk. Thus, the refrigeration can be turned off during the hot cleaning step in the CIP cycle. For cleaning and sanitizing contact surfaces, US researchers have demonstrated, via swab samples, SPCs of $<10\,\text{cfu}/4\,\text{in}^{-2}$ when cleaning temperatures above 65°C were used, followed by recirculation of an acidified final rinse with ambient water at a pH of 5.5–6.5 for 60–90s. The early work of Kaufmann et al. (1960) and Parker et al. (1953) characterize the CIP conditions as applied to the CIP process required to achieve a proper sanitary condition for milk contact surfaces. This work serves as the basis for the US dairy practices and regulations.

Until about 1970 in the United Kingdom, rinsing and swabbing tanks and accessories, whether cleaned manually or mechanically, revealed SPCs on average $<2.5 \times 10^6\,\text{cfu}\,\text{m}^{-2}$, with the greatest SPC frequencies being $<1 \times 10^5\,\text{cfu}\,\text{m}^{-2}$. It was observed that mechanical cleaning appeared to be more effective than manual cleaning (Druce and Thomas, 1972). This observation later was confirmed by Panes et al. (1979), although results indicate a deterioration in the bacteriological cleanliness of farm tanks as shown by approximately 30% of the tanks demonstrating rinse counts of $>5 \times 10^6\,\text{cfu}\,\text{m}^{-2}$ (Table 2.11). Approximately 1% of farms surveyed had serious bacteriological quality problems.

One major source for contaminating raw milk stored in farm bulk tanks is the outlet plug (Figure 2.6). While this plug may appear to be clean, it could be harboring large numbers of bacteria because of the design feature associated with the rubber bung attachment to the metal shaft. Table 2.11 shows that 9% of plug rinses have SPCs of $>1 \times 10^9$ cfu per plug. A precautionary step to minimize development of this bacterial condition is to immerse the bung in boiling water for 2min and conduction-heat the junction of the shaft and the bung at least once a month. Recent plug designs that eliminate the crevice now are commercially available. These outlet plugs are peculiar to UK tanks, but the

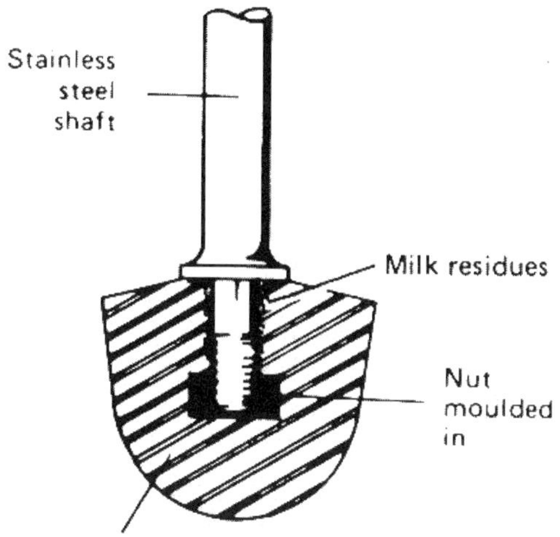

Figure 2.6. Sections of a bulk tank outlet plug, showing site of milk residues. [From Cousins and McKinnon (1979).]

alternative, the outlet valve, also is difficult to clean, and milk residues tend to accumulate around the tap seating.

2.5.1.3 Bucket Milking Equipment. Bucket milking machines have to be cleaned by hand, usually by brushing in a warm detergent–disinfectant solution. The bacterial content can vary just as widely as that of pipeline machines. It is important that the milk buckets are free of open seams, dents, and rust spots. Milk residues, such as biofilms and milkstone, tend to build up because of the difficulty of cleaning buckets with these imperfections. The sanitation process can be complicated further in the cleaning of teat cup clusters and by not allowing adequate time to brush them thoroughly. Consequently, poorly cleaned contact surfaces and the presence of organic material make chemical sanitizing ineffective as reflected in the high rinse SPCs reported by Kaufmann et al. (1960). Hot solution flushing of teat cup clusters, wet storage (or lye storage), immersion cleaning, and boiling water treatment are advocated to improve the bacteriological condition of bucket milking equipment.

2.5.1.4 Milk Cans. As with milk-receiving buckets, milk cans with open seams, dents, or rust spots can be difficult to clean. If the cans are not cleaned effectively or are moist when the lids are placed on them, bacterial multiplication can result in high SPCs. This source of contamination can become even more serious when, in many situations, milk is water-cooled, and bacteria in the milk-filled containers multiply rapidly. Milk cans are suspected as being a source for *Bacillus cereus* spores and other types of thermoduric bacteria found in milk.

2.5.2 Types of Microorganisms on Milking Equipment Surfaces

As might be predicted, the groups and genera of microorganisms found on milk contact surfaces are similar to those found in fresh, raw milk (Table 2.2). However, to be isolated from equipment rinses, colonies must develop on SPC plates and specific types of microorganisms must be present in appreciable proportions (>5%). The survival of bacteria suggests that there is some protection during the cleaning and disinfecting procedures, and subsequently bacteria are able to proliferate during the elapsed time between milkings. Selective methods are available to detect the presence of pathogens, thermodurics, or other microorganisms of special interest.

Certain species, notably mastitis pathogens, have not been reported as forming any appreciable part of the microflora on milking equipment, although large numbers of streptococcal and staphylococcal mastitis organisms can be present in milk passing through the equipment. These microbes can survive long enough to be transferred, via the liner in a teat cup cluster used to milk a cow having an infected quarter, to the noninfected quarters of other cows. These organisms also may be transferred to other cows via udder washing cloths and milk handlers' hands. However, they are unlikely to multiply on the surfaces of milking equipment between milkings.

The temperatures of solutions used for cleaning and disinfecting can influence the microflora remaining on milk contact surfaces. Pipeline milking machines subjected to proper ABW cleaning, in effect, are pasteurized and thus only thermoduric organisms can survive. Application of 70°C detergent solutions has a similar effect. In practice, pasteurization temperatures are not always achieved, and the heat-resistant microbial types such as asporogenous Gram-positive rods (probably *Microbacterium* spp.), micrococci, streptococci, and *Bacillus* spp. usually dominate following hot cleaning treatments. The presence of GNRs including coliforms, which are heat labile, is relatively infrequent. If the microflora recovered by rinsing is predominantly heat labile, it is evident that parts of the machine have not been heated adequately.

Use of lower temperature solutions (40–50°C) permits the development of a heterogeneous-type microflora. Yet, in many pipeline systems, investigations show that the microflora are restricted with one or two of the following groups predominating: micrococci, streptococci, GNRs, and asporogenous Gram-positive rods. In any one milking machine, the microflora may be consistent, or they could vary from time to time. At present, no explanation can be offered for these observations. Druce and Thomas (1972) and Thomas and Thomas (1977a,b,c,d, 1978a,b) reviewed comprehensively the results of many workers on the bacterial content of pipeline milking machines, farm bulk tanks, and bucket milking machines, with particular reference to thermoduric, psychrotrophic, and coliform organisms.

Bucket milking machines with teat cup clusters showing a buildup of milk residues and milkstone often were found to have high thermoduric bacterial counts. These organisms tend to be less prevalent in pipeline milking machines, perhaps because in-place cleaning is more effective in keeping milk contact surfaces clean. A study of the thermoduric and psychrotrophic bacterial content of milking equipment (Mackenzie, 1973) shows that the incidence of thermoduric organisms in pipeline milking machines is slightly lower than that of psychrotrophs. None of the thermoduric counts exceeded 1×10^7 cfu m^{-2}, whereas 7.5% of the systems examined demonstrated psychrotrophic counts above that level.

The SPC of farm bulk tanks is lower than that of milking machines, and the thermoduric bacterial content is very low, $<1 \times 10^5$ cfu m^{-2}. This is because most thermoduric bacteria will not multiply in the cold environment of refrigerated bulk tanks. However, the proportion of GNRs and psychrotrops present in the refrigerated bulk tank appears to be higher than is found in the milking equipment.

In the United Kingdom, there is a trend toward an increase in higher levels of total bacterial contamination in bulk tanks and tank outlet plugs. Mackenzie's (1973) research indicates that with this increase in bacterial numbers, there is also a higher incidence of psychrotrophs present in the tanks.

Based on the comprehensive reviews concerning the significance of psychrotrophic microorganisms in raw milk given by Witter (1961), Thomas et al. (1971), and Cousin (1982), it may be concluded that the sources for these microbes are soil, water, and vegetation. The occurrence of psychrotrophic microorganisms found in raw milk can vary depending on the type and number of cells present, the equipment cleanliness, the temperature, and the length of storage time prior to processing/pasteurization. However, it is fair to assume that where the total bacterial content of equipment is low, the content of specific, undesirable types will be low. The British Standards Institution (BSI, 1975) details cleaning and disinfecting methods for achieving satisfactory bacteriological cleanliness in pipeline milking machines and in bulk tanks. In the United States, the adoption of the PMO by each state provides the guidelines under which raw milk is collected, handled, stored, and picked up at the farm (USDHHS, 1999).

2.6 THE INFLUENCE OF STORAGE AND TRANSPORT ON THE MICROFLORA OF RAW MILK

After production, milk usually is stored in bulk tanks to await collection and subsequent delivery to collection centers or processing plants. The raw milk is delivered to the processing facility in metal milk cans or insulated fiberglass containers in 100-lb quantities or in insulated horizontal tankers in bulk quantities. In most developed countries, construction standards for bulk tankers require that the milk contact surfaces of the tankers be fabricated from stainless steel with a specific polish specification for cleanability. Another requirement is that the tankers be insulated sufficiently to maintain over a 24-h period the milk's temperature to within 1°C from the temperature at time of pickup.

2.6.1 Can Collection

In countries with emerging economies, raw milk usually is collected in metal cans and transported to collection centers within 1 day after milking. Here, the milk is cooled and placed in insulated containers for delivery to processing plants. Under these handling and storage conditions, the raw milk undergoes considerable chemical change due to lipase activity and accelerated microbial growth.

In temperate climates, normally milk is water-cooled to as low a temperature as the cooling medium permits. The method of cooling and the temperature of the available water supply influence the final temperature. In the summer season, it is possible that part of the daily milk yield is stored for 14–18 h at 20–25°C. Where ambient temperatures frequently exceed 25°C, and sometimes 30°C, nonrefrigerated milk often is collected twice daily because of the rapidity of bacterial multiplication and the consequent high risk of spoilage when milk is held at such temperatures for more than about 6 h. On arrival at its destination, the canned milk is used immediately, cooled to ≤5°C, and stored for no more than 24 h before processing or heat treated and then cooled before storage.

2.6.2 Bulk Collection

With bulk collection, milk most commonly is refrigerated immediately after production, either (a) by means of an in-line cooler followed by storage in an insulated tank or (b) in a tank equipped with a refrigeration system. Collection of the milk, which may be within hours, daily, on alternate days, every third day, or, more rarely, at longer intervals, is by means of insulated transport tankers, each picking up milk from several farms. Thus, there is risk that an undetected contaminate from one farm may taint the entire tanker load of milk. Milk haulers normally are authorized to refuse to collect milk from farms where the milk is tainted, appears to be abnormal, or is above a specified temperature (e.g., 7°C). Bulk collection schemes include requirements concerning the design and performance of farm bulk tanks, the rate at which the milk is cooled, the maximum temperature at which it is stored, the frequency of collection, and the cleanliness of the transport tanker (Hoyle, 1979).

On arrival at its destination, if not used immediately, farm refrigerated milk is transferred to insulated storage tanks or silos (see Figure 2.7) where it is stored at 3–5°C until heat treated for processing or manufacture. In some countries, the time delay before processing may be up to 4 days after receipt of the milk. In the Netherlands, where some refrigerated milk is collected every third day, such milk, unless used immediately, is "thermized"—that is, heated to ~63°C for 15 s on arrival at the dairy. This process is sufficient to kill most of the psychrotrophs, and, after cooling to 5–6°C, the milk then can be stored for 2–3 days before use. A study of milk thermilization was published by Griffiths et al. (1986).

2.6.3 Bacterial Multiplication in Stored Milk

The temperature and duration of storage, the numbers and types of bacteria, and, to a lesser extent, the natural inhibitory systems in milk all influence the bacterial number increase that occurs in stored milk. Because of the wide variation in the initial microflora and the conditions under which milk is stored, only generalizations can be made concerning changes in the microflora occurring during storage and transport.

The storage temperature probably is the most important factor in microbial growth. Figure 2.8 illustrates the likely effect of temperature on milk of rather poor quality (having an initial SPC of 50,000 cfu ml^{-1}). If milk is to be kept for more than about 12 h, it is assumed that

Figure 2.7. These insulated storage silos are used to store raw milk prior to processing. The larger silos hold up to 30,000 gal of milk, whereas the smaller ones hold up to 25,000 gal. (Reproduced by courtesy of the Milk Marketing Board, Davidstow.)

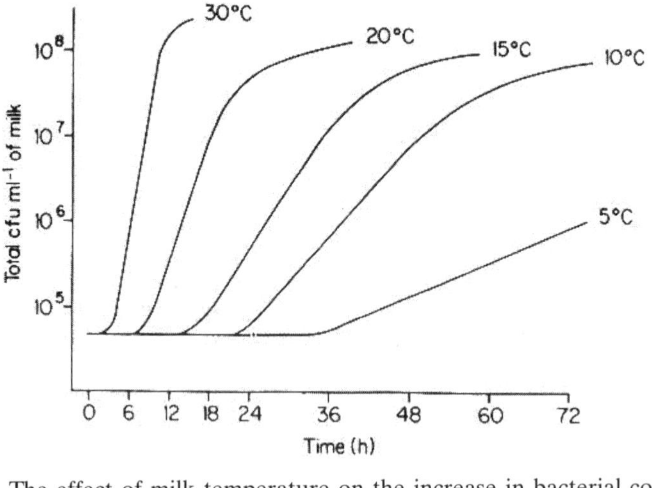

Figure 2.8. The effect of milk temperature on the increase in bacterial count in raw milk having an initial SPC of 50,000 cfu ml^{-1}. [After Druce and Thomas (1968).]

adverse effects will become apparent when the SPC approaches 1×10^7 cfu ml^{-1}. Often, milk delivered in cans is checked upon arrival for these effects and, if tainted or if there is evidence of an acidity above specified limits, the milk is downgraded or rejected.

The spoilage organisms that become predominant at 25–30°C are mainly streptococci and coliforms, and both types increase the acidity of the milk. However, unpleasant tastes/odors produced by variations in the initial microflora can mask the "clean" acid odor normally associated with a predominant streptococci population. GNRs, other than coliforms, and micrococci (including staphylococci) also will multiply unless, or until, any developed acidity becomes inhibitory to them. Between 15°C and 25°C, GNRs could outnumber streptococci. In general, the effects are similar to those at higher temperatures, except that the adaptive phase of microbial growth is extended for several hours. Many species of GNRs can multiply in milk and not produce any noticeable effects even when their numbers reach or exceed 1×10^7 cfu ml^{-1}. Thus, at lower storage temperatures (e.g., 10°C or less), the milk may appear normal for 2–3 days, even though considerable bacterial growth has occurred.

Druce and Thomas (1968) reviewed the effects of preincubation for specified storage periods and at specified temperatures on the results of bacteriological tests applied to milk samples before and after preincubation. In general, saprophytic bacteria multiply most readily under the commonly used preincubation conditions—that is, 12–22°C for 16–24 h. Microorganisms representative of the udder microflora have an optimum growth temperature of 37°C. This microflora will multiply slowly, if at all, during a preincubation period, and an observed increase in SPC (e.g., 100-fold) during storage would indicate a bacterial contamination from sources outside the udder. However, preincubation has its drawbacks in this respect because the saprophytic, thermoduric coryneforms and micrococci originating from milking equipment surfaces do not multiply in raw milk within 24 h at these preincubation temperatures. On the other hand, some types of coliforms from within the udder can grow rapidly.

When compared with heavily contaminated milk, there is no doubt that milk produced with minimal contamination shows, on storage, much smaller increases in bacterial numbers over a wide temperature range. The natural inhibitory properties of milk play some part in this effect, and bacterial multiplication is delayed even in moderately contaminated milk for 2–3 h at 30°C and for longer periods at lower temperatures (Figure 2.8).

2.6.4 Refrigerated Storage of Raw Milk

Refrigeration, by delaying bacterial multiplication, can mask the effects of unsanitary production conditions. In many dairying regions, bulk collection of refrigerated milk accounts for most, if not all, of the milk produced. Because of this, the numbers and types of initial psychrotrophic microflora present in the raw milk and their activities during storage have become an increasingly important area of concern.

Alternate day (AD) collection is the most common practice worldwide. Because four successive additions of milk are made to the farm bulk storage tank, about a quarter of the milk is 2 days old at collection. Studies of AD collection between 1950 and 1970 show no detrimental effect on milk's bacteriological quality up to the point of leaving the farm, provided that the milk added to the tank after each milking was cooled rapidly to ≤4°C (Thomas and Druce, 1972). There is probably a slight increase in psychrotrophic numbers, but it would be insignificant due to a dilution effect resulting from the repeated addition of fresh milk. However, storage at 5–7°C of fresh milk with a high psychrotrophic count most likely would lead to a marked increase in bacterial count at the time of collection. Results of investigations on bacteriological quality of AD collected milk were reviewed by Thomas and Druce (1972).

The initial total microbial count of raw milk is of little value for predicting its count after refrigerated storage. Samples from farm bulk tanks taken shortly after the addition of a second milking were stored at 5°C, and SPCs were determined at daily intervals. The wide range of responses is shown in Table 2.12. Some samples showed only small changes after 4 days, whereas others, showed three- to eightfold increases in just 2 days. As might be expected, the four samples in which bacterial multiplication had been most rapid had similar psychrotrophic and SPCs after 3 days.

TABLE 2.12. SPCs of Milk Stored at 5°C in Individual Farm Bulk Tanks

Farm	Zero Days	Two days	Three days	Four days
A	5,800,	3,300	7,900	14,000
B	14,000	10,000	11,000	70,000
C	14,000	10,000	710,000	15,000,000
D	28,000	83,000	2,800,000	18,000,000
E	62,000	400,000	9,500,000	41,000,000
F	170,000	110,000	110,000	130,000
G	240,000	1,800,000	8,900,000	17,000,000

2.6.5 The Effects of Refrigerated Transport and Subsequent Storage

Because farm collection tankers are insulated, the temperature of the milk remains relatively unchanged from the temperature of the milk at the time of pickup. The majority of the milk collected arrives at its destination at ≤5°C, and often there are no facilities for cooling it to ≤4°C before it is put into storage silos. When stored at temperatures ranging from 5°C to 10°C, samples taken from both transport tankers and storage silos show increases in bacterial counts. For milk collected daily, the results for stored samples having a mean initial psychrotrophic count of 1×10^4 cfu ml^{-1} are shown in Figure 2.9. At 5°C, Cousins et al. (1977) reported counts in excess of 1×10^6 cfu ml^{-1} in 3 days, whereas other workers reported greater psychrotrophic increases. SPCs of commingled milk from storage silos were $\sim 1 \times 10^7$ cfu ml^{-1} after 1 day (LaGrange, 1979), but milk in some of the incoming transport tankers had SPCs above 1×10^6 cfu ml^{-1}. For AD collected milk, Muir et al. (1978) recorded counts ranging from 10^4 to 5×10^6 cfu ml^{-1} for milk in storage silos held for 24 h at the plant. The total psychrotroph and coliform counts of samples from transport tanker milk are higher than

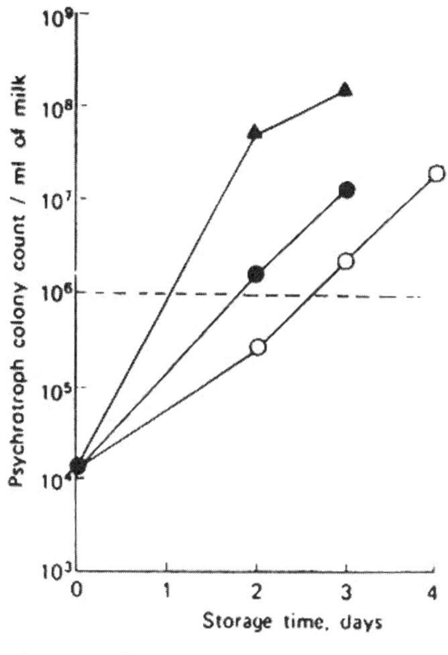

Figure 2.9. Psychrotroph counts (milk, agar, 5°C for 10 days) of transport tanker milk showing increases during storage at 5°C (○), 7°C (●), and 10°C (▲); means of eight samples.

counts of milk samples taken from individual farm tanks, but there is little difference in the thermoduric counts (Thomas, 1974a).

It is not clear to what extent contamination from transport tankers and associated hoses, pumps, meters, and automatic samplers increase the bacterial content of milk. Because of the large volumes of milk involved (e.g., 2500 gal), any significant increase in count per milliliter of milk indicates a heavy contamination of the milk contact surfaces. Some increase in the count may occur from the pumping process that can disrupt clumps and chains of bacteria. However, there is no clear evidence concerning the extent of any such increase. One or two heavily contaminated individual farm supplies will, of course, influence the bacterial content of a whole transport tanker load. For milk that is more than 12 h in transit and has a high content of psychrotrophs, multiplication of these microbes may occur with milk temperatures above 4°C.

During silo storage, multiplication of psychrotrophs derived from bulk handling equipment will contribute to the bacterial content of the milk. If fresh milk is added to a storage silo containing milk stored for 24 h or longer and the mixed milk is held further, bacterial counts are likely to reach undesirable levels more quickly than if the milk were put into a clean, empty silo. For this reason, the Grade A regulations in the United States mandate that milk-handling facilities clean and sanitize milk storage silos immediately after they are emptied (USDHHS, 1999).

Milk inventories should be monitored for bacteriological quality on a daily basis. The use of commercially available rapid methods for enumerating microflora can help in monitoring microbial activity. A balance of raw milk inputs and scheduled product outputs can reduce significantly the storage time required for milk inventories.

2.6.6 Types of Bacteria in Stored Milk

Generally within 2 to 3 days after transfer from transport tankers to storage silos, the microflora of the milk are dominated by psychrotrophs, whereas the thermoduric microflora have not changed significantly in composition. Spores of psychrotrophic *Bacillus* spp. can be present, but germination of these spores and outgrowth to any significant extent is unlikely.

The predominant psychrotrophic genera and species will vary but are derived from those present initially in the raw milk (Table 2.5). *Pseudomonas* spp. are the most frequently reported psychrotrophs in stored milk, with *Pseudomonas fluorescens* being the most common.

However, Law (1979) suggests that this may be due to the ease with which this species can be identified tentatively, rather than due to its true distribution.

Pasteurization will destroy psychrotrophic bacteria in the vegetative state, but not those in the spore state. Pasteurization does not necessarily inactivate the products of psychrotrophs' metabolism or their enzymes. An investigation by Schroeder et al. (1982) indicates that spoilage in commercially pasteurized milk can be associated mainly with GNRs and post-pasteurization contamination. This is in contrast to laboratory pasteurized milk from the same source where the *Bacillus* species were the predominant microflora causing spoilage and no postpasteurization contamination took place. Small-scale experiments show that heat-resistant bacterial protease enzymes can affect adversely the product yield and quality (Cousins, 1978).

Although, on a manufacturing scale, demonstration of the above effects is not clear-cut, dairy industry professionals recognize that the presence of psychrotrophic activity in raw milk will impact both flavor quality and curd yields. However, good handling practices can help to ensure a quality milk supply by preventing bacterial growth and microbial contamination.

ACKNOWLEDGMENTS

This chapter is a rewrite of the "Microbiology of Raw Milk" chapter authored by Drs. A. J. Bramley and C. H. McKinnon of the Milking and Mastitis Centre, AFRA Institute for Animal Health, Newbury, United Kingdom. This author wishes to acknowledge extensive use of their data, figures, and subject content. While retaining the British perspective on the microbiology of raw milk, I have endeavored to incorporate the US perspective, as well.

I also wish to acknowledge Dr. Simon J. Kenyon, Veterinary Clinical Sciences, Purdue University, West Lafayette, Indiana for his counsel and sharing of important literature related to mastitis.

REFERENCES

Akam, D. N. (1979) In *Machine Milking*, C. C. Thiel, and F. H. Dodd, eds., NIRD, Reading, United Kingdom.

AMS (2000) Federal Market 30, Agricultural Marketing Services, U.S. Department of Agriculture, *Upper Midwest Dairy News*, **1**(4), 5.

Anonymous (1979) *Vet. Rec.*, **105**, 294.

APHA (1992) *Standard Methods for the Examination of Dairy Products*, American Public Health Association, R. H. Marshall, ed., Washington, DC.

Armstrong, C., and Parran, T. (1927) *Supplemental 62, Public Health Report*, US Public Health Service, Washington, DC.

Benham, C. L., and Egdell, J. W. (1970) *J. Soc. Dairy Technol.*, **23**, 91.

Bergere, J. L. (1979) *Rev. Lait. Fr.*, **378**, 19.

Booth, J. M. (1988) *Br. Vet. J.*, **144**, 36.

Bramley, A. J. (1975) In *Proceedings of the Seminar on Mastitis Control*, F. H. Dodd, T. K. Griffin, and R. G. Kingwill, eds., Doc. No. 85, International Dairy Federation, Brussels, Belgium.

Bramley, A. J. (1987a) *International Mastitis Symposium*, Ontario College, University of Guelph, Guelph, Ontario, Canada, pp. 142–160.

Bramley, A. J. (1987b) In *Summer Mastitis*, G. Thomas, H. J. Over, U. Vecht, and P. Nansen, eds., Martinus Nijhoff, The Hague, pp. 81–85.

Bramley, J. A., and Dodd, F. H. (1984) *J. Dairy Res.*, **51**, 481.

Bramley, J. A., and Neave, F. K. (1975) *Br. Vet. J.*, **131**, 160.

Bramley, A. J., King, J. S., and Higgs, T. M. (1979) *Br. Vet. J.*, **135**, 262.

Bramley, A. J., McKinnon, C. H., Staker, R. T., and Simpkin, D. L. (1984) *J. Appl. Bacteriol.*, **57**, 317.

BSI (1975) *Recommendations for Cleaning and Sterilization of Pipeline Milking Machine Installation*, BS 5226, British Standards Institution, London.

Bryan, F. L. (1969) In *Food Infections and Intoxications*, H. Reimann, ed., Academic Press, New York.

Carreira, D. F. C., Clegg, L. F. L., Clough, P. A., and Thiel, C. C. (1955) *J. Dairy Res.*, **22**, 166.

Carroll, E. J., Jain, N. C., Schalm, O. W., and Lasmanis, J. (1973) *Am. J. Vete. Res.*, **34**, 1143.

Collins, R. A., Parsons, K. R., Field, T. R., and Bramley, J. A. (1988) *J. Dairy Res.*, **55**, 25.

Cousin, M. A. (1982) *J. Food Prot.*, **45**, 172–207.

Cousins, C. M. (1963) *J. Appl. Bacteriol.*, **26**, 376.

Cousins, C. M. (1972) *J. Soc. Dairy Technol.*, **25**, 200.

Cousins, C. M. (1978) *XXth International Dairy Congress*, Paris, France, Congress paper.

Cousins, C. M., Sharpe, M. E., and Law, B. A. (1977) *Dairy Ind. Int.*, **42**, 12.

Cousins, C. M., and Mckinnon, C. H. (1979) In *Machine Milking*, C. C. Thiel, and F. H. Dodd, eds., NIRD, Reading, UK.

Craven, N., and Williams, M. R. (1985) *Vet. Immunol. Immunopathol.*, **10**, 71–127.

Davies, F. L., and Wilkinson, G. (1973) In *The Microbiological Safety of Foods*, B. C. Hobbs and J. H. B. Christian, eds., Academic Press, London.

Dorner, W. (1930) *Technical Bulletin*, No. 165, Cornell University, New York State Agriculture Experiment Station, Geneva, NY.

Druce, R. G., and Thomas, S. B. (1968) *Dairy Sci. Abstr.*, **30**, 291.

Druce, R. G., and Thomas, S. B. (1972) *J. Appl. Bacteriol.*, **35**, 253.

Ford, J. E. (1974) *Br. J. Nutr.*, **31**, 243.

Gitter, M. (1989) *PHLS Microbiol. Dig.*, **6**, 38–42.

Gitter, M., Bradley, R., and Blampied, P. H. (1980) *Vet. Rec.*, **107**, 390–393.

Griffiths, M. W., Phillips, J. D., and Muir, D. D. (1984) *J. Soc. Dairy Technol.*, **37**, 22.

Griffiths, M. W., Phillips, J. D., and Muir, D. D. (1986) *Dairy Ind. Int.*, **51**, 31.

Goudkov, A. V., and Sharpe, M. E. (1965) *J. Appl. Bacteriol.*, **28**, 63.

Hahn, G., Heeschen, W., and Tolle, A. (1970) *Kieler Milchwirtsch. Forschungsber.*, **22**, 335.

Hoyle, J. B. (1979) In *Machine Milking*, C. C. Thiel and F. H. Dodd, eds., NIRD, Reading, UK.

International Dairy Federation Conference Proceedings (1974).

IDF (1980) In *Factors Influencing the Bacteriological Quality of Raw Milk*, Document No. 120, International Dairy Federation, Brussels, Belgium.

Jackson, E. R., and Bramley, J. A. (1983) *Vet. Rec.*, **113**, 135–146.

Jackson, H., and Clegg, L. F. L. (1966) *Can. J. Microbiol.*, **12**, 429.

Jeffery, D. C., and Wilson, J. (1987) *J. Soc. Dairy Technol.*, **40**(2), 23–26.

Johns, C. K. (1962) *XVIth International Dairy Congress*, Copenhagen, Denmark, Congress paper C, p. 365.

Johns, C. K. (1971) *J. Milk Food Technol.*, **34**, 173.

Juffs, H. S. (1973) *J. Appl. Bacteriol.*, **36**, 585.

Kaufmann, O. W., Hedrick, T. I., Pflug, I. J., Pheil, C. G., and Keppeler, R. A. (1960) *J. Dairy Sci.*, **43**(1), 28–41.

Kingwill, R. G., Dodd, F. H., and Neave, F. K. (1997) In *Machine Milking*, C. C. Theil and F. H. Dodd, eds., NIRD, Reading, UK.

Lagrange, W. S. (1979) *J. Food Prot.*, **42**, 599.

Lander, K. P., and Gill, K. P. W. (1979) *Vet. Rec.*, **105**, 333.

Law, B. A. (1979) *J. Dairy Res.*, **46**, 573.

Luck, H. (1972) *Dairy Sci. Abstr.*, **34**, 101.

Mackenzie, E. (1973) *J. Appl. Bacteriol.*, **36**, 457.

McKinnon, C. H., and Cousins, C. M. (1969) *J. Soc. Dairy Technol.*, **22**, 227.

McKinnon, C. H., and Pettipher, G. L. (1983) *J. Dairy Res.*, **50**, 163.

McKinnon, C. H., Cousins, C. M., and Fulford, R. J. (1973) *J. Dairy Res.*, **40**, 47.

McKinnon, C. H., Higgs, T. M., and Bramley, A. J. (1985) *J. Dairy Res.*, **52**, 355.

McKinnon, C. H., Bramley, A. J., and Morant, S. V. (1988) *J. Dairy Res.*, **55**, 33.

Marshall, J. (1991) *In Practice*, **9**, 198–201.

Marshall, V. M., Cole, W. M., and Bramley, A. J. (1986) *J. Dairy Res.*, **53**, 507.

Mikolajcik, E. M. (1979) *Cul. Dairy Prod. J.*, **14**, 6.

Morin, D. E. (1998) *Milk Quality Issues*, **3**, 2, University of Illinois Extension Publication.

Muir, D. D., Kelly, M. E. Phillips, J. D., and Wilson, A. G. (1978) *J. Soc. Dairy Technol.*, **31**, 137.

Murphy, J. M. (1943) *Cornell Vet.*, **33**, 48.

NMC (1978) *Current Concepts of Bovine Mastitis*, National Mastitis Council, Inc., Washington, DC.

Nickerson, S. C. (1999) *National Mastitis Council Meeting Proceedings*, pp. 76–85.

Oliver, S. P., Lewis, M. J., Gillespie, and Dowlen, H. H. (1992) *J. Dairy Sci.*, **75**, 406–414.

Olsen, J. C., Casman, E. P. Baer, E. F., and Stone, J. E. (1970) *Appl. Microbiol.*, **20**, 605.

Olsen, S. J. (1975) In *Proceedings of the Seminar on Mastitis Control*, F. H. Dodd, T. K. Griffin, and R. G. Kingwill, eds., Doc. 85, International Dairy Federation, Brussels, Belgium.

Olson, J. C., Jr., Macy, H., and Halverson, H. O. (1952) *Agric. Exp. Station Bull.*, **202**, University of Minnesota.

Palmer, J. (1977) *Ir. J. Food Sci. Technol.*, **1**, 57.

Panes, J. J., Parry, D. R., and Leech, F. B. (1979) *Report of a Survey of the Quality of Farm Ilk an England and Wales in Relation to EEC Proposals*, Ministry of Agriculture, Fisheries and Food, London, UK.

Pankey, J. W., Boddie, N. T., Watts, J. L., and Nickerson, S. C. (1985) *J. Dairy Sci.*, **68**, 726–731.

Parker, R. B., Ellicker, P. R., Nelson, G. T., Richardson, G. A., and Wilster, G. H. (1953) *Food Eng.*, **25**(1), 82–86, 176–178.

Peterson, P. K., and Quis, P. G. (1981) *Annu. Rev. Med.*, **32**, 29–43.

Pettipher, G. L., Mansell, R., McKinnon, C. H., and Cousins, C. M. (1980) *Appl. Environ. Microbiol.*, **38**, 423.

Phillips, J. D., and Griffiths, M. W. (1985) *Food Microbiol.*, **2**, 39.

Phillips, J. D., Griffiths, M. W., and Muir, D. D. (1981) *J. Soc. Dairy Technol.*, **34**, 109.

Read, R. B., and Bradshaw, J. G. (1966) *J. Dairy Sci.*, **49**, 202.

Reiter, B. (1978) *J. Dairy Res.*, **45**, 131.

Reiter, B., and Bramley, A. J. (1975) In *Proceedings of the Seminar on Mastitis Control*, F. H. Dodd, T. K. Griffin, and R. G. Kingwill, eds., Document 85, International Dairy Federation, Brussels, Belgium.

Ridgeway, J. D. (1955) *J. Appl. Bacteriol.*, **18**, 374.

Robinson, D. A., Edgar, W. J., Matchett, A. A., and Robertson, L. (1979) *Br. Med. J.*, **1**, 1171.

Runnels, R. J. (1988) *J. Soc. Dairy Technol.*, **41**, 14.

Russell, M. W., Brookerr, B. E., and Reiter, B. (1977) *Res. Vet. Sci.*, **20**, 30.

Schalm, O. W., Lasmanis, J., and Carroll, E. J. (1964) *Am. J. Vet. Res.*, **25**, 83.

Schalm, O. W., Carroll, E. J., and Jain, N. C. (1971) *Bovine Mastitis*, Lea & Febiger, Philadelphia.

Schalm, O. W., Lasmanis, J., and Jain, N. C. (1976) *Am. J. Vet. Res.*, **37**, 885.

Schroder, M. J. A., Cousins, C. M., and McKinnon, C. H. (1982) *J. Dairy Res.*, **49**, 619.

Scott, W. M., and Minett, F. C. (1947) *J. Hyg.*, **45**, 159.

Sinelll, H. J. (1973) In *The Microbiological Safety of Food*, B. C. Hobbs and J. H. B. Christian, eds., Academic Press, London.

Smith, H. (1977) *Bacteriol. Rev.*, **41**(2), 475–500.

Smith, J. (1934) *J. Comp. Pathol.*, **47**, 125.

Swithinbank, H., and Newton, G. (1903) *Bacteriology of Milk*, John Murray Publishers, London, p. 260.

Taylor, P. R., Weinstein, W. M., and Bryner, J. H. (1979) *Am. J. Med.*, **66**, 779.

Theil, C. C., Cousins, C. L., Westgarth, D. R., and Neave, F. K. (1973) *J. Dairy Res.*, **40**, 117.

Thomas, S. B. (1974a) *J. Soc. Dairy Technol.*, **27**, 180.

Thomas, S. B. (1974b) *Dairy Ind.*, **39**, 237, 279.

Thomas, S. B., and Druce, R. G. (1971) *Dairy Sci. Abstr.*, **33**, 339.

Thomas, S. B., and Druce, R. G. (1972) *Dairy Ind.*, **37**, 593.

Thomas, S. B., and Thomas, B. F. (1975) *Dairy Ind.*, **40**, 338.

Thomas, S. B., and Thomas, B. F. (1977a) *Dairy Ind. Int.*, **42**(4), 7.

Thomas, S. B., and Thomas, B. F. (1977b) *Dairy Ind. Int.*, **42**(5), 18.

Thomas, S. B., and Thomas, B. F. (1977c) *Dairy Ind. Int.*, **42**(7), 19.

Thomas, S. B., and Thomas, B. F. (1977d) *Dairy Ind. Int.*, **42**(11), 25.

Thomas, S. B., and Thomas, B. F. (1978a) *Dairy Ind. Int.*, **43**(5), 17.

Thomas, S. B., and Thomas, B. F. (1978b) *Dairy Ind. Int.*, **43**(10), 5.

Thomas, S. B., Hobson, P. M., Bird, E. R., King, K. P., Druce, R. G., and Cox, D. R. (1962) *J. Appl. Bacteriol.*, **25**, 107.

Thomas, S. B., Druce, R. G., and King, K. P. (1966) *J. Appl. Bacteriol.*, **29**, 409.

Thomas, S. B., Druce, R. G., Peters, G. J., and Griffiths, D. G. (1967) *J. Appl. Bacteriol.*, **30**, 265.

Thomas, S. B., Druce, R. G., and Jones, M. (1971) *J. Appl. Bacteriol.*, **34**, 659.

Underwood, H. M., McKinnon, C. H. Davies, F. L., and Cousins, C. M. (1974) *XIXth International Dairy Congress*, Congress paper 1E, p. 373.

USDHHS (1999) *Grade "A" Pasteurized Milk Ordinance*, US Department of Health and Human Services, Public Health Service, Food and Drug Administration, Washington, D.C., USA.

Vakil, J. R., Chandan, R. C., Parry, R. M., and Shahani, K. M. (1969) *J. Dairy Sci.*, **52**, 1192.

Weir, J., and Barbour, D. (1950) *Vet. Rec.*, **62**, 239.

Werner, S. B., Humphrey, C. L., and Kamel, I. (1979) *Br. Med. J.*, **2**, 238.

Westgarth, D. R. (1975) In *Proceedings of the Seminar on Mastitis Control*, F. H. Dodd, T. K. Griffin, and R. G. Kingwill, eds., Document 85, International Dairy Federation, Brussels, Belgium.

Wheelock, J. V., Rook, J. A. F., Neave, F. K., and Dodd, F. H. (1966) *J. Dairy Res.*, **33**, 199.

Wilson, C. D., and Kingwill, R. G. (1975) In *Proceedings of the Seminar on Mastitis Control*, F. H. Dodd, T. K. Griffin, and R. G. Kingwill, eds., Document 85, International Dairy Federation, Brussels, Belgium.

Witter, L. D. (1961) *J. Dairy Sci.*, **44**, 983.

Zarkower, A., and Scheuchenzuber, W. J. (1977) *Cornell Vet.*, **67**, 404.

CHAPTER 3

MICROBIOLOGY OF MARKET MILKS

KATHRYN J. BOOR and STEVEN C. MURPHY
Milk Quality Improvement Program, Department of Food Science,
Cornell University, Ithaca, New York

3.1 INTRODUCTION

Historically, market milk has been defined as fluid milk products sold for direct human consumption (Eckles et al., 1936). Although the collection and use of animal milk for human consumption can be traced back to some of the earliest of recorded histories, the concept of market milk has developed in more recent times. In general, the market milk industry has grown in coincidence with the establishment of major metropolitan areas, which typically are not amenable to individual ownership of cows for personal milk collection. Until the end of the nineteenth century, market milk was primarily raw milk, collected from the farm, distributed fresh to the consumer, and consumed fresh. In a typical pre-1900 US milk distribution system, milk was collected at the farm in 5- to 10-gal cans, transported to the plant, commingled in a vat or can-fill machine, and then filled into 3-gal cans. This product was then distributed by measure into containers left on consumers' porches (Roadhouse and Henderson, 1941). Currently, these practices would be considered unacceptable in the United States because of public health concerns as well as the perishability of raw milk. The development of pasteurization and other heat processes for dairy products has transformed the market milk industry and dramatically expanded product distribution capabilities in the United States and much of the world.

Pasteurization is a process based on the 1860s experiments of Louis Pasteur, who discovered that by heating wine to 50–60°C for a few minutes, microbial spoilage could be prevented or delayed (Hammer,

Dairy Microbiology Handbook, Third Edition, Edited by Richard K. Robinson
ISBN 0-471-38596-4

1948). Soon thereafter, the pasteurization concept was introduced to the dairy industry, primarily to extend the keeping quality of milk. Although milk quality was the primary concern at this time, the association of disease-causing organisms with raw milk was also becoming more apparent. To illustrate, typhoid fever, diphtheria, brucellosis, and tuberculosis were among the human diseases that had been recognized as being transmitted through raw milk consumption. In 1875 and 1886, respectively, scientists in the United States and in Germany began to recommend that milk be heated prior to consumption, especially when intended for infant feeding. Upon entry into the twentieth century, heat treatment of milk was slowly adopted in many segments of the market milk industry. However, the extent and duration of heat treatments lacked uniformity among processing operations, and the efficacy of bacterial destruction for the various treatments was largely undocumented. To address these recognized gaps, extensive research was conducted to determine the heat treatment required to kill *Mycobacterium tuberculosis*, which, at the time, was considered to be the most heat-resistant pathogen associated with milk (Hammer, 1948). This work led to the widespread recognition of the public health significance of thermal milk processing and formed the basis for modern pasteurization processes (Hammer, 1948).

In addition to the widespread application of pasteurization, the microbiological characteristics of modern market milks, which include whole milk, reduced fat milks, nonfat milk, creams, and numerous flavored products, have been influenced by evolving processing technologies and regulatory interventions. Such factors include increased use and efficiencies of mechanical refrigeration, improved processing and sanitation procedures, and enforcement of public health regulations and inspection procedures. This chapter will discuss the types of heat treatments used and current regulations for market milks, as well as the microbiology of these products from the perspectives of quality, shelf life, and safety.

3.2 CURRENT HEAT TREATMENTS FOR MARKET MILKS

The scientific basis for the specific conditions used for heat treating milk products is derived from the concepts of microbial thermal death times in which destruction of the organisms is assumed to be dependent on a linear, semilogarithmic relationship of temperature to time (Jay, 1996). The overall objectives of the heat treatments developed for market milks are to (1) eliminate or reduce the risk of disease associated with pathogens common to raw milk and (2) extend the keeping quality of milk by inactivation of potential spoilage organisms and degradative enzymes. The specific objectives are to ensure that all particles of milk are heated to the desired temperature for the desired time and to avoid significantly damaging the flavor or other quality characteristics of the milk. Rapid cooling after the heat process is also essential to minimize product heat damage and to control the growth of microorganisms that may survive the heat process. In general, the higher the heat treatment, the greater the efficiency of microbial destruction, but also the greater the chance that the quality of the milk will be affected. Heat-associated quality defects depend on the intensity of the heat, the time of exposure, rates of heating and cooling, product composition, and "burn-on" of heat exchangers (Bodyfelt et al., 1988). Defects include cooked flavors that range from slight with mild sulfur notes, which are considered acceptable to most consumers, to "scorched" or "caramelized," which generally are considered unacceptable. Caramel-like flavors are often associated with the Maillard browning reaction, which also may contribute off-colors to highly heated milks.

Microbiological characteristics of market milks are influenced by the design of the overall milk production and handling systems in addition to the temperature and time combinations used to heat the milk. The heat treatments and processing designs generally used for market milks can be categorized as (1) *pasteurization* for limited product shelf life under refrigerated storage, (2) *ultra-pasteurization* or ultra-high-temperature (UHT) pasteurization for extended product shelf life under refrigerated storage, (3) *ultra-high temperature* for shelf-stable products, and (4) *in-container sterilization* for shelf-stable products.

3.2.1 Pasteurization

The public health objective of milk pasteurization is to eliminate all non-spore-forming pathogens commonly associated with milk. Pasteurization also effectively destroys a majority of potential spoilage organisms and contributes to product keeping quality under required refrigeration storage. Pasteurization, as first adopted in the United States, was defined in the 1939 Milk Ordinance and Code as "the process of heating every particle of milk to at least 143°F (61.7°C) and holding at such temperature for at least 30 minutes, or to at least 160°F (71.1°C) and holding at such temperature for at least 15 seconds, in approved and properly operated equipment" (Public Health Service, 1940). These heat treatments were referred to, respectively, as the "holding method" or vat/batch pasteurization and the "flash method" or high-temperature short-time pasteurization. These strategies were considered equivalent for destroying pathogenic microorganisms in milk. In 1956, the vat pasteurization temperature was raised to 63°C (145°F) to ensure destruction of *Coxiella burnetti*, the organism associated with Q-fever, which was found to be more heat-resistant than *Mycobacterium tuberculosis*. Processing requirements from various nations are listed in Table 3.1. Current pasteurization temperature–time combinations that are considered equivalent are defined by the United States Food and Drug Administration (FDA) in the 1999 Grade "A" Pasteurized Milk Ordinance, as shown in Table 3.2.

TABLE 3.1. Minimum Temperature and Times for Fluid Milk Heat Treatments

Country	HTST Pasteurization	Ultra-Pasteurization	UHT Processing
US[a]	72°C for 15s	138°C for 2s	Not defined[d]
EEC[b]	71.7°C for 15s	Not defined	135°C for 1s
Australia/ New Zealand[c]	72°C for 15s	132°C for 1s	132°C for 1s

[a]Public Health Service. US Food and Drug Administration (1999). *Grade "A" Pasteurized Milk Ordinance*, 1999 revision, Publication No. 229.
[b]European Economic Community: 5. Health Rules—Raw milk, heat-treated milk and milk based products. Council Directive 92/46/EEC of June 16, 1992. Laying down the health rules for the production and placing on the market of raw milk, heat-treated milk, and milk based products as amended by Directive EEC No. 92/118 of March 15, 1993; Commission Decision 94/330/E.C. of June 11, 1994; Directive E.C. No. 94171 of December 13, 1994.
[c]Australian New Zealand Food Authority, Food Standards Code. Standard H1—Milk and Liquid Milk Products. http:/anzfa.gov.au/FoodStandardCode/code/parth/H1.htm.
[d]The thermal process and procedures for manufacturing UHT aseptically processed milk and milk products must comply with US Food and Drug Administration requirements for sterilizing low acid foods (Code of Federal Regulations Title 21, Part 113).

TABLE 3.2. Equivalent Temperature and Time Combinations for Milk Pasteurization by US Regulations[a]

Temperature	Time	Temperature	Time
63°C (145°F)[b]	30min	94°C (201°F)	0.1s
72°C (161°F)[b]	15s	96°C (204°F)	0.05s
89°C (191°F)	1.0s	100°C (212°F)	0.01s
90°C (194°F)	0.5s		

[a]Public Health Service. US Food and Drug Administration (1999). *Grade "A" Pasteurized Milk Ordinance*, 1999 revision, Publication No. 229.
[b]If the fat content of the milk is 10% or more or if it contains added sweeteners, the required minimum temperature must be increased by at least 3°C (5°F).

The most commonly used method of pasteurization in the United States is the high-temperature short-time process of 72°C for a minimum of 15s. This is normally accomplished by indirect heating of the milk through a heat-conducting barrier, usually with a plate heat exchanger that separates the heating medium from the milk. A typical design for a pasteurization system utilizing a plate heat exchanger includes (1) a "regeneration" section in which cold raw milk is on the opposite side of the plate from, and flowing counter to, the hot pasteurized milk (this strategy is designed to simultaneously pre-warm the raw milk prior to its entry into the heating section and cool the pasteurized milk), (2) a heating section in which hot water runs on the opposite side of the plate from, and flowing counter to, the pre-warmed milk to heat the milk to the desired temperature, (3) a holding tube designed to maintain the milk at the desired pasteurization temperature for the specified period of time, and (4) a cooling section in which pasteurized milk flows on the opposite side of the plate from, and counter to, a cooling medium. Batch or vat pasteurization is most commonly used in small dairies or for processing small batches of specialty products. Typical pasteurized fluid milk products currently manufactured in the United States have expected shelf lives of 10–21 days at refrigeration temperatures, depending on product manufacturing and distributing conditions.

3.2.2 Ultra-Pasteurization (Extended Shelf-Life Milk)

Ultra-pasteurization is a process recognized in North America in which milk is "thermally processed at or above 138°C for at least 2s, so as to produce a product that has an extended shelf life under refrigerated conditions" (Public Health Service, 1999). Mehta (1980) referred to this process as ultra-high-temperature (UHT) pasteurization because the heating process generally uses the same equipment as UHT sterilized milk, which is described in the next section. Indirect heating by plate or tube heat exchangers or direct heating by steam injection or infusion are applicable to ultra-pasteurized milk. This heat treatment is designed to kill virtually all microorganisms found in raw milk, including vegetative cells and bacterial endospores. Although the packaged product is not considered to be aseptically filled or hermetically sealed, ultra-pasteurization is usually coupled with milk handling and processing procedures that are specifically designed to prevent bacterial recontamination of the milk. Commercially manufactured ultra-pasteurized fluid milk products have been demonstrated to remain free of bacterial contaminants for up to 10 weeks of refrigerated storage

(Boor and Nakimbugwe, 1998). Ultra-pasteurized products are often referred to as "extended shelf life" (ESL) products with shelf lives of 30–90 days or more at refrigeration temperatures. ESL may also be used to describe conventionally pasteurized products that are processed under stringent conditions designed to prevent postpasteurization contamination with the goal of extending the shelf life or keeping quality of these products.

3.2.3 Ultra-High-Temperature Sterilization

Ultra-high-temperature (UHT) sterilization is a process that combines rapid heating of milk to very high temperatures followed by aseptic handling and packaging to produce a shelf-stable, commercially sterile product. The goal of UHT sterilization is to achieve a 9-log reduction of thermophilic endospores that would naturally occur in raw milk (Burton, 1988; Hinrichs and Kessler, 1995). This treatment is considered adequate to ensure a 12-log reduction of *Clostridium botulinum* spores, as required for canning low acid foods (Burton, 1988; Jay, 1996).

Although heat treatments for UHT milk vary from country to country, temperatures of 130–150°C with holding times of 1 s or more are prescribed, with holding times of 2–8 s commonly applied. Temperatures of less than 135°C are considered insufficient for sterilization (Mehta, 1980). Burton (1988) defined the UHT process as one in which the product is heated in continuous flow to a temperature of 135–150°C and maintained at that temperature for a length of time sufficient to ensure commercial sterility with an acceptable change in the product (e.g., flavor, color, and nutrient composition). The product after sterilization is aseptically filled into appropriate sterile containers for nonrefrigerated distribution and sale.

In the United States, the thermal process and procedures for manufacturing UHT aseptically processed milk and milk products must comply with the FDA requirements for sterilizing low acid foods (US Code of Federal Regulations Title 21, Part 113) and the requirements of the Pasteurized Milk Ordinance (PMO; Public Health Service, 1999). The products must be hermetically sealed in a container, free of microorganisms capable of reproducing under unrefrigerated conditions and free of viable microorganisms of public health significance, including bacterial endospores. Heat treatments at or above those commonly used for ultra-pasteurization are generally considered acceptable for UHT processing.

UHT processing first emerged in the late 1940s, with the development of a tubular indirect heating system (Stork Netherlands) and the Uperization steam-into-milk system (Alpura AG & Sulzer AG, Switzerland). The aseptic canning systems that were available at this time made the product difficult and expensive to produce. In 1961, the development of an aseptic form-fill-seal packaging system (Tetra Pak, Sweden) was instrumental in making UHT milk a commercially viable product (Burton, 1988). Today, UHT milk is processed either by indirect heating or by direct heating (Mehta, 1980; Burton, 1988; Bylund, 1995) coupled with aseptic handling and packaging. Indirect processing involves a heat-conducting barrier, such as a stainless steel plate (plate heat exchanger) or tube (tube heat exchanger) that separates the heating medium from the milk. After the required time at the required temperature, milk is generally cooled by successive indirect cooling heat exchangers. Direct processing of UHT milk involves heating milk by mixing it with culinary steam. Direct heating systems for UHT milk include steam injection, steam infusion, and falling film designs. With steam injection systems, steam is injected directly into the milk flow. With steam infusion and falling film systems, the milk is sprayed or dropped as a thin film into a vessel of pressurized steam. The milk is maintained at the required temperature for the required time in the holding area and is then immediately cooled. Cooling is typically accomplished in a condenser equipped with a vacuum/expansion chamber, which also removes water that was added to the milk through the steam. Milk is further cooled, transported to aseptic holding tanks, and then packaged in an aseptic filling machine. Milk products are typically packaged using form, fill, and seal equipment or aseptic bottle fillers. UHT-sterilized milk products are considered shelf-stable, with expected shelf lives of 3 months to 1 year with no refrigeration required.

3.2.4 In-Container Sterilization

Conditions specified for in-container sterilization of milk include temperatures from 105°C to 120°C for 20–40 min (Burton, 1984; Hinrichs and Kessler, 1995). With this strategy, milk is prefilled into cans or bottles that are hermetically sealed, and then the milk is heated in an autoclave or a batch or continuous retort. As with UHT processing, the goal of the in-container sterilization process is to hold the product at a sufficient temperature for a sufficient length of time to ensure a 9-log reduction of thermophilic bacterial endospores (Hinrichs and Kessler, 1995) or a 12-log reduction of *Clostridium botulinum*. In-container sterilized milk products are less commonly manufactured than UHT products. The most common application of the process is in the manufacture of milk-based flavored (e.g., coffee) beverages. In general, milk products processed in this manner are subject to more pronounced defects related to cooked flavors and the Maillard browning reaction than are products processed by other UHT strategies. In-container sterilized milk products have expected shelf lives of a year or more with no refrigeration required.

3.3 THE MICROFLORA AND ENZYMATIC ACTIVITY OF HEAT-TREATED MARKET MILKS—INFLUENCE ON QUALITY AND SHELF LIFE

Food product shelf life is the length of time that a food can be kept under practical storage conditions and still retain acceptable quality characteristics. These characteristics include sensory appeal (odor, flavor, texture, and physical appearance) and product safety (Muir, 1996). The initial quality and shelf-life stability of heat-treated market milks is dependent on (1) the microbiological and enzymatic quality of the raw milk and other ingredients used in manufacturing, (2) the heat resistance and activity of intrinsic and microbial enzymes associated with the raw milk, (3) the types, initial numbers, and potential growth of microorganisms that survive the heat treatment processes, and (4) the incidence, types, and growth of microorganisms that recontaminate the product after the heat treatment (postpasteurization contamination). In general, obvious sensory defects related to microbial growth do not occur until bacterial populations exceed 10^6–10^7 cfu/ml, although the actual defects encountered are dependent on the types and activity of the microflora present (Patel and Blankenagel, 1972; Muir, 1996).

3.3.1 Pasteurized Market Milks

3.3.1.1 The Influence of Raw Milk Quality. Raw milk quality will influence the quality of the processed products. High-quality raw milk is essential for high-quality, long-lasting market milk products. Raw milk can deteriorate prior to processing as a consequence of milk production and handling procedures that result in contamination and growth of microorganisms that degrade milk components. In many countries, raw milk is required to be held under refrigerated conditions prior to processing; however, lack of electrification and appropriate technology prevents on-farm milk refrigeration in many parts of the world. Poor cooling conditions may lead to the rapid growth of mesophilic bacteria such as lactic acid bacteria. When these bacteria are present in large numbers (>10^6 cfu/ml), defects associated with the presence of lactic acid, other fermentation by-products, and milk protein degradation may be detected. Certain strains of *Lactococcus lactis* (formerly *Streptococcus lactis* var. *maltigenes*) are responsible for a "malty" defect that has been associated with inadequate cooling of raw milk. This defect may develop further in products processed from raw milk bearing this defect. A number of compounds have been found to contribute to the "malty" aroma in addition to 2-methylpropanol and 3-methylbutanol, which are considered to be the principal components of this defect (Bodyfelt et al., 1988).

Under refrigerated storage, Gram-negative rods capable of reproducing under refrigeration temperatures (termed psychrotrophic), principally *Pseudomonas* spp., often dominate the microflora of raw milk, especially when the milk is held for extended periods prior to processing (Cousin, 1982; Muir, 1996; Shah, 1994). Defects due to excessive numbers of psychrotrophic microorganisms in the raw milk also can negatively affect the quality of the finished product. Defects are generally the result of the action of exo-enzymes, specifically proteinases, which destabilize milk proteins to release bitter peptides as well as lipases, phospholipases, and esterases that act on milk lipids to yield rancid, unclean, and fruity off-flavors. In addition, some bacterial strains, particularly Gram-negative psychrotrophs, produce heat-stable enzymes that can survive the pasteurization process and continue to degrade product quality, even though the original organism may have been destroyed by the thermal treatment. Such heat-stable enzymes include proteinases that can create bitter compounds, unclean flavor components, and gelation of UHT milks (Fairbairn and Law, 1986; Mottar, 1989). Heat-stable lipases can contribute to the development of rancid and unclean flavor notes (Stead, 1986; Mottar, 1989). Extensive reviews are available on this subject (Fairbairn and Law, 1986; Stead, 1986; Mottar, 1989).

Raw milk may also bear thermoduric bacteria that survive pasteurization and other heat treatments (Olson and Mocquat, 1980; Hull et al., 1992; Kikuchi et al., 1996). Some of these organisms are capable of reproducing under refrigerated conditions (Boor et al., 1998; Ralyea et al., 1998), which can lead to bacterial numbers that exceed the regulatory limits for the finished products and create product defects, thereby shortening product shelf life. Psychrotrophic spore-forming bacteria that survive pasteurization are often implicated in milk spoilage despite the fact that the numbers of these organisms in the raw milk supply are often very low, that is, <1/ml (Boor et al., 1998).

Patel and Blankenagel (1972) found that batch pasteurized milk made from raw milks with bacteria numbers $>10^6$ cfu/ml had a greater tendency to develop defects after 1–2 weeks of storage at 7°C than did products manufactured from milk with lower bacterial counts. Bitterness was the most common defect described. They also found that some pasteurized samples made from raw milk with bacterial numbers in excess of 10^8 cfu/ml had acceptable flavor after 2 weeks of storage, illustrating the concept that the type of bacteria present and their associated enzymatic activities are critical determinants of product stability. While it has been suggested that pasteurized milk products should not be manufactured from raw milks that exceed 5×10^6 cfu/ml at the time of processing (Muir, 1996), regulations and industry standards dictate lower count raw milk to ensure high product quality (Table 3.3).

Enzymes inherent to the raw milk can also affect milk quality. Although many enzymes, such as lipoprotein lipase and alkaline phosphatase (ALP), are inactivated by pasteurization, others may withstand this heat treatment. ALP, which has been shown to be more heat-resistant than *C. burnetii*, is commonly used as an indicator of pasteurization efficacy for bovine milk. Detection of ALP in pasteurized milk suggests either inadequate pasteurization or recontamination with raw milk. Some recent work suggests the possibility that enzymes associated with high somatic cell counts due to mastitis can survive pasteur-

TABLE 3.3. Microbial and Somatic Cell Count Standards for Raw Milk for Pasteurized Milk Products

Country	Producer Raw Milk	Plant Raw Milk
USA[a]	100,000 cfu/ml 750,000 SCC	300,000 cfu/ml
Canada[b]	50,000 cfu/ml 500,000 SCC	50,000 cfu/ml
EEC[c]	100,000 cfu/ml 400,000 SCC	300,000 cfu/ml
Australia/New Zealand[d]	150,000 cfu/ml	150,000 cfu/ml

[a]Public Health Service. US Food and Drug Administration (1999). *Grade "A" Pasteurized Milk Ordinance*, 1999 revision, Publication No. 229.
[b]Canadian Food Inspection System, National Dairy Regulation and Code, October 1, 1997. *Production and Processing Regulations, Part 2: Processing.* http://cfis.agr.ca/codedairyreg2.htm.
[c]European Economic Community. 5. Health Rules—Raw milk, heat-treated milk, and milk based products. Council Directive 92/46/EEC of June 16, 1992. Laying down the health rules for the production and placing on the market of raw milk, heat-treated milk, and milk based products as amended by Directive EEC No. 92/118 of March 15, 1993; Commission Decision 94/330/E.C. of June 11, 1994; Directive E.C. No. 94171 of December 13, 1994.
[d]Australian New Zealand Food Authority, Food Standards Code. Standard H1—Milk and Liquid Milk Products. http:/anzfa.gov.au/FoodStandardCode/code/parth/H1.htm.

ization and cause an increase in bitter and rancid off-flavors in HTST pasteurized products stored beyond 14 days (Ma et al., 2000). Stage of lactation may also affect the quality of finished product. According to Auldist et al. (1996), early lactation milk was more prone to gelation than late lactation milk, whereas milk with higher somatic cell counts tended to gel first in both categories. This study suggests that age gelation may be influenced by milk compositional factors and may not be always directly linked to proteolysis.

3.3.1.2 Microflora of Pasteurized Milk. The microflora of pasteurized milk is primarily bacterial in nature, although contamination with yeasts and molds can occur. Bacteria commonly isolated from pasteurized milk are usually the same types of bacteria that are found in raw milk. Microorganisms found in pasteurized milk originate from (1) thermoduric organisms present in the raw milk supply, (2) raw milk contact with contaminated handling and processing equipment, and (3) entry after the pasteurization process. Typical total bacterial numbers in freshly pasteurized milk are less than 1000 cfu/ml. Acceptable limits for bacterial numbers in pasteurized milk are listed in Table 3.4.

The initial microflora of freshly pasteurized milk usually reflects the Gram-positive thermoduric microflora present in the raw milk. Although the raw milk supply is generally considered to be the principal source of thermoduric species (including *Bacillus* spp.) that are present in pasteurized milk, an improperly cleaned dairy plant processing system may also contribute large numbers of these organisms. In-plant contamination of milk with these types of organism can occur before, during, or after the pasteurization process (Te Giffel et al., 1997). Thermoduric strains of *Bacillus*, *Microbacterium*, *Micrococcus*, *Enterococcus*, *Streptococcus*, *Arthrobacter*, *Lactobacillus*, and *Clostridium* have been isolated from processed milk products (Olson and Mocquat, 1980; Martin, 1981; Hull et al., 1992; Kikuchi et al., 1996; Ralyea et al., 1998). Heat processing characteristics affect the relative proportions of bacterial types that survive pasteurization (Cromie et al., 1989). Gram-negative bacteria generally do not survive pasteurization (Cousin, 1982), unless total bacterial numbers in the raw milk exceed the thermal destruction capability of the pasteurization process (Patel and Blankenagel, 1972). The *Bacillus* species that have been most commonly isolated from raw milk are listed in Table 3.5. *B. licheniformis*, *B. cereus*, *B. circulans*, and *B. subtilis* are the species that have been most commonly isolated from freshly pasteurized products (Hull et al., 1992; Ternström et al., 1993; Crielly et al., 1994).

TABLE 3.4. Microbiological Standards for Pasteurized Milk Products

Country[a]	Total Bacteria[b]	Coliform Bacteria[b]
USA	20,000	10
Canada	$m = 10{,}000$ $M = 25{,}000$ $n = 5$ $c = 2$	$m = 1$ $M = 10$ $n = 5$ $c = 2$
EEC	After 5 days at 6°C $m = 50{,}000$ $M = 500{,}000$ $n = 5$ $c = 1$	 $m = 0$ $M = 5$ $n = 5$ $c = 1$
Australia/ New Zealand	$m = 50{,}000$ $M = 100{,}000$ $n = 5$ $c = 1$	$m = 1$ $M = 10$ $n = 5$ $c = 1$

[a]Sources are Same as those for Table 3.3.
[b]Total bacteria and coliform bacteria counts given as the upper limit of cfu/ml for the United States. For Canada, EEC, and Australia/New Zealand, two-tiered limits are given, with allowable results based on *n* number of samples as described below:
n = number of sample units (subsamples) to be examined per lot.
m = maximum number of bacteria per gram or milliliter of product that is of no concern (acceptable level of contamination).
M = maximum number of bacteria per gram or milliliter of product, which, if exceeded by any one sample unit (subsamples), renders the lot in violation of the Regulations.
c = maximum number of sample units (subsamples) per lot that may have a bacterial concentration higher than the value for m but less than value for M without violation of the regulations.

TABLE 3.5. *Bacillus* Species Isolated from Raw and Pasteurized Milk

B. brevis[a]	*B. lentus*[a]
B. carotarum[a]	*B. licheniformis*
B. cereus[a]	*B. megatarium*
B. cereus var. *mycoides*[a]	*B. polymyxa*[a]
B. circulans[a]	*B. pumilus*[a]
B. coagulans	*B. sphaericus*
B. firmus[a]	*B. stearothermophilus*
B. lateropsorus	*B. subtilis*
	B. thuringiensis[a]

[a]Indicates psychrotrophic strains isolated by Griffiths and Phillips (1990).

Sources: Martin (1981), Matta and Punj (1999), Meer et al. (1991), and Griffiths and Phillips (1990).

Not all thermoduric strains of bacteria are capable of reproducing in pasteurized milk under conditions of refrigerated storage. Small numbers of bacterial contaminants of this nature are unlikely to cause product spoilage within a typical fluid milk product shelf life. However, psychrotrophic strains of *Bacillus*, *Micrococcus*, *Enterococcus*, *Corynebacterium*, *Microbacterium*, *Arthrobacter*, and *Lactobacillus* have been identified (Cousin, 1982; Meer et al., 1991). Strains of *B. cereus* (and the closely related *B. mycoides*) have been implicated as the cause of "sweet-curdling" of milk and "bitty" cream (Overcast and Atmaram, 1974; Meer et al., 1991). Other milk defects that have been associated with *Bacillus* include bitter, yeasty, unclean, and rancid off-flavors as well as coagulation of the milk proteins. Although commonly present in milk, *B. cereus*, a potential foodborne pathogen, may not grow as well at lower refrigeration temperatures (<5°C) as other *Bacillus* spp. such as *B. polymyxa* and *B. circulans* (Langeveld and Cuperus, 1980; Ternström et al., 1993). In a comparison of microbial numbers present in the same milks held for 3 weeks at 5°C and for 2 weeks at 7°C, Ternström et al. (1993) found that 66% and 86% had bacteria counts of $>10^7$ cfu/ml, respectively. In product held at 5°C, *B. polymyxa* was isolated from 17% of the milks that had reached 10^7 cfu/ml, whereas *B. cereus* was not isolated. In product held at 7°C, *B. cereus* was isolated from 18% of the samples, suggesting that the increased temperature influenced the relative abilities of the organisms to reproduce.

Psychrotrophic Gram-positive organisms other than *Bacillus* spp. also may be responsible for limiting the shelf life of pasteurized milk. Kozlowski et al. (1993) evaluated the dominant microflora of 106 commercial pasteurized milk samples with total bacterial counts exceeding 10^6 cfu/ml after 14 days of storage at 6.1°C. Although the predominant organisms were Gram-negative rods, Gram-positive cocci were isolated from five of the samples tested. *Bacillus* strains were not found in this study. In approximately 7% of samples tested, Ternström et al. (1993) found strains of *Leuconostoc*, *Lactobacillus*, *Enterococcus*, and other unidentified Gram-positive bacteria at numbers of 10^7 cfu/ml in milks stored at 7°C for 2 weeks.

In general, postpasteurization contamination contributes the majority of microorganisms that contaminate and spoil pasteurized milk (Griffiths et al., 1984); thus, elimination of contamination sources can dramatically influence processed product shelf life (Gruetzmacher and Bradley, 1999). To illustrate, HTST product shelf life was extended from <7 days to >21 days following identification and elimination of a single point source of *Pseudomonas* spp. contamination in a product filling unit (Ralyea et al., 1998). Rapid trouble-shooting of postpasteurization contamination sites can be accomplished through application of tools such as adenosine triphosphate-bioluminescence hygiene monitoring kits (Murphy et al., 1998).

Postpasteurization contamination occurs when microbes are re-introduced into the pasteurized product as a consequence of product contact with contaminated processing or packaging equipment or workers. Postpasteurization contamination of milk is usually due to deficiencies in a plant's cleaning and sanitation program or to malfunctioning equipment. Typical processed milk contamination sources include improperly cleaned pasteurizers and filling units (Gruetzmacher and Bradley, 1999), plant water, airborne contaminants, and dairy personnel who may mishandle equipment components that are milk contact surfaces. The presence of milk residues on inadequately cleaned milk contact surfaces can support the growth of contaminating microorganisms. If not removed or inactivated, these microorganisms can subsequently contaminate milk during the next processing. Biofilms, which consist of microorganisms and cellular by-products that are attached to surfaces, have been implicated as persistent sources of contamination in dairy processing operations. For example, *P. fragi* has been shown to attach to stainless steel surfaces, even under the dynamic conditions of constant milk flow (Stone and Zottola, 1985). Once biofilm attachment occurs, equipment surfaces become more difficult to clean and sanitize (Frank and Koffi, 1990; Mosteller and Bishop, 1993). Thus, cleaning and sanitizing programs designed to control biofilm formation should help to reduce the likelihood of product contamination through contact with processing equipment. Specific strategies for reducing postpasteurization contamination and for extending the shelf life of pasteurized milk have been described by Barnard et al. (1992).

The presence of Gram-negative spoilage organisms, particularly *Pseudomonas* spp., currently are the primary contributors to the psychrotrophic spoilage of pasteurized milk (Cousin, 1982; Craven and Macauley, 1992; Kozlowski et al., 1993; Ternström et al., 1993; Shah, 1994). Although milk spoilage is most commonly attributed to the presence of high numbers of *P. fluorescens*, strains of *P. fragi*, *P. putida*, and *P. maltophilia* are also common (Wiedmann et al., 2000). Defects associated with *Pseudomonas* growth include fruity, bitter, rancid, and unclean off-flavors as well as coagulation of the milk proteins. Other psychrotrophic Gram-negative bacteria that grow and cause spoilage in pasteurized milk include strains of *Alcaligenes*, *Flavobacterium*, and Enterobacteriaceae. Psychrotrophic Enterobacteriaceae that have been isolated from milk include *Enterobacter cloacae*, *E. agglomerans*, *E. zakazakii*, *Citrobacter freundii*, *Klebsiella oxytoca*, and *Hafnia alvei* (Wessels et al., 1989; Kozlowski et al., 1993).

Because Gram-negative coliform bacteria do not survive pasteurization treatments, the presence of these organisms in pasteurized products is used by the dairy industry as an indication of the possibility of postpasteurization contamination. Coliforms are defined as aerobic and facultatively anaerobic, Gram-negative, non-spore-forming rods that are able to ferment lactose with the production of acid and gas at 32°C or 35°C within 48 hours (Christen et al., 1993). Coliform bacteria isolated from milk and dairy products include species of *Escherichia*, *Enterobacter*, and *Klebsiella* and selected strains of other fermentative bacteria. Currently, in the United States, the legal limit for coliforms in milk is 10 per ml (Public Health Service, 1999). The absence of coliforms does not guarantee the absence of other bacterial contaminants that might cause deterioration of the milk, nor does the presence of coliforms guarantee that other spoilage organisms are present.

Numbers of psychrotrophic contaminants occurring in freshly pasteurized milk can be very low, perhaps below detection limits by conventional enumeration strategies—that is, <10 cfu/ml (Douglas et al., 2000). However, the ability of these organisms to reproduce at refrigeration temperatures can allow them to flourish during storage. Cousin (1982) describes generation times for some psychrotrophic milk contaminants of less than 6 h at temperatures less that 7°C (45°F). Given these kinetics, if one bacterium were to reliably double every 6 h, bacterial numbers could exceed 10 million after 7 days. Thus, even low-levels of postpasteurization contamination can dramatically influence product shelf-life.

A summary of the shelf life characteristics of samples collected over a 2-year period from 25 dairy processors is presented in Table 3.6. After 14 days of refrigeration storage, 23% of the samples had bacterial numbers exceeding 10^6 cfu/ml after only 7 days of refrigeration storage, suggesting a significant level of postpasteurization contamination among many of the manufacturers' products. However, 26% of the HTST pasteurized milk samples tested had <20,000 cfu/ml, which, in general, was due to the absence of Gram-negative psychrotrophic post-processing contaminants in these samples. These data demonstrate the potential for extending fluid milk shelf life beyond 14 days for HTST fluid milk products. The shelf life actually attained by a product depends on a number of variables. Some of these factors include (1) the types and initial numbers of psychrotrophs present in the processed

TABLE 3.6. Percent of Pasteurized Milk Samples Meeting Shelf-Life Criteria During 14-Day Storage at 6.1°C[a]

Test Criteria[b]	Initial	Day 7	Day 10	Day 14
Total count ≤20,000	99	60	43	26
Total count $\leq 10^6$	100	77	51	41
Coliform count ≤10	98	85	79	79
Acceptable flavor[c]	99	96	82	62

[a]From Boor and Bandler (1998) and Boor and Bandler (1999). $n = 371$. Products included homogenized full-fat, reduced fat (2% BF), lowfat (1% BF), and nonfat milk from four samplings at each of 25 processing plants.
[b]Test criteria: ≤20,000 based on the US regulatory limit; $\leq 10^6$ due to likelihood of detection of sensory defects at and above this limit; ≤10 coliform based on the US regulatory limit.
[c]Flavor acceptability determined by six to eight trained panelists following guidelines as described by Bodyfelt et al. (1988).

product, (2) the extent of microbial injury due to heat or exposure to sanitizing chemicals and repair time necessary prior to bacterial reproduction, (3) the length of the lag phase before bacterial growth commences, and (4) bacterial generation times at refrigeration temperatures, because different holding temperatures will influence microbial ecology (Ternström et al., 1993). As processors succeed in preventing postpasteurization contamination by Gram-negative bacteria that are present in the processing environment, product shelf life will be influenced primarily by the psychrotrophic nature of the residual thermoduric microflora (Ralyea et al., 1998).

3.3.2 Ultra-pasteurized Milk

3.3.2.1 Influence of Raw Milk Quality. The primary objective of ultra-pasteurization processing is the elimination by heating of virtually all microorganisms present in raw milk. The heating process may not inactivate heat-stable enzymes, however, and these enzymes can contribute to product deterioration during storage. Sources of heat-stable enzymes in raw milk include Gram-negative psychrotrophic organisms (Adams et al., 1975; Adams and Brawley, 1981; Fairbairn and Law, 1986; Stead, 1986) and indigenous enzyme systems, including plasminogen and plasmin (Manji et al., 1986). To date, the influence of heat-stable enzymes on ultra-pasterized (UP) milk quality has not been well documented; however, refrigerated storage of these products may help reduce enzymatic degradation of the products during product shelf life. For example, in one study of commercial UP milks held at 7°C, stable acid degree values throughout 10 weeks of storage suggested limited product lipolysis. An increase in tyrosine values in the same products suggested some proteolysis; however, no detectable sensory defects were observed within the 10-week storage period (Boor and Nakimbugwe, 1998). In general, the optimal activities of the majority of heat-stable enzymes that have been studied are above refrigeration temperatures, generally ranging from 30°C to 45°C, with reduced or limited activities below 7°C (Mottar, 1989).

3.3.2.2 Microflora of UP Milks. Microbial numbers in UP milk products are controlled by (1) microbial destruction from ultra-pasteurization heat treatment, (2) sterilization of milk contact surfaces, and (3) application of disinfection strategies to UP milk packaging materials. In general, UP milk packaging equipment is enclosed to protect the product from environmental contamination. The air within the filling equipment is usually treated by high-efficiency particulate air (HEPA) filtration to reduce the presence of airborne contaminants. Immediately prior to filling, packaging materials are typically disinfected with hydrogen peroxide, heat, ultraviolet light, or a combination thereof. Effective application of these combined strategies can render the milk product virtually free from microbial recontamination (Henyon, 1999). Prevention of product contamination by potential spoilage microorganisms can allow refrigerated product shelf life of 30–90 days and more. However, any deficiencies in the sanitation process could allow postprocessing contamination, which could result in product spoilage similar to that seen in conventionally pasteurized milk.

Although UP milk packaging material is disinfected, it is not necessarily sterile; thus residual microflora associated with packaging materials may be a limiting factor in UP product shelf life. Paperboard used for liquid packaging may bear various species of fungi such as *Penicillium* (Narciso and Parish, 1997). Bacteria that have been isolated from paperboard include aerobic spore-forming bacteria such as species of *Bacillus* and *Paenibacillus* (Pirttijavi et al., 1996). Product exposure to unlaminated paperboard edges may allow milk to wick into the material, which may promote growth of organisms that may be present in the paperboard (Sammons et al., 2000). Although uncommon, mold growth has been reported in packaged UP milks, especially when storage has been extended beyond 45 days. Gable-top paperboard packaging materials used for UP milks may be treated with gamma irradiation to eliminate microbial contamination.

3.3.3 UHT Sterilized Milk

3.3.3.1 Influence of Raw Milk Quality. UHT milk is heat-processed and packaged to render the product commercially sterile. The objective of UHT processing as defined by the European Economic Community (EEC) Health Rules is to destroy all spoilage organisms and their spores. Burton (1988) gives the following example to calculate the theoretical spoilage rate of a UHT product. If milk contains 100 spores per milliliter and is heat-treated at a temperature and time that results in a sterilization efficiency of 8 for that organism (10^8 reduced to 1) and 10,000 liters are processed, the number of surviving spores would be $10{,}000 \times 1000 \times 100/10^8 = 10$. If this milk was packaged in 1-liter containers, then 1/1000 packages could contain a surviving spore and, thus, have an increased likelihood of bacterial spoilage. Burton (1988) recommends heating raw milk to 100°C for 30 min to obtain an estimate of heat-resistant spores that might survive UHT processing. In one study, the mean number of mesophilic aerobic spores detected after heating raw milk from various sources to 80°C for 12 min was less than 100 cfu/ml, suggesting that numbers of heat-resistant organisms in the raw milk supply may be quite low (Boor et al., 1998).

As with pasteurized and UP milks, raw milk defects due to extensive bacterial growth can negatively affect packaged UHT milk quality. Milk casein exposed to microbial proteinases can become unstable upon exposure to heat, which may result in product coagulation during UHT processing (Mottar, 1989). Heat instability of this nature has been linked to the presence of raw milk mixed flora of 7×10^6 cfu/ml as well as to 10^7 to 10^8 cfu/ml of specific strains of *Pseudomonas*.

Heat-stable enzymes that can survive UHT heat processing and negatively affect processed product quality are produced by a number of Gram-negative bacteria that are common in raw milk, including strains of *Pseudomonas*, *Acinetobacter*, *Achromobacter*, *Aeromonas*, *Flavobacterium*, and *Serratia* (Mottar, 1989). Enzymes that affect UHT milk quality include extracellular proteinases and lipases. To minimize the risk of processed product defects associated with heat stable microbial enzymes, Burton (1988) suggests maintaining raw milk bacterial numbers below 5×10^5 cfu/ml prior to processing.

Heat-stable proteinases of microbial origin have been implicated in the development of bitter peptides and in the acceleration of age gelation in UHT milk. These proteinases have been found to be most active against κ-casein and β-casein, with minor activity against α-casein and little or no activity against the whey proteins (Adams et al., 1976; Law et al., 1977, Lopez-Fandino et al., 1993). The development of bitterness has been shown to correlate with the protease activity (Adams et al., 1976) and the extent of proteolysis in UHT milks (Collins et al., 1993). The proteolytic activity against κ-casein has been likened to chymosin coagulation of milk and is thought to be responsible, in part, for age gelation of UHT milks. Age gelation is characterized by the formation of a soft curd and loss of fluidity (Manji et al., 1986).

Adams et al. (1975) identified multiple *Pseudomonas* isolates capable of producing proteinases that retain >70% activity after a heat treatment of 149°C for 4s. The optimal activities of these *Pseudomonas* proteinases are reported at temperatures ranging from 30°C to 45°C (Fairbairn and Law, 1986; Mottar, 1989), with significant reductions in activity at refrigeration temperatures. In UHT milks processed from 21 batches of naturally contaminated raw milk, proteolysis was most extensive when the milk was held at 30°C in comparison to products stored at other temperatures ranging from 2°C to 50°C (Kocak and Zadow, 1985). Proteolysis was negligible in milks held at or below 10°C. Preprocessing bacterial numbers in this milk ranged from 9.1×10^4 to 7.2×10^7 cfu/ml. In this study, no correlation was found between numbers of microorganisms, extent of proteolysis, or age gelation. In another study, onset of UHT milk gelation was measured in products made from raw milk in which a heat-resistant proteinase-producing strain, *Pseudomonas fluorescens* AR11, was grown to various concentrations (Law et al., 1977). Gelation occurred after 10–14 days or 8–10 weeks at 20°C when raw milk bacterial numbers had been 5×10^7 or 8×10^6 cfu/ml, respectively. Although age gelation has been associated with heat-stable proteolytic enzymes, the extent of milk proteolysis at the onset of gelation can vary considerably, suggesting that proteolytic activity may not be a direct predictor of gelation (Kocak and Zadow, 1985; Manji et al., 1986). Variability in the onset of age gelation has been hypothesized to result from variations in (1) the specific microflora of the raw milk, (2) the specificity, activity, and heat resistance of both microbial and endogenous milk enzymes, and (3) the rate of composition- and temperature-dependent physiochemical reactions (Kocak and Zadow, 1985).

Heat-stable lipases also have been shown to influence the quality of UHT milk, although to a lesser extent than proteinases (Adams and Brawley, 1981; Stead, 1986; Mottar, 1989). Choi and Jeon (1993) found that levels of short-chain fatty acids associated with rancid flavors in milk increased at during UHT product storage at 35°C but not during storage at 25°C.

3.3.3.2 Microflora of UHT Milk. Microbial control strategies for UHT milks are similar to those described for ultra-pasteurization processing, except that the products are aseptically packaged in sterilized packaging materials in a sterile environment to yield commercially sterile products. Product spoilage due to postprocessing contamination is prevented by adherence to and documentation of scientifically established processing parameters with properly operated equipment and reliably sealed containers (Dunkley and Stevenson, 1987). If the sterility of the system is compromised at any point during processing, the process must be shut down and the equipment cleaned and resterilized before processing is resumed.

3.4 PATHOGENIC MICROORGANISMS ASSOCIATED WITH HEAT-TREATED MARKET MILKS

3.4.1 Pasteurized Milk: Pathogenic Agents not Destroyed by Pasteurization

Currently, the most common method for destroying pathogenic organisms and for reducing or eliminating spoilage organisms in US dairy products is through pasteurization by the HTST method. These thermal treatments are designed to destroy the most heat-resistant of the non-spore-forming pathogens, specifically, *Coxiella burnetii.* Some microbes can survive pasteurization (Hammer et al., 1995). Spore-forming bacteria, including those of the *Bacillus* and *Clostridium* genera (e.g., *Bacillus cereus*, *Clostridium botulinum*, *Clostridium perfringens*), are among the heat-resistant pathogens that occasionally can be isolated from pasteurized milk. Some studies also suggest the possibility that *Mycobacterium paratuberculosis*, a bacterium that causes Johne's disease in cattle and that has been hypothetically linked to Crohn's disease in humans, may also survive pasteurization (Mechor, 1997). For example, in one study designed to examine commercially pasteurized milk for the presence of *M. paratuberculosis*, Millar et al. (1996) found that at least 15 of 312 milk samples collected at retail stores contained viable *M. paratuberculosis*. These results do not necessarily prove that *M. paratuberculosis* survives commercial pasteurization procedures, but could also suggest the possibility of postpasteurization contamination with this organism.

Mycobacterium paratuberculosis is the causative agent of Johne's disease in dairy cows. This transmissible disease, which currently affects approximately 33% of US dairy herds (Collins, 1997), dramatically reduces milk production, reproductive performance, and animal condition (Stabel, 1997) and thus has a significant negative economic impact on the dairy industry. This organism, which is excreted in feces and milk, is reportedly not as easily inactivated by pasteurization and thermal treatments as other bacteria infecting humans and animals (Sung and Collins, 1998). Determination of definitive thermal destruction characteristics of *M. paratuberculosis* does not represent a trivial undertaking because culturing techniques for this organism are labor-intensive and very slow, taking up to 16 weeks to produce visible colonies on appropriate microbiological media (Collins, 1997).

The association between *M. paratuberculosis* and human Crohn's disease is highly controversial. In 1984, Chiodini et al. (1986) reported the first isolation of *M. paratuberculosis* from a Crohn's patient. These isolates were subsequently shown to be genetically identical to bovine *M. paratuberculosis* strains and were shown to be able to cause Johne's disease by oral inoculation in goats (Van Kruiningen et al., 1986; Collins, 1997). Since then, a variety of studies have been conducted to determine whether there is a correlation between the presence of *M. paratuberculosis* or other *Mycobacterium* species and human Crohn's disease (reviewed by Chiodini, 1989). These studies used culturing techniques to isolate *Mycobacterium* species from Crohn's patients or DNA-based approaches to screen for the presence of mycobacterial DNA in tissues obtained from Crohn's patients. Many studies have shown that a variety of *Mycobacterium* species, including *M. paratuberculosis*, can be isolated from patients with Crohn's disease. For example, Chiodini et al. (as cited in Chiodini, 1989) isolated *M. paratuberculosis* from 4 of 26 Crohn's patients but from none of 26 control samples. Additionally, Sanderson et al. (1992) found that 65% of the intestinal samples from Crohn's disease patients (n = 40) tested positive for the presence of *M. paratuberculosis* DNA by PCR as compared to 10% of the samples from controls (no inflammatory bowel disease or ulcerative colitis; n = 63). Fidler et al. (1994) reported that 4 of 31 Crohn's disease tissues, but none of 30 control and ulcerative colitis-derived samples, were positive for *M. paratuberculosis* DNA by PCR. Additional studies also found a higher incidence of *M. paratuberculosis* DNA in tissue from Crohn's patients as compared to control samples (e.g., Dell'Isola et al., 1994). However, other studies did not find any evidence for the presence of mycobacterial DNA in tissues from Crohn's disease patients using PCR (Wu et al., 1991; Rowbotham et al., 1995; Frank and Cook, 1996). In summary, although some studies have described an association between the presence of *M. paratuberculosis* and Crohn's disease, the role of *Mycobacterium* species and

M. paratuberculosis in the etiology of this human disease remains unestablished. However, to enable consumer assurance of the microbial safety of their pasteurized dairy products, determination of thermal destruction characteristics of *M. paratuberculosis* is a very high priority for the dairy industry.

Conventional thermal treatments for milk products will not inactivate the causative agent of bovine spongiform encephalitis (BSE), also known as mad cow disease. The agent responsible for this disease, which is not a microbe but rather an infectious protein, shows little loss of infectivity, even after prolonged exposure to temperatures up to 176°F (80°C) (Asher et al., 1986). Fortunately, no evidence exists linking transmission of this disease to consumption of milk from cows with BSE. For example, mice injected with milk from BSE-infected cattle did not develop this disease, nor have epidemiological analyses suggested transmission of BSE to calves via milk (Hillerton, 1997).

3.4.2 Pasteurized Milk: Postpasteurization Contamination

Although some pathogens can survive pasteurization, as described above, the presence of most pathogenic microbes (e.g., *Salmonella* spp. or *L. monocytogenes*) in processed dairy products implies either failure of the pasteurization process or postpasteurization contamination. In fact, the largest single salmonellosis outbreak in US history—over 23,000 culture-confirmed cases from March to April 1985—resulted from consumption of *Salmonella*-contaminated whole and 2% milk that had been produced by a suburban Chicago processor (Lecos, 1986). Other outbreaks linked to fluid milk products are listed in Table 3.7. The importance of protecting processed products from bacterial recontamination is further illustrated by the demonstration that the pathogen *Escherichia coli* O157:H7 can persist for at least 35 days as a postpasteurization contaminant in refrigerated buttermilk with a pH of 4.1 (Dineen et al., 1998).

Multiple routes exist for the entry of contaminating microbes into the dairy processing environment. For example, microbes, including the pathogenic organisms listed in Table 3.8, can be introduced into a dairy processing environment along with the raw milk. Other possible sources include contaminated workers, packaging materials, and distribution equipment (forklifts, pallets, etc.). The abundance of nutrients and moisture in a processing plant can facilitate the survival and growth of many of these contaminants. *Listeria innocua*, *Listeria monocytogenes*, and *Yersinia enterocolitica* are commonly isolated from dairy processing plant locations that involve "wet traffic" including the floors

TABLE 3.7. Examples of Food-Borne Outbreaks Associated with Market Milk Products

Organism	Year	Implicated Food	Extent, Location
Yersinia enterocolitica	1976	Chocolate milk	220 children, New York, USA
Yersinia enterocolitica	1981	Reconstituted nonfat dry milk	239 young adults, New York, USA
Yersinia enterocolitica	1982 (two outbreaks)	Pasteurized milk	>472 people; Connecticut, Tennessee, Arkansas, Mississippi, USA
Listeria monocytogenes	1983	Pasteurized milk	49 illnesses, 14 deaths; Massachusetts, USA
Campylobacter jejuni	1984	Certified raw milk	12 cases; California, USA
Staphylococcus aureus enterotoxin	1985	Chocolate milk	860 children; Kentucky, USA
Listeria monocytogenes	1994	Chocolate milk	45 people; Illinois, USA
Staphylococcus aureus enterotoxin	2000	Pasteurized milk	>14,000 people, Osaka, Japan

of coolers, freezers, and processing rooms; cases and case washers; floor mats and foot baths; and beds of paper fillers (Donnelly, 1990). The possible presence of human pathogenic microorganisms in the dairy processing environment highlights the need to protect processed product from recontamination.

3.5 INFLUENCE OF ADDED INGREDIENTS

Although unflavored products comprise the largest fraction of fluid milk products currently sold in the United States (93%), with flavored milk products constituting only about 6% of total volume, flavored milk sales have shown substantial growth since 1995, with a nearly 8% increase in volume in 1998 and an additional 7% increase in 1999 (Milk Industry Foundation, 1999). Thus, timely attention to improvements in flavored milk quality and shelf life could provide a growth opportunity for the US fluid milk industry.

TABLE 3.8. Microbial Pathogens Isolated from Raw Milk

Bacterial pathogens	**Rickettsia**
Escherichia coli	*Coxiella burnetii*
Salmonella spp.	
Shigella spp.	**Viral**
Yersinia enterocolitica	Enterovirus, including polioviruses, Rotaviruses, and Coxsackie virus
Campylobacter jejuni	
Aeromonas hydrophila	Hepatitis viruses
Vibrio cholera	
Pseudomonas aeruginosa	**Protozoa**
Brucella spp.	
Bacillus cereus	*Entamoeba histolytica*
Bacillus anthracis	*Giardia lamblia*
Clostridium perfringens	*Toxoplasma gondii*
Clostridium botulinum	*Cryptosporidium parvum*
Staphylococcus aureus	
Streptococcus agalactiae	
Streptococcus pyogenes	
Streptococcus zooepidemicus	
Listeria monocytogenes	
Corynebacterium spp.	
Mycobacterium bovis	
Mycobacterium tuberculosis	
Mycobacterium paratuberculosis	

To evaluate the microbiology of flavored milk products, Douglas et al. (2000) measured total bacterial counts and psychrotrophic plate counts in matched flavored (chocolate) and unflavored fluid milk samples at 1, 7, 10, and 14 days postprocessing. Bacterial numbers within 24 h of processing were not significantly different in the unflavored and in the chocolate milk samples ($P > 0.001$). Total bacterial numbers and psychrotrophic plate counts were less than 1000 cfu/ml and 10 cfu/ml, respectively, for all products. However, both total bacterial numbers and psychrotrophic plate counts were higher in chocolate milk samples than in unflavored milk samples after 14 days of storage at 6°C ($P < 0.001$) in products collected from all four processing plants (Douglas et al., 2000).

This study sought to identify the source(s) of the bacterial contaminants in the flavored products. Bacteria present in the cocoa formulations were below the detection limits of the analytical procedures (<10 cfu/g) for products manufactured in all four plants, suggesting that these cocoa powders did not serve as major sources of contamination for these particular products. With one exception, isolation and identification of bacteria present in the various products showed that the profiles of the predominant organisms were the same for flavored and unflavored products processed in the same plants, suggesting that the spoilage organisms were not unique to either set of products. The predominant organisms present in 73% of the samples examined were identified as *Bacillus* spp. (Douglas et al., 2000). These organisms were found to be present in raw and in processed milk samples and were capable of surviving lab pasteurization procedures. These findings are consistent with previous observations of the widespread presence in the raw milk supply (albeit in low numbers) of heat-resistant spore-forming psychrotrophic bacteria that are capable of increasing in number when held at 7°C for 10 days (Boor et al., 1998).

To further investigate the effects of chocolate milk components on bacterial numbers in processed products, total bacterial numbers were monitored in experimentally prepared and aseptically collected unflavored milk, milk with chocolate powder and sucrose (chocolate milk), milk with sucrose only, and milk containing chocolate powder only on the initial day and after storage at 6°C on days 7, 14, and 21 post-processing. At days 14 and 21, total bacterial numbers were higher ($P < 0.001$) in both chocolate milk and in milk with chocolate powder only than in either milk with sucrose or in unflavored milk (Douglas et al., 2000), suggesting that the chocolate powder stimulates increases in bacterial numbers among the residual bacteria surviving HTST pasteurization. From these studies, it appears that extension of flavored (chocolate) milk shelf life and improved milk quality will require removal of bacterial spores from raw milk (e.g., through microfiltration strategies, such as those described below) or by application of processing conditions that utilize increased time or temperature treatments relative to those used for unflavored milk products or through a combination of these strategies.

The influence of chocolate milk ingredients on the growth of *Listeria monocytogenes* has also been examined (Rosenow and Marth, 1987a,b, Pearson and Marth, 1990). In a comparison of final microbial numbers in skim, whole and chocolate milk incubated at 4°C, 8°C, 13°C, 21°C, and 35°C, *Listeria monocytogenes* reached nearly 10-fold higher final numbers in chocolate milk than in the white milk at all temperatures (Rosenow and Marth, 1987a). Further investigation of the influence of the major ingredients of chocolate milk suggested that the combination of cocoa, sucrose, and carrageenan provided the greatest enhancement of growth at 13°C while cocoa and sucrose alone showed a marginal influence (Rosenow and Marth, 1987b). When similar studies were done at an incubation temperature of 30°C, three factors were found to increase numbers of *Listeria monocytogenes*: the addition of cocoa, the addition of sucrose, and agitation (Pearson and Marth, 1990).

Fortification of skim milk with nonfat dried milk has also been shown to stimulate the growth of some psychrotrophic bacteria (Jeong and Frank, 1988), but this effect appeared to be a minor contributor to differences in bacterial numbers.

3.6 POTENTIAL APPLICATIONS OF ALTERNATIVES TO HEAT FOR MARKET MILKS

3.6.1 Microfiltration

US consumers accustomed to flavors associated with HTST pasteurized milk products may report as objectionable the presence of the distinct cooked flavors in high-temperature processed milks such as UHT and ultra-pasteurized products (Hill, 1988). Microfiltration may provide an alternative approach for the production of dairy products with extended shelf lives (Olesen and Jensen, 1989; Eckner and Zottola, 1991; Madec et al., 1992). Microfiltration (MF) is the passage of product under relatively low pressure (approximately 1 bar) through a semipermeable membrane with pore sizes ranging from 0.2 μm to 5 μm (Olesen and Jensen, 1989). Bacause bacteria generally range from 1 μm to 3 μm in size, under some operating parameters, MF should be able to completely remove bacteria from the fluid permeate. MF might provide a lower temperature option, and thus a less pronounced cooked flavor, than UHT processing for production of extended shelf-life dairy products.

Very little published work exists on microfiltration for milk processing; and among the existing reports, the results are sometimes contradictory. Olesen and Jensen (1989) found that the initial content of *Bacillus cereus* spores in milk had a significant effect on the content of spores in the microfiltered milk, but that concentration ratio and circulation pressure had no effects under the conditions studied. Varying operating temperature between 30°C and 50°C did not appear to affect bacterial retention in one study (Eckner and Zottola, 1991); however, increasing microfiltration temperatures from 35°C to 50°C was reported to significantly increase *Salmonella*, but not *Listeria*, retention in another (Madec et al., 1992). Eckner and Zottola (1991) concluded that different membranes with the same molecular weight cutoff have different bacterial retention characteristics, but that bacterial morphology did not affect the ability of the organism to pass through the membrane. In contrast, Madec et al. (1992) reported that *Listeria* and *Salmonella* had differing retention characteristics, but that these characteristics were not influenced by initial contamination numbers, suggesting that bacterial size and shape may play a role in the passage of a particular organism through a membrane. This report also found retention of *Listeria* cells to be much lower when cells were inoculated into milk that had been previously microfiltered. In summary, because generalizations have yet to be established for this process, microfiltration operating parameters must be optimized within each processing establishment to meet the specific goals of a given manufacturing operation.

3.6.2 Carbon Dioxide Addition

The application of carbon dioxide as a bacteriostatic agent in foods and in modified atmosphere packaging has been reviewed by Daniels et al. (1985). CO_2 has been explored as a shelf-life extender for milk and dairy products (Hotchkiss et al., 1999). In milks inoculated with psychrotrophic *Pseudomonas* and *Enterobacter* spp., the presence of carbon dioxide was found to result in lower relative bacterial numbers at given postprocessing time points as a consequence of increasing the bacterial lag phase and reducing the rate of bacterial growth in milk stored at 6.1°C. Packaging contributed to these effects. CO_2-treated milks packaged in high barrier pouches exhibited the greatest inhibition in bacterial growth. The addition of CO_2 at 8.7 mM and 21.5 mM in products packaged in these pouches extended the time needed for bacterial numbers to reach 10^6 cfu/ml from 6.4 days (control) to 9.7 and 13.4 days, respectively. However, both levels of CO_2 were detectable by a trained sensory panel; thus, this strategy is unlikely to be a successful postpasteurization treatment for controlling bacterial numbers in processed milk.

3.7 SUMMARY

The fluid milk processing strategies that are currently most commonly and most successfully applied worldwide to preserve product quality and ensure public health involve the application of heat to destroy microbes that are present in raw milk along with hygienic practices to reduce or eliminate postprocessing reentry of microbes into the products. The time and temperature conditions of the thermal processing treatments, along with the efficacy of the sanitary measures for prevention of product recontamination, dramatically affect product shelf life. Some emerging new nonthermal technologies that show promise for the fluid milk industry include high-pressure processing (Garcia-Risco et al., 1998; Balci and Wilbey, 1999; Datta and Deeth, 1999) and pulsed electric field processing (Reina et al., 1998; Jeyamkondan et al., 1999). The objectives of these and other emerging strategies are to provide the bacterial destruction capabilities equivalent to those attainable by thermal processing while simultaneously yielding products with quality characteristics that are appealing to consumers.

REFERENCES

Adams, D. M., and Brawley, T. G. (1981) *J. Dairy Sci.*, **64**, 1951–1957.

Adams, D. M., Barach, J. T., and Speck, M. L. (1975) *J. Dairy Sci.*, **58**, 828–834.

Adams, D. M., Barach, J. T., and Speck, M. L. (1976) *J. Dairy Sci.*, **59**, 823–827.

Asher, D. M., Gibbs, C. J., Jr., and Gajdusek, D. C. (1986) In *Laboratory Safety: Principles and Practices*, B. M. Miller, ed., American Society of Microbiology, Washington, DC, 59–71.

Auldist, M. J., Coats, S. J., Sutherland, B. J., Hardham, J. F., McDowell, G. H., and Rogers, G. L. (1996) *J. Dairy Res.*, **63**, 377–386.

Balci, A. T., and Wilbey, R. A. (1999) *Int. J. Dairy Technol.*, **52**, 149–155.

Barnard, S. E., Ivkovich, M., and Cauller, P. (1992) *Dairy, Food Environ. Sanit.*, **12**, 66–68.

Bodyfelt, F. W., Tobias, J., and Trout, G. M. (1988) *The Sensory Evaluation of Dairy Products*, Van Nostrand Reinhold, New York.

Boor, K. J., and Bandler, D. K. (1998) *Milk Quality Improvement Program Annual Report*, Cornell University, Ithaca, NY.

Boor, K. J., and Bandler, D. K. (1999) *Milk Quality Improvement Program Annual Report*, Cornell University, Ithaca, NY.

Boor, K. J., and Nakimbugwe, D. N. (1998) *Dairy, Food Environ. Sanit.*, **18**, 78–82.

Boor, K. J., Brown, D. P., Murphy, S. C., Kozlowski, S. M., and Bandler, D. K. (1998) *J. Dairy Sci.*, **81**, 1743–1748.

Burton, H. (1984) *J. Dairy Res.*, **51**, 341–363.

Burton, H. (1988) *Ultra-High-Temperature Processing of Milk and Milk Products*. Elsevier Applied Science, New York.

Bylund, G. (1995) *Dairy Processing Handbook*, Tetra Pak Processing Systems AB, Lund, Sweden.

Chiodini, R. J. (1989) *Clini. Microbiol. Rev.*, **2**, 90–117.

Chiodini, R. J., Van Kruiningen, H. J., Thayer, W. R., and Coutu, J. A. (1986) *J. Clini. Microbiol.*, **24**, 357–363.

Choi, I. W., and Jeon, I. J. (1993) *J. Dairy Sci.*, **76**, 78–85.

Christen, G. L., Davidson, P. M., McAllister, J. S., and Roth, L. A. (1993) In *Standard Methods for the Evaluation of Dairy Products*, 16th ed., R. T. Marshall, ed., American Public Health Association, Washington, DC, 247–269.

Collins, M. T. (1997) *J. Dairy Sci.*, **80**, 3445–3448.

Collins, S. J., Bester, B. H., and McGill, A. E. (1993) *J. Food Prot.*, **56**, 418–425.

Cousin, M. A. (1982) *J. Food Prot.*, **45**, 172–207.

Craven, H. M., and Macauley, B. J. (1992) *Aust. J. Dairy Technol.*, **47**, 38–49.

Crielly, E. M., Logan, N. A., and Anderton, A. (1994) *J. Appl. Bacteriol.*, **77**, 256–263.

Cromie, S. J., Dommett, T. W., and Schmidt, D. (1989) *Aust. J. Dairy Technol.*, **44**, 74–77.

Daniels, J. A., Krishnamurthi, R., and Rizvi, S. S. H. (1985) *J. Food Prot.*, **48**, 532–537.

Datta, N., and Deeth, H. C. (1999) *Aust. J. Dairy Technol.*, **54**, 41–48.

Dell'Isola, B., Poyart, C., Goulet, O., Mougenot, J. F., Sadoun-Journo, E., Brousse, N., Schmitz, J., Ricour, C., and Berche, P. (1994) *J. Infect. Dis.*, **169**, 449–451.

Dineen, S. S., Takeuchi, K., Soudah, J. E., and Boor, K. J. (1998) *J. Food Prot.*, **61**, 1602–1608.

Donnelly, C. W. (1990) *J. Dairy Sci.*, **73**, 1656–1661.

Douglas, S. A., Gray, M. J., Crandall, A. D., and Boor, K. J. (2000) *J. Food Prot.*, **63**, 516–521.

Dunkley, W. L., and Stevenson, K. E. (1987) *J. Dairy Sci.*, **70**, 2192–2202.

Eckles, C. H., Combs, W. B., and Macy, H. (1936) *Milk and Milk Products*, McGraw-Hill, New York.

Eckner, K. F., and Zottola, E. A. (1991) *J. Food Prot.*, **54**, 793–797.

Fairbairn, D. J., and Law, B. A. (1986) *J. Dairy Res.*, **53**, 139–177.

Fidler, H. M., Thurell, W., Johnson, N. M., Rook, G. A. W., and McFadden, J. J. (1994) *Gut*, **35**, 506–510.

Frank, J. F., and Koffi, R. A. (1990) *J. Food Prot.*, **53**, 550–554.

Frank, T. S., and Cook, S. M. (1996) *Mod. Pathol.*, **9**, 32–35.

Garcia-Risco, M. R., Cortes, E., Carrascosa, A. V., and Lopez-Fandino, R. (1998) *J. Food Prot.*, **61**, 735–737.

Griffiths, M. W., and Phillips, J. D. (1990) *J. Soc. Dairy Technol.* **43**, 62–66.

Griffiths, M. W., Phillips, J. D., and Muir, D. D. (1984) *Hannah Res.*, **1984**, 77–87.

Gruetzmacher, T. J., and Bradley, R. L. (1999) *J. Food Prot.*, **62**, 625–631.

Hammer, B. W. (1948) *Dairy Bacteriology*, 3rd ed., John Wiley & Sons. New York.

Hammer, P., Lembke, F., Suhren, G., and Heeschen, W. (1995) In *Heat Treatment & Alternative Methods*. Proceedings of the IDF Symposium held in Vienna (Austria). International Dairy Federation. pp. 9–16.

Henyon, D. K. (1999) *Int. J. Dairy Technol.*, **52**(3), 95–101.

Hillerton, J. E. (1997) In *National Mastitis Council Annual Meeting Proceedings 1997*, 33–41.

Hinrichs, J., and Kessler, H. G. (1995) In *Heat-Induced Changes in Milk*, 2nd ed., International Dairy Federation, Brussels, Belgium, pp. 9–21.

Hotchkiss, J. H., Chen, J. H., and Lawless, H. T. (1999) *J. Dairy Sci.*, **82**, 690–695.

Hull, R., Toyne, S., Haynes, I., and Lehmann, F. (1992) *Aust. J. Dairy Technol.*, **47**, 91–94.

Jay, J. (1996) *Modern Food Microbiology*, 5th ed. Chapman and Hall, New York.

Jeyamkondan, S., Jayas, D. S., and Holley, R. A. (1999) *J. Food Prot.*, **62**, 1088–1096.

Jeong, D. K., and Frank, J. F. (1988) *J. Food Prot.*, **51**, 643–647.

Kikuchi, M., Matsumoto, Y., Sun, X. M., and Takao, S. (1996) *Anim. Sci. Technol.*, **67**, 265–272.

Kocak, H. R., and Zadow, J. G. (1985) *Aust. J. Dairy Technol.*, **39**(4), 14–21.

Kozlowski, S. M., Murphy, S. C., and Bandler, D. K. (1993) *J. Dairy Sci.*, **76**(suppl. 1), 97.

Langeveld, L. P. M., and Cuperus, F. (1980) *Netherlands Milk Dairy J.*, **34**, 106–125.

Law, B. A., Andrews, A. T., and Sharpe, M. E. (1977) *J. Dairy Res.*, **44**, 145–148.

Lecos, C. W. (1986) *FDA Consum.*, **20**, 14–17.

Lopez-Fandino, R., Olano, A., Corzo, N., and Ramos, M. (1993) *J. Dairy Res.*, **60**, 339–347.

Ma, Y., Ryan, C., Barbano, D. M., Rudan, M. A., and Boor, K. J. (2000) *J. Dairy Sci.*, 83, 264–274.

Madec, M. N., Mejean, S., and Maubois, J. L. (1992) *Lait*, **72**, 327–332.

Manji, B., Kayuda, Y., and Arnott, D. R. (1986) *J. Dairy Sci.*, **69**, 2994–3001.

Martin, J. H. (1981) *J. Dairy Sci.*, **64**, 149–156.

Matta, H., and Punj, V. (1999) *J. Dairy Technol.*, **52**, 59–62.

Mechor, G. D. (1997) In *National Mastitis Council Annual Meeting Proceedings 1997*, pp. 50–55.

Meer, R. R., Baker, J., Bodyfelt, F. W., and Griffiths, M. W. (1991) *J. Food Prot.*, **54**, 969–979.

Mehta, R. S. (1980) *J. Food Prot.*, **43**, 212–225.

Milk Industry Foundation (1999) *Milk Facts*. International Dairy Foods Association, Washington, DC, Publication #F-22200, 64 pages.

Millar, D. S., Ford, J., Sanderson, J. D., Withey, S., Tizard, M., Doran, T., and Hermon-Taylor, J. (1996) *Appl. Environ. Microbiol.*, **62**, 2446–3452.

Mosteller, T. M., and Bishop, J. R. (1993) *J. Food Prot.*, **56**(1), 34–41.

Mottar, J. F. (1989) In *Enzymes of Psychrotrophs in Raw Food*. R. C. McKellar, ed., CRC Press, Boca Raton, FL, pp. 227–243.

Muir, D. D. (1996) *J. Soc. Dairy Technol.*, **49**(1), 24–32.

Murphy, S. C., Kozlowski, S. M., Bandler, D. K., and Boor, K. J. (1998) *J. Dairy Sci.*, **81**, 817–820.

Narciso, J. A., and Parish, M. E. (1997) *J. Food Sci.*, **62**, 1223–1225, 1239.

Olesen, N., and Jensen, F. (1989) *Milchwissenschaft*, **44**, 476–479.

Olson, J. C., Jr., and Mocquat, G. (1980) In *Microbial Ecology of Foods*, Vol. II, J. H. Silliker, R. P. Elliott, A. C. Baird-Parker, F. L. Bryan, J. H. Christion, D. S. Clark, J. C. Olson, and T. A. Roberts, eds., Academic Press, New York, pp. 470–485.

Overcast, W. W., and Atmaram, K. (1974) *J. Food Technol.*, **37**, 233–236.

Patel, G. B., and Blankenagel, G. (1972) *J. Milk Food Technol.*, **35**, 203–206.

Pearson, L. J., and Marth, E. H. (1990) *J. Food Prot.*, **53**, 30–37.

Pirttijärvi, T. S. M., Graeffe, T. H., and Salkinoja, M. S. (1996) *J. Appl. Bacteriol.*, **81**, 445–458.

Public Health Service (1940) In *Milk Ordinance and Code, Recommended by the US Public Health Service, 1939*, Public Health Bulletin No. 220, US Government Printing Office, Washington, DC, p. 4.

Public Health Service. US Food and Drug Administration (1999) *Grade "A" Pasteurized Milk Ordinance*, 1999 revision, Publication No. 229.

Ralyea, R. D., Wiedmann, M., and Boor, K. J. (1998) *J. Food Prot.*, **61**, 1336–1340.

Reina, L. D., Jin, Z. T., Zhang, Q. H., and Yousef, A. E. (1998) *J. Food Prot.*, **61**, 1203–1206.

Roadhouse, C. L., and Henderson, J. L. (1941) *The Market Milk Industry*, McGraw-Hill, New York.

Rosenow, E. M., and Marth, E. H. (1987a) *J. Food Prot.*, **50**, 452–459.

Rosenow, E. M., and Marth, E. H. (1987b) *J. Food Prot.*, **50**, 726–729.

Rowbotham, D. S., Mapstone, N. P., Trejdosiewicz, L. K., Howdle, P. D., and Quirke, P. (1995) *Gut*, **37**, 660–667.

Sammons, L., Sumner, S. E., Hackney, C. R., Marcy, J., Duncan, S. E., and Eigel, W. (2000) Program and Abstract Book of the International Association of Food Protection Annual Meeting, August 6–9, 2000, p. 107.

Sanderson, J. D., Moss, M. T., Tizard, M. L. V., and Hermon-Taylor, J. (1992) *Gut*, **33**, 890–896.

Shah, N. P. (1994) *Milchwissenschaft*, **49**, 432–437.

Stabel, J. R. (1997) *J. Dairy Sci.*, **81**, 283–288.

Stead, D. (1986) *J. Dairy Res.*, **53**, 481–505.

Stone, L. S., and Zottola, E. A. (1985) *J. Food Sci.*, **50**, 957–960.

Sung, N., and Collins, M. T. (1998) *Appl. Environ. Microbiol.*, **64**, 999–1005.

Te Giffel, M. C., Beumer, R. R., Langeveld, L. P. M., and Rombouts, F. M. (1997) *Int. J. Dairy Technol.*, **50**, 43–47.

Ternström, A., Lindberg, A. M., and Molin, G. (1993) *J. Appl. Bacteriol.*, **75**, 25–34.

Van Kruiningen, H. J., Chiodini, R. J., Thayer, W. R., Coutu, J. A., Merkal, R. S., and Runnels, P. (1986) *Dig. Dis. Sci.*, **31**, 1351–1360.

Wessels, D., Jooste, P. J., and Mostert, J. F. (1989) *Int. J. Food Microbiol.*, **9**, 79–83.

Wiedmann, M., Weilmeier, D., Dineen, S. S., Ralyea, R., and Boor, K. J. (2000) *Appl. Environ. Microbiol.*, **66**, 2085–2095.

Wu, S. W. P., Pao, C. C., Chan, J., and Yen, T. S. B. (1991) *Lancet*, **337**, 174–175.

CHAPTER 4

MICROBIOLOGY OF CREAM AND BUTTER

R. ANDREW WILBEY
School of Food Biosciences, The University of Reading, Reading, England

Cream is effectively a selective milk concentrate containing an elevated level of milk fat globules dispersed in a continuous phase of skim milk. As such, it provides both a valued range of products and the raw material for butter production.

Historically, cream has been regarded as a luxury food, and production has risen slowly despite concerns over excessive fat intake in the national diet. Unlike milk, cream tends to be treated as an indulgence food and is consumed more at weekends and holiday periods, which puts additional pressure on the manufacturer in terms of shelf life. Butter production is not linked directly to consumption, and UK production data reflects more the surplus milk fat than immediate demand or the competitive position in the market of home produced against imported butters. Production data is summarized in Figure 4.1; the discontinuity in the data is due to changes in the methods of collation of the data, particularly in making allowance for surplus milk fat from production of reduced-fat milks.

Within the European Union (EU), the United Kingdom has a relatively low proportion of its milk going into butter, as illustrated in Table 4.1.

4.1 CREAM

In the United Kingdom, cream is defined as "... that part of milk rich in fat which has been separated by skimming or otherwise ..." (UK

Dairy Microbiology Handbook, Third Edition, Edited by Richard K. Robinson
ISBN 0-471-38596-4

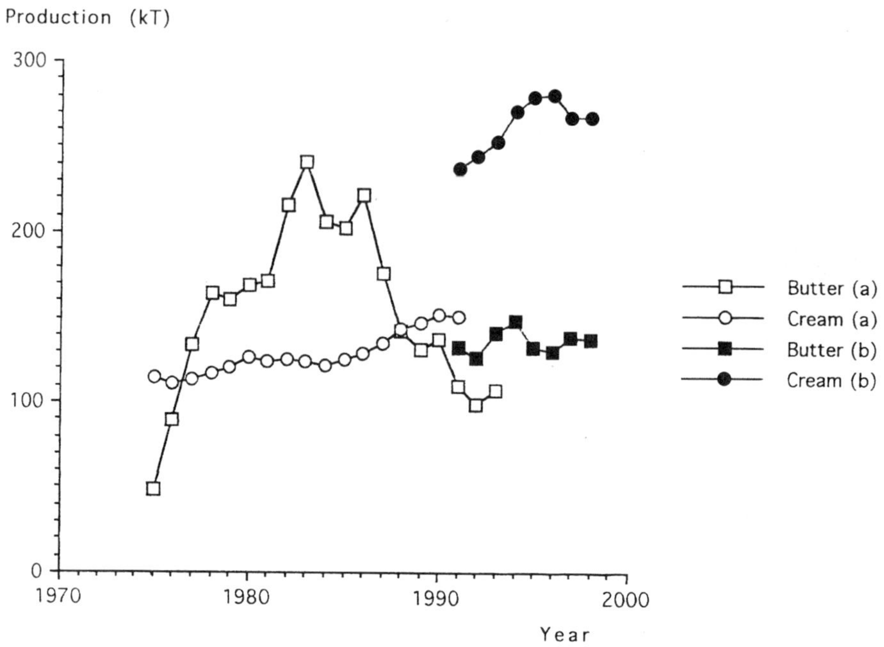

Figure 4.1. Utilization of milk for the production of cream ○ and butter □ in the United Kingdom during the years (April to March) 1975 to 1997. Solid symbols (●, ■) indicate allowance for cream production from low-fat milk processing. (After MMB, 1980, 1985, 1990, 1992, 1994; NDC, 1996, 1997, 1998.)

Regulations, 1995a, 1996). Creams have been defined in terms of the heat treatment that they have been subjected to and in terms of their fat content. Minimum fat contents for UK creams are listed in Table 4.2. Unless qualified by the name of the species, cream is assumed to be prepared from cow's milk.

Definitions for heat treatments of creams are as follows:

Untreated cream that "cream that has not been treated by heat or in any manner likely to affect its nature and qualities and has been derived from milk which has not been so treated." "Pasteurised cream" shall be either (a) heated to a temperature not less than 63°C and retained at that temperature for not less than 30 min, (b) heated to a temperature not less than 72°C and retained at that temperature for not less than 15 s, or (c) heated to any other temperature for such other period of time as has equivalent effect for the elimination of vegetative pathogenic organisms in the cream. Such pasteurized creams must be cooled as soon as practicable after pasteurization and show a negative reaction to the phosphatase test (UK Regulations, 1995a).

TABLE 4.1. Butter Production in 1998 in Terms of Weight and Percent Conversion of Milk Fat

Country	Milk Production (kT)	Butter (kT)	Conversion (%)
Austria	3,256	39	24
Belgium	3,418	109	64
Denmark	4,664	49	21
Finland	2,447	50	41
France	24,793	464	37
Germany	28,500	428	30
Greece	755	3	8
Irish Republic	5,210	141	54
Italy	10,821	98	18
Luxembourg	264	3	23
Netherlands	10,995	149	27
Portugal	1,831	19	21
Spain	5,980	31	10
Sweden	3,331	53	32
United Kingdom	14,635	137	19
European Union total	120,905	1,773	29
Australia	9,731	161	33
New Zealand[a]	11,288	376	67
United States	71,375	522	15

[a]Including anhydrous milk fat.

Source: NDC (2000).

TABLE 4.2. Fat Standards for Creams

United Kingdom	
Half cream (including sterilized)	≥12
Cream or single cream	≥18
Sterilized cream	≥23
Whipping or whipped cream	≥35
Double cream	≥48
Clotted cream	≥55
Australia	
Cream	≥35
New Zealand	
Cream	≥40
United States	
Half and half	≥10.5, <18
Cream	≥18
Light cream	≥18, <30
Light whipping cream	≥30–36
Heavy cream	≥36

Sources: UK Regulations (1996), Australia (1993), S. Carolina (1999).

Clotted cream means "cream which has been produced and separated by the scalding, cooling, and skimming of milk or cream" (UK Regulations, 1996).

Sterilised cream means "cream which has been subjected to a process of sterilisation by heat treatment in the container in which it is to be supplied to the consumer" (UK Regulations, 1996). The heat treatment shall be to not less than 108°C and retained at that temperature for not less than 45 min; or to such a temperature and time that has an equivalent effect for the elimination of pathogenic organisms (UK Regulations, 1995a).

Ultra-high-temperature (UHT) cream means "cream which has been subjected in continuous flow to an appropriate heat treatment and has been packaged aseptically". The heat treatment shall be to not less than 140°C and the cream retained at that temperature for not less than 2 s; or the heat treatment shall be to such a temperature and time that has an equivalent effect for the elimination of pathogenic organisms. For both sterilized and UHT creams there is a further requirement that, after incubation at 30°C for 15 days, the cream should have a plate count ≤ 100 per milliliter and that the product should be organoleptically normal (Regulations, 1995a).

There is a general requirement that "on removal from the processing establishment, milk-based products shall not contain pathogenic organisms and toxins from pathogenic organisms in such quantity as to affect the health of the ultimate consumer." Specific standards are set for *Listeria monocytogenes* and *Salmonella* spp. These regulations gave effect to Council Directive 92/46/EEC, which applies throughout the EU.

Certain additives are permitted in some types of cream, but, with the exception of nisin, these primarily affect the physical rather than the microbiological stability (UK Regulations, 1995b).

4.1.1 Fresh Cream Produced on the Farm

Cream production is essentially a dairy operation, but cream may still be produced on farms and sold direct to the public, just as untreated milk is sold. Hygiene is very variable, and high counts of bacteria, yeasts, and molds may be found in 2 or 3 days, so the shelf life is correspondingly short.

A simple farmhouse pasteurization (heating to 65°C and holding for 30 min) will reduce the bacterial count to about 1% or less of the original, but unless followed by rapid cooling to 5°C, it will have little ultimate effect on keeping quality. The treatment may be repeated the following day with advantage, provided that the cream is hygienically cooled immediately after heating.

Common taints found in farm-produced cream include sour, rancid, cheesy, stale, bitter, putrid, and yeasty; a slight, ill-defined taint may be described as stale or unclean. These problems are always associated with high microbial counts, with a predominant organism (such as *Pseudomonas*, *Micrococcus*, or yeast) being responsible for the dominant taint. Ropiness or sliminess may be caused by some coliforms or lactococci. In bad cases of spoilage, gas may be formed, usually by lactose-fermenting yeasts; and mold growth (e.g., *Geotrichum candidum*) may be visible on the surface of the cream. Souring by lactic acid bacteria may repress putrefactive organisms, but their activity will stimulate yeasts and molds. Sweet curdling may be caused by proteolytic enzymes produced by aerobic spore-formers, which can also be responsible for bitterness.

However, as with other types of dairy processing, such as cheesemaking, the gap in technological proficiency between farms and dairies has been steadily diminishing, except in scale of operation. Hence, the microbiological quality of farm-produced cream should not differ significantly from that retailed by a creamery.

4.1.2 Manufacture on an Industrial Scale

Apart from cultured or soured cream, the entire process of manufacturing, packaging, and distributing cream is, from the microbiological viewpoint, a matter of preventing contamination and keeping the growth of the few organisms that are present to a minimum.

The overall system includes the following stages:

1. Production of milk on the farm
2. Transport to the dairy
3. Storage in the dairy
4. Separation and standardization of cream
5. Homogenization
6. Heat treatments of the cream
7. Cooling and storage after heat treatment
8. Packaging
9. Further cooling, storage, and distribution of cartoned cream
10. Sale—possibly a multistage operation

Thus, cream processing technology is more complex than for milk and there are more opportunities for problems to arise. Different processes may require changes in the sequence of operations, some of the most commonly used are outlined in Figure 4.2.

To produce the best possible cream, each of the above stages must be carried out as efficiently as possible; particularly important are adequate heat treatment, storage, and distribution at 5°C or lower, with excellent hygiene throughout. The old saying that the strength of a chain is the strength of the weakest link is particularly applicable to cream. The most important aspects are contamination after heat treatment, and the product temperature during storage and distribution.

4.1.3 The Collection and Storage of Raw Milk

The hygienic production of milk is of the greatest importance for cream, because although most vegetative cells are easily killed by heat treatment, spores are not; types such as *B. cereus* can be a cause of spoilage [as well as failure in the methylene blue (MB) test]. If the

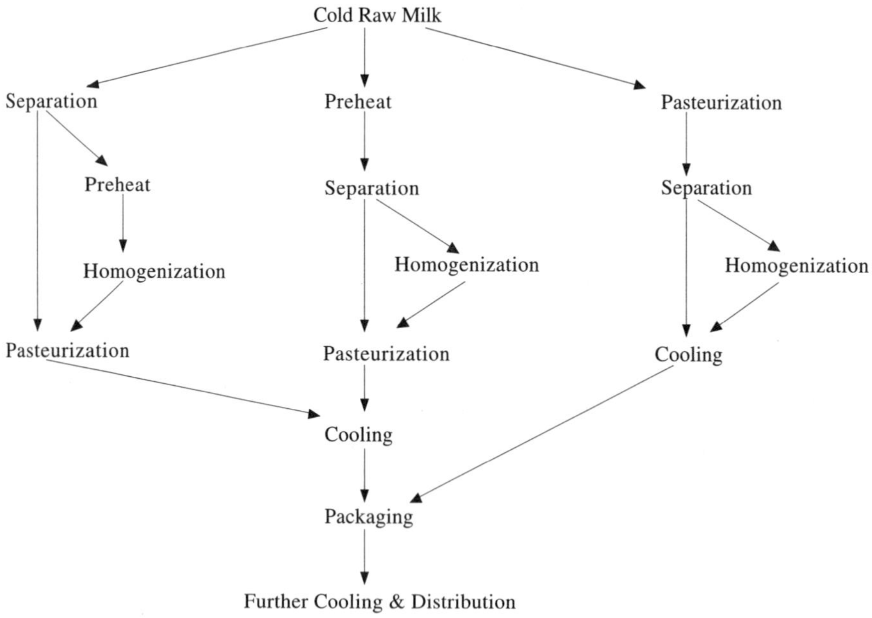

Figure 4.2. Alternative sequences of operations to produce cream.

spore count of the milk is high (over 100 ml^{-1}), it may be worthwhile reducing this by high-speed centrifugal methods because, as aerobic spore-formers tend to form chains, these are more easily removed than single cells.

Bulk collection of milk in Britain became universal in 1979, so that all milk should arrive at distributors' and manufacturers' premises at ≤5°C. The bacteriological quality of ex-farm milk in England and Wales improved with the introduction in 1982 of a quality payment scheme, where a bonus was paid if the average monthly total bacterial count was less than 20,000 cfu ml^{-1}. Subsequent changes in quality payments resulted in further improvements in raw milk quality. However, it is common practice for milk to be stored in creameries at 5°C for up to 48 h and sometimes longer. The change from churn to bulk milk collection has resulted in a change in the microflora of raw milk, with an increase in the level of psychrotrophs. There is normally no problem with good-quality milk, although psychrotrophs can grow very slowly at below 5°C. Psychrotrophs do not usually ferment lactose to lactic acid, but high levels may result in production of proteases and lipases. Thus cold-stored milk does not sour but can develop taints; and although the organisms, typically *Pseudomonas* spp., are killed by pasteurization, their enzymes can survive heat treatment and will continue to produce changes in the pasteurized or even UHT-treated product.

4.1.4 Separation and Standardization

Cream was traditionally made by separating milk in a mechanical centrifugal separator at a temperature between 40°C and 50°C. This is an ideal temperature for bacterial growth, so higher temperatures (e.g., 63°C) may be used which may also reduce fat loss into the skim and denature the lipoprotein lipase. These higher temperatures may be associated with more viscous creams. Some separators are designed specifically to run with cold milk (~5°C) to produce creams with up to 40% fat content. Separation also helps to purify both the skimmed milk and cream by removing dirt, somatic or body cells, and any other foreign matter in the form of slime that is removed mechanically from the separator. Some bacteria, especially clumps of large organisms, including spore-formers, are also removed in this slime. The times between preheating, separation, homogenization (if performed), standardization, and heat treatment should be as short as possible. For a discussion of separation see Towler (1994).

4.1.5 Microbiological Problems of Standardisation

Ideally the milk should be separated to give a cream of the desired fat content, but in practice this is rarely possible. For small-scale producers the normal method is to separate at a slightly higher fat content and then standardize with milk or separated milk. This procedure constitutes a microbiological hazard where warm raw cream is held at about the separation temperature (e.g., 40°C) during the standardization. Furthermore, the cream may be contaminated by inadequately disinfected process plant, or the skim milk may not be of good quality.

With rationalization in the dairy industry leading to larger milk processing plants, it is common to find in-line standardizing systems immediately after separation so that this risk is minimized.

In practice it is a question of balancing the disadvantages of the various methods; but irrespective of the method adopted, it is always preferable to heat treat the final standardized cream and cool and package it immediately. As far as practicable, milk and cream should be held, during processing, at about 5°C or above 60°C. It is expensive to cool warm milk or cream to 5°C, but in hot weather, failure to do so may be disastrous.

4.1.6 Homogenization

Homogenization is the operation of reducing the size of fat globules so that separation due to gravity will be minimized; for further details see Wilbey (1992). Half cream (or coffee cream) and single cream are normally homogenized to increase viscosity, and this treatment may have deleterious effects.

1. Homogenization necessitates an extra treatment and thus adds to the risk of contamination of the cream. It may also involve holding the cream warm for a short time, and always results in breaking up clumps of bacteria. These three factors are usually held to be responsible for any fall in keeping quality which may occur. Homogenization should, therefore, take place immediately before the final stage of heat treatment.
2. The splitting of the fat globules modifies the fat globule membrane and greatly increases the surface area of the fat, thus favoring the action of any lipase that may be present.
3. Homogenizers operate at high pressure and, particularly with older machines, may have seals that are not easily cleaned and disinfected.

There was a general belief in the dairy trade that homogenized milk (and perhaps cream) did not keep as well as the unhomogenized products. If true, this may be because of the extra contamination, the splitting of bacterial clumps, or the oxidized flavor and other lipolytic taints promoted by homogenization.

4.1.7 Heat Treatment

Some form of heat treatment has been used for cream for many years, and cream behaves similarly to milk in all respects microbiologically. Heat penetration is slightly slower, but the chief difference in microbiological considerations is caused by the special demands of retail distribution and keeping quality.

In the early years of the twentieth century, a crude form of batch or holder pasteurization was practiced. Since 1940 the high-temperature short-time (HTST) continuous-flow method, with or without a holding section, has been widely used. The rapid increase in the popularity of semi-skimmed milk over the last 20 years, to take 50% of the liquid milk market (i.e., 25% of UK milk production) by 2000, has produced large quantities of cream as a by-product of milk standardization. Fluctuations in the quantity of cream produced as the separation and standardization provide the different standardized milks make it difficult to run an HTST process efficiently, so high-temperature holding has come back into mainstream processing.

Whatever the heat treatment method, the treatment should be sufficiently severe to denature the native alkaline phosphatase (ALP) in the cream. Destruction of ALP is achieved at temperature and time combinations in excess of those required for the destruction of those vegetative pathogens normally associated with milk products. Because this enzyme is largely associated with the fat globule membrane, the activity in untreated cream is higher than in milk. In addition, a special problem arises in applying the ALP test to pasteurized cream because reactivation of ALP may take place on storage (Wright and Tramer, 1953, 1954, 1956; Peeseboom, 1969, 1970; Kleyn and Ho, 1977; Serebrennikova et al., 1978; Fox and Morrissey, 1981). (Any sample that shows evidence of taint or souring should not be tested.) If more than 10 μg of *p*-nitrophenol are liberated from the buffered substrate (disodium *p*-nitrophenyl phosphate in a sodium carbonate–bicarbonate buffer) at 37 ± 0.5°C for 120-min, then the test must be repeated as follows:

> Ten grams of the cream are added to each of two test tubes and specified quantities of magnesium chloride ($40\,gl^{-1}$) solution added to one

> tube according to the fat content of the cream. After mixing, both tubes are incubated at 37 ± 0.5°C for 60 min with occasional inversion. Two grams of the liquids are then removed and tested for phosphatase. If the color of the sample treated with magnesium chloride is more intense than that of the control, the filtrate is diluted 1 in 4 with the buffer solution and again compared with that of the control. If the color is equal to or more intense than that of the undiluted control, it is concluded that reactivation has taken place and the original positive result is declared void.

Magnesium ions appear to be essential for activity of the reactivated phosphatase but not for the native phosphatase. It has been demonstrated with butter (McKellar et al., 1988) that chelation of the magnesium in the test solution by incorporation of EDTA can avoid interference by reactivated phosphatase. These methods may be carried out more rapidly and sensitively using fluorimetric techniques (IDF, 1999b).

The heat treatment used for pasteurization of cream has typically been more severe than the statutory minimum, being established by experimentation with that processing plant to meet the following criteria:

1. Destruction of all pathogens.
2. Achievement of desired shelf life by reduction in spoilage microflora.
3. Avoidance of "cooked taints" that result from the production of volatile sulfur compounds when milk and cream are heated above 80°C; these usually disappear in 1 or 2 days.
4. Destruction of milk enzymes—particularly lipases, which may cause rancidity.

The normal HTST method used for cream should achieve all these objectives, provided that hygiene and subsequent refrigeration are adequate.

The heat treatment forms the critical control step in the manufacturing process. Optimizing the heat treatment is useless unless the cream is cooled and packed under hygienic conditions, with the cool chain being maintained throughout the shelf life. With poor cooling, the surviving bacteria, particularly the spores and postpasteurization contaminants, will grow and produce taints and physical defects.

4.1.7.1 Holder Pasteurization. This method, involving heating cream to at least 63°C (typically about 65°C) for 30 min, was regarded as obsolete as far as ordinary dairies are concerned, but has advantages for farms and very small dairies processing up to 200 liters a day. Equipment and control are very simple, and heating and cooling in the same vessel eliminates one cause of contamination. However, because cooling is inevitably slower, all sources of contamination, particularly when filling into cartons, become potentially more dangerous.

With cream production becoming a by-product of liquid milk standardization, a form of holder pasteurization has been used as a deaeration treatment, holding the freshly separated cream for ~30 min in a top-filled tank. This technique will not guarantee adequate heat treatment but will contribute to the overall destruction of the raw milk microflora. Typically, this system would operate with a pair of tanks on a 4.5- to 5.5-h cycle, with the cleaning ensuring that there would be no buildup of thermophilic organisms. The heat treatment would then be completed using an HTST process.

4.1.7.2 The High-Temperature Short-Time Method. The HTST method is now universal for cream in virtually all dairies. Its main advantages are in compactness, thermodynamic efficiency, speed of operation, total enclosure, and overall cost. It can give a cream of very good keeping quality, provided that it is properly operated. The entire sequence of operations—reception of raw milk, warming to separation temperature, homogenization (if done), pasteurization, cooling, storage, and filling—can be carried out in one continuous process in an entirely enclosed system, with obvious advantages. The equipment used for the cream should be designed to treat the cream as gently as is practicable.

Minimum temperature–time conditions for pasteurization of cream were approved at the 64th Annual Sessions of the International Dairy Federation (IDF) in 1980. Examples of these temperature–time combinations are:

1. 18% cream, 75°C for 15 s
2. 35% fat or more, 80°C for 15 s

These are more severe than the minimum heat treatment permitted in the EU, which is 72°C for 15 s or an equivalent temperature–time combination of equivalent lethality to vegetative pathogens in the cream. In the United States, a minimum of 74.4°C for 15 s is recommended for dairy products with more than 10% fat.

The holding time is must be the minimum residence time for that type of cream in the holding tube of that heat exchanger. Changing the viscosity of the cream—for example, by changing the fat content or homogenization conditions—may bring about a change in the residence time. The presence of interconnecting pipe work will usually increase the total minimum residence time to 18–20 s. The average holding time, estimated by dividing the volume of the holding tube by the flow rate, will always be greater than the minimum holding time that governs the bacteriological safety of the plant. (The average holding time is more relevant to the extent of the chemical reactions taking place during the heat treatment.) The relationship between the average and minimum residence times depends on the flow pattern within the holding tube; with turbulent flow the minimum residence time may be up to 0.83 of the average residence time, whereas under streamline flow conditions the minimum residence time drops to only half the average time. This has a significant effect on the potential for survival of pathogenic organisms. The lethality of the heat treatment process will be a function not only of the holding conditions but also of the whole temperature–time profile of the process. Typical HTST holding conditions will be 75–85°C for 15–20 s.

When the temperature of pasteurization is raised or the holding time is increased, there is not a simple relationship between the survival of microorganisms and the heat treatment. This is because the cream generally contains a mixed microflora of differing thermolability. Eibel and Kessler (1984) reported a stepped destruction of microorganisms when monitored on a total count basis. The first step took place at 57–63°C with the extensive destruction of heat-sensitive, mainly gram-negative, organisms. Between 63°C and 68°C the further destruction of Gram-negative organisms was largely masked by the death of the more heat-labile lactococci and by the more heat-resistant organisms which in turn were destroyed over the range 68–100°C. There appeared to be little benefit in increasing the holding time beyond 20 s. Mesophilic spores were not destroyed until temperatures in excess of 100°C were used.

The use of higher temperatures may not result in an increased shelf life. The germination of spores is a complex subject, but one of the most important factors is the stimulus given to germination by heat. In the present context, because of the requirement for a long shelf life, the possible protective effect of fat on bacteria, and the slower heat transfer in cream, it is usual to heat cream appreciably above the statutory minimum, either by flash heating or with a holding period. However, this higher heat treatment may, in fact, have an adverse effect on keeping quality because this more drastic exposure may increase the germination rate of spores which survive. Effects on spore germination may be exacerbated by destruction of the natural biostatic systems in the original milk.

This anomalous effect is well illustrated by the results of Brown et al. (1980). Little difference was found between heat-treated creams held for up to 9 days at 7°C; but after this, appreciably higher counts were found when temperatures of 76.5°C, 79.0°C, and 81.5°C (with a holding time of 15s) were used rather than 74°C. Similarly with a holding time of 1s, a temperature of 80°C gave lower counts after 6 days storage than 82.5°C, 85.0°C, 87.5°C, and 90°C. Shelf-life tests on these creams confirmed these results; the lowest temperatures give values of 20 days, and 87.5°C for 1s give values of only 9 days. Eibel and Kessler (1984) concluded that lowering the storage temperature was more effective for increasing shelf life than was raising the pasteurizing temperature.

Concern has been expressed that some pathogens, particularly *Listeria monocytogenes*, may survive the minimum pasteurization conditions. The concern with *L. monocytogenes* is increased by its ability to grow at chill chain temperatures. Rosenow and Marth (1987) reported generation times in whipping cream (33–36% milk fat) of 29–46h for different strains at 4°C with a maximum population on the order of 107 cells per milliliter achieved in 30 days. D values at 68.9°C for the strain *Scott A* were given by Bradshaw et al. (1987) as 6s in raw 38% milk fat cream, increasing to 7.8s in inoculated "sterile" cream. Z values of 6.8°C and 7.1°C, respectively, were calculated.

4.1.7.3 "In-bottle" Pasteurization. It was established many years ago that if milk was pasteurized in the bottle, an excellent keeping quality could be obtained because subsequent contamination became impossible. Unfortunately, the cost of the process rendered it uneconomic. However, a higher profit margin is obtainable with cream, and some dairies have utilized this principle to obtain creams with very low levels of survivors. Thus the milk would be pasteurized and separated, and the bottled cream would be heated to 65°C or higher for 30min.

4.1.7.4 Clotted Creams. This is the most popular holiday type of cream in the United Kingdom; it is made in Devon and Cornwall and often sent away. In the traditional or farmhouse method, milk (from Channel Island or South Devon breeds) was put into pans, 30cm in diameter and 20cm deep, and held 12h for the cream to rise. The pan was then put in a steamer until a layer of solidified cream formed round the edge. After cooling, the cream was ladled off with a perforated dipper and packed as layers in the containers.

The traditional method has been superseded by large-scale creamery methods, principally:

1. The float process closely resembles the traditional process, in that it involves heating double cream over a layer of skim or whole milk in a large, shallow, jacketed tray till a crust is formed. The tray is then cooled overnight to set the cream before it is removed.
2. The scald process where a thin layer of high-fat cream, 54–59% milk fat, is heated in a tray to 77–85°C to form a crust before cooling.

These processes were reported in more detail by Wilbey and Young (1989).

The more severe heat treatment that clotted cream receives results in a different microflora, in which aerobic spore-formers of the *Bacillus subtilis* type are usually prominent. In the United Kingdom, nisin may be added to cream for clotting to control spore germination. In some processes, draughts may result in marginal heat treatment of the surface layer. However, the major microbial problems are more likely to be associated with the slow cooling in open trays and in the handling of the cooled cream. Poor air quality and/or poor process hygiene can lead to mould spoilage, coliforms, and other postprocess contaminants.

4.1.7.5 The Ultra-High-Temperature (UHT) Process. Cream that has been subjected to this treatment is known as "ultra-heat-treated cream" and must have been heated to at least 140°C for at least 2 s, or an equivalent temperature/time in order to render the cream free of both viable microorganisms and their spores (UK Regulations, 1995a). It must be immediately put into a sterile container with aseptic precautions. There are three important advantages of this process over the traditional heat treatments.

1. The cream is sterilised, i.e. all forms of life are destroyed.
2. The process is very rapid.
3. The treatment induces very little cooked flavour in the cream.

The method has obvious advantages where a long keeping quality is required because, microbiologically, the cream will keep indefinitely without refrigeration. Shelf-life is limited by biochemical considerations. Calcium-casein interactions will destabilize the emulsion, and proteases surviving the heat treatment will bring about gelation. The original slightly sulfurous flavor soon passes off, and the remaining cooked flavor is usually augmented ultimately by a stale, "cardboardy" or oxidized flavor. These independent effects limit the shelf life to 3–6 months according to circumstances.

The UHT method has been used extensively for retail cream, especially for small units in the catering industry. Single or half cream is generally used, packed in very small cartons. The UHT method becomes progressively more difficult to control as the fat content of the cream rises. UHT processing of cream has been reviewed recently, including methods for avoiding recontamination of UHT-treated products (IDF, 1996; Lewis and Heppell, 2000). Hydrogen peroxide, ethanol, and ethylene oxide are used commercially to sterilize the paper and plastics used in aseptic cartoning.

The ATAD friction process provides an alternative method for the rapid sterilization of cream and other liquids. In this process, the liquid is preheated to 70°C and then heated at 140°C for 0.54 s, and it can be applied successfully to 12% and 33% fat creams (Alais et al., 1978).

4.1.7.6 Tyndallization. In the early days of bacteriology, considerable trouble was experienced through the inability of the available processes to kill spores in a medium. Tyndall in 1877 suggested that if a medium was heated at 100°C for 30 min on three successive days, first the vegetative cells would be killed, and then the spores would germinate and the new vegetative cells would be killed on the second and third days. The idea has been resurrected from time to time for foods that would be affected organoleptically by autoclaving. The method fails if the medium does not permit the spores to germinate, and it is unreliable for anaerobic and thermophilic aerobic spores. Various forms of the method have been suggested, a recent one by Pien (1977) is based on double HTST pasteurization. The main reason for the failure of all these methods is the unpredictability of the germination of spores (Franklin, 1969; Jayne-Williams and Franklin, 1960).

Brown et al. (1979) investigated a double pasteurization treatment separated by periods of aerobic and anaerobic incubation at 30°C. None of the treatments had any effect on the spore load or the storage life of the creams. The organoleptic spores were not influenced by the different types of treatment.

A method formerly used for making "long-life" cream was to heat the cream in bottles for 30 min at 105°C on the first, second, and fourth days, leaving the bottles at room temperature for the third day. This period allowed surviving spores to germinate, and it allowed the cells to be killed on the fourth day. The method was reported to give a very long keeping quality, but serum began to separate after a month.

4.1.7.7 Sterilized (or Canned) Cream and Half Cream. Sterilized cream must contain not less than 23% fat, and sterilized half cream not less than 12%; but otherwise, and from the microbiological point of view, they are identical.

After standardization, the cream is heated to homogenizing temperature, usually about 65°C. After homogenization at about 17 MPa, the cream is filled into cans at a temperature of 30–50°C. As filling is a well-recognized source of contamination, some makers fill at 75–80°C, though viscosity may be affected. Sterilization of the containers must be at not less than 108°C for a minimum of 45 min or a temperature–time combination of equivalent lethality. Konietzko and Reuter (1980), working with spores of *B. stearothermophilus*, found a linear thermal death curve in the range 130–145°C; but for homogenized 25% fat cream, the best temperature–time combination for sterilization is 15 min at 121°C or 10 min at 122°C (Dhamangaonkar and Brave, 1978).

The sterilization process may use a batch rotary retort or the continuous method; the temperature–time conditions vary according to size of the can (Smith, 1989). Continuous systems are being increasingly used for large-scale production. In the three-stage cooker, the cans are transferred by a conveyor-valve system through a preheating, sterilizing, and cooling process of the type long used for evaporated milk. The trend is now toward the hydrostatic system in which the pressure required to give the necessary temperature is maintained by columns of water at the entrance and the exit of the equipment. Cans are heated at 116–121°C for 30 min; but if a UHT treatment is applied first to the cream, then a lower temperature and shorter time could be used to overcome the postprocessing and container contaminants.

Cooling must be carried out as soon as possible after the sterilization, typically using hyperchlorinated water to minimize the risk of postprocess contamination should a seam leak.

The microbiological problems of sterilized cream are somewhat different from those of pasteurized cream. The cream is theoretically sterile, but problems similar to those of evaporated and sterilized milk can occur. If the cream becomes contaminated with very heat-resistant spores of *B. subtilis*, these can germinate and produce a bitter taint and thinning of the cream by the production of lipolytic and proteolytic enzymes. This fault is usually associated with poor-quality raw milk high in spores and/or dirty equipment. Such spores can sometimes survive heating at 120°C for 40 min (Nichols, 1939). Improvement in the hygienic production and handling of the raw milk is essential in these circumstances, though as an interim measure it may be necessary to raise the sterilization temperature.

Sterilized cream must satisfy a colony count test in the same manner as for UHT cream (UK Regulations, 1995a). The test is deemed to be satisfied if the number of colonies is found to be less than 100 cfu/ml after incubation.

If any microbiological defect does arise, then the causative organism should be identified, because this information can provide an important clue to the source of infection. Thus, if a non-spore-forming organism is responsible for spoilage in a sterilized product, this indicates contamination after sterilization; in a canned product, this indicates a defective can or a "leaker." In the latter case, bacteria can then enter the cream from the cooling water, or elsewhere, and cause various faults. For example, common water-borne organisms, such as *Proteus*, can cause bitterness and thinning, coliforms can form gas, and lactococci give rise to acid curdling. It is important to remember that sterilized cream is just as good a medium for microbiological growth as raw milk or cream, if it becomes contaminated.

Careful control is essential at all stages in the manufacture of canned foods. Ordinary microbiological control methods are of little use because one surviving spore in a can may germinate weeks or months after manufacture and thus produce a defect. A few cans from each batch should be subjected to accelerated storage tests by incubation at 37°C and 55°C for at least 7 days. Growth can often be detected by external inspection (swelling of cans or shaking them), as well as by internal examination of others. Retention of the whole batch at ambient factory temperature for a month followed by inspection before dispatch is a wise precaution.

In order to obtain a satisfactory sterilization in the manufacture of canned cream, the heating process must be evaluated by a suitable method (Hersom and Hulland, 1980). Using heat penetration data, Board and Steel (1978) have compared sterilizing values (F_0) obtained using an automated version of Gillespy's method with those obtained by the general method. The former usually gave lower F_0 values, but the differences were not significant.

4.1.8 Packaging of Cream

The packaging of any perishable food is technically demanding, because the greatest efficiency in processing is invalidated if the product is unhygienically filled into unclean containers. At one time, retail cream was filled into bottles or jars and sold like milk, but cartons are now used almost universally for pasteurized cream. In the modern dairy, the carton is the last item coming under control in the sequence of operations.

Cartons, whether of paper or plastics, are virtually always of good hygienic quality because of their method of manufacture, provided that they are stored in a clean atmosphere. The manufacturer normally packs the cartons into polyethylene bags within cardboard boxes. This will protect the cartons from environmental contamination, but the outer box can become contaminated. Hence it is necessary to separate the cartons from the outer box under hygienic conditions before they are brought into the cream packing area. Where ultraclean or aseptic packaging is to be practiced, the low level of microbial contaminants should be eliminated. For instance, this may be by fumigation or γ-irradiation of the packaging for ultraclean systems, or by hydrogen peroxide and heat in the case of aseptic packaging systems. It is critical that recontamination of the packaging and the filling system be avoided.

Semi-bulk containers, typically bag-in-box systems holding 5–10 liters, are commonly used for small bakery and catering outlets. Blow-molded 2- to 5-liter polyethylene bottles may also be used. As with cartons, these are virtually sterile from their method of manufacture. Bacteriologically, all that is required is to store them under clean conditions and avoid contamination in handling, filling, and sealing.

Intermediate quantities of 25–50 liters may be distributed in aluminium or plastic cans or churns. These must be cleaned and disinfected adequately and will normally be used with a polyethylene liner.

Bulk quantities of cream (e.g., 250–1000 liters), may be transported in stainless containers, whereas larger quantities may be pumped direct into road tankers. In either case, the cleaning and disinfection will employ CIP systems, as with the other equipment in the dairy.

4.1.9 Cooling of Heat-Treated Creams

From the microbiological viewpoint, all creams should be cooled as quickly as possible within a closed system to 5°C or less. However, for many retail creams the shear forces associated with such cooling would lead to an unacceptably thin cream and to the risk of damage to the milk fat globules. An old "trick of the trade" known as "rebodying" was to cool the cream quickly to 30°C or less and then to continue the cooling slowly down to 5°C. This procedure can give a cream of much higher viscosity. The thickest creams would be packed warm and cooled in the container. The microbial quality will be at risk from growth of survivors of the heat treatment and from postpasteurisation contaminants; hence such procedures should be used with caution and with the greatest emphasis on good hygiene.

Where cream is packed warm, then the packaging and handling systems should be set up to maximize the cooling rates—for example, by using containers no greater than 500-ml capacity and trays with air vents. Blast cooling should be used, but the air temperature should not cause the cream to freeze (the freezing point is the same as for milk, approximately –0.52°C).

4.1.10 Whipped Cream

Whipped cream often requires the addition of stabilizers to minimize serum loss. Sugar may also be added. Small-scale bakers may add stabilizer mixes, usually including some sugar, to the cream during whipping. Thus the microbiological quality of that stabilizer will be critical in governing the quality of the final product. Stabilized creams for large-scale whipped cream manufacture should be prepared by pasteurizing the stabilized cream. Whipping cream should be held at or below 5°C for 24h before use to give the best whipping properties.

The whipping process will introduce an equal volume of air to the mix. Carrying out this operation on a small scale with a planetary mixer will expose the cream to environmental contamination. Larger-scale operations using continuous aerators in a closed system can avoid such environmental contamination; the oil-free compressed air or nitrogen supplied must be passed through a high-efficiency filter before use.

Whipped cream, unless whipped using nitrogen, will provide a highly aerobic medium for microbial growth, and spoilage can be expected to occur more rapidly than for liquid cream. Addition of sugar will not reduce the water activity (a_w) sufficiently to inhibit growth. Storage should be below 5°C, because low temperatures also inhibit serum separation and collapse of the foam.

4.1.11 Frozen Cream

Freezing is arguably the least objectionable method for preserving perishable foods. Organisms cannot grow, and nutritional value is maintained virtually unaltered. Textural and flavor problems may occur, depending on the method of freezing and type of product.

Cream for freezing should first be pasteurized at ≥75°C for 15s, and then it should be cooled to 1°C as quickly as possible before freezing.

Freezing may be carried out within containers, on a band as a sheet or as pellets, or by direct contact with liquid nitrogen. Bulk storage of frozen blocks of cream in polyethylene bags may be used for surplus cream for later conversion into butter or anhydrous milk fat. The cream should be frozen as quickly as possible and stored at –18 to –26°C, the lower the better. A keeping quality of 2–18 months with an average of about 6 months can be expected. The shelf life of the cream will be limited by physical and chemical constraints, the main problem being destabilization of the emulsion.

Freezing cannot be used to reduce bacterial numbers, although a few organisms may be killed mechanically by the formation of ice crystals and some may die during storage (e.g., coliforms). Normally pasteurized cream would be used, and the microbiology would be the same as for the nonfrozen product.

4.1.12 Cream-Based Desserts

There is no legal definition of dairy desserts in the United Kingdom, though a convenient description would be a dessert product where milk ingredients make up at least 40% of the dry matter. Most products are sweetened, though they may contain fermented ingredients. (Ice cream forms a major sector of the cream-based desserts market and is covered separately in Chapter 6.) Cream-based desserts may be single products or multicomponent. Typical examples of single-component products can be found in the puddings produced by heat treating blends of skim milk, cream, sugar, and jelling agents, such as starches, carrageenan, locust bean gum, and/or gelatin, to give a range of gelled and semisolid products. Typical heat treatment temperatures are well in excess of minimum pasteurization, particularly when native starches have to be cooked, so that the main microbiological problems are likely to arise from thermoduric organisms, especially spore-formers, and from postpasteurization contaminants. The hot filling of carrageenan-based desserts to get a set product (e.g., creme caramels) can reduce the risk of postpasteurization contamination. Where permitted, the addition of nisin to creme caramel can increase the shelf life at 7°C (Anonymous, 1985).

The inclusion of sugars in the mix widens the range of contaminants that can grow in cream-based desserts, for instance, a wider range of yeasts and molds can be supported.

Incorporation of fruit conserves will lower the pH and can favor yeasts and molds over most bacterial contaminants. Suggested internal standards for dairy desserts are given in Table 4.3.

TABLE 4.3. Example of Internal Standards for Dairy Desserts (cfu/g)

	Target	Acceptable	Doubtful	Reject
Total viable count	<1,000	1,000–5,000	5,000–20,000	>20,000
Coliforms	absent in 5 × 1 g	<5 in 1 of 5 × 1 g	<5 in 2 of 5 × 1 g	<5 in 3 of 5 × 1 g
Yeasts	<10	10–50	50–100	>100
Molds	<10	10–50	50–100	>100

Multicomponent desserts will provide microbiological problems characteristic of both the individual components and of the blends generated from their mixing. Should problems be suspected, then supplementary examination of the individual components is needed.

4.1.13 Hygienic Control in Cream Processing

The keeping quality or shelf life of cream, as with all perishable heat-treated foods, is largely determined by the extent of post-heat-treatment contamination. The control of the cleaning and disinfection of the equipment (sometimes referred to as "sanitizing," though this term is not recommended by the IDF) is thus second to none in its importance in the operation of any dairy. Control can be considered in two parts: the methods of cleaning and disinfection (Romney, 1990) and the laboratory examination of the disinfected plant to check for "sterility." True microbiological sterility of the process plant is only necessary for a UHT plant.

"Good manufacturing practice" is a term used to embrace all operations in the dairy, including hygiene, whereas "quality control" refers to laboratory and ancillary tests. Good manufacturing practice may also be considered to embrace the quality control aspects (IFST, 1998). These operations assure the manufacturer and the customer of a satisfactory product and may be referred to as quality assurance. Ideally, this should extend from the production of milk on the farm to the delivery to the customer; it is this concept which underlies the EU Milk Hygiene Directive (EEC, 1992).

Essentially, hygienic preparation of cream means the prevention of contamination of the cream at all stages. The greatest hazard is usually from dirty equipment. All items that come into contact with the cream in any way at any stage must be cleaned by removing all soil, and any residual organisms should be killed by heat or by chemical disinfectants such as chlorine compounds. The requirements are exactly the same as for milk. Once good process hygiene has been established, storage temperature is the second most critical factor. In the United Kingdom, cream comes within the general requirement to hold perishable foods below 8°C (UK Regulations 1995c), but, in practice, temperatures below 5°C would be needed to maximize the shelf life of pasteurized creams.

It is essential that the plant environment be maintained in a satisfactory state in order to minimize postprocess contamination. This should include noncontact plant surfaces, as well as air and water quality.

4.1.13.1 In-Line Testing of Cream Equipment and Product. The modern trend toward ever-larger processing units has resulted in complex computer-controlled systems. The cleaning and disinfection of the equipment is carried out by a cleaning-in-place (CIP) system involving typically a water rinse, circulation of hot caustic-based detergent solution [e.g., containing NaOH ($5\,gl^{-1}$) at 70°C], a rinse, then disinfection by hot water or cold hypochlorite treatment and a final rinse. With a properly planned and operated system using equipment in good condition, there are normally no microbiological problems. However, if there is a defect anywhere, product faults can arise and these can sometimes be very elusive. Problems occurred with early automated systems because failures of valves and other plant items were not being recognized; this was overcome by incorporating feedback signals into the control system. Planned maintenance of the plant is an essential component in maintaining quality because, for instance, worn seals can readily become sources of contamination.

If there is an equipment or operational problem, then keeping quality can fall drastically and, even if taints do not develop, the cream may fail rapid screening tests such as the methylene blue test. Follow-up tests may reveal a high count and sometimes the presence of coliforms.

The first step is then to check all the operations from the initial treatment of the raw milk onwards and to examine the condition of all items in the processing line. If no apparent fault can be found, the next procedure is to take samples at various points in the processing line during a typical run and examine them by a suitable microbiological test, a method commonly described as in-line testing. It is essential that the sampling be above reproach. A satisfactory way is to take the samples by hypodermic needle through an Astell seal using the usual precautions. When a severe heat treatment is used (e.g., 85–90°C with 15-s holding), the surviving colony count is, or should be, very low—for example, about $10\,cfu\,ml^{-1}$. However, because normally only 0.1 ml is

TABLE 4.4. Some Typical Results Obtained by Taking In-Line Samples of Cream During a Series of Processing Runs[a]

Sampling Point	Counts (MPN per milliliter) No. 1	No. 2	No. 3
Exit from pasteurizer	4	3	<3
Entry to storage tank	7	4	11
Exit from storage tank	9	7	93
Entry to filler	39	9	120
Exit from filler	43	120	150
Filled carton	64	120	210

[a]Sources of infection are shown by marked differences in counts. For example, in test No. 1 the pipework was clearly responsible for contamination, in No. 2 the filler, and in No. 3 the storage tank. All samples must be tested under the same conditions.

plated, a reliable count cannot be obtained by this method unless contamination is severe, and hence an alternative approach such as the most probable number (MPN) plate counts or preincubation of samples should be used.

The resulting counts are recorded against each sampling point, and this operation should be repeated twice before any conclusions are drawn. Some typical results are given in Table 4.4.

4.1.14 Keeping Quality or Shelf Life

The keeping quality (KQ) of retail cream is more critical than that of milk because although cream usually receives a more severe heat treatment than does milk, the distribution system and pattern of consumption is different. Thus, while milk is usually pasteurized one day, delivered the next, and consumed a day or two later, cream has to be separated, standardized, cartoned, and distributed, often through a retail dairy organization, and then sold through supermarkets or shops. Temperature control may not be complete during this procedure, and because cream sales tend to be concentrated at weekends and on special occasions, and the carton may be opened and used more than once over a period of days, the KQ requirements are severe.

4.1.14.1 Causes of Poor Keeping Quality in Cream. The following are the main factors responsible for a short shelf life in retail cream:

1. Poor-quality raw milk, particularly with high spore and thermoduric counts
2. Poor hygiene in separation
3. Wrong choice of temperature for heat treatment
4. Poor hygiene in processing
5. Poor hygiene in packaging
6. Too high a temperature for storage and distribution

Of these factors, the quality of the raw milk, hygienic handling, and proper control of the temperature of the end product are, perhaps, the more important.

4.1.14.2 Raw Milk Quality. Bacteria found in milk may be placed in three groups from the point of view of keeping quality in cream:

1. Thermolabile types that are killed by ordinary pasteurization (72°C with 15-s holding)—for example, *Pseudomonas* spp., lactococci, and coliforms
2. Thermoduric types that do not form spores but survive ordinary pasteurization—for example, *Enterococcus faecalis* and *Micrococcus* spp.
3. Aerobic spore-forming bacteria—for example, *B. cereus*

High levels of thermolabile organisms can give rise to taints in the milk which are carried over into the cream. High levels of *Pseudomonas* spp. are also undesirable in milk to be separated for UHT creams, since lipases and proteases produced by these organisms are extremely heat-resistant and will cause spoilage of the product during storage. The presence of *Pseudomonas* or coliforms indicate postpasteurization contamination in a cream that has been adequately heat-treated. Griffiths et al. (1984) suggested a rapid detection method for postpasteurization contamination of cream by Gram-negative psychrotrophs.

With control of postpasteurization contamination, then the heat-resistant groups are of the greatest importance for cream. Brown and Prentice (1982) reported *Bacillus* species as the main psychrotrophic spore-formers in pasteurized single cream stored at 7°C.

It was shown in the 1960s that an initially good (i.e., low count) cream will last for 6 days at 5°C but develops a high count (though no coliforms) and fails the methylene blue (MB) test after 2 days at 15°C. An initially poor (high count) cream becomes sour, coliform contamination becomes evident, and the product fails the MB test after 6 days at

TABLE 4.5. Comparison of Microbial Growth in Good and Poor Cream Held at 5°C and 15°C

	Held at 5°C			Held at 15°C		
Age (Days)	Total Count at 30°C	Coliforms[a]	MB[b] (h)	Total Count at 30°C	Coliforms[a]	MB[b] (h)
			Good Cream			
0	1,900	– – –	>4.5	1,900	– – –	>4.5
1	2,100	– – –	4.5	12,000	– – –	1
2	3,200	– – –	4.5	>1,600,000	– – –	0
3	2,900	– – –	4.5			
4	2,100	– – –	4.5			
6	2,500 (still acceptable)	– – –	4.5			
			Poor Cream			
1	20,000	– – –	>4.5	24,000	+ + +	1
2	20,000	+ – –	4.5	>1,600,000	+ + +	0
3	25,000	+ + –	4.5	(unclean odor and taste)		
4	67,000	+ + +	4.5			
6	560,000 (sour)	+ + +	0			

[a]Coliforms: The three results indicate presence or absence in 1, 0.1, and 0.01 g.
[b]MB stands for methylene blue test (PHLS, 1971).

5°C; or it develops an unclean odor and taste and coli, and it fails the MB test after 2 days at 15°C (Davis, 1969) (Table 4.5).

Pasteurized creams would be expected to last longer than 5 days, and microbiological standards are set correspondingly tighter.

The bacterial population (P) of cream, after holding for a known time at a specified temperature, may be estimated by using the formula

$$P = \text{initial number} \times 2^N$$

where

$$\mathrm{N} = \frac{\text{Age in hours}}{\text{Generation time in hours}}$$

Because the generation time in cream for any one organism is roughly constant for a considerable time at a low temperature, this emphasizes the importance of obtaining a low count cream at the end of processing/filling—that is, at the start of the shelf life of the cream.

4.1.14.3 The Effect of Temperature. From the microbiological point of view, cream is a remarkably standard product, in that the pH value, water activity, and nutrient content are virtually constant. It follows, therefore, that for a given degree of microbial contamination, the shelf life is entirely a question of temperature. The rate of growth of all microorganisms is controlled mainly by temperature, and each organism has a range over which growth occurs and an optimum. As far as bacterial contaminants of cream are concerned, the psychrotrophs may grow in the range 0–45°C, mesophiles grow in the range 10–45°C, and thermophiles grow in the range 35–63°C. These are not rigidly defined limits but serve as a guide, and there are also a large number of organisms that cannot be so readily classified. With cream and other refrigerated perishable foods, the problem is to minimise the rate of growth of the psychrotrophs. The generation times are inversely related to the storage temperature (up to the optimum growth temperature), and they vary from several hours to about 15 min. For the psychrotrophic bacteria, there are two ranges of importance: 0–4°C and 5–10°C. While the rate of growth is very slow up to 4°C, it increases progressively from 5°C to 10°C and particularly above 6°C. This observation is important in that reported temperatures in the dairy industry are not always accurate, so that although refrigeration temperatures during the night may not exceed 4°C, during working hours they can easily reach 7°C or even higher (Airey, 1978; Jackson, 1978; Muir et al., 1978).

It is also relevant that, although modern instrumentation is very efficient, temperatures should *always* be checked with a thermometer that has been calibrated against a standard thermometer kept in the laboratory. The accuracy of portable electronic thermometers may deteriorate as their batteries discharge.

4.1.14.4 Microbiological Problems in the Distribution of Cream. Cream presents more problems than milk because of the methods of distribution and the requirements for a longer keeping quality. Sales may be erratic depending on the weather, holiday seasons, local activities, and many other factors. Broadly speaking, cream sales and desserts (such as cream cakes) peak at Christmas. There is a lesser peak at Easter and other public holidays. Any increase in consumption of foods traditionally associated with cream will increase the demand, so there is a period of increased sales during the soft fruit season (in the United Kingdom usually during June and July). Strawberries and cream is one of the most popular desserts, and this accounts for a considerable consumption of cream. Unfortunately, this increased demand often results in problems in methods of distribution, reduced control, and longer storage (to cover erratic consumption); and, because these adverse factors may occur during periods of warm weather, microbiological problems are enhanced and keeping quality may be seriously affected. There seems to have been a belief that cream keeps better than milk, other things being equal, but this is very doubtful. Bacteria may be more static in cream because of the greater viscosity, which incidentally results in larger clumps and therefore lower colony counts, though this effect would not restrain metabolic activities, such as souring, or affect the results of the MB test.

All but sterilized and UHT creams must be dispatched from the manufacturing dairies by chilled distribution vehicles. Larger retailers may be supplied direct, either to their shops or to central distribution depots for subsequent delivery. Smaller retailers may be supplied via wholesalers. Once cream has been dispatched by the manufacturer, he has no further control over the way it is handled; and the longer the distribution chain, the greater the potential for the cream to be mishandled before it finally reaches the customer. Some suggested and statutory standards are included in Table 4.6. Even in the best of circumstances, the chill chain will be broken when the consumer purchases the cream and transports it home at ambient.

In terms of shelf life, the crucial temperatures are 5°C and 13°C. Below 5°C, growth is so slow that an adequate shelf life is usually

TABLE 4.6. Some Suggested Microbiological Standards for Cream at the Point of Sale to the Consumer

	Colony Counts per Gram		
	Satisfactory	Doubtful	Unsatisfactory
Total Colony Count			
SDT (1975)	<10,000	10,000–100,000	>100,000
Davis (1969)	<10,000	10,000–100,000	>100,000
Jackson (1978)	<50,000	50,000–250,000	>250,000
EEC (1992)	<50,000	≤2 @ 50,000–100,000	>2 @ 50,000–100,000
(5 × 1-ml samples)			
New Zealand	<50,000		>50,000
MB Test			
Jackson (1978)	2.5–4 h	≤2 h	0 h
Coliform Bacteria			
SDT (1975)	<10	10–100	>100
Davis (1969)	<10	10–100	>100
Jackson (1978)	Absent in 0.1 g	Present in 0.1 g	Present in 0.01 g
EEC (1992)	Absent in 5 × 1 ml	<5 in 2 of 5 × 1 ml	>5 per ml

assured, other factors being satisfactory. The nearer the temperature is to 0°C, the better, but such refrigeration is costly. Between 5°C and 13°C the bacteria grow at an increasing rate, and deterioration of quality accelerates; though 8°C is the upper limit permitted in distribution and storage, the temperature will exceed this after purchase as the product is being taken home. Above 13°C, bacterial growth rapidly produces souring or other taints, or at least a failure in the total count, coliform, or MB tests. Refrigeration is the only legally permitted method for controlling the shelf life of cartoned cream during distribution and storage.

The above considerations do not apply to sterilized or UHT-treated creams. However, if the former is contaminated through a leaking seam, temperature will affect it; also, those enzymes that survive UHT treatment will cause changes that proceed faster at higher temperatures.

4.1.15 Taints in Cream

Taints in cream may be absorbed as chemical entities, or they may develop as the result of the growth and metabolism of microorganisms.

Because the milk fat in cream is emulsified, it has an enormous surface area (>$0.3\,m^2\,ml^{-1}$ in unhomogenized cream alone) and readily absorbs odors from the atmosphere. It is therefore of the greatest importance that cream should not be stored where any odoriferous material, such as disinfectants, paint, varnish, scents, or "strong-smelling" foods, are stored. The taint is usually so characteristic that it is unmistakable and the tainted cream will be inedible. An absorbed odor is readily distinguished from a microbiological taint because the former passes-off on standing open to the atmosphere, whereas the latter steadily increases in intensity.

Another type of chemical taint can be caused by cows eating certain plants, such as garlic or decayed fruits (e.g., apples). Oxidized taints can occur in cream of very good microbiological quality held at low temperatures. Ultraviolet light plus minute concentrations of copper can promote oxidation of milk fat, rendering cream so unpleasant as to be inedible in a few hours. The higher the bacterial count, the slower the development of oxidized taint, because bacteria consume oxygen and thereby lower the E_h of the cream.

Occasionally an absorbed type of taint can be imitated by the growth of certain organisms. Thus yeasts can produce fruity flavors (sometimes quite specific), and some bacteria can produce taints resembling apples and other fruits.

The origin of a taint in cream may be difficult to elucidate, but some fundamental causes of taints are as follows:

1. Abnormalities in the milk inside the udder due to mastitis, late lactation, method of feeding, or weeds in the pasture.
2. Failure to cool the milk immediately after milking, thereby permitting lipase and other enzymes to act on the milk; aeration and agitation may accelerate the chemical changes involved.
3. High count milk and/or cream (lack of hygiene in production).
4. Use of stale milk.
5. Dirty separators and other equipment (this gives a characteristic dirty taint).
6. Failure to cool the cream.
7. High temperature of holding during distribution and sale.
8. Cream stale when sold.

Asking the following simple questions may be helpful.

1. Was the taint in the original milk?
2. How old was the milk when separated?
3. Was the taint in the cream after separation?
4. What time elapsed between pasteurisation and sale?
5. At what temperature was the cream held?
6. What was the bacterial count at the time of sale?

A critical survey of the answers to these questions will usually afford some clue as to the nature of the problem. If the taint has developed only after pasteurization, it may be due to dirty equipment, thermoduric bacteria, and/or holding at too high a temperature. If present in the cream immediately after separating, it may be due to enzyme action in the raw milk, or it may be inherent in the milk itself.

A more sophisticated approach is to test the milk, raw cream, and pasteurized cream at the point of sale for the following:

1. Thermoduric bacteria
2. Lipolytic bacteria using tributyrin agar (5 days at 22°C)
3. Caseinolytic bacteria using caseinate agar (5 days at 22°C)

Usually, 1 and 3 give similar results. If large numbers are found, the taint may be of bacterial origin, but, if not it, may be due to lipase or oxidase action.

The coliform and MB tests are usually of little use for this purpose, although a high coliform count suggests dirty equipment. Bitterness and other taints in cream may be caused by contaminated water infecting the milk or cream with *Proteus, Pseudomonas, Achromobacter*, or other proteolytic and lipolytic organisms able to grow at low temperatures.

In general, all bacteriological tests for faults in cream should be made at 22°C, and never at above 30°C. In all quality control work for milk and cream, odor and taste tests should always be made immediately after processing, and at a time corresponding to sale. The product should be adjusted to 20°C for this purpose.

4.1.15.1 Bitterness. A bitter taste can be developed in cream by a number of microorganisms. It usually results from proteolysis giving rise to hydrophobic peptides. Partial glycerides resulting from lipolysis may also be responsible. In addition to *Proteus* and other Gram-negative rods, some yeasts and molds can produce bitterness, although associated growth may be necessary. For example, souring by a lactic organism, such as *Lactococcus lactis*, may be necessary to allow *Rhodotorula mucilaginosa* to produce a bitter flavor.

4.1.16 Microorganisms Causing Defects in Cream

All milk and products made from it, such as cream, become contaminated by microorganisms from the udder, or from the cow, or during the milking process. The "original" flora in this sense consists mainly of lactococci, streptococci, micrococci, corynebacteria, and aerobic and anaerobic spore-forming bacteria (Crossley, 1948; Davis, 1971; Tekinsen and Rothwell, 1974). Thus, if cream is sold raw, all these organisms will generally be present as well as cow-derived pathogens. From the time the milk is put through the various stages necessary for the production of cream, a variety of organisms accumulate until the final heat-treatment destroys most of them. Staphylococci and lactobacilli, Gram-negative rods from watery environments (many of them psychrotrophic), are usually prominent, but the proportions depend not only on the level of hygiene but also on the temperature of the cream. The last two types and *B. cereus* are favored by failure to cool the cream rapidly to 5°C or under.

4.1.16.1 Types of Organisms Found in Cream. In their investigation of changes in the microflora in retail creams held at 5°C for 5 days, Tekinsen and Rothwell (1974) found that initial counts were low and that the proportion of psychrotrophs were small. After storage, the psychrotrophic count (at 5°C) varied from about 10^2 to over 10^7 cfu ml^{-1}, and the mesophilic count (at 30°C) various from about 10^3 to 10^8 cfu ml^{-1}. In fresh cream, the predominating organisms at 5°C were *Pseudomonas, Alcaligenes, Acinetobacter, Aeromonas*, and *Achromobacter*, and at 30°C, they were *Corynebacterium, Bacillus, Micrococcus, Lactobacillus*, and *Staphylococcus*. The distribution of types varied greatly with the source of the cream. After holding for 5 days at 5°C, nonfluorescent *Pseudomonas* tended to become the predominant type and *Corynebacterium* and *Micrococcus* were reduced in number, although there were still differences between samples.

Similar results were reported by Phillips et al. (1981a,b), with post-heat-treatment contamination by Gram-negative organisms being the major factor in limiting the shelf life. If this contamination were eliminated, then survival of bacterial spores would then be a limiting factor; increasing the storage temperature from 6° to 10°C halved the shelf life of the cream. The mean shelf-life of the samples varied from 5 to 23 days (a considerable difference). At the end of shelf life, *Pseudomonas* spp., and especially the nonfluorescent types, were very definitely the predominant flora.

In their study of the bacteriological condition of retail cream in Worcestershire, Colenso et al. (1966) found that 223 out of 540 samples failed the MB test, and only 181 were satisfactory. Counts were fantastically high, with 70 samples having counts of 2–50 × 10^8 g^{-1} and 137 having counts of over 10^6 g^{-1}; large numbers of coliforms and *Bacillus* spp. were recorded. However, Barrow et al. (1966) reported that cream could have a count of 5 × 10^7 g^{-1} without the flavor being affected.

4.1.16.2 Identification of Bacteria Causing Taints in Cream. It is not necessary to identify organisms causing taints in cream with academic precision. A broad typing sufficient to identify the genus and probable species is quite adequate to indicate the source (e.g., dirty equipment) or the reason for development of the taint (e.g., storage at too high a temperature) and the method that should be adopted to eliminate the contamination.

4.1.16.3 Psychrotrophic Organisms. All perishable foods have to be held at low temperatures, usually not above 5°C, and this treatment constitutes a form of selective enrichment. Psychrotrophic organisms are defined as those *capable* of growing at low temperatures, although they may in some cases have a high maximum temperature—for examples, 50°C and an optimum in the range 30–40°C (Harrigan, 1998).

While a temperature of about 30°C is generally suitable for making total counts on dairy products, an examination of long shelf-life products (e.g., keeping for 7–14 days at 5°C) should include counts made at 20–22°C as well as at 30°C in order to check the numbers of organisms capable of growing at lower temperatures. The correlation between counts at 5–10°C and those at 22°C is usually high.

The term "pseudomonads" is sometimes used to embrace all Gram-negative, non-spore-forming, oxidase-positive rods that commonly contaminate dairy products from dirty water. They do not sour cream but produce a variety of taints when the count approaches 10^8 cfu ml^{-1}.

A count of oxidase-positive bacteria can be made by flooding the plate with a reagent that develops a color under oxidase reaction, such as freshly prepared 1% tetramethyl-*p*-phenylenediamine hydrochloride, which turns successively pink, purple, and finally black. A parallel flooding of another plate with 1% hydrogen peroxide solution will identify catalase-positive colonies by a continuous evolution of tiny bubbles of oxygen.

The pseudomonads may be broadly distinguished from the Enterobacteriaceae by being oxidase- and catalase-positive and by not fermenting lactose. Most Enterobacteriaceae ferment lactose and are catalase-positive but oxidase-negative.

4.1.16.4 Pseudomonas. This genus consists of Gram-negative non-spore-forming rods that attack proteins and fats strongly, but have little or no effect on sugars. They require air for growth and often produce greenish pigments and various taints. They are associated with watery environments and can grow at low temperatures, but are easily killed by pasteurization. The most common in dairy products is *Pseudomonas fluorescens*, which attacks fat and produces rancidity. *P. fragi* develops an apple-like ester taint before rancidity is detected by taste, and *P. putrefaciens* produces a putrid odor. *P. nigrifaciens* can give a blackish discoloration on the surface of dairy products.

In general the genus is harmless, but *P. aeruginosa* can grow at 42°C, is very resistant to antibacterial chemicals (antibiotics and quaternary ammonium compounds), and is now recognized as an opportunist pathogen.

4.1.16.5 Yeasts. Yeasts are not commonly the cause of defects in dairy products, because (with a few exceptions such as *Kluyveromyces lactis* which may be selected by consistently poor hygiene practices) they do not ferment lactose and grow comparatively slowly. However, if organisms (such as lactococci) that are capable of hydrolyzing lactose are present, or if sugar is added (in whipped cream, for example), then yeasts can grow rapidly and produce a characteristic yeasty or fruity flavor and obvious gas. *Torula cremoris* or *Candida pseudotropicalis* and *Torulopsis sphaerica* have been responsible for outbreaks of this defect in cream.

4.1.16.6 Anaerobic Spores. The presence of bacterial spores in milk is of considerable importance for certain dairy products, particularly those that are sterilized by retorting or UHT processes. Destruction of anaerobic spores is essential in sterilized and UHT products, because cream is a low acid product. Thus all creams and similar long-shelf-life products should be subjected to a "botulinum cook"—that is, a heat treatment in excess of 121.1°C for 3 min or equivalent. Assuming a Z value of 10°C, this treatment should give a 12-decimal reduction in the probability of *Clostridium botulinum* surviving the heat treatment (IFST, 1998).

4.1.16.7 Aerobic Spores. Aerobic spores can be a serious source of trouble and may be classified into groups according to their heat resistance.

The spores of *B. cereus* are only moderately heat-resistant, but can easily survive pasteurization and somewhat higher temperatures. This and other *Bacillus* species can grow at low temperatures (Cox, 1975), although a temperature of 5°C will retard their growth for some days (Davis, 1969, 1971). *B. cereus* is of particular importance for cream, because it can not only induce sweet curdling (or "bitty cream" in milk) and proteolytic spoilage in cream, but can also reduce methylene blue and thereby lead to failure in the official PHLS test.

Other species, such as *B. licheniformis* and *B. coagulans*, can also become prominent in cream (Cox, 1975; Tatzel, 1994). *Bacillus subtilis* and its variants is also important in terms of producing spores that are markedly heat-resistant, and they may be responsible for bitterness and thinning in sterilized creams. More recently, *B. pumilus* has been recognized as a potential contaminant carried over from the raw milk (Lewis, 1999), and *B. sporothermodurans* has been recovered from UHT cream (Herman et al., 2000). *B. sporothermodurans* is far more heat-resistant than *B. stearothermophilus* under UHT conditions, with D_{140} values of 3.4–7.9 s compared to 0.9 s for the latter with z values of 13.1–14.2°C and <10°C, respectively (Klijn et al., 2000).

The survival of spores is also a significant factor with clotted cream. There are two reasons for this. The process kills all vegetative cells and thus leaves a clear field for the growth of the spores, plus the heating gives a shock to the spores which stimulates germination (Davies, 1975).

4.1.17 Microbiological Associations

Organisms seldom occur in pure culture in nature, and the importance of interconnecting activities or associated action is not always appreciated. An organism may exert a powerful influence on the shelf life by releasing a source of energy (e.g., by hydrolyzing lactose) for another organism to utilize, or by creating favorable or unfavorable conditions (acidic or anaerobic conditions). For example, souring of milk by *Lac. lactis* may inhibit growth of *Bacillus* spp. and other types sensitive to acid (Harman and Nelson, 1955).

4.1.18 Food Poisoning from Cream

Untreated cream poses similar risks to untreated milk, possibly increased by the additional handling in its preparation. Untreated cream was implicated in an outbreak of *E. coli* 0157 phage type 2 food poisoning in the United Kingdom (Anonymous, 1998).

Heat-treated cream may differ from milk for two reasons:

1. Cream is normally more severely heat-treated than retail milk.
2. Cream is expected to have a longer shelf-life than retail milk.

Thus, in practice, while there is less probability of food-poisoning organisms from milk passing into cream, any which do pass into cream have more chance of proliferating and thereby causing trouble. Any contamination *after* heat treatment may be more serious for this reason, although if the cream is held below 5°C, the chance of food poisoning organisms growing is considerably reduced.

The most common cause of deterioration of cream is water-borne contamination, the most common source of *Pseudomonas*; and if the water is contaminated with sewage, *Salmonella* and other fecal types may be present. *Salmonella* spp. have been found in cream and chocolate cakes (Serra et al., 1989), and in the past this was the usual cause of food poisoning from cream, but such incidents are now rare.

Jensen (1990) reported holding temperatures of 69.1–72.5°C to give a 1-s decimal reduction time for different strains of *L. monocytogenes* when using a plate heat exchanger, indicating a low probability of survivors. The generation times for *L. monocytogenes* strains varied from 22.2–28.8h at 4°C to 8.6–14.4h at 7°C.

EU standards (EEC, 1992) for pasteurized cream include:

L. monocytogenes	Absence in 1g
Salmonella spp.	Absence in 25g where $n = 5, c = 0$

Other species are covered by the general requirement that the products should not contain pathogenic microorganisms in such quantity as to affect the health of the ultimate consumer.

4.2 BUTTER

Cream is the primary ingredient for butter, and its microbiology will have a major influence on the butter made from it. However, butter is a water-in-oil emulsion with at least 80% fat, so the microflora will be primarily concentrated within the dispersed aqueous phase and subject to both steric and compositional limitations, varying with the butter-making method used.

4.2.1 Butter Making

Butter making is essentially the controlled destabilization of the oil-in-water emulsion of cream, selective concentration of the lipid components by removal of the aqueous buttermilk fraction, and subsequent formation of a stable water-in-oil emulsion. The products may be divided into sweet cream and cultured cream butters, while the processes employed may be divided between:

1. Traditional batch churning
2. Continuous churning, often referred to as the Fritz process
3. High-fat processes

In both the traditional and continuous churning processes, the pasteurization of the cream is in effect the critical control point, with all steps thereafter potentially leading to contamination of the final product. With the high-fat processes there is less potential for contamination. An outline of the various processes is given in Figure 4.3.

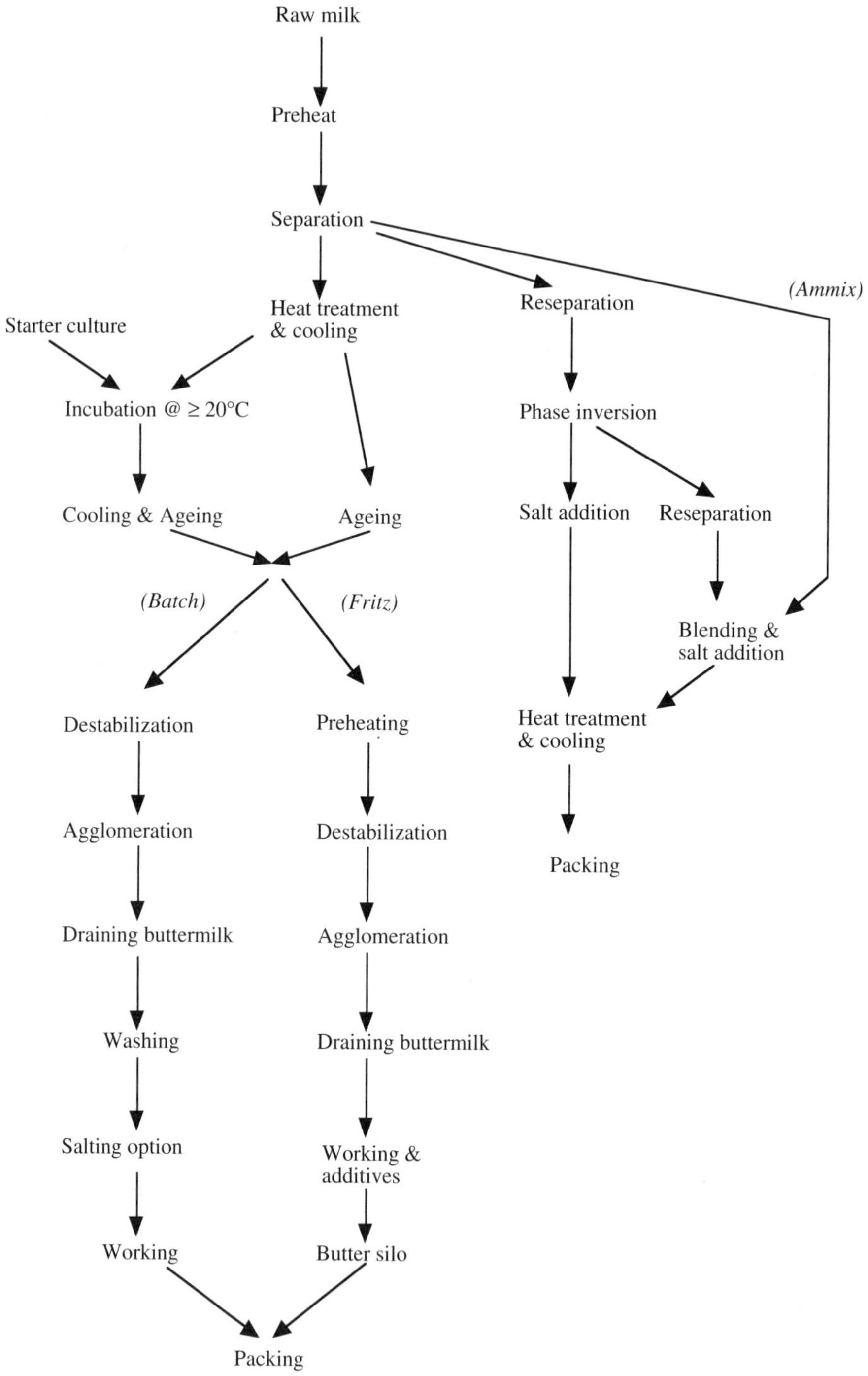

Figure 4.3. Alternative sequences of operations to produce butter.

The microbiological considerations in preparing cream for butter making are similar to those for other forms of cream—in particular, the risks of sour milk introducing off-flavors and high psychrotroph counts, which, in turn, lead to the presence of heat-resistant lipases. However, butter may be used as a sink product for surplus and downgraded milk or cream, and microbial quality problems may be expected in these circumstances.

4.2.2 Traditional Batch Churning

4.2.2.1 Original Farmhouse Methods. Up to the late nineteenth century, cream was separated by standing raw milk in bowls overnight and then decanting off the top cream layer. Extensive growth of the endogeneous microflora was normal, with lactic acid bacteria predominating. The cream was not heat-treated but transferred to a wooden churn, and the emulsion was broken by manual churning or beating. Washing the butter grains with unchlorinated, often unfiltered water would remove much of the milk solids-not-fat (MSNF), thereby reducing the substrate available to support microbial growth but introducing additional water-borne organisms. The addition of salt and/or storage in brine was needed as a preservative.

Similar microbiological problems may still be found where very simple production methods are being used.

4.2.2.2 Early Creamery Production. The introduction of mechanical separators led to centralization of separation in "creameries," with butter making then being scaled up and mechanized. Further progress was made at the beginning of the twentieth century with the introduction of cream pasteurization. Hygiene was also improved considerably, although until the middle of the twentieth century the majority of churns were still made of wood, with consequent problems in cleaning and disinfection. Wooden churns were replaced by stainless steel churns (and some aluminum churns initially), permitting more rigorous cleaning methods to be applied.

Reduction of the natural souring microflora from the cream gave a much blander product and gave rise to a division in manufacturing methods. The United Kingdom followed New Zealand practice in churning the sweet cream, then adding salt to the butter grains and working the mass to form a salted butter. Other European countries tended to add a starter culture to sour the cream to regain the traditional flavor before butter making.

4.2.2.3 Current Practice. Cream for traditional batch butter making is normally separated at 35% fat but may be up to 40% fat, using temperatures of 50–63°C for separation. The cream is then pasteurized at 74–80°C for 15s and cooled to 5–8°C using a plate heat exchanger, the treatment being sufficient to eliminate vegetative pathogens and virtually all lactic acid bacteria and other contaminants. The cooled cream contains a mixture of crystalline, supercooled, and liquid fats and should be held cold for at least 4h, preferably overnight, to allow further crystallization of the milk fat to take place. Most of this additional crystallization takes place within the first two hours and is accompanied by the release of latent heat of crystallization. The energy release brings about a rise in the cream temperature, typically about 2°C, which must be compensated for by either greater initial cooling or by cooling the cream in the aging tank. Cream aging tanks are typically 5000- to 10,000-liter capacity, equipped with a slow-moving stirrer and chilled water jacket. Such tanks, like the pasteurizer, will be cleaned-in-place and have minimal effect on the cream microflora.

Following aging, the cream at ~8°C will be pumped over to the churn, with no more than half filling it. (The churn will previously have been disinfected—for example, by either hypochlorite or peracetic acid treatment.) Rotation of the churn introduces air into the cream, creating an unstable coarse aeration which, in conjunction with the mechanical impacts between globules, leads to damage of some of the fat globules, release of free fat, and agglomeration of both the damaged and undamaged milk fat globules to form butter grains. The temperature of the mass also rises to ≥12°C. The microflora of the cream will be divided between the free serum (known as buttermilk) and the butter grains, with most being retained within the buttermilk.

Buttermilk is drained off from the butter grains and pumped away. Chilled, preferably pasteurized, water is then added to the butter grains, and the churn is rotated slowly for a short time to ensure adequate mixing. Pasteurization or an equivalent process will minimize contamination from the water, because mains water normally contains psychrotrophic organisms, which, while they should not be a public health problem, do pose a spoilage risk. The wash water is then drained off, removing approximately 90% of the MSNF and hence providing a less attractive medium for microbial growth. The cooling of the butter grains also aids subsequent handling.

Salt may then be added to the drained butter grains. In the United Kingdom up to 2% finely milled salt is normally added for salted butter, though in Wales there is also a market for extra salted butter with >3% salt. The salt dissolves in the residual serum, a maximum of 16% moisture being permitted in butter virtually worldwide. Though some moisture droplets already dispersed in the fatty matrix may not become salted, most of the serum becomes salted, with a 2% addition being equivalent to >11% in the serum phase. This is equivalent to an a_w of 0.94 and is sufficient to inhibit the growth of many bacteria.

Restarting the churn causes the mass of butter grains to tumble, consolidating the butter grains into a continuous phase of liquid fat containing dispersed phases of damaged and undamaged fat globules, milk fat crystals, and serum. The shear forces generated in the tumbling action break up the pools of serum into progressively finer droplets, the ideal state being when the droplets have been reduced to ≤10 μm (Richards, 1982). At a mean particle size of 10 μm, there would be over 38 million droplets per milliliter of butter so that even with a poor cream containing 10,000 cfu/ml less than 0.1% of the droplets should contain any organisms, the growth of which would be inhibited by the relative lack of nutrients and space.

Once the butter has been worked sufficiently, it is dropped from the churn into a butter trolley. Smaller-scale production may use simple wheeled trolleys from which the butter is dug using stainless or plastic shovels, whereas larger-scale operations have an auger and butter pump built into the trolley to minimize the handling.

Packaging is either into consumer packs, typically 250-g parchment or foil laminate, or into 25-kg polyethylene-lined cardboard boxes for storage.

4.2.3 Continuous Butter Making

Prototype continuous butter makers were evolved in the 1930s, but it was not until the 1950s that the methods had been developed sufficiently for the traditional batch techniques to be displaced. The great advantage of these continuous methods was the possibility of improving the hygienic aspects of production while reducing labor costs.

4.2.3.1 Fritz-Type Processes. The Fritz process was developed in Germany in the 1930s as a method for rapidly destabilizing small quantities of cream to produce a high-moisture butter, thus giving an hourly output similar to the larger batch churns. Subsequent development of the process by a number of manufacturers led to this technique becoming the most popular butter making method in Western Europe. These processes are described in more detail by Wilbey (1994).

In most modern production lines, the aged cream is fed to the butter maker via a preheater, using a small temperature differential between the cream and the warm water. The cream, now at the optimum for destabilizing the emulsion (about 12°C), is fed into the first cylindrical chamber where a beater rotating at ~1000 rpm beats and aerates the thin layer of cream to break the emulsion and promote aggregation of surviving and damaged fat globules. The mixture of aggregates and serum from the cream drops down a chute into a second cylindrical chamber that rotates slowly at up to 35 rpm. The rolling action of the particles in the first part of this cylinder promotes further aggregation of the butter grains with some recovery of fat from the serum. The second part of the cylinder is perforated so that the serum, now referred to as buttermilk, can drain away while the butter grains increase further in size and consolidate. A significant proportion of the microflora in the cream will be lost in the buttermilk. Early designs of butter makers and those used for poor-quality cream may introduce a chilled wash water at this stage, as with batch butter making. Some butter makers cool and recycle buttermilk within this stage to cool the butter grains.

Butter grains fall from the second cylinder into the first working stage, where a pair of contra-rotating augers consolidate the butter grains and squeeze out some of the buttermilk. The compacted butter grains are transported up a slight incline through a series of orifice plates and cutter blades, similar to a mincing machine. The effect of this action is to consolidate the butter grains into a continuous lipid phase of liquid milk fat in which are distributed the moisture droplets, milk fat crystals, and fat globules. Ideally these moisture droplets should have a diameter ≤10 μm; but because this is difficult to achieve in a single working stage, most butter makers now have a second stage worker, linked to the first worker unit by a vacuum chamber that is used to standardize the air content in the butter.

As described above, the butter maker would only be able to produce unsalted butter. However, salt may be introduced by dosing at the first orifice plates in the first working section. Dry salt is difficult to handle, so either a saturated brine (26% salt) for slightly salted butter or a 50% suspension for salted butter is used. The higher shear forces employed in the working sections should give a finer dispersion of moisture in the butter, thus giving superior keeping qualities to that from the traditional batch butter makers.

4.2.3.2 High Fat Processes. Several different processes based on reseparation before reversal of the emulsion had evolved and found niches in the world market for butter making. These are essentially hot cream processes that do not require cream to be aged but employed reseparation of the cream at temperatures varying from 52°C to 90°C to achieve the desired 80% minimum milk fat content (McDowall, 1953). Successful commercial exploitation of the processes required efficient scraped surface heat exchangers in which the cooling, crystallization, and formation of the water-in-oil emulsion took place. Salted butters could be produced by blending salt with the high-fat cream before cooling, but ripened cream butters were less easily produced by his technique.

Most recently, the Ammix process was developed in New Zealand for continuous butter production. In this process, part of the cream is reseparated, the emulsion is broken, and a third stage separation is applied to produce an anhydrous milk fat that is then mixed with the remainder of the cream and possibly salt to give a mixture with >80% fat, <16% moisture. This mixture is then cooled and worked by passage through scraped surface heat exchangers and worker units to produce the desired plastic product that would pass directly from the cooler exit to the filler for packing into either retail or bulk packs for storage.

Production plant for the high-fat process should always be cleaned in place. With a closed system and a shorter product pathway downstream of the heat treatment, the risks of recontamination are less than those for the Fritz-type processes. Adequate working of the butter in the scraped surface units should ensure that moisture droplet sizes are kept below 10 μm so that the product should be stable in microbiological terms.

4.2.3.3 Recombined Butters. Recombined butters may be produced by the recombination of the fatty and nonfat components of butter, selected for their long shelf life so that the recombined butter may be produced in areas with insufficient milk supply to meet their dairy product demands. Anhydrous milk fat, through its extremely low moisture content (<0.1%), will not support microbial growth and, providing that it is correctly packaged and stored, has its shelf life limited by the development of oxidative rancidity. Milk powder, typically skim milk powder (SMP), will carry a microbial load of thermoduric organisms and spores but will also be stable, provided that the moisture content remains below 4%. To make up the aqueous phase, SMP is dispersed in potable water and salt is added (if desired). Flavorings may also be added. The lipid phase is primarily milk fat, but emulsifiers may be added where the product is intended for baking or frying. The emulsion may be created either by a batch blending or by in-line dosing and mixing systems. Though the ingredients should be free of pathogens, it is desirable that either the individual phases or preferably the emulsion be heat treated. Heat treatment may use plate, tubular, or scraped surface heat exchangers but all cooling below ~40°C must use scraped surface heat exchangers as the viscosity rises quickly once the milk fat begins to crystallise. As with the high fat butter making processes, the high shear forces during cooling promote the production of a fine, plastic emulsion.

4.2.4 Ripened Butters

Improved hygiene by the end of the nineteenth century and heat treatment of the cream removed much of the natural microflora associated with the cream ripening. It was then necessary to produce the ripened creams by adding a starter culture to the heat treated cream. The starter cultures were originally extracted from souring milks and cream, selected on the basis of their lactic acid or flavor production.

4.2.4.1 Cream Fermentation. Strains of *Lac. lactis* subspp. *lactis* and *cremoris* were selected for their rapid production of lactic acid by fermentation of lactose. *Lac. lactis* subsp. *lactis* biovariant *diacetylactis* is a much weaker producer of lactic acid but carries the plasmid for metabolism of citrate, yielding a number of metabolites including diacetyl, which is a major contributor to the flavor of soured creams and butters as well as having some biostatic properties. The fourth organism normally present is *Leuconostoc mesenteroides* subsp. *cremoris*, which, in addition to its ability to metabolize citrate, can reduce acetaldehyde to ethanol, thus avoiding the green flavor note that is associated with yoghurt. Hugenholtz and Starrenburg (1992) looked at eight strains of Leuconostoc, finding that while citrate lyase and alpha-acetolactate synthetase were induced by citrate, the citrate appeared to be converted to D-lactate. These lactic acid bacteria are mesophiles and are normally grown at 20–30°C.

Cream for ripened butter production is frequently subjected to a relatively severe heat treatment—for example, 90–95°C for 15s or flash treatment at >100°C. This should be sufficient to remove thermoduric enterococci, lactobacilli, and micrococci, and the additional protein denaturation will release sulfhydryl compounds, thus reducing the E_h in the cream and providing a better medium for starter growth. Fermentation at 20°C will take 12–20hs, depending upon the activity of the starter culture and the desired acidity. Flavor development is controlled by pH, not just in terms of the amount of lactic acid produced but by the control that pH exerts on the biosynthesis of diacetyl. Above pH 5.4, the enzyme acetolactate synthetase is inhibited, thus inhibiting production of acetolactate, a precursor to diacetyl (Cogan and Hill, 1993). Below that threshold, citrate is converted to acetoin, which may then be oxidized to diacetyl.

The final acidity of creams for the manufacture of ripened butters may vary from pH 5.3 for mild ripened butters to pH 4.5 for the more strongly flavored varieties, which are more common in Scandinavia. The lower pH products will normally contain higher levels of diacetyl and lactic acid. The final pH is achieved by cooling the fermenting cream to ≤10°C, at which point metabolism will virtually stop. This cooling may be achieved by batch cooling using a chilled water jacket while stirring slowly, or particularly for larger quantities, by pumping from the fermentation tank through a plate heat exchanger to a holding tank. In either case, cooling must start before the target pH is reached because fermentation will continue in the uncooled cream.

Traces of short-chain fatty acids may contribute to the flavors of butters, the contribution being highly dependent on concentration for the particular fatty acids. Excessive levels can lead to cheesy and rancid off-notes. The starter microfloras are weakly lipolytic and thus unlikely to be a cause of such off-flavors.

Ripened creams may be churned to butter using either the traditional batch churning or Fritz type of process. The combination of a low pH and a microflora rich in lactococci is usually sufficient, as with salted butter, to inhibit growth of microbial contaminants within a correctly produced fine emulsion.

4.2.4.2 NIZO Process. Though the use of ripening of cream produces a butter that is preferred by a large proportion of European consumers, the process creates approximately equal quantities of lactic buttermilk. This lactic buttermilk is difficult to heat treat and process into products for which there is consumer demand; thus, much of the lactic buttermilk has to go for animal feed, which gives a minimal return.

One way to get around this economic problem was invented at the Netherlands Institute voor Zuivelondazoek (NIZO), where the starter culture and lactic acid were added to the butter grains made from sweet cream using a Fritz-type continuous butter maker (Veringa et al., 1976). The *Lac. lactis* subspecies, including the *diacetylactis* biovariant, were grown up and the cultured medium aerated to oxidize acetoin to diacetyl, raising the diacetyl level to ~40 ppm. The *Leu. mesenteroides* culture was grown separately. Addition of these cultures could provide the characteristic microflora to the butter, but there was not sufficient acidity from this source to reduce the pH to below 5.3. Additional lactic acid was provided by blending the *Lactococcus* culture with a "culture concentrate," prepared by fermenting whey with *Lactobacillus helveticus* and then recovering the dilute lactic acid by ultrafiltration and evaporating to 11% acid. Total addition rates corresponded to 2% starter culture and 0.7% concentrate.

Other inventions have sought to simplify the process further—for instance, by using lactic acid, cultures, and a flavor distillate prepared off-site (Wiles, 1978).

4.2.5 Reworking

Worldwide, most butter production is based on the surplus milk fat that accumulates during seasonal peak milk production, whereas consumer demand is more constant. The majority of butter is thus packed as "bulk butter" into 25-kg card boxes lined with polyethylene (which has largely replaced parchment). This butter will be stored frozen (–10°C to –27°C) until required and will be microbiologically stable. Indeed, the total colony counts will tend to decline with time, along with coliforms (Jensen et al., 1983).

For consumer sales, this frozen bulk butter must first be thawed out, chopped up, and reworked to return it, albeit temporarily, to a plastic state and packed into retail portions, typically 250 g. The thawing process can take up to a week in a cool room for large quantities of butter, providing an opportunity for any yeast, mold, or bacterial contaminants on the surface to grow. Both floor space, time, and the potential for deterioration can be reduced by deboxing and then using a microwave tunnel attemperation system. The 25-kg blocks are then stripped of their polyethylene covering and comminuted to flakes using a shiver. These flakes must then be converted into a plastic mass by reworking.

Smaller reworking facilities may still use batch sigma or "Z" blenders, so named after the shape of the mixing arms. These are essentially open mixing vessels, and the blended butter will be either dug out by hand or tipped into an open auger system. Hand cleaning is the norm, and there is plenty of scope for recontamination.

Most butter reworking plants now use continuous reworkers linked directly to the shiver. These are similar to the working units on the continuous butter makers and work in exactly the same way. Continuous reworking systems are essentially closed systems and can be installed with CIP so that cleaning and disinfection may be closely controlled. Contamination of the butter during reworking is much less of an problem than with the open batch units.

TABLE 4.7. European Union Milk Fat Standards for Spreadable Fats

Sales Description	% Milk Fat	
Butter	≥80	≤90
Three-quarter-fat butter	≥60	≤62
Half-fat butter	≥39	≤41
Dairy spread *X* %	<39% or	
	>41	<60
	>62	<80
Reduced-fat butter	>41	<62
Low-fat or light butter	<41	

Additional water and/or salt may be added during the reworking to optimise their levels. Any water added must be at least of potable quality.

4.2.6 Reduced-Fat Butters

Compositional standards for reduced-fat butters and similar products are covered by EU regulations, summarized in Table 4.7 (EC, 1994; UK Regulations, 1995d).

Reducing the fat content of butters and similar fat spreads will result in an increase in the moisture content. In the case of salted products, the salt content cannot be increased so that the level of salt in the aqueous phase will decrease pro rata with the moisture increase. The consequent increase in a_w will make the product more susceptible to spoilage. Substitution of milk fat by vegetable fats will have no direct effect on microbiological stability, but may alter the pattern of free fatty acids liberated by any lipolytic activity.

The increased moisture content increases the dependence upon a fine particle size distribution for microbial stability. However, fine emulsions tend to destabilize more slowly on the palate and give poorer sensory properties. In some low-fat products a biphasic dispersion may be formed (Charteris et al., 1992), which, while giving satisfactory sensory properties, may be less stable in microbiological terms. Reduction in pH and/or the inclusion of potassium sorbate may inhibit spoilage (Charteris, 1995).

Production methods for reduced- and low-fat spread products rely mainly on scraped surface technology (Wilbey, 1994). The aqueous phase may contain milk solids, salt, flavoring, stabilizers, preservatives, and other permitted additives. In most cases the ingredients will have been pasteurized. The lipid phase may contain vegetable fats (in the case of blends), emulsifiers, color, vitamins, and flavor. The emulsion may be formed in a batch process or by continuous blending. If the aqueous phase or emulsion are at temperatures of 20–60°C, then there is potential for growth of thermoduric survivors such as *Enterococcus* spp., *B. cereus*, and/or thermophilic contaminants. Milk proteins will provide a source of amino acids to support microbial growth (Bullock and Kenny, 1969), whereas lactose will encourage the growth of lactic acid bacteria; this lactic acid may help to inhibit many bacterial species.

Either the individual phases or, preferably, the emulsion should be heat-treated. Cooling of the emulsion within the scraped surface heat exchangers should avoid recontamination, with the remaining risk being at the packaging stage. Thus the main threats to product quality will come from failures in the CIP of the plant and from contamination during the filling operation.

In the final product, gelling agents such as milk proteins and stabilizers will reduce the diffusion rates of nutrients, thus inhibiting microbial growth (Robins et al., 1994). Gelation has an additional benefit in minimizing coalescence of the dispersed moisture droplets.

4.2.7 Microbiological Associations

In well-made salted butter the surviving microflora will be tolerant of low temperature and relatively high salt concentrations, allowing micrococci to predominate (Shehata et al., 1978). Environmental contaminants are likely to be inhibited and may slowly die out if the moisture is finely dispersed.

Where the storage environment has a high humidity, particularly if a permeable packaging such as vegetable parchment is used, then psychrotrophic molds may grow on the surface of the butter. Mold species associated with fungal spoilage of butter include *Cladosporium, Alternaria, Aspergillus, Mucor, Rhizopus, Penicillium*, and *Geotrichium*, especially *G. candidum* (Jay, 1996). The color of the discoloration may be a good clue to the type of spoilage, but a black discoloration may be attributable to *Cladosporium* spp., yeasts such as *Torula* spp., or *P. nigrifaciens*.

Mold growth on butter may be inhibited by the addition of potassium sorbate, with an additional benefit being the inhibition of coliforms (Kaul et al., 1979). However, most authorities regard the combination of good process hygiene and correct storage as the preferred method of avoiding mold and coliform contamination.

Unsalted sweet cream butter is the most susceptible to microbial contaminants because there is only the fine dispersion of the aqueous phase to inhibit growth. In ripened butters the lactic acid bacteria and reduced pH will inhibit many contaminants, including *Pseudomonas* spp., but will allow yeasts to grow if free moisture is available.

As with all fresh products, contamination of wash water with lipolytic psychrotrophs will present a risk to product quality. *P. putrefaciens, P. fluorescens, P. mephitica*, and *P. fragi* have all been associated with defects in butter. Surface growth of *P. putrefaciens* has been associated with surface taint resulting from release of free fatty acids, particularly isovaleric acid. *P. fluorescens* and *P. fragi* are also associated with lipolytic rancidity while a skunklike odor may be caused by *P. mephitica* (Jay, 1996). The organisms may be eliminated from the water supply by hyperchlorination to ≥20 ppm available chlorine without causing taints in the butter.

Poor quality may also be associated with the presence of coliforms, an indicator of poor hygiene and potential risk of food poisoning.

4.2.8 Food Poisoning from Butter

Historically, butter has not been regarded as a high-risk product. However, the presence of a potentially pathogenic genus gives an indication of a postprocess contamination problem, and there is a risk that any pathogenic contaminants could be transferred to a more amenable substrate and hence reach numbers that may cause problems for the consumer of that food.

EU regulations (EEC, 1992) require *L. monocytogenes* to be absent in 1 g butter, *Salmonella* spp. to be absent in 25 g ($n = 5, c = 0$), and coliforms in butter made from pasteurized cream to be $<10^5$ ($m = 10^4, n = 5, c = 2$). As with other foods, there is the overall caveat that pathogenic microorganisms and their toxins must not be present in quantities such as to affect the health of consumers.

Sims et al. (1969) contaminated salted and unsalted sweet cream butters with *S. typhimurium* var. *copenhagen*. *Salmonella* counts declined over a 10-week period when the butter was held ≤4.5°C, particularly ≤–18°C. At 25°C, counts increased approximately a thousandfold in the first 5 weeks before declining to a similar level to the original inoculum by the tenth week. The salt did not have a protective effect.

One cluster of listeriosis cases in the United States has been attributed to butter contaminated by *L. monocytogenes*, and there have been at least four recalls of butter contaminated with this organism.

Experimental manufacture of salted butter deliberately made from cream containing high numbers of *L. monocytogenes* indicated that the majority of the organisms were lost in the buttermilk but that at 1.2% salt there was an increase of 1.9–2.7 orders of magnitude at 4–6°C and 13°C storage, respectively, while survivors in deep frozen butter dropped by less than one order (Ryser, 1999).

Butter was also implicated in an outbreak of listeriosis in a hospital in Finland in 1998–1999 in which 25 patients were affected and 6 died. The outbreak strain was found in both the packaged butter and the manufacturing dairy (Lyytikainen et al., 2000).

These reports emphasize the need for good hygiene practices in avoiding postpasteurization contamination.

REFERENCES

Airey, F. K. (1978) *J. Soc. Dairy Technol.*, **31**, 148.

Alais, C., Lorient, D., and Humbert, G. (1978) *Ann. Nutr. Aliment.*, **32**, 511.

Anonymous. (1985) *Dairy Ind. Int.* **50**(6), 41.

Australia (1993) *Dairy Industry Regulation 1993* Section 5. State of Queensland.

Barrow, G. I., Milier, D. C., and Johnson, D. L. (1966) *Lancet*, **2**, 802.

Blankenagel, G. (1976) *J. Milk Food Technol.*, **39**, 301.

Board, P. W., and Steel, R. J. (1978) *Food Technol. Aust.*, **30**, 169.

Bradshaw, J. G., Peeler, J. T., Corwin, J. J., Hunt, J. M., and Twedt, R. M. (1987) *J. Food Prot.*, **50**, 543.

Brown, J. V., Wiles, R., and Prentice, G. A. (1979) *J. Soc. Dairy Technol.*, **32**, 109.

Brown, J. V., Wiles, R., and Prentice, G. A. (1980) *J. Soc. Dairy Technol.*, **33**, 78.

Brown, J. V., and Prentice, G. A. (1982) *XXI International Dairy Congress* Vol. 1, book 2, Mir Publishers, Moscow.

Bullock, D. H., and Kenny, A. R. (1969) *J. Dairy Sci.*, **52**, 625–628.

Burton, H. (1988) *Ultra-High-Temperature Processing of Milk and Milk Products*, Elsevier Applied Science, London.

Cassedi, M. A., de Matos, R. E., Harrison, S. T., and Gaze, J. E. (1998) Heat resistance of *Listeria monocytogenes* in dairy products as affected by the growth medium. *J. Appl. Microbiol.*, **84**(2), 234–239.

Charteris, W. (1995) *J. Soc. Dairy Technol.*, **48**(3), 87–96.

Charteris, W., Kennedy P., Heapes M., and Rerille W. (1992) *Farm & Food* **2**(1), 18–19.

Cogan, T. N., and Hill, C. (1993) Cheese starter cultures. In Cheese: chemistry, physics and microbiology Vol 1. Ed Fox P. F., Chapman & Hall, London.

Colenso, R., Court, G., and Henderson, R. J. (1966) *Monthly Bull. Min. Health Public Health Lab. Service*, **25**, 153.

Cox, W. A. (1975) *J. Soc. Dairy Technol.*, **28**, 59.

Crossley, E. L. (1948) *J. Dairy Res.*, **15**, 261.

Davies, F. L.(1975) *J. Soc. Dairy Technol.*, **28**(2), 69–78.

Davis, J. G. (1969) *Dairy Industries*, **34**, 555.

Davis, J. G. (1971) *Dairy Ind.*, **36**, 267.

Dhamangaonkar, A. D., and Brave, O. (1978) *XXth Int. Dairy Congr.*, **E**, 846.

EC (1994) *Council Regulation (EC) No. 2291/94 of 5 December 1994 laying down standards for spreadable fats.* Council of the European Communities, Brussels.

EEC (1992) *Council Directive 92/46/EEC of 16 July 1992 laying down the health rules for the production of raw milk, heat treated milk and milk-based products.* Council of the European Communities, Brussels.

Eibel H., and Kessler H. G. (1984) *Milschwissenschaft*, **39**(11), 648–651.

Felmingham, D. M., and Juffs, H. S. (1977) *Aust. J. Dairy Technol.*, **32**, 158.

Fox, P. F., and Morrissey, P. A. (1981) In *Enzymes and Food Processing*, Applied Science, London, p. 213.

Franklin, J. G. (1969) *J. Soc. Dairy Technol.*, **22**, 100.

Griffiths, M. W., Phillips, J. D., and Muir, D. D. (1984) *J. Soc. Dairy Technol.*, **37**(1), 22–26.

Harmon, L. G., and Nelson, F. E. (1955) *J. Dairy Sci.*, **38**, 1189.

Harrigan, W. F. (1996) *Laboratory Methods in Food Microbiology*, 3rd ed. Academic Press, London.

Herman, L., Heyndrickx, M., Vaerewijck, M., and Klijn, N. (2000) *Bulletin of the International Dairy Federation No 357/2000*, International Dairy Federation, Brussels, pp. 9–13.

Hersom, A. C., and Hulland, E. D. (1980) *Canned Foods*, Churchill Livingstone, Edinburgh.

Holt J. G. (editor-in-chief) (1984–1989) *Bergey's Manual of Systematic Bacteriology*, Williams & Wilkins, Baltimore.

Hugenholtz, J., and Starrenburg, M. J. C. (1992) *Applied Microbiology and Biotechnology* **38**(1), 17–22.

Hughes, D. (1979) *J. Appl. Bacteriol.*, **46**, 125.

IDF (1996) *Bulletin 315: UHT Cream*, International Dairy Federation, Brussels.

IDF (1999a) *Bulletin 339: The World Dairy Situation 1999*, International Dairy Federation, Brussels.

IDF (1999b)International Standard 155A: 1999—Milk and milk-based drinks: Determination of alkaline phosphatase activity using a fluorimetric method. International Dairy Federation, Brussels.

IFST (1982) *Guidelines for the Handling of Chilled Foods*, Institite of Food Science and Technology, London.

IFST (1998) *Food and Drink Manufacture—Good Manufacturing Practice*, 4th ed, Institute of Food Science and Technology, London.

Jackson, A. C. (1978) *J. Soc. Dairy Technol.*, **31**, 80.

Jay, J. M. (1996) *Modern Food Microbiology*, 5th ed., Chapman & Hall, New York.

Jayne-Williams, D. J., and Franklin, J. G. (1960) *Dairy Sci. Abstr.*, **22**, 215–221.

Jensen, H. (1990) Forsøgrapport-Statens Mejerifor søg No 47 52pp. Cited in Dairy Science Abstracts **53**(10), no 7027.

Jensen, H., Danmark, H., and Mogensen, G. (1983) *Milschwissenschaft*, **38**(8), 482–484.

Kaul, A., Singh, J., and Kuila, R. K. (1979) *J. Food Prot.*, **42**(8), 656–657.

Kleyn, D. H., and Ho, C. L. (1977) *J. Assoc. Offic. Agric. Chem.*, **60**, 1389.

Klijn, N., Wagendorp, A., Huemer, I., Langeveld, L., and de Jong, P. (2000) *Bulletin of the International Dairy Federation No 357/2000*, International Dairy Federation, Brussels, pp. 19–20.

Konietzko, M., and Reuter, H. (1980) *Milchwissenschaft*, **35**, 274.

Lewis, M. J. (1999) *Int. J. Dairy Technol.* **52**(4), 121–125.

Lewis, M. J., and Heppell, N. (2000) *Continuous Thermal Processing of Foods*, Aspen, Gaithersburg.

Lyytikainen, O., Autio, T., Maijala, R., Ruutu, P., Honkanen-Buzalski, T., Miettenen, M, Hatakka, M., Mikkola, J., Anttila, V. J., Johansson, T., Rantala, L., Aalto, T., Korkeala, H., and Siitonen, A. (2000) *J. Infectious Diseases* **181**(5), 1838–1841.

McDowall, F. H. (1953) The Buttermaker's Manual Vols 1 & 2. New Zealand University Press, Wellington.

McKellar, R. C., Cholette, H., and Emmons, D. B. (1988) *Can. Inst. Food Sci. Technol. J.*, **21**(1), 97–101.

MMB (1980) *Dairy Facts and Figures 1980*, Milk Marketing Board of England and Wales, Thames Ditton.

MMB (1985) *Dairy Facts and Figures 1985*, Milk Marketing Board of England and Wales, Thames Ditton.

MMB (1990) *Dairy Facts and Figures 1990*, Milk Marketing Board of England and Wales, Thames Ditton.

MMB (1992) *Dairy Facts and Figures 1992*, Milk Marketing Board of England and Wales, Thames Ditton.

MMB (1994) *Dairy Facts and Figures 1994*, Milk Marketing Board of England and Wales, Thames Ditton.

Muir, D. D., Kelly, M. E., and Phillips, J. D. (1978) *J. Soc. Dairy Technol.*, **31**, 203.

NDC (1996) *Dairy Facts and Figures 1996*, National Dairy Council, London.

NDC (1997) *Dairy Facts and Figures 1997*, National Dairy Council, London.

NDC (1998) *Dairy Facts and Figures 1998*, National Dairy Council, London.

NDC (2000) *Dairy Facts and Figures 1999*, National Dairy Council, London.

Nichols, A. A. (1939) *J. Dairy Res.*, **10**, 231.

Peeseboom, J. W. C. (1969) *Milchwissenschaft*, **24**, 266.

Peeseboom, J. W. C. (1970) *Erhakrungsindustrie*, **72**, 299.

Phillips, J. D., Griffiths, M. W., and Muir, D. D. (1981a) *J. Soc. Dairy Technol.*, **34**(3), 109–113.

Phillips, J. D., Griffiths, M. W., and Muir, D. D. (1981b) *J. Soc. Dairy Technol.*, **34**(3), 113–118.

PHLS (1971) *J. Hyg. Camb.*, **69**, 155.

Pien, J. (1977) *Technique Lait.*, **904/905**, 7.

Richards, E. (1982) *J. Soc. Dairy Technol.*, **35**(4) 149–153.

Robins et al. (1994)

Romney, A. J. D. (ed.) (1990) *CIP: Cleaning in Place*, 2nd ed., Society of Dairy Technology, Huntingdon.

Rosenow, E. M., and Marth, E. H. (1987) *J. Food Prot.*, **50**(6), 452–459.

Rothwell, J. (1969) *J. Soc. Dairy Technol.*, **22**, 26.

Ryan, J. J., and Gough, R. H. (1982) *Food Prot.*, **45**(3), 279.

Ryser (1999)

SDT (1975)

Serebrennikova, V. A., Patratii, A. P., Rashkina, N. A., and Kravtsova, A. M. (1978) *Molochnaya Promyshlennost*, No. 10, p. 23.

Serra M. de S., Ferrer-Escobar, M. D., Pericas-Bosch, E., DeSimom-Serra, M., Escobar, D. F., and Bosch, E. P. (1989) *Anales Bromatol.*, **41**(1), 81–86.

Shehata, A. E., Mogdoub, M. N. I., El-Samagry, Y. A. A., and Hassan, A. A. (1978) *Milschwissenschaft*, **33**(5), 292–294.

Sims et al. (1969)

Smith, G. (1989) Sterilised Cream. In *Cream Processing Manual* 2nd edn, Ed Rothwell J. Society of Dairy Technology, Huntingdon.

Carolina, S. (1999) South Carolina Code of Regulations Chapter 61, in *State Regulations* **23** issue 10, South Carolina.

Tatzel (1994)

Tekinsen, O. C., and Rothwell, J. (1974) *J. Soc. Dairy Technol.*, **27**, 57.

Towler, C. (1994) Developments in cream separation and processing. In *Modern Dairy Technology*, Vol. 1, 2nd ed., R. K. Robinson ed., Chapman & Hall, London, pp. 61–106.

UK Regulations (1995a) *The Dairy Products (Hygiene) Regulations 1995*, SI 1995, No. 1086, HMSO, London.

UK Regulations (1995b) *Miscellaneous Food Additives Regulations 1995*, SI 1995, No. 3187, HMSO, London.

UK Regulations (1995c) *Food Safety (Temperature Control) Regulations 1995*, SI1995, No. 2200, HMSO, London.

UK Regulations (1995d) *Spreadable Fats (Marketing Standards) Regulations 1995*, SI1995, No. 3116, HMSO, London.

UK Regulations (1996) *Food Labelling Regulations 1996*, SI 1996, No. 1499, HMSO, London.

Veringa, H. A., van den Berg, G., and Stathouders, J. (1976) Milschwissenschaft **31**, 658–662.

Wilbey, R. A. (1992) *J. Soc. Dairy Technol.*, **45**(2), 31–32.

Wilbey, R. A.(1994) Production of butter and dairy based spreads. In *Modern Dairy Technology*, Vol. 1, 2nd ed. R. K. Robinson, ed., Chapman & Hall, London. pp. 107–158.

Wilbey, R. A., and Young, P. (1989) Clotted cream. In *Cream Processing Manual*, J. Rothwell, ed., Society of Dairy Technology, Huntingdon.

Wiles, R. (1978) UK Patent 1 579 068.

Wright, R. C., and Tramer, J. (1953) *J. Dairy Res.*, **20**, 177, 258.

Wright, R. C., and Tramer, J. (1954) *J. Dairy Res.*, **21**, 37.

Wright, R. C., and Tramer, J. (1956) *J. Dairy Res.*, **23**, 248.

CHAPTER 5

THE MICROBIOLOGY OF CONCENTRATED AND DRIED MILKS

RICHARD K. ROBINSON and PARIYAPORN ITSARANUWAT
School of Food Biosciences, The University of Reading, Reading, England

Dairy products of reduced moisture content may be produced to achieve savings in transportation and merchandising costs related to the reduced volume and weight; and these products, with their greater concentration of milk solids, are useful in the manufacture of ice cream, candies, and a variety of other food items. In some instances, desirable special properties result from one or more of the processing operations. In this chapter the concern will be with the microbiology of a number of important groups of product: (1) condensed and evaporated milks, (2) sweetened condensed milk, (3) the "retentates" obtained by reverse osmosis and ultrafiltration, and (4) dried milk powders.

Some of these products are little different from pasteurized milk from a microbiological standpoint; but heat, at the level of commercial sterilization, confers excellent microbiological keeping quality on "canned" evaporated milk. Sugar at a level inhibitory to most microbial growth retards spoilage in sweetened condensed milk, whereas in dried milk products the low available water acts as the inhibitory factor with respect to any bacterial spores or vegetative cells that have survived the drying process.

Dairy Microbiology Handbook, Third Edition, Edited by Richard K. Robinson
ISBN 0-471-38596-4

5.1 CONDENSED AND EVAPORATED MILKS

5.1.1 Concentrated Milk

This name is commonly used for a condensed milk prepared for human consumption, after appropriate dilution, as fluid milk, but without further processing. It is ordinarily prepared from Grade A milk, usually with a 3:1 concentration. The processing is done under conditions that satisfy Grade A standards, including a final pasteurization. One of the advantages is the reduced space required in the refrigerator, but this can be offset by the need to dilute before use; in addition, the keeping quality needs to be superior if advantage is taken of less frequent purchase. While considerable scientific and commercial interest was shown in this product in the 1950s, it has not become an important retail product.

The raw milk supply for the concentrate is given a heat treatment approximating pasteurization, and it is then concentrated at low temperature to minimize changes in flavor or physical characteristics. The concentrate is then standardized, homogenized, and pasteurized before packaging. Pasteurization is usually at a somewhat elevated temperature, such as 79.4°C for 25s, because this compensates for the slight protective effect of the greater solids concentration. The product has essentially the same keeping quality as pasteurized milk. The increased solids level is not great enough to inhibit microbial growth, so that storage at 10°C will permit growth of both thermoduric bacteria and any postpasteurization contaminants; the defects encountered are essentially the same as those in pasteurized milk.

5.1.2 Bulk Concentrated Milk

This product is usually made from manufacturing-grade raw milk, and it is used as a source of milk solids for candy, bakery products, ice cream, concentrated yogurt, and a number of other manufactured foods. Some concentrated milk may be made from Grade A raw milk for use in the standardization of market milk; this product must be processed and handled according to the provisions of the recommended sanitation ordinance. Condensing is usually in the range of 2.5:1 to 4:1, depending upon the use for which it is being prepared. The product is not sterilized during or after processing, so it contains a number of viable microorganisms, and the concentration of milk solids is not great enough to inhibit microbial development. Keeping quality is limited, particularly when great care is not used to control postheating contamination.

The methylene blue reduction test, with a reduction time of not less than 2.5h, has been used for evaluation of the raw milk prior to concentration; but with better cooling on the farm, the shortcomings of this procedure for the detection of psychrotrophic bacteria limit is usefulness (Lück and Andrew, 1975); other tests for microbial quality have now replaced dye reduction tests in most countries. If the milk is to be held for any period prior to processing, it should be cooled to 4.6°C or below to avoid a potential buildup of the microbial population and the attendant chance that acidity or other microbially produced defects could become a problem.

If a skim-milk condensed product is to be made, the milk is preheated and separated before further processing. In making concentrated whole milk, the product is customarily homogenized. Standardization for the desired fat:solids ratio may precede condensing, or may be done at the same time as the product is standardized for total solids content following condensing. Prior to condensing, the milk is heated in a continuous preheater, or in a "hot well," and heating is commonly to 65.6–76.7°C. This preheating temperature may be increased to 82.2–93.3°C for as much as 15min to increase viscosity and to impart other desirable characteristics to the product for use in special applications.

The preheating is usually not controlled to the same degree as pasteurization would be, but the microbiological results are much the same when temperatures in the upper part of the range are employed. However, the products should not be labeled as "pasteurised" because of the lack of proper safeguards. In the lower range of preheating temperatures, the product is usually not held at the forewarning temperature for any significant period of time, particularly if the process is continuous, and thus the microbiological destruction would be less than with pasteurization. The heater, as well as the "hot wells" or surge tanks, can serve as incubators for thermophilic bacteria under these latter conditions, and when such equipment is operated for long periods without intermediate clean-up, or when the milk supply contains excessive numbers of thermophilic bacteria, the numbers may build up to a point where acid and unclean flavors result.

The preheated milk is then concentrated in a vacuum pan or in a multiple-effect evaporator. The exact temperatures and times at which it is processed depend greatly on the type of equipment employed and the desired characteristics of the final product. A temperature range of 54.4–57.2°C is not uncommon, and this is very suitable for the growth of thermophilic bacteria. Operation over an extended period provides an opportunity for development of a considerable thermophilic population, so that proper sanitation and control in the preceding phases of the operation become essential.

The heat treatments prior to, and during, condensing are appreciably less than adequate to provide a sterile product. The cooling, standardization, and packaging operations provide opportunities for contamination, particularly from improperly cleaned and microbicidally treated equipment and containers. In addition, the material(s) used for standardization must be of good microbiological quality, because it must be emphasized that heat treatments, effective as they can be on microorganisms present at the time of treatment, leave no residual microbiocidal effect to act upon subsequent contaminants or to limit the growth of survivors of the heat treatment; again, the solute concentration in a condensed product is not sufficient to inhibit the growth of microorganisms. Because the product does contain microorganisms that have survived processing, and usually contains microorganisms that have contaminated the product subsequent to heat treatment, the time over which the product can be held without serious microbial spoilage is frequently very limited, especially at ambient temperatures.

5.1.3 Evaporated Milk

This product has much in common with bulk condensed milk through the condensing stage. However, it is given long-term keeping quality at room temperature by commercial sterilization, either before or after placing in a hermetically sealed container to protect against subsequent contamination. Most of the product is made from whole milk. The Federal Standards in the United States require at least 7.9% milk fat and 25.9% total solids in the final product, and the addition of limited amounts of salts for stabilization is permitted. Some products of this type are made from skim milk, and limited amounts of "filled milk" are made with other fats replacing the milk fat. The Evaporated Milk Association, a trade group of manufacturers of the product, has established sanitary and other standards. These standards involve microbiological tests for the incoming milk, as well as procedures for the inspection of producing farms and their operations; some aspects of the processing plant and its operation also are specified.

The milk is frequently clarified centrifugally and, while this process removes somatic cells (leucocytes) and some bacteria, the change in total bacterial count is not significant. The materials used to standardize the fat:solids ratio, whether cream or skim milk, should be of the same or better quality than the raw milk, and this means that proper provisions must be made for the production and handling of these products.

Stability of the milk to the sterilizing temperatures used later is essential, and the milk proteins are ordinarily stabilized against subsequent coagulation by heating to above 93.4°C with holding for some minutes. This may be done in a "hot well" with holding for as long as 20–25 min; but in larger continuous operations, an enclosed system may permit the use of temperatures as high as 121°C for a few minutes. Any of these temperatures destroys all of the nonsporulating bacteria present and many of the less resistant sporulating types; any bacteria that might cause infectious disease would also be killed. However, the enterotoxin produced by enterotoxigenic staphylococci would not be inactivated, making it essential that proper precautions are taken to prevent extensive growth of these organisms in the incoming milk (Anonymous, 2000). The temperatures employed are too high to permit the growth of thermophilic bacteria, so these organisms are not a problem.

In the actual condensing operation, whether it be batch or continuous, temperatures seldom exceed 54.5°C, and they may be significantly lower under some circumstances; these temperatures will not kill those bacteria that may have survived preheating. Thermophilic bacteria may grow under these conditions, and they may become a factor in limiting the length of time the equipment can be operated before a shutdown for cleaning becomes necessary. Operation for extended periods may also result in such extensive buildup of solids on the heating surfaces that satisfactory cleaning will be made quite difficult.

Following condensing, the product is homogenized, and the usual precautions concerning homogenizer care must be observed to avoid excessive contamination from this source. The product is then cooled and placed in storage, where the final standardization of composition takes place. Holding under good refrigeration until packaging and sterilization take place is essential, because the product is not sterile and can thus spoil if conditions permit appreciable microbial growth.

The cans for evaporated are usually fabricated in an adjacent plant, and the heat used in fabrication is enough to ensure that the can will contribute few, if any, bacteria to the canned product. Because the milk is cold when placed in the can, sufficient head space must be allowed for the expansion that occurs during the sterilization process; otherwise the internal pressure may open one or more of the seams, and the resulting "leak" will permit the entry of bacteria when the residual heat is no longer adequate for their destruction. The closure must be hermetic; that is, it must not permit the passage of air or fluid in either direction. Automatic weighing may be used to check the amount of fill and detect spillage before, during, or after closure, or through leaks during processing and cooling; any cans that are beyond the "normal" range should be discarded.

To have adequate keeping quality at room temperature, evaporated milk must be "commercially sterile." This means that it must not contain organisms that will grow, and probably produce defects, under the normal storage conditions. If a residual organism is prevented from growing by the lack of oxygen in the environment (a highly aerobic organism) or if it is an obligately thermophilic organism that will grow only at elevated temperatures, such as 45°C, the product may be still be "commercially sterile." The presence of thermophilic bacteria that survive the usual heat treatment can become a problem when the product is held at unusually warm temperatures, such as in a poorly ventilated warehouse in a tropical climate. Product going to such an environment may need to be processed at a slightly elevated temperature to provide the necessary keeping quality.

Treating evaporated milk with a time–temperature combination that would provide "absolute sterility" would not be a problem, but such absolute sterility would be associated with an unacceptable level of "cooked" flavor, a dark color, and probably some modified physical characteristics. Therefore, the heat exposure customarily chosen is the lowest that will provide "commercial sterility" under that particular set of conditions, thus keeping the modifications of flavor and physical characteristics to a minimum. However, as Galesloot (1962) has pointed out, an increase in temperature from 120°C to 150°C will increase the relative spore destruction rate about 1000-fold, whereas the relative rate of browning will be increased by only about 15.7-fold; with the markedly reduced time of exposure required at the higher temperatures to give an equivalent bacteriocidal effect, the retention of flavor and color can be impressive.

5.1.3.1 Available Heat Treatments. The historical procedure for sterilization was to hold the canned milk for 15–20 min at 115°C or slightly higher, with both the heating and cooling periods contributing some microbiocidal action, as well as affecting the flavor and physical properties. Both batch and continuous sterilizers have been used, the former usually only in smaller operations. More recently, the tendency has been to increase the temperature and reduce the time of holding to gain some of the advantages of an ultra-high-temperature (UHT) treatment, while still avoiding the problems of aseptic packaging.

However, as the technology of the process has improved, interest in the UHT heat treatment has increased, particularly when used in combination with aseptic packaging. The heating can be with tubular or plate heat exchangers using steam as the indirect heating medium, or by the direct injection of steam into the product, followed by a vacuum treatment to remove the water added by the steam. Two of the methods employed to permit very quick heating are (a) the use of falling films of product and (b) its dispersal as droplets. The temperatures that are effective can only be achieved with both steam and product under pressure, so that the closed system must be engineered to withstand the necessary pressures; treatments from 130°C for 30s to 150°C for less than 1s have been employed. The time–temperature combination selected must be such that it will effectively kill a "normal" load of mesophilic spore-forming bacteria, and *Bacillus subtilis* is probably the most resistant of this group. Highly heat-resistant, obligately thermophilic strains of the species *Bacillus stearothermophilus* might survive such a treatment, probably in small numbers—as might "flat-sour" bacilli (Kalogridou-Vassiliadou, 1990)—but would not be expected to grow or be responsible for defects under the normal holding conditions for canned evaporated milk.

It may be relevant, however, that *B. stearothermophilus* is capable of forming biofilms in dairy plants, and hence numbers could be higher than expected unless the cleaning regime is adequate (Flint et al., 2001).

When the product is to be held at higher temperatures and, similarly, when the load of any type of spore is excessive, a treatment more stringent than usual might need to be employed. Spores that develop at the optimum growth temperature for the organism have been reported to be the most resistant (Theophilus and Hammer, 1938), but the fore-warming temperatures apparently constitute a sublethal heat shock that makes the spores more sensitive to the final heat treatment (Curran and Evans, 1945). For a more detailed discussion of factors that influence spore resistance, one should consult the classic treatise by Stumbo (1972) or the text by Burton (1988).

Aseptic packaging of UHT products poses numerous microbiological problems. The container and closure used must be sterile, as must the equipment through which the product passes, and contamination by microorganisms from the air must be avoided. The container must be hermetically scaled, so that air, water, and other sources of contaminants cannot gain access to the product. Avoidance of rough handling subsequent to filling is essential, because damage to the package or even temporary weakening of the closure may permit contamination. When metal cans are used, they may be flame-heated or autoclaved with superheated steam to be made sterile. Glass containers must normally be cleaned very thoroughly and then autoclaved. When composite paper–plastic–foil containers are used, the blanks are customarily treated, just before forming and filling under aseptic conditions, by a combination of hydrogen peroxide and dry heat; the air used for the latter purpose is heated to about 200°C to effectively remove any residual peroxide.

All equipment and enclosures must be treated to destroy contaminating microorganisms prior to use. Superheated steam is the usual treating agent, although hot air could be used for some parts under appropriate circumstances. One of the problems is that in the case of jamming or other malfunction that requires entry to the system, the sterilization procedure must be repeated. An atmosphere of superheated steam or hot gas must be maintained around the filler, closing machine, and interconnecting conveyor system to preserve sterility.

5.1.3.2 Microbiological Examination. Examination of evaporated milk by customary procedures immediately after packaging will usually reveal no viable organisms, and microorganisms seldom develop even after prolonged holding at room temperature. In past years, much canned evaporated milk was held for 2–3 weeks in a warm room to detect spoilage before shipping; but, with improved technology and laboratory control, holding of the entire lot is seldom done. Representative cans will be held out for incubation and examination. Where the product is to be shipped to warmer areas, incubation of the cans at 37°C or 55°C is frequently used to detect facultative or obligate, respectively, thermophilic bacteria that may have survived the heat treatment. If a laboratory test is needed, small amounts (1 ml) of the milk can be removed, with great attention to completely aseptic conditions, and tranferred to tubes of sterile bromocresol purple milk (aerobic spore-formers) and reinforced clostridial medium (anaerobic spore-formers); the latter tubes are stratified with wax after inoculation (BSI, 1968). Attempts to obtain quantitative results are seldom made, because the presence of a viable organism is the criterion employed. The presence of non-spore-forming organisms indicates poststerilization contamination, frequently as the result of a leaky container, but the presence of spore-formers is usually associated with inadequate heat treatment of the product.

5.1.3.3 Defects. Many of the defects reported in the literature and summarized by Hammer and Babel (1957) are almost never found at the present time because of improvements in process technology and laboratory control by the industry. Microbial defects can be divided into (a) those that are due to organisms of high heat resistance that survive a slightly inadequate heat treatment and (b) those that gain entrance after heat treatment, and which are usually of low heat resistance.

Most of the heat-resistant forms are species of the genus *Bacillus*, although an occasional species of the genus *Clostridium* has been encountered. *Bacillus coagulans and B. stearothermophilus* may cause an acid coagulation and a slight cheesy odor and flavor, and "flat-sour" bacilli can survive as well (Kalogridou-Vassiliadou, 1992). These organisms grow best at 37°C and above, and high storage temperatures and/or inadequate cooling are factors in this type of spoilage. *B. subtilis* causes a nonacid curd, which may then be digested to a brownish liquid with a bitter taste. The coagulum formed by *Bacillus megaterium* is accompanied by some gas and a cheesy odor. Gas production associated with putrefaction and a smell of hydrogen sulfide have been reported as caused by a *Clostridium* sp., but this type of defect is very rare. Bulged cans are caused much more commonly by a chemical action on the metal of the can or by overfilling of the cans with cold milk, which then expands on heating.

Contamination subsequent to heating may result in a greater variety of defects. Leaks in the hermetic seal of the container may be due to improper closure, subsequent corrosion, or mechanical damage during subsequent handling; even a momentary leak may permit microbial entry. Such problems should be rare; but if they do occur, then the consequences can be serious.

The consumer must be made aware of the perishability of evaporated milk once the can is opened. The heat treatment used to kill the organisms and provide keeping quality in the unopened can has no residual effect that will control the growth of subsequent contaminants, and opening the can under kitchen conditions is almost certain to lead to some contamination. Nevertheless, the product will keep for several days under refrigeration, if reasonable care is used to minimize contamination, but holding without refrigeration is almost certain to permit spoilage in as little as 24 h.

5.2 SWEETENED CONDENSED MILKS

Sweetened condensed milk may be made either from whole milk or from skim milk. The product made from whole milk must have at least 8% fat and 28% milk solids, whereas the standard for the type made from skim milk is 24% milk solids and less than 0.5% milk fat. Water is evaporated and sugar added to yield a product with a sufficiently high solute concentration to prevent the growth of most microorganisms. For the retail trade, sweetened condensed milk is packaged in hermetically sealed, metal containers, but for industrial purposes it is packaged in milk cans, barrels, steel drums, and bulk tanks. Bakers, confectioners, and the prepared food industry use considerable amounts of these products.

The keeping quality of sweetened condensed milk is largely the result of (a) the increase in osmotic pressure (reduction in water activity) and (b) the "binding" of water by the added sugar. The increased concentration of milk solids brought about by the removal of water by evaporation also contributes to the increase in osmotic pressure, but this is relatively minor compared to the effect of added sugar. The absence of significant amounts of air in the hermetically sealed package also contributes considerably to the keeping quality of the canned product, inhibiting the growth of a number of aerobic microorganisms, particularly molds, but a few yeasts and some micrococci can tolerate the high osmotic pressure.

The concentration of sugar-in-water of sweetened condensed milk is known as the "sugar number" or "sugar index" (Carić, 1992), and the index is usually 63.5–64.5 for the canned product and about 42 for the bulk, whole milk product. The lower index for the latter product is permissible because the storage time is fairly short, and refrigeration is used. The use of sugar to extend the shelf life should not be considered a substitute for good-quality raw milk, proper sanitation, or adequate processing and holding practices. The heat treatments employed in processing are insufficient to "sterilize" the product, so residual organisms are always present to cause problems if the product is not handled satisfactorily.

The raw milk used in production usually is of "manufacturing grade," and the manufacture of sweetened condensed milk can be divided into forewarming, superheating, sugar addition, condensing, cooling, forced crystallization, and packaging. The only truly microbiocidal heat used is during the forewarming and superheating (if used) phases, so these must be depended upon to destroy all pathogenic microorganisms and also most of the potential spoilage microorganisms. The most common temperatures and times are 82–100°C for 10–30 min, with occasional use of higher temperatures and shorter times in an enclosed chamber. All but the most heat-resistant types of organism will be destroyed, including potential spoilage agents, and hence the presence of the latter in the end product is almost invariably due to contamination subsequent to forewarming. The natural enzymes of milk are also inactivated, but the proteolytic and lipolytic enzymes resulting from excessive microbial growth may not be affected and may cause problems in the final product.

The sugar normally added is sucrose, but other sugars may be used, at least in part, for special purposes. The sugar is normally an unimportant source of microorganisms, but under unfavorable conditions, it may be contaminated with mold spores, osmophilic yeasts, or bacteria that will produce acid and gas. Sugar storage should be in a dry place, free from dust, insect, and rodent contamination. The sugar may be added to the forewarmed milk prior to entry to the vacuum pan, or as an approximately 65% solution late in the condensing operation. Addition prior to forewarming does reduce the microbiocidal effectiveness of the heating, but addition to the forewarmed milk may contribute to age thickening, so this latter procedure is used primarily for bulk product to be used quickly.

The condensing operation is carried out in a vacuum pan at approximately 57.2°C, but the temperature may be permitted to drop at 48–9°C late in the cycle. The vacuum pan and its associated equipment must be cleaned very well, and the microbiocidal treatment must be carried out very thoroughly to avoid having the pan become a source of contamination. The sticky nature of the product increases the difficulty of cleaning, and "cooked on" material may become a real problem. An acid detergent treatment following the usual alkali detergent wash has been suggested to help control the formation of milkstone.

The forced crystallization step consists of seeding the partially cooled milk with very fine lactose crystals, usually when the temperature is approximately 30°C, to induce the formation of numerous small lactose crystals, rather than fewer large ones during subsequent cooling (Carić, 1992). The added crystals are usually not heavily contaminated, but conversion of the lactose to the α-anhydride form by heating under vacuum to 93.3°C, followed by fine grinding and autoclaving in sealed cans at 130°C for 1–2 h, has been suggested for sterilization.

The cans for the retail trade (usually 12 oz for the skimmed, and 14 oz for the whole milk product), as well as the lids, should be microbiocidally treated by gas flames, superheated steam, or ultraviolet radiation. The fillers are usually of the plunger type, equipped with a cutoff to prevent dripping, and these fillers are quite complex and difficult to clean adequately. They can, therefore, be sources of heavy contamination and the cause of serious outbreaks of spoilage in the packaged product. Fillers should be disassembled after each day's run, thoroughly washed, steamed, and stored dry; otherwise, this equipment can be a major source of micrococci, yeasts, and molds in sweetened condensed milk. The cans must be filled as full as possible without causing later bulging, because a minimum of air space plus a hermetic seal will help restrict the growth of the aerobic organisms that may cause defects. Imperfect seals, or can damage subsequent to closure, may permit the entry not only of microorganisms, but also of the air that is necessary for the growth of a number of spoilage organisms.

Bulk product may be stored and shipped in containers varying in size from 45-liter cans to railroad tank cars. The industrial product is usually used within a few days; but if it is not used quickly, it should have something approaching the high sugar content of the retail canned product. Surface growth of aerobic organisms may be restricted by having the containers full; and ultraviolet lights over the storage tanks, along with protection from atmospheric contamination, may be used to combat surface mold. Provision must also be made to avoid any surface dilution, such as by condensate from above the liquid line.

5.2.1 Microbiological Examination

Viable microorganisms are commonly found in the final product, and some reports place the numbers from a few hundred to 1.0×10^6 cfu g^{-1}. The heat treatments used are not adequate to kill spore-forming bacteria, and further processing and handling usually contribute a variety of microorganisms; the sugar levels employed permit some types to grow if other conditions are favorable. Enough oxygen may be present in the head space of an incompletely filled, or poorly sealed, container to permit the growth of organisms able to tolerate the reduced water activity of the product.

Sampling for laboratory analysis requires considerable care. Cans must be thoroughly cleaned, the area of opening must be treated adequately with heat and/or microbiocidal chemicals, the opening instrument must be sterilized, and great care must be used in the withdrawal of the product. In bulk product, representative sampling of the viscous material is a problem because adequate mixing is difficult; and, in some instances where aerobic microorganisms are involved, a sample from the product surface may be more revealing than a mixed sample.

Three types of count are made routinely on sweetened condensed milk—namely the total colony count, a coliform count, and a yeast and mold count—and the procedures are outlined in the Standard Methods for the Examination of Dairy Products (APHA, 1992). High total colony counts and the presence of coliform bacteria or yeasts and molds in recently processed product are considered indices of contamination following preheating. As might be expected, postheating contamination remains a constant threat, and the work of Farrag et al. (1990) on the survival of *Listeria monocytogenes* in sweetened condensed milk should serve as a useful reminder. Some organisms, such as the coliforms, may die off with holding at room temperature (the counts of *L. monocytogenes* declined at room temperature but not at 7°C), but micrococci, yeasts, and molds may proliferate.

5.2.2 Defects

With current improved technology, defects in sweetened condensed milk are relatively uncommon, but defects of microbiological origin are still of some concern. Gas formation may be caused by yeasts of the genus *Torulopsis*, although coliform bacteria have been implicated occasionally when the sugar ratio has been in the 40–45 range. Because neither yeasts nor coliforms are resistant to forewarming temperatures, contamination during subsequent processing is indicated; the defect is more common during warmer months. The cans may be bulged and blown by carbon dioxide, or a mixture of carbon dioxide and hydrogen, depending upon the organism involved. The defect usually develops slowly because of the slow growth rate of the responsible organisms in the high sugar concentrations encountered (Bhale et al., 1989).

Thickening is usually accompanied by some acidity and cheesy odors, and this defect is encountered primarily in the bulk product of lower sugar content. Many species of bacteria have been mentioned as responsible, with the micrococci and spore-formers being encountered most frequently. Lower storage temperatures and improved plant sanitation have been found helpful in combating this defect.

Small masses of mold mycelium and coagulated casein (buttons), usually colored white to brown, may be found on the surface or in the subsurface layers, especially of the canned product, because this is held for longer times; a disagreeable taste is associated with the defect. Species of *Aspergillus* and *Penicillium* have been implicated. The molds will grow until all the available oxygen in the head space is exhausted, although the buttons may continue to increase in size because of continued enzymic activity. Poor plant sanitation is a major factor, because these molds are not heat resistant and do not survive forewarming. Underfill of the cans increases available oxygen and thus favors this defect, but storage below 16°C may be helpful in reducing the incidence of faulty cans.

5.3 RETENTATES

Retentates are the materials produced when selective membranes in combination with pressure are used to concentrate desirable components. Reverse osmosis is the term applied to the process separating low-molecular-weight components from their solvents, usually water; and the membranes employed permit passage into the discarded portion of only low-molecular-weight materials, such as water and some salts. Ultrafiltration is essentially a sieving process in which molecules such as proteins are retained, while the membrane is permeable not only to water but also to solutes such as sugars. By selection of appropriate structural materials and use of various preparation techniques, a range of membranes having considerable variation of characteristics has been prepared. Cellulose acetate and its derivatives have been used for preparation of membranes for both procedures. Polysulfone and several related polymers have been used for membranes for ultrafiltration.

Glover et al. (1978) and Grandison and Glover (1994) prepared extensive reviews on the uses of reverse osmosis and ultrafiltration in the production of dairy products. These processes have been used to concentrate whole milk, skim milk, and whey, and they are alleged to be cheaper than evaporation in the 2× to 4× range of concentration. The retentates have been used as concentrated sources of solids for such products as ice cream and yogurt, and some manufacturers have attained a "better texture and flavor" in their products than when milk powder or bulk condensed milk was used (Lankes et al., 1998). Ultrafiltration has been used for making soft cheeses, such as Camembert, by concentrating the skim milk by as much as a factor of 6×, with resultant savings in rennet required and an increase of significant magnitude in the yield of cheese. Concentration of whey proteins, without subjecting them to the partial denaturation associated with concentration in the vacuum pan, has been achieved and ultrafiltration has been suggested as a way to produce low-lactose or lactose-free milks for lactose-intolerant people.

Certain microbiological problems are associated with these processes, because all microorganisms will be rejected by the membranes and will, therefore, increase in numbers in line with the concentration factor being employed. The retention time, as well as the temperature during the processing, must be controlled to limit microbial growth, and the feed may need to be thermized or pasteurized prior to filtration (Lewis, 1996). The higher the temperature, within the normal range used to minimize denaturation, the shorter the time that can be tolerated without excessive microbial growth in the product; in a few instances, temperatures high enough to inhibit microbial growth have been used. Once concentration has been achieved, the product should be used or processed further with minimum delay because the concentrates are not sterile and further heating, which could be employed to reduce the microbial populations, would negate some of the process advantages. Freezing might be advantageous under some circumstances to permit holding.

The membranes, which are the heart of the processes, require special cleaning and microbiocidal treatment. The cellulose acetate membranes must be kept moist; and because they will not tolerate high temperatures, such as above 50°C, these membranes cannot tolerate the methods customarily used for cleaning of dairy equipment. One solution is to remove them from the system and treat them separately, but backwashing is helpful if the mounting of the somewhat delicate membrane will support the necessary backpressure. Some workers have suggested that the detergents used should contain proteolytic enzymes that will assist in the removal of substances fouling the membrane surface. Lodophors are among the most effective microbiocidal agents in use, and a level of 10 ppm has been recommended. Some of the polysulfone flat sheet membranes withstand a wide pH range and temperatures up to 100°C, and traditional dairy cleaning methods may be used on these. A microbiocidal treatment with 0.1% hydrogen peroxide has been suggested as an alternative.

Greater use of reverse osmosis and ultrafiltration by the food industry in the future appears highly probable; and with further experience will come a better understanding of the microbiological considerations involved, permitting more specific recommendations to be made.

5.4 PRODUCTION OF DRIED MILK POWDERS

The production of dried milk powders has been influenced since the 1980s by a demand for products with a range of performance characteristics. Although skim-milk powder and whole milk powder were once the basic commodities produced by the milk drying industry, they have been rapidly overtaken by powders with tailor-made qualities. Instant products are commonplace; but lipase-free, calcium-reduced, or high-heat powders are now widely available.

The development of these specific products, coupled to an energy-conscious industry, has resulted in new and innovative processing plants. These new developments have not lessened the need for vigilance in microbiological terms, but have enhanced the requirement for speedier and even more reliable test procedures.

5.5 MANUFACTURING PROCESSES

Two basic processes are commonly employed for the drying of milk and milk products: roller drying and spray drying. Roller drying has given way, over the last two decades, to spray drying. The roller drying process is plant- and energy-intensive and can result in considerable heat damage to the product. This product damage results in poor solubility characteristics in comparison with the spray-dried product. However, roller drying does impart some desirable flavor characteristics as a result of the high beat levels employed in the process. The same or similar characteristics can be obtained in spray-dried milk powder by subjecting the concentrate to higher temperatures before drying. It is unlikely that roller drying will find favor in the future, and its continuing demise can be anticipated.

5.5.1 Spray Drying

Raw milk is received, bulked, and standardized for fat content or, alternatively, is separated into skim milk and cream as in the case of the manufacture of skim-milk powder. The milk is then passed to the evaporation plant for concentration. Modern evaporation plants are complex and highly energy-efficient, and the dairy industry has generally accepted the falling film design as its standard. The falling film evaporator is simple in comparison with other evaporators, retains a small amount of liquid product during operation, and has a low product residence time. The evaporator operates under a vacuum; this allows the use of low boiling temperatures, which, in conjunction with short residence times, results in minimal damage to heat-sensitive components.

Evaporation plants can be fitted with heating units to heat products to 140–150°C as in UHT plants. This results in a final product of excellent microbiological quality, which may be an essential requirement where the use is as an ingredient in compounded baby foods.

Following evaporation, the concentrated product is passed to the dryer. Current trends are to link the evaporation plant and the dryer as a single integrated unit. Two basic systems of spray drying are used: jet or nozzle dryers and rotary atomizer dryers. Historically, jet or nozzle dryers have featured predominantly in plants in the United States, with rotary atomizers finding greater favor in Europe. Within the last 10 years, there has been a diffusion of these two systems throughout the world. Each system produces powders of different basic characteristics, and the small nuances of performance can be important in specific applications. Bulk density tends to be higher in powders produced by the jet system of atomisation, for example (Knipschildt and Andersen, 1994).

Milk to be dried by the jet atomizing system is fed to a high-pressure pump, similar in design to the ubiquitous homogenizer. It is pumped under pressure to a series of specially designed jets within the drying chamber. The jet, which incorporates a swirl chamber, sprays the concentrate in a thin cone. The hot air stream impacts with the liquid cone, causing it to break up into fine droplets that form the powder. The heavier powder falls to the base of the drying chamber, normally rectangular in shape, where it is removed by a series of mechanical scrapers. The lighter powder (fines) is carried out in the moisture-laden exhaust air stream. The fines are removed either by a cyclone or series of cyclones, or by means of cloth filters.

In the case of spray drying using a rotary atomizer, the concentrated milk is fed by a pump to the atomizer. The rotary atomizer consists of a high-speed, electric motor driving a shaft on which is fixed a circular atomizer disc. The design of atomizer discs varies between manufacturers, but, essentially, around the periphery of the disc are a number of rectangular slots or circular holes. The concentrated milk is directed to the exit points by curved vanes. The product exiting the disc is broken up into droplets into a hot air stream inside a large drying chamber, usually conical in shape. The droplets dry rapidly and form a powder that exits from the chamber with the now moisture-laden air. The powder is separated initially in a primary cyclone from the main air stream, while finer powder (fines) is separated in a secondary cyclone system; the combustible nature of these materials means that the process plant must be well-designed (Skov, 1994).

Over the last decade, the trend in spray-drying systems has been toward the incorporation of fluid beds. Three such systems have been described by Masters (1987): a two-stage dryer with external fluid bed;

a concurrent, integrated fluid-bed spray dryer; and a mixed-flow, integrated fluid-bed spray dryer. In the case of the two-stage dryer, the fines from the fluid bed are recycled back into the main chamber where they come into contact with the bulk of the powder, thereby forming simple agglomerates. This results in a dust-free powder with improved wettability characteristics.

The developments in evaporation and spray-drying plants have resulted in an increasing range of dried products; and the co-current, integrated fluid-bed dryer and the mixed-flow, integrated fluid-bed dryer have provided the means whereby hygroscopic and tenacious products can be handled more easily.

The packaging of dried milks has changed very little, with bulk packaging largely confined to multiwall paper sacks with loose inner liners of plastic. Retail units have centered mainly on board containers with and without bag inserts. Dried milk as a commodity product is not in the forefront of demand for innovative packaging, but is reliant on a functional, low-cost container.

5.5.2 Alternative Drying Procedures

A number of specialized drying procedures are still in use for a small number of products, such as malted milks and chocolate-based beverages.

These involve modified concentration plants and drying procedures utilizing vacuum ovens. The process imparts a particular flavor and textural profile to products that have become established in the marketplace. There is little doubt that such products could be made in the newer, spray-drying systems that will process hygroscopic and "sticky" materials, but the particular texture of a product produced from a vacuum oven, for example, is difficult to reproduce in a spray dryer plus fluid-bed combination. In the traditional product, the honeycomb structure of the final particle is a continuum of the product, whereas in the case of the agglomerated product, it is a fusion of particle surfaces.

Microbiologically, these products have not attracted attention, and this may be due to their long, food-poisoning-free history. Such information as is available suggests that the heat treatment throughout the process, which imparts the unique flavor, is a contributory factor to their excellent record.

Another alternative procedure, the Filtermat drying system, employs a three-stage drying process that encompasses agglomeration all in a single unit. The end product is porous, as with agglomerated products generally, and has good dispersability and wettability.

5.6 MICROBIOLOGICAL ASPECTS OF PROCESSING

The microbiological aspects of processing are, from a routine standpoint, of direct concern to factory managers, quality controllers, and designers of plant. The new developments, combined with the increasing trend toward processing conditions aimed at minimizing nutrient damage, place greater emphasis on reliable microbiological control.

The earliest large-scale examination of the microbiology of the spray drying of milk was the Staplemead experiment. The principles established in that trial, carried out in 1942, are still valid. The work by Mattick, Crossley, and colleagues was basically directed to extending the shelf life of whole milk powder, but it also identified the basic microbiological factors critical to the operation of evaporating and drying plants (Crossley and Johnson, 1942; Mattick et al., 1945).

5.6.1 Raw Milk Quality

As with all milk products, the quality of the raw milk is paramount. Unfortunately, the commonly used term "manufacturing milk" has all too frequently given the impression that quality is less important for making milk products than it is for bottling as market milk for direct consumption. Nothing could be further from the truth, because the nature of the evaporation process provides ideal conditions for the multiplication of microorganisms in those stages where the temperature is low.

The relationship between the manufacturing process and raw milk quality has been examined in Germany. Otte (1980) suggested that raw milk for use in infant feeding formulations should not contain in excess of 1.0×10^4 organisms ml^{-1} as an initial count and, furthermore, that raw milk with counts in excess of 1.0×10^5 organisms ml^{-1} could result in counts of over 1.0×10^4 organisms g^{-1} in the dried product.

However, it is not simply a question of numbers but also of types, and the presence of large numbers of thermoduric organisms, such as *Streptococcus thermophilus* or *Alcaligenes tolerans*, together with heat-resistant spore-formers, can have a deleterious effect on end-product quality. Quite clearly the thrust for improved nutrient retention must be accompanied by a maintenance of, and, if possible, improvement in, raw milk quality.

5.6.2 Process Microbiology

5.6.2.1 Evaporation. In any consideration of the microbiological facets of processing, it is necessary to understand the process and the influencing factors. Dried milk manufacture has a number of requirements. First, the process must be economic; second, it should have a minimal effect on nutrients; third, the product must be free from harmful or deleterious organisms; and finally it must have the properties of ease of reconstitution and acceptable flavor. Clearly, there have to be some compromises made in achieving the utopian ideal.

Commencing with the standardization of whole and formulated milks, blending of cream or other components is not likely to result in any diminution of microorganisms. Thus, the importance of the raw material cannot be overemphasized.

The current design of evaporators means that the preheating stages are an integral part of the plant. In many cases, the preheating stages are in a duplex arrangement that allows them to be changed over during extended runs; this need arises from product fouling and permits long uninterrupted production runs. In addition, many evaporators have a feedback or recirculating circuit for the start-up phase; this facilitates the buildup of the solids level before the milk goes forward into the drying phase.

The preheating stage, quite apart from its effect in reducing the microbial content, also has the effect of providing a vapor phase when it enters the tubes in the falling film section. Preheating to high temperatures (i.e., in excess of 100°C) gives rise to fouling problems. These high temperatures can also create problems when the milk enters the main body of the evaporator, due to the amount of flash vapor produced, which, in turn, can become difficult to remove at the appropriate rate. However, the use of lower temperatures can also lead to problems from the growth of thermophilic spore-forming bacilli (Murphy et al., 1999).

Entering the evaporation side, three, five, or even seven effects (stages) may be involved, but the commonest format seems to be three effects plus a finisher stage. First-effect temperatures are within the region of 70°C, but this drops down to a final-effect temperature of 43–46°C. The vapor produced from each effect is used to supply some of the heat for the previous stage. The make-up heat is supplied by either thermal vapor recompression (TYR) or mechanical vapor recompression (MYR) systems. Clearly, this use of waste heat has resulted in significant economics that can be realized in energy costs (Knipschildt and Andersen, 1994).

The tubes in the falling film sections range up to 12 m in length, and they are located in compact bundles in each effect. The number of tubes is based on the amount of concentrate destined for each effect, the thickness of the film, and its viscosity and surface tension. With skim milk, the viscosity increases extremely rapidly in the range of 45–55% total solids and, if the total solids are pushed too high, gelation can occur; if this occurs in the tubes, then the problem becomes extremely serious. If the gelation goes unchecked and a plug is formed, then it becomes difficult, if not impossible, to remove and the only solution may be tube replacement. The significance for the microbiologist is that any form of blockage presents a potential hazard. In the case of tube blockage, the problem would be best attended to before detergents are applied, but blockages are not easy to detect. If allowed to go unattended, such blockages become serious microbiological hazards. Visual inspection of tubes still has a role to play, and plant microbiologists would be well-advised to incorporate this procedure as a routine. Early signs will be an increase in solids or a temporary reduction in feed to the atomizer. In-line density controllers are now commonplace for controlling evaporation plants, but they need to be calibrated at regular intervals.

The evaporation stage may be called upon to deal with a wide range of conditions, such as concentrate for low-, medium- and high-heat skim-milk powders; and unless this flexibility was specified in the original design, problems of operation may arise. Clearly the desire for maximum economy of operation invariably means less tolerance in operational parameters and an increased risk of problems arising. The thrust toward multiple-effect plus finisher systems, encouraged by the economies in energy of incorporating thermal vapor or mechanical recompression, has resulted in highly complex plant configurations. These "plumbing nightmares," while a tribute to processing engineering design, nevertheless increase the microbiological hazards. It means that the design of valves and bends and the quality of welding must be faultless, and it is to their credit that process engineers and designers have acknowledged this fact.

5.6.2.1.1 Cleaning and Sanitization. The second area of importance is that of cleaning and sanitization. Undoubtedly, a "closed" plant permits the use of higher cleaning and sanitizing temperatures and permits a move toward stronger detergent/disinfectant solutions. However, a great deal of reliance has to be placed on the cleaning and sanitizing procedure, because those parts of the plant that may not be cleaned properly are also likely to prove inaccessible.

The cleaning and sanitizing of evaporators has to be carried out by CIP procedures, and this adds to the complexity of the pipework. Concentrated residues, especially if not dealt with expeditiously, can prove to be tenacious and difficult to remove; the location of the CIP jets or spray balls is also of the utmost importance. Every part of a soiled surface must be contacted with cleaning liquids, but the increasing complexity of plants has meant that there are areas that are difficult to access; regular visual inspection, accompanied by swabbing, still has a role to play. Inspection of tubes for blockages is recommended, as well as valves and gaskets, and jets and sprays need to be examined to ensure the free flow of liquids. Problems may not arise with new or near-new plants, and it is easy to become complacent until a problem arises.

In relation to the cleaning operation, it is not unknown for some plants to recover the initial rinsings from a plant for recycling. Clearly, this appears to make sense. Water is, after all, a valuable raw material, and only clean water has been used to remove the residues. It is, however, a false economy and exceedingly dangerous and is a good way of "seeding" successive batches in microbiological terms. Such residues can usually be disposed of as animal feed, but, in any event, they should not be recycled.

5.6.2.2 Spray Drying. The spray-drying operation has become equally as complex as that for evaporation. The basic spray dryer has been superseded by the spray dryer plus fluid bed in a number of configurations. The evaporative capacity of spray dryers has increased together with the range of dried products. The introduction of instantizing in the 1950s gave a new impetus to drying technology, and the cumbersome earlier designs of re-wet plants have given way to straight-through units. The success of these plants has, in turn, enabled manufacturers to produce specially formulated products, high-fat powders, whey-based powders, and baby foods, and developments in reverse osmosis and ultrafiltration will undoubtedly stimulate further changes.

The increasing complexity of design means additional potential microbiological hazards. However, once the initial drying phase has taken place, the moisture level in the product is not likely to allow unrestricted microbial growth. Two-stage drying gives rise to moisture levels of around 5–7% from the first stage and, even with the newer compact dryers with integrated fluid beds, the moisture content is only approximately 10%. Thus in terms of moisture, the danger of microbial proliferation is not high, but particles entrained in areas where the humidity may be high can still present a risk.

Turning to the operation of the spray dryer, concentrate is nowadays pumped direct, to the atomizer or nozzle with, perhaps, a small balance tank in between. However, some older plants still use twin balance tanks as a means of ensuring continuity of operation. This practice is frequently adopted where evaporators or finishers may be cleaned independently. Hawley and Benjamin (1955) and Keogh (1965, 1966) have indicated that this is a microbiological hazard. The temperature of the concentrate on exiting from evaporators can be in the region of 42–45°C, and it will drop further when it is in the balance tanks. The trend to high total solids, thus increasing the economics of spray drying, is also likely to result in thickening, thus giving rise to a sludge or, at worst, gelation. Dual balance tanks may be used to overcome this problem, and they can be switched over during the drying run. Mesophilic bacteria, including pathogens, can and do grow in such environments, and it is important to ensure that the tanks are fitted with covers and that these are kept in place during operation. The newer drying systems, as stated previously, should not present any serious hazards. Spray drying per se has been well-researched, and designers as well as dairy plant management are well aware of the potential dangers (Keceli and Robinson, 1997).

There are no inherent microbiological hazards within the drying chamber or, indeed, through the system to the powder outlet. Inlet air temperatures have settled into the range 190–250°C, and the filtration of air is now extremely efficient with the use of highly retentive filters. It is important that the air inlet is located at a distance from the exhaust air outlet.

The main chambers of dryers are washed using high-pressure rotating jet systems. The frequency varies from plant to plant—some as frequently as once per week, others once every season. The atomizing unit is cleaned more frequently, maybe two or three times in a production run, but this is usually a hand-washing operation due to the size of the unit. Dry cleaning of drying plants is carried out more frequently using vacuum cleaning units, which may be built-in or mobile units. An important feature of the drying process is to keep the main chamber as free as possible of powder. Hammers, vibratory units in the side walls of the dryer, have some effect, but the main control is exercised through feed rate. A sudden increase in feed will give rise to damp powder that will cling to the walls or ceiling of the chamber and provide a surface for further buildup. On the positive side, as drying temperatures increase, microbial counts in the finished powder decrease (Galesloot and Stadhouders, 1968; Chopin et al., 1977, 1978).

The microbiological risks from aerial contamination were raised by Hawley and Benjamin (1955) and confirmed by Crossley and Campling (1957). It is important that the manufacturing area should be kept free of dust and particulate matter; the use of ring-main vacuum systems is one reliable method for the removal of such residues.

5.7 MICROFLORA OF DRIED MILKS

The microflora of dried milks has been examined by a number of workers, including Crossley and Johnson (1942), Higginbottom (1944), and Keogh (1966, 1971), and Otte (1980) and El-Bassiony and Aboul-Khier (1983) have confirmed these earlier findings. Attention to specific groups, such as clostridia, has been given by the New Zealand Dairy Research Institute (1978) and Appuswamy and Ranganathan (1981), while facultative thermophiles are featured in the work of Arun et al. (1978) and Chopra and Mathur (1984).

Thermophilic actinomycetes in dried milk products have been examined by Falkowski (1978), who described the isolates as *Thermoactinomycetes vulgaris* and *Micromonospora* sp. Molds and yeasts may also be present in dried milk products, and *Aspergillus*, *Penicillium*, and *Mucor* spp. were isolated by Aboul-Khier et al. (1985).

The range of microorganisms found in dried milks appears to contradict the advances made in processing technology. However, improvements in recovery media and methodology, together with the geograpical location of some of the isolations, are factors that should not be overlooked.

5.7.1 Examination of Dried Milks

The routine microbiological control of dried milk manufacture has evolved over many years. Sampling points are usually chosen at well-established stages—for example, raw milk, milk after preheating, concentrated milk prior to drying, and powder ex-final discharge. In the case of systems incorporating fluid beds, additional points may be incorporated—for example, powder ex-main chamber and powder ex-fluid bed.

One of the problems facing the microbiologist is the difficulty of taking samples at stages in the process. It is a facility that is overlooked by plant designers, possibly because they are never asked to provide ease of access for the purpose of sampling. If sampling does prove impossible at intermediate points, then there is little choice but to access the product at the first opportunity. The well-established "Hazard Associated (Analysis) Critical Control Point (HACCP)" procedure is widely practiced (see Chapter 13 for full details) as an alternative to intermediate sampling and, while it is argued that this can make microbiological criteria on the end product unnecessary, end-product specifications still have to be met (Lück and Gavron, 1989; also see Chapter 14).

The tests normally carried out on in-line samples include a total colony count, a count for thermoduric organisms, a count for *Staphylococcus aureus*, a coliform count, and, of increasing importance, an examination for *Salmonella*. This latter development has come about as detection techniques have improved, as well as to meet the requirements of bulk users increasingly sensitive of the public health risk.

Sampling techniques for the final product have been specified by a number of authorities, but invariably depend on the total number of containers being known before determining how many should be examined.

As a practical guide, samples should be taken hourly or on the basis of every 50 sacks (25 kg), particularly in the case of a new plant starting-up. Once the microbial pattern for the plant emerges, say after 3 months' results, the sampling frequency can be changed. Sacks or other containers need to be numbered sequentially, so that if further sampling is necessary, it can be carried out in a logical manner. The process of withdrawing samples before sacks are stitched is easy and far less disruptive than having to open sacks and then have them restitched. The latter problem is further aggravated if the stock, as per usual, is located in a store far away from the laboratory and the sacks to be sampled are at the base of a stack. These practical difficulties can frequently prove troublesome in terms of time, additional labor, and the reclosing procedure. Thus, there is a good argument to be made for initially taking more samples than are required; in any event, there is no reason why these samples should not be taken to provide a series of bulk samples for microbiological analysis.

5.7.1.2 Specific Aspects. The procedures for determining counts has always been dogged by the problems of reproducibility and the pressure to adopt newer procedures—always on the grounds of recovering even more organisms than previous techniques. This demand for improved accuracy has led to comparisons between laboratories and individual workers, but it tends to be overlooked that microbiologists are constantly faced with the vagaries of techniques seeking to assess equally variable microorganisms. Kilsby and Baird-Parker (1983) discussed the difficulties associated with the enumeration of microorganisms in relation to sampling programs and concluded that "the application of microbiological criteria will always be an inefficient and imprecise method of exercising microbiological control over batches."

Nevertheless, numerical standards are still used, and a total colony count of 50,000 cfu g^{-1} is still accepted as the bench mark for Extra Grade Dried Milk by the American Dairy Products Institute (ADPI, 1990)—even though modern plants have exceptionally low results well inside this standard.

A valuable contribution to the significance of counts in spray-dried milk powders was made by Kwee et al. (1986). The work focused on the total bacterial load, thermodurics, thermophiles, spore-formers, psychrotrophs, coliforms, yeasts, molds, and *Salmonella*. These were measured in low-, medium-, and high-heat powders and at various intermediate stages during production, including before and after preheating of the milk and after concentration. They established that preheating had the most significant effect and that with the exception of thermoduric, thermophilic, and spore-forming organisms, counts were reduced to negligible proportions in this stage of the process. This work is important in that it suggests that manufacturers can evaluate the destructive effect(s) of a process from the spectrum of microorganisms found in the product emerging from the stage in question.

One of the major problems associated with microbiological analysis is that of the time needed to obtain a colony count; and, over recent years, there has been a concerted effort toward the development of rapid methods for the estimation of microbial numbers. McMurdo and Whyard (1984) examined the application of a number of new rapid methods including the direct epifluorescent filter technique (DEFT), ATP bioluminescence, and impediometry; impediometry measures changes in impedance or conductivity brought about by microbial activity in a defined medium (Pettifer, 1994). It was concluded that, of the techniques then available, the conductivity method was the best available system, but the same workers were quick to point out, however, that the technique is still subject to limitations. However, the evaluation of any new method is not without difficulties, and Wood and Gibbs (1982) drew attention to the difficulties of comparing different bases of measurement. It is quite understandable to want to compare measurements in terms of colony count equivalents, but how, they argue, can one relate the measurement of adenosine triphosphate in the bioluminescence assay technique to a colony count?

However, while acceptance of indirect parameters may not prove to be an insurmountable hurdle, the colony count technique still has one important advantage in the hands of a skilled microbiologist, namely that the appearance of colonies can provide an invaluable guide as to type or species. Experienced microbiologists in dairy plants can frequently pinpoint areas for further investigation as a result of examining a relevant series of plates, thus saving much valuable time in isolating problems. It seems likely though that rapid methods will emerge as the means of providing a microbiological audit of the process, with the traditional colony count being used for specification requirements (Hall, 1994). Further discussion of this issue can be found in Chapter 14, and details of the techniques that are being currently considered for dairy products can be found in Robinson et al. (2000).

The low A_w of milk powder has a bearing on the microbial levels that can result from storage. Crossley (1962) confirmed that the microbial population in a milk powder does decrease during storage, but that the rate varies with different powders. Mair-Waldburgh and Lubenau-Nestle (1974) observed decreases in microbial counts in both spray and roller-dried powders after storage for 6 months. In the case of instant, dried skim-milk powder, Miercurio and Tadjall (1979) found that, after storage in well-sealed tins for 20 years, the count was comparable to that of a freshly made product. While this result implies a fair degree of microbiological stability, it also suggests that the microflora was probably made up of spore-formers.

5.7.1.3 Specific Organisms. The first concern of all food microbiologists is with food-borne disease organisms; and because milk is a naturally occurring food, it is the ideal medium for microbial growth. Given the right temperature, microorganisms will multiply in milk at a prolific rate, and the dairy industry has become well-attuned to this problem.

Gilbert (1983) reviewed current trends and future prospects for food-borne infections, and he observed that these are likely to increase in the United Kingdom for the next few years. The occurrence of important food–organism relationships continues to be a cause for concern: For example, the increase in milk-borne *Campylobacter* infections (Robinson, 1981; Robinson and Jones, 1981; Blankenship and Hoffman, 1986), as well as problems with the so-called emerging pathogens, such as *Mycobacterium avium* sub-sp. *paratuberculosis* (Grant et al., 2001), and *Escherichia coli* 0157 (Keceli and Robinson, 1997; McKillip et al., 2000), are likely to increase as well.

Nevertheless, salmonellosis still remains one of the dominant types of food-borne disease, and the occurrence of salmonellae in dried milk is well-documented. Schroeder (1967) examined 3315 samples from 200 factories in 19 American states, and he claimed that 1% were contaminated with salmonellae. Collins (1968) has drawn attention to the presence of *Salmonella newbrunswick* in instant milk, and he suggested that the "instantizing" process was at fault. Craven (1978) noted a link between salmonellosis in human infants and a particular brand of infant food produced in one factory in Victoria, Australia. As a result, all dried milk factories in Victoria have, since that date, had samples routinely examined for salmonellae.

The effects of the spray drying process on the survival of salmonellae have been well-researched (McDonough and Hargrove, 1968; Licari and Potter, 1970). It has been suggested that product temperature, particle density, and fat content are the main factors that influence survival. Licari and Potter (1970) demonstrated that *Salmonella* spp. are not eradicated completely from milk powder by spray drying. They also reported that, while storage at 45°C and 55°C had a lethal effect on the test organisms—*Salmonella typhimurium* and *Salmonella thompson*—the numbers of salmonellae were only reduced, but not eliminated, at temperatures of 25°C and 35°C.

The question of where the salmonellae come from in the first place has also to be addressed. Heldman et al. (1968) identified a number of areas where there was contact between the air and the product, implying that salmonellae could be an airborne contamination, and Otte (1980) also suggested that contamination takes place during manufacture. Clearly there is no simple answer as to the mode of ingress of salmonellae into dried milk powder, and quality control procedures must be trusted to trace any likely contamination.

Methods for the detection of *Salmonella* spp. in dried milks continues to be an area of continuing research. The ICMSF (1986) has outlined proposals for sampling and the derivation of sampling plans for dried milk powders and, with powders targeted at vulnerable groups of consumers, it is widely assumed that salmonellae should be "*absent*" in 500-g samples. The ICMSF (1986) points out that few outbreaks of food-borne disease have been attributed to dried milks in recent years, and consider that this is due to the rigorous testing and control procedures employed within the industry (Mettler, 1994). The detection of salmonellae, both in terms of recovering scant numbers and then rapid confirmation, is another area of ongoing research. Andrews et al. (1983) and Rappold et al. (1984) have proposed rapid cultural methods, while Ibrahim et al. (1985) have evaluated rapid detection methods based on radio-immunometric and enzyme immunometric assays; the latter authors claim that selective enrichment cultures can be examined within 8h. In a more recent test, Karpiskova and Holasova (1999) reported that the use of the immunomagnetic separation technique for detecting salmonellae in milk eliminated the problem of "false negatives" associated with conventional methods, and it saved at least 1 day by eliminating the selective enrichment step. The immunomagnetic separation technique was recommended previously by Parmar et al. (1992), but the AOAC International appear to have opted for enrichment on modified semisolid Rappaport–Vassiliadis medium as the most appropritae method for detecting salmonellae in milk-based powders (Bolderdijk and Milas, 1996).

However, one of the problems with milk powders is that the cells of *Salmonella* may be heat-stressed, and Baylis et al. (2000) have alerted workers to the fact that commercial brands of buffered peptone water differ greatly in their ability to recover injured cells.

Unquestionably the effort to find simpler, faster, and more reliable methods for the detection of salmonellae will, due to its public health significance, continue. The fact should not be overlooked, however, that dried milk has an excellent record, which can be attributed to increased attention to equipment design and process control, coupled with improved quality assurance procedures.

The next most important organism with respect to dried milk powders is *S. aureus*. Crossley and Campling (1957) investigated the spray-drying process and concluded that a small proportion of cells of *S. aureus* could survive the drying conditions. However, Crossley (1962) observed that, while *S. aureus* occurred in raw milk supplies, there was no connection with the strains isolated from milk powder, and he concluded that the most likely source was in-plant contamination. Otte (1980) supported this proposition and concluded that the presence of *S. aureus* is clear evidence of contamination during manufacture; only 2% of *S. aureus* survived spray drying. However, if *S. aureus* is allowed to grow in the milk prior to heat treatment, then the enterotoxins can be carried through into the milk powder produced subsequently and at levels that can cause food poisoning (Anonymous, 2000). Hill (1983) studied the occurrence of enterotoxin-producing strains of *S. aureus*, and he found them to be present in both dried milk and sodium caseinate; however, phage-typing could not determine whether the isolated strains were of human or bovine origin.

The enumeration of *S. aureus* in milk powder was the subject of a large ICMSF study involving 14 laboratories in 12 countries (Chopin et al., 1985). In the trial, *S. aureus* was added before and after drying using seven strains of the organism, and direct inoculation of serial dilutions onto Baird–Parker medium or into Giolitti and Cantoni's enrichment broth (MPN technique) gave similar results. Such large-scale, cooperative evaluations of methods have been a growing feature in microbiology in recent years, and the initiative is to be applauded. Thus, while the nature of microbiological techniques makes them still dependent on operative skill, a good measure of agreement between different laboratories enhances confidence in the procedures.

Other pathogens have joined the salmonellae and staphylococci in recent years as important contaminants of milk powder, and notable amongst these has been *Bacillus cereus*. It gives rise to food poisoning in two clinical forms caused by at least two enterotoxins (Gilbert, 1983). Holmes et al. (1981) described a food poisoning outbreak affecting eight people who had eaten macaroni cheese which was found to contain 10^8–10^9 *B. cereus* organisms g^{-1}; *B. cereus* was recovered from the milk powder used in the preparation of the macaroni cheese. They concluded that the temperature profile of the meal-production system could have contributed to the multiplication of the organism.

Johnson (1984) researched food-borne illnesses and confirmed that dried skim milk and malted milk figured amongst those foods that contained *B. cereus*, while Becker et al. (1994) confirmed that infant foods could also contain the organism. Helmy et al. (1984) examined milk and milk products in Egypt and reported that 7 out of 10 samples of dried milk contained *B. cereus* in numbers ranging from 100 to 1×10^6 cfu g^{-1}, but in Australia the counts in local samples of milk powder did not exceed 900 cfu g^{-1} (Rangasamy et al., 1993). Although *B. cereus* can survive severe heat treatments, in many cases, contamination from in-plant sources can be as important carryover from the raw milk (teGriffel et al., 1996). Elimination of the organism places demands on raw milk quality, plant design, and plant hygiene (Hammer et al., 2001). Methods for the enumeration of *B. cereus* have recently been evaluated by Schulten et al. (2000) and Torkar and Mozina (2001).

Enterococci have also featured prominently in recent years, and Batish et al. (1982, 1984) found *Enterococcus faecium* and *Enterococcus faecalis* sub-sp. *zymogenes* in the high numbers in dried milk. El-Bassiony (1985) also examined dairy products for enterococci and found *Ent. faecalis* and *Ent. faecium*.

Other microbial species appearing in association with dried milk have included *Yersinia enterocolitica* (Morse et al., 1984), which has been implicated in food-related disease outbreaks. While dried milk may be the principal vehicle by which consumers can be infected with *Yersinia*, its reinforces the need to produce milk powder to strict standards, because it is a very widely used ingredient and a primary target for laboratory examination. It may be for this reason that a number of studies of *L. monocytogenes* have included isolation from "spiked" samples of skim-milk powder (Twedt et al., 1994); and although an outbreak of listeriosis has never been linked with dry milk powders, simple MPN methods could also be adapted for routine control purposes (Gohil et al., 1996; Tran and Hitchins, 1996).

Debate about the significance of coliforms, primarily as indicator organisms, is undiminished, and Otte (1980) regards their presence as clear evidence of contamination. However, Law and Mabbitt (1983) pointed out that *Escherichia coli* is an important pathogen causing mastitis and that its presence in milk invalidates the proposition that it is an indicator of faecal contamination. Nevertheless, the presence of *E. coli* after the heat treatment stage in the drying process can be regarded as adventitious contamination; and testing of milk powders for coliforms, as an adjunct to the testing for *Salmonella*, still has a valid role.

The need to be able to enumerate coliforms resulted in another collaborative study involving 14 laboratories and reported by Entis (1983). The examination involved nonfat dry milk and canned custard using the hydrophobic grid membrane filter technique that is now a recommended official first method in North America (APHA, 1992). However, the ease of long-established procedures for determining coliforms, such as MPN counts in MacConkey Broth and confirmation with Brilliant Green Bile Broth or similar medium (BSI, 1968), along with the understanding that dairy microbiologists have of this group of organisms, means that testing for coliforms will be around for a long time to come.

Other research into the microbiology of dried milk has been concerned with yeasts and molds. Aboul-Khier et al. (1985) have isolated *Aspergillus*, *Penicillium*, and *Mucor* spp. from a range of dried products including whole milk, skim-milk, and ice cream mixes, but there was no suggestion that these fungi posed any risk to consumers.

5.8 PRODUCT SPECIFICATIONS AND STANDARD METHODS

The question of standards in respect to dried milks has been the subject of continuous debate. The American Dry Milk Institute Standards (now the Dairy Products Institute Standards) predominated for many years and are still widely quoted today. More recently, the Codex Alimentarius Commission has focused on a Code of Hygienic Practice for Dried Milk, but indicated that microbiological guidelines should not be included. In terms of microbial quality, however, they proposed end-product specifications and made specific reference to *Salmonella*.

In relation to specifications, each major milk-powder-producing country appears to have its own; and while these specifications are designed to provide a common minimum level that is pertinent to general public health, they may not deal with the specific needs of individual bulk users. These latter specifications may form a contractual obligation and need careful preparation. Quite frequently they are directed to some particular need, such as the incorporation of a specially processed dried milk into a baby food formulation, and it is not unknown for special requirements in relation to thermophilic or thermoduric organisms to be stated. It is fundamental that such specifications should be definitive and exact, and vague statements, such as "absent," mean nothing without some qualification as to specific parameters. All too frequently these situations arise because contractual arrangements are made between nontechnical persons or, even worse, between individuals with a vague knowledge of the subject.

The same situation can be found with respect to standard methods and, while each major country has standard methods for total counts (coliforms and *Salmonella*), so do international bodies, such as the International Commission on Microbiological Specifications for Foods (ICMSF), the International Dairy Federation, the International Standards Organisation, and the Association of Official Analytical Chemists. Clearly there is a degree of overlap and, with many of the same people contributing to each of these organizations, a level of agreement does arise. Agreed procedures with some authoritative backing are important, they provide a reference point against which other methods can be assessed, and, in the rare event of litigation, can prove invaluable.

In contracts, standards must be quoted clearly, and they define both the mass or volume of the sample under examination and the number of analyses to be performed—whether this be in duplicate or triplicate. Next there needs to be an agreement on the method of analysis or, at the very least, the reference method to be employed in case of dispute. The agreement should also include the requirement for independent assessment by an officially recognized laboratory, again using a reference procedure. It has always been appropriate in reference procedures to use those prescribed by one of the international bodies cited earlier. The procedures are well proven and enjoy the highest levels of credibility as a result; their wider use is to be encouraged particularly within the contractual situation. Microbiology is dependent on reproducible and accurate techniques, and it is fastidious and demanding. The need to confront these facts in supplier–customer situations cannot be underestimated.

ACKNOWLEDGMENTS

The authors acknowledge without reservation that this chapter has been based on the contributions of Professors F. E. Nelson and H. R. Lovell to the second edition of the book *Dairy Microbiology*, published in 1990 by Chapman & Hall, London. Because this present text is envisaged as the "natural" successor to the earlier book, it was decided to utilize the expertise of the previous authors rather than allow their knowledge to go "out of print." Any errors in the present text can be ascribed to the authors, but all credit must be assigned to the "experts" who supplied the essential framework.

REFERENCES

Aboul-Khier, F., El-Bassiony, T., Abd-El Hamid, A., and Moustafa, M. K. (1985) *Assiut Vet. Med. J.*, **14**(28), 71–79.

ADPI (1990) *Standards for Grades of Dry Milks*, American Dairy Products Institute, Chicago.

Andrews, W. H., Wilson, C. R., and Poelm, P. L. (1983) *J. Food Sci.*, **48**(4), 1162–1165.

Anonymous (2000) *Food Sci Technol.*, **30**, 1–2.

APHA (1992) *Standard Methods for the Examination of Dairy Products*, 16th ed., American Public Health Association, Washington, DC.

Appuswamy, S., and Ranganathan, B. (1981) *J. Food Sci. Technol.*, **18**(6), 258–259.

Arun, A. P. S., Prasad, C. R., and Sinha, B. K. (1978) *Orissa Vet. J.*, **12**(1), 19–23.

Batish, V. K., and Ranganathan, B. (1984) *NZ J. Dairy Sci. Technol.*, **19**(3), 189–196.

Batish, V. K., Chandler, H., and Ranganathan, B. (1982) *J. Food Prod.*, **45**(4), 348–352.

Baylis, C. L., MacPhee, S., and Betts, R. P. (2000) *J. Appl. Microbiol.*, **89**(3), 501–510.

Becker, H., Schaller, G., Vonwiese, W., and Terplan, G. (1994) *Int. J. Food Microbiol.*, **23**(1), 1–15.

Bhale, P., Sharma, S., and Sinha, R. N. (1989) *J. Food Sci. Technol. Mysore*, **26**(1), 46–48.

Blankenship, L. C., and Hoffman, P. S. (1986) In *Developments in Food Microbiology—2*, R. K. Robinson, ed., Elsevier Applied Science Publishers, pp. 91–122.

Bolderdijk, R. F., and Milas, J. E. (1996) *J. AOAC Int.*, **79**(2), 441–450.

BSI (1968) British Standard 4285—Examination of Milk Products, British Standards Institute, Milton Keynes, Bucks, England.

Burton, H. (1988) *Ultra-High-Temperature Processing of Milk and Milk Products*, Elsevier Applied Science, London.

Carić, M. (1992) Concentrated and dried dairy products. In *Dairy Science and Technology Handbook*, Y. H. Hui, ed., VCH Publishers, New York.

Chopin, A., Mocquot, G., and Legraet, Y. (1977) *Can. J. Microbiol.*, **23**, 716–762.

Chopin, A., Tessone, S., Vila, J.-P., and Legraet, Y. (1978) *Can. J. Microbiol.*, **24**, 1371–1381.

Chopin, A., Malcolm, S., Jarvis, G., Asperger, H., Beckers, H. J., Bertona, A. M., Cominazzini, C., Carini, S., LoDi, R., Hahn, G., Heeschen, W., Jans, J. A., Jerns, D. I., Lanier, J. M., O'Connor, F., Rea, M., Rossi, J., Seligmann, R., Tessone, S., Waes, G., Mocquot, G., and Pirnik, H. (1985) *J. Food Prot.*, **48**(1), 21–27, 34.

Chopra, A. K., and Mathur, D. K. (1984) *J. Appl. Bacteriol.*, **57**(2), 263–271.

Collins, R. N. (1968) *J. Am. Med. Assoc.*, **203**(10), 838–844.

Craven, J. A. (1978) *Victorian Vet. Proc.*, **36**, 56–57.

Crossley, E. L. (1962) *Milk Hygiene*, WHO Monograph No. 48, Geneva.

Crossley, E. L., and Campling, M. (1957) *J. Appl. Bact.*, **20**(1), 65–70.

Crossley, E. L., and Johnson, W. A. (1942) *J. Dairy Res.*, **13**(1), 5–44.

Curran, H. R., and Evans, F. R. (1945) *J. Bacteriol.*, **39**, 335–346.

El-Bassiony, T. A. (1985) *Assuit Vet. Med. J.*, **15**(29), 107–111.

El-Bassiony, T. A., and Aboul-Khier, F. (1983) *Assuit Vet. Med. J.*, **11**(21), 159–163.

Entis, P. (1983) *J. Assoc. Off. Analyt. Chemists*, **66**(4), 897–904.

Falkowski, J. (1978) *Zentralbl. Bakteriol. Parasitenk. Infektionskr. Hyg. 1B*, **167**(1/2), 165–170.

Farrag, S. A., El-Gazzar, F. E., and Marth, E. H. (1990) *J. Food Prot.*, **53**(9), 747–750.

Flint, S., Palmer, J., Bloemen, K., Brooks, J., and Crawford, R. (2001) *J. Appl. Microbiol.*, **90**(2), 151–157.

Galesloot, T. E. (1962) The sterilization of milk. In *Milk Hygiene* World Health Organization, Geneva, Switzerland.

Galesloot, T. E., and Stadhouders, B. (1968) *Netherlands Milk Dairy, J.*, **22**, 158–172.

Gilbert, R. J. (1983) In *Food Microbiology: Advances and Prospects*, T. A. Roberts and F. A. Skinner, ed., Academic Press, London.

Glover, F. A., Skudder, P. J., Stothart, P. H., and Evans, E. W. (1978) *J. Dairy Res.*, **45**, 291–318.

Gohil, V. S., Ahmed, M. A., Davies, R., and Robinson, R. K. (1996) *Food Control*, **6**(6), 365–369.

Grandison, A. S., and Glover, F. A. (1994) Membrane processing of milk. In *Modern Dairy Technology*, Vol. 1, R. K. Robinson, ed., Chapman & Hall, London, pp. 273–312.

Grant, I. R., Rowe, M. T., Dundee, L., and Hitchings, E. (2001) *Int. J. Dairy Technol.*, **54**(1), 1–13.

Hall, P. A. (1994) Scope for rapid microbiological methods in modern food production. In *Rapid Analysis Techniques in Food Microbiology*, P. Patel, ed., Blackie Academic & Professional, London.

Hammer, B. W., and Babel, F. J. (1957) *Dairy Bacteriology*, 4th ed., John Wiley & Sons, New York.

Hammer, P., Wiebe, C., Walte, H. G., and Teufel, P. (2001) *Kieler Milchwirtsch. Forschungber.*, **53**(2), 123–146.

Hawley, H. B., and Benjamin, M. I.-W. (1955) *J. Appl. Bacteriol.*, **18**(3), 495–502.

Heldman, D. R., Hall, C. W., and Hedrick, T. J. (1968) *J. Dairy Sci.*, **51**(3),466–470.

Helmy, Z. A., Abd-El-Bakey, A., and Mohamed, E. L. (1984) *Zentralbl. Mikrobiol.*, **139**(2), 129–133.

Higginbottom, C. (1944) *J. Dairy Res.*, **14**(1,2), 184–194.

Hill, B. M. (1983) *NZ J. Dairy Sci. Technol.*, **18**(1), 59–62.

Hobbs, B. C. (1955) *J. Appl. Bacteriol.*, **18**(3), 484–492.

Holmes, J. R., Plunkett, T., Pate, P., Roper, W., and Alexander, W. J. (1981) *Arch. Int. Med.*, **141**(6), 766–767.

Ibrahim, G. F., Lyons, M. J., Walker, R. A., and Fleet, G. H. (1985) *Appl. Environ. Microbiol.*, **50**(3), 670–675.

ICMSF (1986) In *Micro-organisms in Foods*, Vol. 2, University of Toronto Press, Toronto, Canada.

Johnson, K. M. (1984) *J. Food Prot.*, **47**(2), 145–153.

Kalogridou-Vassiliadou, D. (1990) *Lebensm. Wiss. Technol.*, **23**(4), 285–288.

Kalogridou-Vassiliadou, D. (1992) *J. Dairy Sci.*, **75**(10), 2681–2686.

Karpiskova, R., and Holasova, M. (1999) *Vet. Med.*, **44**(8), 225–228.

Keceli, T., and Robinson, R. K. (1997) *Dairy Industries International*, **62**(4), 29–33.

Keogh, B. P. (1965) *Aus. J. Dairy Technol. Spray Drying of Milk*, Technical Publication. No. 16.

Keogh, B. P. (1966) *Food Technol. Aust.*, **18**(3), 126–133.

Keogh, B. P. (1971) *J. Dairy Res.*, **38**(1), 91–111.

Kilsby, D. C., and Baird-Parker, A. C. (1983) In *Food Microbiology: Advances and Prospects*, T. A. Roberts and F. A. Skinner, eds., Academic Press, London.

Knipschildt, M. E., and Andersen, G. G. (1994) Drying of milk and milk products. In *Modern Dairy Technology*, Vol. 1, R. K. Robinson, ed., Chapman & Hall, London, pp. 159–254.

Kwee, W. S., Dommett, T. W., Giles, J. E., Roberts, R., and Smith, R. A. D. (1986) *Aust. J. Dairy Technol.*, **41**(1), 3–8.

Lankes, H., Ozer, B., and Robinson, R. K. (1998) *Milchwissenschaft*, **53**(9), 510–513.

Law, B. A., and Mabbitt, L. A. (1983) In *Food Microbiology: Advances and Prospects*, T. A. Roberts and F. A. Skinner, eds., Academic Press, London.

Lewis, M. J. (1996) Pressure-activated membrane processes. In *Separation Processes in the Food and Biotechnology Industries*, A. S. Grandison and M. J. Lewis, eds., Woodhead Publishing Ld., Cambridge, England.

Licari, J. J., and Potter, N. N. (1970) *J. Dairy Sci.*, **53**(7), Part 1, 865–870; Part 11, 871–876; Part Ill, 877–882.

Lück, H., and Andrew, M. J. A. (1975) *S. Afr. J. Dairy Technol.*, **7**, 39–42.

Lück H., and Gavron, H. (1989) In *Dairy Microbiology*, Vol. 2, 2nd ed., R. K. Robinson, ed., Elsevier Science Publishers, London.

Mair-Waldburgh, H., and Lubenau-Nestle, R. (1974) *XIXth Int. Dairy Congress*, **IE**, 553–554.

Marth, E. M., and Steele, J. L. (2001) *Applied Dairy Microbiology*, Marcel Dekker, New York.

Masters, K. (1987) In *Food Technology International Europe*, Sterling Publications Ltd., London.

Mattick, A. T. R., Hiscox, E. R., Crossley, E. L., Lea, C. H., Findlay, J. D., Smith, J. A. B., Thompson, S. Y., Kon, S. K., and Edgell, J. W. (1945) *J. Dairy Res.*, **14**(1,2), 116–159.

McDonough, F. E., and Hargrove, R. E. (1968) *J. Dairy Sci.*, **51**(10), 1587–1591.

McKillip, J. L., Jaykus, L. A., and Drake, M. A. (2000) *J. Appl. Microbiol.*, **89**(1), 49–55.

McMurdo, I. H., and Whyard, S. (1984) *J. Soc. Dairy Technol.*, **37**(1), 4–9.

Mettler, A. E. (1994) *J. Soc. Dairy Tcehnol.*, **47**(3), 95–107.

Miercurio, K. C., and Tadjall, V. A. (1979) *J. Dairy Sci.*, **26**(4), 633–636.

Morse, D. L., Shayegani, M., and Gallo, R. J. (1984) *Am. J. Public Health*, **74**(6), 589–592.

Murphy, P. M., Lynch, D., and Kelly, P. M. (1999) *Int. J. Dairy Technol.* **52**(2), 45–50.

New Zealand Dairy Research Institute (1978) Annual Report.

Otte, I. (1980) *Mikrobiologie der Trockenprodukte*, Institut fur Hygiene Bundesanstalt fur Milchforschung, Kiel, Germany.

Parmar, N., Easter, M. C., and Forsythe, S. J. (1992) *Lett. Appl. Microbiol.*, **15**(4), 175–178.

Pettifer, G. (1994) Microbiological analysis. In *Modern Dairy Technology*, Vol. 2, R. K. Robinson, ed., Chapman & Hall, London, pp. 417–454.

Rangasamy, P. N., Iyer, M., and Roginski, H. (1993) *Aust. J. Dairy Technol.*, **48**(2), 93–95.

Rappold, H., Bolderdijk, R. F., and de Smedt, J. M. (1984) *J. Food Prot.*, **47**(1), 46–48.

Robinson, D. A. (1981) *Br. Med. J.*, **282**, 1584.

Robinson, D. A., and Jones, D. M. (1981) *Br. Med. J.*, **282**, 1374–1376.

Robinson, R. K., Batt, C. A., and Patel, P. D. (2000) *Encyclopedia of Food Microbiology*, Academic Press, London.

Schroeder, S. A. (1967) *J. Milk Food Technol.*, **30**, 376.

Schulten, S. M., in't Veld, P. H., Nagelkerke, N. J. D., Scotter, S., de Buyser, M. L., Rollier, P., and Lahellec, C. (2000) *Int. J. Food Microbiol.*, **57**, 53–61.

Shayegani, M., Morse, D, De Forge, I., Root, T., Parsons, L. M., and Maupin, P. S. (1983) *J. Clin. Microbiol.*, **17**(1), 35–40.

Skov, O. (1994) Protection against fire and explosion in spray dryers. In *Modern Dairy Technology*, Vol. 1. R. K. Robinson, ed., Chapman & Hall, London. pp. 255–272.

Stumbo, C. R. (1972) *Thermobacteriology in Food Processing*, 2nd ed., Academic Press, New York.

teGriffel, M. C., Beumer, R. R., Bonestroo, M. H., and Rombouts, F. M. (1996) *Netherlands Milk Dairy J.*, **50**(4), 479–492.

Theophilus, D. R., and Hammer, B. W. (1938) Influence of growth temperature on the thermal resistance of some bacteria from evaporated milk, *Iowa Agric. Exp. Stn. Res. Bull.*, 244.

Torkar, K. G., and Mozina, S. S. (2001) *Period. Biol.*, **103**(2), 169–173.

Tran, T. T., and Hitchins, A. D. (1996) *J. Food Prot.*, **59**(6), 928–931.

Twedt, R. M., Hitchins, A. D., and Prentice, G. A. (1994) *J. AOAC Int.*, **77**(2), 395–402.

Wood, J. M., and Gibbs, P. A. (1982) In *Developments in Food Microbiology—1*, R. Davies, ed., Elsevier Applied Science Publishers, New York, pp. 183–214.

CHAPTER 6

MICROBIOLOGY OF ICE CREAM AND RELATED PRODUCTS

PHOTIS PAPADEMAS
P. Roussounides Enterprises Ltd., Pralina, Nicosia, Cyprus

THOMAS BINTSIS
Laboratory of Food Microbiology and Hygiene, Department of Food Science and Technology, Faculty of Agriculture, Aristotelian University of Thessaloniki, Thessaloniki, Greece

"I doubt whether the world holds for anyone a more soul-stirring surprise than the first adventure with an ice cream."
—Heywood Hale Broun (1888–1939), American journalist

6.1 INTRODUCTION

It is possible that the origins of ice cream and other similar frozen confections lie in the "milk and honey" mentioned in the Old Testament. It has also been suggested that Abraham and Isaac had frozen or chilled drinks. It is certain that ices were known in ancient times, because the Chinese mixed snow and fruit juices to make an iced sweet and records exist which indicate that, during the time of Confucius, ice cellars were used to keep foods cool.

In the first century A.D., the Romans were using ice from the mountains to chill mixtures of honey and fruit pulp or juice, but it is then not until about 1292, when Marco Polo returned from his journey to Asia and China, that further definite information is available. He brought back recipes for water ices said to have been used in Asia for thousands of years.

Dairy Microbiology Handbook, Third Edition, Edited by Richard K. Robinson
ISBN 0-471-38596-4

Henrietta Maria most probably introduced ice cream to Great Britain when she became the wife of Charles I in 1630. The first manuscript on ice cream (The Art of Preparing Ice Cream) was written in French, but the author was never discovered.

Ice cream was introduced to America by Europeans—in particular, English colonists around 1700. Nancy Johnson created the first manually operated freezer in 1846. Jacob Fussell, the father of the wholesale ice cream industry in America, opened the first ice cream factory in Baltimore in 1856. Two more factories closely followed in Washington, DC and New York. In 1859 the total production of ice cream in America was 4000 gallons, but by 1899 this had risen to more than five million gallons; it is now in excess of 1000 million gallons (Marshall and Arbuckle, 1996).

Two major inventions, in addition to the introduction of mechanical refrigeration, were largely responsible for this enormous increase. They were, in 1899, the homogenizer and, in 1929, the continuous freezer. In addition, high-temperature short-time (HTST) pasteurization of ice cream was approved by the US authorities (1953), and this development gave a degree of automation and better product quality. Moreover, an important period in the history of ice cream was between 1965 and 1981 when highly automated and high-volume processing equipment was introduced, and the definitions and standards for frozen desserts were revised.

At the present time, many different types of frozen dessert are being made. It is necessary to classify them, and although the descriptions do vary from country to country, the list given below is typical, and legislation usually follows a similar pattern.

6.2 CLASSIFICATION OF FROZEN DESSERTS

6.2.1 Cream Ices

Creme glacée, eiskrem, crema di gelato, and roomijs are ice creams made with a statutory minimum of butterfat (milk fat). Flavoring materials, including fruit, nuts, and chocolate, may be added to produce the corresponding flavored cream ice. Many countries (e.g., France and Germany) insist on the use of milk fat only in any ice cream and prohibit the use of any substitute fats.

6.2.2 Ice Cream

Ijs (The Netherlands), *mellorine* (in some states of the United States, but not all), *glaces de consommation* (Belgium), and *margarin-is*

(Norway) are ice creams with a statutory minimum of fat, some or all of which may be fat other than milk fat.

6.2.3 Milk Ices

Glace au lait, milcheis, gelato al latte, and milkijs are ices based on milk. Only milk fat may be used, and minimum standards of 2.5–3% are established in most countries.

6.2.4 Custards

The traditional milk ice, which was made by boiling 4.5 liters of whole milk with 170 g of corn flour and 0.68 kg of sugar, is sometimes called custard. A true custard is probably made with fresh whole eggs, or egg yolk solids, and in the United States it must contain at least 1.4% by weight of egg solids. It may be called French ice cream or French custard ice cream. In France this product is called glace aux oeufs and must contain at least 7% egg yolk solids.

6.2.5 Ices or Water Ices

These are made from fruit juices and/or pulp diluted with water and with additions that may include sugar, citric, malic or tartaric acids, a stabilizer such as gelatin or pectin, color, and flavor. They may be frozen with or without agitation and the incorporation of air. If sold in a "slushy" condition, they may be called "frappe"; and if made with an alcoholic liquid in place of the water, they may be known as a "punch." Water ices frozen without agitation and usually on a stick are called "ice lollies" in the United Kingdom.

6.2.6 Sherbet

This is made from ingredients similar to those of water ices, but it incorporates some ice cream, liquid milk, cream, or milk powder. It is usually frozen with agitation and thus contains some air.

6.2.7 Sorbets

The composition of sorbets is similar to that of ices. Sorbets have a high sugar and fruit and fruit juice content (30 and 30–50%, respectively). Stabilizers and egg white are also added, and the product has an overrun of 20% or less. Exotic flavors are often included in sorbets.

6.2.8 Mousse

Originally a mousse was a well-whipped mixture of cream, fruit, sugar, and egg white. It had a large content of air; and to prevent complete collapse, it was necessary to include high level of total solids. Present-day mousse in many ways resembles an ordinary ice cream, but does not melt when taken from deep-freeze storage. This is because it contains a relatively high quantity of stabilizer, which produces a gel-like consistency.

In addition, over the years more complex products have been developed, and this is still continuing. Two of these are detailed below (see Sections 6.10 and 6.11).

6.2.9 Cassata

This is made in a round mold, hinged so that it may be filled with ice cream and other frozen products. The confection is built up in layers of rich, variously flavored ice cream, some with fruits, some with liqueurs, and sometimes with chocolate or nuts. Fingers or slices of sponge cake, sometimes soaked in liqueur, may be added. The cassata is frozen for several hours, then turned out of the mold for serving.

6.2.10 "Splits"

These are made on a stick, and they consist of a central section of ice cream and an outer layer of fruit water ice. They may be dipped in chocolate, broken nuts, or biscuit crumbs.

6.2.11 Frozen Yogurt

This product is prepared by freezing while stirring a pasteurized mix of milk fat, MSNF, sweetener, stabilizer, and yogurt, and it is usually flavoured with fruit puree. The amount of yogurt added could range from 10% to 20% of the total weight of the mix.

Frozen yogurt products are low in fat, and two variations—nonfat frozen yogurt and low-fat frozen yogurt—are commercially available. Recently, Davidson et al. (2000) has reported that frozen yogurt could serve as an excellent vehicle for the incorporation of probiotic bacteria.

TABLE 6.1. Ice Cream Consumption in Europe (Liters per Capita in 1995)

United Kingdom	9	Germany	7
Ireland	10	Netherlands	7
France	6	Italy	6
Greece	5	Norway	13
Spain	4	Sweden	14
Portugal	3	Finland	14
Austria	7	Denmark	10
United States	22		

Source: After Doxanakis (1998).

6.3 ICE CREAM AND FROZEN DESSERT SALES

In the United Kingdom, ice cream and frozen desserts represent one of the largest packaged grocery sectors. The market, though, is extremely concentrated and is dominated by three major food companies.

The total retail sales have risen considerably over the past 5–6 years and the estimated sales for 1999 reached £1460 million, with ice cream representing £1040 million of the total. Market experts forecast that by 2004 the figure of total sales will reach £1605 million (Anonymous, 2000).

The UK ice cream market seems to be affected by the season; for example, in 1998 the poor summer was a reason for the decrease in total sales of ice cream (£984 million) when compared to £1030 mil in 1997. The manufacturers have tried to deseasonalize the market by encouraging year-round sales and introducing winter-themed products.

On the other hand, season seems not to play a very important role in the consumption of ice cream by the Scandinavian countries because their climate is characterized as cold, but they are still the top ice cream consumers in Europe (see Table 6.1).

The United States remains the world's leader in ice cream manufacture and development with annual ice cream consumption at 22 liters per capita (Doxanakis, 1998). The United States is also one of the top ice cream and edible ice exporters (see Table 6.2), whereas imports remain very low (see Table 6.3).

6.4 LEGISLATION

6.4.1 Composition

Ice cream and other whipped frozen desserts are foams made up of air cells surrounded by a partially frozen emulsion. Ice crystals and

TABLE 6.2. Total Exports of Ice Cream and Edible Ice (Quantity in Metric Tons)

	1993	1994	1995	1996	1997	1998
European Union (15 countries)	257,724	303,056	308,263	312,140	321,928	382,957
Austria	3,690	2,080	5,984	3,800	4,220	1,381
Belgium–Luxembourg	62,589	69,324	67,872	68,596	75,677	90,455
Denmark	32,411	24,557	26,183	21,507	19,766	3,270
Finland	9,141	16,463	7,717	5,952	4,059	2,598
France	43,069	56,176	47,247	50,963	59,503	60,517
Germany	26,953	30,745	33,059	38,797	35,456	67,401
Greece	5,432	6,042	6,162	5,343	4,852	5,090
Ireland	4,240	2,561	2,167	3,199	3,922	6,122
Italy	19,313	23,746	28,707	31,342	28,776	43,614
Netherlands	12,379	17,211	16,579	19,680	26,873	24,087
Portugal	455	182	206	1,126	498	1,692
Spain	22,716	21,225	24,062	28,673	23,406	30,354
Sweden	4,500	13,848	14,412	9,955	9,162	14,358
United Kingdom	10,836	18,896	27,906	23,207	25,758	32,018
Russian Federation	172	3,743	2,431	1,692	1,692	5,768
United States	37,684	41,974	40,117	42,118	41,037	41,781

Source: FAOSTAT (2000).

TABLE 6.3. Total Imports of Ice Cream and Edible Ice (Quantity in Metric Tons)

	1993	1994	1995	1996	1997	1998
European Union (15 countries)	215,277	225,855	237,975	237,312	264,803	341,352
Austria	3,180	3,526	5,819	6,046	8,313	26,103
Belgium–Luxembourg	26,958	22,647	24,342	22,121	21,087	23,339
Denmark	2,797	2,227	2,481	3,881	6,158	19,096
Finland	2,176	5,635	7,425	6,563	7,468	7,538
France	33,937	42,551	46,723	47,885	55,941	72,118
Germany	28,275	31,501	29,960	33,002	37,657	37,334
Greece	2,993	3,193	4,714	4,263	4,831	5,552
Ireland	3,984	3,599	6,152	4,614	5,016	6,220
Italy	14,446	14,263	13,425	11,947	14,469	17,995
Netherlands	32,927	34,705	34,386	32,770	31,804	27,913
Portugal	5,620	6,497	9,023	10,117	10,235	10,791
Spain	13,346	16,855	18,516	21,674	19,903	25,312
Sweden	7,226	4,714	4,871	5,224	3,789	7,962
United Kingdom	37,412	33,942	30,138	27,205	38,132	54,079
Russian Federation	6,700	39,771	26,903	33,278	21,800	19,033
United States	267	248	706	1,041	1,365	3,243

Source: FAOSTAT (2000).

TABLE 6.4. Typical Chemical Composition of Ice Cream

Constituent	Concentration (% w/w)
Milkfat	10
MNFS	8.8
Sucrose	14
Emulsifier	0.2
Stabilizer	0.2
Overrun (50%)	0.06
Water (50–90% frozen)	66.74
Total	100

Source: After Walstra and Jonkman (1998).

solidified fat globules are embedded in the continuous unfrozen liquid phase that contains proteins, carbohydrates, salts, and gums. It is well recognized that each ingredient in ice cream can play a significant role in texture, flavor, and stability (Huang and Platt, 1995). Ice cream has water, fat, emulsifiers, milk solids non-fat (MSNF), sugars, stabilizers, air, and ice as its major components. A typical composition of ice cream is given below (Table 6.4).

The legal compositional requirements vary considerably between countries (Pappas, 1988). In the United States, for example, *ice cream* is a product that has at least 10% milk fat, *premium ice cream* has to contain at least 12% milk fat, and *superpremium ice cream* contains at least 14% milk fat (Stogo, 1998).

The various regulations for ice cream composition that were introduced after the 1939 War in the United Kingdom were finally revoked and replaced by the Food Standards (Ice Cream) Regulations 1967 (SI No. 1866). More recently, the latter has been replaced, and now the composition of ice cream is included in the Food Labelling Regulations, 1996 (SI No. 1499).

The description of "ice cream" states that it should apply to the frozen product containing not less than 5% fat and not less than 2.5% milk protein. This frozen product is obtained by subjecting an emulsion of fat, milk solids, and sugar [including any permitted sweeteners, Sweeteners in Food Regulations 1995(a)] to heat treatment and subsequent freezing.

The description of "dairy ice cream" is essentially the same as for "ice cream," with the difference that the fat used (minimum 5%) must consist exclusively of milk fat.

6.4.2 Labeling

The correct labeling of food products is of paramount importance in order to protect both the consumer and the manufacturer. Special legislation is now in force, and in the United Kingdom, this is covered by the Food Labelling Regulations, 1996 (SI No. 1499). The requirements for ice cream are given, briefly, below, and the major points that must be covered by the label include:

1. The correct name of the product as described in the Food Labelling Regulations, 1996 (see Section 6.4.1). That is, "ice cream" or "dairy ice cream," and in the case of "ice cream" it is necessary for the words "contains non-milk fat" or "contains vegetable fat" as applicable to be included near the name "ice cream."
2. The pack must include a list of the ingredients given in weight descending order. It is permitted to list the stabilizer, emulsifier, and color using their "E" numbers. It is not sufficient to use the term "milk solids non-fat," because the actual source itself has to be named (e.g., liquid skim milk, skim milk powder, condensed milk, or whey solids). For example, dairy ice cream made from full cream milk, cream, skim milk powder, sugar, dextrose, stabilizer and emulsifier, flavor, and color would probably be labeled: Ingredients: full cream milk, sugar, cream (48% milk fat), skim milk powder, dextrose, stabilizer, emulsifier, permitted flavor and color.
3. In the United Kingdom, it is not at present necessary to include a durability indication on edible ices sold in individual portions. In the case of ice cream a "best before" indication is included together with the appropriate storage conditions (e.g., keep at temperature below –18°C).
4. The name and address of the manufacturer or packer, or the name and address of the seller established in the EU.
5. The size of the pack, either in fluid ounces, pints, or gallons or in metric sizes for which there is a set of specified metric volumes.

6.4.3 Heat Treatment

In the United Kingdom, under the Dairy Products (Hygiene) Regulations of 1995 (SI No. 1086), dairy ice cream mix (see Section 6.4.1) has to be heat-treated in order to produce a safe food product. The regulations are summarized below:

Pasteurized ice cream is obtained by the mixture being heated to:

(a) a temperature of not less than 65.6°C and retained at that temperature for not less than 30 min;
(b) a temperature of not less than 71.7°C and retained at that temperature for not less than 10 min; or
(c) a temperature of not less than 79.4°C and retained at that temperature for not less than 15 s.

The temperature of the pasteurized dairy ice cream mix must be reduced to no more than 7.2°C within 90 min and kept at such a temperature until the freezing process is begun. The temperature of the frozen dairy ice cream must remain at a temperature below –2.2°C at all times after freezing. If the temperature rises above –2.2°C, it must be heat-treated again.

Sterilized dairy ice cream is obtained by the mixture being heated to a temperature of not less than 148.9°C for at least 2 s. If the sterilized mix is immediately transferred in sterile airtight containers, under sterile conditions, and the containers remain unopened then, there is no need for the temperature of the sterilized mix to be reduced to no more than 7.2°C within 90 min prior to freezing. The temperature of the frozen sterile dairy ice cream has to remain at all times below –2.2°C. If the temperature rises above –2.2°C, it must be heat-treated again.

Although dairy ice cream and ice cream (contains non-milk fat) is differentiated in the United Kingdom, their heat treatment requirements remain the same (Anonymous, 1995).

Moreover, heat treatment is not required in the case of a mix that has a pH of 4.5 or less (e.g., some lolly mixes). In addition, mixes made by adding a heat-treated powdered mix to cold potable water need not be further treated. Only sugar may be added as well as water, and the mixes must be frozen within 1 h of being reconstituted (Jukes, 1997).

In the case of HTST pasteurization or sterilization, full thermostatic control is required, and a positive drive pump must be used to ensure the correct holding time. There must also be a device for automatically diverting mix, or for stopping the sterilization process, if the mixture is not heated to the correct temperature.

After such heat treatment, and given careful low temperature storage of the mix, it is to be expected that the mixture will have a very low bacterial count. Heat treatment regulations vary considerably from country to country. In the United States there is legislation that applies to individual states; in addition, there is Federal legislation.

6.5 INGREDIENTS

6.5.1 Milk Solids Non-fat (MSNF)

MSNF ingredients include sugars (lactose), proteins, and minerals. The function of MSNF is to increase the viscosity and melting resistance of the ice cream, with the proteins absorbing free water and preventing the growth of large ice crystals (Fearon and Moruzzi, 1999).

Liquid whole milk, skim milk, and cream are all good sources of MSNF but none of them contributes sufficient MSNF to satisfy either the legal requirements for MSNF or the technical demands. So, although they are, when fresh and of good bacteriological quality, excellent sources of MSNF, it is necessary to add more MSNF in a concentrated form.

A satisfactory source of this is concentrated liquid skim milk, obtained by vacuum evaporation to give a total solids content of 25–35% (Hamilton, 1990). Because this is not normally a sterilized product, it has to be used rapidly and requires refrigerated storage.

Sweetened condensed milk can also be used as to increase the MSNF of ice cream, but the possible formation of large lactose crystals may result to a texture defect known as "sandiness" (Huang and Platt, 1995).

Spray-dried skim milk powder is the most widely used milk product, and this excellent ingredient is used to either provide the whole of the MSNF in the mix or increase the MSNF content of liquid whole milk or skim milk. The use of membranes to concentrate skim milk is also a very satisfactory method of increasing the MSNF. Spray-dried, full cream milk powder is another source of MSNF, but the shelf-life of this product is much less than skim milk powder due to the presence of milk fat; this may become off-flavored and eventually rancid on prolonged storage. If the full cream milk powder is properly used, though, it is a very satisfactory source of MSNF and also of some milk fat.

Another good alternative may be the use of buttermilk powder, which can replace 50% of the skim milk powder. Buttermilk powder can be kept for several weeks to a few months, depending on the storage conditions, moisture content, and, of course, initial quality (Marshall and Arbuckle, 1996).

There is some use now being made of various whey products. The quantity of whey powder in the ice cream mix has to be restricted to 25% of the total MSNF (Stogo, 1998; Hamilton, 1990). The problem with ordinary whey powder is that it contains appreciably more lactose than skim milk powder (SMP), about 74% compared to 50% in SMP, and this may cause a "sandy-mouth feel."

Modified whey powders and liquid concentrates may be obtained in which some, at least, of the lactose has been hydrolyzed to its constituent sugars, glucose, and galactose. This is an ingredient that is appreciably sweeter and may, in some cases, be used as a partial source of sweetener for the mix.

Ice cream manufactured with low-lactose or delactosed milk products can also be more easily consumed by individuals that suffer from lactose malabsorption.

6.5.2 Milk Fat

Fat is essential because it provides ice cream with its rich, mellow, full, and creamy flavor. Fat will also increase the viscosity of the mix, lower the tendency to melt, and provide a smoother ice cream. The best source of milk fat, and the most costly, is cream. This must be freshly produced, and it will produce an excellent "cream ice." Other sources are frozen cream, unsalted butter, and anhydrous milk fat (AMF) or butteroil (Hamilton, 1990).

6.5.3 Other Fats

The use of fats other than milk fat is prohibited by law in a large number of countries. However, they are permitted for use in the United Kingdom, Sweden, Belgium, Denmark, and the Netherlands; and to be suitable for use in ice cream, they should have a melting point below 37°C, with a slip melting point of about 30°C. This is to avoid any "clinging fatty" sensation being left in the mouth, which fats with higher melting points might produce.

The fats that are most commonly used include partially hydrogenated palm oil, palm kernel oil, and coconut oil, suitably blended to give a satisfactory melting range. Because they are usually bland, suitable flavorings must be added as required. Moreover, the addition of fats and oils not of milk origin should be mentioned on the label (Pappas, 1988).

6.5.4 Sugars and Sweeteners

Apart from providing the required sweetness, which balances the "fattiness" of the product, sugars play other more important roles in ice cream. These roles include the following; (a) adding to the total solids, thereby providing texture and body as well as "creaminess," and (b) controlling the freezing point and hence the hardness and "scoopability" of ice cream (Rothwell, 1998).

The sugar most commonly employed is sucrose, obtained either from cane or beet. Sucrose is highly soluble, thereby making it an ideal ingredient. Glucose (dextrose) syrups made by the acid and/or enzyme hydrolysis of corn flour, wheat, and other sources of starch are also employed to a considerable extent. The amount and type of hydrolysis governs both the sweetening power and the amount of freezing point depression, and thus these syrups may be used to produce different types of ice cream as required. The various glucose syrups are differentiated by their dextrose equivalent (DE) number. The higher the DE number, the higher the amount of dextrose present, which leads to increased relative sweetness; by increasing the DE number, the freezing point depression is also increased (Fearon and Moruzzi, 1999).

On a large scale, sugar/glucose syrup mixtures are used, according to the ice cream manufacturers' requirements. They contain about 70% solids, and they are handled in tanker loads. Dextrose is also used, particularly in smaller-scale manufacture, because it is a powder and is much more easily handled than the glucose syrups. Finally, spray-dried glucose syrups are also available, but they cost more than dextrose.

There is a small, but growing, demand for ice cream suitable for diabetics; for this, the sucrose has to be replaced by suitable sweeteners, of which fructose is one. It is appreciably more sweet than sucrose, but bulking agents may be required to produce an acceptable texture. Among these is polydextrose, which provides only 1 kcal g^{-1}. Polydextrose is tasteless, is easy to use, and dissolves quickly. Diabetic ice may also be produced using polydextrose and an artificial sweetener (i.e., Aspartame). The total caloric value of the final ice is about half of that of a comparable standard ice cream. Another experimental ice for diabetics uses sorbitol, which provides both sweetness and bulk. The disadvantage of using sorbitol is that intake has to be restricted due to its laxative properties (Rothwell, 1998).

6.5.5 Stabilizing Agents

The primary function of stabilizers is to prevent heat shock (Fearon and Moruzzi, 1999). This is a process by which fluctuating temperatures cause the ice crystals in the frozen product to melt and then refreeze. During this process, ice crystals can become mobile and combine (and refreeze) into larger ice crystals. The formation of large ice crystals will cause undesirable texture defects. The stabilizers will absorb free moisture in the frozen ice cream and reduce the amount of water available to participate in the phase changes ice to water and water to ice (Marshall and Arbuckle, 1996).

Stabilizers also give the ice cream "body" without making it too "heavy or gummy" and allow the ice cream to melt down satisfactorily, but relatively slowly. They can also improve the whippability of the mix, giving a drier ice cream, and probably minimize shrinkage problems (Dubey and White, 1997).

To a limited extent, milk proteins act in this way, but additional stabilization is necessary to give a satisfactory product. Gelatin, which is an animal protein, was the first substance to be used in this way. It produces a gel network during a storage or "aging" period of about 4–5 h at 5°C. Milk proteins from added milk powders also require some time to hydrate fully.

However, for modern, high-speed production methods, other stabilizers, which do not require this time for satisfactory hydration, have been introduced. The ice cream manufacturer now has a wide range to select from, and usually a mixture of two or more are employed. Those most commonly used now include sodium alginates, locust bean and guar gums, sodium carboxymethyl cellulose (CMC), carrageenan, xanthan gum, and pectin. Most of these will dissolve reasonably well, but locust bean gum requires a temperature of about 70°C for up to 15 min. It is, therefore, not very suitable for ultra-high-temperature (UHT) or HTST mix processing methods; but guar gum, which is in other ways very similar to locust bean gum, may be used in its place because it dissolves in the cold quite satisfactorily.

Stabilizer manufacturers and suppliers now make mixtures, which not only include a blend of several different stabilisers, but also incorporate an emulsifying agent made in a form that is easy to handle and which can be rapidly incorporated into the mix.

6.5.6 Emulsifying Agents

As will be mentioned later, ice cream mix has to be submitted to mechanical emulsification. This is to ensure that the fat globules are sufficiently small to give a homogeneous and smooth mix, but, at the same time, it greatly increases the surface area of the fat globules. To prevent the fat globules from agglomerating after homogenization, it is necessary to "protect" their surfaces either by a natural membrane (as is present in milk and cream) or by adding some surface-active material.

An emulsifying agent is used for this purpose, and it acts by contributing to the formation of a different membrane. After cooling and aging of the homogenized mix at 4°C the membranes rearrange and, with the emulsifier being adsorbed more strongly, it replaces proteins on the surface. The extent of protein displacement from the membrane, and hence the extent of dryness achieved, is a function of the type and concentration of the emulsifier used (Goff, 1997). Fat globules with adsorbed emulsifiers have an increased susceptibility to destabilization forces existing in the freezer.

The emulsifying agent is thus required to help to maintain the fat globules as individual units. For optimal structure and texture of regular ice cream, approximately 60% of the fat in the emulsion should be destabilized (Marshall and Arbuckle, 1996). The emulsifier also assists in the formation of larger numbers of smaller and more uniform air cells, which helps to produce a smoother ice cream and aid whippability.

However, it is now accepted that emulsifier action is more complex and that during the freezing process in the ice cream freezer, possible stabilization (flocculation) of fat globules may lead to fat churning. To control this it is necessary to use the correct amount of a suitable emulsifier. If too much agglomeration of the fat occurs, this may lead to a very fatty, greasy-tasting product, which will not be fully acceptable; the melt-down characteristics of the product will also be impaired.

Egg yolk was the first emulsifying agent used (having lecithin as the active ingredient), and this was followed by glyceryl monostearate (GMS). Normal GMS has a monoglyceride content of about 40%, but it is possible to obtain specially processed types that contain up to 80% monoglyceride. Other emulsifying agents include polysorbitol esters, but these have never been generally allowed by the health authorities in Europe (Krog, 1998).

6.5.7 Flavoring Materials

Flavorings of natural or synthetic origins are available mainly in mixtures, for the proper flavoring of foods (Marshall and Arbuckle, 1996).

Natural flavorings are derived from citrus and noncitrus fruits, spices, cocoa and chocolate, coffee, and natural flavorings from vanilla beans and nuts. Liqueur flavorings include whiskey and distilled beverages, fruit brandy distillates, and fruit liqueurs.

Vanilla is one of the most popular flavorings used in the US ice cream industry, and it is produced from the beans that are present in the pods of the vine *Vanilla planifolia*. Another widely used flavoring is chocolate and cocoa, which are obtained from the cocoa bean, the fruit of the *Theobroma cacao* tree. Chocolate ice cream, made either from flavoring syrup or from a special mix incorporating cocoa, is quite popular as are chocolate coatings on ice cream cones and bars.

Fruits are also used as flavorings in both ice creams and frozen yogurts. They are available in fresh, frozen, and heat-processed forms. Fresh or fresh-frozen fruit is often the best source of flavor; but fruits high in sugar content, such as strawberries, raspberries, and black currants, which are very commonly employed, may cause a problem with yeast and mold contamination.

A health-conscious public has pushed the industry to reduce the calorific value of ice cream and ingredients such as blueberry chunks, raspberry and strawberry purees, and chocolate flakes, are now sweetened with aspartame.

Nuts are also used as a coating or included in the ice cream, and the most popular are pecans, walnuts, almonds, pistachios, filberts, and peanuts. They must be clean, be in a sound condition, have a low count of microorganisms, and be free of pathogens. Finally, spices used in the frozen dessert industry include cinnamon, nutmeg, cloves, allspice, and ginger.

6.5.8 Colorings

Colorings that are permitted for use in ice cream are "certified" and must be declared on the ingredient labels. The coloring used in ice cream depends on the type of ice cream produced. For example, in chocolate ice creams, extra coloring is not added because cocoa imparts a high color at the concentration of use. On the other hand, fruit ice creams need to be colored because the usual amounts of fruits added do not impart a desirable color to the end product.

6.6 OTHER TYPES OF ICE CREAM

6.6.1 Ice Cream with Different Fat Contents

The ever-increasing demand of health-conscious consumers to lower the daily fat intake has prompted changes in many food products, and a recent trend in ice cream formulation has been to reduce the calorific contribution of fat in the mix. Ice cream made with 25% less fat than the reference ice cream is labeled reduced fat, whereas a light version is also available (50% less fat than the reference ice cream). A reduced-fat, sugar-free ice cream is also available.

Two more products with even lower fat contents are also marketed: (a) low-fat with not more than 3.0g of fat per serving and (b) non-fat with less than 0.5g of fat per serving (Marshall and Arbuckle, 1996). Substitutes for fat in ice cream may include protein mimetics (e.g., Simplesse), while Polyestra is a carbohydrate (sucrose polyester) used to replace all or part of the fat in ice cream.

The challenge for the lower fat products is that they must possess the same or comparable organoleptic and flavor characteristics as ice cream.

6.6.2 Probiotic Ice Cream

The generally widespread use of probiotic bacteria in the production of yogurt has triggered the interest of researchers in the possible addition of such bacteria to ice cream, as well.

Hagen and Narvhus (1999) have recently described the production of ice cream containing probiotic bacteria. Four different probiotic strains (*Lactobacillus reuteri*, *Lactobacillus acidophilus*, *Lactobacillus rhamnosus* "*GG*," and *Bifidobacterium bifidum*) were used for ice cream production, and the end products were evaluated for several sensory attributes. The authors concluded that viable numbers of the probiotic cultures remained above 10^6 cfu g^{-1} during storage of 52 weeks at –20°C, and the ice cream obtained high scores during the sensory evaluation. In a similar study, Hekmat and McMahon (1992) used *L. acidophilus* and *B. Bifidum* to ferment a standard ice cream mix. Their results showed high levels of viable organisms after 17 weeks of frozen storage. Marshall (1998) also reports that Bifidobacteria and *L. acidophilus* could be added as dietary adjuncts to frozen desserts.

The above studies show that ice cream could be used as a good source for delivering probiotic bacteria to the consumers, without affecting the sensory properties of the end product.

6.6.3 Ice Cream Novelties

An emerging sector, prominent during the last 15–20 years, within the wider ice cream market is the "novelties." An ice cream novelty is defined by Stogo (1998), as "a unique single-serve portion-controlled product." Novelties include special combinations of ice cream with flavors and confections, cup items, and fancy molded items (Marshall and Arbuckle, 1996).

They are usually produced by either extrusion or molding, and examples include coated ice cream bars (e.g., Mars), coated ice cream bars on a stick (e.g., Magnum), ice cream cakes, and ice cream logs (e.g., Vienetta). The production of "novelties" is illustrated in Figure 6.1.

6.7 MANUFACTURE OF ICE CREAM

The previous sections highlighted that mix composition and ingredient quality, coupled with compliance to legal requirements, are important prerequisites for the production of high-quality ice cream.

The manufacture of ice cream is certainly a complex operation, and a number of steps are involved (Figure 6.1) which to some extent may affect the microbiological quality of the end product. Although these steps are only briefly described below, Marshall and Arbuckle (1996) provide excellent and detailed information on the subject.

1. Once the formulation of the mix has been decided, the ingredients will be weighted or measured into the mixing vessels. In large enterprises, continuous operations under computer control are employed, and the mix is pumped through a closed system. This approach reduces the costs of handling, decreases the risks of contamination, and makes possible the automated cleaning-in-place (CIP) of the equipment.

2. After mixing, the ingredients are either heat-treated in a batch system (using the same mixing vessel) for small-scale enterprises or pumped to a pasteurization plant for HTST continuous-flow pasteurization (larger-scale enterprises). UHT processing is mainly carried out by steam injection, or in scraped-surface heat exchangers, because the mix may foul the surfaces of plates if heated to temperatures in excess of about 110°C. Actual operating conditions depend on the legal requirements of individual countries.

3. The homogenizer in a batch system is placed right after the pasteurizer, whereas if a continuous system is employed, it is located between the heating section of the pasteurizer and the raw product regenerator. The pressures employed vary with the type and quantity of fat used (Marshall and Arbuckle, 1996). Usually, vegetable fats are treated at lower pressures around 1600 psi, whereas mixes with milk fat are homogenized at around 2200 psi. If two-stage homogenization is used, the pressures for both types of fat are usually 2000 psi, followed by about 500 psi. The formulation of the mix is also important in

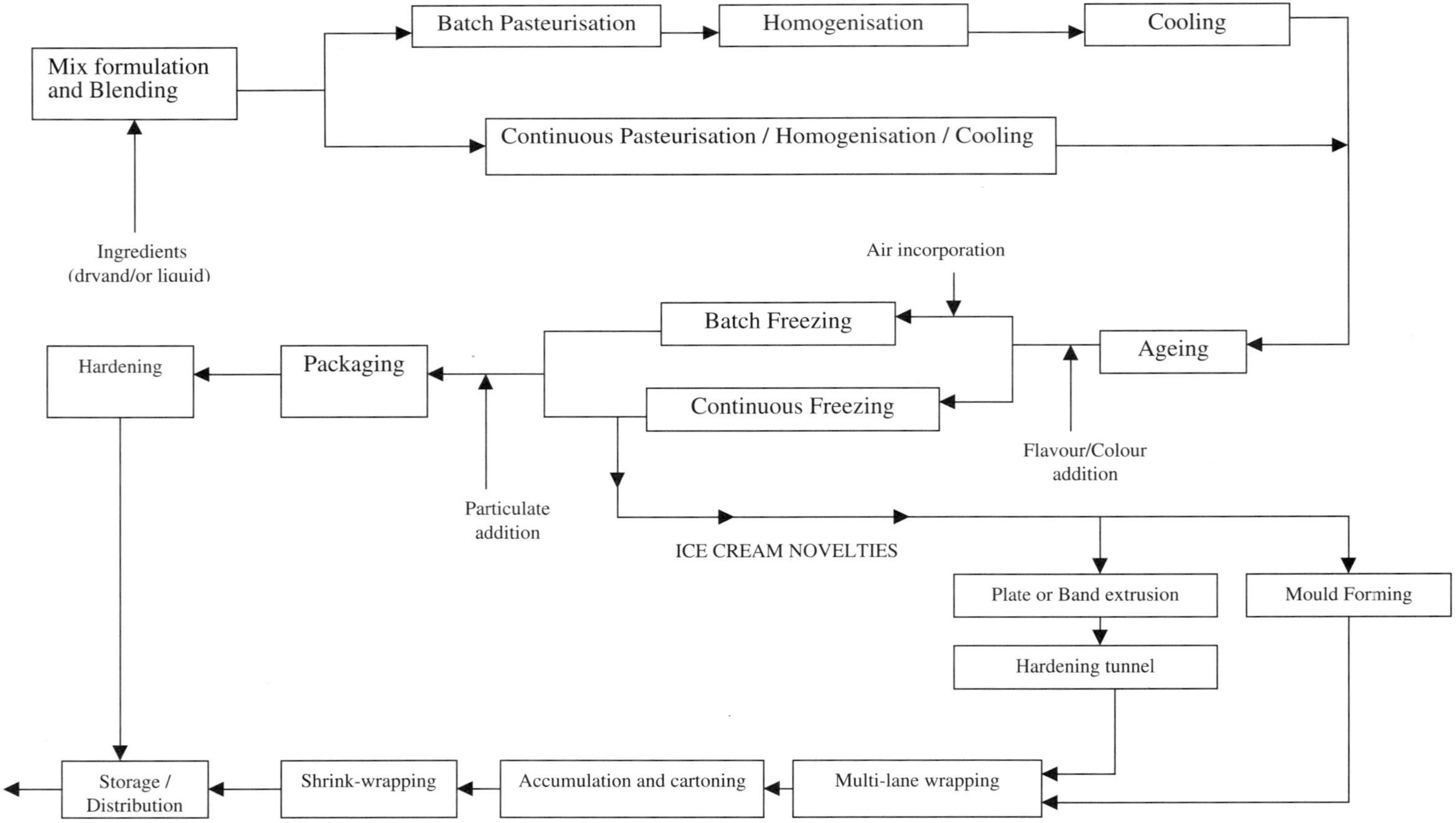

Figure 6.1. Schematic drawing of the major operations involved in the manufacture of ice cream (batch or continuous process) and ice cream novelties.

determining the correct pressure during homogenization. For example, ice cream mixes that contain chocolate require pressures that are about 500 psi higher than those used for plain mixes containing the same amount of fat. Homogenization will reduce the size of the fat globules to less than 2 μm, producing a uniform and stable suspension of fat in the mix. Therefore, homogenization prevents fat separation and reduces fat churning during the freezing process. As a result, the organoleptic characteristics of the finished product are improved by providing the correct body and texture, mouth feel, and appearance. The homogenizer is a complex piece of equipment and must be carefully cleaned and disinfected each time it is used, or the mix may be considerably contaminated. Therefore, it is suggested that homogenization of the ice cream mix is carried out before the mix is finally heat treated wherever this is possible.

4. After pasteurization and homogenization is completed, the mix is rapidly cooled to 2–4°C in order to preserve the bacteriological quality of the mix. In batch systems the mix is pumped over a cooler, whereas in a continuous system, cooling is achieved by contact of the mix with the cold part of a plate-type or scraped-surface heat exchanger (Stogo, 1998). The mixture is then stored for approximately 24 h, and during this period a process known as "aging" takes place; the milk proteins hydrate, the fats begin to crystallize, and any added hydrocolloids absorb quantities of water. A non-aged mix is very wet at extrusion and exhibits variable whipping qualities (Goff, 1997). Normally the mix should be frozen within 24 h of being heat-treated because undue prolongation of storage may lead to proliferation of psychrotrophic organisms with a serious risk of spoilage of the mix.

5. The mix is then passed to the ice cream freezer where its temperature is reduced rapidly, and at the same time the mix is subjected to considerable agitation. Air is incorporated to give an aerated product with overrun (the amount of expansion caused by this incorporation of air) of up to 100%, depending on the freezer and the type of ice cream being produced. Small vertical freezers (batch) (Figure 6.2) normally give overruns of up to about 50%, but a large continuous freezer can produce overruns of up to 100% by the incorporation of air under pressure. Batch-frozen products have larger ice crystals and air cells than the continuous-frozen products because the drawing temperature of ice cream from the batch freezer is slightly higher than that used for continuous freezers.

6. On leaving the freezer, the ice cream will normally be packaged, either in bulk or in smaller-sized "family" packs of 1 liter or less, or in

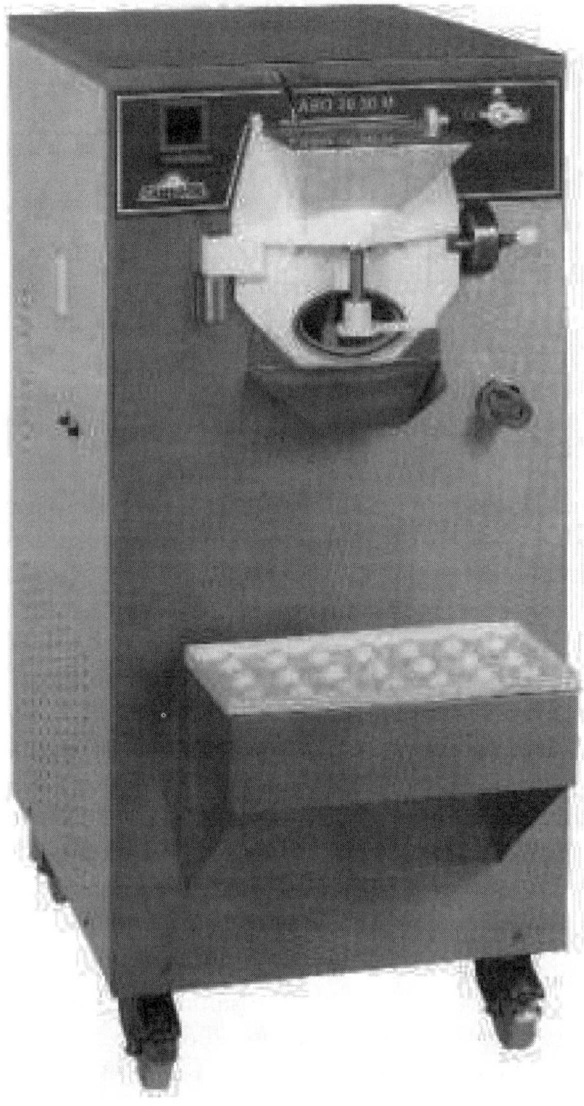

Figure 6.2. A batch freezer. (Reproduced by courtesy of Ingrams Ltd. and Caprigiani SPA.)

individual retail packs. All these are quick-hardened (to avoid the formation of large ice crystals) in wind cabinets at –40°C, in hardening tunnels (Figure 6.3), or in contact plate freezers, packaged as suitable, and then kept at a temperature of about –30°C until, and during, dis-

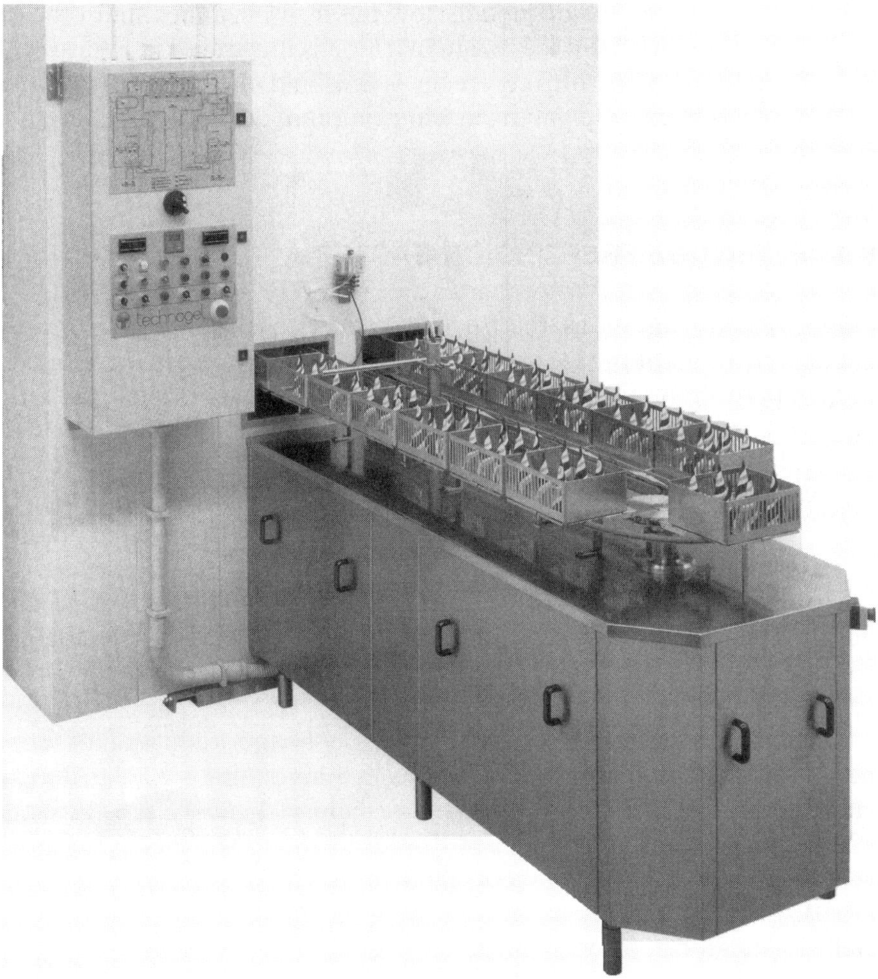

Figure 6.3. The Technogel hardening tunnel. (Reproduced by courtesy of Alfred & Co. Ltd.)

tribution. Hardening of ice cream has been accomplished when the temperature in the center of the package drops to −18°C. The process may take from 30 min to 24 h according to the package quantity and type of hardening facility.

Some ice cream is sold direct from a dispensing freezer as "soft-serve" ice cream either on cones or in cups. There are several types of soft-serve freezers available—that is, floor and counter-top models, single- and multiple-flavor models, low-overrun machines, and others. A "self-pasteurizing" soft ice cream machine is illustrated in Figure 6.4. The fat content of a soft ice cream is 7% and the MSNF is raised to 12% to prevent the fat from churning-out during whipping and freezing.

6.8 EFFECT OF FREEZING ON BACTERIA

Under certain conditions, the freezing process has some effect on the bacterial content of the ice cream. In the freezer, as already mentioned, the temperature of the mix is rapidly reduced while air is incorporated, accompanied by vigorous whipping of the ice cream. Both intra- and extracellular ice crystals can mechanically injure frozen cells. Ice crystals that form outside the cell reduces the amount of free water in which solute can be dissolved. Those that form inside the cells have the potential to puncture cell membranes.

However, in a continuous freezer, where very rapid freezing occurs during a short period of agitation, the numbers of survivors fluctuated considerably, but were generally much greater than in the other types of freezer.

Once the ice cream leaves the freezer, any further reduction of temperature occurs in quiescent conditions in a hardening tunnel or hardening room. The ice cream will then be held at temperatures around −20°C or lower until it is sold, and this period has been shown to cause very little change in the numbers of bacteria, even after periods of many months. Pathogenic organisms, amongst many different types artificially inoculated before the mix was frozen, have been shown to survive for years.

Dean and Zottola (1996) investigated the survival characteristics of *Listeria monocytogenes* in full fat (10% fat) and reduced fat (3% fat) ice cream. They found that the viable cell count of *L. monocytogenes* remained constant throughout manufacture and 3 months storage at −18°C. They concluded that the lactose and sucrose in the ice cream acted as cryoprotectants by depressing the freezing point of the suspension media, in addition to protection by other milk components and the viscous nature of ice cream.

Therefore, it is essential that the bacteriological content of the ice cream from the freezer be as low as possible, because neither the final hardening process nor the low-temperature storage can be relied upon to reduce the numbers appreciably; pathogenic organisms should be absent.

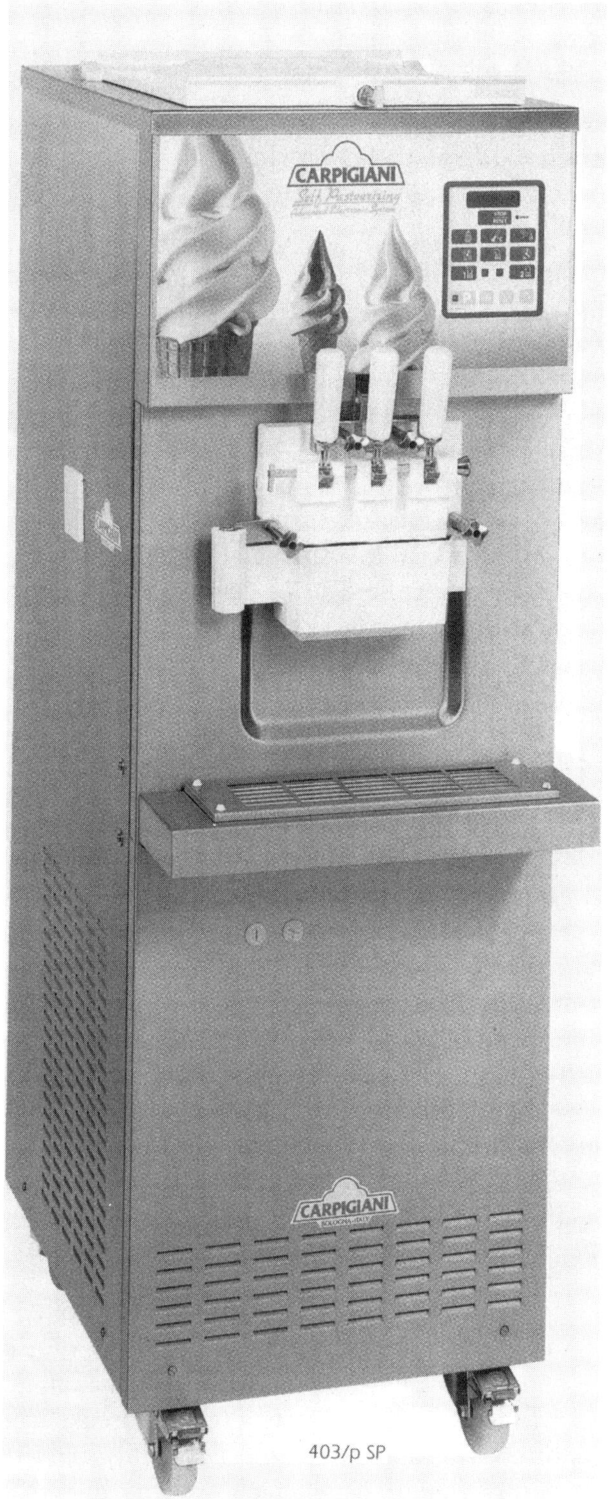

Figure 6.4. A self-pasteurizing soft-serve ice cream freezer. (Reproduced by courtesy of Ingrams Ltd. and Caprigiani SPA.)

6.9 ICE CREAM AS A CAUSE OF FOOD-BORNE DISEASES

6.9.1 Disease Outbreaks—Historical

In the United Kingdom there were several outbreaks of food poisoning before the Second World War, but, because there were no heat treatment requirements at that time, the significant cases are those that occurred later. Hobbs and Gilbert (1978) give examples of two outbreaks of food poisoning in the United Kingdom which have been directly attributed to the consumption of ice cream. One, in 1945, was due to staphylococci, carried by a worker in the cookhouse of an army hospital, which were introduced into batches of ice cream mix after the ingredients had been cooked. The mix was allowed to cool slowly overnight and was frozen 20–30h later. Around 700 people were affected by a staphylococcal toxin, which developed during this period.

A major outbreak of typhoid fever occurred during the summer of 1947 in Aberystwyth, soon after ice cream manufacture was allowed again after several years of being prohibited during the war (Evans, 1947). About 210 cases were reported, including four deaths, and ice cream from one particular source appeared to be the common factor. It was found that the ice cream manufacturer was an active urinary excretor of typhoid bacteria (*Salmonella typhi*), and specimens from him and from patients were shown to be of the same phage type. The manufacturer had been ill with typhoid fever in 1938, but on recovery he had been declared free of the infection.

Other cases reported by Hobbs and Gilbert (1978) include a paratyphoid incident at a North Devon holiday resort, where the cause was found to be small doses of bacteria (*S. paratyphi*) on the hands of the ice cream vendor and his wife. A case of *Shigellae* dysentery is also mentioned, caused by a monkey (in the pets' corner of a large store) that touched an ice cream cornet being eaten by a child. *Shigella flexneri* 103Z was isolated from a stool specimen after the child became ill, and it was also found in a specimen from the monkey. The infective dose in this case was probably that transferred by the animal's paw, and thus there had been little or no time for an increase in the number of organisms.

Between 1950 and 1955, there were 11 outbreaks of food-borne disease in the United Kingdom due to the consumption of ice cream and ice lollies involving salmonellae, two involving staphylococci and six of unknown cause. Since 1955, only a few outbreaks from ice cream have been reported. This decline is almost certainly due to the fact that the heat treatment regulations, which were first introduced in 1947,

TABLE 6.5. Recent Outbreaks of Food-Borne Disease Caused by the Consumption of Ice Cream

Type of Frozen Dairy Product	Year	Country of Manufacture	Pathogen
Ice cream	1989	Belgium	*Listeria monocytogenes*
Ice cream	1992	South West England	*Salmonella enteritidis*
Ice cream	1993	Wales	*Salmonella enteritidis*
Ice cream products	1994	Minnesota, United States	*Salmonella enteritidis*

became really effective from 1950 onwards. In addition, inspections by local health authorities have played a major role in improving the hygienic quality of ice cream and related products.

6.9.2 Recent Disease Outbreaks

One of the most serious outbreaks of food-borne infection during the last 20 years was attributed to *Salmonella enteritidis* contamination of ice cream products (see Table 6.5) manufactured in the United States. Epidemiologists from the Minnesota Department of Health calculated that 29,100 Minnesotans became ill after eating ice cream and up to approximately 224,000 people in the United States were infected (Vought and Tatini, 1998). Investigations revealed that cross-contamination of a pasteurized ice cream premix occurred during transport in tanker trailers that had previously hauled nonpasteurized liquid eggs containing *S. enteritidis*.

According to Djuretic et al. (1997), 2 out of 20 general outbreaks associated with the consumption of milk and dairy products between 1992 and 1996 in England and Wales were reported as associated with ice cream. *Salmonella enteretidis* PT4 infection was the cause in both cases. In the first case (September 4, 1992 in South West England), a commercial ice cream served in a hotel was probably contaminated and stored at the wrong temperature; the second one (February 6, 1993 in Wales) was the result of using raw shell eggs in the manufacture of ice cream. For the second outbreak, it has been suggested that because the attack rate was high and the incubation period relatively short, a large inoculum of *Salmonella* was the cause (Dodhia et al., 1998).

The higher severity of illness from the multistate outbreak in the United States as compared with the English outbreak, because both cases involved *Salmonella enteritidis* in ice cream, was possibly due to the virulence characteristics of the different strains. The phage type involved in the US outbreak was type 8, while type 4 was found in the homemade ice cream in England. Vought and Tatini (1998) found that the former page type caused illness with a very low number of cells (no more than 25).

Finally, a sporadic case of listeriosis has been reported in Belgium (Andre et al., 1990). A 62-year-old man, apparently immunocompromised, had consumed ice cream containing 10^4 cfu g^{-1} of *Listeria monocytogenes* serovar 4b.

6.10 OCCURRENCE OF PATHOGENS IN ICE CREAM

Although the reported cases of food-borne diseases appear to have been limited over the last 20 years, the presence of potentially pathogenic bacteria in ice cream and related products is not rare.

During the FDA Dairy Initiative Microbiological Surveillance Program (1986–1987), *Listeria monocytogenes* was isolated from 3.03%, 3.03% and 8.55% of samples of ice milk, ice cream, and ice cream novelties, respectively (Ryser, 1999).

Consequently, a large number of voluntary recalls of dairy products possibly contaminated with *Listeria monocytogenes* were issued in the United States after the FDA DIMS Program. More than 10,200 tonnes (48 recalls) of ice creams, ice milks, sherbets, ice cream novelties, and frozen yogurts were recalled from July 1986 to March 1998 (see Table 6.6).

A national survey in England and Wales for the presence of *Listeria* spp., carried out by the Public Health Laboratory Service (PHLS), showed that the occurrence of *L. monocytogenes* in ice cream was 2%, whereas *Listeria* spp. other than *L. monocytogenes* were found in 1.3% of the ice cream samples (Greenwood et al., 1991); all isolations were obtained by enrichment cultures only.

In another study, a total of 2612 edible ice samples were examined by PHLS laboratories under the 1993 European Community Coordinated Food Control Programme (ECCFCP) (Nichols and de Louvois, 1995). The results showed that edible ices sold in England and Wales were generally of good microbiological quality; 100% of the branded ice cream bars and cones and over 99% of the ice lolly and sorbet samples met the microbiological targets for hygienic manufacture (Ice Cream Alliance and Ice Cream Federation, 1995). In addition, 90% of hard ice cream and 77% of soft ice cream had satisfactory aerobic plate counts (APCs). Both hard and soft ice cream samples were found to be *Listeria monocytogenes* positive (0.2%); the detection of *Salmonella* spp. in one of the 1000 hard ice cream samples (0.1%) suggests that

TABLE 6.6. Reported Recalls Issued in the United States of Frozen Dairy Products Contaminated with *L. monocytogenes*

Type of Frozen Dairy Product	Year of Recall	State of Manufacture	Quantity
Ice cream bars	1986	Virginia	Large but unknown
Ice cream, sherbet	1986	Wisconsin	30,280 liters
Ice cream, ice milk, sherbet and gelati products	1986	Minnesota	~3.7 million liters
Ice cream and ice cream novelties	1986	Iowa	Unknown
Ice cream	1986	New York	~3,160 liters
Ice cream	1986	West Virginia	1,703 liters
Ice cream, ice milk, sherbet	1987	Iowa	~3.7 million liters
Ice cream	1987	New York	~1,196 liters
Ice cream, ice milk, sherbet	1987	California	~227,100 liters
Ice cream nuggets	1987	Maryland	20,400 boxes
Ice cream, ice milk	1987	Nebraska	~113,550 liters
Chocolate ice cream	1987	Kentucky	~3,652 liters
Ice cream nuggets	1987	Maryland	Unknown
Ice cream novelties	1987	Ohio	Unknown
Ice cream bars	1987	Ohio	51,780 bars
Ice cream, sherbet, ice milk	1988	Ohio	>4,099 liters
Ice cream pies	1988	Connecticut	1,700 pies
Ice cream	1988	Pennsylvania	813 liters
Ice cream bars	1988	New York	~465 liters
Ice cream bars	1989	Wisconsin	~1,381 liters
Ice cream bars	1990	New Mexico	~3,785 liters
Ice cream	1990	Ohio	~2,945 liters
Ice cream novelties	1991	Tennessee	Unknown
Ice cream, ice milk	1991	Illinois	>8,611 liters
Ice cream, ice milk	1992	Wisconsin	~4,258 liters
Ice cream bars	1993	Ohio	34,752 bars
Ice cream bars	1994	Wisconsin	632 liters
Ice cream, frozen yogurt, sherbet, sorbet, ice cream mix	1995	Ohio	~1,589,700 liters
Ice cream	1995	California	>1,892 liters
Frozen yogurt	1995	Ohio	~46,862 liters
Ice cream, sherbet	1996	Ohio	~10,840 liters
Frozen strawberry yogurt	1997	Pennsylvania	~159 liters
Ice cream bars	1997	Texas	29,814 cases
Ice cream	1998	North Carolina	492,050 liters

Source: After Ryser (1999).

commercially produced ice cream may still occasionally be a potential cause of food-borne disease (Nichols and de Louvois, 1995).

More recently, 1246 soft ice creams from fixed premises and mobile vendors were examined by 39 PHLS and 11 non-PHLS laboratories in England, Wales, Scotland, and Northern Ireland between August and October 1997 (Little and de Louvois, 1999). This study showed that 74% of the soft ice cream samples were of acceptable microbiological quality. However, 26% of the samples, mainly from mobile vendors, were classified, using published guidelines, as unsatisfactory (Little and de Louvois, 1999). *Salmonella* spp. were not found in any of the 1219 samples examined, but *L. monocytogenes* was detected in 10 out of 1204 soft ice cream samples. It should be noted that no recall of ice cream and other frozen dairy products has been reported in the United Kingdom.

In a surveillance of the incidence of pathogenic psychrotrophs in ice creams in Mumbai, India, all samples were found to be free from *Salmonella*, whereas 53% of the packed and 100% of the open ice creams exhibited *Listeria* contamination (Warke et al., 2000). In Australia, 23 of 166 samples (13.9%) of ice cream samples were found to be positives for *L. monocytogenes*; the incidence was 2% in Costa Rica and 10% in Turkey (Ryser, 1999).

6.11 MICROBIOLOGICAL STANDARDS

Due to the frequent presence of *Listeria* spp., a considerably high number of recalls of ice cream and related products has been reported in the United States during recent years. This is probably due to the fact that the US Government has adopted the most stringent policy regarding the presence of *L. monocytogenes* in ice cream: zero tolerance.

France, Germany, and the Netherlands accept up to 100 cfu g^{-1}. Denmark has established a zero tolerance for foods that have received a listericidal treatment after packaging: If >10 cfu g^{-1} are found, corrective HACCP actions are required; if the level is >100 cfu g^{-1}, the product is recalled. Italy has a zero tolerance policy, whereas in the United Kingdom, <100 cfu g^{-1} of *L. monocytogenes* is considered fairly satisfactory, whereas 100–1000 cfu g^{-1} is unsatisfactory and >1000 cfu g^{-1} is unacceptable (Gravani, 1999). In Canada, the criteria for *L. monocytogenes* in ready-to-eat foods are composed of three categories: Category 1, ready-to-eat foods causally linked to listeriosis (Action level >0 cfu 50 g^{-1}); Category 2, all other ready-to-eat foods supporting growth of *L. monocytogenes* with refrigerated shelf life of >10 days (Action level >0 cfu $25\,g^{-1}$); Category 3, ready-to-eat foods supporting growth of *L. monocytogenes* with refrigerated shelf life of <10 days and all ready-to-eat foods not supporting growth (Action level ≤100 cfu g^{-1}), depending on the plants Good Manufacture Practice (Farber and Harwig, 1996).

Recently, the EC have adopted microbiological criteria for milk-based products, including ice cream, in the Milk and Milk Products Directive (see Table 6.7).

A microbiological criterion for foodstuffs defines the acceptability of the process, product, or food lot based on the absence or presence, or number of microorganisms and/or quantity of their toxins/metabolites, per unit(s) of mass, volume, or area (Anonymous, 1992). Two main categories of microbiological criteria are generally applied:

1. *Microbiological Standards.* Mandatory criteria included in the legislation or regulations where failure to comply with them can result in rejection of the food.

TABLE 6.7. Microbiological Criteria for Ice Cream and Related Products Suggested by the EU Committee

	Compulsory Criteria: Pathogenic Microorganisms			
Listeria monocytogenes	Absent in 1 g			
Salmonella spp.	Absent in 25 g[a]/$n = 10$, $c = 0$			
	Analytical Criteria: Organisms Indicating Poor Hygiene			
	m	*M*	*n*	*c*
Staphylococcus aureus	10	100	5	2
	Indicator Organisms: Guidelines			
	m	*M*	*n*	*c*
Coliforms 30°C	10	100	5	2
Plate count	10^5	5×10^5	5	2

[a]The 25-g sample to consist of five specimens of 5 g taken from different parts of the same product.
n = number of sample units in the sample.
c = number of sample units where the bacterial count may be between m and M.
m = threshold value; the result is satisfactory if the number of bacteria in all sample units does not exceed m cfu g^{-1}.
M = maximum value for the number of bacteria; the result is unsatisfactory if the number of bacteria in one or more units exceeds M cfu g^{-1}.
Source: After Anonymous (1992).

2. *Microbiological Guidelines.* Criteria included in legislation or regulations which are intended to guide the manufacturer and help to ensure good hygienic practice.

Members of the EC have been required to incorporate this legislation into their national regulations since January 1, 1994.

The Scientific Committee on Veterinary Measures Relating to Public Health (1999) examined the EC microbiological criteria and whether they should be retained, modified, or deleted.

The Committee suggested that there should be:

1. Separate mandatory (e.g., pathogens) and guideline (e.g., indicator) criteria
2. A limit on the number of criteria that are not relevant to consumer protection
3. Better uniformity

For the frozen dairy products, including ice cream, the Committee proposed deletion of the criteria for pathogenic bacteria. Furthermore, they proposed deletion of the plate count at 30°C and replacement of the coliform test with a test for total *Enterobacteriaceae*.

Accordingly, the IFST Professional Food Microbiology Group has published an extended discussion of the development of microbiological criteria for foods (Anonymous, 1997); a division into two categories is proposed (see Table 6.8).

TABLE 6.8. Microbiological Criteria Suggested by the IFST for Dairy Products, Including Ice Cream

Organism	GMP	Maximum
	Pathogens	
Salmonella spp.	ND in 25 ml/g	ND in 25 ml/g
Listeria monocytogenes	ND in 25 ml/g	10^3 cfu g^{-1}
Staphylococcus aureus	<20 cfu g^{-1}	10^3 cfu g^{-1}
	Indicator and Spoilage Organisms	
Coliforms/Enterobacteriaceae	<10 cfu g^{-1}	10^4 cfu g^{-1}
Escherichia coli	<10 cfu g^{-1}	10^3 cfu g^{-1}

GMP, target expected immediately following production under good manufacture practice.
ND, not detected.

Source: After Anonymous (1997).

In the United States, there are no federal standards for the APC, but some health authorities set a maximum of 50,000 cfu g^{-1} (Marshall and Arbuckle, 1996). For coliform counts, the standards are almost uniformly at a maximum of 10 cfu g^{-1} for both pasteurized ingredients and the finished ice cream product.

In countries where a maximum count is applied, problems can arise if a sample contains more than the "permitted number" of organisms, particularly in borderline cases, as to whether the sample should be condemned or not. Normally, such ice cream should be subjected to further investigation.

In addition to the microbiological criteria for the final product, criteria for surfaces in contact with the product (using swabs) have been published by Ice Cream Alliance and Ice Cream Federation (1995) as shown in Table 6.9.

6.12 MICROBIOLOGICAL QUALITY OF FROZEN DAIRY PRODUCTS

Published reports on the bacteriological conditions of ice cream in some countries show that there is a wide variation both in the microbiological quality of the products examined and in the overall standards of hygiene. In addition, wide variation exists in the microbial populations of different frozen dairy products and also within a single product.

TABLE 6.9. Microbiological Criteria for Surfaces in Contact with Food Products

	m	*M*	*n*	*c*
APC	10	100	5	2
Coliforms	0	10	5	2
Escherichia coli	0	0	5	0
Staphylococcus aureus	0	0	5	0
Listeria monocytogenes	0	0	5	0

n = number of sample units in the sample.
c = number of sample units where the bacterial count may be between *m* and *M*.
m = threshold value; the result is satisfactory if the number of bacteria in all sample units does not exceed *m* cfu 100 cm^{-2}.
M = maximum value for the number of bacteria; the result is unsatisfactory if the number of bacteria in one or more units exceeds *M* cfu 100 cm^{-2}.

Source: After Ice Cream Alliance and Ice Cream Federation (1995).

A study of 100 samples of ice cream in India showed poor bacteriological quality with a high incidence of *Staphylococcus aureus* contamination (Ramakrishnan et al., 1986). Only 10% conformed to the standard for total count (25×10^5 cfu g^{-1}) and coliforms (less than 90 cfu g^{-1}), and all samples had more than the limit of *S. aureus* (100 cfu g^{-1}). Also in India, Shrestha and Sinha (1987) found that 33 out of 43 samples of ice cream had coliform counts in excess of the Indian specification, and they stressed the need for greater care to prevent contamination with coliforms during processing. Similarly, the results of a more recent study showed that a considerable number of ice cream samples (55–81%), marketed in Tirupati, India, exceeded the standard of 90 cfu g^{-1} for coliform counts (Reddy et al., 1994).

In addition, the coliform counts of ice cream samples in Mumbai, India were found to be 10- to 100-fold higher than the safety limits prescribed by the Indian Standards Institute (90 cfu g^{-1} for coliforms) (Warke et al., 2000). While *Salmonella* spp. were not detected in any of the 30 samples tested, 53% of the packed and 100% of the open ice creams exhibited *Listeria* contamination.

The bacteriological quality of soft-serve mixes and ice cream in Louisiana was investigated by Ryan and Gough (1982) over a 21-month period. Two hundred and fifty-two samples of mix were tested, and 10.71% were found to contain more than 50,000 cfu g^{-1}, and 7.54% contained more than 10 coliform cfu g^{-1}. The ice creams tested numbered 817, out of which 38.55% contained more than 50,000 bacteria g^{-1}, and 51.22% contained more than 10 coliforms g^{-1}. It would appear from this investigation that the soft-serve or dispensing freezer is relatively difficult to maintain in a satisfactory hygienic condition.

In a more recent study, carried out by the PHLS, a total of 2612 edible ice cream and ices were examined. It was found that 6% had APC $\geq 10^6$ cfu g^{-1}, while less than 1% of samples contained *Staphylococcus aureus* at between 10^2 and 10^4 cfu g^{-1} and *Bacillus cereus* at between 10^2 and 10^3 cfu g^{-1}. *Escherichia coli* was detected in 3% of the samples and was present in excess of 10^2 cfu g^{-1} in 1% (Nichols and de Louvois, 1995).

In addition, Little and de Louvois (1999) reported the microbiological quality of soft ice cream (1246 samples) in the United Kingdom. Some soft ice creams (41%) had APC $<10^4$ cfu g^{-1}, *Enterobacteriaceae* <10 cfu g^{-1}, and *Staphylococcus aureus* <20 cfu g^{-1}, 33% had APC at 10^4–10^5 cfu g^{-1}, *Enterobacteriaceae* at 10–10^4 cfu g^{-1}, and *S. aureus* at 20–10^2 cfu g^{-1}, and 26% had APC $\geq 10^5$ cfu g^{-1}, *Enterobacteriaceae* $\geq 10^4$ cfu g^{-1}, and *S. aureus* at 10^2–10^4 cfu g^{-1}; 1% were found to be unacceptable and/or potentially hazardous due to the presence of *S. aureus* in excess of 10^4 cfu g^{-1}.

Although the microbiological quality appeared to vary widely between the samples, a significant proportion of the samples tested were found to be unsatisfactory.

Therefore, it is indicated that the hygienic practices applied to the manufacture of these products continue to give some cause of concern, despite the fact that outbreaks of food poisoning or food-borne infection caused by ice cream and related products are nowadays very rare.

6.13 FACTORS THAT AFFECT THE MICROBIOLOGICAL QUALITY OF ICE CREAM

Many factors affect the microbiological quality of the finished product, including the relatively complex manufacturing process and the equipment that are used for the processing. These factors include those described in the following three sections.

6.13.1 The Raw Materials

There should be very few problems caused by high numbers of organisms being introduced into the mix by the raw materials, provided that they have been handled and stored under satisfactory conditions—for example, frozen or refrigerated as appropriate. Liquid milk, cream, and skim-milk concentrate should have been subjected to adequate heat treatment by the supplier and, if kept under refrigeration and used promptly, should be quite satisfactory. If processing is being carried out on a farm, the milk should be that produced at the latest milking. Skim-milk powder may, on occasion, contain numbers of *Bacillus cereus*; and, although this is not a major health hazard, it is preferable that the numbers are as low as possible. Bacilli reduce methylene blue rapidly and can produce clotting in milk; and because they can grow at relatively low temperatures, they may, in extreme cases, cause spoilage of the mix. In addition, *Listeria monocytogenes* may survive the typical spray-drying process (Doyle et al., 1985), and thus dried milk may serve as a source of *Listeria*.

Granulated sugar, as well as other dry sugars such as dextrose, should be almost sterile, and the only organisms that may normally be present are small numbers of yeasts. Sugar syrups, whether sucrose, corn syrups, or mixtures of these, or lactose and whey syrups, again may contain some yeasts, especially when the concentration of solids in the syrup is low. It must be remembered, however, that osmophilic yeasts may be able to grow in these syrups, and molds may grow on the surface if contamination should occur. It is suggested that tests for yeasts should be made on bulk deliveries of sugar and sugar syrups. Typical manufacture's maximal standards for microorganisms in syrups are: APC, 100 cfu g^{-1}; yeasts, 20 cfu g^{-1}; molds, 20 cfu g^{-1}; *Escherichia coli* absent in 30 g and *Salmonella* absent in 100 g (Marshall, 1998).

Butter and anhydrous milk fat (butter oil) are products made under careful control, from cream that has been heat treated at a relatively high temperature. Therefore, a very high microbiological quality is to be expected, and spoilage is usually the result of chemical changes producing rancid and other off-flavors. Tests for yeasts, molds, mesophilic bacteria, coliforms, and the presence of lipophylic organisms should be curried out, however. In particular, the presence or absence of *Pseudomonas fragi* should be noted, as this organism can cause unpleasant taints in butter.

Butter should preferably be stored at a temperature of no more than –20°C, and, as for all the other ingredients, careful stock control should ensure the proper rotation.

Vegetable fats are normally made by processes that involve high temperatures, particularly during the refining and deodorizing processes. They contain almost no moisture and, therefore, should contain very few organisms indeed.

Stabilizers and emulsifiers should not present any problems; but gelatin, as an animal product, may be a hazard and, like all the other ingredients that are used in ice cream manufacture, should be obtained from a reputable supplier and be kept quite cool and dry. In addition, it is likely that certain stabilizers increase the heat resistance of thermotolerant bacteria (Holsinger et al., 1992).

Many other foodstuffs are added to ice cream, either mixed with it or as coatings. These include fruits [either canned, fresh, or frozen (in concentrated sugar syrup)], nuts, chocolate, broken biscuit, colors, and flavors. Most of these should be of a satisfactory microbiological standard, particularly canned fruits, but fresh and frozen fruits may contain yeasts; nuts may be infected with molds (with the risk of aflatoxin formation); and desiccated coconut may be a hazard because it can be contaminated by salmonellae and should be heat-treated. The examination of these materials should include a visual inspection and the enumeration of mesophilic bacteria, coliforms, yeasts, and molds.

Eggs are usually added for their emulsifying and stabilizing properties; but because raw eggs can be contaminated with salmonellae and other pathogens, only pasteurized eggs must be used if they are added after the heat treatment of the ice cream mix. The pasteurized egg products should meet the following standards: APC <10,000 cfu g^{-1}, coliforms <10 cfu g^{-1}, yeast and molds <10 cfu g^{-1}, and *Salmonella* spp. absent in 25 g.

Colors may become infected by careless handling, and this must be avoided by maintaining good management control. Flavors are normally added to the mix after it has been heat treated, and so they must also be handled with great care to avoid contamination. In common with all the other ingredients, they should be stored in cool, dry conditions. Some additives (e.g., cinnamon) may have antimicrobial properties against food-borne pathogens.

Air that incorporated into the ice cream may be a source of contamination. Therefore, it is of crucial importance that the air be processed (i.e., filtered) so as to ensure that no contamination can be attributed to the introduction of air.

Finally, the packaging material should be obtained from a reputable supplier and subsequently handled and stored under appropriate conditions.

6.13.2 The Efficiency of Processing

Because ice cream mix is a heat-treated product, its quality will depend on the manner in which the processing equipment is operated and maintained. The heat treatment applied to the ice cream mix (e.g., 71.7°C for at least 15 min) is more severe than the pasteurization of the milk (at least 71.7°C for 15 s); as a consequence, the destruction of the microflora in the mix is more severe too. Most countries have legal requirements for heat treatment conditions (see Table 6.10), and these must, obviously, be adhered to. Holder processing in vats has a greater safety margin than short-time processing, but it is essential that the time of holding be strictly observed in all cases. In many designs of equipment this is under manual control, so management must ensure that the time and temperature are being monitored. In the case of continuous-operation plate heat exchangers, the time depends on the design of the plant and the rate of flow of the mix. This should be set on installation and must not be altered.

The whole process, including the rate of cooling of the mix to below 7.2°C and keeping it below that temperature until it is frozen, must be carefully monitored by management. Recording thermometers must be installed not only on the processing equipment but also on the storage vats and must be regularly checked for accuracy.

TABLE 6.10. Microbiological Tests for the Microbiological Examination of Ice Cream and Related Products

Microbiological Test	Method
Aerobic Plate Count at 30°C	IDF Standard 100B, (IDF, 1991) BS 4285, Section 2.1 (BSI, 1990b)
Enumeration of coliforms	IDF Standard 73A (IDF, 1985b) BS 4285, Section 3.7 (BSI, 1987a)
Enumeration of *Escherichia coli*	IDF Standard 170A, Part 1 (Most Probable Number Technique) (IDF, 1999) IDF Standard 170A, Part 2 [Most Probable Number Technique Using 4-Methylumbelliferyl-β-D-glucoronide (MUG)] (IDF, 1999) IDF Standard 170A, Part 3 (Colony Count Technique at 44°C using membranes) (IDF, 1999)
Enumeration of yeasts and molds	IDF Standard 94B (IDF, 1990a) BS 4285, Section 3.6 (BSI, 1990b)
Enumeration of *Staphylococcus aureus*	IDF Standard 145 (IDF, 1990c) BS 4285, Subsection 3.10.1 (BSI, 1990e)
Detection of *Salmonella* spp.	IDF Standard 93A (IDF, 1985a) BS 4285, Subsection 3.9.1 (BSI, 1987b) BS 4285, Subsection 3.9.2 (BSI, 1992) (Screening method using electrical conductance)
Detection of *Listeria monocytogenes*	IDF Standard 143 (IDF, 1990b)
Methylene Blue Test	BS 4285, Section 5.2 (BSI, 1989)

Ice cream mix has been shown (e.g., Holsinger et al., 1992) to give greater protection to microorganisms than milk. These authors found that the *D* values at 54.4°C of *Listeria monocytogenes* in ice cream mix were approximately four to six times (depending on the concentration of the stabilizer) those obtained in milk. This further stresses the importance of following the legal requirements covering the time and temperature of the heat treatment process used.

6.13.3 Hygiene

In addition to the careful operation of the equipment, it is most important that the plant be properly cleaned and disinfected, in accordance with the relevant documentation. This is essential, not only to fulfill the legal requirements but also to increase the efficiency of the equipment and reduce the risk of contamination of the product. Not only is the processing equipment important, but ancillary equipment, in particular at the final sales point, must be kept in a satisfactorily hygienic condition.

Any contamination that may occur during the handling of ingredients, packing materials, and the product during processing and distribution must be eliminated, or kept to the very minimum. Occasional sources of trouble are the packaging materials, but these are initially produced to a very high standard; if problems arise, it is usually because they have not been handled and stored properly.

Probably the most important and dangerous source of contamination is the operative. Ice cream mix will be heat-treated in enclosed equipment and will be frozen with the very minimum of handling. Then, during the final stages of packaging and at the retail selling point, it will be subjected to many opportunities for human contamination. The personal hygiene and habits of all in the factory (whether it is large or small) and at sales points must be above reproach. Education, in addition to medical inspection, is necessary; and if an operator is unwell for any reason, he or she must not be allowed to work without full medical clearance. Many major food poisoning outbreaks have been caused by human contamination. Similarly, all animals, such as birds—and, in particular, rodents and insects—must be rigorously excluded. A contract for the control and eradication of these by a reputable firm is strongly recommended. "Pet" animals, such as dogs and cats, similarly have no place in a food production factory.

The method of sale has a major bearing on the amount of contamination to which the product is subjected. Ice cream which is sold prepacked in single retail portions, and which has only to be handled by the consumer during unwrapping, should have the least contamination of all. In complete contrast to this is the sale of ice cream in cones, as well as other individual portions scooped from bulk ice cream. Here there is a possibility of considerable contamination, unless all the equipment used (servers, etc.), the method of dispensing, and the personal hygiene of the operator are all of a very high standard. Wherever possible, the scoops and other serving equipment should be allowed to stand in a stream of running, pure, cold water, so that any ice cream residues are quickly removed. In the case of running water not being available, the scoop water and disinfectant solution should be changed at least once every hour, as recommended by the Milk Marketing Board (United Kingdom) (Wilson et al., 1997).

Soft-serve ice cream, which is sold directly from a special dispensing freezer, may also prove a hazard unless the most stringent precautions are taken. However, there are now special self-pasteurizing dispensing freezers available, which eliminate the need for daily cleaning and disinfecting (see Section 6.7). In these, at the end of each day's operation, or other convenient time, the freezer is placed in the "self-pasteurize" mode, and all the mix and every part of the freezer that can come into contact with ice cream or mix is raised to a temperature above that required for pasteurization of the mix and is held at that temperature for the legally required time. The freezer and its contents are then cooled rapidly to about 4°C and held at that temperature. Tests have shown that there is little or no increase in the bacterial content of the product over a period of more than a week, and that there is little or no discernible flavor change over this period. Various safeguards are incorporated; for example, if power is cut off for any reason during the "self-pasteurization" cycle, the whole cycle will recommence from the very beginning.

The equipment that is most difficult to clean and disinfect includes: (a) the freezers, although the latest and largest are now designed for cleaning-in-place (CIP); (b) homogenizers, which have many parts and intricate passages for the mix to flow through, though here again the modern types are suitable for CIP; and (c) mix holding tanks and the tanks used for the storage of raw materials.

A strict cleaning and disinfecting regime for the freezer must be instituted and performed thoroughly each day. The sequence of operations for the cleaning (removal of residues of fat, sugar, milk solids, etc.) and disinfecting (removal of microorganisms) of other equipment will differ in detail according to the type and size of the equipment. One sequence that is often adopted on large plants is as follows:

1. The plant is thoroughly rinsed with warm water until no more ice cream appears to be being flushed out.
2. The plant is then dismantled.
3. Washing with detergents and physical methods of cleaning as appropriate.
4. Reassembling.
5. The plant is rinsed with clean water; the temperature depends on the type of equipment.
6. Disinfecting.
7. Final rinsing, draining, and then leaving the equipment to dry.

If large-scale equipment is used, the method of cleaning may involve CIP, provided that the whole of the plant to be cleaned by this method has been designed specially for this system. In this case, dismantling of the plant will not be required, the whole operation will be automated, and, in most cases, the results obtained are better than when hand-washing is carried out—partly because the risks of contamination are reduced, and partly because the whole sequence of operations will have been designed to give the best results with the most efficient use of materials and energy.

In the case of raw material storage tanks, it is suggested that the tanks holding milk or milk concentrates should be thoroughly cleaned and disinfected each day; sugar syrup tanks at least once every 14 days, or each time the tank is emptied; fat tanks as often as possible to avoid the mixing of old fat (which may rapidly oxidize as the fat is kept at a relatively high temperature so that it is liquid) with new fat (these tanks must be completely dry before the new fat is allowed in); and chocolate tanks, where probably only hand-washing may be possible in order to remove the chocolate completely. Here again, problems may arise if the tank is not properly dried before it is used again.

Batch heat treatment vats and HTST and UHT plants are cleaned in the same way as milk or cream processing plants, and a suitable approach is described in the BS 5305, Section 7 (1984).

Detergents and disinfecting agents suitable for use in the dairy and ice cream industry have been described, together with recommendations for their use, in the British Standards Code of practice for cleaning and disinfecting of plants and equipment used in the dairying industry [BS 5305, Section 3: Disinfection and Chemical Agents (BSI, 1984)].

Moreover, the quality of the water used for cleaning the equipment is crucial. The use of soft or softened water is a must, because high levels of salts (calcium and magnesium) form films on the equipment; these films can absorb organic material from the milk and thus form an organic deposit.

Aerosols in the manufacture of frozen dairy products provide another possible contamination point. It is suggested that high-pressure sprays should not be used in the processing area; these are usually used for floors and drains, and the resulting aerosols may crosscontaminate the final product or some food-contact surfaces. Air filters are strongly recommended, and these should always be clean, dry, and regularly inspected. It should be noted that the maintenance of positive air pressure in the manufacturing area is highly desirable.

The main problem with contaminants—such as *Listeria monocytogenes*, which can survive and/or grow on organic soil in the moist environment of the dairy plant—is that the organism can adhere to the food-contact surfaces and form biofilms (Gravani, 1999). Therefore, it is crucial that all areas within the processing line should be kept dry and free from any food residues.

6.14 BACTERIOLOGICAL CONTROL

The obvious intention of any bacteriological control is to ensure that the product reaches the consumer in a satisfactory condition. In addition, the manufacturer needs to meet the regulations in existence and, furthermore, minimize spoilage losses. This is particularly difficult in modern factories, where products are becoming more and more complex—for example, with coatings or layers that contain nuts or confectionery items.

In those countries where standards do exist, bacteriological control will involve the use of tests, as established by regulations in existence, to check on the ice cream.

In the United Kingdom, the methylene blue test (MBT) [BS 4285: Section 5.2 (BSI, 1989)] has been used by local health authorities for many years. It is a very useful laboratory test for monitoring ice cream hygiene for public health purposes. However, it cannot be used as a measure of quality, but only as a simple and cheap test for monitoring the safety of ice cream in a country using statutory heat treatment regulations (Barton, 1981). However, this test is no longer considered appropriate to monitor the safety and quality of ice cream, due to several problems arising from the presence of coloring additives (particularly chocolate and chocolate products) in ice creams which may interfere with the determination of the end point of the MBT. In addition, reliance on the MBT does not permit bacterial numbers to be determined, nor the absence of specific pathogens (Casemore et al., 1992). It must be pointed out, though, that very few, if any, general tests can distinguish pathogenic organisms, and certainly the standard total colony count does not.

Methods for the microbiological examination of frozen dairy products are described in the British Standard 4285 (which consists of five parts), in the Standard Methods for the Examination of Dairy Products (American Public Health Association, 1993), and in the IDF Standards (see Table 6.11).

TABLE 6.11. Pasteurization Requirements in the Countries Indicated

Country	Heat Treatment Requirements
Austria	70°C for 10 min or equivalent within 1 h of preparation
China	78°C for 20 min, then cooled to 60°C to homogenize
Denmark	At least 65°C for 30 min within 1 h of mixing ingredients
Finland	72°C for 15 s
France	60–65°C for at least 30 min
Hong Kong	66°C for not less than 30 min 71°C for not less than 10 min 79°C for not less than 15 s
India	68.3°C for at least 30 min 79.5°C for at least 25 s
Italy	64°C for 30 min or equivalent
Japan	68°C for 30 min or equivalent
Portugal	65°C for at least 30 min
United Kingdom	65.6°C for at least 30 min 71.7°C for at least 10 min 79.4°C for at least 15 s 148.8°C for 2 s
United States	68°C for 30 min 79°C for 25 s

Source: After ICMSF (1998).

The colony count method for coliform organisms is of considerable importance as a control test, because it can produce results within 24 h of sampling the product. For a full confirmation of the presence of coliform organisms, further testing is necessary. However, using petrifilm plates as an alternative to the traditional agar plates, the results for coliforms can be achieved within 4 h of incubation. The confirmed coliform colonies appear after 8 h of incubation, but at 4 h there is a very good indication of the presence of coliforms; if high numbers of coliforms are detected, then postpasteurization contamination has occurred and it is beneficial for the company that such contamination is detected at 4 h rather than at 24 h (Stogo, 1998). In addition, the petrifilm technique can handle larger volumes that the classical plate count techniques, that is 5 or 10 ml can be plated, whereas 5 or 10 plates respectively are required with the classical technique.

Coliforms, particularly *E. coli*, are regarded as an indication of recent fecal pollution if they are found in water (they die out rapidly in water);

but, in ice cream, as in many other foods, they do not die out, and conditions, except in frozen ice cream, are favorable to their growth. Almost always, the presence of coliform organisms in ice cream is due to contamination from equipment that has not been properly cleaned and disinfected, or due to incorrect operation of the heat treatment process. Even where the initial infection may have been low, the cells will probably have multiplied due to reasonably favorable conditions (e.g., they will grow at temperatures above 7°C). In all cases where the presence of coliforms is indicated in the final product, this must be considered to be unsatisfactory and is a very bad failing of the plant management in its widest sense.

The test for *Enterobacteriaceae* instead of coliforms is a more sensitive test for postpasteurization contamination, because the test detects all of the heat-sensitive, nonsporing Gram-negative rods and provides good evidence that contamination has occurred. In this case, the media used for the coliform test must contain glucose instead of lactose [e.g., violet red bile glucose agar (VRBGA)]. For coliform counts, the direct plating of an ice cream sample on media like violet red bile agar (VRBA) may cause some false-positive results, because non-lactose-fermenting bacteria may ferment sugars contained in the undiluted sample (Matushek et al., 1992). It should be remembered that contamination with *Enterobacteriaceae* shows that serious pathogens may have contaminated the product as well.

For the detection of pathogenic organisms, the conventional methods involve enrichment by culturing in a selective medium, selective isolation on diagnostic media, and subsequent identification—that is, confirmation by biochemical and/or serological tests; the approach is time-consuming. Gooding and Choudary (1997) have developed a rapid and sensitive method for the detection of enterohaemorrhagic *E. coli* O157:H7 by combining polymerase chain reaction (PCR) with both culture enrichment and immunomagnetic separation; the method can detect a single bacterial cell present in an ice cream sample in less than one working day.

Whatever method of testing is adopted, it is essential that sampling be carried out in the proper manner. Sampling procedures are described in detail in the IDF Standard 122C (IDF, 1996), British Standards 4285, Section 1.1 (BSI, 1984a), and American Public Health Association (1993). The frozen products must be kept frozen during sampling and shipment to the laboratory, while soft-serve products should be maintained at 0–4°C. A representative sample (at least 50 g) must be aseptically transferred to a sterile wide-mouthed sample container. Problems can arise with composite ice creams, where fruits, nuts, chocolate, and other types of confection are added to the ice cream or layered with the ice cream.

In addition to the assessment of the microbiological quality of the raw materials and final product, the efficiency of disinfection of the processing plant must be tested. The microbiological state of the dairy plant should be routinely monitored. Any variation from the normal, regularly obtained results for the APC and coliform tests could indicate an inefficiency of the cleaning-disinfecting process. The sanitary condition of the equipment is determined using the swab test procedure. A description of the swab test procedure in included in the Standard Methods for the Examination of Dairy Products (American Public Health Association, 1993). It is suggested, in addition, that swabs of particular areas of equipment are carried out at intervals, and unless particular points of contamination are identified, the swabs should be taken at different points on successive occasions. Wherever possible, the use of rinses instead of, or in addition to, swabs can be very helpful in the examination of equipment.

Even with the very wide variation that is reported from total counts, it is usually possible to state whether an ice cream sample is satisfactory, doubtful, or unsatisfactory, and the intelligent use and interpretation of the tests mentioned should enable proper microbiological control to be carried out.

6.15 HACCP SYSTEM IN THE MANUFACTURE OF ICE CREAM

Several recent reports have been published concerning the implementation of HACCP in the manufacture of ice cream and related products. With such an approach, potential hazards are identified, using a completely documented procedure, and control measures are decided. A detailed application of HACCP system in the manufacture of a flavored ice cream is described by the Ice Cream Alliance and Ice Cream Federation (1995). Also, Sandrou and Arvanitoyannis (2000) and ICMSF (1998) have identified the critical control points for the manufacture of a simple ice cream.

Although the implementation of a HACCP system is considered essential by the retailing companies (Sandrou and Arvanitoyannis, 2000), it should be emphasized that the HACCP approach cannot replace the traditional hygienic practices. However, it can reduce the importance of traditional microbiological control based on testing the final product.

In fact, HACCP is an internationally recognized quality control system that focuses on prevention rather than relying on end-product testing (Martimore and Wallace, 1998). Implementation of HACCP system should, therefore, give confidence that the product is safe. This is particularly important for the ice cream market, which is characterized by globalization, because increasing numbers of US products are being exported throughout the world, and European products are imported in the US market.

6.16 HYGIENE AT THE FINAL SELLING POINT

The final selling point is one of the weakest links in the whole chain so far as hygiene is concerned. Soft-serve ice cream and ice cream scooped from a bulk container is handled, to a greater or lesser extent, by necessity. It is here, therefore, that the personal hygiene of the server becomes of the utmost importance. This applies even if pre-wrapped ice cream is being sold. Cleanliness of hands, clothing, and habits must be above reproach, and the operators must be trained in the best ways of maintaining this, and in the distribution of the individual portions of the ice cream, as suggested by the dairy industry guidelines (Ice Cream Alliance and Ice Cream Federation, 1995).

The equipment servers, wafer holders, and so on, have to be kept free of all residues of ice cream, which may melt and allow the growth of bacteria to recommence. Wherever possible, these items of equipment should be kept in running cold water. If they have to be kept in a jug of water, this water must be regularly changed to avoid it becoming a source of contaminating bacteria. Wilson et al. (1997) reported that the opened ice cream served by scooping had the microbial load increased by one order of magnitude when compared with the unopened ice cream, which had APC of around 10^3–10^4 cfu ml^{-1}.

6.17 CONCLUSION

Despite the fact that several factors are important in the production of high-quality ice cream, special emphasis should be placed on certain points in the manufacture, namely:

- Selection of ingredients of high microbiological quality
- Use of the appropriate heat treatment

- Avoidance of postpasteurization contamination
- Maintenance of hygienic condition during production, storage, and distribution and up to the final selling point.

Because the freezing process will cause little reduction in the microbial load of the product, a total quality management program is needed so that the final product is of the highest microbiological quality.

It is generally recognized that ice cream and related products are nutritious and delicious dairy products, which are now consumed worldwide. The limited number of disease outbreaks caused by ice cream shows that it is a safe product if certain quality assurance steps are followed.

ACKNOWLEDGMENT

The authors wish to acknowledge the work of the late J. Rothwell, which formed the basis of this chapter.

REFERENCES

American Public Health Association (1993) *Standard Methods for the Examination of Dairy Products*, 16th ed., R. T. Marshall, ed., APHA, Washington, DC.

Anonymous (1992) *Off. J. Eur. Communities*, **L268**, 24–30.

Anonymous (1995) Code of Practise for the Hygienic Manufacture of Ice Cream, Ice Cream Alliance, Nottingham, United Kingdom.

Anonymous (1997) Development and use of microbiological criteria for foods. *Food Science and Technology Today*, **11**(3), 137–177.

Anonymous (2000) *Ice Cream (Magazine of the Ice Cream Alliance)*, **51**(6), 248–249.

Andre, P., Roose, H., Van Noyen, R., Dejaegher, L., Uyttendaele, I., and de Schrijver, K. (1990) *Med. Mal. Infect.*, **20**, 570–572.

Barton, B. W. (1981) *Ice Cream Frozen Confect.*, **33**(8), 496.

BSI (1984a) In *Microbiological Examination for Dairy Purposes*, BS 4285, Section 1.1, British Standards Institution, London.

BSI (1984b) In *Cleaning and Disinfecting of Plant and Equipment Used in the Dairying Industry*, BS 5305, British Standards Institution, London.

BSI (1987a)

BSI (1987b) In *Microbiological Examination for Dairy Purposes*, BS 4285, Subsection 3.9.1, British Standards Institute.

BSI (1989) In *Microbiological Examination for Dairy Purposes*, BS 4285, Section 5.2, British Standards Institution, London.

BSI (1990a) In *Microbiological Examination for Dairy Purposes*, BS 4285, Part 0, British Standards Institution, London.

BSI (1990b) In *Microbiological Examination for Dairy Purposes*, BS 4285, Section 2.1, British Standards Institution, London.

BSI (1990c) In *Microbiological Examination for Dairy Purposes*, BS 4285, Section 3.7, British Standards Institution, London.

BSI (1990d) In *Microbiological Examination for Dairy Purposes*, BS 4285, Section 3.6, British Standards Institution, London.

BSI (1990e) In *Microbiological Examination for Dairy Purposes*, BS 4285, Section 3.10, British Standards Institution, London.

BSI (1992) In Microbiological Examination for Dairy Purposes, BS 4285, Subsection 3.9.2.

Casemore, D. P., Richardson, K., Sands, R. L., and Stevens, G. (1992) *PHLS Microbiol. Dig.*, **9**(4), 166–171.

Davidson, R. H., Duncan, S. E., Hackney, C. R., Eigel, W. N., and Boling, J. W. (2000) *J. Dairy Sci.*, **83**, 666–673.

Dean, J. P., and Zottola, E. A. (1996) *J. Food Prot.*, **59**(5), 476–480.

Djuretic, T., Wall, P. G., and Nichols, G. (1997) *Commun. Dis. Rep. CDR Rev.*, **7**, R41–R45.

Dodhia, H., Kearney, J., and Warburton, F. (1998) *Commun. Dis. Public Health*, **1**(1), 31–34.

Doxanakis, V. (1998) In *Ice Cream: Proceedings of the International Symposium (Greece)*, International Dairy Federation, Brussels, Belgium, pp. 10–16.

Doyle, M. P., Meske, L. M., and Marth, E. H. (1985) *J. Food Prot.*, **48**, 740–742.

Dubey, U. K., and White, C. H. (1997) *J. Dairy Sci.*, **80**, 3439–3444.

Evans, D. I. (1947) The Medical Officer, 25th January, 39.

FAOSTAT (2000) *www.fao.org*

Farber, J. M., and Harwig, J. (1996) *Food Control*, **7**(4/5), 253–258.

Fearon, A., and Moruzzi, T. (1999) *Food Ingredients Anal.*, **21**(6), 17–23.

Goff, D. (1997) *Int. Dairy J.*, **7**, 363–373.

Gooding, C. M., and Choudary, P. V. (1997) *J. Dairy Res.*, **64**, 87–93.

Gravani, R. (1999) In *Listeria, Listeriosis and Food Safety*, 2nd ed., revised and expanded, E. T. Ryser and E. H. Marth, eds., Marcel Dekker, New York, Chapter 17, pp. 657–709.

Greenwood, M. H., Roberts, D., and Burden, P. (1991) *Int. J. Food Microbiol.*, **12**, 197–206.

Hagen, M,. and Narvhus, J. A. (1999) *Milchwissenscaft*, **54**(5), 265–268.

Hamilton, M. P. (1990) *J. Soc. Dairy Technol*, **43**(1), 17–20.

Hekmat, S., and McMahon, D. J. (1992) *J. Dairy Sci.*, **75**, 1415–1422.

Hobbs, B. C., and Gilbert, R. J. (1978) In *Food Poisoning and Food Hygiene*, 4th ed., Edward Arnold, London, pp. 91, 131–135.

Holsinger, V. H., Smith, P. W., Smith, J. L., and Palumbo, S. A. (1992) *J. Food Prot.*, **55**(4), 234–237.

Huang, V., and Platt, S. (1995) *Chem. Ind.*, **2**, 51–54.

Ice Cream Alliance and Ice Cream Federation (1992) Code of practice for the hygienic manufacture of ice cream.

ICMSF (1998) In *Micro-organisms in Foods 6—Microbial Ecology of Food Commodities*, Blackie Academic & Professional, London, 559–576.

IDF (1985a) In *Milk and Milk Products, Detection of Salmonella spp.*, IDF Standard 93A, Brussels, Belgium.

IDF (1985b) In *Milk and Milk Products, Enumeration of Coliforms*, IDF Standard 73A, Brussels, Belgium.

IDF (1990a) In *Milk and Milk Products, Enumeration of Yeasts and Moulds*, IDF Standard 94B, Brussels, Belgium.

IDF (1990b) In *Milk and Milk Products, Detection of Listeria monocytogenes*, IDF Standard 143, Brussels, Belgium.

IDF (1990c) In *Milk and Milk Products, Enumeration of Staphylococcus aureus*, IDF Standard 145, Brussels, Belgium.

IDF (1991) In *Milk and Milk Products, Aerobic Plate Count at 30°C*, IDF Standard 100B, Brussels, Belgium.

IDF (1996) In *Preparation of Samples and Dilutions for Microbiological Examinations*, IDF Standard 122C, Brussels, Belgium.

IDF (1999) In *Milk and Milk Products, Enumeration of presumptive Esherichia coli*, IDF Standard 170A, Brussels, Belgium.

Jukes, D. J. (1997) *Food Legislation of the UK-A Concise Guide*, 4th ed., Butterworth-Heinemann, Oxford, England.

Krog, N. (1998) In *Ice Cream: Proceedings of the International Symposium (Greece)*, International Dairy Federation, Brussels, Belgium, pp. 37–44.

Little, C. L., and de Louvois, J. (1999) *Int. J. Environ. Health Res.*, **9**, 223–232.

Marshall, R. T. (1998) In *Applied Dairy Microbiology*, E. H. Marth and J. L. Steele, eds., Marcel Dekker, New York, 1998, pp. 81–108.

Marshall, R. T., and Arbuckle, W. S. (1996) *Ice Cream*, 5th ed., International Thomson Publishing, New York.

Martimore, S., and Wallace, C. (1998) *HACCP: A Practical Approach*, 2nd ed., Aspen, Gaithersburg, MD.

Matushek, M. G., Curiale, M. S., McAllister, J. S., and Fox, T. L. (1992) *J. Food Prot.*, **55**(2), 113–115.

Nichols, G., and de Louvois, J. (1995) *PHLS Microbiol. Dig.*, **12**(1), 11–15.

Pappas. C. P. (1988) *Br. Food J.*, **90**(6), 250–254.

Ramakrishnan, S., Selvaraj, K., and Devi, R. M. (1986) *Indian Dairyman*, **38**(5), 225.

Reddy, B. B., Reddy, Y. K., Ranganadham, M., and Reddy, V. P. (1994) *J. Food Sci. Technol.*, **31**(2), 151–152.

Rothwell, J. (1998) In *Ice Cream: Proceedings of the International Symposium (Greece)*, International Dairy Federation, Brussels, Belgium, pp. 46–53.

Ryan, J. J., and Gough, R. H. (1982) *J. Food Prot.*, **45**(3), 279.

Ryser, E. T. (1999) In *Listeria, Listeriosis and Food Safety*, 2nd ed., revised and expanded, E. T. Ryser and E. H. Marth, Marcel Dekker, New York, Chapter 11, pp. 359–410.

Sandrou, D. K., and Arvanitoyannis, I. S. (2000) *Food Rev. Int.*, **16**(1), 77–111.

Scientific Committee on Veterinary Measures Relating to Public Health (1999) European Commission Unit B3—Management of Scientific Committees SC4.

Shrestha, K. G., and Sinha, R. N. (1987) *Indian J. Dairy Sci.*, **40**(1), 121.

Statutory Instrument (SI) (1995) Food Milk and Dairies—The Dairy Products (Hygiene) Regulations, No. 1086.

Statutory Instrument (SI) (1996) Food—The Food Labelling Regulations, No. 1499.

Stogo, M. (1998) *Ice Cream and Frozen Desserts—A Commercial Guide to Production and Marketing*, John Wiley & Sons, New York.

Vought, K. J., and Tatini, S. R. (1998) *J. Food Prot.*, **61**(1), 5–10.

Walstra, P., and Jonkman, M. (1998) In *Ice Cream: Proceedings of the International Symposium (Greece)*, International Dairy Federation, Brussels, Belgium, pp. 17–24.

Warke, R., Kamat, A., Kamat, M., and Thomas, P. (2000) *Food Control*, **11**, 77–83.

Wilson, I. G., Heaney, J. C. N., and Weatherup, S. T. C. (1997) *Epidemiol. Infect.*, **119**, 35–40.

CHAPTER 7

MICROBIOLOGY OF STARTER CULTURES

ADNAN Y. TAMIME
Scottish Agricultural College, Ayr, Scotland

7.1 INTRODUCTION

The major functions of microbial starter cultures in food and dairy products may be summarized as follows:

- To biopreserve the product due to a fermentation that results in an extended shelf life and enhanced safety; the production of bacteriocins may also have potential uses as food preservatives.
- To enhance the perceived sensory properties of the product due, for example, to the production of organic acids, carbonyl compounds and partial hydrolysis of the proteins and/or fats.
- To improve the rheological properties (i.e., viscosity and firmness) of the product, and in some instances encourage gas production (i.e., eye formation in cheese) or color (white and blue mold or red smear).
- To contribute dietetic/functional properties to food, such as occurs with the use of probiotic microfloras.

However, the preservation of food by fermentation is one of the oldest methods known to humankind (see the reviews by Hammes, 1990; Mittal, 1992; Stiles, 1996). Throughout the world, around 400 and 3500 names are applied to traditional and industrialized fermented

Dairy Microbiology Handbook, Third Edition, Edited by Richard K. Robinson
ISBN 0-471-38596-4

milk products and cheeses, respectively (Kurmann et al., 1992; Masui and Yammada, 1996; Robinson and Wilbey, 1998). These products may have different local names, but they are practically the same. Nevertheless, when taking into account the microbial species that dominate(s) the microflora, a more accurate list would only include a few varieties. Tamime and Robinson (1999) have reported that fermented milks can be divided into three broad categories based on the metabolic products (e.g., lactic fermentations, yeast–lactic fermentations and mold–lactic fermentations; see Chapter 8 in this volume). Such an approach could also be applied to wide ranges of cheeses, and Table 7.1 illustrates a possible scheme of classification of fermented milks and cheeses. Such fermentation processes are the result of the presence of microorganisms (bacteria, molds, yeasts, or combinations of these) and their enzymes in milk. In the dairy industry, these organisms are known as starter cultures; however, in some cheese varieties other microfloras

TABLE 7.1. Proposed Scheme of Classification for Fermented Milks and Cheese Products

Category	Fermented milks	Cheeses
Lactic fermentations		
Mesophilic	Cultured buttermilk	Cheddar
	Långofil	Cheshire
	Täetmjolk	Gouda
	Ymer[a]	Edam
Thermophilic	Yoghrt, Zabadi	Emmenthal
	Labneh[a]	Sbrinz, Mozzarella
	Skyr[a]	Parmesan
Therapeutic[b]	Acidophilus milk	Gouda[b]
	ABT[c], Labneh[a]	Quarg[b]
	Yakult	Cottage[b]
Mold–lactic fermentations	Viili	Camembert, Brie, Roquefort
Yeast–lactic fermentations	Kefir	Mold-ripened cheeses (see text)
	Koumiss	
Miscellaneous microflora		
Acetobacteria	Kefir	
Propionibacteria		Wide range of cheeses
Brevibacteria		Wide range of cheeses
Secondary cheese cultures		Wide range of cheeses

[a]Concentrated fermented milk products.

[b]For further information refer to Gomes et al. (1995), Tamime et al. (1995), and Tamime and Marshall (1997).

[c]ABT: *Lactobacillus acidophilus*, *Bifidobacterium lactis*, and *Streptococcus thermophilus*.

could be present, and these are referred to as *secondary cultures* (see later). The essential roles of starter cultures are summarized as follows. First, the production of lactic acid as a result of lactose fermentation; the lactic acid helps to form the gel and imparts a distinctive and fresh, acidic flavor during the manufacture of fermented milks; however, in cheesemaking, lactic acid is important during the coagulation and texturizing of the curd. Second, the production of volatile compounds (e.g., diacetyl and acetaldehyde) that contribute toward the flavor of these dairy products. Third, the starter cultures may possess a proteolytic or lipolytic activity that may be desirable, especially during the maturation of some types of cheese. Fourth, other compounds may be produced—for example, alcohol, which is essential during the manufacture of Kefir and Koumiss. Fifth, the acidic condition in these dairy products, and in some instances the production of bacteriocins, prevents the growth of pathogens, as well as many spoilage organisms.

Traditionally, milk was left to sour naturally prior to the manufacture of cheese and fermented milks. Such a method of production is not reliable, is prone to failure, and may promote undesirable side effects, and the quality of the end product can vary tremendously. These drawbacks were basically due to a lack of scientific knowledge in the field of microbiology (i.e., starter cultures); but since the turn of the twentieth century, starter cultures have been widely studied, and their behavior and metabolism are well established. Hence, the selection of starter cultures has become feasible, and their activity has become more predictable. Such an approach is important in factories handling large volumes of milk; furthermore, greater uniformity in the quality of the end product can be expected.

Although the traditional method of manufacture was not scientifically controlled, it has provided the industry with the basic technology required today. At present, the manufacture of fermented dairy products including cheese is more centralized. Hence, starter cultures have become an integral part of a successful industry; their relevance is reflected by the world output of these products. In view of the production outputs of cheese and fermented milks, the classification, maintenance, preservation, and propagation of starter cultures is of paramount importance, and these aspects will be reviewed in this chapter; however, the problem associated with starter culture bacteriophages and the genetic developments of these micro-organisms will be discussed separately. The following are recommended for further reading about different aspects of starter cultures (Zourari et al., 1992; Poolman, 1993; Tan, 1994; Roussis, 1994; Bruinenberg

TABLE 7.2. World Production Figures of All Types of Cheeses (Thousand Tonnes)

Region	1970	1980	1990	1998
Africa	294	364	466	579
America (North and Central)	1,574	2,471	3,519	4,207
America (South)	377	462	535	677
Asia	478	661	803	936
Europe	4,768[a]	7,165[a]	8,904[a]	7,537[b]
Oceania	177	254	306	552
World	7,668	11,797	14,533	14,488

[a]Data include production figures in the former USSR.
[b]Production figures for former USSR was not reported.
Source: After FAO (1972, 1981, 1996, 1999).

and Limsowtin, 1995; Geisen and Holzapfel, 1996; Kunji et al., 1996; Davidson et al., 1996; de Vos, 1996; Law and Haandrikman, 1997; Grossiord et al., 1998; de Vos et al., 1998; Venema et al., 1999; Siezen, 1999; Delcour et al., 1999; Macedo et al., 1999; Christensen et al., 1999).

7.2 ANNUAL UTILIZATION OF STARTER CULTURES

There are no data available in the dairy industry concerning the actual amounts or types of starter cultures utilized and/or produced every year. Since the mid-1980s, there has been greater reliance in the industry on the use of direct-to-vat inoculation (DVI) starter cultures for (a) the manufacture of cheese and fermented milk products or (b) the production of bulk starter cultures. It is possible, however, to estimate the annual volume of starter cultures from world production figures of cheese and fermented milks based on the assumption that all these products are made using bulk starter cultures only.

According to the FAO (1999), the world production of all cheese varieties was in the region of 14.5 million tonnes in 1998 (see Table 7.2), but the actual figure is higher because production data from the former USSR since 1990 was not reported. Nevertheless, if it is accepted that 10 liters of milk is required to produce 1 kg of cheese, then a total of 145 million tonnes of milk was used for the manufacture of cheese. On average, cheese starter cultures are inoculated at a rate of 1–1.5 ml $100\,ml^{-1}$ of milk, and hence in 1998 the world production of cheese bulk starter cultures was in the region of 1.5–2.2 million tonnes. However, according to Christian Hansen (i.e., one of the main suppliers of starter

TABLE 7.3. Total Production Figures of Different Types of Fermented Milks in Some Selected Countries (Thousand Tonnes)

	Buttermilk		Yogurt		Others	
Country/Region	1982	1992	1982	1992	1982	1992
Australia	—[a]	—	30.0	76.4	—	—
Europe	486.8	626.2	2,123.4[b]	3,670.4[b]	2,172.4	2,011.9
North America	438.3	381.3	323.9	583.8	—	51.0
Chile	—	—	22.6	54.3	—	—
India	NR	—[c]	2,574.0	3,950.0	—	—
Iceland	1.8	4.7	1.3	2.4	3.2	4.2
Israel	—	—	29.1	55.9	37.0	54.0
Japan	—	—	283.3	533.0	148.2	476.0
Total	926.9	1,012.2	5,387.6	8,926.2	2,360.8	2,597.1

[a]Indicate no figures or product is not manufactured.
[b]In some countries, the yogurt figures include other types of fermented milk.
[c]Refer to text for details.
Source: Adapted from Tamime and Marshall (1997).

cultures in the world), it was estimated that in 1998, 80% of sales of DVI starter cultures were for the production of bulk starter culture, while 20% were added directly into the cheese milk (Nagle, personal communication).

World production figures of fermented milks and other related products are only available in some selected countries (see Table 7.3), and the last data published by the International Dairy Federation (IDF) was for 1992 (IDF, 1994; see also Tamime and Marshall, 1997).

World production figures of fermented milks and other related products are not available. Table 7.3 shows that the production of cultured buttermilk, yogurt, and fermented products in some member countries of the IDF had increased from 8.7 million tonnes in 1982 to 12.5 million tonnes in 1992. However, under commercial practice, starter cultures are used at an average rate of 2–3 ml $100\,ml^{-1}$; hence 251,000–376,000 tonnes of bulk starters were required.

It could be argued, however, that the production data shown in Table 7.3 does not include countries like Bulgaria, Turkey, or the Arab countries that are high producers of fermented milk products. Also, over the past decade there has been increased sales of fermented milk products in southeast Asia and South America, but no production figures are available. Lastly, the production data of buttermilk in India in 1991 amounted to 19.5 million tonnes (IDF, 1993), with no indication whether it was "sweet" or fermented using the traditional process (i.e., the by-product of the manufacture of ripened cream butter). If it was fermented buttermilk, the starter culture required at an average rate of 1.5 ml 100 ml^{-1} would have been 292,500 tonnes.

7.3 CLASSIFICATION OF STARTER ORGANISMS

As mentioned earlier, several microorganisms (bacteria, yeasts, molds, or combinations of these) are employed in the manufacture of cheese and other fermented milk products. However, detailed characterizations of these microorganisms have been reported by Sneath et al. (1986), Bezkorovainy and Miller-Catchpole (1989), Spencer and Spencer (1990), Balows et al. (1992), Holt et al. (1994), Hui and Khachatourians (1995), Wood and Holzapfel (1995), Cogan and Accolas (1996), Tamime and Marshall (1997), and Salminen and von Wright (1998). The relevance of these microorganisms to the dairy industry is as follows.

7.3.1 Traditional Microflora

***7.3.1.1 The Genus* Lactococcus.** The starter cultures of the genus *Lactococcus* consist of *Lactococcus lactis* subsp. *lactis*, subsp. *cremoris* and subsp. *lactis* biovar *diacetylactis*. The latter microorganism is closely related to *L. lactis* subsp. *lactis*, but differs by being citrate-positive. These cultures previously belonged to the genus *Streptococcus* (i.e., mesophilic streptococci belonging to Lancefield group N; Lancefield, 1933). However, Sneath et al. (1986) and Holt et al. (1994) (i.e., in *Bergey's Manuals*) have combined all these lactic acid bacteria to form a single species of *L. lactis*. This approach is based on closely related characteristics, which include (a) possession of identical isoprenoid quinones and the enzyme, β-phosphogalactase, (b) indistinguishable lactic dehydrogenase, and (c) the fact that all have identical percentages of guanine and cytosine. The only properties that distinguish them are plasmid-controlled, and they show high DNA homology. However, these subspecies are of great importance in the cheese industry for their fermentation characteristics. For example, *L. lactis* subsp. *lactis* biovar *diacetylactis* is a flavor producer with the production of diacetyl and CO_2 from citrate, while *L. lactis* subsp. *lactis* and subsp. *cremoris* are classified as non-flavor producers.

The overall characteristics of these microorganisms are as follows:

- Spherical or ovoid cell morphology, 0.5–1.0 μm in diameter and forming short to long chains or occurring in pairs (see Figure 7.1A); they are often elongated in the direction of the chain.
- Gram-positive, microaerophilic, homofermentative lactic acid bacteria that produce L(+)-lactate from lactose; only *L. lactis* subsp. *lactis* biovar *diacetylactis* produces diacetyl.
- Some strains produce exopolysaccharide (EPS) material and/or bacteriocins; all the species do not have flagella, nor do they form endospores; and for optimum growth they require B vitamins.
- The peptidoglycan in the cell wall is similar to that of *Streptococcus pyogenes* except that the cross-bridges consist of D-isoasparagine.
- Absence of growth at 45°C, but able to grow in 0.3 g $100\,g^{-1}$ methylene blue and hydrolyze arginine.
- The differentiating characteristics of *Lactococcus* spp. are shown in Table 7.4 (see also Bissonnette et al., 2000).

***7.3.1.2 The Genus* Leuconostoc.** In the genus *Leuconostoc*, the organisms that are associated with dairy starter cultures are *Leuconostoc mesenteroides* subsp. *cremoris* (previously known as *L. cremoris* or *L. citrovorum*), *L. mesenteroides* subsp. *dextranicum*, and in some instances *L. lactis*. These microorganisms are related phenotypically to the genera *Lactobacillus* and *Pediococcus*, and they share many features with heterofermentative lactobacilli. The *Leuconostoc* spp. have complex nutritional requirements, and some strains produce EPS material. They are capable of producing D(−) lactate, CO_2, and aroma compounds (e.g., ethanol, diacetyl and acetic acid). These organisms are normally used in multiple or mixed-strain cheese/fermented milks starter cultures that contain flavor producers (refer to Section 7.4). Incidentally, Milliere et al. (1989), Cogan and Jordan (1994), Vedamuthu (1994), and Thunell (1995) have recently provided a thorough description of the genus *Leuconostoc*, including their metabolites and their use in dairy products (see also Table 7.4).

Leuconostoc mesenteroides subsp. *cremoris* has a long generation time (48 h at 22–30°C) vis-à-vis (24 h at 30°C) for *L. mesenteroides* subsp. *mesenteroides*. Other characteristics may include the following:

- The cell morphology varies with growth medium; and in the presence of glucose the cells are elongated, resembling lactobacilli.

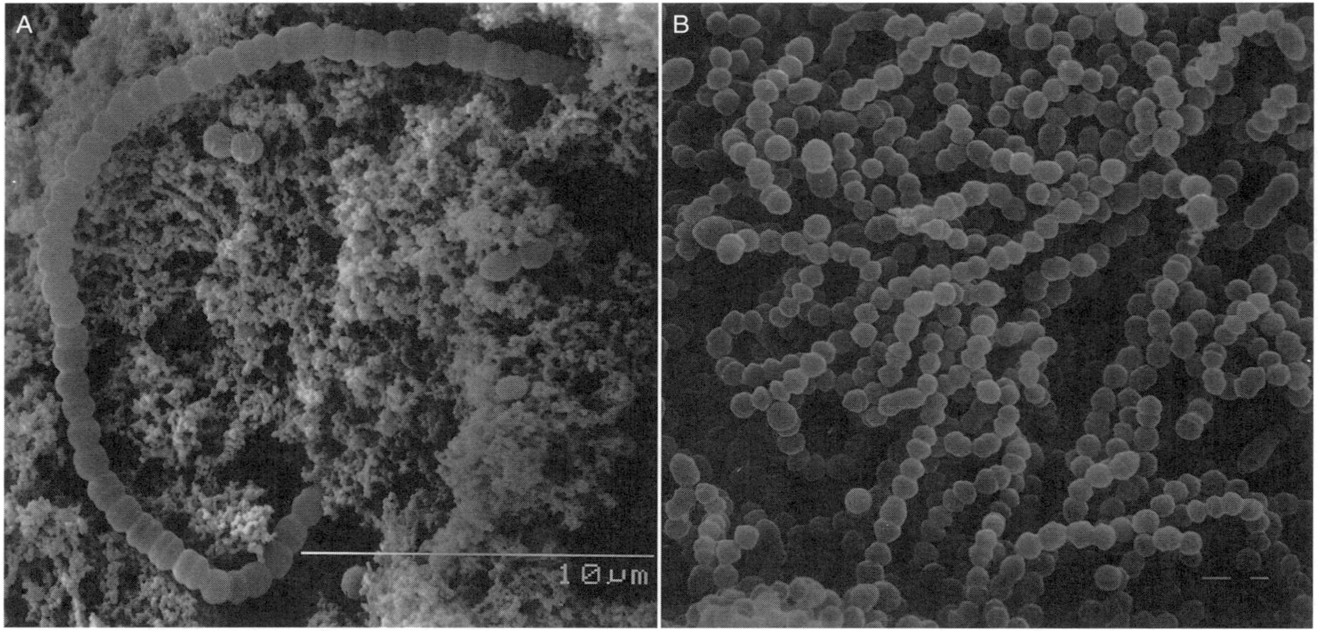

Figure 7.1. A small microcolony of (A) an aroma producer *L. lactis* subsp. *lactis* biovar *diacetylactis* and (B) a thermophilic lactic acid bacteria (*S. thermophilus*). (Reproduced by courtesy of V. Bottazzi, F. Bianchi, and C. Garabazza.)

TABLE 7.4. Selected Characteristics of Some Lactic Acid Bacteria Associated with Fermented Milks and Cheeses

	Lactococcus lactis subsp.			*Leuconostoc mesenteroides* subsp.				
Characteristic	*lactis*	*lactis* biovar *diacetylactis*	*cremoris*	*mesenteroides*	*dextranicum*	*cremoris*	*Pediococcus acidilactici*	*Streptococcus thermophilus*
G + C[a] mean %	33.8–36.9	33.8–34.8	35.0–36.2	37.0–39.0	37.0–40.0	38.0–40.0	38.0–40.0	37.0–40.0
Lactic acid isomer(s)	L(+)	L(+)	L(+)	D(−)	D(−)	D(−)	DL	L(+)
Growth at 10/45°C	+/−	+/−	+/−	+/−	+/−	+/−	−/+	−/+
Carbohydrate utilization[b]								
Aesculin	d	d	d	d	d	−		+
Amygdalin				d	d	−	d	−
Arabinose	d	d	−	+	−	−	d	d
Cellobiose				d	d	−	+	−
Fructose				+	+	−	+	+
Galactose	+		+	+	d	+	+	(d)
Lactose	+	+	+	(d)	+	+	d	+
Maltose	+	+	−	+	+	d	−	−
Mannitol				d	d	−	d	−
Mannose				+	d	−	+	
Melezitose	−	−	−				−	
Melibiose	−	−	−	d	d	−	−	(d)
Raffinose	−	−	−	d	d	−	d	d
Rhamnose	−	−	−				d	
Ribose	+	+	+	+		−	+	−
Salicin	d	d	d	d	d	−	d	−
Sorbitol	−	−	−				−	
Saccharose	d	−	−	+	+	−	(d)	+
Trehalose	d	d	d	+	+	−	d	−
Xylose	d	d	−	d	d	−	+	d

[a]Mean % of guanine and cytosine in deoxyribonucleic acid (DNA).

[b]+, >90% strain positive; −, >90% strain negative; d, positive reaction by 11–89% of strains; (d), delayed reaction; empty spaces data not reported.

Source: Adapted from Marshall and Tamime (1997).

- The cells are coccoid in shape, Gram-positive, asporogenus, and nonmotile, and they occur singly or in pairs that may form short- to medium-length chain.
- The cell wall peptidoglycan type of most species is L-Lys-L-Ser-L-Ala_2, but L-Lys-L-Ala_2 for *L. mesenteroides* subsp. *mesenteroides* and *L. lactis*.

***7.3.1.3 The Genus* Pediococcus.** The only strain of this genus to be used in dairy starter cultures is *Pediococcus acidilactici*. However, this organism had other synonyms such as *Pediococcus lindneri*, *Pediococcus cerevisiae*, and *Streptococcus lindneri*. The pediococci divide alternatively in two perpendicular directions to form tetrads that differentiate them morphologically from other lactic acid bacteria. The cells do not form spores or capsules, but are Gram-positive cocci of uniform size, produce DL-lactate, and are nonmotile (see Table 7.4).

Pediococcus spp. have been found in cheese, but represented only a small proportion of the total lactic acid bacteria in the product; their precise role is not fully understood (Fox et al., 1990; Olson, 1990, Bhowmik and Marth, 1990). However, Biokys®, which is produced in the Czech Republic, is a fermented milk with therapeutic properties because the starter culture consists of *Pediococcus aciditactici*, *Lactobacillus acidophilus*, and *Bifidobacterium bifidum* in a ratio 1:0.1:1 (Tamime and Marshall, 1997).

***7.3.1.4 The Genus* Streptococcus.** The taxonomic status of *Streptococcus thermophilus* reported by Orla-Jensen (1931) has fluctuated since the 1980s due to the close relationship between this organism and *Streptococcus salivarius* (i.e., nucleic acid hybridization and the long-chain fatty acid profiles); as a consequence, it was denoted as a subspecies (e.g., *S. salivarius* subsp. *thermophilus*) in the mid-1980s (Farrow and Collins, 1984). In the early 1990s, a separate species status was reproposed on the basis of genetic and phenetic criteria (Schleifer et al., 1991). This microorganism is predominantly used in combination with other starter cultures for the manufacture of cheese (Swiss and Italian varieties), yogurt, and "bio" fermented milk products. Some selected characteristics of *S. thermophilus* may include the following:

- The cells are spherical or ovoid, are $<1\,\mu m$ in diameter, and form chains or occur in pairs (see Figure 7.1B).

- The organism does not grow at 15°C, but most strains are able to grow between 40°C and 50°C, and they require B vitamins and some amino acids for maximum growth.
- Such streptococci are anaerobic homofermentative lactic acid, are Gram-positive, and produce L(+)-lactate, acetaldehyde, and diacetyl from lactose in milk (see Table 7.4).
- Some strains produce EPS, and the cell wall peptidoglycan type is Lys-Ala_{2-3}, while the cell wall polysaccharide component have not been determined.
- A group antigen for seriological identification has not been demonstrated.

***7.3.1.5 The Genus* Lactobacillus.** Orla-Jensen (1931) classified the lactobacilli (i.e., rod-shaped and catalase-negative species) into three genera: *Thermobacterium*, *Streptobacterium*, and *Betabacterium*. In the 1980s, the genus *Lactobacillus* was still divided into three main groups (I, II, and III) resembling the classification of Orla-Jensen (1931), but without designating them as subgeneric taxa (Sneath et al., 1986). The organisms that are employed in the dairy industry are as follows: (a) **Group I, obligately homofermentative lactobacilli**. The species are *L. delbrueckii* subsp. *lactis*, *L. delbrueckii* subsp. *bulgaricus*, *L. helveticus*, *L. kefiranofaciens*, *L. acidophilus*, *L. gasseri* and *L. johnsonii.* The latter three organisms are used in "bio" products, while *L. kefiranofaciens* is present in Kefir grains. It is now accepted that *L. jugurti* is a biotype of *L. helveticus*, only lacking the ability to ferment maltose (see Manachini and Parini, 1983; also see Figure 7.2). However, *L. caucasicus*, which is sometimes reported to be among the organisms in a Kefir grain, is now proposed as a rejected name. (b) **Group II, facultative heterofermentative lactobacilli**. Some examples are *L. casei*, *L. paracasei* subsp. *paracasei*, *L. paracasei* subsp. *tolerans*, *L. rhamnosus*, and *L. plantarum*; most of these organisms are used in "bio" products. (c) **Group III, obligately heterofermentative lactobacilli**. These are not important as dairy starter cultures except for Kefir production, and some examples are *L. brevis*, *L. fermentum*, *L. kefir*, *L. viridescens*, and *L. reuteri* (this organism is used in "bio" fermented milks).

At present, the lactobacilli have been regrouped based on phylogenetic relationship into Groups A, B, and C that correspond to the respective groups or genera mentioned earlier. Furthermore, within each group, the phylogenic groupings is subgrouped as **a**, **b**, and **c** where species belong to the *L. delbrueckii*, *L. casei-Pediococcus*, and

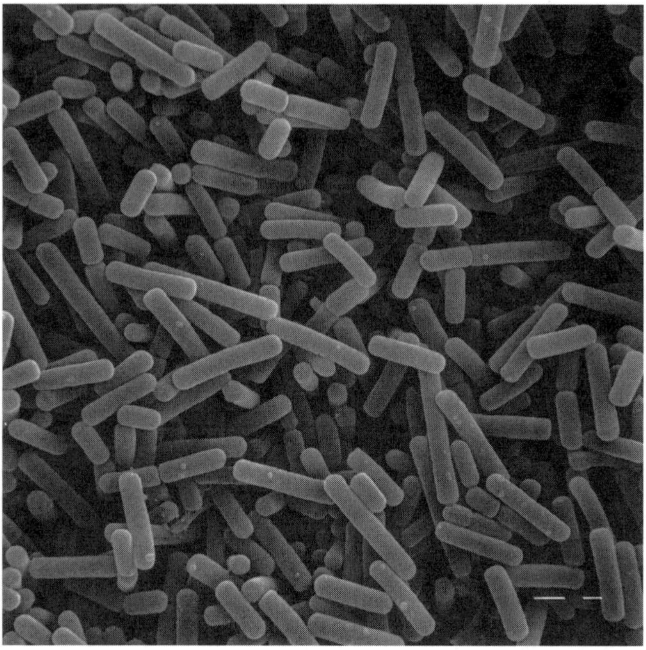

Figure 7.2. Scanning electron micrograph illustrating the cell morphology of *L. helveticus* starter culture grown in whey. (Reproduced by courtesy of V. Bottazzi, F. Bianchi, and C. Garabazza.)

Leuconostoc group, respectively (Hammes and Vogel, 1995). The history of the lactobacilli has been recently detailed in a review by Pot et al. (1994).

Recently, Collins et al. (1991) and Klein et al. (1998) reviewed the taxonomy and physiology of probiotic lactic acid bacteria (the genera *Lactobacillus*, *Bifidobacterium*, and *Enterococcus*) and reported that the lactobacilli species belonged to (a) the *L. acidophilus* group, (b) the *L. casei* group, and (c) the *L. reuteri/L. fermentum* group. However, most *L. acidophilus* strains used in probiotic fermented milk products have been identified as *L. gasseri* or *L. johnsonii*; both species are members of the *L. acidophilus* group.

In view of the large number of species within the genus of *Lactobacillus* that can be used as starter cultures, some detailed characteristics are shown in Table 7.5. However, the reviews by Sneath et al. (1986) and Hammes and Vogel (1995) are recommended for further reading regarding the specific characterisation of individual microbial isolates.

It can be observed, however, that many different species of lactic acid bacteria can be used as starter cultures in the dairy industry, and the overall classification is illustrated schematically in Figure 7.3.

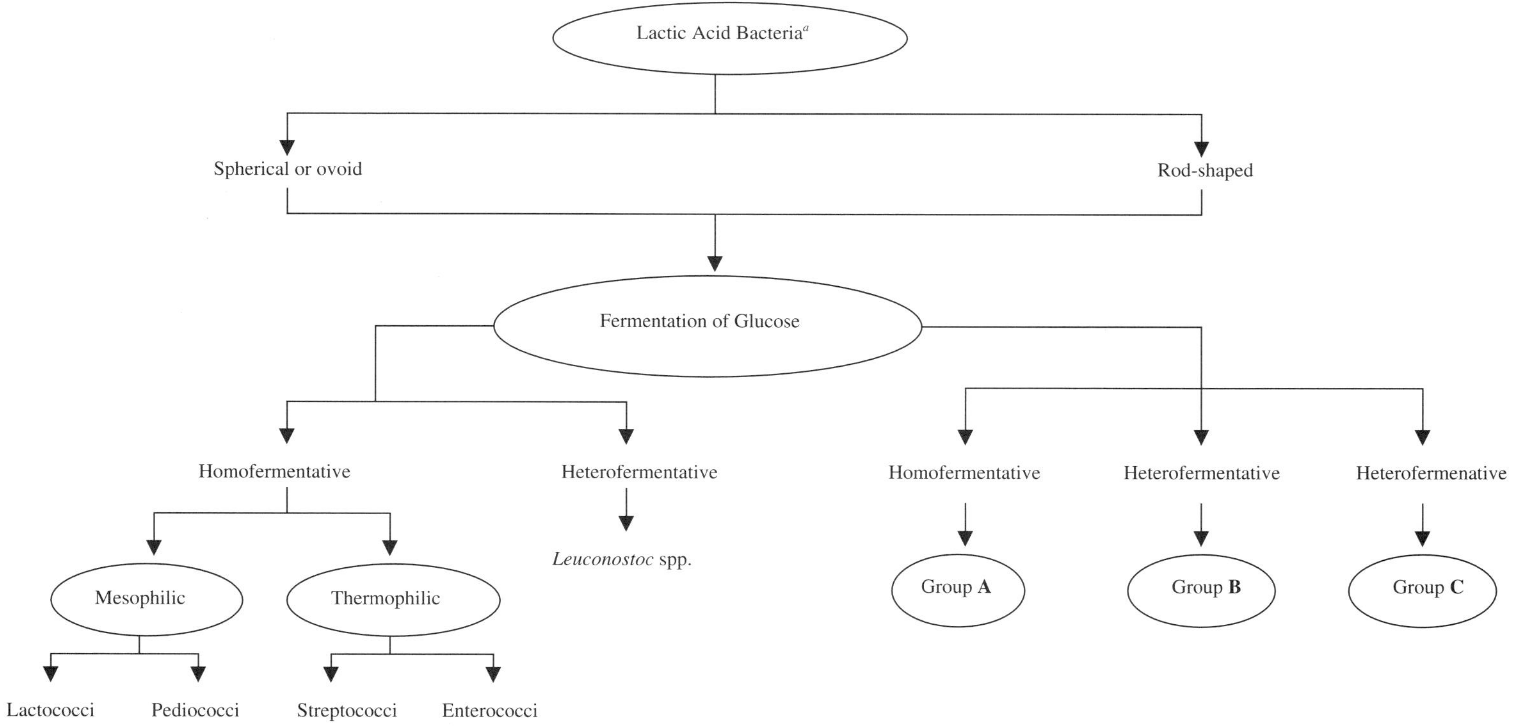

Figure 7.3. Classification and differentiation of main dairy starter cultures. [a]For further details or examples of species, refer to Tables 7.4, 7.5, and 7.6. [After Sharpe (1979) and Sneath et al. (1986).]

TABLE 7.5. Selected Characteristics of Some *Lactobacillus* spp. Associated with Fermented Milks and Cheeses

Starter Organisms/ Group[a]	Phylogenic Grouping[b]	G + C (Mean %)	Lactic Acid Isomer(s)	Growth 15/45°C	Carbohydrate Utilization[a,c]			
					ESC	AMY	ARA	CEL
Group **A**—obligately homofermentative								
L. acidophilus	**a**	34–37	DL	–/+	+	+	–	+
L. delbrueckii								
subsp. *delbrueckii*	**a**	49–51	D(–)	–/+	–	–	–	d
subsp. *bulgaricus*	**a**	49–51	D(–)	–/+	–	–	–	–
subsp. *lactis*	**a**	49–51	D(–)	–/+	+	+	–	d
L. gasserie	**a**	33–35	DL	–/+	+	+	–	+
L. helveticus	**a**	38–40	DL	–/+	–	–	–	–
L. johnsonii	**a**	33–35	DL	+/+		+		+
L. kefiranofaciens	**a**	34–35	D(L)	–/–		–		–
Group **B**—facultatively heterofermentative								
L. paracasei								
subsp. *paracasei*	**b**	45–47	L/DL[d]	+/d	+	+	–	+
subsp. *tolerans*	**b**	45–47	L(+)	+/–	–	–	–	–
L. rhamnosus	**b**	45–47	L(+)	+/+	+	+	d	+
L. plantarum	**b**	44–46	DL	+/–	+	d	+	+
Group **C**—obligately heterofermentative								
L. brevis	**b**	44–47	DL	+/–	d	–	+	–
L. fermentum	**b**	52–54	DL	–/+	–	–	d	d
L. kefir	**b**	41–42	DL	+/–	–	–	d	–
L. reuteri	**b**	40–42	DL	–/+	–	–	+	–
L. viridescens	**c**	41–44	DL	+/–	–	–	–	–

[a]For symbols refer to Table 7.4.
[b]Group **a**, **b**, and **c** species belong to the *L. delbrueckii* group, the *L. casei–Pediococcus* group, and the *Leuconostoc* group, respectively.
[c]ESC, aesculin; AMY, amygdalin; ARA, arabinose; CEL, cellobiose; FRU, fructose; GAL, galactose; LAC, lactose; MAL, maltose; MAN, mannitol; MNE, mannose; MLZ, melizitose; MEL, melibose; RAF, raffinose; RIB, ribose; SAL, salicin; SOR, sorbitol; SAC, saccharose/sucrose; TRE, trehalose.
[d]Strains designated *L. casei* subsp. *pseudoplantarum* produce DL lactic acid.
Source: Data compiled from Sneath et al. (1986) and Hammes and Vogel (1995).

7.3.2 Bacterial Species Incorporated into Dairy Starter Cultures

A wide range of microorganisms are sometimes added to milk (besides dairy starter cultures) for specific applications, and some examples follow.

***7.3.2.1 The Genus* Bifidobacterium.** The taxonomic situation and nomenclature of the bifidobacteria have changed since the turn of the twentieth century, when they were classified as *Lactobacillus* spp., while in the latest edition of *Bergey's Manual* (Sneath et al., 1986; Holt et al., 1994) the same organisms are grouped in a separate genus, *Bifidobacterium* (see also Klein et al., 1998). Currently, 30 different

Carbohydrate Utilization[a,c]													
FRU	GAL	LAC	MAL	MAN	MNE	MLZ	MEL	RAF	RIB	SAL	SOR	SAC	TRE
+	+	+	+	−	+	−	d	d	−	+	−	+	d
+	−	−	d	−	+	−	−	−	−	−	−	+	d
+	−	+	−	−	−	−	−	−	−	−	−	−	−
+	d	+	+	−	+	−	−	−	−	+	−	+	+
+	+	d	d	−	+	−	d	d	−	+	−	+	d
d	+	+	d	−	d	−	−	−	−	−	−	−	d
	+	d	+	−	+		d	d		+		+	d
	+	+	+	−			+	+		−		+	−
+	+	d	+	+	+	+	−	−	+	+	+	+	+
+	+	+	−	−	−	−	−	−	−	−	−	−	−
+	+	+	+	+	+	+	−	−	+	+	+	+	+
+	+	+	+	+	+	d	+	+	+	+	+	+	+
+	d	d	+	−	−	−	+	d	+	−	−	d	−
+	+	+	+	−	+	−	+	+	+	−	−	+	d
+	−	+	+	−	−	−	+	−	+	−	−	−	−
+	+	+	+	−	−	−	+	+	+	−	−	+	−
+	−	−	+	−	+	−	−	−	−	−	−	d	d

species of bifidobacteria isolated from humans, animals, insects, and the environment have been identified. Only six species (*Bifidobacterium adolescentis*, *breve*, *bifidum*, *infantis*, *lactis*, and *longum*) have attracted attention in the dairy industry, primarily for the manufacture of "bio" fermented milk products.

The differentiating characteristics (see Table 7.6) have been reviewed by many researchers (Bezkorovainy and Miller-Catchpole, 1989; Barlows et al., 1992; Tamime et al., 1995; Wood and Holzapfel, 1995; Kok et al., 1996; Meile et al., 1997; Salminen and von Wright, 1998; Cai et al., 2000), and other characteristics may include the following:

- The bacteria are Gram-positive, anaerobic heterofermentative, nonmotile, and non-spore-forming rods (0.5–1.5 × 1.5–8 μm).
- When grown aerobically in a trypticase–peptone–yeast medium, the cells of these bifidobacteria have distinctive shapes and arrangements (e.g., amphora-like, specific epithet, thin and short; very irregular contours and rare branching), while the rods on some occasions may have a slightly concave central region and

swollen ends (see Figure 7.4). However, it is not unusual to encounter cells that are coccoid or appear as very long or short bacilli of varying widths; or the cells may be V-, Y-, or X-shaped, depending on the constituents of the medium on which the colony is growing.

- The cell wall peptidoglycan varies among the species, and the guanine plus cytosine molecular percentage of the deoxyribonucleic acid (DNA) of the genus ranges between 54 and 67 (see Table 7.6).
- A wide range of components have been identified as bifidogenic growth stimulators (see Tamime et al., 1995; Tamime, 1997).
- Different species utilize different types of carbohydrate, and Table 7.6 illustrates some examples. Such fermentations are used for identification purposes, and one key enzyme involved is fructose-6-phosphate phosphoketolase (F-6-PPK), known as the "bifidum shunt"; this enzyme can be used to identify the genus. It should be noted that not all strains possess enough F-6-PPK for it to be identified. The fermentation of two molecules of glucose yield two molecules of lactate and three molecules of acetate.

***7.3.2.2 The Genus* Enterococcus.** Over the past 100 years, the classification of the streptococci (and enterococci) has changed. By the 1930s, the streptococci were divided into four groups: (a) the enterococci, (b) the lactic streptococci (at present is known as lactococci; see Section 7.3.1.1), (c) the viridance streptococci, and (d) the pyogenic streptococci. This classification relied for a long time on the seriological groups introduced by Lancefield (1933); however, 16S rNA sequencing within the genus *Entercoccus* has revealed the presence of three species groups, and some strains form individual lines of descent (Devriese and Pot, 1995). The enterococci that have been used in some cheese varieties and "bio"-fermented milk products (Tamime and Marshall, 1997; Klein et al., 1998) are as follows:

- *Enterococcus faecium* and *Entercoccus durans* are found in the first group based on 16S rNA sequencing.
- *Enterococcus faecalis* forms an individual line of descent based on 16S rNA sequencing.

Some selected characteristics of the above *Enterococcus* spp. are shown in Table 7.6. These bacteria are Gram-positive; they may be motile, non-spore-forming cocci or anaerobic homofermentative lactic acid; and they produce L(+)-lactate from glucose. The cell wall

TABLE 7.6. Some Selected Characteristics of Bifidobacteria and Enterococci Used for the Manufacture of "Bio"-Fermented Milks[a]

	Bifidobacterium spp.						*Enterococcus* spp.		
Characteristics	*adolescentis*	*bifidum*	*breve*	*infantis*	*lactis*	*longum*	*faecium*	*durans*	*faecalis*
G + C (mean %)	58.9	60.8	58.4	60.5	61.9	60.8	37–40	38–40	37–40
Type of peptidoglycan	Lys (Orn)-D-Asp	Orn (Lys)-D-Ser-D-Asp	Lys-Gly	Orn (Lys)-Ser-Ala-Thr-Ala	Lys (Orn)-Ala-(Ser)-Ala_2	Orn (Lys)-Ser-Ala-Thr-Ala	Lys-D-Asp	Lys-D-Asp	Lys-Ala_{2-3}
Carbohydrate utilization									
Arabinose	+	–	–	–	d	+	+	–	–
Cellobiose	+	–	d	–	–	–			
Fructose	+	+	+	+	+	+			
Galactose	+	+	+	+	+	+			
Gluconate	+	–	–	–	–	–	d	–	+
Inulin	d	–	d	d	+	–			–
Lactose	+	+	+	+	+	+			
Maltose	+	–	+	+	+	+			
Mannitol	d	–	d	–	–	–	d	–	+
Mannose	d	–	+	d	–	d			
Melezitose	+	–	d	–	–	+			+
Melibiose	–	d	–	+	+	+	+	–	–
Raffinose	+	–	+	+	+	+	–	–	–
Ribose	+	–	+	+	+	+			
Salicin	+	–	+	–	–	–			
Sorbitol	d	–	d	–	+	–	–	–	+
Starch	+	–	–	–	–	–			
Sucrose	+	d	+	+	+	+	+	–	
Trehalose	d	–	d	–	+	–			
Xylose	+	–	–	d	d	d	–	–	–

[a]For identification of symbols see Table 7.4.

Source: Data compiled from Devriese and Pot (1995) and Tamime and Robinson (1999).

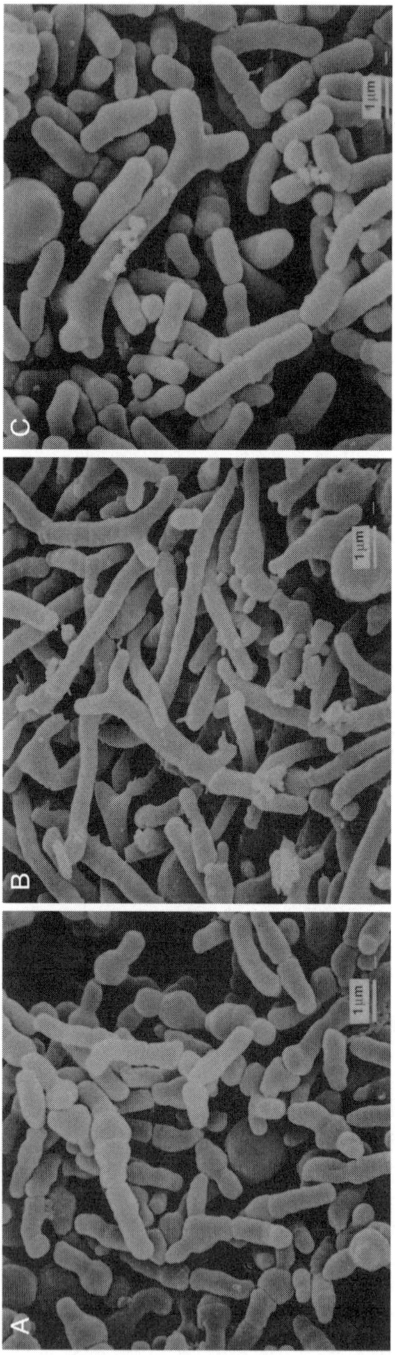

Figure 7.4. Illustrations of different cellular morphology of (A) *Bifidobacterium infantis*, (B) *Bifidobacterium animalis* (this organism has been reclassified as *Bifidobacterium lactis*; Meile et al., 1997, but recently it has been renamed as *Bifidobacterium animalis*; Cai et al., 2000). (Reproduced by courtesy of V. Bottazzi, F. Bianchi, and C. Garabazza.)

peptodoglycan is Lys-D-Asp in all species, except *E. faecalis* (Lys-Ala_{2-3}). The minimal nutritional requirements of these microorganisms are complex.

***7.3.2.3 The Genus* Propionibacterium.** The genus *Propionibacterium* has two principal groups of organisms: (a) strains associated with dairy products and (b) others which are found on human skin and in the intestine. The former strains are known as "classic propionibacteria" or "dairy propionibacteria" (Sneath et al., 1986; Britz and Riedel, 1991). The most important species is *P. freudenreichii*, which is widely used in Swiss cheese varieties (Emmenthal and Gruyère), mainly for its ability to produce large gas holes in the cheese during the maturation period. However, other propionibacteria have been isolated from dairy products—for example, *P. jensenii*, *P. thoenii* (isolated from red spots in Emmenthal cheese), and *P. acidipropionici*.

The cell of all species of propionibacteria are irregularly staining Gram-positive, nonmotile and non-spore-forming rods (see Figure 7.5).

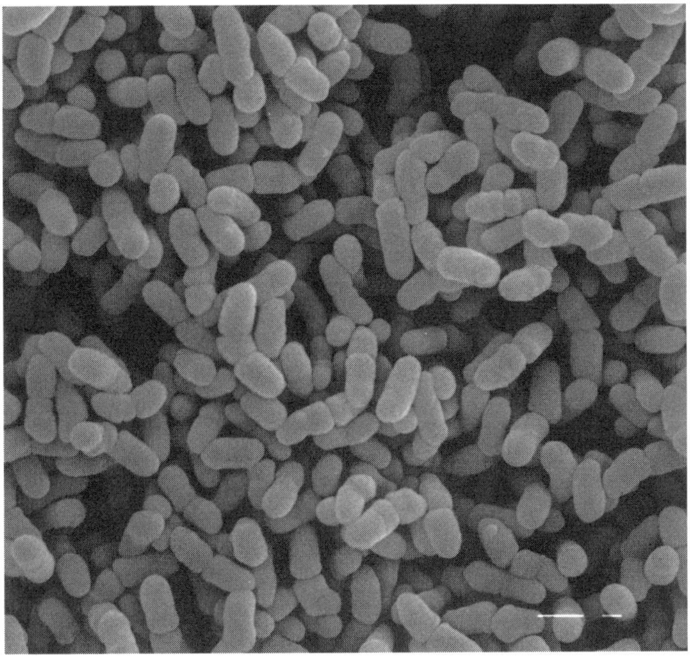

Figure 7.5. An illustration showing the microbial cell morphology of *Propionibacterium* spp. grown on lactate agar. (Reproduced by courtesy of V. Bottazzi, F. Bianchi, and C. Garabazza.)

The "classic" *Propionibacterium* spp. are short and rather thick, but variation in morphology is related to strain and phase of growth (e.g., early log phase). *Proprionibacterium thoenii* and *P. jensenii* are capsulated in wet mount India ink preparations, and some characteristics of *Propionibacterium* spp. are shown in Table 7.7; others may include the following:

- Some strains produce EPS material.
- The cell wall peptidoglycan consists of L-DAP (i.e., diamino acid such as Ala, Glu, or Gly; however, Gly is not present in strains with *meso*-DAP) and polysaccharides (e.g., mannose, galactose, glucose, rhamnose).

TABLE 7.7. Some Selected Characteristics of Propionibacteria and Brevibacteria[a]

	Propionibacterium spp.			*Brevibacterium* spp.	
Characteristic	*freudenreichii*	*jensenii*	*thoenii*	*linens*	*casei*
G + C (mean %)	65	67	66		
Type of peptidoglycan	*meso*-DAP	L-DAP	L-DAP		
Carbohydrate utilisation					
Amygdalin	–	d+	d+	All *Brevibacterium*	
Arabinose	+	–	–	spp., including *B.*	
Cellobiose	–	d-	–	*linens* and *B. casei,*	
Erythritol	+	+	d+	produce slight or	
Aesculin	–	–	+	no acid from	
Fructose	+	+	+	glucose or other	
Galactose	+	+	+	carbohydrates in a	
Glucose	+	+	+	peptone medium.	
Lactose	d–	d+	d-		
Maltose	–	d+	d+		
Mannitol	–	+	–		
Mannose	+	+	+		
Melezitose	–	d+	d+		
Melibiose	d-	+	d+		
Raffinose	–	+	d+		
Ribose	d+	+	+		
Sorbitol	–	–	d+		
Starch	–	–	+		
Sucrose	–	+	d+		
Trehalose	–	+	+		
Xylose	–	d+	d+		

[a]Symbols: +, positive reaction in 90–100% of strains or pH < 5.7 in 90–100% of strains; –, negative reaction in 90–100% or pH > 5.7 in 90–100% of strains; d–, reaction positive in 10–40% of strains; d+, reaction positive in 40–90% of strains.

Source: Data compiled from Sneath et al. (1986).

- In nutritional requirements, all strains of propionibacteria seem similar (for example, in their B-vitamin requirement), and oleate stimulates the growth of some strains; however, *P. freudenreichii* produces large quantities of vitamin B_{12}.

Recently, the review by Mantere-Alhonen (1995) has reported that *P. freudenreichii* can be considered as a potential probiotic microorganism, a view based on (a) the production of propionic acid, (b) bacteriocins, (c) sythesis of vitamin B_{12}, (d) better exploitation of fodder, (e) growth stimulation of other beneficial bacteria, and (f) survival during gastric digestion.

***7.3.2.4 The Genus* Brevibacterium.** The *Brevibacterium* spp. (*B. linen* and *B. casei*) impart a distinctive reddish-orange color to the rind, or cause the formation of a smear, on Brick and Limburger cheese (Olson, 1969; Reps, 1993); they also produce a gray white color on traditional Camembert (Gripon, 1993). The microorganisms are characterized by having coccal-rod cells of variable length, but generally 0.6–1.0 μm in diameter (Sneath et al., 1986). The bacteria are Gram-positive, obligate aerobes, and the optimum growth temperatures are 20–25°C (*B. linens*) or 30–37°C (*B. casei*). Some selected characteristics are shown in Table 7.7; in addition, these organisms are salt-tolerant, nonmotile, and not acid-fast, and they produce no endospores. The cell wall peptidoglycan contains *meso*-DAP without arabinose. The nutritional requirements is complex and not fully established, but most strains require amino acids and B vitamins (Sneath et al., 1986). Recently, the purification and characterization of serine proteinases produced by *B. linens* has been reported by Hayashi et al. (1990).

Their primary function in cheesemaking, besides their pigment production, is the formation of aromatic sulfur compounds. The commercial freeze-dried cultures available on the market are nonlipolytic and have varying proteolytic activities—that is, medium or high (Bockelmann, 1999).

7.3.2.5 Miscellaneous Microorganisms. At present, there are several commercial preparations of either single or mixed cultures that are primarily used as cheese milk adjuncts. These can be added into the cheese milk or brushed onto the smear cheese varieties. Some examples include coryneform bacteria, yeasts (*Debaromyces hansenii*, *Candida utilis*, and *Saccharomyces cervisiae*; see Figure 7.6A), and micrococci (e.g., *Staphylococcus xylosus* and *Staphylococcus carnosus*). The latter microorganism helps the texture and aroma development in

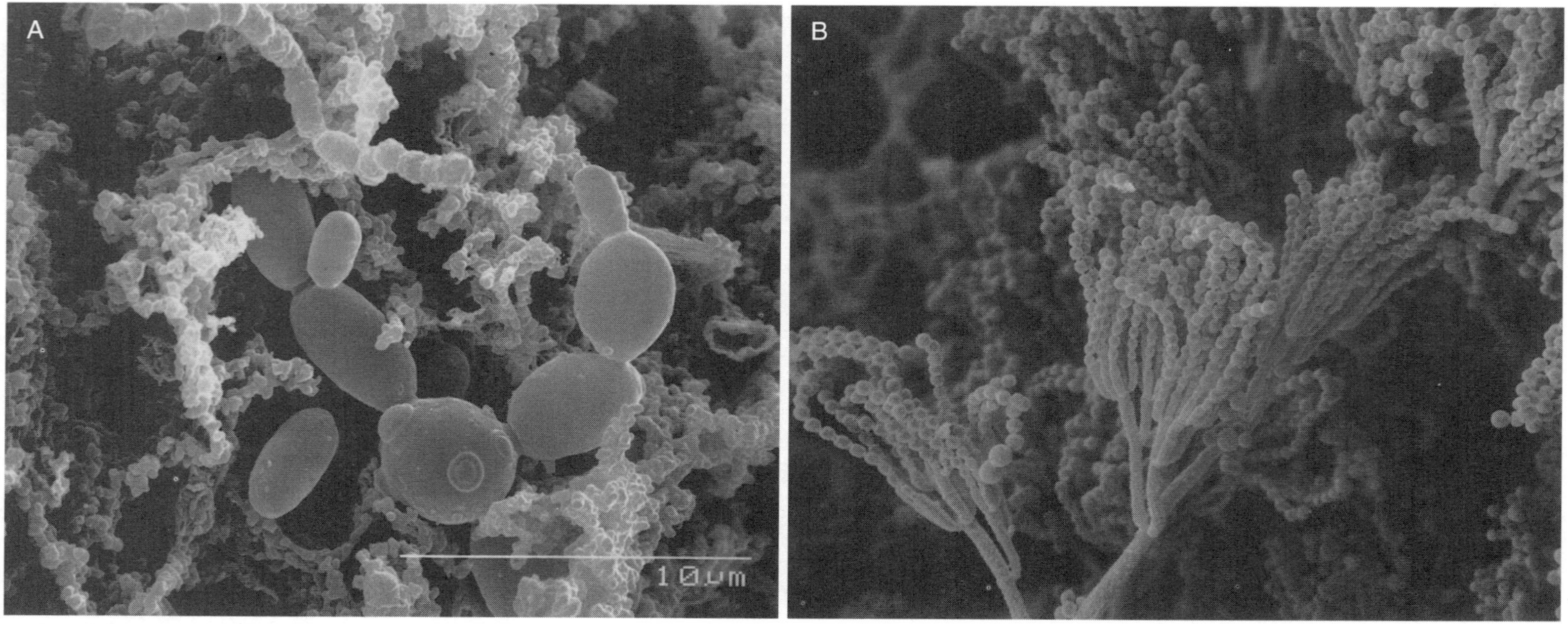

Figure 7.6. The microstructure of Gorgonzola cheese showing the presence of yeast and lactic acid bacteria (A), and typical compact branching of *P. roqueforti* (B). (Reproduced by courtesy of V. Bottazzi, F. Bianchi, and C. Garabazza.)

the cheese, while the yeast strains help to replace the indigenous yeast flora in raw milk that has been inactivated by pasteurization (Bockelmann, 1999; Choisey et al., 2000).

7.3.2.6 Molds. Molds are mainly used in the cheese industry for the manufacture of some semisoft cheese varieties. Their major role is to enhance the flavor and aroma and to modify, slightly, the body and texture of the curd (Cerning et al., 1987; Leistner, 1990; Gripon, 1993). The molds can be divided into two types, taking into consideration their color and growth characteristics, namely the white and blue molds (see Figure 7.7). The former type, which grows externally on the cheese (e.g., Camembert and Brie) is known as *Penicillium camemberti;* however, in some older dairy textbooks/literature or commercial culture suppliers, the following *Penicillium* species have been reported (*P. caseicolum*, *P. caseicola*, *P. candidum*, and *P. album*), but these are now considered biotypes of, or synonyms for, *P. camemberti* (Pitt, 1979; Ramirez, 1982).

The blue mold, *P. roqueforti*, grows internally in the cheese, and examples of "blue cheeses" are Roquefort, Blue Stilton, Danish Blue, Gorgonzola (see Figure 7.6B), and Mycella; other synonyms for this blue mold may include *P. gorgonzolae* and *P. stilton* (Pitt, 1979; Ramirez, 1982). However, Boysen et al. (1996) have reported that although the current classification of *P. roqueforti* is into two varieties (i.e., *P. roqueforti* var. *roqueforti*, which is used in cheesemaking, and the ubiquitous patulin-producing variety known as *P. roqueforti* var. *carneum*), it should be reclassified into three species (*P. roqueforti*, *P. carneum*, and *Penicillium paneum*) on the basis of molecular genetic and biochemical profiles. Some characteristics of blue mold cultures are shown in Figure 7.7.

Other genera of molds which have very limited application, or are traditionally used in some parts of the world, are *Mucor rasmusen*, used in Norway for the manufacture of ripened skimmed milk cheese, and *Aspergillus oryzae*, used in Japan for the production of soya milk cheese varieties (Kosikowski and Mistry, 1997).

Geotrichum candidum grows on the surface of the milk to form the white velvet layer on Viili, which is a cultured milk product from Finland (Tamime and Robinson, 1988a; Tamime and Marshall, 1997; see also Chapter 8). For further information regarding the systematics of *Penicillium* and *Aspergillus* species, refer to the latest textbooks by Samson and Pitt (1985) and Peberdy (1987).

7.3.2.7 Yeasts. In general, the presence of yeasts in dairy products is considered as contamination and has negative aspects (IDF, 1998;

MOLDS	White molds	*P. camemberti*	Pale-gray-green colonies; it is highly proteolytic and produces a soft product. Grows on agar showing abundant conidial heads and also showing conidiophores arising from aerial hyphae as well as substrate. Conidia are uncolored.
	Blue molds	*P. roqueforti*	Dark green, velvety molds; simple monoverticillate, penicilli or verticils of metula and sterigmata, or compact branching systems; conidorphores are short and ascend from aerial loops or submerged sections of vegetative hyphae.
	Miscellaneous	{ *M. rasmusen* { *A. oryzae* { *G. candidum*	
YEASTS		*Candida kefyr*	Morphological characteristics: Growth at 25°C for three days in glucose/yeast extract/peptone broth produces short to long ovoid cells. Growth on agar medium (as above) for one month; the culture is creamy-yellowish in color, soft, semi-dull and finely reticulate. Growth on corn meal agar; the pseudomycelium appears as coarse and slightly curved branches. G & C mean % in DNA is 41.3 and is able to ferment glucose, galactose, and sucrose (lactose fermentation is variable).

Figure 7.7. Classification and differentiation of yeasts and molds used as dairy starter cultures. [After Raper and Thom (1968), Pitt (1979), Kreger-van Rij (1984), and Kurtzman and Fell (1998).]

Loureiro and Querol, 1999). However, as mentioned elsewhere (see Section 7.3.2.5), certain species of yeast are added to cheese milk; and in specific application(s) in the dairy industry, the addition of yeasts to milk, besides the lactic acid bacteria, results in a yeast-lactic fermentations (see Table 7.1). This type of fermentation is limited to the manufacture of Kefir and Koumiss. In the Kefir starter culture, often known as Kefir grains, the exact population of microorganisms is still somewhat controversial, and the latest reviews by Tamime and Robinson (1988a) and Tamime and Marshall (1997) reported the yeast species that have been identified in Kefir (see Table 7.8). It was suggested by many researchers in the past that *Candida kefyr* (old name *Candida kefir*) and *L. kefir* are the only organisms that are intimately associated with Kefir grains, but other researchers have included more yeast species, acetic acid bacteria and lactic acid bacteria (Tamime and Marshall, 1997); Figure 7.7 shows some characteristics of *C. kefyr*. With regard to the yeast species, the confusion may have arisen because of the following aspects: First, the same yeast species may be known by two generic names depending on spore formation (i.e., imperfect or perfect stage; *C. kefyr* and *Candida valida* (imperfect) are synonymous with *Kluyveromyces marxianus* var. *marxianus* and *Pichia membranaefaciens* (perfect), respectively. Second, synonymous names are evident between old and new nomenclature (see Table 7.8). The reviews by Kreger-van Rij (1984), Koroleva (1991), Kurtzman and Fell (1998), Tamime and Marshall (1997), and Valderrama et al. (1999) are recommended for further reading. Furthermore, *G. candidum* and acetic acid bacteria are considered as contaminants in Kefir grains in some countries (see Tamime and Marshall, 1997; see also Chapter 8).

In Japan, Toba et al. (1987, 1990, 1991) have isolated a capsular polysaccharide producing *L. kefiranofaciens* K_1 from Kefir; and a fermented milk was prepared from the isolated strain, which had a ropy consistency and was resistant to syneresis.

Again, the microflora of Koumiss is not well-defined, but consists mainly of lactobacilli, lactose-fermenting (*Saccharomyces lactis*, *Torula koumiss*) and non-lactose-fermenting (*Saccharomyces cartilaginosus*) yeasts, and the non-carbohydrate-fermenting yeast (*Mycoderma* spp.) (Koroleva, 1991; Tamime and Marshall, 1997; Oberman and Libudzisz, 1998).

It can be observed that many different types of yeasts and molds are used as starter cultures in the dairy industry, and some differentiating characteristics are shown in Figure 7.7.

TABLE 7.8. The Yeast Microflora of Kefir Grains

Microflora	Latest Nomenclature
Saccharomyces cerevisiae	
delbrueckii	*Torulaspora delbrueckii*
florentinus	*Zygosaccharomyces florentinus*
exiguus	
fragilis	*Kluyveromyces marxianus* var. *fragilis*
carlbergensis	*Saccharomyces cerevisiae*
globus	*Saccharomyces cerevisiae*
dairensis	
unisporus	
kefyr[a]	*Kluyveromyces marxianus* var. *marxianus*
Kluyveromyces marxianus	*Kluyveromyces marxianus* var. *marxianus*
lactis	*Kluyveromyces marxianus* var. *lactis*
Candida kefyr[a]	
pseudotropicalis	
tenuis	
holmii	
valida	
friedrichii	
Mycotorula kefyr	*Kluyveromyces marxianus* var. *marxianus*
lactis	
lactosa	
Torulopsis kefyr	
holmii	*Candida holmii*
Cryptococcus kefyr	*Kluyveromyces marxianus* var. *marxianus*
Torulaspora delbrueckii	
Pichia fermentans	
membranaefaciens	

[a]Previously known as *kefir*.

Source: Adapted from Tamime and Marshall (1997).

7.4 TERMINOLOGY OF STARTER CULTURES

7.4.1 Background

The microorganisms, which are employed in the dairy industry, are used either *singly*, *in pairs*, or *in a mixture*, thus giving the industry the opportunity to manufacture different types of cheeses, fermented milks, and cultured cream products. Examples of these applications include the following:

- Cheese (Sellars, 1967; Lücke et al., 1990; Robinson, 1995; Cogan et al., 1997; Tamime, 2000)

- Fermented milks including "bio" products (Tamime and Marshall, 1997; Tamime, 1998; Tamime and Robinson, 1999)
- Cultured cream products (Tamime and Marshall, 1997)
- Alcoholic/lactic beverages (Tamime and Marshall, 1997)

Traditionally, the fermentation of milk was achieved either by leaving the milk at room temperature for a time or, alternatively, by seeding the fresh milk with small quantities from a previous day's good-quality product (i.e., inoculum). In both instances, the indigenous microorganisms (e.g., lactic acid bacteria) utilize the lactose in milk as an energy source for growth and, subsequently, yield lactic acid [see the reviews by Cogan (1995) and by Marshall and Tamime (1997)]. The adoption of the latter approach meant that the fermentation process was, to some extent, brought under control; and this technique, which was passed from one generation of producers to the next, may be considered the "primitive" origin of the starter cultures known today. This craft has survived for centuries, because it is only recently that humans discovered microorganisms and their essential role of the fermentation of milk.

Mesophilic lactic starter cultures are widely used in the dairy industry (see Table 7.4); and in the case of cheese, for example, they are divided into these categories: single, pairs, multiple, or mixed strains; these starter systems have been developed in order to achieve the following objectives during cheesemaking:

- Better control of starter culture activity (i.e., level of acidity)
- Development of flavor
- Intolerance of salt and cooking temperature
- Safeguard against bacteriophage (phage) attack
- Minimize variation in the quality of end product
- Control of the level of bitterness and/or gas production in certain cheeses, such as Cheddar

7.4.2 Defined Starter Cultures

A few decades ago, mixed-strain cultures of *unknown* composition were used without rotation for the production of Cheddar cheese in many different countries. Certain faults, such as open texture, were identified as being due to the presence of gas-producing strains, while bacteriophage attack on these mixed-strain cultures was not well-identified, possibly due to the development of mutant strains that were insensitive to phages.

At present, cheese starter cultures are categorized on the basis of flavor and/or gas production (Cogan, 1983; Gilliland, 1985; Cogan and Hill, 1993; Cogan and Accolas, 1996) as follows:

Type	Microorganism	Flavor Producer
B or L	*Leuconostoc* spp.	Positive
D	*Lactococcus lactis* subsp. *lactis* biovar *diacetylactis*	Positive
BD or DL	Mixture of the above two species	Positive
O	*Lactococcus lactis* subsp. *lactis* and subsp. *cremoris*	Negative

Incidentally, Crawford (1972) has referred to the pairing of *L. lactis* subsp. *lactis* and *cremoris* as "the non-gas- and non-aroma-producing cheese lactic starters." However, the flavor-producing species are mainly citrate utilizers that yield diacetyl and CO_2 (Cox et al., 1978; Cogan and Hill, 1993). The aroma-producing lactic starters are essential for the production of buttermilk, sour cream, cultured butter, Ymer, Filmjölk, and other fermented milk products (see Tamime and Robinson, 1988a; Tamime and Marshall, 1997).

Defined-strain, starter culture systems (i.e., single, pairs or multiple) were pioneered in New Zealand at the Dairy Research Institute, which dates back over the past five decades. The detailed evolution of such systems has been recently reviewed by Lawrence and Heap (1986) and Heap and Lawrence (1988), and a brief summary of these developments is given below:

In the 1930s: Commercial, mixed-strain cultures in powder form were imported from Europe for the production of Cheddar cheese and apparently caused "open texture" defects.

By 1935: Whitehead isolated pure lactic lactococci from these cultures (e.g., gas- and non-gas producers), and the latter type was introduced to cheese factories as *single-strain* cultures that eventually failed due to infection by phage.

In the 1940s: Rotation of *pairs of single-strain* starter cultures was introduced to cheese factories up to 1950, and the bulk starter was produced in a mechanically protected tank to keep out phage (see section on production systems of bulk starter cultures).

In the 1960s: Centralization of cheese production meant that vats were filled two or three times a day, but eventually it was difficult to keep phage under control despite the availability of a large number of starters—including up to 14 days' rotation; furthermore, the defects in the cheese were identified as being "acid" and bitter flavor.

In the 1970s: Slow and fast acid producers in combinations of single strains (e.g., a pair of AM_2 and ML_8) were introduced to overcome the above defects in the cheese and, in view of the knowledge gathered, for selecting and screening strains for their sensitivity/insensitivity to phage, temperature, and salt. A ratio of 2:1 (slow:fast) was adopted by the industry, and a 4-day rotation was used. However, the use of *triplets* (i.e., one strain is temperature-insensitive and two strains are temperature-sensitive) has also been recommended by the Institute, but in practice it was observed that better control of the desirable ratio of these strains could be maintained by using "pairs" of starter strains that gave a typical Cheddar cheese flavor. By the end of the 1970s, a *multiple* starter system of six strains was introduced (e.g., MS_6), which can be used without rotation for an extended period of time, and it was recommended that at least one of the phage-sensitive strains should be replaced each year. A similar approach has been adopted in the United States, Australia, and Ireland (Cogan and Accolas, 1996).

By the 1980s: Work at the NZ Dairy Resarch Institute continued to differentiate many types of phage capable of attacking the *Lactococcus* spp. (Cogan and Accolas, 1990); but two new isolates, which were temperature-insensitive, appeared to be free from all phage. Thus, since 1980 most cheese factories have been using this new "*single-pair*" starter culture system without rotation (Lawrence et al., 1984); more recently, *new* phages have begun to attack this culture, and the fight against phage is still going on.

By 1990s: DVI cultures have been widely used in the industry, including genetic manipulation of lactic acid bacteria to combat bacteriophage attack. Furthermore, currently *S. thermophilus* is blended with DVI mesophilic lactococci for the manufacture of Cheddar-type cheeses because such starter cultures have a high degree of bacteriophage durability due to the biphasic growth of the microfloras during cheesemaking (Stanley, 1998).

In The Netherlands, a different approach has been adopted using a phage-insensitive, *mixed-strain* starter culture system that is highly dependent on the method of handling and storage of the microorganisms. The work was first pioneered by Galesloot et al. (1966), who observed that the propagation of mixed-strain starter cultures under factory conditions (i.e., not aseptic) involved both phage-sensitive and phage-insensitive strains. The latter strains become predominant and, while they may show some fluctuation in their activity, there is never complete failure. These cultures are referred to as "P" cultures (i.e., practice); however, the same cultures propagated in the laboratory under aseptic conditions are called "L" cultures, which are found to be more susceptible to phage vis-à-vis the "P" cultures. Hence, screening of the best "P" cultures at The Netherlands Dairy Research Institute (NIZO) brought about the best strain isolates in terms of (a) ability to produce good-quality cheese, (b) phage-insensitivity, (c) appropriate rate of acid development, and (d) best flavor profile. The "L" cultures are maintained at –196°C with the minimum number of transfers, and they are dispatched to the industry, when required, to be used in multifill cheese vats without rotation. The performance of these cultures is ensured by producing the bulk starter culture in a tank protected against bacteriophage and by minimizing the buildup of phage in the factory. The mechanism(s) of phage-insensitivity in these cultures is very complex and could be due to (a) cell wall resistance, (b) resistance due to restriction/modification, (c) resistance due to lysogeny, or (d) resistance due to other reasons (Stadhouders and Leenders, 1984; Stadhouders, 1986); however, some strains are not attacked by any phage, and the starter has to be contaminated by a great number of different phages before even all the sensitive strains are attacked (see also Hugenholtz, 1986).

Other types of mesophilic lactic starter cultures are proteinase-deficient strains (Prt^-) and whey-derived starters. The latter type has been developed in Australia (Hull, 1977, 1983); and, in principle, it could be considered as a modification of the *single-strain* system that was developed in New Zealand. The moment phage is detected in the starter culture, and the culture is replaced by a resistant derivative that has been challenged in a growth medium containing filtered whey—that is, free from bacteria but not bacteriophage. It has been reported that more than 50% of Australian Cheddar cheese is manufactured by using "whey-derived, single-strain, phage-resistant starters"; however, the use of such starters alone may cause a "bitter flavor" defect in the cheese (Heap and Lawrence, 1988).

Prt^- starter strains have been used in cheesemaking for some time, and they overcome the problem of phages due to their slow growth.

Prt^- are used in conjunction with some Prt^+ strains, because the former type depends on a supply of amino acids for growth. These organisms have other advantages, which can be summarized as follows: (a) They are more resistant to inhibition due to the presence of antibiotics in milk; (b) they are less proteolytic, thereby minimizing "bitter flavor" production in cheese; (c) by increasing the inoculation rate of Prt^- heat-sensitive strains, more acid development occurs at cooking temperature as compared with Prt^+ strains; (d) cell growth and acid production take place in the bulk starter tank and, to a lesser degree, in the cheese vat due to the longer generation time; and (e) they eliminate bacteriophages (see Gilliland, 1985; Hugenholtz, 1986; Oberg et al., 1986; Hugenholtz et al., 1987).

Thermophilic lactic starter cultures are used for the manufacture of yogurt, acidophilus milk, and high-temperature scalded cheeses (e.g., Swiss and Italian varieties). These thermophilic cultures are classified into two main types: (a) natural starters of undefined strains (see Section 7.4.3) and (b) the defined starters. Examples of the defined, thermophilic starter culture systems are as follows: (a) *S. thermophilus* and *Lactobacillus delbrueckii* subsp. *bulgaricus*, where acid production is enhanced due to the growth association between these species (Marshall and Tamime, 1997; Tamime and Robinson, 1999; see also Figure 7.8); (b) a single-strain starter culture of *Lactobacillus acidophilus* is mainly used for the production of acidophilus milk, and, because this microorganism tends to grow slowly in milk, the manufacture of different therapeutic products involves *L. acidophilus*, yogurt starter cultures, and bifidobacteria in a multitude of combinations (Tamime and Robinson, 1988b, 1999; Tamime and Marshall, 1997); (c) *L. paracasei* biovar *shirota* or a mixture with *L. acidophilus*, *Bifidobacterium* spp., and mesophilic and/or thermophilic lactic acid bacteria are used as a defined starter culture for the production of "bio"-fermented milk products.

In some instances, certain strains of cheese and yogurt starter cultures are EPS producers (Cerning, 1990; Tamime and Robinson, 1999; Kitazawa et al., 2000; see also Chapter 8), and the use of such strains in yogurt making can modify the consistency/viscosity of the coagulum. However, few phage problems have ever been reported in the yogurt industry, and Lawrence and Heap (1986) suggested the following possible reasons:

- The phage of thermophilic streptococci are larger vis-à-vis the phage that can attack mesophilic strains, and they replicate more slowly.

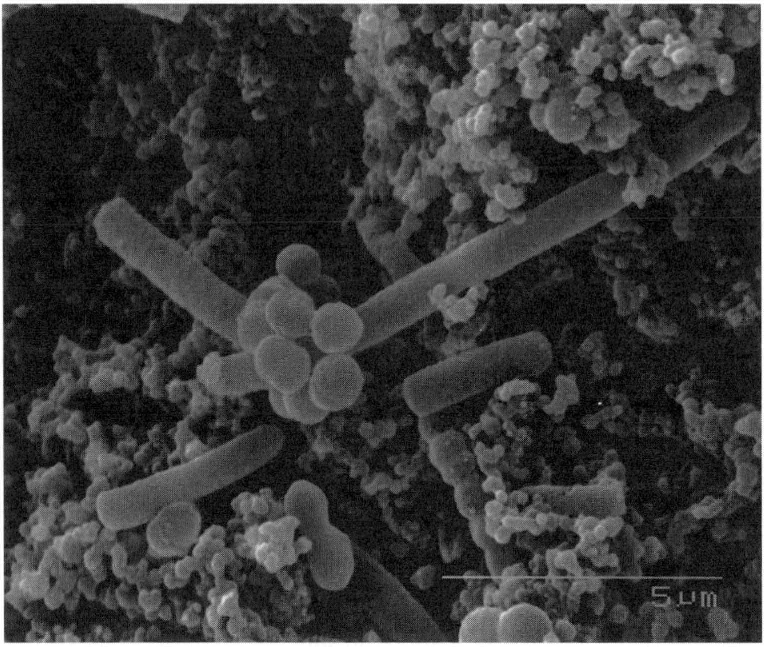

Figure 7.8. Structural characteristics of the casein micelles in yogurt and typical microbial yogurt cells of *S. thermophilus* and *L. delbrueckii* subsp. *bulgaricus*. (Reproduced by courtesy of V. Bottazzi, F. Bianchi, and C. Garabazza.)

- The phage that can attack the lactobacilli and streptococci are unrelated, and the opportunity for the phage of the lactobacilli to replicate is limited; the lactobacilli begin to grow at pH < 5.2 (i.e., the milk has become viscous), and the spread of phage in the milk is reduced.
- No whey separation occurs during the manufacturing stages, and thus the spread of phage in the factory is minimized.
- In Swiss and Italian cheeses, the lactobacilli start to grow after the cooking stage, which limits the propagation of phage.

7.4.3 Undefined Starter Cultures

The undefined, thermophilic, lactic starter cultures are sometimes known as "artisanal" or traditional starters, and a typical application is their use during the production of Swiss and Italian cheeses. Recently, Limsowtin et al. (1996) have reviewed these starter cultures in detail and have classified them into three main groups: (a) artisanal or "natural" milk cultures, (b) whey cultures, and (c) whey cultures plus rennet.

Natural milk starter cultures are made from heat-treated milk, followed by incubation at high temperature, which creates a selective condition for the growth of *S. thermophilus*. In some instances, *Enterococcus* spp. may be present.

Whey starter cultures (i.e., freshly drawn whey from the cheese vat) have been used in Italy since 1890 (Wood, 1981). The microflora is undefined, but is used successfully for the production of Grana and other cheeses using raw milk. Over the years, *S. thermophilus* has become predominant in low-acid, whey cultures, whereas the lactobacilli have become predominant in high-acid, whey cultures consisting mainly of *L. helveticus*, *L. delbrueckii* subsp. *lactis*, *Lactococcus* spp., and enterococci; however, *L. delbrueckii* subsp. *bulgaricus* and *L. fermentum* may present in lower numbers or occasionally (Coppola et al., 1988, 1990). These cultures are used during the manufacture of water buffalo Mozzarella and other Italian cheese; in the latter products, the lactobacilli mentioned earlier are predominant. These cultures are still widely used, and in 1974, laws were passed in Italy to prohibit the use of reconstituted milk powder or phosphate-buffered medium for the production of bulk starters, which can only be produced from fresh milk.

Italian-type cheeses that are manufactured in the United States are made from heat-treated milk (possibly a few days old) and the use of defined cultures that are preserved in liquid nitrogen and propagated in phosphate-buffered medium.

Whey cultures plus rennet are another type of artisanal whey starter culture of undefined strains, and they are widely used by the traditional Swiss cheesemakers. The starter is prepared by macerating air-dried calf stomachs in fresh or deproteinized whey. According to Auclair and Accolas (1983) and Accolas and Auclair (1983), such artisanal starters may consist of the following thermophilic, lactic acid bacteria: the streptococci (*S. thermophilus* and, in some instances, fecal enterococci, possibly *E. faecium*) and the lactobacilli (*L. fermentum*, *L. helveticus*, *L. delbrueckii* subsp. *lactis*, and, to a lesser extent, *L. delbrueckii* subsp. *bulgaricus* and *L. acidophilus*). Such types of culture contain both the milk-coagulating enzymes and phage-insensitive, thermophilic starter strains.

In Switzerland, the Federal Dairy Research Institute has screened large numbers of artisanal starters obtained from factories producing Emmenthal cheese, with a view toward minimizing some of the faults in the cheese. The isolates were screened for (a) production of acid,

(b) proteolytic activity, (c) storage stability, (d) cheesemaking performance, and (e) presence of phage. However, the quality of the cheese is always monitored, and if a starter shows any irregularities at the factory level, the starter is quickly modified (Auclair and Accolas, 1983). Incidentally, the role of the propionic acid bacteria should not be overlooked, because they are important, during the secondary fermentation/maturation period of cheese, for the production of CO_2 giving rise to the characteristic "eyes" in Emmenthal and Gruyère cheeses.

The differentiating characteristics of certain species of lactobacilli are shown in Table 7.5. *L. delbrueckii* subsp. *bulgaricus* and *L. delbrueckii* subsp. *lactis* are closely associated and mainly produce D(−)-lactic acid, are galactose-negative (gal^-), and show the presence of metachromatic granules. *Lactobacillus helveticus* produces DL-lactic acid, is galactose-positive (gal^+), and lacks metachromatic granules. The accumulation of galactose in Swiss cheese, due to the activity of *S. thermophilus*, is not desirable, and it may cause "off-flavor" development by the nonstarter culture bacteria. Gal^+ lactobacilli should also be used (i.e., *L. helveticus*), which is the case for the manufacture of Swiss cheese in Europe. However, in the United States, *L. delbrueckii* subsp. *bulgaricus* is used instead, but, according to Turner and Martley (1983), such lactobacilli, which are gal^+, are incorrectly classified and should be redesignated as *L. helveticus*. This confusion may explain why different types of lactobacilli are apparently used during the manufacture of Swiss-type cheeses in different parts of the world.

7.4.4 Lactic/Yeast Starter Cultures

The combined activity of mesophilic and thermophilic lactic acid bacteria and yeasts yields a lactic acid/alcohol fermentation in milk (e.g., Kefir and Koumiss). Ethyl alcohol is mainly produced, and the level can reach as high as $2\,g\ 100\,ml^{-1}$; the flavor components are acetaldehyde, diacetyl, and lactic acid. The Kefir grains are irregular and whitish in color, and the bacteria and yeasts are held together due to the formation of a glucose–galactose polymer. Chapter 8 is recommended for further reading regarding the microflora of such types of starter culture.

7.4.5 Lactic/Mold Starter Cultures

Finally, mesophilic lactic acid bacteria are mixed with *Penicillium* spp. for the production of mold-ripened cheeses (Marth, 1987) or with *G. candidum* for the production of Viili (Tamime and Marshall, 1997;

see also Chapter 8). It is important to note that these organisms are added separately to the processed milk rather than being combined as in the case, for example, of Kefir grains.

7.5 STARTER CULTURE TECHNOLOGY

The fermentation process of any cultured dairy product relies entirely on the purity and activity of the starter culture, provided that the milk or growth medium is free from any inhibitory agent (e.g., antibiotics or bacteriophage). The traditional method for the production of bulk starter is illustrated in Figure 7.9 (system 1); and although the propagation procedure is time-consuming, requiring skilled operators and may lead to contamination by bacteriophage (which is one of the major

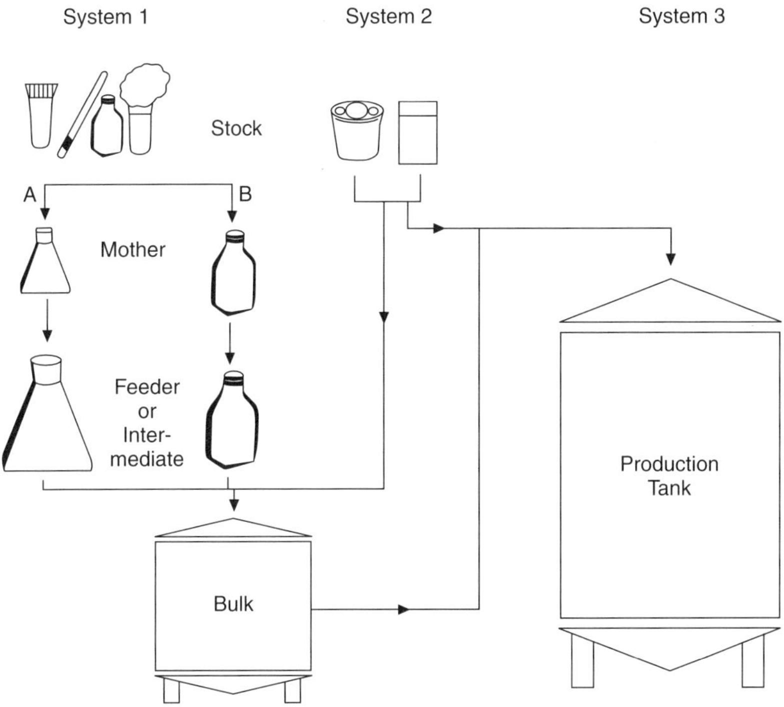

Figure 7.9. Starter culture preparations. *Note*: In System 1, stock culture may be liquid, freeze-dried, or frozen at –196°C for the production of bulk starter and cheese or fermented milks, respectively. In Systems 2 and 3, stock culture may be concentrated freeze-dried or frozen at –60°C to –196°C for the production of bulk starter and cheese or fermented milks, respectively. [After Tamime and Robinson (1999). Reproduced by courtesy of Woodhead Publishing, Cambridge, England.]

hazards in the industry), it is still widely used. Nevertheless, research work has been intensified in this area to overcome various problems, and developments in this field have focused on the areas of starter preservation and concentration.

The starter culture must contain the maximum number of viable organisms, must be highly active under production conditions in the dairy, and must be free from contaminants. Provided that culture inoculation is carried out under aseptic conditions and growth is initiated in a sterile medium, Foster (1962) suggested that one of the following principles must be adopted in order to maintain activity:

- Reducing and/or controlling the metabolic activity of the microorganisms.
- Separating the organisms from their waste products.

The former principle is evident in refrigeration, while the latter approach is mainly used during the concentration and/or preservation of starters—that is, during the production of a concentrated, active bulk starter, either in a continuous fermentor or in a batch process, for direct-to-vat inoculation (DVI) of the milk (Figure 7.9, systems 2 and 3).

7.5.1 Methods of Preservation of Lactic Acid Bacteria

It is essential that starter cultures be preserved in order to maintain an available stock of organisms, especially in the case of a starter failure. Also, successive subculturing can induce mutant strains that may alter the overall behavior and general characteristics of the starter. Dairy cultures may be obtained from research establishments, educational colleges, culture bank organizations, or commercial manufacturers, and starter culture bacteria may be preserved by one of the following methods:

- Liquid starter (mother → bulk).
- Dried starter: (a) spray-dried (unconcentrated), (b) freeze-dried or lyophilized (unconcentrated), and (c) concentrated freeze-dried.
- Frozen starter: (a) frozen at –20°C (unconcentrated), (b) deep frozen at –40 to –80°C (concentrated), and (c) ultra-low-temperature freezing at –196°C in liquid nitrogen (concentrated).

Thus, at present, cheesemakers and fermented milk manufacturers can choose from various forms of starter culture, depending on the size and complexity of operation. It can be observed from the above methods of preservation that the organisms may have been subjected to stress conditions during growth (e.g., adverse pH and/or concentration), refrigeration shock, freezing and thawing, freezing and drying, or drying. Hence, some of the starter bacteria may die or be injured; and, in view of the economic importance of starter cultures in the dairy industry, the general aim of scientists in this field has been to minimize the death rate of, or injury to, the preserved cultures. For further information about the factors affecting the survival of microorganisms, including dairy starter cultures, refer to the following texts: Ray (1984), and Andrew and Russell (1984).

7.5.1.1 Liquid Starter Cultures. This is the most popular and widely used form in which starter cultures are handled in the dairy. Starters are normally preserved in small quantities, but to meet the required volume for any production line, a scale-up system of propagation is required.

Stock culture → Mother → Feeder or Intermediate → Bulk

For example, processing 10,000 liters of milk into cheese per day with a rate of inoculation of 2 ml $100\,ml^{-1}$ would require a scale-up propagation as follows:

$$\underset{0.4\,\text{ml}}{\text{Stock}} \xrightarrow{1\,\text{ml}\ 100\,\text{ml}^{-1}} \underset{40\,\text{ml}}{\text{Mother}} \xrightarrow{1\,\text{ml}\ 100\,\text{ml}^{-1}} \underset{4\ \text{liters}}{\text{Feeder}} \xrightarrow{2\,\text{ml}\ 100\,\text{ml}^{-1}} \underset{200\ \text{liters}}{\text{Bulk}}$$

The working stock cultures are maintained in autoclaved (0.1 MPa for 15 min) reconstituted, antibiotic-free, skimmed milk powder [10–12 g $100\,g^{-1}$ solids non-fat (SNF)], with either weekly or daily subculturing. Cheese starter cultures (*L. lactis* subsp. *lactis* and subsp. *cremoris*, *L. mesenteroides* subsp. *cremoris*) can be propagated up to 50 times without any fear of mutation, and the sterilized medium is inoculated at a rate of 1 ml $100\,ml^{-1}$ and incubated at 22°C or 30°C for 18 h or 6 h, respectively (Walker, personal communication). However, recent work in this field (Lawrence and Heap, 1986) suggests that repeated subculturing of certain strains of starter bacteria may lead to a loss of plasmid material that, consequently, can affect the characteristics of the organism (i.e., phage-resistant becomes phage-sensitive). The yogurt starter cultures (*S. thermophilus* and *L. delbrueckii* subsp. *bulgaricus*) are normally subcultured only 15–20 times as a safeguard against imbalance and to retain the ratio of cocci:rods as 1:1 (Sellars, personal communication). Incubation is carried out at 42°C for 3–4 h, or at 30°C for 16–18 h using 2 or 1 ml 100 ml^{-1} inocula, respectively.

Starter culture activity is affected by the rate of cooling after incubation, level of acidity at the end of the incubation period, and the temperature and duration of storage. Cooling is important to control the metabolic activity of the starter; in practice, however, a warm starter (freshly incubated and uncooled) is sometimes used in cheese factories and, to some extent, in the yogurt industry.

The reserve stock culture can be maintained in a liquid form, and a slightly extended preservation of liquid cultures of most lactic acid bacteria can be achieved using litmus milk [(g 100 g^{-1}) reconstituted SMP 10–12, litmus solution 2, yeast extract 0.3, dextrose/lactose 1, enough calcium carbonate to cover the bottom of the test tube, panmede 0.25, and lecithin 1 (the latter two components were adjusted to pH 7)]. The medium is autoclaved at 69 kPa for 10 min, and it is incubated for a week to check sterility (Shankar, personal communication). The inoculated medium is incubated for a short period of time, and it is stored under ordinary refrigeration. Reactivation is only necessary once every 3 months.

However, alternative techniques for maintaining cheese cultures in the laboratory may include the following methods: *First*, inoculate 0.1 ml of a fresh, active culture into 10 ml cold, sterile milk and store at 4°C until required; *second*, incubate culture until a visible clot occurs and store at 4°C. The latter approach can increase the lag phase during subsequent use if the storage period is extended for several weeks, so that subculturing once per week is recommended (Ross, 1982).

7.5.1.2 Dried Starter Cultures. The preservation of starter cultures by drying is an alternative method for culture retention. The development of such processes seeks to overcome the work involved in maintaining liquid stock cultures; it also facilitates the dispatch of dried cultures by post without any loss in activity. The different methods used are as follows:

- Vacuum- and spray-drying (i.e., old methods not used at present time).
- Lyophilization or freeze-drying (this method is widely used in laboratories).
- Freeze-drying of concentrated cultures (widely used commercially).

Prior to the 1950s, vacuum drying was the normal practice (Tofte-Jespersen, 1974a,b, 1976). The process consists of mixing a liquid culture with lactose and then neutralizing the excess acid with calcium carbonate. The mixture is partially concentrated by separation and/or expressing the whey, thereby yielding granules that are dried under vacuum. The dried starter contains only 1–2% viable bacteria, and they may require several subcultures before regaining maximum activity.

Higher survival rates in dried starters can be achieved by growing the culture with pH control using $Ca(OH)_2$ as a neutralizing agent, followed by evaporation at 27°C to 22 g 100 g^{-1} total solids (TS), spraying drying at 70°C to 9 g 100 g^{-1} moisture with the powder temperature below 42°C, and finally vacuum drying at 27°C and 1–2 mm Hg until the dried culture has ~5 g 100 g^{-1} moisture. This method was developed in Holland by Stadhouders et al. (1969). The dried starter is claimed to be as active as a 24-h liquid starter. Although this development in starter technology proved promising, the system has not been developed commercially. The reason could be the usually low survival rates of the dried cultures—that is, 10% for the majority of mesophilic lactic acid bacteria and 44% for *L. lactis* subsp. *lactis* biovar *diacetylactis*. However, the addition of mono-Na-glutamate and ascorbic acid to a starter culture propagated in a buffered medium did protect the bacterial cells to some extent, and the spray-dried culture retained its activity after storage for 6 months at 21°C (Porubcan and Sellars, 1975a). Anderson (1975) claimed in a Swedish Patent No. 369470 in 1974 that yogurt starter cultures (ratio of cocci : rods –40 : 60 to 60 : 40) can be obtained in the spray-dried form when the starter is propagated in concentrated skimmed milk (18–24 g 100^{-1} TS) fortified with cyanocobalamin, lysine, and cystine. The drying temperature can be as high as 75–80°C without causing any bacterial damage. Despite the advantages claimed for spray-dried cultures, it seems that this system of preservation is not widely used.

Teixeira et al. (1994, 1995a,b,c) reported that the death kinetics of *L. delbrueckii* subsp. *bulgaricus* during spray-drying were influenced by many factors such as the following:

- The logarithmic survival ratio decreased with increased outlet air temperature with first-order kinetics, and pseudo-z for the organism was ~17°C.
- The calculated activation energy (E_a) above 70°C and below 70°C were 33.5 and 86 kJ mol^{-1}, respectively.

- The relationship between the entropy and enthalpy of activation for both spray-drying and heating in liquid medium was linear; the data for drying fell in the range of negative entropy.
- High storage temperature and water activity (A_w) reduced the survival rate of the dried microbial cells.
- The survival rate of *L. delbrueckii* subsp. *bulgaricus* was higher in the presence of mono-na-glutamate and ascorbic acid during storage at 4°C and 20°C, respectively.
- The preserved cells were sensitive to NaCl, antibiotics, and lysozyme due to damage of the cell membrane, DNA, and cell wall.
- The ratio of unsaturated:saturated fatty acids of the microbial cell membrane decreased after spray-drying and during storage in air-tight packages, indicating lipid oxidation (Teixeira et al., 1996).

However, highest survival of rate of spray-dried *L. lactis* subsp. *lactis* was obtained using a cell concentration of ~20% in the feed solution to the dryer, along with an outlet air temperature at 77°C (Fu et al., 1994, 1995; Fu and Etzel, 1995). Similar results were reported for *L. acidophilus* by Johnson and Etzel (1993, 1994). In a separate study, To and Etzel (1997) found that the reduction in survival rate of some starter cultures was dependent on the species and method of preservation; the survival after spray-drying was greatest for *S. thermophilus* > *L. paracasei* subsp. *paracasei* > *L. lactis* subsp. *cremoris*. Egyptian researchers (Abd El-Gawad et al., 1989; Metwally et al., 1989) recommended the following for optimal survival after spray-drying of the yogurt microflora and *L. lactis* subsp. *lactis*: (a) addition of dextrin or neturalization of the culture to pH 6.8 before drying, (b) growth of the culture for 24h only, and (c) storage of the preserved culture at 5–7°C or –15°C; the latter storage temperature was more effective, especially for *Lactobacillus* spp. Another effective method of protecting the cells of starter cultures against drying damages is microencapsulation (Desmons et al., 1998). The reader is referred to Hill (1987) and Boyaval and Schuck (1994) for general aspects regarding the spray-drying of starter cultures or the effect of spray-drying on the cell viability and bacteriocin activity of dried lactic acid bacteria stored for 60 days at 4°C (Mauriello et al., 1991).

The addition of mannitol to washed pellets of *L. lactis* subsp. *lactis* before drying at 20°C in Petri dishes by exposure to air (31 g $100\,g^{-1}$ relative humidity (RH) for 72h) and subsequent drying by desiccation resulted in survival rate of the dried cells at a level equaling the viable cell count of the culture before drying (Efiuvwevwere et al., 1999). However, the lag phase of the rehydrated culture was extended by 4h, and the same authors hypothesized that the radical activity of mannitol could, in part, explain the protection of the culture during drying.

In contrast, freeze-dried cultures are produced when the starter is dried in the frozen state. This method of starter preservation improves the survival rate of the dried culture, and good results have been achieved as compared with spray-dried starters (To and Etzel, 1997). It has been observed that the process of freezing and drying can damage the bacterial cell membrane, but the damage is minimized by the addition of certain cryogenic agents/compounds prior to freezing and drying (Nastaj, 1996). Many different media, additives, or techniques have been studied by many researchers to determine the optimum conditions for the production of freeze-dried cultures, and some examples include milk solids, Na-glutamate, gelation, Tween 80, β-glycerophosphate, malt extract, soya plus casein, sheep or horse serum, sucrose in chopped meat carbohydrate broth, and vitamin E (Tamime and Robinson, 1976; Kilara et al., 1976; Hup and Stadhouders, 1977; Yang and Sandine, 1979; Ozlap and Ozlap, 1979; Ishibashi et al., 1985; Staab and Ely, 1987; Kim et al., 1988). In addition, an American patent was filed in the mid-1970s for the production of freeze-dried starter cultures (Porubcan and Sellars, 1975b); in brief, the process can be described as follows: Propagate the starter culture in a milk-based medium (pH adjusted to 6.0–6.5) plus additives (e.g., ascorbic acid, mono-Na-glutamate, aspartate compound) plus cryoprotective agents (e.g., inositol, sorbitol, mannitol, glucose, sucrose, corn syrup, DMSO, PVP, maltose, mono- or disaccharides; see also Nazzaro et al., 1999). The early commercial freeze-dried starter cultures were not suitable for DVI application, and it was necessary to propagate these cultures a few times to reestablish their activity prior to fermentation (Porubcan and Sellars, 1979).

In other patents (Amen and Cabau, 1984, 1986), cheese starter cultures are grown in a special medium containing a "nutritive substrate," and the pH is maintained >5.5 by the addition of a neutralizing agent (i.e., ammonium hydroxide). The removal of the inhibitory ammonium lactate is carried out by ultrafiltration and the addition of water. The concentrated culture is then freeze-dried, and it is suitable for DVI applications in the manufacture of cheese.

It is evident that a wide range of compounds could be used as cryoprotective agents during the freeze-drying of starter cultures, and Morichi (1972, 1974) has studied the effect of different protective solutes on the survival rates of lactic acid bacteria (see Table 7.9). He

TABLE 7.9. Effect of Selected Cryogenic Agents[a] on the Survival Rate of Freeze-Dried Lactic Acid Bacteria (All Figures as % of Original Cell Number)[b]

Microorganisms	L-Glutaric acid	L-Arginine	L-Lysine	DL-Threonine	DL-Pyrolidine carboxylic acid	Acetyl glycine	DL-Malic acid
Lactococcus lactis subsp. *cremoris*	40–60	42–60	1–16	20–30	53–67	48–59	23–57
subsp. *lactis*	31–70	36–53	0–7	8–39	19–81	10–56	26–44
subsp. *lactis* biovar *diacetylactis*	44–53	48–54	4	14–37	47–60	38–43	20–33
Streptococcus thermophilus	35–40	21–40	6–7	7–11	24–48	29–44	52–59
Lactobacillus delbrueckii subsp. *bulgaricus*	16–21	20–35	1–10	6–10	9–11	7–33	6–15
Lactobacillus acidophilus	42–63	39–57	4–38	6–21	24–56	3–35	28–66
Lactobacillus helveticus	48	35	23	14	23	32	35
Lactobacillus casei subsp. *casei*	52–69	28–40	6–10	6–9	13–29	4–18	6–22
Lactobacillus plantarum	57	44	5	10	48	20	46

[a]Suspending medium (0.06 M solution of agent adjusted to pH 7).
[b]The range of survival (%) is due to different strains tested.

Source: Adapted from Morichi (1972).

TABLE 7.10. Effect of Different Protective Compounds in Preserving Maximum Viability (>50%) of Lactic Acid Bacteria Subjected to Freeze-Drying

Microorganisms	Protective Agent	Survival (%)
Lactobacillus casei subsp. *casei* ATCC 393	Dextran-W[a] or -M[b]	50 or 74
	Na-glutamate-M	60
	β-Glycerophosphate-W or -M	70 or 54
Lactobacillus plantarum ATCC 8014	PEG 1000-W	50
	Dextran-M	58
	Na-glutamate-M	70
	β-Glycerophosphate-W or -M	85 or 58
Lactobacillus delbrueckii subsp. *bulgaricus* ATCC 11842	Cysteine-M	78
Lactococcus lactis subsp. *cremoris* ATCC 19257	Na-glutamate-W or -M	60 or 57
Lactococcus lactis subsp. *lactis* T164	PEG 1000-M	85
	Na-glutamate-W or -M	95 or 86
	Glycerol-W or -M	75 or 53
	β-Glycerophosphate-W or -M	93 or 80
Streptococcus thermophilus ATCC 19258	PEG 1000-W or -M	84 or 52
	Na-glutamate-W or -M	97 or 82
	Asparagine-M	78
	β-Glycerophosphate-W or -M	62 or 94
Leuconostoc mesenteroides subsp. *cremoris* ATCC 19254	Na-glutamate-W	100
	β-Glycerophosphate-W	78

[a]W = Additive suspended in sterile distilled water.
[b]M = Additive suspended in sterile reconstituted skimmed milk powder (lactobacilli and lactococci in 20 and $10\,g\,100\,g^{-1}$ solids, respectively).
PEG 1000 = polyethylene glycol 1000.
ATCC = American Type Culture Collection.
Source: Adapted from Font de Valdez et al. (1983a).

concluded that these protective solutes are hydrogen-bonding and/or ionizing groups that help to prevent cellular injury by stabilizing the cell membrane constituents during the preservation procedures. However, in a separate study, Font de Valdez et al. (1983a,b) have studied the efficacy of some additives in protecting 13 different strains of lactic acid bacteria against freeze-drying, and a summary of their work is shown in Table 7.10.

Up to around 1990 [see the reviews by Tamime (1990) and Champagne et al. (1991b)], it was possible to summarize the factors that can affect the survival rate of freeze-dried dairy starter cultures as follows:

- Most lactic acid bacteria preserve well—with the exception of *L. delbrueckii* subsp. *bulgaricus* and *L. helveticus*, which are sensitive to freezing and drying. Propagation of the starter culture in milk fortified with yeast extract and hydrolyzed protein improves the survival rate, and raising the cell concentration of a culture e.g., $>10^{10}$ cfu ml^{-1}) can increase the viable number of bacteria in the dried culture.

- Starter cultures are less sensitive to freezing and drying if the cells are harvested toward the latter part of the exponential phase—with the exception of *L. delbrueckii* subsp. *bulgaricus* and *Lactococcus lactis* subsp. *cremoris*, where the cells are harvested in the early stages of the stationary phase.
- Media in a pH range of 5–6 are more favorable to high survival rates, and neutralization of the growth and the suspending medium is essential (e.g., skimmed milk plus Na-malate proved suitable for *S. thermophilus;* a solution of lactose and arginine hydrochloride gave protection to *L. delbrueckii* subsp. *bulgaricus*; and glutamic acid gave protection to *Leuconostoc* spp. (see also Pereda Alardin et al., 1990).
- The moisture content of the dried culture must be less than 3 g 100 g^{-1}.
- Dried cultures stored at 5–10°C showed higher rates of survival during prolonged storage than did those stored at room temperature. Vacuum or modified atmosphere packaging of the dried cultures is highly recommended because the preserved organisms are sensitive to oxygen; however, the most popular type of packaging material for dried cultures is the glass vial, followed by the laminated, aluminum foil sachet (see Figure 7.10A; Heiner, 1990).
- Freezing cultures at –20 to –30°C and drying at temperatures between –10 and 30°C results in high bacterial activity of the dried culture.
- Carbonyl compounds, such as pyruvate and diacetyl, which can react with the amino groups within the preserved cells, can accelerate their death. It is recommended that these compounds should be separated from the harvested cells. For the long-term preservation of freeze-dried cultures, the suspending medium must be fortified with nonreducing sugars, amino acids, and/or semicarbazide. The presence of adonitol and glycerol in the suspending medium protected different strains of lactic acid bacteria which were subjected to freeze-drying (Font de Valdez et al., 1983b, 1985a).
- Freeze-drying of *L. acidophilus* adversely affected the H bonds involved in binding the surface layer protein to the cell wall, and

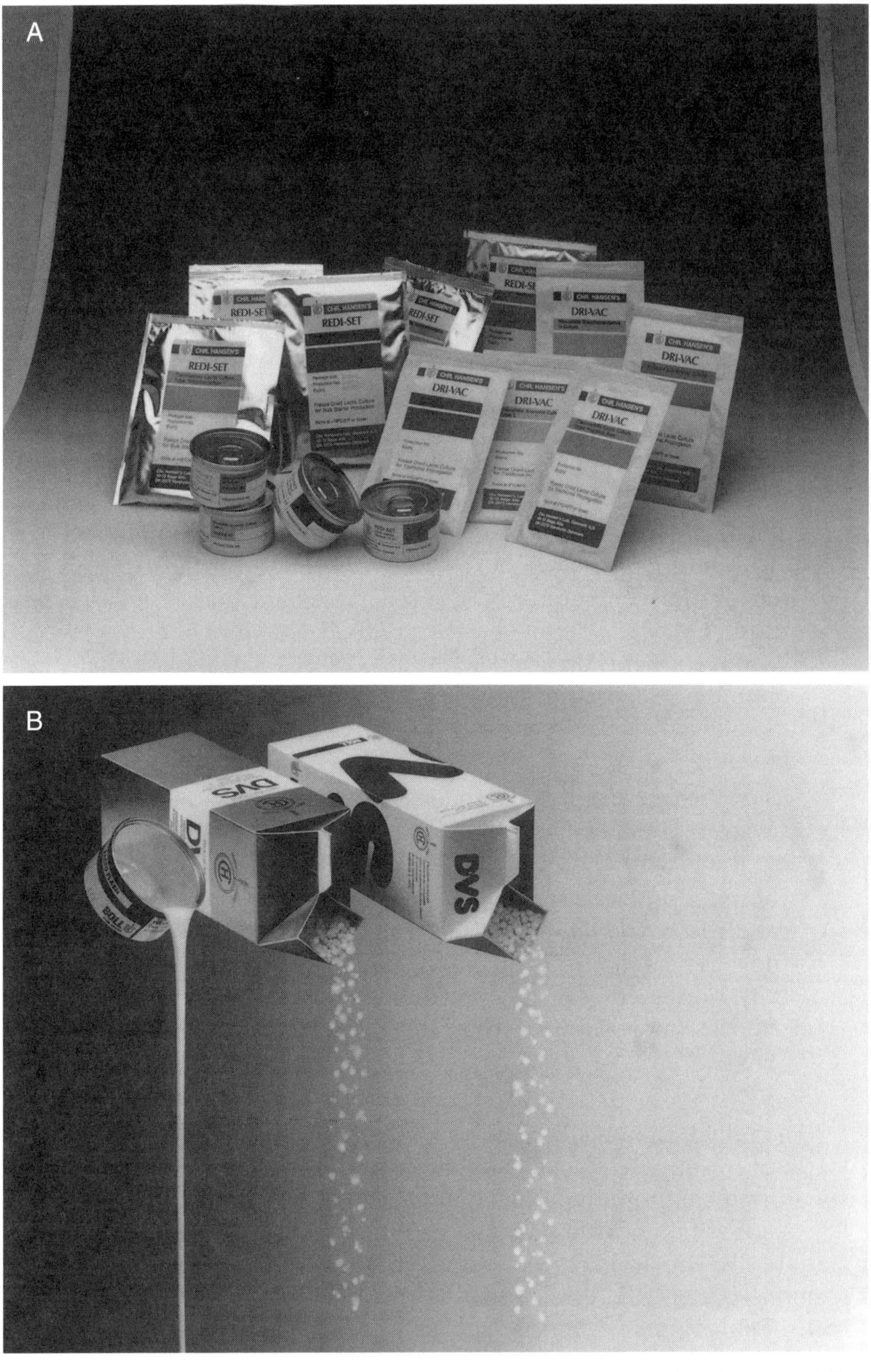

Figure 7.10. Some packaging systems used for starter cultures. (A) aluminum foil sachet and (B) laminated cartons and metal cans. [Reproduced by courtesy of Chr. Hansen (UK) Ltd., Hungerford, England.]

it was lost from the dried cells; glycerol protected such bonds (Brennan et al., 1986; Ray and Johnson, 1986).

- Factors that increased the survival rate of *L. casei* subsp. *casei* and *B. bifidum* after freeze-drying were as follows: (a) Addition of β-glycerophosphate (8.1 g $100\,g^{-1}$) had greater cryoprotection effect than adonitol, polyethylene glycol, or Na-glutamate, (b) immobilization of the cells on Na-alginate increased survival rates during the freeze-drying of concentrated cultures, (c) addition of Ca^{2+} to growth media enhanced the survival rate of the microorganims compared to a Ca^{2+}-deficient medium, and (d) viability of the cells was affected by the pH of the growth medium, and optimum pH 4 and 6 were recommended for the lactobacilli and bifidobacteria, respectively (Kim and Yoon, 1995).
- Rehydration temperature can affect the leakage of cellular ribonucleotides from damaged cells. Although *S. thermophilus* and *L. lactis* subsp. *lactis* and subsp. *cremoris* showed little response to rehydration temperatures, it is recommended that *L. delbrueckii* subsp. *bulgaricus* is rehydrated at 20–25°C (Morichi et al., 1967). However, Font de Valdez et al. (1985b) concluded that the optimum rehydration temperature of both mesophilic and thermophilic lactic acid bacteria is 20°C.
- Up-to-date studies on the preservation of dairy starter cultures using the freeze-dried method are shown in Table 7.11 [see also the review by Tamime and Robinson (1999) for specific details regarding the preservation of the yogurt microflora].
- Recently, probiotic starter cultures such as *L. paracasei* subsp. *paracasei* F19, *L. rhamnosus* GG, *L. johnsonii* La1, and *B. lactis* BB12 (now renamed as *B. animalic*; Cai et al., 2000) were successfully preserved (i.e., freeze-dried and frozen) by Chr. Hansen, and the survival rate was $\geq 10\log_{10}$ cfu ml^{-1} or g^{-1} (Saxelin et al., 1999).

The rehydration medium plays a major role in the recovery or survival of freeze-dried lactic acid bacteria; and because the rehydration of a dried culture in liquid medium is complete within a few seconds, the composition of such diluents can either protect or damage the bacterial cells. Thus, certain compounds in the rehydration medium act as osmotic buffers, thereby regulating the entry of water into the dried cells. Published work in this field is somewhat limited; and although different rehydration media have been studied, definite conclusions cannot be drawn because the wide range of starter organisms available,

TABLE 7.11. A Selection of Different Growth and Cryogenic Compounds ($g\,100\,g^{-1}$) Employed for Optimizing the Survival Rate of Freeze-Dried Dairy Starter Cultures

Microorganisms	Growth Media and/or Cryogenic Compounds	References
Lactococcus lactis subsp. *lactis*	Skimmed milk solids 28, sucrose 4, and ascorbic acid 0.3 or mannitol 1–2 increased survival rate during drying and maintained high viability during storage.	Coutour et al. (1991)
Streptococcus thermophilus	Reconstituted SMP provided good protection of the cells; better survival rate was obtained using vacuum rather than atmospheric freeze-dryer.	Wolff et al. (1990)
Lactobacillus delbrueckii subsp. *Bulgaricus*	Survial rate of freeze-dried culture was increased by a factor of 10 in the presence of Tween 80 before drying. Viability of freeze-dried culture grown in skimmed milk was greatest when stored at 11% RH and 5°C; the cell membrane damage was due to a decrease membrane bound proton-translocating ATPase activity.	Champagne et al. (1991a), Rumain et al. (1993), Castro et al. (1995, 1996, 1997), Riis et al. (1995)
Lactobacillus plantarum	The inactivation of the starter culture during freeze-drying was due to thermal and dehydration inactivation; addition of sorbitol before drying increasd the residual glucose fermenting activity of the cell after drying due to the reduced rate of DNA hydrolysis. The use of fluidized bed/spray drying was reported by Fuchs (1995).	Lievense et al. (1994), Linders et al. (1994)
Lactobacillus acidophilus	The cells were grown in MRS broth or whey supplemented with yeast extract 0.5 and glucose as a cryoprotective agent; whey media gave slightly higher counts, and the dried culture is used as probiotic preparation.	Gandhi and Shahani (1994)
	The addition of glycerol 5 in the suspended media before freeze-drying improved the survival rate of the organism and decreased injury to the cells.	Bozoğlu and Gurakan (1989), see also Espina and Packard (1979), King and Su (1993), Desmons et al. (1997)

TABLE 7.11. *Continued*

Microorganisms	Growth Media and/or Cryogenic Compounds	References
Lactobacillus reuteri	Cells treated with 0.15 oxgall showed increased activity of β-galactosidase, but showed low survival rate possibly due to the presence of membrane structures containing simple folds and buds of the cell membrane.	Font de Valdez et al. (1997)
Bifidobacterium infantis	Suspending the culture in UF retentate with or without lactose before freeze-drying did not enhance the survival rate; prolonged storage (i.e., 8 months at 4°C) reduced β-galactosidase activity by 18%, and hence such media could act as cryoprotective agent during long-term storage of freeze-dried culture of bifidobacteria.	Blanchette et al. (1995)
Bifidobacterium longum	Commercial freeze-dried cultures were spray coated with gelatin, xanthan gum or milk fat and by co-crystalisation of lactose; prolonged storage at 20°C decreased microbial activity except the milk fat-coated cultures; the crystalization process did not increase cell count before drying, and freeze-dried culture containing lactose crystals was ~20%.	Champagne et al. (1995)
Different species	Similar to the media described by Coutour et al. (1991) plus maltodextrin 10, gelatin 2, or xanthan gum; gelatin improved survival and storage stability at ~20°C of freeze-dried *L. rhamnosus, B. longum*, and *S. thermophilus*.	Champagne et al. (1996a)
	Lactobacillus delbrueckii subsp. *bulgaricus*, *E. faecium, L. plantarum*, and *L. halotolerance* were subjected to salt or osmotic stress before drying; betaine was the major solute accumulated by these organisms which helped to enhance considerably the survival rate of cultures.	Kets and de Bont (1994), Kets et al. (1996, 1997)
	Streptococcus thermophilus was found more resistant to freeze-drying compared with *L. delbrueckii* subsp. *bulgaricus*; storage under vacuum or nitrogen was superior than under air because oxygen diffuse into the dry cell through the interfacial area.	Bozoğlu et al. (1987), Qiao et al. (1999), (see also Béal and Corrieu, 1994; Béal et al., 1994)

and sometimes different strains within the same bacterial species, respond in different ways. The reader should consult the following publications regarding the behavior of dairy starter cultures (Ray, 1984; Font de Valdez et al., 1985c,d, 1986).

Starter cultures preserved by freeze-drying tend to have a prolonged lag phase, and they are mainly used as inoculants for the propagation of mother cultures (see Figure 7.9, system 1). Larger quantities are needed for direct inoculation of the bulk starter, and an extended incubation time may be required (Sellars and Babel, 1985). Developments in the last few years have made it feasible to produce concentrated, freeze-dried cultures (CFDC) for direct inoculation of the bulk starter vessel or for DVI of milk for the manufacture of cheese and other fermented dairy products (see Figure 7.9, systems 2 and 3, respectively).

Since 1980, there have been great technological developments in the production and preservation of direct-to-vat freeze-dried cultures, and Porubcan (1991) has detailed the preparation and commerical production of freeze-dried cultures. The activity of such cultures is primarily dependent on the following factors:

- Percentage survival rate—the average count for most commercial applications is $\sim 10^{11}$–10^{12} cfu g^{-1}.
- The moisture content—a target figure is <3 g 100 g^{-1}.
- Gas flushing the package with N_2.
- Storage at refrigeration temperature—however, prolonged shelf-life is achieved by storing the dried cultures at ~–20°C.

Another technique for the preservation of starter cultures uses milk-based powders to absorb the moisture from the growth medium, and Harju et al. (1984) achieved a 50% survival rate for *L. helveticus* and propionibacteria species when mixed with SMP; the process of freeze-drying was more successful.

7.5.1.3 Frozen Starter Cultures. Starter cultures can also be preserved in the frozen form, and such cultures are produced by two different routes:

- Deep or subzero freezing (–20 to –80°C).
- Ultra-low-temperature freezing at –196°C in liquid nitrogen.

Sterile milk freshly inoculated with an active starter culture is deep frozen at –30 to –40°C for preservation as a mother or feeder/intermediate culture. Such frozen cultures can retain their activity for several months when stored at –40°C, and this method of culture preservation became popular in the UK cheese industry because deep frozen cultures were produced in centralized laboratories and could be dispatched to a dairy in dry ice whenever required. Such cultures have now been replaced by the concentrated, frozen type for direct inoculation of bulk starter tanks or DVI of milk for the manufacture of cheese or fermented milks (see Figure 7.9, systems 2 and 3, respectively).

Freezing and prolonged storage at –40°C can lead to a deterioration in starter culture activity, and can damage certain lactobacilli, but the use of a medium containing (g $100\,g^{-1}$) skimmed milk 10, sucrose 5, fresh cream, $NaCl_2$ 0.9, or gelatin 1 can improve survival rates (Imai and Kato, 1975). In addition, concentrated cells (10^{10}–10^{12} cfu ml^{-1}) frozen at –30°C and in the presence of certain mixtures of cryogenic compounds (Na-citrate, glycerol, Na-β-glycerophosphate, yeast extract, sucrose, cream, sterile skimmed milk, peptone, trehalose, or lactose) have been retained as active as the original cultures in the case of mesophilic organisms, *Lactobacillus* spp. or propionic acid bacteria (Barbour and Priest, 1986; Oberman et al., 1986; Toyoda et al., 1988; de Antoni et al., 1989; Tamime, 1990; Abraham et al., 1990; Zlotowska and Ilnicka-Olejniczak, 1993; Fonseca et al., 2000). However, Weerkamp et al. (1996) managed to preserve pure cultures of mesophilic lactic acid bacteria at –82°C in cryovials containing M17 medium supplemented with 15 ml $100\,ml^{-1}$ glycerol. Recently, Wouters et al. (1999) reported a ~100-fold increase in the survival rate of *L. lactis* subsp. *lactis* MG1363 after freezing when the cells were temperature-shocked at 10°C for 4 h compared to cells taken midlogarithmic phase during growth at 30°C; this protection was attributed to the induced 7-kDa cold-shock proteins (CSP) generated at 10°C by the lactococci.

Nevertheless, earlier studies by researchers have suggested that although freezing and storage at –40°C has proved to be a successful process for preserving starter cultures, storage at –80 to –100°C in liquid nitrogen vapor improves the survival rate of the frozen organisms during storage. *L. lactis* subsp. *cremoris* strains have been grown individually in PHASE 4 medium (see Section 7.7) and then frozen unconcentrated; glycerol is required to preserve cultures stored at –20°C, but no cryoprotectants were needed to preserve the viability and activity of cultures stored at –40 and –80°C (Thunell et al., 1984a,b). However, freezing in liquid nitrogen at –196°C is by far the best method, and one of the earliest commercial applications has been patented in the United States (Christensen, 1969). The review by Gilliland (1985) shows the extent of the work that has been carried out in this field, and the freezing and thawing cycle is still regarded as an important factor in the successful use of frozen cultures in the dairy industry. An organism that is highly susceptible to damage during freezing is *L. delbrueckii* subsp. *bulgaricus*, but it was found that the presence of Tween 80 and Na-oleate improved cell stability. *L. acidophilus* is also susceptible to freezing and thawing, and the injury is associated with cell wall components other than peptidoglycan; such injury is reversible by repair of the cell wall components (Johnston et al., 1984). However, the type of growth medium, neutralizing agent used, and/or cryoprotective compound(s) can play a major role in the activity of the preserved culture, and the review by Gilliland (1985) highlights these factors in relation to different species of lactic acid bacteria.

Because cryoprotective compounds may not be suitable for the preservation of certain species of lactic acid bacteria by freezing, it is possible to grow cultures in, for example, pepsinized sweet whey. Mitchell and Gilliland (1983) managed to grow *L. acidophilus* in such a medium, and they maintained at pH 6.0 using a neutralizer consisting of sodium carbonate and ammonium hydroxide. The cell count was around 1×10^9 cfu ml^{-1}; and after freezing in liquid nitrogen, the stability of the culture after 28 days of storage was excellent.

Freezing cultures in liquid nitrogen has made possible the DVI of milk for cheese and yogurt production, or direct inoculation of the bulk starter (see Figure 7.9, systems 2 and 3). The advantages of this approach are as follows: convenience, culture reliability, improved daily performance and strain balance, greater flexibility, better control of phage, and possible improvement in quality. However, the disadvantages are as follows: difficulties in providing liquid nitrogen facilities, higher cost, greater dependence on starter suppliers, and apportioning of responsibility in case of starter failure (Tamime and Robinson, 1976; Wigley, 1977; Maruejouls and Caigniet, 1983). However, Wigley (1980) has discussed the successful use of liquid nitrogen frozen cultures for the manufacture of Cheddar cheese.

It is of the utmost importance that the thawing and handling of frozen cultures is carried out according to the supplier's recommendations, and a typical procedure is as follows:

- Remove can from liquid nitrogen storage.
- Thaw in water containing 100–200 μg g^{-1} hypochlorite solution at 20°C for 10 min.

- When culture is partially thawed (i.e., contents are just loosened), remove can from water, open lid, and add directly to bulk starter milk or milk for processing.

Rapid thawing in a water bath at 20–45°C rather than slow thawing at 4°C has been recommended, while other researchers have suggested rapid defrosting at 50–55°C for 3 min (Tamime, 1990). Thawing of frozen cultures in pellet form is not required at all, and the culture is added directly to milk.

Current research work on frozen starter cultures suggest that the use of skimmed milk as a growth media preserved *L. delbrueckii* subsp. *bulgaricus* and *L. lactis* subsp. *lactis* better at –20°C compared with MRS broth (Kim and Yu, 1990). The same authors also reported the following: (a) Washed cells treatment before freezing enhanced lactic acid production after thawing the cultures that had been stored for 8 months at –20°C, and (b) the survival rate (cfu ml^{-1}) of the lactococci (i.e., stored for 7 months at –20°C) grown in skimmed milk supplemented with glycerol, SMP, or lactose after freezing was 2.7 × and 0.79 × 10^8 cfu ml^{-1}, respectively. Incidentally, the count before freezing averaged 9.12 × 10^8 cfu ml^{-1}. However, supplementation of the growth media (skimmed milk 10 g 100 g^{-1} SNF) with $CaCO_3$ (0.1 g 100 ml^{-1}) provided more stable cells of *L. acidophilus* frozen at –20°C (Bozoğlu and Gurakan, 1989). The addition of lactose or sucrose (5 g 100 g^{-1}) to centrifuged pellets of cells stored at –40 or –70°C resulted in the highest viability of *L. lactis* subsp. *lactis* (Chavarri et al., 1988; see also McIntyre and Harlander, 1989).

Recently, Morice et al. (1992) suggested the following recommendations to optimize the survival rate of *S. thermophilus* during freezing and storage:

- Freezing the cells in the stationary phase rather than the exponential phase enabled the cells to resist the processing conditions better when the organism was grown in synthetic media, but no difference was observed if milk was used as the growth medium.
- The survival rate of the starter culture was influenced by the freezing rate and the strain.
- Cryoprotectants (e.g., dimethyl-sulfoxide, glycerol, xylitol, adonitol, and raffinose) also improved resistance to cellular damage during freezing.
- Prolongation of the freeze-thawing stage altered the survival rate of the preserved cells, and the effect was also strain-related.

Hence, it is evident that the rate of freezing and the process of freeze-thawing are important if damage to the microbial cells is to be minimized. Oberman et al. (1995) and Libudzisz et al. (1991) recommended that freezing rate should be at 2.3°Cs^{-1} when using a vessel ~7mm in diameter in an ethanol/CO_2 bath at –78°C; such a freezing rate is suitable for processing the biomass of species belonging to the genera *Lactococcus Leuconostoc*, and *Lactobacillus.* However, two-phase freezing was found to be suitable for *L. acidophilus*—for example, first phase to –10°C and second phase to –70°C at a rate of –2.3 to –9.0°C and –8°Cs^{-1}, respectively (Lidudzisz and Mokrosinska, 1995). Nevertheless, the thawing rate for frozen cells of lactococci, leuconostocs, and lactobacilli should be ~3°Cs^{-1} in order to achieve 86–94% survival of the frozen cultures (Piatkiewicz and Mokrosinska, 1995). The kinetics of the freezing and thawing processes for lactic acid bacterial biomass was detailed by Walczak et al. (1995). According to Font de Valdez and de Giori (1993), freezing and thawing render *L. delbrueckii* subsp. *bulgaricus* more sensitive to NaCl and liver extract, and the amino acid transport system is depressed. The injured cells could be partially repaired with a solution of pyruvate, $MgSO_4$, and KH_2 PO_4 (see also Foschino et al., 1992).

Over the past decade, more knowledge has become available on cryotolerance and cold stress in starter cultures. Panoff et al. (2000) have suggested that the connection between the stress generated by transferring lactic acid bacteria to a low-temperature environment and the correlated response may be evaluated using the following approaches: (a) the physiological response of the microbial cell (i.e., growth and cryotolerance), (b) the generated biochemical modifications (e.g., degree of fatty acids desaturation and protein profile), and (c) the control of the cold shock response (see also Panoff et al., 1994, 1995, 1998; Thammavongs et al., 1996; Mayo et al., 1997). However, Gómez-Zavaglia et al. (2000) studied the fatty acid composition and freeze–thaw resistance in lactobacilli (*L. delbrueckii* subsp. *delbrueckii*, *lactis* and *bulgaricus*, *L. acidophilus*, and *L. helveticus*), and they reported the following observations:

- Five fatty acids [C14:0 to C19:0 (cyclopropane)] made up to 90% of the cellular pool.
- Strains containing a high level of unsaturated fatty acids (66–70%) had decreased freeze–thaw resistance with increasing concentration of cyclopropane (cyc 19:0).
- Increased freeze–thaw resistance was observed in strains of lactobacilli with low concentrations of unsaturated fatty acids

(42–49%) and with increasing levels of cyc 19:0 (see also Johnsson et al., 1995).

Commercial frozen starter cultures are packaged in metal cans, or the pelleted frozen type is packaged in laminated cartons (see Figure 7.10B).

7.5.1.4 Immobilized Starter Cultures. Immobilized cell technology (ICT) has many potential applications in the dairy industry, but it has not been applied in large-scale operations (see the reviews by Champagne et al., 1994a; Sodini-Gallot et al., 1998). According to the same authors, the applications of ICT in the dairy industry are as follows:

- Culture processing
- Milk and whey treatments

Some of the dairy applications of ICT include (a) the continuous fermentation of milk and UF milk by starter cultures (Kim et al., 1985a,b; Sodini-Gallot et al., 1995, 1997a), (b) the production of fresh cheese (Prevost and Divies, 1987; Sodini-Gallot et al., 1997b), (c) the production of yogurt (Prevost and Divies, 1988a,b), and (d) the growth of starter cultures in whey (Steenson et al., 1987; Champagne et al., 1986, 1988, 1993; Norton et al., 1994; Passos et al., 1994; Lapointe et al., 1996; Lamboley et al., 1997). However, other applications of ICT include the production of bacteriocins, cheese flavors, and starter cultures (Sodini-Gallot et al., 1998); the latter aspect will be reviewed in detail.

ICT of starter cultures can be used to store cultures at dairy factories or to transport cultures without the need of freezing or drying (Champagne et al., 1994b). According to Sodini-Gallot et al. (1998), the advantages of ICT in industrial fermentations are as follows:

- High productivity.
- Protection of cultures against biological or chemical contamination.
- Continuous and stable operation of fermentation processes.
- More adaptable to starter culture production than milk processing/treatments in dairy factories.

Some of these aspects have been evaluated on specific starter cultures, and some examples include (a) *Lactococcus* spp. (Passos and Swaisgood, 1993; Cachon et al., 1998), (b) yogurt microflora

(Büyükgüngor and Caĝlar, 1990; Audet et al., 1991b; Ragout et al., 1996), and (c) *L. casei* subsp. *casei* (Arnaud et al., 1992). Current studies on the production and preservation of starter cultures using ICT technology are shown in Table 7.12.

7.5.1.5 Miscellaneous Starters. Different methods for the preservation of starter cultures have been reported in the literature, (see Tamime and Robinson, 1999), and a summary of those techniques is provided below:

- Mesophilic and thermophilic lactic acid bacteria have been preserved on anhydrous silica gels under vacuum for 3 years; only the yogurt culture showed reduced activity after storage for 2 years (de Silva et al., 1983).
- The addition of 0.5–2 g 100 g^{-1} of calcium carbonate to starter cultures increased the survival rate even at elevated temperature—for example, 20°C or 30°C (Kang et al., 1985).
- Na-citrate and potassium phosphate buffer solutions have been reported to conserve the activity of dairy starter cultures, but the survival rate was rather low, which affected the rate of acid development and flavor production by some cultures (Sultan et al., 1987).
- Lactic acid bacteria (i.e., single strains) have been suspended in MRS broth or 15 g 100 g^{-1} reconstituted skimmed milk powder and dried onto 5- to 7-mm porcelain beads; in general, the percentage survival was low, but the rate was influenced by the suspending medium; the potential to produce lactate and carbonyl compounds decreased after 12 months' storage at 4°C (Magdoub et al., 1987).
- *Lactococcus lactis* subsp. *lactis* C_2 and subsp. *cremoris* HP have been immobilized in calcium alginate beads; these cultures were immune to bacteriophage attack because they were embedded in the gel matrix away from the phage particles, but the rate of acid production, vis-à-vis cells freely suspended in milk, was lower due to the limited diffusion of nutrients into the beads (Steenson et al., 1987; see also Section 7.5.1.4).

7.5.2 Concentration of Cells

It can be observed from the information above that the survival rate of the preserved starter culture is dependent on the processing

TABLE 7.12. Some Examples of Immobilized Cell Technology for the Preservation of Starter Cultures

Microorganisms	Comments	References
Lactococcus lactis subsp. *lactis*	Concentrated cultures were immobilized in Ca-alginate beads, added to glycerol/soytone solution (i.e., to maintain A_w at 0.93), and stored for 30 days at 4°C lost only 22% of its viability.	Champagne et al. (1994b)
	Lactococci cells were immobilized in Ca-alginate; and before freeze-drying, the cells were mixed with milk-based protective solution; cell survival rate ranged between 62% and 79%.	Champagne et al. (1992)
	Comparative behavior of lactococci in free and immobilized cell technology was evaluated, and the specificity of the latter technique was characterized.	
Bifidobacterium longum	Low survival rate (i.e., reduction by 2 $\log_{10}$ cfu g^{-1}) was observed for such microorganism when immobilized in Ca-alginate gels, freeze-dried and stored at 20°C, but cell mortality was reduced by exposing the cells to whey-based media before drying; such feature was strain/species related because this method of preservation was not detrimental to *Lactococcus* spp. or *L. rhamnosus*, but had negative effect on the survival of *S. thermophilus*.	Champagne et al. (1996b)
	One batch fermentation of bifidobacteria in MRS broth and whey permeate was required to achieve a count of 3×10^{10} cfu g^{-1} in freeze-dried immobilised cells in κ-carrageenan-locust bean gum gel.	Maitrot et al. (1997)

Streptococcus thermophilus	The culture was successfully preserved in alginate-immobilized freeze-dried beads; the culture grew in alginate beads and reached a populatoin of 10^{10} cfu g^{-1} after 6 h incubation, and the rehydrated cells contained 13% free cells.	Champagne et al. (2000)
Lactobacillus delbrueckii subsp. *bulgaricus*	Immobilization of such organims in κ-carrageenan gel carrier had higher mechanical and chemical stability than alginate cells (see also Klein et al., 1983).	Büyükgüngör (1992)
Lactobacillus helveticus	Cells were entrapped in Ca-alginate and later preserved using fluidized-bed dryer; best survival rate was experienced with adonitol and SMP (71% and 57%, respectively—compared with intitial cell count) immediately after dehydration.	Selemer-Olsen et al. (1999)
Yogurt cultuers	The immobilized strains (i.e., entrapped in κ-carrageenan-locust bean gum mixed gel beads) stored well during storage at 4 and 25°C; the most effective storage solutions to retain cell viability were NaCl, glycerol, and sorbitol for *S. thermophilus* or phosphate buffer and sorbitol for *L. delbrueckii* subsp. *bulgaricus* stored at 4°C; storage at 25°C for 14 days could be used for the streptococci in all solutions except glycerol, while the lactobacilli only stored for 4 days in sorbitol.	Audet et al. (1988, 1991a)
	In classic free cell fermentation, the counts of streptococci and lactobacilli were 1.5×10^9 and 6.0×10^8 cfu ml^{-1}; respectively; and when immobilized in Ca-alginate beads, cell densities were 5 times higher than under free cell conditions.	Champagne et al. (1996c)

conditions (growth medium, presence of cryogenic compounds, freezing, and drying) and on the method of cell concentration. A few decades ago, the cell concentration systems that were widely used were (a) mechanical separators (e.g., Sharples at 5,500 × *g*, or ultracentrifuge at 15,000–20,000 × *g*), which can cause some physical damage to the bacterial cells, and (b) diffussion culture techniques or dialysis [see the reviews by Tamime (1990) and Pörtner and Märkl, (1998)]. However, the current method of concentration used by commercial culture manufactures consists of two-stage concentration before freeze-drying or freezing (Høier et al., 1999). The process line consists of the following sections:

- Tank for growth media preparation.
- Heat treatment equipment [i.e., ultra-high temperature (UHT)] to sterilize the growth media.
- Aseptic fermentation tank fitted with (a) pH control including continuous neutralization of the growth medium to maintain the pH at 6.0–6.3 or 5.5–6.0 for mesophilic and thermophilic lactic acid bacteria, respectively, by the addition of NaOH or NH_4OH; this system produces a high cell biomass, which is then concentrated (see later); the formation of Na-lactate can be inhibitory, and hence separation increases the degree of concentration that is possible (Anonymous, 1970, 1972; Barach and Kamara, 1986; Parente and Zottola, 1991; Borzani et al., 1993), (b) an inoculation port to inoculate the media with starter culture, (c) a temperature control system to maintain the optimum growth conditions of the culture, and (d) control of the speed of agitation and composition of head space gases which are optimized for each bacterial strain.
- Cooling to stop the metabolic activity of the starter culture for storage in an intermediate buffer tank.
- Concentration of the microbial cell biomass by centrifugation, ultrafiltration, or cross-flow filtration (Ferras et al., 1986; Taniguchi et al., 1987; Boyaval et al., 1987, 1988; Prigent et al., 1988; Hayakawa et al., 1990; Roy et al., 1992; Corre et al., 1992; Gagne et al., 1993; Suzuki, 1996), along with storage in an intermediate tank for further processing.
- Production of commerical starter cultures by (a) freeze-drying followed by packaging or (b) formation of pellets followed by freezing and packaging or freeze-drying and later packaging (see Figure 7.10).

It is evident that rapid development has taken place in the technology of starter cultures since the 1980s, and a typical cell count in commercial cultures may be 100–200 billion cfu g^{-1} in a freeze-dried culture, or 100–300 billion cfu ml^{-1} in frozen cultures which can be used for the direct production of bulk starter or product (Hansen, 1980; Tamime and Robinson, 1999; Høier et al., 1999). However, Martin (1983) described the production of freeze-dried type by Rhodia in France (formerly known as IZAL or Eurozyme). An illustration of a fermentor used to produce high number of cell biomass is shown in Figure 7.11.

Another method used to concentrate starter culture (i.e., *L. acidophilus* and *B. bifidum*) cells continuously involves two successive cell-recycle bioreactors (Boyaval et al., 1992). The growth medium was prepared by recombination [(g $liter^{-1}$) of sweet cheese whey powder 65, autolyzed yeast extract 10, ascorbic acid 1, caseinates 1] in deionized

Figure 7.11. Equipment for the production of starter culture concentrate prior to freeze-drying. *Note*: On-site view of a processing tank for production of starter culture; ammonium compound is used to neutralize the acid produced. (Reproduced by courtesy of Rhodia Food UK Ltd., Stockport, England.)

water. *Lactobacillus acidophilus* was propagated in a second bioreactor, which was connected to the permeate outflow of a microfiltration plant fitted onto the first bioreactor. Also, the second bioreactor was fitted with another microfiltration plant, and both units are used to concentrate the microbial cells. The cell productivities of *L. acidophilus* and *B. bifidum* were 7.6×10^7 and 3.2×10^7 cfu ml^{-1} h^{-1}, respectively. This method of starter culture production also allows a study to be made of the relationship between the two microfloras without mixing them together (Boyaval et al., 1992).

7.5.3 Preservation of Molds

Blue and white molds are normally preserved by freeze-drying, or with a fluid-bed drier as reported by Hylmar and Teply (1970) for the preservation *of P. roqueforti.* The dried cultures are resuspended in sterile or boiled water, and the resultant preparation is referred to as the "working culture." The rehydrated spores can remain active for a week at 5°C. The application of these molds in the dairy industry depends on the end product; and according to Galloway (personal communication), a culture of white mold can be used in one of the following ways during the manufacture of Camembert or Brie:

- Direct inoculation of the milk, along with the lactic starter, prior to the addition of rennet.
- Spraying the mold "solution" onto the cheese curd before salting.
- Coating the surface of cheese with a special mixture of salt and dried mold spores.

However, the application of blue molds is as follows: (a) using two of the methods mentioned above, (b) adding the mold culture to the curd immediately after filling the cheese molds, and (c) growing the mold on brown bread crumbs, shaping the mycelial mass into balls, which are then wrapped in muslin cloth and rubbed by hand into the milk after the addition of the lactic starter.

Ottogalli and Rondinini (1976) reported on the preservation of the molds and lactic acid bacteria used in the production of Gorgonzola cheese. The method comprised growing the cultures in milk at pH 5.5–6.5 to give 10^8–10^9 cfu ml^{-1}, followed by freezing at −18°C. The survival rates for the particular organisms were as follows: *S. thermophilus.* 18%; *L. lactis* subsp. *lactis*, 38%; *L. mesenteroides* subsp. *cremoris*,

80%; *L. delbrueckii* subsp. *bulgaricus*, 47%; *Torula* spp., 50%; and *P. roqueforti*, 100%. From such results it is safe to assume that blue molds can be preserved by freezing without any loss in their activity. However, Larroche and Gros (1986) managed to cultivate *P. roqueforti* in a fermentor filled with buckwheat seeds for the production of spores; and, in a semicontinuous fermentation, the average productivity was 9.2×10^6 external spores g^{-1} dry matter h^{-1}. In addition, Godíndez and Calderón (2000) freeze-dried three and four strains of *P. camemberti* and *P. roqueforti*, respectively, and they reported that the viability of both species remained high over 12 months' storage. However, it was observed by the same authors that the lipolytic and proteoltic activities of the molds were reduced with increase in the duration of the storage period.

The growth kinetics for *G. candidum* and *P. camemberti* cultivated on complex liquid media and the sporulation of *P. camemberti* in submerged culture were reported by Amrane et al. (1999) and Bockelmann et al. (1996), respectively. The cell biomass concentration (g liter^{-1}) reached to 1.5 and 2.5 for *G. candidum* and *P. camemberti*, respectively, while in the submerged culture the spore count reached 1.5×10^8 spores ml^{-1}. However, freezing of these cultures at –80°C followed by storage at –196°C (i.e., irrespective of cryoprotective agent used) ensured cell viability and stability of the mold strains (Schmidt et al., 1991; see also Moebus and Teuber, 1986; Desfarges et al., 1987).

7.5.4 Preservation of Kefir Grains

Kefir grains are preserved either dry or wet. One simple method consists of washing the excess grains with water, followed by drying at room temperature. This crude method of preservation can lead to contamination of the dried grains, and it is possible that the associative microflora may be altered. The preservation of the grains by freeze-drying is a better system. The preserved culture is in the form of a powder or small crystals, and after two to three subcultures the grains start to form in the growth medium. The production of freeze-dried Kefir grains is discussed in a Russian patent described by Lagoda et al. (1979). Alternatively, the washed grains are suspended in a sterile medium that can be stored for a few months at ordinary refrigeration temperature without any appreciable loss in activity; the grains retain their normal shape and form.

The preservation of Kefir starter cultures by freezing at –50°C, or by freeze-drying techniques, caused losses in lactobacilli and yeasts (i.e.,

>80%); the survival rates of the acid- and aroma-producing bacteria were <70% in frozen cultures and 50% in freeze-dried. However, in order to maintain an active level of yeast in these preserved cultures, the yeasts are grown on wort agar, stored with sterilized starch, and added to the starter during the production of Kefir (Kramkowska et al., 1986). Recently, Garrote et al. (1997) successfully preserved grains by suspending them in milk with freezing at –20 and –80°C; Kefir produced from the preserved grains was similar to the control (i.e., similar microflora, rheological characteristics, and acid and CO_2 contents). However, Pettersson et al. (1985) claimed that a freeze-dried Kefir culture (i.e., 1.4×10^{11} cfu g^{-1}) consisting of 90% lactic lactococci, 9% citric acid-fermenting lactococci, <0.5% lactobacilli, and <0.3% yeasts was suitable for the production of a bulk starter culture, and that the preserved culture maintained its activity at –80°C for 1–2 years. Nonetheless, Brialy et al. (1995) recommended glycerol as a cryoprotective agent during the lyophilization of the Kefir grains.

ICT was used for lactic acid bacteria, *Enterococcus* spp. and yeasts on Ca-alginate beads for the continuous production of Kefir (Gobbetti and Rossi, 1993, 1994). The fermented product contained higher microbial counts (~2 $\log_{10}$ cycles) than did the control Kefir.

7.5.5 Preservation of Koumiss Starter

Little data are available on Kumiss starter cultures, but in a Russian patent (Stoyanova and Pushkareva, 1986) the production of the starter concentrate is as follows:

- Centrifuge Koumiss starter culture consisting of *L. delbrueckii* subsp. *bulgaricus* and *K. marxianus* var. *lactis.*
- Dilute the concentrate in a ratio of 1:10 with a solution containing lactose, NaCl, and $(NH_4)_2HPO_4$.
- Freeze-dry at 30°C for 10–12 h, which ensures a survival rate of 90–95% and 82–86% for the lactobacilli and yeasts, respectively.
- Such a culture is suitable for the production of Koumiss, and the processing time is reduced by 0.5–1.0 h.

7.5.6 Implementation of HACCP System During the Preservation of Starter Cultures

The hazard analysis critical control points (HACCP) system was devised years ago in the United States to produce safe food for astronauts. In principle, the system focuses on safety and quality of the manufactured products, and it had identified seven aspects of production that merit constant attention (Mortimore and Wallace, 1994). This approach can be easily applied during the production and preservation of dairy starter cultures. Although the primary consideration of HACCP is to produce a "safe" starter culture, its suitability to produce acid during the manufacture of cheese and fermented milks is also important. Shapton (1989) suggested that these objectives can be achieved by having sufficient active cells and freedom from bacteriophage, other contaminants, and/or pathogens. In addition, the critical control points (CCPs) in starter handling in factories include (a) condition at reception, (b) starter storage (e.g., freeze-dried at <5°C or –18°C and frozen at –40°C or lower), (c) transfer to inoculation point, and (d) production of starter culture, incubation and transfer into vat or DVI into vat.

The purity and activity of DVI starter cultures are highly important, and the application of HACCP is normally adopted during the production of these cultures. A typical illustration of the manufacturing stages of freeze-dried cultures, including the CCPs, is shown in Figure 7.12.

7.6 FACTORS CAUSING INHIBITION OF STARTER CULTURES

There are many factors that can cause an inhibition of, and/or a reduction in, starter culture activity, and either event can lead to (a) poor-quality fermented dairy products reaching the consumer and (b) financial loss to the manufacturer. It is recommended, therefore, that milk intended for bulk starter culture production, or the manufacture of dairy products, should be free from these factors. The causes of inhibition of starter cultures are summarized in the following sections.

7.6.1 Compounds that Are Naturally Present in Milk

There are different antimicrobial systems that are present naturally in milk, and their main function is to protect the young suckling animal against infection and disease. The inhibitory compounds, which have been identified as inhibiting the growth of starter culture bacteria, are lactenins and the lactoperoxidase–thiocyanate–hydrogen peroxide (LPS) system (Tamime and Robinson, 1999). The former compounds are heat-sensitive, and they are destroyed during the preparation of the starter culture milk. In a typical investigation, Roginski et al. (1984a,b)

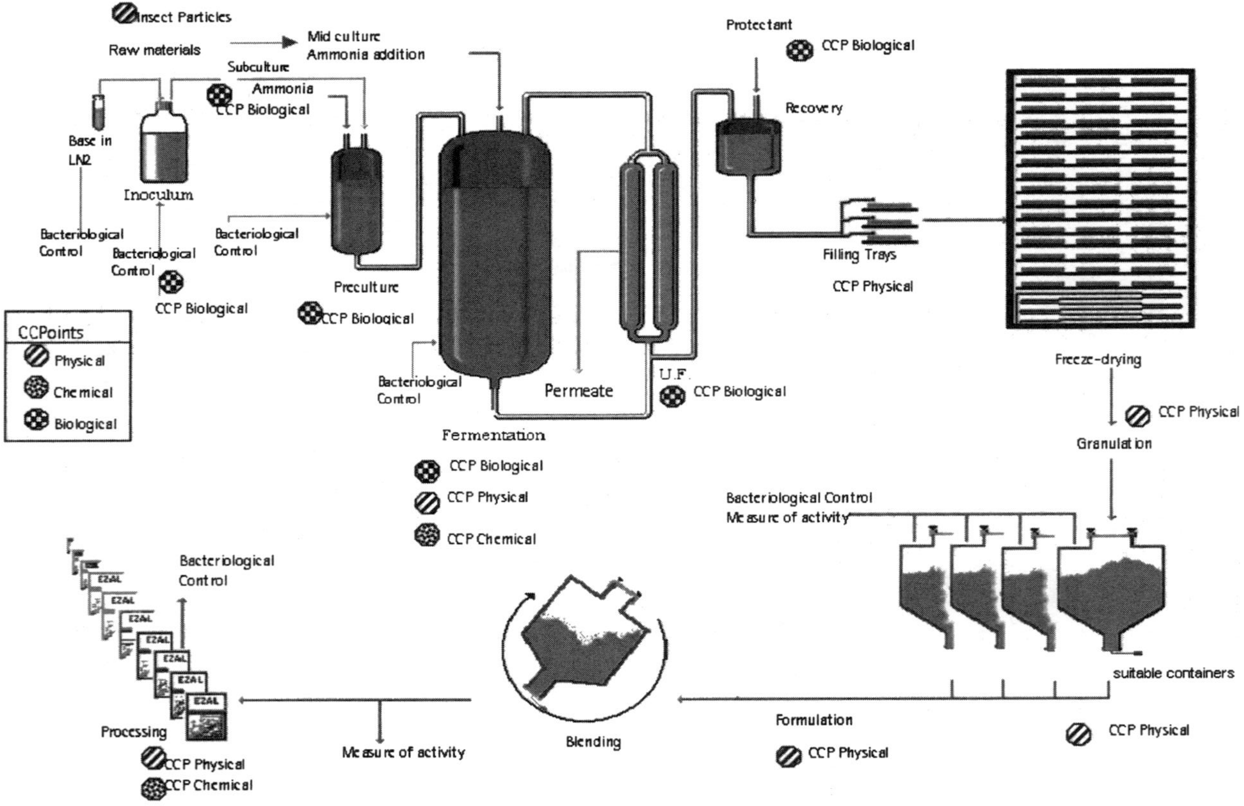

Figure 7.12. HACCP flow diagram including CCPs during the production of freeze-dried DVI starter culture. (Reproduced by courtesy of Rhodia Food UK Ltd., Stockport, England.)

observed inhibition of certain mesophilic lactic acid starter cultures in bulk milk and suggested that the inhibition could be attributed to the LPS system and the presence of inhibitory compounds (e.g., quaternary ammonium compounds).

The susceptibility of dairy starter cultures to LPS inhibition is dependent on the following:

- Strain sensitivity,
- Ability of the strain to generate hydrogen peroxide (H_2O_2), which activates the LPS system (Reiter, 1978; Reiter and Harnulv, 1984; Roginski et al., 1984a,b, 1991; Thomas, 1985; Guirguis and Hickey, 1987a; Font de Valdez et al., 1988; Piard and Desmazeaud, 1991; Pruit and Kamau, 1991; Nichol et al., 1995; Kot et al., 1996), or
- The presence of nonspecific enzymes (e.g., xanthine oxidase), and/or hypoxanthine to generate significant amounts of H_2O_2.

Hydrogen peroxide is added to raw milk produced in hot countries to improve the shelf-life during storage. According to IDF (1988) the recommended rate to activate the LPS system in milk is 3 mg $100\,g^{-1}$ of Na-precarbonate and 1.4 mg $100\,g^{-1}$ of Na-thiocynate. Thus, the natural presence of H_2O_2 in milk and activation of LPS, which can inhibit the growth of starter cultures, is the result of normal bacterial metabolism; a wide range of reactions and catalyzing enzymes are involved, and these have been reviewed by Piard and Desmazeaud (1991).

7.6.2 Antibiotic Residues

Residues of antibiotics in milk result from mastitis therapy in the dairy cow. There are various types of antibiotics used, and starter cultures are susceptible to very low concentrations (Cogan, 1972). Reinbold and Reddy (1974) surveyed 30 different antibiotics and antimicrobial agents, and they concluded that lactic streptococci and yogurt starter cultures were very sensitive to all the antibiotics tested; however, strains of *Leuconostoc* spp., *E. faecium*, and *B. linens* were less sensitive.

It is evident that dairy starter cultures are very sensitive to the presence of antibiotics in milk, and the minimum inhibitory concentration is dependent on differences between species and/or strains. For further information, refer to the published work by Park et al. (1984), Ramakrishna et al. (1985), Orberg and Sandine (1985), Narayanan and

Dhanalakshmi (1999), and Tamime and Robinson (1999). However, Teply and Cerminova (1974) tested the effect of eight antibiotics under processing conditions, and their results are illustrated in Table 7.13. It can be observed that the inhibitory levels of streptomycin, chloramphenicol, oxytetracycline, and tetracycline seem rather high, and this apparent resistance could be attributed to strain variation, variation in the commercial preparations of antibiotics used, or some feature of the test method used to detect the level of antibiotics.

7.6.3 Bacteriophage

The occurrence of bacteriophage (phage) in dairy starter cultures was first reported in the 1930s, and the existence of phage is a recognized problem in the dairy industry (Walker et al., 1981); the following precautionary measures may be practiced:

- The propagation of the starter culture must be carried out employing aseptic techniques (refer to later section 7.7.2).
- Heat treatment of bulk starter milk ensures the destruction of these viruses, but it is vital that the starter tank should be filled to its maximum capacity: otherwise, prolonged heat treatment is necessary to eliminate the phages in the airspace.
- Daily rotations of phage-unrelated starter strains, or phage-resistant strains, must be used in the dairy.
- Effective filtration of the air in the starter room and the production area can help to control the phage problem (see Chapter 14).
- The equipment must be properly sanitized (i.e., by heat or chemicals).
- Location of the starter room far away from the cheese production area and whey handling department reduces the possibility of airborne infection.
- Ensure that plant personnel, particularly those from the cheese room, are not allowed into the starter handling area.
- Propagate the starter culture in a phage-inhibitory medium (refer to Section 7.7.3).
- Avoid starter culture strains sensitive to those phages categorized as having short latent periods and large burst sizes.
- "Fog" the air in the starter preparation room with a solution of hypochlorite; alternatively, the use of UV light can control

TABLE 7.13. Sensitivity of Dairy Starter Cultures to Various Antibiotics (IU ml^{-1}) under Processing Conditions

Starter Cultures	Penicillin	Strepto-mycin	Chloram-phenicol	Erythro-mycin	Chlortetra-cycline	Bacitracin	Oxytetra-cycline	Tetra-cycline
Yogurt	0.01	1.0	0.5	0.10	0.01	0.04	0.4	1.0
Cream	0.20	2.5	1.0	0.50	0.05	0.10	0.4	1.0
Acidophilus	0.07	2.5	2.0	0.07	0.50	—	2.5	2.0
Emmenthal cheese	0.20	2.5	2.0	0.08	0.10	0.03	—	—
Kefir	2.50	2.5	3.0	0.50	2.00	—	1.5	1.0
Mesophilic lactic starter	0.01	2.5	2.0	0.05	0.50	0.12	2.0	—

Note: For further information refer to Adamse (1955), Berridge (1956), Cogan and Fitzgerald (1980), Hsu et al. (1987), and IDF (1995).
Source: Adapted from Teply and Cerminova (1974).

phage in the atmosphere (see Chapter 12 in this volume; IDF, 1991).

- "Fog" the cheese production area, although this approach is regarded by some as an extremely dubious practice.
- Use mixed-strain starter cultures, and possibly phage-resistant strains.

7.6.4 Detergent and Disinfectant Residues

Detergents and disinfectants are used for cleaning and sanitation purposes for dairy equipment. Residues of these compounds [alkaline detergents, chlorine-based materials, iodophors, quaternary ammonium compounds (QAC), and ampholytes] can affect starter culture activity. Pearce (1978) reported that strains of *L. lactis* subsp. *lactis* and subsp. *cremoris* showed inhibition at 2–4 mg QAC l^{-1}, except for one strain that tolerated up to 12 mg QAC l^{-1}. Yogurt cultures are more tolerant; and the inhibitory levels (mg liter^{-1}) of chlorine compounds, QAC, and iodophors for *S. thermophilus* and *L. delbrueckii* subsp. *bulgaricus* are 100, 100–500 and 60, and 100, 50–100 and 60, respectively (Bester and Lombard, 1974), but in a recent study, Guirguis and Hickey (1987b) have observed that some *Lactobacillus* spp. are sensitive to 0.5 mg QAC l^{-1}.

The effect of nine different detergents and sanitizers (0.1 g 100 ml^{-1} concentration in skimmed milk) on the activity of starter cultures was determined by El-Zayat (1987), and only one culture was completely inhibited by Solvay (caustic detergent) and Henkel P3-Oxonia (acidic disinfectant). The effect of disinfectants and sanitizers, as well as threshold of inhibition concentration (partial or total), on Cheddar cheese starter cultures have been well documented (Dunsmore, 1984; Liewen and Marth, 1984; Dunsmore et al., 1985; Petrova, 1990; Mäkelä et al., 1991; Vallado and Sandine, 1994).

Contamination of starter milk with these compounds is due to human error or due to a breakdown in the automatic cleaning cycle. In practice, it is necessary to ensure that the rinsing cycle is long enough to wash these chemicals from the bulk starter tank.

7.6.5 Bacteriocins

Antibacterial substances, which are usually segregated from antibiotics, are produced by a wide range of bacteria including starter cultures. These compounds are proteinaceous in nature, and they are known as bacteriocins. Some examples of bacteriocins produced by starter culture are (a) nisin, lacticin, lactococin (*Lactococcus* spp.), (b) mesenteroicin (*Leuconostoc* spp.), (c) thermophilin, STB (*S. thermophilus*), (d) bulgarican, lactobacillin, acidophilin, helveticin, caseicin, reutericin (*Lactobacillus* spp.), (e) bifidocin B (*Bifidobacterium* spp.), and (f) pediocin (*Pediococcus* spp.). These compounds have been found to inhibit the growth of a wide range of microorganisms, including pathogens and bacteria of the same genera. For further information refer to the following reviews: Piard and Desmazeaud (1992), de Vuyst and Vandamme (1994), Nes et al. (1996), Marshall and Tamime (1997), Weinbrenner et al. (1997); Yildirim and Johnson (1998); Tamime and Robinson (1999); and Moll et al. (1999).

7.6.6 Miscellaneous Inhibitors

7.6.6.1 Mastitis Milk and Somatic Cell Count. Leucocytes in mastitis milk can absorb starter organisms by phagocytosis, and heating results in no significant improvement (Sellars and Babel, 1985). However, Gajdusek and Sebela (1973) observed that while yogurt starter cultures are inhibited by up to 35% in milk containing high somatic cell counts, boiling the milk for 2 min (or heating it at 90°C for 20 min) removed the inhibition and allowed normal acid production.

However, somatic counts in milk of 4.0×10^5 cells ml^{-1} caused some inhibition of the yogurt microflora, *L. acidophilus* and *L. paracasei* subsp. *paracasei* (Fang et al., 1993; see also Tamime and Robinson, 1999). The quality of yogurt and other dairy products is influenced by the somatic cell count, and it is recommended that the count should be low (Mitchell et al., 1986; Kosikowski and Mistry, 1988; Mistry and Kosikowski, 1988; Rogers and Mitchell, 1994; Auldist and Hubble, 1998).

7.6.6.2 Seasonality of Milk. Late lactation milk and spring milk have some effect on starter culture activity. The reason(s) has not been elucidated yet, but it is possible to speculate that thiocyanates may be responsible.

7.6.6.3 Environmental Pollution. Other inhibitors in milk could be due to environmental pollution, such as insecticides, which can inhibit the starter organisms (Deane and van Patten, 1971; Dean and Jenkins, 1971; Gajduskova and Lat, 1972). The presence of these compounds in milk can inhibit the growth of the starter culture or can affect cell morphology (e.g., increase in cell size and the formation of longer chains).

However, many Egyptian scientists (see the review by Tamime and Robinson, 1999) concluded that the presence of pesticides in milk affected the activity of starter cultures and the quality of the dairy products. The results of these studies could be summarized as follows:

- Longer coagulation times were required, and the cheeses had many holes.
- The level of pesticide decreased in freshly made Zabadi (*i.e.*, Egyptian fermented milk) as a result of the fermentation.
- The growth rates of the yogurt organisms was reduced in the presence of DDT and malathion, while the cells of *L. delbrueckii* subsp. *bulgaricus* flocculated into clumps in milk containing aldicarb and the cell count was lower than the control (see Dhanalakshmi et al., 1998).
- Heating and fermentation of pesticide-contaminated milk contributed toward the degradation of pesticides.

7.6.6.4 Miscellaneous Compounds. Volatile and nonvolatile compounds (fatty acids, formic acid, formaldehyde, acetonitrile, chloroform, ether) in concentrations up to 100 mg $liter^{-1}$ inhibit the growth of the *Lactococcus* spp. and *L. mesenteroides* subsp. *cremoris* (Kulshrestha and Marth, 1974a–c, 1975), while Umanskii and Borovkova (1985) have reported that 3 mg ml^{-1} of a mixture of free fatty acids in sterilized milk significantly reduced the growth of *S. thermophilus*; a 5-mg ml^{-1} concentration inhibited the growth of *L. lactis* subsp. *lactis* and subsp. *lactis* biovar *diacetylactis*, *L. helveticus*, and *L. delbrueckii* subsp. *lactis*.

The review by Tamime and Robinson (1999) suggested that many compounds can affect starter culture growth or reduce acid development during the manufacture of fermented milk, and some examples include the following:

- Addition of flavors (coffee and garlic extract, ginseng saponins) affected the growth of *L. acidophilus*, yogurt cultures, and single strains of mesophilic lactic acid bacteria.
- Most strains of *L. helveticus* and a strain of *L. delbrueckii* subsp. *bulgaricus* were found sensitive to lysozyme (10 or 20 μg ml^{-1}) when added to cheese milk to control or inhibit the growth of clostridia.
- The addition of nitrates or nitrites to milk reduced the rate of acid development by the yogurt cultures.

- Increasing the concentrations (0.5–8 g $liter^{-1}$) of herbs (oregano, rosemary, sage, and thyme) progressively reduced acid production of *L. plantarum* and *P. acidilactici*; the former culture was more resistant (Zaika et al., 1983).
- Contamination of milk with radioactive iodine ^{131}I (i.e., equivalent to 6–12 kβq kg^{-1}) reduced the counts of yogurt organisms by 49% and reduced lactococcal species by 30% in cheese and buttermilk and 26% in ripened butter.
- Addition of sugar ≥9 g 100 g^{-1} to the milk may cause some inhibition or delay the fermentation period.
- The use of UF milk or phosphated bulk starter media may cause some inhibition of starter culture strains (see Section 7.7.3.1, discussion of bacteriophage inhibitory media).

7.6.6.5 Processing Conditions. The growth behavior of any starter culture (i.e., single, multiple, or mixed strain) is influenced by the accurate inoculation rate and incubation at the optimum temperature. The accuracy during processing (e.g., production of bulk starter culture or manufacture of fermented milks and cheese) ensures maximal rate of acid development within a short period of time.

7.7 PRODUCTION SYSTEMS FOR BULK STARTER CULTURES

The production of a bulk starter culture—that is, the culture used directly in the fermentation process—necessitates several stages of subculturing in order to meet the quantity required, or direct inoculation of the bulk starter tank using concentrated cultures (see systems 1 and 2 in Figure 7.9). The most important aspect of starter production is the preparation of the growth medium, along with the protection of the culture from phage attack. The methods used may be divided into the following systems: first, the use of simple microbiological techniques; second, the use of mechanically protected tanks; and third, the propagation of the starter culture in a special medium that is inhibitory to phage. The mechanically protected approach is widely used in the United Kingdom, Australia, and New Zealand, whereas the latter system is popular in the United States.

7.7.1 Simple Microbiological Techniques

In this system, the equipment/materials are basically laboratory utensils and a starter tank, which may be of a simple design—that is, batch pasteurizer/starter incubator. A full description of this method has been reported by Tamime and Robinson (1999).

7.7.2 Mechanically-Protected Systems

Different techniques have been developed in the dairy industry for the production of starter cultures in mechanically protected systems; and in these systems, the following aspects are important: First, both the processing of the growth medium and starter growth take place in a completely enclosed tank; and second, the inoculation of the starter takes place through a barrier which prevents the entry of unclean air. Some examples of mechanically protected systems follow.

7.7.2.1 The Lewis System. The development of the Lewis technique is well documented by Lewis (1956) and by Cox and Lewis (1972); recently, Lewis (1987) has published a book which illustrates the progress made in the field of handling, production and supply of cultures from 1948 to the present time. The technique consists of using reusable polythene bottles (115- and 850-g capacity) for mother and feeder cultures, respectively. These bottles are fitted with Astell rubber seals, and the growth medium (i.e., 10–12 g $100\,g^{-1}$ reconstituted, antibiotic-free, skimmed milk powder) is sterilized in the same bottles. The starter culture transfers are carried out by means of two-way hypodermic needles, and the overall technique is illustrated schematically in Figure 7.13. The Lewis system requires a pasturized bulk starter tank so that the growth medium can be heat-treated inside the sealed vessel. It is worthwhile pointing out that, during the heating or cooling of the milk, no air escapes from or enters the tank. The top of the tank is flooded with sodium hypochlorite solution ($100\,mg\,liter^{-1}$), so that the transfer of the starter inoculum (feeder) to the bulk starter medium is through a sterile barrier. Incidentally, special lids can he fitted to 22.7- to 45.5-liter churns, so that the Lewis system can be used successfully for the production of smaller volumes of starter culture. On-site illustrations of such tanks have been reported by Lewis (1987) and by Tamime and Robinson (1999).

7.7.2.2 The Jones System. The other protected method is the Jones system and, in this case, the tank is not a pressurized starter culture vessel. A detailed illustration of this bulk starter system, including a combined Lewis–Jones system, has been reported by Tamime (1990). However, during the heating or cooling of the milk, air leaves

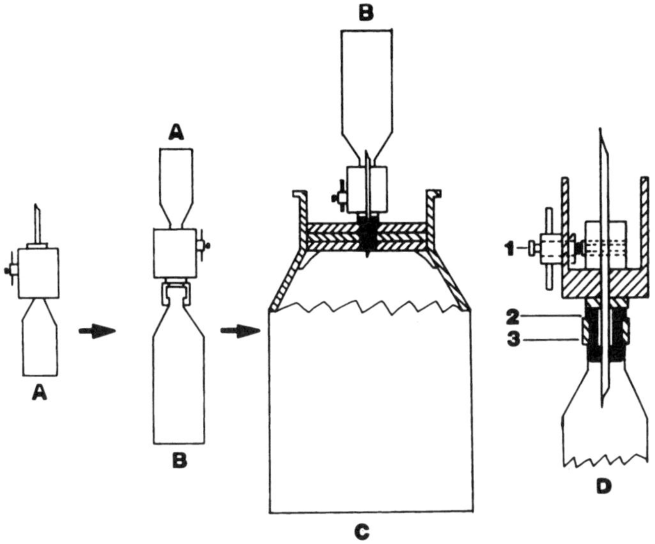

Figure 7.13. Schematic illustration of the Lewis system for starter culture transfers. *Note*: A, mother culture; B, feeder/intermediate culture; C, bulk culture; D, detail of meedle assembly (1, tap; 2, Astell rubber seal; 3, hypochlorite solution).

or enters the starter vessel; that is, the tank is not pressurized. The manway cover, the inoculation port, and the drive shaft of the agitator system (top-driven type) are specially designed to act as mechanical barriers that are referred to as water seals. During the heat treatment of the milk, the air in the head space of the tank is forced out, and it reenters during the cooling stages. The air is sterilized using a combination of heat and cotton wool filtration. The starter inoculum is poured into the tank through a special narrow opening, and a "ring of flame" or steam is used to provide a sterile point of entry. It is important to point out that the air filtration unit is activated during the heating and cooling of the milk and during the cooling of the active culture, thus ensuring that air entering the tank is always sterile.

The historical background to, and development of, this system has been reported by Whitehead (1956) and Robertson (1966a,b). Recently, Heap and Lawrence (1988) have described an up-to-date version of this starter tank which is widely used in New Zealand, and the tank is fitted with a compressed air system incorporating a special filter to sterilize the air.

The conventional procedure for culture transfer is to remove the inoculation port and pour the culture through a "ring of flame" or a "blanket of steam." In either method, the combined effect of heat and the slight positive pressure in the head space of the starter vessel ensures that the flow of air is always in one direction—that is, from inside the tank to the atmosphere. This minimizes the possibility of airborne contamination.

7.7.2.3 The Tetra Pak System. In the Lewis system, the starter culture transfer from one container to another relies on squeezing the polythene bottle to eject the culture. However, the Tetra Pak system (Bylund, 1995) uses filter-sterilized air, under pressure, for transferring the culture. The mother and feeder/intermediate cultures are prepared in a special unit called a "Viscubator," and the capacity of these containers is 0.5 and 20 liters, respectively. The former unit is a glass bottle that is sealed with a rubber stopper and a metal screw-cap with an annular space, and the feeder container is a stainless steel canister which has two fittings: one is for compressed air, and the other is in the form of a length of stainless steel pipe that connects to the bulk starter tank during culture transfer.

During culture transfer (i.e., from mother → feeder) two disposable sterile syringes are used: One is short and the other is long enough to reach the bottom of the bottle. Through the short syringe, compressed, sterile air causes a buildup of pressure in the head space of the bottle which displaces and forces the culture through the long needle to the feeder vessel. The same principle is applied for culture transfer from the feeder to the bulk starter tank using special fitments that are equipped with special valves for quick release/coupling. These units are fully illustrated by Bylund (1995) and by Tamime and Robinson (1999).

For the manufacture of starter cultures under aseptic conditions, Tetra Pak has developed a tank of aseptic design. According to Bylund (1995), the specifications of this bulk starter tank can be summarized as follows:

- The tank is constructed using stainless steel, and it is capable of withstanding negative and positive pressures up to 30 and 100 kPa, respectively.
- It has an aseptic design (i.e., hermetically sealed and triple jacketed).
- The growth medium is heated in the tank, and the tank can be fitted with a stationary integrated pH meter that is designed to withstand the great temperature differences that occur between the heat

treatment of the medium or cleaning-in-place (CIP) and growth or storage of the starter culture.

- The tank is fitted with high-efficiency particulate air (HEPA) filters that can be sterilized by steam at 140°C.
- A two-speed motor is used to operate the agitator of the tank, and the shaft of the agitator is double sealed.

It is recommended that two tanks should be used in rotation. One tank contains a ready-made starter for use on a production day, whereas the other is for preparing starter culture for the following day (Bylund, 1995).

7.7.2.4 Miscellaneous Systems. Different types of bulk starter tanks using filtered, sterile air (under pressure) are available to dairy processors in many countries. Illustrations and descriptions of these tanks have been published by Tamime (1990). The application of HEPA filtration systems to bulk starter tanks was studied at NIZO in the 1970s (see Stadhouders et al., 1976; Tofte-Jesperson, 1979: Stadhouders, 1986). Leenders et al. (1984) have evaluated the effect of a HEPA filter (ultrapolymembrane PF-PP 30/3, 0.2 μm HF) on air entering bulk starter tanks in factories, and they have observed that less than one phage out of 1.9×10^8 phage passed through—this is a *priority* requirement.

An example of such a tank is that manufactured by GEA Tuchenhagen (see Figure 7.14); and according to Anderson (personal communication), the processing stages of the bulk starter are as follows:

- Sterilize the milk pipeline.
- Steam sterilize the bulk starter tank; afterwards the tank should be filled with sterile air under pressure, and the pressure should be maintained until the starter is used.
- Fill the tank with skimmed milk (i.e., fresh or reconstituted), heat to 90–95°C for 30–45 min, and cool to 20–45°C depending on the starter culture used (mesophilic or thermophilic).
- Inoculate the starter culture (for example, freeze-dried or frozen) through the manhole, mix the culture into the milk, and incubate to the desired pH.
- Stir and cool the culture in-tank to stop the metabolic activity of the microorganisms; the culture is ready to be used for the manufacture of cheese and/or fermented milks.

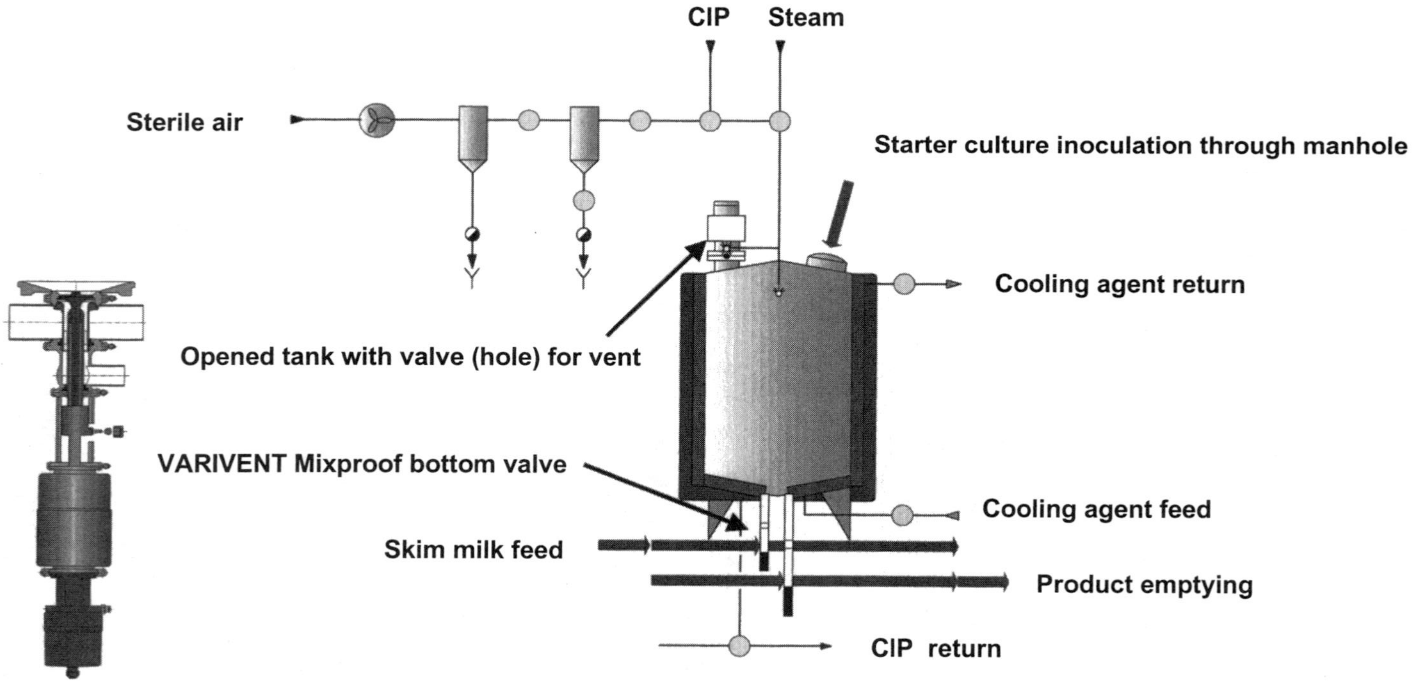

Figure 7.14. Cultivation tank for bulk starter with overpressure of sterile air. (Reproduced by courtesy of GEA Tuchenhagen UK Ltd., Warrington, England.)

7.7.3 Bacteriophage Control Systems

7.7.3.1 Phage-Resistant/Inhibitory Medium (PRM/PIM). The proliferation of phage in a dairy starter culture is dependent on the presence of free calcium ions in the growth medium. Reiter (1956) observed the inhibition of phage of lactic streptococci in a milk medium lacking in calcium (see Shalaby et al., 1986). The medium was referred to as a "phage-resistant medium" (PRM), but is sometimes also called a "phage-inhibitory medium" (PIM). The use of phosphates to sequester the free calcium ions in bulk starter skimmed milk was reported by Hargrove (1959) and Hargrove et al. (1961), and this initiated the commercial production of PRM/PIM. Since 1960, there have been tremendous developments in PRM/PIM composition, and the existing PRM/PIM on the market consists mainly of milk solids, sugar, growth stimulatory factors, and buffering agents (e.g., phosphates and citrates). However, when PRM/PIM were first developed, their effectiveness to protect and stimulate the growth of starters was limited [see the review by Whitehead et al., (1993)]; a view based on some of the following research results, but at present such media have been shown to be more effective in controlling phage:

- Sozzi (1972) and Sozzi et al. (1978) observed that some phage can still destroy starter cultures despite the absence of calcium, and that phage activity is maximized at the optimum growth temperatures of the starter organisms.
- Henning et al. (1965) reported that *Leuconostoc* spp. do not grow well in PRM/PIM.
- Ledford and Speck (1979) observed that phosphates in PRM/PIM cause metabolic injury to starter cultures, but that this effect was least pronounced in cultures and growth medium obtained from the same commercial source; however, the presence of 1–2 g $100\,g^{-1}$ phosphates in milk adversely affects the growth and morphology of *L. delbrueckii* subsp. *bulgaricus* (Wright and Klaenhammer, 1984).
- Gulstrom et al. (1979) evaluated seven commercial brands of PRM/PIM and concluded the following: First, only two media were effective against the proliferation of phages, and effectiveness was mainly related to buffering capacity. Second, the best results were obtained in media that contained citrate buffers and hydrolyzed cereal solids that stimulated the growth of *Lactococcus* spp. Third, the best medium contained enough nutrients to support starter growth; otherwise, depressed growth is inevitable due to the high

concentration of phosphates and/or citrates. This latter aspect is important because certain strains of cheese starter cultures—for example, the New Zealand blend of two single strains *L. lactis* subsp. *cremoris* AM_2 and *Lactococcus lactis* subsp. *lactis* ML_8—fail to grow in these buffered media (Tamime, unpublished data).

- Roy et al. (1987) observed that PIM appeared to enhance the proteolytic activity of most Cheddar cheese starter cultures studied, but the effect did not persist during the intitial stages of cheesemaking.

The degree of heat treatment of PRM/PIM is important, and dairy processors should follow the guidelines provided by the suppliers; however, maintenance at 85°C for 30 min can be employed with the different starter media manufactured by Rhodia Food (Anonymous, 1986), and the production of bulk starter in a mechanically protected tank will provide a supplementary measure to limit the proliferation of phages.

7.7.3.2 pH Control Systems. The production/development of bulk starter systems using the pH control techniques could be attributed to the following aims:

- To overcome the drawbacks associated with PRM/PIM mentioned above, including the cost of such media.
- To minimize the daily fluctuations in acid development of the conventional bulk starter (i.e., "overripe" or less active) that occur under commercial practice; two surveys in New Zealand and the United Kingdom have illustrated this point (Pearce and Brice, 1973; Walker et al., 1981).

In principle, the pH control systems maintain the level of acidity in the bulk starter medium, thus reducing the cellular damage that can occur to certain starter cultures when the pH drops below 5. In general, when starter cultures are held for a long period at low pH, they become progressively less active. The control of acid development in the growth medium is achieved by neutralization, which consequently helps the starter culture to remain active and increase in cell count. Furthermore, cultures that are propagated in a pH-controlled medium tend to be protected against the inhibitory effect of the phosphate buffer that is present in PRM/PIM (Ausavanodom et al., 1973). Basically, there are two different methods of pH control that can be used (see Thunell, 1988; Rao et al., 1989; Thunell and Reinbold, 1993).

External pH control system was developed in the United States during the late 1970s by G. H. Richardson and his colleagues. Cheese whey was used as the starter culture growth medium, and this was fortified with phosphates (i.e., as PIM/PRM) and yeast extract. The pH is maintained at around 6 during the growth phase of the starter culture by the intermittent injection of anhydrous or aqueous ammonia. The system has been very successful for the production of starter cultures for the manufacture of most American-style cheeses (Richardson et al., 1977), cottage cheese (Chen and Richardson, 1977), and Italian and Swiss cheeses (Reddy and Richardson, 1977).

As mentioned earlier, a phosphate buffer in the growth medium is essential to control bacteriophage; but, by using the external pH system, the amount of buffer can be reduced without affecting the inhibitory benefits (Ausavanodom et al., 1977). Thus, using a pH controlled bulk starter system, it is feasible to produce an active, concentrated starter culture in whey buffered with ammonia (Richardson, 1978, 1983a,b; Richardson et al., 1979; Thomas et al., 1981; Wright and Richardson 1982; Richardson and Ernstrom, 1984), or other media buffered with sodium hydroxide (Reddy, 1984a). Some of the advantages claimed for such starter cultures are as follows:

- They are less expensive as compared with the PIM/PRM method.
- An active, highly concentrated, culture is produced that never sours (i.e., because the lactose is completely utilized), and it requires no ripening time in the cheese vat.
- Daily or weekly propagation is not required.
- The growth medium is cheap, is readily available, and consists of whey water, lactose, and stimulants.
- Acid-producing activity is greater vis-à-vis a culture grown in milk or PIM/PRM.
- The inoculation rate is reduced by 70–80%.
- There is a reduction in labor requirements.

The external pH control system is widely used in the United States, but not in Europe yet, and some of the latest designs incorporate microprocessors to control the instrumentation for pH and temperature. One such system is marketed by Rhodia Food and Debelak Technical System, and an illustration of the layout is shown in Figure 7.15.

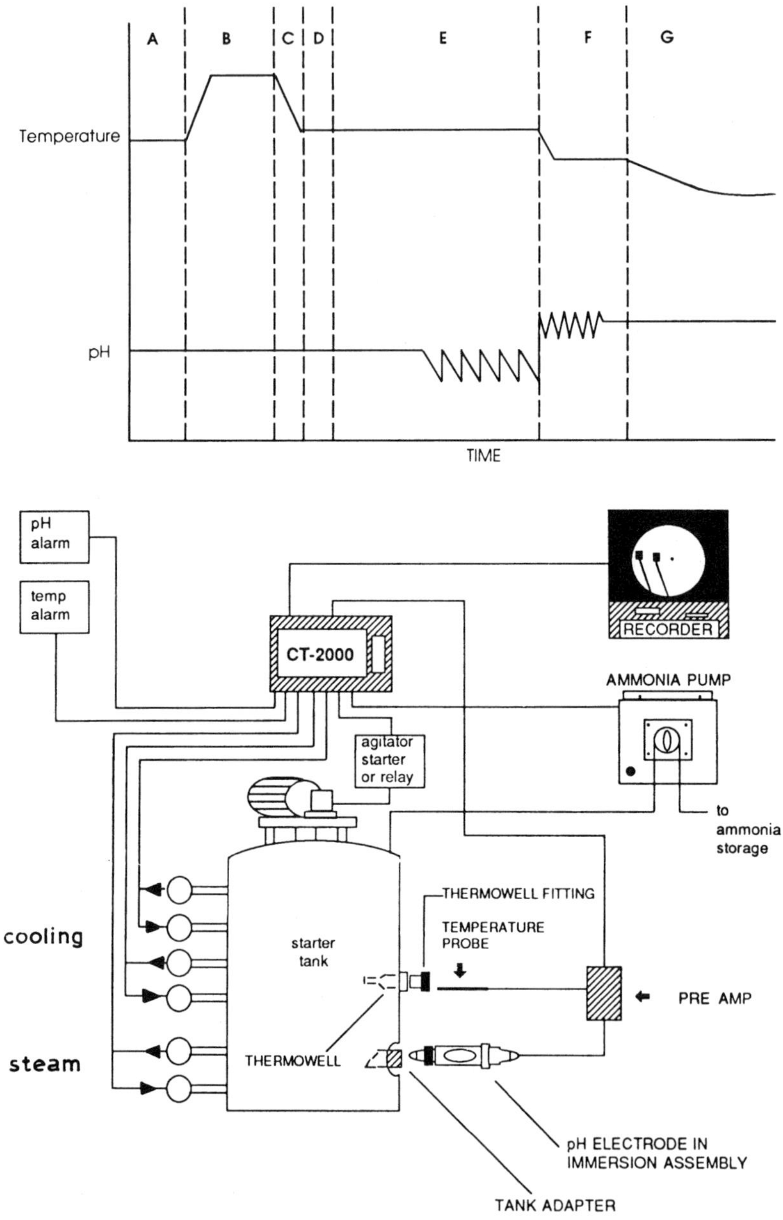

Figure 7.15. Operating modes pH, temperature control, and schematic illustration of the CT-2000 S.S. equipment; curves characteristics of thermophilic fermentations (refer to text). *Note*: A, idle; B, heat treatment; C, cool down; D, inoculation; E, growth I; F, growth II; G, cool. (Reproduced by courtesy of Rhodia Food UK Ltd., Stockport, England.)

When different bacterial strains (*S. thermophilus* and *L. delbrueckii* subsp. *bulgaricus*) are grown together for Italian type cheeses, the pH and incubation temperature can be adjusted to accommodate each strain. For example, the growth of the lactobacilli is maintained at low pH, but raising the pH and lowering the temperature stimulates the growth of *S. thermophilus*. However, the supplementation of whey permeate with 0.1 g $100\,g^{-1}$ SMP and 0.5–1 g $100\,g^{-1}$ yeast extract improved the growth and the activity of *L. helveticus*, *S. thermophilus*, and *L. delbrueckii* subsp. *bulgaricus* using the external pH control system (Parente and Zottola, 1991). Yezzi et al. (1993) reported increased nisin production when *Lactococcus* spp. were grown in pH-controlled bulk starter systems; subsequently, the nisin content in Cheddar cheese increased by 20% because of the higher cell numbers in the bulk starter.

This approach to starter culture production has been manipulated in the United States and New Zealand to increase the working capacity of the bulk starter tank (Limsowtin et al., 1980; Heap and Richardson, 1984). The culture is grown in a whey medium using an external pH control system; and when the culture is fully grown and 50–75% has been used, the balance is mixed with sterile whey to restore the original volume. After 2–3 h of incubation, the *new* culture is ready for use in the cheese vat. A similar approach is used in Australia using conventional bulk starter production in skimmed milk, and without the application of an external pH control system (Hall et al., 1985).

A pH-controlled system using a lactose-limited, milk-based medium is used in New Zealand to provide the industry with concentrated cultures at –40°C for the direct inoculation of bulk starter tanks (Turner et al., 1979). In addition, Champagne et al. (1996a) successfully used whey protein concentrate (WPC) to grow mesophilic and thermophilic starter cultures under an external pH control system; these cultures had higher counts than the same starters grown in milk (i.e., without pH control).

Internal pH control system contains the buffering agent in the growth medium, and such buffers solubilize in response to acid production by the starter culture bacteria. This technique was developed in the late 1970s at Oregon State University (Willrett et al., 1979; Sandine and Ayres, 1981, 1983); and it is marketed commercially under the name of "PHASE 4" (see also Mermelstein, 1982; Waes, 1988; Whitehead et al., 1993), or as another type of medium containing phosphates and free lecithin (Reddy, 1984b). The "PHASE 4" growth medium consists of a whey-based powder that contains an encapsulated citrate–phosphate buffering system that is initially insoluble. The buffer is released gradually during the ripening of the bulk starter, and it maintains the pH around 5.2. Many advantages have been claimed for "PHASE 4," and, in general, they are similar to those mentioned for the external pH system; however, with "PHASE 4" there is no requirement for special equipment for the addition of buffer or the neutralizing chemicals. Incidentally, "PHASE 4" was primarily developed for the production of mesophilic lactic acid starter cultures, and some modifications have been carried out for the starters required for the production of Italian and Swiss cheeses in the United States (Willrett et al., 1982). Evaluations of "PHASE 4" bulk starter medium have been carried out by Birlison and Stanley (1982) with buttermilk, cottage cheese, and Cheddar cheese starter cultures, and the results confirm the claims of the manufacturer. Champagne and Gagné (1987) have reported greater acid-producing activity with two out of three strains of *S. thermophilus* when grown in "PHASE 4" (Rhodia Food) and "In-Sure" (Chr. Hansen). However, mesophilic starter cultures, *S. thermophilus* and *L. delbrueckii* subsp. *bulgaricus* (the latter cultures are used in mozzarella cheesemaking), grown in internal pH control media to produce bulk starters, were more stable, could be held at refrigeration temperature for a few days, and had higher cell counts compared with the same cultures grown in reconstituted SMP media (Ustunol et al., 1986; Khasravi et al., 1991). In addition, Rajagopal et al. (1990) compared six commercially available bulk starter media including whey-based PIM for growing mozzarella starter cultures.

Another bulk starter medium for cheese with internal pH control has been patented by Sinkoff and Bundus (1983), and this contains magnesium ammonium phosphate as the buffering agent and an alkali metal tripolyphosphate for controlling phage (see also Willrett and Comotto, 1989, 1990); another medium, called IPC-55, has been developed in Australia (Ziolkowski and Goonan, 1985).

Changes in the cell counts of *Listeria monocytogenes* in a commercial starter culture medium with internal pH control system during the production of bulk starter using *L. lactis* subsp. *cremoris* were studied by Wenzel and Marth (1990a, 1990b, 1991); the results suggest that despite a reduction in numbers of *Listeria*, a substantial count (10^3–10^5 cfu ml^{-1}) was present when the bulk starter was ready for the manufacture cheese or fermented milk products.

7.7.4 Growth Medium and Costing of Bulk Starter Cultures

It is agreed that high-quality milk is essential as a substrate for the growth of bulk starters. It must be free from antibiotics and other inhibitory substances, and under commercial practice (Walker et al., 1981) the available types of starter culture growth media are:

- Whole or skimmed milk (fortification with SMP is optional)
- Reconstituted SMP
- Whey-based medium (Christopherson and Zottola, 1989a; Rajagopal et al., 1990; Bury et al., 1998)
- pH-controlled medium (Okigbo et al., 1985)
- PRM/PIM
- Mixture of reconstituted SMP and PRM/PIM
- Retentate from ultrafiltration of milk at different concentrations (Mistry and Kosikowski, 1985a,b, 1986a–c; Mistry et al., 1987; Christopherson and Zottola, 1989b; Srilaorkul et al., 1989; Jones et al., 1990; Premaratne and Cousins, 1991; Kosikowski and Mistry, 1992; Ventling and Mistry, 1993; Meijer et al., 1995)

It is evident that there are many different types of growth medium available to cheesemakers for the production of bulk starter, and the choice is primarily governed by the availability of substrate, economics, tradition, and effect on the quality of cheese. However, the cost of bulk starter production plays a major role, and surveys by LaGrange and Reinbold (1968), Dennien (1980), and LaGrange (1987) have shown that cost is dependent on the system used, including direct-to-vat inoculation of bulk starter milk or the cheese vat.

7.7.5 Inoculation Systems for DVI Bulk Starter Cultures and Production Tanks

Over the past 50 years, various methods have been designed to transfer starter cultures from one container to another under aseptic conditions, as well as to eliminate phage contamination during the preparation of the bulk starter. Knuston and Zottola (1989) evaluated an inoculation system for a closed bulk starter tank where the starter culture was aseptically injected through the inoculation port using a sterile needle and syringe. In addition, aerosol, which contained *Escherchia coli* and phage particles, was used at the time of starter inoculation. After the fermentation period, the bulk starter was *free* from phage and coliforms. The inoculation port used was an improvement to the type developed in the 1960s and 1970s (Robertson et al., 1974).

Since the 1980s, there has been greater reliance of fermented milks and cheese processors to use DVI starter cultures (i.e., concentrated frozen or freeze-dried), which are marketed by major culture suppliers in the world. It is evident that in some mechanically protected bulk starter tanks such DVI starter cultures have to be thawed or rehydrated before inoculating the bulk starter tank using a sterile needle and syringe. In addition, care should be exercised in order to minimize cellular damage or contamination of the starter culture.

Under normal commercial practice and during the manufacture of cheese or fermented milks, DVI starter cultures are added to the milk by opening the manhole of each fermentation tank. This method of inoculation may possibly contaminate the milk; and in the case of freeze-dried culture, ineffective rehydration may occur. Recently, an automatic inoculation system (AISY) was developed through collaboration between Chr. Hansen and Tetra Pak (see Figure 7.16). This system is suitable for the in-line inoculation of DVI cultures into the processed milk. The AISY method of inoculation can be summarized as follows:

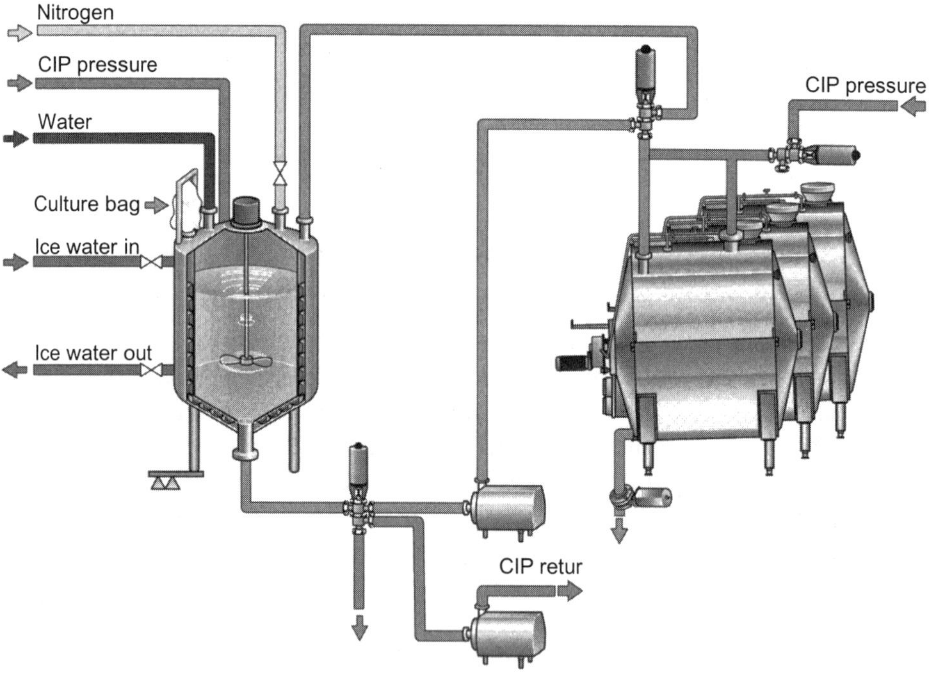

Figure 7.16. An illustration of the Tetra Pak/Chr. Hansen AISY for in-line culture dosing unit. [Reproduced by courtesy of Tetra Pak (Dairy and Beverage System) A/B, Lund, Sweden.]

- Add the required amount of water to a buffer tank to make a 10 g 100 g^{-1} bacterial solution (e.g., 10 kg frozen DVI culture and 90 kg of water).
- It is recommended that (a) the water should be pasteurized at 72°C for 15 s and followed by cooling to 14–18°C, (b) it should be good-quality water, and (c) an active carbon filter should be installed to remove chlorine if it is present >5 μg g^{-1}.
- Remove the culture from the freezer (i.e., at ≤–45°C) and empty into the buffer tank with agitation; the agitators should operate at slow speed to minimize foam formation or the incorporation of air into the solution.
- The dissolved culture is stirred for 10 min before being metered into the processed milk or production tank; the activity of the dissolved culture can be maintained up to 24 h if it is cooled to <10°C.
- The head space of the buffer tank should be purged with nitrogen to minimize frothing or loss in microbial activity.

It is worthwhile reporting at this stage that, over the last decade, "new" fermented milk products containing viable bacteria of human origin (i.e., bifidobacteria and lactobacilli) have been produced and marketed in many countries. Although these cultures are available as DVI cultures for process milk or bulk starter production, the use of one bulk tank starter may lead to changes in the ratio of the several bacterial species used. Hence, special growth media have to be used; as a consequence, many bulk starter tanks are required. Reviews of the technical aspects of the preparation of "bio" starter cultures have been published by Gilliland et al. (1991) and Hunger and Peitersen (1992).

7.8 QUALITY CONTROL

The ability of a starter culture to perform its function during the manufacture of fermented dairy products depends on its purity and activity. Different activity tests have been proposed (Hull and Juffs, 1980; Heap and Lawrence, 1981; Oberg and Richardson, 1984), and some of the recommended routine quality control tests on starter cultures are as follows.

7.8.1 Lactic and Other Bacterial Starter Cultures

- Microscopic examination using Gram's stain and/or Newman's staining method; the starter bacteria are Gram-positive, and the latter method has been used to monitor the ratio of cocci to rods in yogurt cultures.

- Detection of contaminants: (a) purity using the catalase test; LAB are catalase-negative, and a positive reaction is due to contamination; (b) a positive coliform test indicates gross contamination; (c) yeast and molds must not be present in LAB cultures; (d) detection of phage.
- An activity or vitality test helps to determine the rate of acid development by a starter culture prior to its use in the processing vat—for example, a simulated cheesemaking process in the laboratory.
- Tolerance to scalding temperatures is an important test for cheese cultures.
- The Voges–Proskauer or Creatine test is a biochemical test to check aroma-producing cultures, and this test is sometimes used to detect gas-producing organisms in a non-gas-forming culture.
- In some instances, lactococcal agglutination in bulk starter cultures can be minimized by (a) homogenization of the fermentate between 17 and 24 MPa of pressure (Milton et al., 1990; Hicks and Ibrahim, 1992) or (b) the use of whey-based medium treated with enzymes (Ustunol and Hicks, 1994).
- Enumeration of the "bio" species in a starter culture to maintain the appropriate ratio and count; it should be $>10^6$ cfu ml^{-1} in the final product at the end of its shelf life.

7.8.2 Molds

- The inoculation rate is determined by counting the spores in a sterile suspension.
- Sometimes a coliform test is carried out to detect gross contamination.

7.8.3 Kefir Grains

- Activity test on the preserved grains.
- Two possible contaminants—that is, coliforms and white molds (e.g., *G. candidum*)—may be present in the grains.

7.8.4 Swabs

These tests are carried out on starter culture equipment and propagation vessels and are essential to check on the efficiency of cleaning and the sanitary condition of the plant.

7.8.5 Microbiological Specifications

Microbiological specifications of commercial DVI starter cultures (freeze-dried or frozen) are:

Starter organims	10^{10}–10^{12} cfu g^{-1}
Coliforms	Absence in g^{-1}
Enterococci	<20 cfu g^{-1}
Yeasts and molds	Absence in g^{-1}
Staphylococci (coagulase-positive)	Absence in 10 g^{-1}
Listeria	Absence in 25 g^{-1}
Salmonella	Absence in 25 g^{-1}

REFERENCES

Abd El-Gawad, I. A., Metwally, M. M., El-Nockrashy, S. A., and Ahmed, K. E. (1989) *Egypt. J. Dairy Sci.*, **17**, 273–279.

Abraham, A. G., de Antoni, G. L., and Añón, M. C. (1990) *Cryobiology*, **27**, 336–342.

Accolas, J.-P., and Auclair, J. (1983) *Ir. J. Food Sci. Technol.*, **7**, 27–38.

Adamse, A. D. (1955) *Netherlands Milk Dairy J.*, **9**, 121–138.

Amen, J., and Cabau, M. (1984) European Patent Application, **EP** 65895 B1.

Amen, J., and Cabau, M. (1986) French Patent Application, **F** 2 554 125.

Amrane, A., Plihon, F., and Pringent, Y. (1999) *World J. Microbiol.*, **15**, 489–491.

Anderson, L. (1975) *Dairy Sci. Abstr.*, **37**, 310.

Andrew, M. H. E., and Russell, A. D. (eds.) (1984) *The Revival of Injured Microbes*, Academic Press, London.

Anonymous (1970) British Patent Application, **GB** 1 205 733.

Anonymous (1972) British Patent Application, **GB** 1 260 247.

Anonymous (1986) *Making Milk Works Wonders*, Miles Technical Bulletin, Rhodia Food UK Ltd., Stockport, England.

Arnaud, J.-P., Lacroix, C., and Choplin, L. (1992) *Biotechnol. Tech.*, **6**, 265–270.

Auclair, J., and Accolas, J. P. (1983) *Antonie van Leeuwenhoek*, **49**, 313–326.

Audet, P., Paquin, C., and Lacroix, C. (1988) *Appl. Microbiol. Biotechnol.*, **29**, 11–18.

Audet, P., Paquin, C., and Lacroix, C. (1991a) *Biotechnol. Tech.*, **5**, 307–312.

Audet, P., Paquin, C., and Lacroix, C. (1991b) *Int. Dairy J.*, **1**, 1–15.

Auldist, M. J., and Hubble, I. B. (1998) *Aust. J. Dairy Technol.*, **53**, 28–36.

Ausavanodom, N., White, R. S., and Richardson, G. H. (1973) *J. Dairy Sci.*, **56**, 637.

Ausavanodom, N., White, R. S., Young, G., and Richardson, G. H. (1977) *J. Dairy Sci.*, **60**, 1245–1251.

Barach, J. T., and Kamara, B. J. (1986) European Patent Application, **EP** 0 196 593 A2.

Barbour, E. A., and Priest, F. G. (1986) *Lett. Appl. Microbiol.*, **2**, 69–71.

Barlows, A., Trüper, H. G., Dworkin, M., Harder, W., and Schleifer, K.-H. (eds.) (1992) *The Prokaryotes*, Vols. 1 and 2, 2nd ed., Springer-Verlag, New Work.

Béal, C., and Corrieu, G. (1994) *Lebensm. Wiss. Technol.*, **27**, 86–92.

Béal, C., Spinnler, H. E., and Corrieu, G. (1994) *Appl. Microbiol. Biotechnol.*, **41**, 95–98.

Berridge, N. J. (1956) *J. Dairy Res.*, **23**, 348–354.

Bester, B. H., and Lombard, S. H. (1974) *S. Afr. J. Dairy Technol.*, **6**, 47–52.

Bezkorovainy, A., and Miller-Catchpole, R. (1989) *Biochemistry and Physiology of Bifidobacteria*, CRC Press, Boca Raton, FL.

Bhowmik, T., and Marth, E. H. (1990) *J. Dairy Sci.*, **73**, 859–866.

Birlison, F. G. T., and Stanley, G. (1982) *Dairy Ind. Int.*, **47**(11), 27–30.

Bissonnette, F., Labrie, S., Deveau, H., Lamoureux, M., and Moineau, S. (2000) *J. Dairy Sci.*, **83**, 620–627.

Blanchette, L., Roy, D., and Gauthier, S. F. (1995) *Milchwissenschaft*, **50**, 363–367.

Bockelmann, W. (1999) Secondary cheese cultures. In *Technology of Cheesemaking*, B. A. Law, ed., Sheffield Academic Press, Sheffield, pp. 132–162.

Bockelmann, W., Portius, S., Heller, K. J., and Neve, H. (1996) *Milchwissenschaft*, **51**, 306–310.

Borzani, W., Sanches, P. A., Luna, M. F., Jerke, P. R., and Stein, M. A. C. F. (1993) *J. Biotechnol.*, **31**, 61–66.

Boyaval, P., and Schuck, P. (1994) *IAA*, **111**, 807–818.

Boyaval, P., Corre, C., and Terre, S. (1987) *Biotechnol. Lett.*, **9**, 207–212.

Boyaval, P., Terre, S., and Corre, C. (1988) *Lait*, **68**, 65–84.

Boyaval, P., Madec, M.-N., and Corre, C. (1992) *Biotechnol. Lett.*, **14**, 589–592.

Boysen, M., Skouboe, P., Frisvad, J., and Rossen, L. (1996) *Microbiology*, **142**, 541–549.

Bozoğlu, T. F., and Gurakan, G. C. (1989) *J. Food Prot.*, **52**, 259–260.

Bozoğlu, T. F., Özilgen, M., and Bakir, U. (1987) *Enzyme Microb. Technol.*, **9**, 531–537.

Brennan, M., Wanismail, B., Johnson, M. C., and Ray, B. (1986) *J. Food Prot.*, **49**, 47–53.

Brialy, C., Rivalland, P., Coiffard, L., and de Roeck Holtzhauer, Y. (1995) *Folia Microbiol.*, **40**, 198–200.

Britz, T. J., and Riedel, K.-H. J. (1991) *J. Appl. Bacteriol.*, **71**, 407–416.

Bruinenberg, P. G., and Limsowtin, G. K. Y. (1995) *Aust. J. Dairy Technol.*, **50**, 47–50.

Bury, D., Jelen, P., and Kimura, K. (1998) *Int. Dairy J.*, **8**, 149–151.

Büyükgüngör, H. (1992) *J. Chem. Technol. Biotechnol.*, **53**, 173–175.

Büyükgüngör, H., and Caĝlar, A. (1990) *Doĝa—Turk. J. Chem.*, **14**, 316–324.

Bylund, G. (1995) *Dairy Processing Handbook*, Tetra Pak (Processing Systems Division) A/B, Lund, Sweden.

Cachon, R., Anterieux, P., and Divies, C. (1998) *J. Biotechnol.*, **63**, 211–218.

Cai, Y. M., Matsumoto, M., and Benno, Y. (2000) *Microbiol. Immunol.*, **44**, 815–820.

Castro, H. P., Teixeira, P. M., and Kirby, R. (1995) *Appl. Microbiol. Biotechnol.*, **44**, 172–176.

Castro, H. P., Teixeira, P. M., and Kirby, R. (1996) *Biotechnol. Lett.*, **18**, 99–104.

Castro, H. P., Teixeira, P. M., and Kirby, R. (1997) *J. Appl. Microbiol.*, **82**, 87–94.

Cerning, J. (1990) *FEMS Microbiol. Rev.*, **87**, 113–130.

Cerning, J., Gripon, J. C., Lamberet, G., and Lenoir, J. (1987) *Lait*, **67**, 3–39.

Champagne, C. P., and Gagné, D. (1987) *Can. Inst. Food Sci. Technol. J.*, **20**, 34–37.

Champagne, C. P., Baillargeon-Coté, C., and Goulet, J. (1986) *J. Appl. Bacteriol.*, **66**, 175–184.

Champagne, C. P., Girard, F., and Morin, N. (1988) *Biotechnol. Lett.*, **10**, 463–468.

Champagne, C. P., Detournay, H., and Hardy, M.-J. (1991a) *J. Ind. Microbiol.*, **7**, 147–150.

Champagne, C. P., Gardner, N., Brochu, E., and Beaulieu, Y. (1991b) *Can. Inst. Food Sci. Technol. J.*, **24**, 118–128.

Champagne, C. P., Morin, N., Couture, R., Ganon, C., Jelen, P., and Lacroix, C. (1992) *Food Res. Int.*, **25**, 419–427.

Champagne, C. P., Girard, F., and Rodrigue, N. (1993) *Int. Dairy J.*, **3**, 257–275.

Champagne, C. P., Lacroix, C., and Sodini-Gallot, I. (1994a) *Crit. Rev. Biotechnol.*, **14**, 109–134.

Champagne, C. P., Gardner, N., and Dugal, F. (1994b) *J. Ind. Microbiol.*, **13**, 367–371.

Champagne, C. P., Raymond, Y., Mondou, F., and Julien, J. P. (1995) *Bifidobacteria Microflora*, **14**, 7–14.

Champagne, C. P., Mondou, F., Raymond, Y., and Roy, D. (1996a) *Food Res. Int.*, **29**, 555–562.

Champagne, C. P., Mondou, F., Raymond, Y., and Brochu, E. (1996b) *Biosci. Microflora*, **15**, 9–15.

Champagne, C. P., St.-Gelais, D., and Audet, P. (1996c) *Milchwissenschaft*, **51**, 561–564.

Champagne, C. P., Gardner, N., Soulignac, L., and Innocent, J. P. (2000) *J. Appl. Microbiol.*, **88**, 124–131.

Chavarri, F. J., de Paz, M., and Nunez, M. (1988) *Biotechnol. Lett.*, **10**, 11–16.

Chen, Y., and Richardson, G. H. (1977) *J. Dairy Sci.*, **60**, 1252–1255.

Choisey, C., Desmazeaud, M., Gueguen, M., Lenoir, J., Schmidt, J.-L., and Tourneur, C. (2000) Microbial phenomena. In *Cheesemaking*, 2nd edn., A. Eck and J.-C. Gillis, eds., Lavoisier Publishing, Paris, pp. 353–417.

Christensen, V. W. (1969) United States Patent Application, **US** 3 483 087.

Christensen, J. E., Dudley, E. G., Pedersen, J. A., and Steele, J. L. (1999) *Antonie van Leeuwenhoek*, **76**, 217–246.

Christopherson, A. T., and Zottola, E. A. (1989a) *J. Dairy Sci.*, **72**, 2862–2868.

Christopherson, A. T., and Zottola, E. A. (1989b) *J. Dairy Sci.*, **72**, 2856–2861.

Cogan, T. M. (1972) *J. Appl. Microbiol.*, **23**, 960–965.

Cogan, T. M. (1983) *Ir. J. Food Sci. Technol.*, **7**, 1–13.

Cogan, T. M. (1995) *J. Appl. Bacteriol. Symp. Suppl.*, **79**, 49S–69S.

Cogan, T. M., and Accolas, J.-P. (1990) Starter cultures: types, metabolism and bacteriophage. In *Dairy Microbiology*, Vol. 1, 2nd ed., R. K. Robinson, ed., Elsevier Applied Science, London, pp. 77–114.

Cogan, T. M., and Accolas, J.-P. (eds.) (1996) *Dairy Starter Cultures*, VCH Publishers, New York.

Cogan, T. M., and Fitzgerald, G. F. (1980) *Ir. J. Food Sci. Technol.*, **4**, 79–88.

Cogan, T. M., and Hill, C. (1993) Cheese starter cultures. In *Cheese: Chemistry, Physics and Microbiology*, Vol. 1, 2nd ed., P. F. Fox, ed., Chapman & Hall, London, pp. 193–255.

Cogan, T. M., and Jordan, K. N. (1994) *J. Dairy Sci.*, **77**, 2704–2717.

Cogan, T. M., Barbosa, M., Beuvier, E., Bianchi-Salvadori, B., Cocconcelli, P. S., Fernandes, I., Gomez, J., Gomez, R., Kalantzopoulos, G., Ledda, A., Medina, M., Rea, M. C., and Rodriguez, E. (1997) *J. Dairy Res.*, **64**, 409–421.

Collins, M. D., Rodrigues, U., Ash, C., Aguirre, M., Farrow, J. A. E., Martinez-Murcia, A., Phillips, B. A., Williams, A. M., and Wallbanks, S. (1991) *FEMS Microbiol. Lett.*, **77**, 5–12.

Coppola, S., Parente, E., Dumontet, E. S., and La Peccerella, A. (1988) *Lait*, **68**, 295–310.

Coppola, S., Villani, F., Coppola, R., and Parente, E. (1990) *Lait*, **70**, 411–423.

Corre, C., Madec, M.-N., and Boyaval, P. (1992) *J. Chem. Tech. Biotechnol.*, **53**, 189–194.

Coutour, R., Gangée, D., and Champagne, C. P. (1991) *Can. Inst. Food Sci. Technol. J.*, **24**, 224–227.

Cox, W. A., and Lewis, J. E. (1972) Methods of handling and testing starter cultures. In *Safety in Microbiology*, Technical Series No. 6, D. A. Shapton and R. G. Board, eds., Academic Press, London, pp. 133–150.

Cox, W. A, Stanley, G., and Lewis, J. E. (1978) Starters: purpose, production and problems. In *Streptococci*, Technical Series No. 7, F. A. Skinner and L. B. Quensel, eds., Academic Press, London, pp. 279–296.

Crawford, R. J. M. (1972) *Dairy Ind.*, **37**, 648–654.

Davidson, B. E., Kordias, N., Dobos, M., and Hillier, A. J. (1996) *Antonie van Leeuwenhoek*, **70**, 161–183.

Deane, D. D., and Jenkins, R. A. (1971) *J. Dairy Sci.*, **54**, 749.

Deane, D. D., and van Patten, M. M. (1971) *J. Milk Food Technol.*, **34**, 16–22.

de Antoni, G. L., Pérez, P. F., Abraham, A. G., and Añón, M. C. (1989) *Cryobiology*, **26**, 149–153.

Delcour, J., Ferain, T., Deghorain, M., Palumbo, E., and Hols, P. (1999) *Antonie van Leeuwenhoek*, **76**, 159–184.

Dennien, G. J. (1980) *Dairy Products*, **8**(2), 10–14.

Desfarges, C., Larroche, C., and Gros, J. B. (1987) *Biotechnol. Bioeng.*, **29**, 1050–1058.

de Silva, M. C. T., Tessi, M. A., and Moguilevsky, M. A. (1983) *J. Food Prot.*, **46**, 699–701.

Desmons, S., Zgoulli, S., Destain, J., and Thonart, P. (1997) *MFLRBER Universiteit Gent*, **62**, 1713–1716.

Desmons, S., Zgoulli, S., Evrard, P., Roblain, D., Destain, J., and Thonart, P. (1998) *MFLRBER Universiteit Gent*, **63**, 1253–1261.

de Vos, W. M. (1996) *Antonie van Leeuwenhoek*, **70**, 223–242.

de Vos, W. M., Hols, P., van Kanenburg, R., Luesink, E., Kuipers, O. P., van der Oost, J., Kleerebezam, M., and Hugenholtz, J. (1998) *Int. Dairy J.*, **8**, 227–233.

Devriese, L. A., and Pot, B. (1995) The genus *Enterococcus*. In *Lactic Acid Bacteria*, Vol. 2, B. J. B. Wood and W. H. Holzapfel, eds., Blackie Academic & Professional, London, pp. 327–367.

de Vuyst, E., and Vandamme, E. J. (eds.) (1994) *Bacteriocins of Lactic Acid Bacteria*, Blackie Academic & Professional, London.

Dhanalakshmi, B., Khan, M. Md. H., Narasimhan, R., Vijayalkshmi, R., and Ramasamy, D. (1998) *Cheiron*, **27**, 44–49.

Dunsmore, D. G. (1984) Effects of sanitizers on starter performance. In *Proceedings of Dairy Culture Review Conference*, No. 27, Australian Society of Dairy Technology, Glenelg North, pp. 40–49.

Dunsmore, D. G., Makin, D., and Arkins, R. (1985) *J. Dairy Res.*, **52**, 287–297.

Efiuvwevwere, B. J. O., Gorris, L. G. M., Smid, E. J., and Kets, E. P. W. (1999) *Appl. Microbiol. Biotechnol.*, **51**, 100–104.

El-Zayat, A. I. (1987) *Egypt. J. Dairy Sci.*, **15**, 79–86.

Espina, F., and Packard, V. S. (1979) *J. Food Prot.*, **42**, 149–152.

Fang, W., Shi, M., Huang, L., Shao, Q., and Chen, J. (1993) *Vet. Microbiol.*, **37**, 115–125.

FAO (1972) *Production Yearbook*, Vol. 26, Food and Agriculture Organization of the United Nations, Rome, pp. 213–214.

FAO (1981) *Production Yearbook*, Vol. 34, Food and Agriculture Organization of the United Nations, Rome, pp. 229–230.

FAO (1996) *Production Yearbook*, Vol. 49, Food and Agriculture Organization of the United Nations, Rome, pp. 220–221.

FAO (1999) *Production Yearbook*, Vol. 52, Food and Agriculture Organization of the United Nations, Rome, pp. 217–218.

Farrow, J. A. E., and Collins, M. D. (1984) *J. Gen. Microbiol.*, **130**, 357–362.

Ferras, E., Minier, M., and Goura, G. (1986) *Biotechnol. Bioeng.*, **28**, 523–533.

Fonseca, F., Béal, C., and Corrieu, G. (2000) *J. Dairy Res.*, **67**, 83–90.

Font de Valdez, G., and de Giori, G. S. (1993) *Cryobiology*, **30**, 329–334.

Font de Valdez, G., Savoy de Giori, G., Pesce de Ruiz Holgado, A., and Oliver, G. (1983a) *Cryobiology*, **20**, 560–566.

Font de Valdez, G., Savoy de Giori, G., Pesce de Ruiz Holgado, A., and Oliver, G. (1983b) *Appl. Environ. Microbiol.*, **45**, 302–304.

Font de Valdez, G., Savoy de Giori, G., Pesce de Ruiz Holgado, A., and Oliver, G. (1985a) *Appl. Environ. Microbiol.*, **49**, 413–415.

Font de Valdez, G., Savoy de Giori, G., Pesce de Ruiz Holgado, A., and Oliver, G. (1985b) *Milchwissenschaft*, **40**, 147–148.

Font de Valdez, G., Savoy de Giori, G., Pesce de Ruiz Holgado, A., and Oliver, G. (1985c) *Milchwissenschaft*, **40**, 518–520.

Font de Valdez, G., Savoy de Giori, G., Pesce de Ruiz Holgado, A., and Oliver, G. (1985d) *Appl. Environ. Microbiol.*, **50**, 1339–1341.

Font de Valdez, G., Savoy de Giori, G., Pesce de Ruiz Holgado, A., and Oliver, G. (1986) *Milchwissenschaft*, **44**, 286–288.

Font de Valdez, G., Bibi, W., and Bachman, M. R. (1988) *Milchwissenschaft*, **43**, 350–352.

Font de Valdez, G., Mantos, G., Taranto, M. P., Lorca, G. L., and Pesce de Ruiz Holgado, A. (1997) *J. Dairy Sci.*, **80**, 1955–1958.

Foschino, R., Beretta, C., and Ottogali, G. (1992) *Ind. Latte*, **28**(2), 49–67.

Foster, E. M. (1962) *J. Dairy Sci.*, **45**, 1290–1294.

Fox, P. F., Lucey, J. A., and Cogon, T. M. (1990) *Crit. Rev. Food Sci. Nutr.*, **29**, 237–253.

Fu, W. Y., and Etzel, M. R. (1995) *J. Food Sci.*, **60**, 195–200.

Fu, W. Y., Suen, S. Y., and Etzel, M. R. (1994) Injury to *Lactococcus lactis* var. *lactis* C_2 during spray drying. In *Drying 94*, Vol. B, V. Rudolph and R. B.

Kerry, eds., Proceedings 9th International Drying Symposium, Gold Coast, pp. 785–792.

Fu, W. Y., Suen, S. Y., and Etzel, M. R. (1995) *Drying Technol.*, **13**, 1463–1476.

Fuchs, K. (1995) *Dairy Sci. Abstr.*, **57**, 142.

Gagne, J., Roy, D., and Gauthier, S. F. (1993) *Milchwissenschaft*, **48**, 501–505.

Gajdusek, S., and Sebela, F. (1973) *Dairy Sci. Abstr.*, **35**, 364.

Gajduskova, V., and Lat, J. (1972) *Acta Veterinaria-Bruno*, **41**, 447–452.

Galesloot, T. E., Hassing, F., and Stadhouders, J. (1966) *XVll Int. Dairy Congr.*, **D**, 491–498.

Gandhi, D. N., and Shahani, K. M. (1994) *Indian J. Microbiol.*, **34**, 45–47.

Garrote, G. L., Abraham, A. G., and de Antoni, G. L. (1997) *Lebensm. Wiss. Technol.*, **30**, 77–84.

Geisen, R., and Holzapfel, W. H. (1996) *Int. J. Food Microbiol.*, **30**, 315–324.

Gilliland, S. E. (ed.) (1985) *Bacterial Starter Cultures for Foods*, CRC Press, Boca Raton, FL.

Gilliland, S. E., Peitersen, N., and Hunger, W. (1991) *XXIII Int. Dairy Congr.*, **3**, 1925–1936.

Gobbetti, M., and Rossi, J. (1993) *Microbiol., Aliments, Nutr.*, **11**, 119–127.

Gobbetti, M., and Rossi, J. (1994) *Int. Dairy J.*, **4**, 237–249.

Godíndez, S., and Calderón, M. (2000) *Alimentaria*, **37**(313), 107–109.

Gomes, A. M. P., Malcata, F. X., Klaver, F. A. M., and Grande, H. J. (1995) *Netherlands Milk Dairy J.*, **49**, 71–95.

Gómez-Zavaglia, A., Disalvo, E. A., and de Antoni, G. L. (2000) *J. Dairy Res.*, **67**, 241–247.

Gripon, J. C. (1993) Mould-ripened cheeses. In *Cheese: Chemistry, Physics and Microbiology*, Vol. 2, 2nd ed., P. F. Fox, ed., Chapman & Hall, London, pp. 111–136.

Grossoird, B., Vaughan, E. E., Luesink, E., and de VOS, W. M. (1998) *Lait*, **78**, 77–84.

Guirguis, N., and Hickey, M. W. (1987a) *Aust. J. Dairy Technol.*, **42**, 14–16 and 26.

Guirguis, N., and Hickey, M. W. (1987b) *Aust. J. Dairy Technol.*, **42**, 11–13.

Gulstrom, T. J., Pearce, L. E., Sandine, W. E., and Elliker, P. R. (1979) *J. Dairy Sci.*, **62**, 208–221.

Hall, R. J., Linklater, P. M., and Antonucci, W. T. (1985) *N. Z. J. Dairy Sci. Technol.*, **20**, 35–42.

Hammes, W. P. (1990) *Food Biotechnol.*, **4**, 383–397.

Hammes, W. P., and Vogel, R. F. (1995) The genus *Lactobacillus*. In *Lactic Acid Bacteria*, Vol. 2, B. J. B. Wood and W. H. Holzapfel, eds., Blackie Academic & Professional, London, pp. 19–54.

Hansen, R. (1980) *North Eur. Dairy J.*, **46**, 62–69.

Hargrove, R. F. (1959) *J. Dairy Sci.*, **42**, 906.

Hargrove, R. F., Mcdonough, F. E., and Titsler, R. P. (1961) *J. Dairy Sci.*, **67**, 1674–1679.

Harju, M., Mattila, L., Heikonen, M., and Linko, P. (1984) *Dairy Sci. Abstr.*, **46**, 347.

Hayakawa, K., Sansawa, H., Nagumene, T., and Endo, I. (1990) *J. Ferment. Bioengi.*, **70**, 404–408.

Hayashi, K., Cliffe, A. J., and Law, B. A. (1990) *Int. J. Food Sci. Technol.*, **25**, 180–187.

Heap, H. A., and Lawrence, R. C. (1981) *N. Z. J. Dairy Sci. Technol.*, **16**, 91–94.

Heap, H. A., and Lawrence, R. C. (1988) Culture systems for the dairy industry. In *Developments in Food Microbiology*, Vol. 4, R. K. Robinson, ed., Elsevier Applied Science, London, pp. 149–185.

Heap, H. A., and Richardson, G. H. (1984) *N. Z. J. Dairy Sci. Technol.*, **19**, 249–254.

Heiner, H. (1990) *DMZ Lebensm. Milchwirtsch.*, **11**, 1374–1375,

Henning, D. R., Sandine, W. E., Elliker, P. R., and Hays, H. A. (1965) *J. Milk Food Technol.*, **28**, 273–277.

Hicks, C. L., and Ibrahim, S. A. (1992) *J. Food Sci.*, **57**, 1086–1092.

Hill, F. F. (1987) *Alimenta*, **3**, 73–80.

Høier, E., Janzen, T., Henriksen, C. M., Rattray, F., Brockmann, E., and Johansen, E. (1999) The production, application and action of lactic cheese starter cultures. In *Technology of Cheesemaking*, B. A. Law, ed., Sheffield Academic, Sheffield, pp. 99–131.

Holt, J. G., Krieg, N. R., Sneath, P. H. A., Staley, J. T., and Williams, S. T. (eds.) (1994) *Bergey's Manual of Determinative Bacteriology*, 9th ed., Williams & Wilkins, Baltimore.

Hsu, H.-Y., Jewett, Jr., F. F., and Charm, S. E. (1987) *Cult. Dairy Prod. J.*, **22**(4), 18–24.

Hugenholtz, J. (1986) *Netherlands Milk Dairy J.*, **40**, 129–140.

Hugenholtz, J., Splint, R., Konings, W. E., and Weldkamp, H. (1987) *Appl. Environ. Microbiol.*, **53**, 309–314.

Hui, Y. H., and Khachatourians, G. G. (eds.) (1995) *Food Biotechnology Microorganisms*, VCH Publishers, New York.

Hull, R. R. (1977) *Aust. J. Dairy Technol.*, **32**, 65–70.

Hull, R. R. (1983) *Aust. J. Dairy Technol.*, **38**, 149–154.

Hull, R. R., and Juffs, H. (1980) *Dairy Prod.*, **8**(1), 15–19.

Hunger, W., and Peitersen, N. (1992) New technical aspects of the preparation of starter cultures. In *New Technologies for Fermented Milks*, Bulletin No. 277, International Dairy Federation, Brussels, pp. 17–21.

Hup, G., and Stadhouders, J. (1977) United States Patent Application, **US** 4 053 642.

Hylmar, B., and Teply, M. (1970) *XVIII International Dairy Congress*, **IE**, 127.

IDF (1988) *Code of Practice for the Preservation of Raw Milk by the Lactoperoxidase System*, Bulletin No. 234, International Dairy Federation, Brussels.

IDF (1991) *Practical Phage Control*, Bulletin No. 263, International Dairy Federation, Brussels.

IDF (1993) *Consumption Statistics for Milk and Milk Products—1991,* Bulletin No. 282, International Dairy Federation, Brussels, pp. 1–3.

IDF (1994) *Consumption Statistics for Milk and Milk Products—1992,* Bulletin No. 295, International Dairy Federation, Brussels, pp. 1–3.

IDF (1995) *Residues of Antimicrobial Drugs and Other Inhibitors*, Special Issue 9505, International Dairy Federation, Brussels.

IDF (1998) *Yeasts in the Dairy Industry: Positive and Negative Aspects*, Special Issue No. 9801, International Dairy Federation, Brussels.

Imai, M., and Kato, M. (1975) *J. Agric. Chem. Soc. Japan*, **49**, 93–98.

Ishibashi, N., Tatematsu, T., Shimamura, S., Tomita, M., and Okonogi, S. (1985) *Refrig. Sci. Technol.*, **1**, 227–232.

Johnson, J. A. C., and Etzel, M. R. (1993) *AICHe Symp. Ser.*, **89**, 98–107.

Johnson, J. A. C., and Etzel, M. R. (1995) *J. Dairy Sci.*, **78**, 761–768.

Johnston, M., Ray, B., and Speck, M. L. (1984) *Cryo-Lett.*, **5**, 171–176.

Johnsson, T., Nikkilä, P., Toivonen, L., Rosenqvist, H., and Laakso, S. (1995) *Appl. Environ. Microbiol.*, **61**, 4497–4499.

Jones, T. H., Ozimek, L., and Stiles, M. E. (1990) *J. Dairy Sci.*, **73**, 1166–1172.

Kang, K. H., Ahn, S. N., and Lee, J. Y. (1985) *Dairy Sci. Abstr.*, **47**, 555.

Kets, E. P. W., and de Bont, J. A. M. (1994) *FEMS Microbiol. Lett.*, **116**, 251–256.

Kets, E. P. W., Teunissen, P. J. M., and de Bont, J. A. M. (1996) *Appl. Environ. Microbiol.*, **62**, 259–261.

Kets, E. P. W., de Bont, J. A. M., and Gorris, L. G. M. (1997) *Recent Res. Dev. Microbiol.*, **1**, 189–199.

Khasravi, L., Sandine, W. E., and Ayres, J. W. (1991) *Cult. Dairy Prod. J.*, **26**(2), 4–6 and 8–9.

Kilara, A., Shahani, K. M., and Das, N. K. (1976) *Cult. Dairy Prod. J.*, **11**(2), 8–11.

Kim, K. I., and Yoon, Y. H. (1995) *Korean J. Dairy Sci.*, **17**, 129–135.

Kim, K. S., and Yu, J. H. (1990) *Dairy Sci. Abstr.*, **52**, 202.

Kim, N., NaveH, D., and Olsen, N. (1985a) *Milchwissenschaft*, **40**, 605–607.

Kim, N., NaveH, D., and Olsen, N. (1985b) *Milchwissenschaft*, **40**, 645–649.

Kim, J. M., Baek, Y. J., and Kim, H. U. (1988) *Dairy Sci. Abstr.*, **50**, 144.

King, V. A.-E., and Su, J. T. (1993) *Process Biotechnol.*, **28**, 47–52.

Kitazawa, H., Ishii, Y., Uemura, J., Kawai, Y., Saito, T., Kaneko, T., Noda, K., and Itoh, T. (2000) *Food Microbiol.*, **17**, 109–118.

Klein, J., Stock, J., and Vorlop, K. D. (1983) *Eur. J. Appl. Microbiol. Biotechnol.*, **18**, 86–91.

Klein, G., Pack, A., Bonaparte, C., and Reuter, G. (1998) *Int. J. of Food Microbiol.*, **41**, 103–125.

Knuston, K. M., and Zottola, E. A. (1989) *J. Food Prot.*, **52**, 715–718.

Kok, R. G., de Waal, A., Schut, F., Welling, G. W., Weenk, G., and Hollingwerf, K. J. (1996) *Appl. Environ. Microbiol.*, **62**, 3668–3672.

Koroleva, N. S. (1991) Products prepared with lactic acid bacteria and yeasts. In *Therapeutic Properties of Fermented Milks*, R. K. Robinson, ed., Elsevier Applied Science, London, pp. 159–179.

Kosikowski, F. V., and Mistry, V. V. (1988) *Milchwissenschaft*, **43**, 27–30.

Kosikowski, F. V., and Mistry, V. V. (1992) United States Patent Application, **US** 5 098 721.

Kosikowski, F. V., and Mistry, V. V. (1997) *Cheese and Fermented Milk Foods*, Vol. I, F. V. Kosikowski L. L. C., Westport.

Kot, E., Furmanov, S., and Bezkorovainy, A. (1996) *J. Dairy Sci.*, **79**, 758–766.

Kramkowska, A., Fesnak, D., Kornacki, K., and Bauman, B. (1986) *Acta Biotechnol.*, **6**, 167–174.

Kreger-van Rij, N. J. W. (ed.) (1984) *The Yeast—A Taxomonic Study*, 3rd ed., Elsevier Science Publishers BV, Amsterdam.

Kulshrestha, D. C., and Marth, E. H. (1974a) *J. Milk Food Technol.*, **37**, 593–599.

Kulshrestha, D. C., and Marth, E. H. (1974b) *J. Milk Food Technol.*, **37**, 600–605.

Kulshrestha, D. C., and Marth, E. H. (1974c) *J. Milk Food Technol.*, **37**, 606–611.

Kulshrestha, D. C., and Marth, E. H. (1975) *J. Milk Food Technol.*, **38**, 138–141.

Kunji, E. R. S., Mierau, I., Hagting, A., Poolman, B., and Konings, W. N. (1996) *Antonie van Leeuwenhoek*, **70**, 187–221.

Kurmann, J. A., Rasic, J. L. J., and Kroger, M. (1992) *Encyclopedia of Fermented Fresh Milk Products*, Van Nostrand Reinhold, New York.

Kurtzman, C. P., and Fell, J. E. (eds.) (1998) *The Yeast—A Taxonomic Study*, 4th ed., Elsevier Science Publishers BV, Amsterdam.

Lagoda, I. V., Bannikova, L. A., Babina, N. A., Rozhkova, I. V., and Prusakova, N. V. (1979) *Dairy Sci. Abstr.*, **41** 109.

LaGrange, W. S. (1987) *J. Dairy Sci.*, **70**, 367–372.

LaGrange, W. S., and Reinbold, G. W. (1968) *J. Dairy Sci.*, **51**, 1985–1990.

Lancefield, R. C. (1933) *J. Exp. Med.*, **57**, 571–595.

Lamboley, L., Lacroix, C., Champagne, C. P., and Vuillemard, J. C. (1997) *Biotechnol. Bioeng.*, **56**, 502–516.

Lapointe, M., Champagne, C. P., Vuillemard, J. C., and Lacroix, C. (1996) *J. Dairy Sci.*, **79**, 767–774.

Larroche, C., and Gros, J.-B. (1986) *Appl. Microbiol. Biotechnol.*, **24**, 134–139.

Law, J., and Haandrikman, A. (1997) *Int. Dairy J.*, **7**, 1–11.

Lawrence, R. C., and Heap, H. A. (1986) *Special Addresses Given at IDF Annual Sessions*, Bulletin No. 199, International Dairy Federation, Brussels, pp. 14–20.

Lawrence, R. C., Heap, H. A., and Gilles, J. (1984) *J. Dairy Sci.*, **67**, 1632–1645.

Ledford, R. A., and Speck, M. L. (1979) *J. Dairy Sci.*, **62**, 781–784.

Leenders, G. J. M., Bolle, A. C., and Stadhouders, J. (1984) *Dairy Sci. Abstr.*, **46**, 233.

Leistner, L. (1990) *Food Biotechnol.*, **4**, 433–441.

Lewis, J. E. (1956) *J. Soc. Dairy Technol.*, **9**, 123–130.

Lewis, J. E. (1987) *Cheese Starters—Development and Application of the Lewis System*, Elsevier Applied Science, London.

Libudzisz, Z., and Mokrosinska, K. (1995) *Acta Microbiol. Polon.*, **44**, 305–313.

Libudzisz, S., Mokrosinska, K., Piatkiewicz, A., Moneta, J., and Oltuszak, E. (1991) *Pr. Inst. Lab. Badaw. Przem. Spozyw.*, **45**, 144–156.

Lievense, L. C., Verbeek, M. A. M., Noomen, A., and van't Riet, K. (1994) *Appl. Microbiol. Biotechnol.*, **41**, 90–94.

Liewen, M. B., and Marth, E. H. (1984) *J. Food Prot.*, **47**, 197–199.

Limsowtin, G. K. Y., Heap, H. A., and Lawrence, R. C. (1980) *N. Z. J. Dairy Sci. Technol.*, **15**, 219–224.

Limsowtin, G. K. Y., Powell, I. B., and Parente, E. (1996) Types of starters. In *Dairy Starter Cultures*, T. M. Cogan and J.-P. Accolas, eds., VCH Publishers, New York, pp. 101–129.

Linders, L. J. M., de Jong, G. I. W., Meerdink, G., and van't Riet (1994) The effect of disaccharide addition on the dehydration of *Lactobacillus plantarum* during drying and the importance of water activity. In *Drying 94*, Vol. B, V. Rudolph and R. B. J. Keey, eds., Proceedings, 9th International Drying Symposium, Gold Coast, pp. 945–952.

Loureiro, V., and Querol, A. (1999) *Trends Food Sci. Technol.*, **10**, 356–365.

Lücke, F.-K., Brümmer, J.-M., Buckenhüskes, H., Fernandez, A. G., Rodrigo, M., and Smith, J. E. (1990) Starter cultures development. In *Processing and Quality of Foods*, Vol. 2, P. Zeuthen, J. C. Cheftel, C. Eriksson, T. R. Gormley, P. Linko, and K. Paulus, eds., Elsevier Applied Science, London, pp. 2.11–2.36.

Macedo, M. G., Champagne, C. P., Vuillemard, J. C., and Lacroix, C. (1999) *Int. Dairy J.*, **9**, 437–445.

Magdoub, M. N. I., Nawawy, M. A., Sultan, N. I., and Kanany, Y. M. (1987) *Egypt. J. Dairy Sci.*, **15**, 57–63.

Maitrot, H., Paquin, C., Lacroix, C., and Champagne, C. P. (1997) *Biotechnol. Tech.*, **11**, 527–531.

Mäkelä, P. M., Korkeala, H. J., and Sand, E. K. (1991) *J. Food Prot.*, **54**, 632–636.

Manachini, P. L., and Parini, C. (1983) *Antonie van Leeuwenhoek*, **49**, 143–152.

Mantere-Alhonen, S. (1995) *Lait*, **75**, 447–452.

Marshall, V. M. E., and Tamime, A. Y. (1997) Physiology and biochemistry of fermented milk. In *Microbiology and Biochemistry of Cheese and Fermented Milk*, 2nd ed., B. A. Law, ed., Blackie Academic & Professional, London, pp. 153–192.

Marth, E. H. (1987) Dairy products. In *Food and Beverage Mycology*, 2nd ed., L. R. Beuchat, ed., Van Nostrand Reinhold, New York, pp. 175–209.

Martin, M. (1983) *Tech. Lait.*, **976**(5), 45–49.

Maruejouls, A., and Caigniet, A. (1983) *Tech. Lait.*, **976**(5), 37–43.

Masui, K., and Yamada, T. (1996) *French Cheeses*, DK Publishing, New York.

Mauriello, G., Aponte, M., Andolfi, R., Moschetti, G., and Vilani, F. (1991) *J. Food Prot.*, **62**, 773–777.

Mayo, B., Derzelle, S., Fernandez, M., Leonard, C., Ferain, T., Hols, P., Swarez, J. E., and Delcour, J. (1997) *J. Bacteriol.*, **179**, 3039–3042.

McIntyre, D. A., and Harlander, S. K. (1989) *Appl. Environ. Microbiol.*, **55**, 2621–2626.

Meijer, W. C., Tacken, M., Noomen, A., and Hugenholtz, J. (1995) *J. Dairy Sci.*, **78**, 17–23.

Meile, L., Ludwig, W., Reuger, U., Gut, C., Kaufmann, P., Dasen, G., Wenger, S., and Teuber, M. (1997) *Syst. Appl. Microbiol.*, **20**, 57–64.

Mermelstein, N. H. (1982) *J. Food Technol.*, **36**(8), 69–76.

Metwally, M. M., Abd El-Gawad, I. A., El-Nockrashy, S. A., and Ahmed, K. E. (1989) *Egypt. J. Dairy Sci.*, **17**, 35–43.

Milliere, J. B., Mathot, A.-G., Schmitt, P., and Divies, C. (1989) *J. Appl. Bacteriol.*, **67**, 529–542.

Milton, K., Hicks, C. L., O'Leary, J., and Langlois, B. E. (1990) *J. Dairy Sci.*, **73**, 2259–2268.

Mistry, V. V., and Kosikowski, F. V. (1985a) *J. Dairy Sci.*, **68**, 1613–1617.

Mistry, V. V., and Kosikowski, F. V. (1985b) *J. Dairy Sci.*, **68**, 2536–2543.

Mistry, V. V., and Kosikowski, F. V. (1986a) *J. Dairy Sci.*, **69**, 945–950.

Mistry, V. V., and Kosikowski, F. V. (1986b) *J. Dairy Sci.*, **69**, 1484–1490.

Mistry, V. V., and Kosikowski, F. V. (1986c) *J. Dairy Sci.*, **69**, 2577–2582.

Mistry, V. V., and Kosikowski, F. V. (1988) *J. Dairy Sci.*, **71**, 2333–2341.

Mistry, V. V., Kosikowski, F. V., and Bellamy, W. D. (1987) *J. Dairy Sci.*, **70**, 2220–2225.

Mitchell, S. L., and Gilliland, S. E. (1983) *J. Dairy Sci.*, **66**, 712–718.

Mitchell, G. E., Fedrick, I. A., and Rogers, S. A. (1986) *Aust. J. Dairy Technol.*, **41**, 12–14.

Mittal, G. S. (1992) *Food Biotechnology*, Technomic, Lancaster, pp. 295–365.

Moebus, O., and Teuber, M. (1986) *Kieler Milchwirtsch. Forschungsber.*, **38**, 255–264.

Moll, G. N., Konings, W. N., and Driessen, A. J. M. (1999) *Antonie van Leeuwenhoek*, **76**, 185–198.

Morice, M., Bracquart, P., and Linden, G. (1992) *J. Dairy Sci.*, **75**, 1197–1203.

Morichi, T. (1972) *Mechanism and Prevention of Cellular Injury of Lactic Acid Bacteria Subjected to Freezing and Drying*, 56th Annual Session of IDF, Japan.

Morichi, T. (1974) *Jpn. Agric. Res. Q.*, **8**, 171–176.

Morichi, T., Irie, R., Yano, N., and Kembo, H. (1967) *Agric. Biol. Chem.*, **31**, 137–141.

Mortimore, S. E., and Wallace, C. (1994) *HACCP—A Practical Approach*, Chapman & Hall, London.

Narayanan, R., and Dhanalakshmi, B. (1999) *Egypt. J. Dairy Sci.*, **27**, 263–267.

Nastaj, J. F. (1996) *Drying Technol.*, **14**, 1967–2002.

Nazzaro, F., Coppola, R., Marotta, M., Maurelli, L., de Rosa, M., and Addedo, F. (1999) *MFLRBER Univ. Gent*, **64**, 253–256.

Nes, I. F., Diep, D. B., Håvanrstein, L. S., Brurberg, M. B., Eijsink, V., and Holo, H. (1996) *Antonie van Leeuwenhoek*, **70**, 113–128.

Nichol, A. W., Harden, T. J., Dass, C. R., Angel, L., and Louis, J. P. (1995) *Aust. J. Dairy Technol.*, **50**, 41–46.

Norton, S., Lacroix, C., and Vuillemard, J. C. (1994) *J. Dairy Sci.*, **77**, 2494–2508.

Oberg, C. J., and Richardson, G. H. (1984) *J. Dairy Sci.* (Suppl. 1), **67**, 82.

Oberg, C. J., Davis, L. H., Richardson, G. H., and Ernstrom, C. A. (1986) *J. Dairy Sci.*, **69**, 2975–2981.

Oberman, H., and Libudzisz, Z. (1998) Fermented milks. In *Microbiology of Fermented Foods*, Vol. 1, 2nd ed., B. J. B. Wood, ed., Blackie Academic & Professional, London, pp. 308–350.

Oberman, L., Libudzisz, Z., and Piatkiewicz, A. (1986) *Nahrung*, **30**, 147–154.

Oberman, H., Libudzisz, Z., Piatkiewicz, A., Mokrosinska, K., Oltuszak, E., and Moneta, J. (1995) *Pol. J. Food Nutr. Sci.*, **4**(2), 21–31.

Okigbo, L. M., Oberg, C. J., and Richardson, G. H. (1985) *J. Dairy Sci.*, **68**, 2521–2526.

Olson, N. F. (1969) *Ripened Semisoft Cheeses—Pfizer Cheese Monograph*, Vol. IV, Chas. Pfizer & Co., New York.

Olson, N. F. (1990) *FEMS Microbiol. Rev.*, **87**, 131–147.

Orberg, P. K., and Sandine, W. E. (1985) *Appl. Environ. Microbiol.*, **49**, 538–542.

Orla-Jensen, S. (1931) *Dairy Bacteriology*, 2nd ed., translated by P. S. Arup, J. & A. Churchill, London.

Ottogali, G., and Rondinini, G. (1976) *Dairy Sci. Abstr.*, **38**, 109.

Ozlap, E., and Ozlap, G. (1979) *Dairy Sci. Abstr.*, **41**, 871.

Panoff, J.-M., Legrand, S., Thammavongs, B., and Boutibonnes, P. (1994) *Curr. Microbiol.*, **29**, 213–216.

Panoff, J.-M., Thammavongs, B., Laplace, J.-M., Hartke, A., Boutibonnes, P., and Auffray, Y. (1995) *Cryobiology*, **32**, 516–520.

Panoff, J.-M., Thammavongs, B., Guéguen, M., and Boutibonnes, P. (1998) *Cryobiology*, **36**, 75–83.

Panoff, J.-M., Thammavongs, B., and Guéguen, M. (2000) *Sci. Aliments*, **20**, 105–110.

Parente, E., and Zottola, E. A. (1991) *J. Dairy Sci.*, **74**, 20–28.

Park, S. Y., Kim, J. H., Kwon, I. K., and Kim, H. U. (1984) *Korean J. Dairy Sci.*, **6**, 78–84.

Passos, F. M. L., and Swaisgood, H. E. (1993) *J. Dairy Sci.*, **76**, 2856–2867.

Passos, F. M. L., Klaenhammer, T. R., and Swaisgood, H. E. (1994) *J. Dairy Res.*, **61**, 537–544.

Pearce, L. E. (1978) *N. Z. J. Dairy Sci. Technol.*, **13**, 55–58.

Pearce, L. E., and Brice, S. A. (1973) *N. Z. J. Dairy Sci.*, **8**, 17–21.

Peberdy, J. F. (ed.) (1987) *Penicillium and Acremonium, Biotechnology Handbooks*—1, Plenum Press, New York.

Pereda Alardin, A. L., Salgado Vega, A. L., Mota de la Garza, L., Mungui á Perez, J. L., and Galinez, J. (1990) *Rev. Lactinoam. Microbiol.*, **32**, 30–35.

Petrova, N. (1990) *Dairy Sci. Abstr.*, **52**, 55.

Pettersson, H. E., Christiansson, A., and Ekelund, K. (1985) *Scand. J. Dairy Technol.*, **Annual No. 2**; *Nordisk-Mejerindustri*, **12**(8), 58–60.

Piard, J. C., and Desmazeaud, M. (1991) *Lait*, **71**, 525–541.

Piard, J. C., and Desmazeaud, M. (1992) *Lait*, **72**, 113–142.

Piatkiewicz, A., and Mokrosinska, K. (1995) *Pol. J. Food Sci.*, **4**(2), 33–45.

Pitt, J. I. (1979) *The Genus Penicillium and Its Teleomorphic States Eupenicillium and Talaromyces*, Academic Press, London, pp. 344, 358.

Poolman, B. (1993) *Lait*, **73**, 87–96.

Pörtner, R., and Märkl, H. (1998) *Appl. Microbiol. Biotechnol.*, **50**, 403–414.

Porubcan, R. S. (1991) Large-scale freezing drying of bacterial concentrates. In *Granulation Technology for Bioproducts*, K. L. Kadam, ed., CRC Press, Boca Raton, FL, pp. 233–255.

Porubcan, R. S., and Sellars, R. L. (1975a) *J. Dairy Sci.*, **58**, 787.

Porubcan, R. S., and Sellars, R. L. (1975b) United States Patent Application, **US** 3 897 307.

Porubcan, R. S., and Sellars, R. L. (1979) Lactic starter culture concentrates. In *Microbial Technology*, Vol. 1, 2nd ed., H. J. Peppler and D. Perlman, eds., Academic Press, New York, pp. 59–92.

Pot, B., Ludwig, W., Kerster, K., and Schleifer, K. H. (1994) Taxonomy of lactic acid bacteria. In *Bacteriocins of Lactic Acid Backteria*, L. de Vuyst and E. J. Vandamme, eds., Blackie Academic & Professional, London, pp. 13–89.

Premaratne, R. J., and Cousins, M. A. (1991) *J. Dairy Sci.*, **74**, 3284–3292.

Prevost, H., and Divies, C. (1987) *Biotechnol. Lett.*, **9**, 789–794.

Prevost, H., and Divies, C. (1988a) *Milchwissenschaft*, **43**, 621–625.

Prevost, H., and Divies, C. (1988b) *Milchwissenschaft*, **43**, 716–719.

Pringent, C., Corre, C., and Boyaval, P. (1988) *J. Dairy Res.*, **55**, 569–577.

Pruitt, K. M., and Kamau, D. N. (1991) The lactoperoxidase systems of bovine and human milk. In *Oxidative Enzymes in Foods*, D. S. Robinson and N. A. M. Eskin, eds., Elsevier Applied Science, London, pp. 133–174.

Qiao, F. D., Nan, Q. X., and Lan, F. Y. (1999) *Dairy Sci. Abstr.*, **61**, 373.

Ragout, A., Siñeriz, F., Kaul, R., Guoqiang, D., and Mattiasson, B. (1996) *Appl. Microbiol. Biotechnol.*, **46**, 126–131.

Rajagopal, S. N., Sandine, W. E., and Ayres, J. W. (1990) *J. Dairy Sci.*, **73**, 881–886.

Ramakrishna, Y., Singh, R. S., and Anand, S. K. (1985) *Cult. Dairy Prod. J.*, **20**(3), 12–13.

Ramirez, C. (1982) *Manual and Atlas of the Pencillia,* Elsevier Biomedical Press, Amsterdam, pp. 381, 419.

Rao, K. H., Singh, S., and Kanawjia, S. K. (1989) *Indian Dairyman*, **41**, 97–101.

Raper, K. B., and Thom, C. (1968) *A Manual of the Penicillia,* Hafner Publishing, New York.

Ray, B. (1984) Reversible freeze-injury. In *Repairable Lessions in Bacteria*, A. Hurst and A. Nasim, eds., Academic Press, London, pp. 237–271.

Ray, B., and Johnson, M. C. (1986) *Cryo-Lett.*, **7**, 210–217.

Reddy, M. S. (1984a) PCT International Patent Application, **WO** 8 404 106A1.

Reddy, M. S. (1984b) PCT International Patent Application, **WO** 8 404 107A1.

Reddy, K. P., and Richardson, G. H., (1977) *J. Dairy Sci.*, **60**, 1527–1531.

Reinbold, G., and Reddy, M. S. (1974) *J. Milk Food Technol.*, **37**, 517–521.

Reiter, B. (1956) *Dairy Ind.*, **21**, 877–879.

Reiter, B. (1978) *J. Dairy Res.*, **45**, 131–147.

Reiter, B., and Harnulv, G. (1984) *J. Food Prot.*, **47**, 724–732.

Reps, A. (1993) Bacterial surface-ripened cheeses. In *Cheese: Chemistry, Physics and Microbiology*, Vol. 2, 2nd ed., P. F. Fox, ed., Chapman & Hall, London, pp. 137–172.

Richardson, G. H. (1978) *Dairy and Ice Cream Field*, **161**(9), 80A–80D.

Richardson, G. H. (1983a) *Utah Sci.*, **44**, 60–63.

Richardson, G. H. (1983b) *Cult. Dairy Prod. J.*, **18**(4), 27–32.

Richardson, G. H., and Ernstrom, C. A. (1984) The Utah State University lactic culture system. In *Proceedings of a Dairy Culture Review Conference*, Glenelg, pp. 12–18.

Richardson, G. H., Cheng, C. T., and Young, R. (1977) *J. Dairy Sci.*, **60**, 378–386.

Richardson, G. H., Hong, G. L., and Ernstrom, C. A. (1979) *Lactic Cultures*, Research Report No. 42, Utah State University.

Riis, S. B., Pedersen, H. M., Sørensen, N. K., and Jakobsen, M. (1995) *Food Microbiol.*, **12**, 245–250.

Robertson, P. S. (1966a) *Dairy Ind.*, **31**, 805–809.

Robertson, P. S. (1966b) *XVII Int. Dairy Congr.*, **D**, 439–446.

Robertson, P. S., Pearce, L. E., Heap, H. A., and Bysouth, R. (1974) Improved inoculation system for bulk starter. In *New Zealand Dairy Research Institute Annual Report*, No. 46, pp. 55–57.

Robinson, R. K. (ed.) (1995) *A Colour Guide to Cheese and Fermented Milks*, Chapman & Hall, London.

Robinson, R. K., and Wilbey, R. A. (1998) *Cheesemaking Practice—R. Scott*, Aspen Publishers, Gaithersburg.

Rogers, S. A., and Mitchell, G. E. (1994) *Aust. J. Dairy Technol.*, **49**, 70–74.

Roginski, H., Broome, M. C., and Hickey, M. W. (1984a) *Aust. J. Dairy Technol.*, **39**, 23–27.

Roginski, H., Broome, M. C., Hungerford, D., and Hickey, M. W. (1984b) *Aust. J. Dairy Technol.*, **39**, 28–32.

Roginski, H., Hickey, M. W., and Legg, G. A. (1991) *Aust. J. Dairy Technol.*, **46**, 31–35.

Ross, G. (1982) *OATS—Suppl. Aust. J. Dairy Technol.*, **37**, 1.

Roussis, I. G. (1994) *Chim. Chron.*, **23**, 137–153.

Roy, D., Goulet, J., and Provencher, P. (1987) *Lait*, **67**, 41–50.

Roy, D., Gagne, J., and Gauthier, S. F. (1992) *J. Dairy Sci. (*Suppl. 1), **75**, 131.

Rumain, N., Angelov, M., and Tsvetkov, T. S. (1993) *Cryobiology*, **30**, 438–442.

Salminen, S., and von Wright, A. (eds.) (1998) *Lactic Acid Bacteria*, 2nd ed., Marcel Dekker, New York.

Samson, R. A., and Pitt, J. E. (eds.) (1985) *Advances in Penicillium and Aspergillus Systematics*, NATO ASI Series A: Life Sciences, Vol. 102, Plenum Press, New York.

Sandine, W. E., and Ayres, J. W. (1981) United States Patent Application, **US** 4 282 255.

Sandine, W. E., and Ayres, J. W. (1983) United States Patent Application, **US** 4 382 965.

Saxelin, M., Grenov, B., Svensson, U., Fondén, R., Reniero, R., and Mattila-Sandholm, T. (1999) *Trends Food Sci. Technol.*, **10**, 387–392.

Schleifer, K. H., Ehrmann, M., Krusch, U., and Neve, H. (1991) *Syst. Appl. Microbiol.*, **14**, 386–388.

Schmidt, J. L., Diez, M., and Lenoir, J. (1991) *Sci. Aliments*, **11**, 653–672.

Selemer-Olsen, E., Sørhaug, T., Birkeland, S. E., and Pehrson, R. (1999) *J. Ind. Microbiol. Biotechnol.*, **23**(2), 79–85.

Sellars, R. L. (1967) Bacterial starter cultures. In *Microbial Technology*, H. J. Peppler, ed., Reinhold Publishing, New York, pp. 34–75.

Sellars, R. L., and Babel, F. J. (1985) *Cultures for the Manufacture of Dairy Products*, 2nd ed., Chr. Hansen Laboratory Inc., Wisconsin.

Shalaby, S. O., Nour, M. A., Abd-el-Tawab, G., and Mohamed, O. A. (1986) *Egypt. J. Dairy Sci.*, **14**, 143–154.

Shapton, N. (1989) *Dairy Ind. Int.*, **54**(6), 25–29.

Sharpe, M. E. (1979) *J. Soc. Dairy Technol.,* **32**, 9–18.

Siezen, R. J. (1999) *Antonie van Leeuwenhoek*, **76**, 139–155.

Sinkoff, B. A., and Bundus, R. H. (1983) United States Patent Application, **US** 4 404 986.

Sneath, P. H. A., Mair, N. S., Sharpe, M. E., and Holt, J. G. (eds.) (1986) *Bergey's Manual of Systematic Bacteriol.*, Vol. 2, Williams and Wilkins, Baltimore.

Sodini-Gallot, I., Corrieu, G., Boquien, C. Y., Latrille, E., and Lacroix, C. (1995) *J. Dairy Sci.*, **78**, 1407–1420.

Sodini-Gallot, I., Boquien, C. Y., Corrieu, G., and Lacroix, C. (1997a) *Enzyme Microb. Technol.*, **20**, 381–388.

Sodini-Gallot, I., Boquien, C. Y., Corrieu, G., and Lacroix, C. (1997b) *J. Ind. Microbiol. Biotechnol.*, **18**, 56–61.

Sodini-Gallot, I., Corrieu, G., and Lacroix, C. (1998) Potential in using immobilized cell bioreactor. In *Proceedings of 25th International Dairy Congress*, Vol. II, Aarhus, pp. 61–73.

Sozzi, T. (1972) *Milchwissenschaft*, **27**, 503–507.

Sozzi, T., Poulin, J. M., and Maret, R. (1978) *J. Dairy Res.*, **45**, 259–265.

Spencer, J. F. T., and Spencer, D. M. (eds.) (1990) *Yeast Technology*, Springer-Verlag, Berlin.

Srilaorkul, S., Ozimek, L., and Stiles, M. E. (1989) *J. Dairy Sci.*, **72**, 2435–2443.

Staab, J. C., and Ely, J. K. (1987) *Cryobiology*, **24**, 174–178.

Stadhouders, J. (1986) *Netherlands Milk Dairy J.*, **40**, 155–173.

Stadhouders, J., and Leenders, G. J. M. (1984) *Netherlands Milk Dairy J.*, **38**, 157–181.

Stadhouders, J., Jansen, L. A., and Hup, G. (1969) *Netherlands Milk Dairy J.*, **23**, 182–199.

Stadhouders, J., Bangma, A., and Driessen, F. M. (1976) *North Eur. Dairy J.*, **42**, 191–208.

Stanley, G. (1998) Cheeses. In *Microbiology of Fermented Foods*, Vol. 1, 2nd ed., B. J. B. Wood, ed., Blackie Academic & Professional, London, pp. 263–307.

Steenson, L. R., Klaenhammer, T. R., and Swaisgood, H. E. (1987) *J. Dairy Sci.*, **70**, 1121–1127.

Stiles, M. E. (1996) *Antonie van Leeuwenhoek*, **70**, 331–345.

Stoyanova, L. G., and Pushkareva, E. I. (1986) *Dairy Sci. Abstr.*, **48**, 615.

Sultan, N. I., El-Nawawy, M. A., Khalafalla, S. M., and Kenany, M. (1987) *Egypt. J. Dairy Sci.*, **15**, 31–37.

Suzuki, T. (1996) *J. Ferment. Bioeng.*, **82**, 264–271.

Tamime, A. Y. (1990) Microbiology of starter cultures. In *Dairy Microbiology*, Vol. 2, 2nd ed., R. K. Robinson, ed., Elsevier Applied Science, London, pp. 131–201.

Tamime, A. Y. (1997) Bifidobacteria—an overview of physiological, biochemical and technological aspects. In *Non-Digestible Oligosaccharides: Healthy Food for the Colon*, H. Hartemink, ed., Vermeulen Publications, Heteren, pp. 9–24.

Tamime, A. Y. (1998) *Latte*, **23**(1), 52–59.

Tamime, A. Y. (2000) Cheese—in the market place. In *Encyclopedia of Food Microbiology*, Vol. 1, R. K. Robinson, C. A. Batt, and P. D. Patel, eds., Academic Press, London, pp. 372–381.

Tamime, A. Y., and Marshall, V. M. E. (1997) Microbiology and technology of fermented milks. In *Microbiology and Biochemistry of Cheese and Fermented Milk*, 2nd ed., B. A. Law, ed., Blackie Academic & Professional, London, pp. 57–152.

Tamime, A. Y., and Robinson, R. K. (1976) *Dairy Ind.*, **41**, 408–411.

Tamime, A. Y., and Robinson, R. K. (1988a) *J. Dairy Res.*, **55**, 281–307.

Tamime, A. Y., and Robinson, R. K. (1988b) Technology of manufacture of thermophilic fermented milks. In *Fermented Milks: Science and Technology*, Bulletin No. 227, International Dairy Federation, Brussels, pp. 82–95.

Tamime, A. Y., and Robinson, R. K. (1999) *Yoghurt Science and Technology*, 2nd ed., Woodhead Publishing, Cambridge.

Tamime, A. Y., and Marshall, V. M. E., and Robinson, R. K. (1995) *J. Dairy Res.*, **62**, 151–187.

Tan, P. S. T. (1994) *Snow Brand R&D Rep.*, **102**, 1–16.

Taniguchi, M., Kotani, N., and Kobayashi, T. (1987) *Appl. Microbiol. Biotechnol.*, **25**, 438–441.

Teixeira, P. C., Castro, M. H., and Kirkby, R. M. (1994) *Lett. Appl. Microbiol.*, **18**, 218–221.

Teixeira, P. C., Castro, M. H., and Kirkby, R. M. (1995a) *J. Food Prot.*, **57**, 934–936.

Teixeira, P. C., Castro, M. H., and Kirkby, R. M. (1995b) *J. Food Microbiol.*, **78**, 456–462.

Teixeira, P. C., Castro, M. H., Malcata, F. X., and Kirkby, R. M. (1995c) *J. Dairy Sci.*, **78**, 1025–1031.

Teixeira, P. C., Castro, M. H., and Kirkby, R. M. (1996) *Lett. Appl. Microbiol.*, **22**, 34–38.

Teply, M., and Cerminova, N. (1974) *XIX Int. Dairy Congr.*, **1E**, 428–429.

Thammavongs, B., Corroler, D., Panoff, M.-M., Auffray, Y., and Boutibonnes, P. (1996) *Lett. Appl. Microbiol.*, **23**, 398–402.

Thomas, E. L. (1985) Bacterial hydrogen peroxide production. In *The Lactoperoxidase System*, K. M. Pruitt and J. O. Tenovuo, eds., Marcel Dekker, New York, pp. 179–202.

Thomas, E. L., Hong, G. L., Ernstrom, C. A., and Richardson, G. H. (1981) *Dairy Food Sanit.*, **1**, 236–237.

Thunell, R. K. (1988) *Cult. Dairy Prod. J.* **23**(3), 10–14, 16.

Thunell, R. K. (1995) *J. Dairy Sci.*, **78**, 2514–2522.

Thunell, R. K., and Reinbold, R. S. (1993) *Mod. Dairy*, **72**(3), 17–19.

Thunell, R. K., Sandine, W. E., and Bodyfelt, F. W. (1984a) *J. Dairy Sci.*, **67**, 1175–1180.

Thunell, R. K., Sandine, W. E., and Bodyfelt, F. W. (1984b) *J. Dairy Sci.*, **67**, 24–36.

To, B. C. S., and Etzel, M. R. (1997) *J. Food Sci.*, **62**, 576–578.

Toba, K., Arihara, K., and Adachi, S. (1987) *Milchwissenschaft*, **42**, 268–565.

Toba, T., Arihara, K., and Adachi, S. (1990) *Int. J. Food Microbiol.*, **10**, 219–224.

Toba, T., Uemura, H., Mukai, T., Fuji, T., Itoh, T., and Adachi, S. (1991) *J. Dairy Res.*, **58**, 497–502.

Tofte-Jespersen, N. J. (1974a) *S. Afr. J. Dairy Technol.*, **6**, 63–68.

Tofte-Jespersen, N. J. (1974b) Concentration of cultures. In *A New View of Int. Cheese Production*, Chr. Hansen Lab. A/S, Copenhagen, pp. 77–86.

Tofte-Jespersen, N. J. (1976) *Dairy Ice Cream Field*, **159**(5), 58A–58G.

Tofte-Jespersen, N. J. (1979) *J. Soc. Dairy Technol.*, **32**, 190–194.

Toyoda, S., Oki, Y., and Yoshioka, Y. (1988) *Dairy Sci. Abstr.*, **50**, 284.

Turner, K. W., and Martley, F. G. (1983) *Appl. Environ. Microbiol.*, **45**, 1932–1934.

Turner, K. W., Davey, G. P., Richardson, G. H., and Pearce, L. E. (1979) *N. Z. J. Dairy Sci. Technol.*, **14**, 16–22.

Umanskii, M. S., and Borovkova, Yu, A. (1985) *Dairy Sci. Abstr.*, **47**, 638.

Ustunol, Z., and Hicks, C. L. (1994) *J. Dairy Sci.*, **77**, 1479–1486.

Ustunol, Z., Hicks, C. L., and O'Leary, J. (1986) *J. Dairy Sci.*, **69**, 15–21.

Valderrama, M. J., Siloniz, M. I., Gonzalo, P., and Peinado, J. M. (1999) *J. Food Prot.*, **62**, 189–193.

Vallado, M., and Sandine, W. E. (1994) *J. Dairy Sci.*, **77**, 1509–1514.

Vedamuthu, E. R. (1994) *J. Dairy Sci.*, **77**, 2725–2735.

Venema, G., Kok, J., and Van Sinderen, D. (1999) *Antonie van Leeuwenhoek*, **76**, 3–23.

Ventling, B. L., and Mistry, V. V. (1993) *J. Dairy Sci.*, **76**, 962–971.

Waes, G. (1988) *Dairy Sci. Abstr.*, **50**, 451.

Walczak, P., Mokrosinska, K., and Libudzisz, Z. (1995) *Pol. J. Food Nutr. Sci.*, **4**(2), 9–20.

Walker, A. L., Mullan, W. M. A., and Muir, M. E. (1981) *J. Soc. Dairy Technol.*, **34**, 78–84.

Weerkamp, A. H., Klijn, N., Neeter, R., and Smit, G. (1996) *Netherlands Milk Dairy J.*, **50**, 319–332.

Weinbrenner, D. R., Barefoot, S. F., and Grinstead, D. A. (1997) *J. Dairy Sci.*, **80**, 1246–1253.

Wenzel, J. M., and Marth, E. H. (1990a) *J. Dairy Sci.*, **73**, 3357–3365.

Wenzel, J. M., and Marth, E. H. (1990b) *J. Food Prot.*, **53**, 918–923.

Wenzel, J. M., and Marth, E. H. (1991) *J. Food Prot.*, **54**, 183–188.

Whitehead, H. R. (1956) *Dairy Eng.*, **73**, 159–163.

Whitehead, W. E., Ayres, J. W., and Sandine, W. E. (1993) *J. Dairy Sci.*, **76**, 2344–2353.

Wigley, R. C. (1977) *J. Soc. Dairy Technol.*, **30**, 45–47.

Wigley, R. C. (1980) *J. Soc. Dairy Technol.*, **33**, 24–30.

Willret, D. L., and Comotto, M. (1989) United States Patent Application, **US** 4 851 347.

Willret, D. L., and Comotto, M. (1990) United States Patent Application, **US** 4 919 942.

Willret, D. L., Sandine, W. E., and Ayres, J. W. (1979) *Cheese Rep.*, **103**(18), 8–9.

Willret, D. L., Sandine, W. E., and Ayres, J. W. (1982) *Cultured Dairy Prod. J.*, **17**(3), 5–9.

Wolff, E., Delisle, B., Corrieu, G., and Gibert, H. (1990) *Cryobiology*, **27**, 569–575.

Wood, N. J. (1981) *Dairy Ind. Int.*, **46**(12), 14–15, 18.

Wood, B. J. B., and Holzapfel, W. H. (eds.) (1995) *The Genera of Lactic Acid Bacteria*, Vol. 2., Blackie Academic & Professional, London.

Wouters, J. A., Jeynov, B., Rombouts, F. M., de Vos, W. M., Kuipers, O. P., and Abee, T. (1999) *Microbiology (Reading)*, **145**, 3185–3194.

Wright, C. T., and Klaenhammer, T. R. (1984) *J. Dairy Sci.*, **67**, 44–51.

Wright, S. L., and Richardson, G. H. (1982) *J. Dairy Sci.*, **65**, 1882–1889.

Yang, N. L., and Sandine, W. E. (1979) *J. Dairy Sci.*, **62**, 908–915.

Yezzi, T. L., Ajao, A. B., and Zottola, E. A. (1993) *J. Dairy Sci.*, **76**, 2827–2831.

Yildirim, Z., and Johnson, M. G. (1998) *J. Food Prot.*, **61**, 47–51.

Zaika, L. L., Kissinger, J. C., and Wasserman, A. E. (1983) *J. Food Sci.*, **48**, 1455–1459.

Ziolkowski, R., and Goonan, J. (1985) *Dairy Prod.*, **13**(2), 15–17.

Zlotowska, H., and Ilnicka-Olejniczak, O. (1993) *Dairy Sci. Abstr.*, **55**, 52.

Zourari, A., Accolas, J.-P., and Desmazeaud, M. J. (1992) *Lait*, **72**, 1–34.

CHAPTER 8

MICROBIOLOGY OF FERMENTED MILKS

RICHARD K. ROBINSON
School of Food Biosciences, The University of Reading, Reading, England

ADNAN Y. TAMIME
Scottish Agricultural College, Ayr, Scotland

MONIKA WSZOLEK
University of Agriculture, Animal Products Technology Department, Kraków, Poland

8.1 INTRODUCTION

The origins of fermentations involving the production of lactic acid are lost in antiquity, but it is not difficult to imagine how nomadic communities gradually acquired the art of preserving their meager supplies of milk by storing them in animal skins or crude earthenware pots. Initially, the intention could well have been simply to keep the milk cool through the evaporation of whey from the porous surface, but the chance transformation of the raw milk into a refreshing, slightly viscous foodstuff would soon have been recognized as a desirable innovation. The introduction of refinements, such as concentrating the raw milk over an open fire to give a thicker coagulum, must also have evolved over a considerable period of time, but the end result was that many communities in the Middle East and along the Eastern seaboard of the Mediterranean gradually acquired the skills of making yogurt and related products.

In some places, the basic character of the yogurt was changed to give a product with improved keeping qualities. The derivation of con-

Dairy Microbiology Handbook, Third Edition, Edited by Richard K. Robinson
ISBN 0-471-38596-4

densed yogurt (for example, in which some of the whey is expressed from the coagulum) or Kishk, a product made by sun-drying a mixture of yogurt and a cereal (for example, cracked wheat known as Burghol or flour), could well have been instigated by the desire for some method of storing yogurt in the absence of refrigeration, and numerous local variations on these themes can still be found (Tamime and O'Connor, 1995; Tamime and Robinson, 1999; Tamime et al., 1999a–c; Tamime and McNulty, 1999). Occasionally, yeasts would figure in the fermentation as well, and mildly alcoholic beverages like Kefir and Koumiss made their appearance among the multitude of fermented milks.

Nevertheless, despite the popularity of these products in certain societies, communities in northern Europe and elsewhere paid scant attention to their properties until the classic text by Metchnikoff (1910) stirred the imagination of those attracted by his hypothesis. Thus, his proposal that the apparent longevity of the hill tribesmen of Bulgaria was a direct result of their lifelong consumption of yogurt inspired an interest in the nutritional characteristics of the product that has never abated. Even today the controversy smoulders on; and although modern commercial yogurt bears little resemblance to its Balkan counterpart, there are still many consumers who believe that it is more than "just another foodstuff" (Robinson, 1989).

The validity, or otherwise, of Metchnikoff's views will be debated for many years to come, but one indisputable effect of his work was a marked increase in the popularity of yogurt throughout Europe. The almost "mystical" properties of natural yogurt were not, however, enough to sustain the interest of the market for very long, and it was not until the introduction of fruit/flavored varieties in the late 1950s that yogurt became a major dairy product. Since that time, its popularity as an inexpensive and convenient dessert has increased almost annually, so that, at the present time, fruit yogurts represent a major source of income to the dairy industry. This massive growth has, of course, been accompanied by considerable changes in the process plant employed for production, while, in a modern factory, control of the microbiological aspects of the fermentation has reduced the chances of unacceptable variation between batches. Nevertheless, the principles underlying the various aspects of manufacture have altered little with time, and an understanding of these basic tenets is essential for efficient control at plant level.

In addition to yogurt, of course, there are now numerous types of fermented milk manufactured in different parts of the world, and the International Dairy Federation (IDF, 1988) has recently published a monograph dealing with the major types (see also Tamime and

Robinson, 1999). In general, the different types tend to be classified on the method of fermentation and/or processing, and a simple scheme based on these criteria is shown in Figure 8.1.

Although many of these fermentations involve specific microfloras (Marshall, 1987; Tamime and Marshall, 1997; Marshall and Tamime, 1997), there is a considerable degree of similarity in respect of the technological aspects (Tamime and Robinson, 1988, 1999), especially with the lactic fermentations. Obviously, the fine details of manufacture differ from product to product, and to some extent from plant to plant even for the same product, but certainly many of the technical aspects have much in common. It is entirely appropriate, therefore, to consider these aspects under just one heading, and the universal popularity of yogurt makes it the obvious choice for detailed appraisal. The production of the remaining groups in Figure 8.2 will, as a consequence, be discussed in outline only.

8.2 LACTIC FERMENTATIONS

The majority of the products mentioned in Figure 8.2 are now manufactured with starter cultures composed solely of lactic acid bacteria, and careful selection of species is important in maintaining the integrity of each product. The manifestation of these desirable features depends, in turn, on the provision of optimum growth conditions for the organisms, and one of the essentials in this respect is temperature. It is convenient, therefore, to consider the various products in relation to the temperature employed for incubation, because, as can be concluded from Table 8.1, this factor will be a direct reflection of the selected microflora.

The basic requirements in terms of plant and equipment are similar to those described in relation to yogurt, and hence the important differences between products arise as a result of the following:

- The specific starter organism associated with the product.
- The precise conditions that are employed during the fermentation.
- The composition and treatment of the milk base.

In practice, however, the essential characteristics of any given milk are usually the result of microbial activity, and the basic identity of each product results from the process conditions providing the correct environment for the bacteria or molds or yeasts concerned. In this way, the

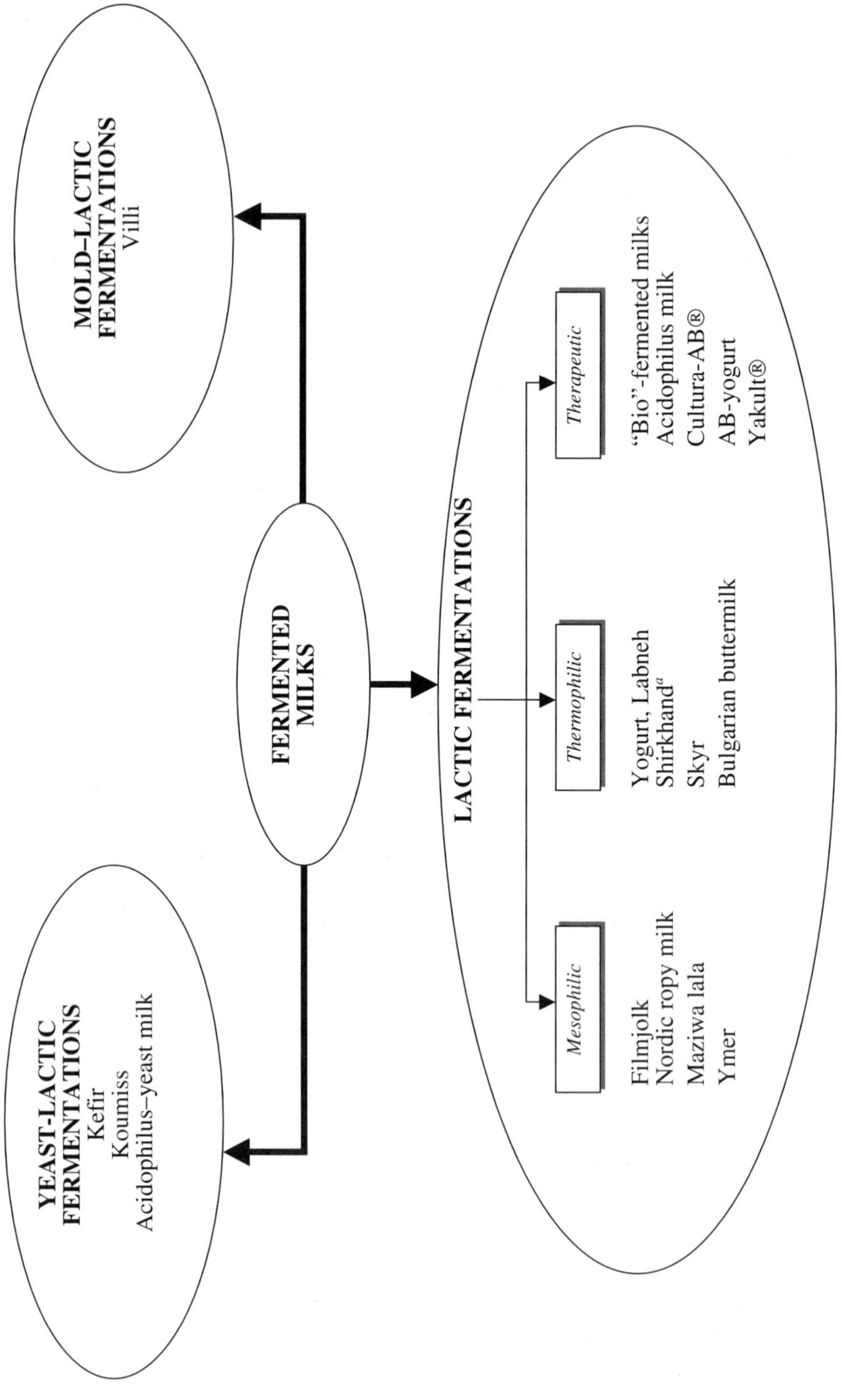

Figure 8.1. Scheme of classification for fermented milk products. [a]In some instances, mesophilic cultures are employed.

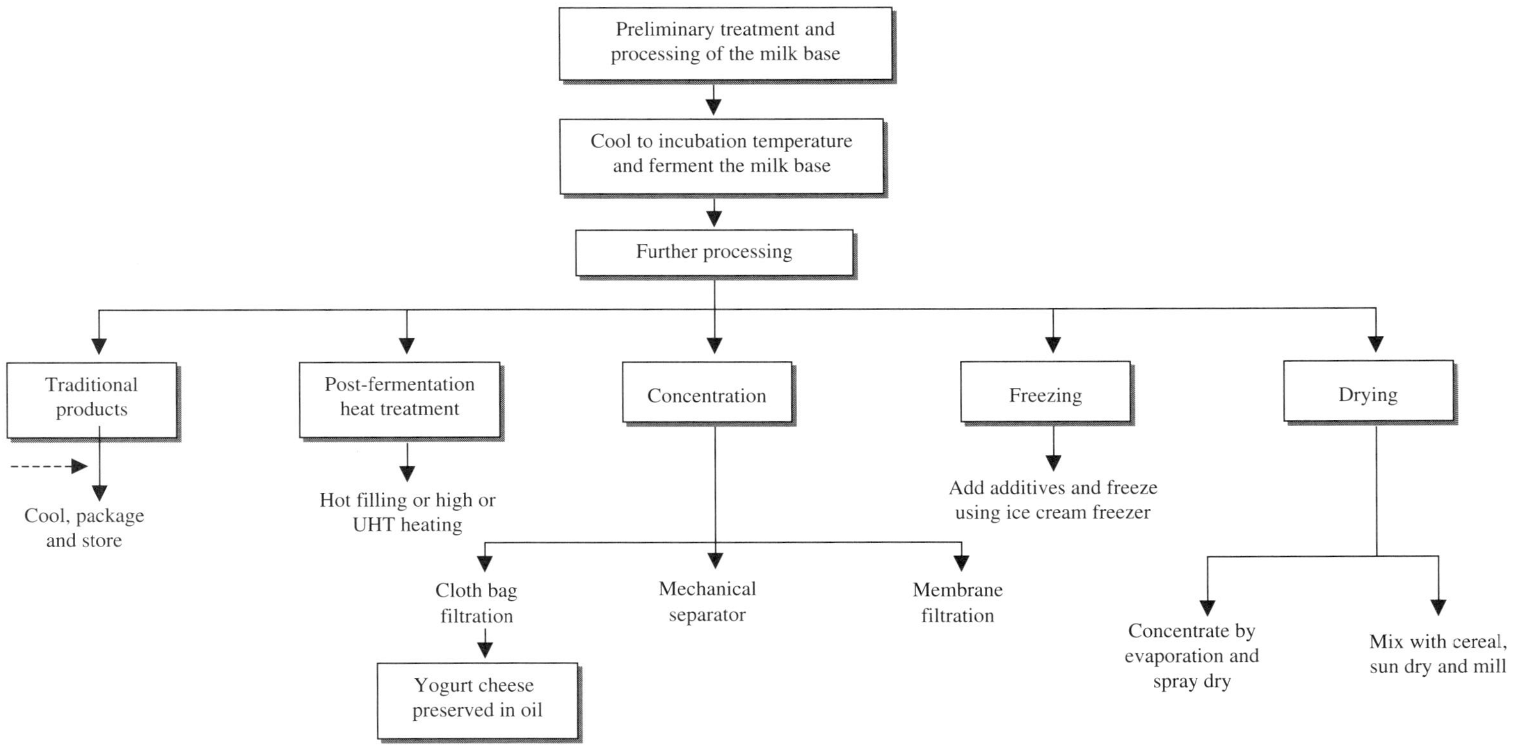

Figure 8.2. An outline indicating the manufacturing stages of fermented milks and related products. The dashed arrow (-----▶) denotes that addition of fruit flavors is optional.

TABLE 8.1. Some Established Categories of Fermented Milk Products and an Indication of the Microflora Responsible for Their Production[a]

Category	Genera of Microorganisms[b]	Fermentation Temperature (°C)
Lactic fermentations		
Mesophilic	*Lactococcus* *Leuconostoc* *Pediococcus*[c]	≤30
Thermophilic	*Streptococcus* *Lactobacillus*	42–45
Therapeutic[c]	*Bifidobacterium* *Streptococcus* *Lactobacillus* *Enterococcus*	~37
Yeast–lactic fermentations	Mesophilic LAB[d] Thermophilic LAB[d] Yeasts Contaminants	~22
Mould–lactic fermentations	Mesophilic LAB[d] Mold	18–20

[a]Refer to text.
[b]For detailed characterization of these microorganisms, refer to Chapter 7.
[c]Refer to Chapter 9 for further detail.
[d]LAB, lactic acid bacteria.

basic metabolism of the species or groups of species stamps its unique character on the milk; and although the final flavors will result from a complex range of chemical interactions, it is the microflora which paves the way. This dependence on microbial activity means that care over starter culture preparation is extremely important, and much of what has been said elsewhere in this volume about starter production and maintenance is equally applicable to the range of fermented milks that are available in various parts of the world.

It must be admitted, however, that with the exception of yogurt and acidophilus milk, the microflora of fermented milks has, until recently, been subject to little systematic investigation. Thus, the role of strain differences and/or incompatibility in relation to the organoleptic properties of the various products is still largely a matter for speculation, and this restraint applies even where interspecific relationships are of major importance. A prime example of this latter situation is the granular starter employed for Kefir, where the nature of the inoculum represents, perhaps, the ultimate complexity.

In some instances, this absence of information about a microflora and its activity has clearly inhibited the development of new or improved products; but in other cases, traditional expertise more than compensates for any deficiencies in fundamental knowledge. As a consequence, a number of fermented milks are manufactured on a considerable scale, and a brief outline of some of the more popular ones will serve to indicate how variation between the various products has been achieved.

In general, the stages of manufacture consist of a preliminary treatment of the milk base [fat standardization and fortification of the solids-not-fat (SNF)], followed by homogenization, heat treatment, cooling, fermentation, cooling, fruit addition, packaging, and cold storage. The only differences associated with the sequence of the manufacturing stages of fermented milks are (a) homogenization of the milk base before or after heat treatment and (b) homogenization of the product after the fermentation stage (Tamime et al., 2001).

8.2.1 Mesophilic Fermentations

8.2.1.1 Traditional or Natural Buttermilk. Traditional buttermilk is the by-product of buttermaking that arises after churning ripened cream, while cultured buttermilk is a fermented skimmed milk. The starter culture is composed of a mixture of *Lactococcus* spp. (*Lactococcus lactis* subsp. *lactis* and *Lactococcus lactis* subsp. *cremoris*, which are the main lactic acid producers, and *Lactococcus lactis* subsp. *lactis* biovar *diacetylactis*) and *Leuconostoc mesenteroides* subsp. *cremoris*. Incidentally, the latter two species are responsible for the aroma or flavor production (Tamime and Marshall, 1997).

During production of the butter, the cream is heated before ripening at 90–95°C for 15 s or at 105–110°C with no holding. Vacuum treatment of the cream during cooling is optional, but it is recommended to remove any cooked flavor due to heating. The perceived sensory characters of the resultant buttermilk are (a) acid or sour taste, (b) buttery flavor due to diacetyl, and (c) the presence of some residual butter granules.

8.2.1.2 Cultured Buttermilk. In contrast, cultured buttermilk is made from skimmed milk (9.5 g $100\,g^{-1}$ SNF; optional 7.5 to 11.4 g $100\,g^{-1}$ SNF and 0.3–2.0 g $100\,g^{-1}$ fat) which is heated to 95°C for 5 min, cooled to 22°C and inoculated with mesophilic mixed starter cultures (Tamime and Marshall, 1997). After fermentation, the product is warmed to 40°C, homogenized (7.5–10.0 MPa), cooled, and packaged;

this latter process improves the water holding capacity and stability of the gel. A similar approach is used in Ymer and Ylette making (e.g., Nordic products), but the fermented milk is homogenized before cooling and packaging stages (Tamime et al., 2001).

To mimic the characteristics of traditional buttermilk, incorporation of butter granules into cultured buttermilk can be achieved by (a) the addition of freeze-dried butter flakes or granules, (b) the use of a churning method, or (c) dripping melted cream or anhydrous milk fat (AMF) into the cold butter milk (Tamime and Marshall, 1997).

8.2.1.3 Nordic Sour Milks. The traditional Nordic or Scandinavian buttermilk products (Là6ngfil, Langmjölk, Taetmjölk, Filmjölk) are slimy or ropy due to (a) rubbing the interior of the milk pails with leaves of Butterwort (*Pinguicula vulgaris*) and (b) the presence of the slime-producing *Bacterium lacticus longi*, which has been identified as a variant of *Lactococcus* spp.; some researchers have proposed the name *Lactococcus lactis* biovar *longi* (Macura and Townsley, 1984).

In some countries, the fat content is standardized within a range of 0.5–3.0 g $100\,g^{-1}$, and the Scandinavian "sour milk" (Filmjolk) is a typical variant. However, whatever the national differences in respect of the end product, manufacturing procedures have much in common. In the most direct process, the milk is simply pre-heated to >75°C, deaerated in a vacuum chamber (the temperature drops to ~70°C; this step of the process is regarded as desirable to impart a smooth consistency to the product), and the fat is separated for continuous in-line standardisation of the fat level required in the end product. Then, the milk is homogenised (17.5–20.0 MPa), heated to 90–95°C for 3 min, cooled to 20°C and inoculated with a buttermilk starter culture (1–2 ml $100\,ml^{-1}$ bulk starter) including slime-forming strains. The inoculated milk is agitated for 10 min and fermented at 20°C for 20 h or until acidity reaches ~0.9% lactic acid. The coagulum is cooled, packaged and moved to the cold store (Tamime and Marshall, 1997). However, insufficient flavor in buttermilk and related products is often associated with deficiencies in the starter culture, but the most frequent complaint concerns physical separation. Some manufacturers avoid this latter problem by incorporating gelatin or additional fat into the process milk, but the danger of consumer complaints can usually be avoided by close attention to process details.

8.2.1.4 Cultured Cream. Sour cream is an extremely viscous product with the flavor and aroma of buttermilk, but with a fat content of 10–12 or 20–30 g 100 g^{-1} fat; its method of consumption is more akin to that of normal cream. A typical schedule for production involves fortifying whole milk with cream to give the desired fat content and, optionally, adding citric acid or Na-citrate to enhance the metabolic activity of the starter culture i.e., cit^{+} *Lactococcus* and *Leuconostoc* species). The cream (10–12 and 20–30 g 100 g^{-1} fat) is prewarmed to ~70°C and homogenized at 15–20 and 10–12 MPa for low- and high-fat creams, respectively; this process improves the consistency of the final product. The homogenized cream is heated to 90°C for 30 min or 90°C for 5 min, cooled to ~20°C, inoculated with 1–2 ml 100 ml^{-1} bulk starter culture (three *Lactococcus* species and *L. mesenteriodes* subsp. *cremoris*), and fermented for 18–20 h or until acidity reaches 0.8 g 100 g^{-1} of lactic acid.

On cooling, the sour cream is then ready for packaging in cartons prior to dispatch, but care must be taken at this stage to avoid any serious deterioration in viscosity. It is for this reason that some manufacturers incubate the cream in the retail cartons, and certainly this process can give rise to a markedly thicker material.

8.2.1.5 Miscellaneous Products. A wide range of indigenous fermented milk products are traditionally made in rural areas of many countries. Most of these products rely on spontaneous fermentation due to the indigenous lactic acid bacteria present in the milk. The following reviews are recommended for further reading in relation to the processing and microbiological aspects of such products (Kurmann et al., 1992; Dirar, 1993; Tamime and Marshall, 1997). In some countries, the spontaneous fermentation of the milk takes place in gourds made out of the fruits of plants (e.g., *Lagenaria peucantha*). Although the mechanism is not understood, a degree of control of the fermentation can be achieved by smoothing the inner side of the gourd with glowing splints; in Kenya, for example, wood from the tree *Olea africana* is used (Kimonye and Robinson, 1991). Some examples of other fermented products are as follows:

Maziwa Lal is a traditional fermented milk made in Kenya, and the commercial product is called Mala. A buttermilk starter culture is used to ferment the milk, and optionally the coagulum is flavored with fruit juices. In addition, the product is sweetened with the addition of sugar and may be stabilized (e.g., pectin, gelatin, or Na-caseinate).

Susa is made from camel's milk in Kenya, and the fermentation is carried out using a heterofermentative mesophilic starter culture [Farah et al. (1990); see also Tamime and Robinson (1999) for examples of thermophilic-type fermentations of camel's milk]. However, when using homofermentative type starter cultures were used, consumers did not rate the product highly.

Lben is a Moroccan fermented product which is equivalent to buttermilk. The milk is fermented spontaneously at 18–24°C for 24–48 h, and then it is churned to remove the butter granules. The microflora of the Lben is mainly composed of *L. lactis* subsp. *lactis* biovar *diacetylactis*, *Leuconostoc lactis*, and *L. mesenteroides* subsp. *cremoris* and subsp. *dextranicum*; lactobacilli, yeasts, molds and coliforms are also present (Tamime and Marshall, 1997).

8.2.2 Thermophilic Fermentations

The term "thermophilic" is reserved for starter cultures whose optimum growth temperature lies between 37°C and 45°C; and the genera *Streptococcus*, *Lactobacillus*, *Bifidobacterium*, and *Enterococcus* are often used for the production of fermented milk. Only the products made with *Streptococcus thermophilus*, *Lactobacillus delbrueckii* subsp. *bulgaricus* and subsp. *lactis*, and in some instances *Lactobacillus helveticus* will be reviewed, whilst the "bio"-fermented milk products are detailed in Chapter 9 of this volume.

8.2.2.1 Bulgarian Buttermilk. This fermented milk product is made in Bulgaria using *L. delbrueckii* subsp. *bulgaricus* alone, and other synonyms for the product are Bulgaricus acid or cultured buttermilk, Bulgarian milk, or Bulgaricus milk (Tamime and Marshall, 1997). The manufacturing stages are similar to yogurt making, and the processed milk base is fermented overnight at ~41°C to achieve the desired level of acidity (e.g., 1.4 g $100 g^{-1}$). On some occasions the starter culture may also contain *S. thermophilus* or a cream culture (Gyosheva, 1985; van den Berg, 1988; Kurmann et al., 1992; Tamime and Marshall, 1997).

It is reported by Marshall (1986) and Tamime and Marshall (1997) that the product has a "clean" flavor reminiscent of yogurt, and this reaction suggests that *L. delbrueckii* subsp. *bulgaricus* metabolizes some of the components in milk to acetaldehyde. Furthermore, this product is made into pharmaceutical tablets containing viable cells [2.5×10^9 colony forming unit (cfu) g^{-1}]; when fed to babies, their feces contain 1.0×10^9 cfu g^{-1} of viable lactobacilli (Kurmann et al., 1992).

8.2.2.2 Yogurt. The types of yogurt that are produced worldwide can be divided into various categories, and the subdivisions are usually created on the basis of the following:

TABLE 8.2. Scheme for the Classification of Yogurt

Group	Physical State	Product Type
I	Liquid/viscous	Yogurt including drinking products
II	Semisolid	Concentrated/strained yogurt
III	Solid	Soft and hard frozen yogurt and mousse yogurt
IV	Powder	Dried yogurt

Source: Adapted from Tamime and Deeth (1980).

- Existing or proposed legal standards (full, medium, or low-fat)
- Method of production (set or stirred)
- Flavors (natural, fruit, or flavored)
- Post-incubation processing (heat treatment, freezing, drying, or concentration)

It can be observed from this brief review that it is difficult, if not impossible, to find a common definition for all those products at present termed "yogurt." However, Tamime and Deeth (1980) (see also Tamime and Robinson, 1999) have proposed a scheme of classification that does cover all types of yogurt, with the groupings based primarily on the physical state of the product (Table 8.2).

Some of the steps involved in the manufacture of the different types of yogurt are illustrated in Figure 8.3 (see also Table 8.3), and it can be observed that the stages of production for set or stirred or drinking yogurts are similar until after the addition of the starter culture; illustrations of yogurt production lines have been detailed by Bylund (1995), Tamime and Robinson (1999), and Tamime et al. (2001). However, Figure 8.4 shows the processing equipment/installations required for the production of set, stirred, and extended shelf life of stirred yogurt. The primary difference in the plant design are governed mainly by the options used for the preliminary treatment of the milk base and the type of yogurt that is produced (i.e., set or stirred). Some examples of the former aspect include (a) the use of powders, (b) online concentration of the milk base (e.g., vacuum evaporator or membrane processing), and (c) the method used to standardize the fat content.

8.2.2.2.1 Preparation of the Milk Base. The preparation of the milk base involves the fortification and/or standardization of milk, and fortification of the yogurt milk means, in this context, increasing the level of the SNF in the milk in order to achieve the desired

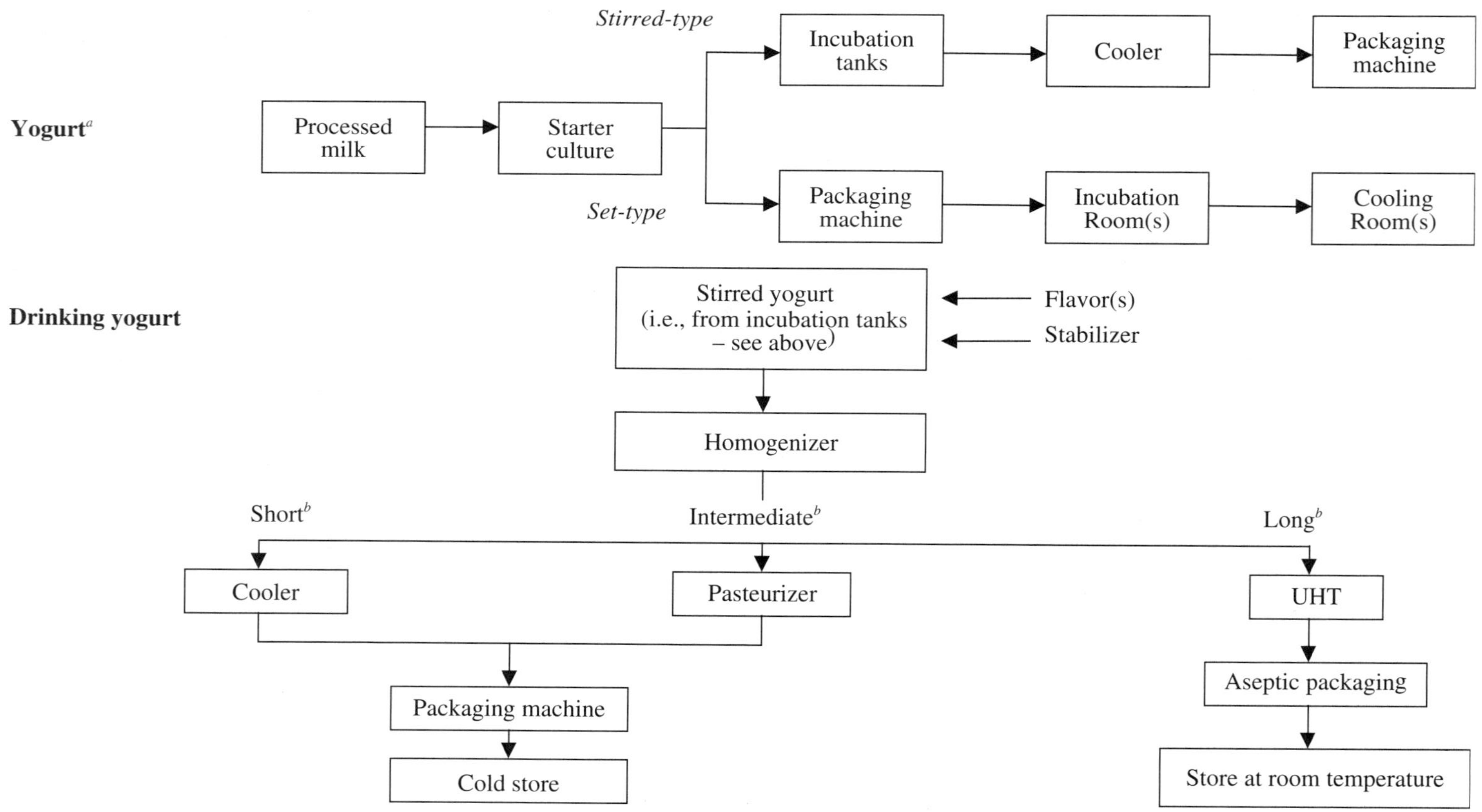

Figure 8.3. The principal stages of manufacture associated with the main types of yogurt [[a]Group I: liquid or viscous products (refer to Table 8.2)]. [b]Shelf life 2–3 weeks, 1–2 months, and several months, respectively. [Adapted from Tamime and Robinson (1999).]

TABLE 8.3. Outline of the Important Stages in the Manufacture of Set or Stirred Yogurt

Process	Materials	Comments
Standardization of fat, addition of SMP[a] or vacuum/membrane concentration of the milk base	Whole or skimmed milk at 14–16 g $100\,g^{-1}$ total solids (0.1–4.5 g^{-1} 100 fat)	Sucrose (7–10 g $100\,g^{-1}$) and stabilizers may be added to the milk base for stirred, fruit yogurt
Homogenization at ~17 MPa and 60–70°C	Process the milk base	Reduces fat globules to <2 μm and improves the texture of end product
Heat treatment at 80–85°C for 30 min or 90–95°C for 5–10 min	Process the milk base	Reduces bacterial load and oxygen content of milk; denatures whey proteins that interact with κ-casein and improves texture of end product
Cooling to 30°C or 42°C and inoculation with starter cultures	Inoculate the processed milk base	Set yogurt will be packaged at this point, optional addition of fruits or flavors
Incubation for 16–18 h at 27–30°C or 3.5–4.5 h at 42°C	Milk coagulated by lactic acid, and flavor compounds released by starter cultures	Incubation rooms for set yogurt or in-tank incubation for stirred yogurt
Cooling to <5°C for set yogurt or 20°C for stirred yogurt	Addition of fruit (10–15 g $100\,g^{-1}$) or flavors for stirred yogurt	The coagulum must be handled gently to avoid structural damage of the gel
Packaging of stirred yogurt and in-container cooling to <5°C	Retail products	Individual carton (120–150 ml) and family packs (500 g) are normal

[a]SMP, skimmed milk powder.

rheological properties of the manufactured product, while standardization means adjusting the fat level in the milk. The different methods for fortification/standardization of the milk base are summarized in Table 8.4, which also indicates the possibilities that exist for either increasing or decreasing the various milk constituents. However, the choice of any particular system is primarily governed by the following factors:

- Cost and availability of the raw materials
- Scale of production
- Capital investment in the processing equipment

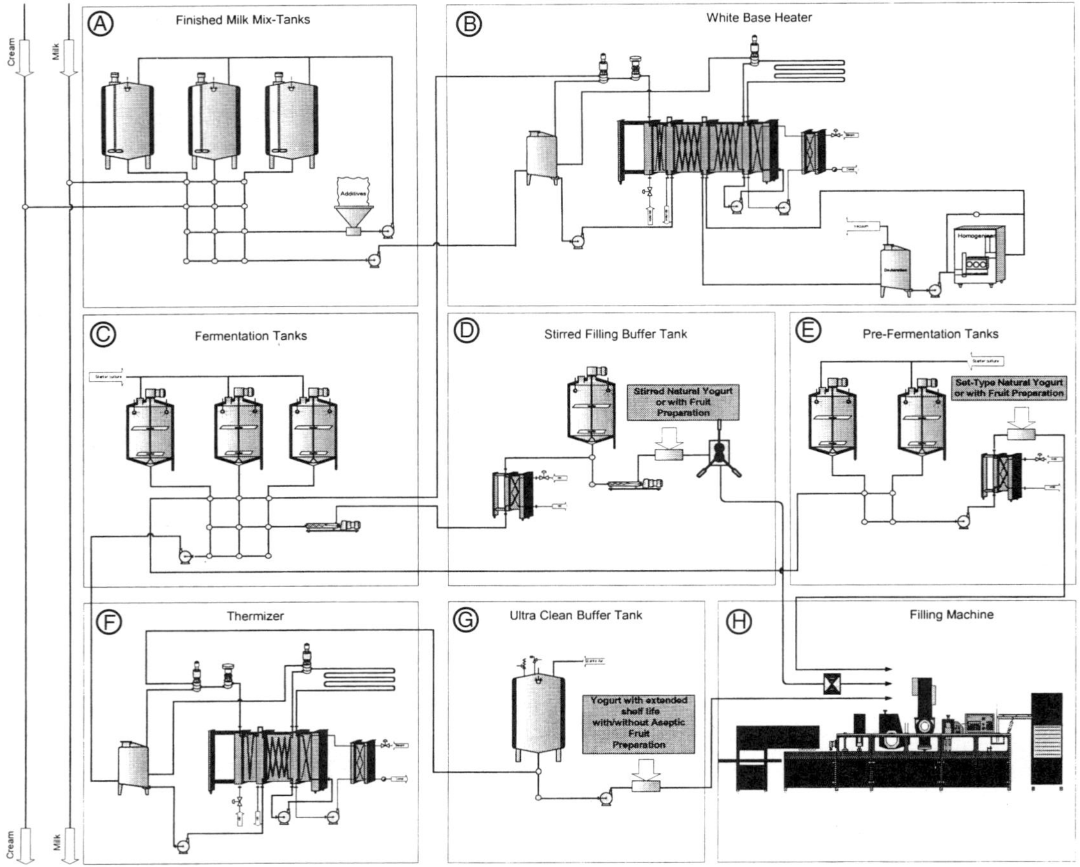

Figure 8.4. Large-scale yogurt production lines. (A) = Preparation of the milk base; (B) = heat treatment, homogenization, and cooling of the milk base; (C, D, and H) fermentation tanks, pre-cooling and mixing with fruit, and filling machine, respectively, of stirred yogurt; (E and H) pre-fermentation tanks and filling machine during the manufacture of set yogurt (incidentally, the incubation rooms are not shown); (F, G, and H) thermization plant, aseptic buffer tank, and aseptic filling machine, respectively, for the production of extended shelf-life yogurt. (Reproduced by courtesy of GEA Tuchenhagen UK Ltd., Warrington, England.)

TABLE 8.4. Some Methods of Fortification/Standardization of the Milk Base Employed for the Manufacture of Yogurt

Raw Materials	Process	Addition[a]	Principal Effect on Milk Base
Liquid whole milk	Evaporation (EV)	EV whole milk	Increases all constituents
		EV skimmed milk	Increases SNF[b]
	Ultrafiltration (UF)	UF whole milk	Increases all constituents except lactose
		UF skimmed milk	Increases protein
	Reverse osmosis (RO)	RO whole milk	Increases all constituents
		RO skimmed milk	Increase SNF
		Whole milk powder (WMP)	Increases all constituents
		Retentate WMP	Increases protein and fat
		Skimmed milk powder (SMP)	Increases SNF
		Retentate SMP	Increases protein
		Whey powder	Increases whey protein and lactose
		Whey powder concentrate	Increases whey protein
		Buttermilk	Increases SNF and phospholipids
		Caseinate	Increases casein
		Cream	Increases fat and slightly SNF[c]
		Anhydrous milk fat	Increases fat
Liquid skimmed milk	As above[d]	As above[d]	Increases SNF and/or protein
WMP and/or SMP	Either powder can be reconstituted to the desired total solids, or can be made equivalent to liquid milks and amended as above.		

[a]If mixtures of these products are added, the algebraic method of calculation could be used to calculate accurately the SNF and fat contents of the milk base (Tamime and Robinson, 1999).
[b]SNF, solids non-fat.
[c]The increase in the SNF content is governed by the fat content of the cream used (e.g., low-fat creams slightly increase SNF, while high-fat cream does not increase SNF).
[d]The increases of SNF and/or protein is only applicable when using skimmed milk products.

The composition of yogurt (percentage total solids) varies considerably (IDF, 1969, 1992a,b), and the available information is well-documented (Robinson and Tamime, 1975; Tamime and Deeth, 1980; Tamime and Robinson, 1999). However, commercial yogurts (natural) in the United Kingdom tend to contain around 15 g $100\,g^{-1}$ milk solids (Tamime et al., 1987a), and the most popular methods of fortification are the addition of milk powder to liquid milk (see Figure 8.4A) or the use of evaporation techniques. Alternatively, the solids content of the yogurt milk can be increased by the addition of an ultrafiltration (UF) concentrate instead of milk powder; the equipment remains basically the same, but the milk powder mixing unit shown in Figure 8.4A would not be required (see Tamime et al., 2001). However, it is feasible to concentrate the milk base on-site, and a description of such plants has been reported by Bylund (1995) and Tamime and Robinson (1999); additional information on the use of membrane filtration has been evaluated by Grandison and Glover (1994) and Cheryan (1998).

It is also at this stage of pretreatment of the yogurt milk that stabilizers, such as pregelatinized starch (up to 1 g $100\,g^{-1}$) or plant gums (up to 0.5 g $100\,g^{-1}$), may be added to the process milk, because syneresis from the coagulum of stirred yogurts is most easily avoided through the judicious use of hydrocolloids (see Tamime and Robinson, 1999; Fiszman et al., 1999).

8.2.2.2.2 Homogenization and Heat Treatment. The importance of total solids in this context stems from the improved consistency imparted to the yogurt coagulum, an improvement that is carried further by the homogenization stage that follows fortification (see Figure 8.4B). The effect of this treatment on the milk, and particularly on the lipid fraction, has been discussed by Tamime and Robinson (1999), and these changes tend to impart a rather "smooth" texture to the coagulum. It is also reported that homogenization can reduce the incidence of "pips" in yogurt—that is, the white flecks that appear in some fruit varieties (Robinson, 1981)—because although the origin of the "pips" is now known for certain, diminishing their visual impact goes some way toward reducing consumer complaints.

Although homogenization (~17 MPa at 60–70°C) is widely practiced by the industry, its effects are not so apparent as those associated with the subsequent heat treatment. The optimum conditions for this treatment are 80–85°C with a holding time of 30 min; but, in many factories, time and temperature are dictated by the available plant. However, although the impact of a given heating regime will vary with the conditions employed, the effects of heat treatment can be summarized as follows:

- There is a denaturation of the whey proteins (albumins and globulins), as well as an interaction with the κ-casein. Such interaction, along with the yogurt coagulum produced subsequently, is rendered more viscous (Walstra and Jenness, 1984).
- The bacterial load in the milk is reduced, and hence the starter culture has less competition from adventitious organisms.
- There is a reduction in the amount of oxygen in the milk, and because the normal yogurt cultures are micro-aerophilic, the lowered oxygen tension encourages their growth.
- Some limited damage to the milk proteins may occur during heating, and the breakdown products can either stimulate or inhibit starter activity (see Tamime and Marshall, 1997; Tamime and Robinson, 1999).

The temperature of heat treatment of the yogurt milk can vary from as low as ordinary pasteurization (72°C for 15s) to as high as 133°C for 1s (UHT). However, convenient industrial practice involves preheating the milk to 85°C for 30min (batch process) or 90–95°C for 5–10min (continuous process). In the batch system, heating of the milk is carried out in a vessel known as a "multipurpose processing tank," where all the stages of yogurt manufacture are carried out in the one holding unit. These tanks are normally jacketed, and a typical cycle of operations might be as follows:

- Fill the tank with milk and add the necessary ingredients.
- Circulate hot water in the jacket to raise the temperature of the milk base.
- Hold the milk at the desired temperature for a standard period of time.
- Circulate chilled water to cool the milk to the incubation temperature (i.e., 42°C).
- Maintain this temperature during the entire period of fermentation/coagulation.
- Circulate chilled water just before the yogurt reaches the desired acidity, and continue the cooling stage until the temperature of the yogurt is less than 10°C.

It should be noted that while the quality of the end product can be extremely good, the time required to manufacture, say, 5000 liters of yogurt can be as long as 12h.

The use of multipurpose processing tanks for the manufacture of yogurt can, therefore, prove inconvenient, and it is for this reason that plate or tubular heat exchangers are often used for continuous heat treatment of the milk base. Tamime and Robinson (1999) have reviewed in detail the use of such systems for the manufacture of yogurt, and the advantages of a continuous process as compared with the batch system are as follows:

- A smaller floor area is required for a given volume of production.
- Less energy is required.
- Productivity can be increased by utilizing the fermentation tanks more than once per day.
- The milk base can be homogenized more easily.
- The yogurt is cooled outside the fermentation tanks.

8.2.2.2.3 Inoculation with Starter Cultures. The processed milk is cooled to 30°C or 40°C and inoculated with starter cultures to ferment the milk (i.e., long or short incubation time, respectively).

8.2.2.2.3.1 MICROBIOLOGICAL EXAMINATION. The type of starters available have been discussed in Chapter 7 of this volume, but one popular material for inoculation of the production vessels is still a liquid bulk culture containing *S. thermophilus* and *L. delbrueckii* subsp. *bulgaricus*, for example, in the ratio of 1:1 (chain:chain). In practice, this requirement means checking the balance by direct microscopic examination; and, if the count is made quantitative as well (i.e., with a Breed Smear technique), then the total count for each species should confirm that the culture is suitable for use (Robinson and Tamime, 1976). Staining with Newman's stain—or, after defatting, with methylene blue (Cooper and Broomfield, 1974) or Gram's stain (Davis et al., 1971)—is a useful aid to differentiation, and for routine purposes the number of fields to be examined need not be excessive. Thus, Tamime (1977) found that counting 10 fields in a five-by-five cross-pattern overcame uneven spreading, and a reasonable estimate of the relative numbers of each species in the starter culture could be obtained.

An alternative technique for obtaining information about the ratio between the two organisms in a starter culture, or in the retail product for that matter, is the total colony count using a medium that selects for one or other species, or differentiates between them on the same plate. A selection of possible media is shown in Table 8.5, but it is

TABLE 8.5. Some of the Culture Differentiating Media[a] that Can Be Employed to Enumerate the Yogurt Starter Cultures; Responses Are Based on Colony Morphology or Detectable Growth

	Microorganisms	
Culture Media	*S. thermophilus*	*L. delbrueckii* subsp. *bulgaricus*
TGV agar	Smooth colonies	Hairy edge colonies
Hansen's agar	High mass colonies (1–3 mm)	Low mass colonies (2–10 mm)
LAB	Smooth colonies	Irregular, hairy or rough
Lee's medium	Yellow colonies	White colonies
L-S medium[b]	Round, red colonies with clear zone (<0.5 mm)	Irregular, red colonies with opaque zone (>1.0 mm)
Lactic agar	No growth	Growth
Modified lactic agar	Small, red colonies	Large, white colonies
Reinforced clostridial medium with Prussian blue	Pale blue colonies with thin, blue halo	Pale blue colonies with wide, royal blue halo
Tryptose proteose peptone yeast (TPPY) with Eriochrome dye	Oval, convex colonies (1–3 mm), opaque white/violet often with a dark centre	Transparent, diffuse colonies (4–6 mm), unidentified shape with irregular edge
TPPY agar with Prussian blue	Pale blue colonies with thin, blue halo	Small, shiny white colonies with wide royal blue halo
M17	Growth	No growth
Acidified MRS	No growth	Growth
Bromocresol green whey agar	Green lenticular colonies with entire edges	Light in color, irregular mass with twisted, filament Projections
Microassay	Growth	No growth
Streptosel agar	Growth	No growth
TYP-HGME agar	Small, light blue colonies	Large, dark blue colonies
YGLP-YL agar	Small, brilliant white colonies	Large, white colonies
Hydrophobic grid membrane filter with erioglaucine supplemented with TPYE agar	Light blue colonies	Dark blue colonies

[a]These media may *not* be selective against other thermophilic lactic acid bacteria, and not all strains of *S. thermophilus* or *L. delbrueckii* subsp. *bulgaricus* will give typical reactions.
[b]This medium is no longer available from Oxoid Ltd. (see text).
Source: Data compiled from Robinson and Tamime (1976), Reuter (1985), Matalon and Sandine (1986), Millard et al. (1990), Sanchez-Banuelos et al. (1992), Onggo and Fleet (1993), Pirovano et al. (1995), Yamani and Ibrahim (1996) and Tamime and Robinson (1999).

important that different strains of *S. thermophilus* and *L. delbrueckii* subsp. *bulgaricus* will behave differently on the same medium, and the performance of Lee's medium is a case in point (Lee et al., 1974; Ghoddusi and Robinson, 1996). Thus, while some strains of *L. delbrueckii* subsp. *bulgaricus* will give rise to white colonies on the medium, others produce yellow colonies that are identical to those of *S. thermophilus*, and it appears that the acid-producing capacity of *L. delbrueckii* subsp. *bulgaricus* is the critical factor. L-S differential medium agar [Brisdon (1998); currently, this agar (CM 495) is no longer commercially available from Oxoid Ltd.] and modified lactic agar (Matalon and Sandine, 1986) are other media that appear to give different responses according to the strains of bacteria under examination; hence a medium should only be employed for monitoring a starter culture once its performance had been tested. Tryptose proteose peptone yeast (TPPY) agar with Eriochrome black or Prussian blue has been reported to give good differentiation, as has reinforced clostridial Prussian blue (RCPB) agar (Ghoddusi and Robinson, 1996; Rybka and Kailasapathy, 1996).

While a single differentiating medium may be preferred for visual counts, the introduction of automatic colony counting may necessitate a change to the use of a medium selective for only one species—for example, M17 agar for *S. thermophilus* (Jordano et al., 1992). However, it should be noted that selective media are not always entirely inhibitory to other organisms. For example, acidified MRS agar can support the growth of both yeasts and *L. delbrueckii* subsp. *bulgaricus*; and although the difference in colony morphology is evident to the human eye, the electronic system will record just one total count.

8.2.2.2.3.2 ACTIVITY TESTS. The essential characteristic of a good starter for yogurt is that it should produce the desired level of lactic acid within a given time, and a simple test that can be completed on liquid bulk cultures involves placing a standard volume of process milk in a sterile flask, adding a specific amount of starter culture ($\sim$2 ml 100 ml^{-1}) and incubating the inoculated milk for 4 h at 42°C.

At the end of this time, the acidity of the milk should be around 0.85–0.95 g 100 g^{-1} lactic acid, and any cultures that fall to achieve these figures should be regarded with suspicion.

8.2.2.2.3.3 ABSENCE OF CONTAMINATION. The presence of gas bubbles in a starter culture or an "unclean" smell is often a clear indications of gross bacterial contamination, and a useful confirmatory test is the catalase reaction. Thus, the starter organisms are catalase-negative, so that if 5 ml of a culture are added to 1 ml of hydrogen peroxide (10 ml 100 ml^{-1}), the formation of gas bubbles indicates a considerable infection by nonstarter bacteria (Harrigan, 1998).

Although an examination for bacterial contamination can be helpful, the presence of yeasts or molds at >10 cfu ml^{-1} of starter can be of much greater concern, partly because many bacteria will be inactivated at the low pH of yogurt and partly because the presence of fungi is likely to lead to spoilage during the shelf life of the retail product. Contamination of this magnitude can be readily monitored using malt extract agar acidified with lactic acid or chloramphenicol agar (IDF, 1990a), and a 10^{-1} dilution of the starter is convenient for incorporation into pour plates (1 ml/Petri dish). This approach should, at least, indicate if yeasts are present; but, if the original counts are below 100 cfu ml^{-1}, it may be necessary to dispense 1 ml of undiluted culture into three standard Petri dishes (9.0-cm diameter) or one large dish (14 cm). Particular attention should be paid to any signs of infection by species capable of utilizing lactose (e.g., *Kluyveromyces marxianus* var. *marxianus* or *Kluyveromyces marxianus* var. *lactis*), and their presence must be regarded as a stimulus for immediate action.

These routine examinations of bulk starters are essential where culture maintenance is carried out on-site, but commercial freeze-dried or deep-frozen cultures for direct-to-vat inoculation (DVI) of the bulk starter/process milk have an excellent record with regard to freedom from contamination and overall performance, and yogurt manufacturer can normally be excused further examinations.

8.2.2.2.4 Incubators/Fermentation Tanks. The acidification of milk during the manufacture of yogurt is a biological process that must be carried out, under controlled conditions, in special incubators and/or fermentation tanks. In the case of set yogurt, the processed milk, which is inoculated with the starter cultures, is packaged and fermentation takes place in the retail container or the milk may be pre-fermented in large tanks to pH >5.7 before packaging and the final stages of fermentation in the container (see Figure 8.4E). Nevertheless, the final incubation system for set-type yogurt may be selected from among the following:

Water Baths. This is an old process for fermentation which is still used in some small-scale production units. The inoculated milk is bottled in glass retail containers; and, after packing in metal trays, the bottles are immersed in shallow water baths or tanks. The temperature of the water is maintained at 40–45°C until the desired acidity is reached, at which point the warm water is replaced with cold water in order to reduce the metabolic activity of the starter culture. Final cooling takes place in the cold store. It is, perhaps, worth pointing out that these types of incubator are labor-intensive, their energy consumption is high, and they require a large floor area.

Cabinets. The cabinet comprises a small insulated room that is divided into compartments, and most incubators of this type are multipurpose chambers capable of circulating hot or cold air. In practice, the retail cartons are "palletized" inside the cabinet, and hot air is circulated during the fermentation period followed by cold air during the cooling stage. Sometimes these cabinets are used as incubators only, and the yogurt is cooled in a refrigerated cold store; however, the warm coagulum may suffer structural damage during transfer, and hence the latter procedure is avoided if possible.

Tunnels. In the previous types of incubator, set yogurt is produced in batches, but in the tunnel system, production can be continuous. The tunnel is divided into two sections, with the first part as a heat chamber and the second part as a "cooler." The trays of yogurt cartons move through the tunnel on a conveyor belt, and the speed and length of the conveyor are governed by the temperature of incubation, the percentage of starter culture used, and the activity of the inoculum; the lower the temperature of incubation (e.g., less than 35°C) and the smaller the inoculation rate (e.g., 1 ml $100 ml^{-1}$), the longer and slower the conveyor belt. Although the warm coagulum is in motion during the fermentation period, minimum structural damage is achieved by fitting smooth rollers beneath the pallets. Incidentally, any type or size of retail container can be used in the case of cabinet or tunnel incubators.

During the manufacture of stirred yogurt, the milk is fermented in bulk in a special incubation tank. The use of these tanks in the manufacture of yogurt has been reviewed by Tamime and Greig (1979), and they are divided into two main types: (a) Fermentation tanks are used as incubators only, and they are usually insulated in order to maintain the appropriate temperature (see Figure 8.4C); the processing of the milk and the cooling of the yogurt is carried out in other equipment on the production line (see Figure 8.4D), and (b) multipurpose tanks (as mentioned earlier, this type of tank is jacketed) can be used for all stages of yogurt production.

8.2.2.2.5 Cooling. After the incubation period, the yogurt is cooled in order to control the level of lactic acid in the product. There are a number of different methods that can be used to cool stirred yogurt, but it is of note that the rate of cooling can affect the structure of the coagulum. Thus, very rapid cooling can lead to whey separation, due to a too rapid contraction of the protein filaments, which, in turn, affects their hydrophilic properties (Rasic and Kurmann, 1978).

The normal industrial practice is, therefore, to cool the yogurt to 15–20°C before mixing it with fruit/flavors and packaging. The final cooling to 5°C takes place in a refrigerated cold store. The methods used for cooling yogurt are as follows:

In-Tank Cooling. The cooling of yogurt in a multipurpose tank requires the circulation of chilled water through the jacket, and a tank of 2500–5000 liters of yogurt can take up to 4h to cool from 45°C to 10°C (Jay, personal communication). Alternatively, by fitting stationary coils in a multipurpose tank (e.g., Goavac yogurt tank—5000 liters), the efficiency of cooling is improved, and it would take less than 30min to cool 5000 liters of yoghurt from 42°C to 20°C before mixing with fruit, packaging, and final cooling in the cold store (Tamime and Robinson, 1985).

Plate or Tubular Cooler. An efficient system of cooling yogurt quickly and continuously involves the use of a plate or tubular cooler; the latter type coolers may cause the least structural damage to the coagulum (Piersma and Steenbergen, 1973).

Although cooling of yogurt is required mainly to control the metabolic activity of the starter cultures, there is no fixed pH value(s) at which to start the agitator of the fermentation tank. However, consumer acceptability of the final product and processing experience to minimize protein lumps or syneresis in yogurt can dictate the time of agitation. According to Anderson (personal communication), good-quality yogurts have been obtained when agitation of the coagulum has started at

- pH 4.6 for creamy yogurt (10g $100g^{-1}$ fat)
- pH 4.4 for whole milk yogurt (3.5g $100g^{-1}$ fat)
- pH 4.25 for low-fat yogurt (1.5g $100g^{-1}$ fat)
- pH 4.1 for skimmed milk yogurt (~0.3g $100g^{-1}$ fat)

8.2.2.2.6 Packaging. Yogurt packaging machines are based on one of the following principles: (a) volumetric level filling (e.g., when fluid yogurt is poured into glass bottles) or (b) volumetric piston filling, as applied to the packaging of stirred yogurt into plastic containers. The latter type is, of course, more widely used, but the piston pump can cause some shearing of the coagulum. It is also extremely important that the design of the filling head should allow for a high standard of hygiene. For further details of the different packaging systems (e.g., cup fillers), form–fill–seal, carton fillers and multiunit packs are available in the publication by Tamime and Robinson (1999).

In general, most filling machines including GEA Finnah GmbH (see Figure 8.5) are available in the following types/categories:

- Standard machine constructed with basic hygiene in mind,
- Machines suitable for "clean-room" operation and fitted with laminar flow equipment,
- Machines suitable for "ultraclean" operation where the equipment sanitizes the aluminum foil and provision for sterile air system, or
- Aseptic types fitted with aluminum foil sterilization and hermetically sealed tunnels and having provision for sterile air system (Anderson, personal communication).

Some other features of the GEA Finnah "form–fill–seal" machine are: (a) pre- and post-filling stations that are suitable for multichamber containers, (b) patch labeling of individual cups, (c) wrap-around labeling, (d) provision for snap-on lids, (e) temperature-controlled contact heating for sealing, (f) sterilization of packaging materials is effected using an ultrasonically produced hydrogen peroxide (H_2O_2) aerosol, and (g) the machines are equipped for cleaning-in-place (CIP). Some filling outputs of the GEA Finnah models are as follows:

4100	4200	4300
6 × 75 mm 15,100 cups h^{-1}	10 × 75 mm 25,200 cups h^{-1}	24 × 75 mm 54,000 cups h^{-1}
5 × 95 mm 12,600 cups h^{-1}	8 × 95 mm 20,200 cups h^{-1}	18 × 95 mm 24,500 cups h^{-1}
10 × 95 mm 24,000 cups h^{-1}	16 × 95 mm 38,000 cups h^{-1}	
4 × 127 mm 10,100 cups h^{-1}	6 × 127 mm 15,100 cups h^{-1}	
3 × 164 mm 7,600 cups h^{-1}	5 × 164 mm 12,600 cups h^{-1}	

Note: The filling capacity specifications apply to 42 strokes min^{-1}.

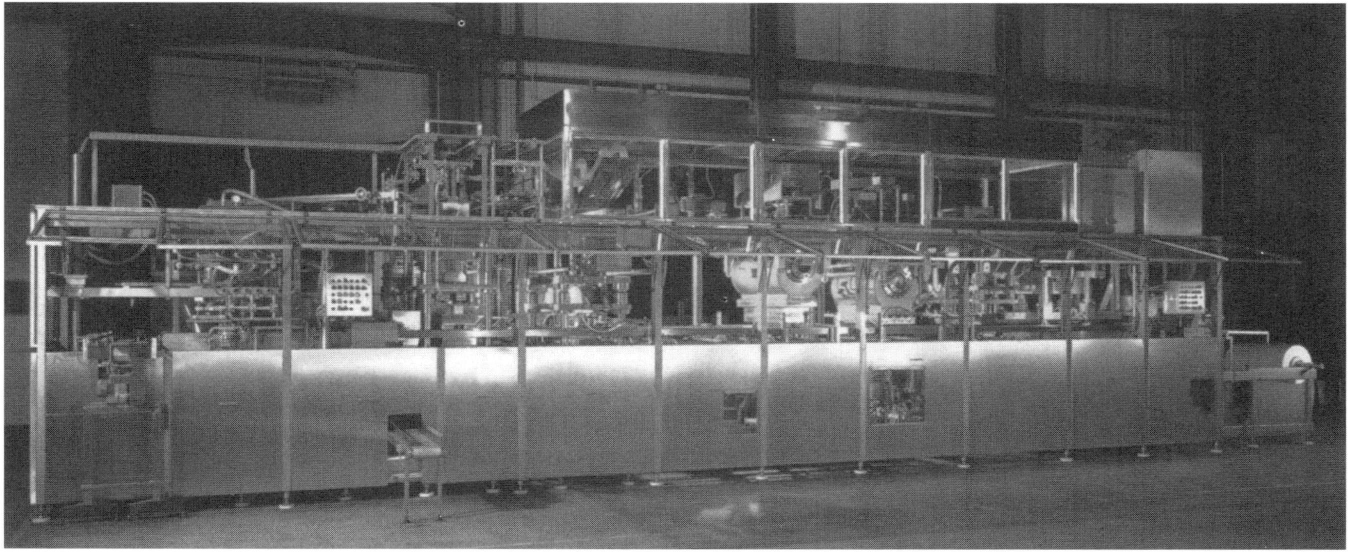

Figure 8.5. A general view of the GEA Finnah form–fill–seal packaging machine. Reproduced by courtesy of GEA Tuchenhagen UK Ltd., Warrington, England.)

8.2.2.2.7 Microbiology of the Fermentation. The crude inoculum that was employed in the production of traditional yogurt consisted of a whole range of lactic acid bacteria belonging to the genera *Lactobacillus* and *Streptococcus*, and it was Metchnikoff (1910) who postulated that the derivation of yogurt depended on the presence of one particular bacterium, *Bulgarian bacillus*. This organism, renamed *Thermobacterium bulgaricum* by Orla-Jensen (1931), is now recognized as being *L. delbrueckii* subsp. *bulgaricus* (Weiss et al., 1983; Kandler and Weiss, 1986).

Although the inclusion of *L. delbrueckii* subsp. *bulgaricus* in starter cultures for yoghurt is now almost obligatory (Winkelmann, 1987), the occasional employment of *L. delbrueckii* subsp. *lactis* or *L. helveticus* in starter cultures is a distinct possibility, because the relationship between these associated organisms is fairly close. However, the organisms associated with the production of yogurt are normally restricted to *L. delbrueckii* subsp. *bulgaricus* and *S. thermophilus* (Kon, 1972). In some sources, the latter organism has been reclassified as *Streptococcus salivarius* subsp. *thermophilus* because the DNA homology between *S. salivarius* and *S. thermophilus* was found to be in the range of 75–97% (Kilpper-Bälz et al., 1982; Farrow and Collins, 1984). Nevertheless, these genetic data have not been supported by numerical taxonomic studies using phenotypic characters (Bridge and Sneath, 1983; Hardie and Wiley, 1995), and hence most authorities have reverted to the old nomenclature of *S. thermophilus*.

8.2.2.2.7.1 THE INDIVIDUAL SPECIES. *S. thermophilus* is a Gram-positive bacterium with spherical/ovoid cells of 0.7- to 0.9-μm diameter, and it is a natural inhabitant of raw milk in many parts of the world. It occurs in milk in long chains of 10–20 cells, and it ferments lactose homofermentatively to give L(+)-lactic acid as the principal product. Above 10g of lactic acid kg^{-1} of yogurt (around pH 4.3–4.5), the growth and metabolism of *S. thermophilus* is normally inhibited, and cell numbers that may have reached $10 \times 10^{7-8}$ cfu ml^{-1} of yogurt tend to stabilize. Glucose, fructose, and mannose can also be metabolized, but the fermentation of galactose, maltose, and sucrose is strain-specific. Thus, as mentioned earlier, numerous strains of *S. thermophilus* have been isolated over the years and/or modified in the laboratory, and hence the loss or gain of alleles for specific aspects of metabolic performance is not uncommon.

The principal sugar in the yogurt base, lactose, is actively transported across the cell membrane of *S. thermophilus* through the mediation of a membrane-located enzyme, galactoside permease; and once inside the cell, the enzyme β-galactosidase hydrolyzes the sugar to glucose and galactose. The glucose is then metabolized to pyruvate via the Embden–Meyerhof–Parnas (EMP) Pathway, and lactic dehydrogenase converts the pyruvate to lactic acid. The galactose and lactic acid usually leave the cell and accumulate in the medium, but some strains of *S. thermophilus* possess a galactokinase that converts the galactose to galactose-1-P. This phosphorylated form of galactose can then be transformed into either glucose-1-P or galactose-6-P, depending on the strain, and further metabolized into lactic acid.

Despite its protein-rich habitat, *S. thermophilus* displays limited proteolytic ability, and hence its source of nitrogen is, at least initially, free amino acids occurring naturally in the milk or released during the heat treatment. However, some amino acids, such as glutamic acid, histidine, cysteine, methionine, valine, or leucine, are not present in milk at levels sufficient to support the essential growth of *S. thermophilus*, so that the increase in cell numbers necessary to complete the yogurt fermentation depends upon the absorption of short-chain peptides released by *L. delbrueckii* subsp. *bulgaricus* and hydrolysis of these to the constituent amino acids (Robinson, 2000).

The optimum growth temperature for *S. thermophilus* is ~37°C, but it is sufficiently thermophilic in nature to grow alongside *L. delbrueckii* subsp. *bulgaricus* during the commercial production of yogurt at 42°C. The growth of *S. thermophilus* ceases at ~10°C. *Lactobacillus delbrueckii* subsp *bulgaricus* is also Gram-positive, but occurs in milk as chains of three to four short rods with rounded ends, 0.5–0.8 × 2.0–9.0 μm. Its basic metabolism is again homofermentative, but in this case the end product is D(−)-lactic acid at levels of around 18 $g\,kg^{-1}$ of yogurt. This form of lactic acid is less readily metabolized by humans than the L(+)-acid isomer, and it is well above the level appreciated by the average consumer at ~10 $g\,kg^{-1}$. The tolerance of *L. delbrueckii* subsp. *bulgaricus* to acidity also contrasts dramatically with that of *S. thermophilus*. Lactose, fructose, glucose, and, in some strains, galactose can all be utilized by *L. delbrueckii* subsp. *bulgaricus*, but, unlike *S. thermophilus*, *L. delbrueckii* subsp. *bulgaricus* can hydrolyze casein—especially β-casein—by means of a wall-bound proteinase to release polypeptides. However, the peptidase activity of *L. delbrueckii* subsp. *bulgaricus* is limited; but, because *S. thermophilus* can readily hydrolyze peptides to free amino acids, it is likely that some of the latter are available to *L. delbrueckii* subsp. *bulgaricus* (Robinson, 2000).

The optimum growth temperature for *L. delbrueckii* subsp. *bulgaricus* is 45°C, and hence the value of 42°C selected for commercial production is an effective compromise between the growth optima of the two essential species.

8.2.2.2.7.2 SYNERGISTIC OR ASSOCIATIVE INTERACTIONS. The trend in favor of *L. delbrueckii* subsp. *bulgaricus* and *S. thermophilus* as the dominant starter organisms in yogurt is valuable in that it helps to give yogurt a distinctive character vis-à-vis other fermented milks, such as acidophilus milk made with *Lactobacillus acidophilus*. There are also sound microbiological reasons for promoting the association, because it has been widely observed that both the cell count and the production of lactic acid during yogurt manufacture is greater when mixed cultures of *S. thermophilus* and *L. delbrueckii* subsp. *bulgaricus* are used, as compared with single species (Tamime and Robinson, 1999). This observation led Pette and Lolkema (1950a,b) to postulate the existence of a synergistic relationship between compatible strains of the two organisms; in particular, it is now widely accepted that:

- The proteolytic activity of *L. delbrueckii* subsp. *bulgaricus* releases peptides and, to a lesser extent, amino acids that can be metabolized by *S. thermophilus*.
- The peptidases formed during the active growth of *S. thermophilus* act on the peptides to liberate amino acids that are utilized by *L. delbrueckii* subsp. *bulgaricus* (Accolas et al., 1980; Radke-Mitchell and Sandine, 1984; Tamime and Robinson, 1999).

The stimulation of *L. delbrueckii* subsp. *bulgaricus* is further enhanced due to the production by *S. thermophilus* of formic acid or the addition of sodium formate (Galesloot et al., 1968; Veringa et al., 1968; Bottazzi et al., 1971; Marshall and Mabbitt, 1980). In addition, Driessen et al. (1982) reported that the growth of *L. delbrueckii* subsp. *bulgaricus* was stimulated by carbon dioxide produced by *S. thermophilus* and, in particular, that the optimum level was around $31\,mg\,CO_2\,kg^{-1}$ of milk. In commercial practice, heat treatment of the milk base reduces the natural level of carbon dioxide substantially, but strains of *S. thermophilus* have been found that produce 30–50 mg $CO_2\,kg^{-1}$ of milk within the first hour of incubation; these levels more than sufficient to stimulate *L. delbrueckii* subsp. *bulgaricus*. The source of this carbon dioxide is the breakdown of urea (Tinson et al., 1982).

The value of the synergism between the species can be easily demonstrated by isolating strains of *S. thermophilus* and *L. delbrueckii* subsp. *bulgaricus* from a commercial starter culture and then inoculating the individual species into milk (20 ml of liquid culture $liter^{-1}$). If the rate of acid development in the separate milks is then compared with the level of lactic acid liberated in a "control" [10 ml of *S. thermophilus* and 10 ml of *L. delbrueckii* subsp. *bulgaricus* (liquid cultures $liter^{-1}$)], the contrast is most dramatic. Thus, while the combined culture may well generate an acidity of >10 g $liter^{-1}$ in 4 h at 42°C, the values in the individual cultures could be ~4 g $liter^{-1}$ for *S. thermophilus* and around 2 g $liter^{-1}$ for *L. delbrueckii* subsp. *bulgaricus* (Robinson, 2000).

The implications of this pattern for the dairy industry are self-evident, and it confirms also that *S. thermophilus* grows more rapidly than *L. delbrueckii* subsp. *bulgaricus* in milk and, at least initially, releases more lactic acid. However, the developing acidity provides an environment that is conducive to the growth and metabolism of *L. delbrueckii* subsp. *bulgaricus* so that, at the end of 4 h, the latter component of a combined culture will be releasing more lactic acid than *S. thermophilus*. Although the growth rates of the two species are markedly different over a typical 4-h fermentation, "abundant and viable" populations of both species should be present in the retail product.

8.2.2.2.7.3 METABOLIC PRODUCTS IMPORTANT FOR YOGURT QUALITY. The production of acetaldehyde in yogurt is most pronounced in mixed cultures (Bottazzi and Vescovo, 1969); and although *S. thermophilus* does form acetaldehyde, the threonine aldolase pathway (threonine → glycine and acetaldehyde) is less active at normal fermentation temperatures (i.e., ~42°C) than the corresponding synthesis by *L. delbrueckii* subsp. *bulgaricus*; the possible role of alternative pathways for the production of acetaldehyde in yogurt has been reviewed by Tamime and Robinson (1999).

In mixed cultures, the final concentrations of acetaldehyde in yogurt can range from 2.4 $\mu g\,g^{-1}$ to 41.0 $\mu g\,g^{-1}$; and these levels, along with lower amounts of acetone (1.0–4.0 $\mu g\,g^{-1}$), acetoin (2.5–4.0 $\mu g\,g^{-1}$), and diacetyl (0.4–13.0 $\mu g\,g^{-1}$), give yogurt its distinctive flavor profile. However, the flavor of milks that arise from natural fermentations are not easy to classify (Ulberth and Kneifel, 1992); this point was confirmed by Ott et al. (1997, 1999, 2000a–c), who identified 21 components of Swiss yogurt that could be having a major impact on flavor.

While lactic acid is the primary product of lactose catabolism (Marshall and Tamime, 1997), the production of extracellular polysaccharides (EPS) is extremely important in those cultures employed for the manufacture of stirred, fruit yogurts. Scanning electron micrographs of yogurt manufactured with appropriate strains show that the gum-like material forms filamentous links between the cell surfaces of the bacteria and the protein matrix (Kalab et al., 1983; Tamime et al., 1984). These findings were confirmed by Schellhaass and Morris (1985), Bottazzi and Bianchi (1986), Kalab (1993), and Skriver et al. (1995), and a typical example of this behavior is shown in Figure 8.6. The precise identity of these "gums" is still under investigation, but Tamime (1977)

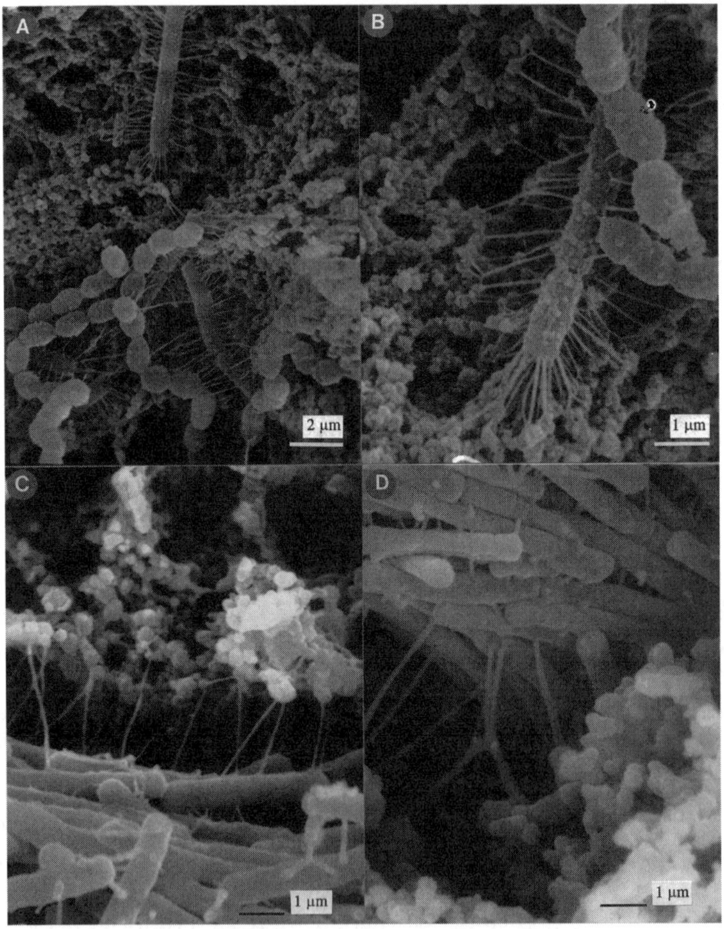

Figure 8.6. A scanning electron micrograph of natural yogurt showing the long chains of cocci (*S. thermophilus*) and the short chains of rods (*L. delbrueckii* subsp. *bulgaricus*); the matrix of coagulated milk proteins is also clearly visible, as are the strands of extracellular polysaccharide linking the bacterial cells to the protein. Micrographs A and B, Tamime and Kalab (Scotland and Canada); micrographs C and D, Bottazzi (Italy). [After Tamime and Robinson (1999). Reproduced by courtesy of Woodhead Publishers, Cambridge, England.)

reported that the material secreted by a so-called "slimy" culture was a glucan. By contrast, Schellhaass and Morris (1985) found that the secretion from another "slimy" strain was formed from glucose and galactose in a ratio of 1:2. This contrast between strains is not unexpected, nor is the ability of different strains of *S. thermophilus* (Cerning et al., 1988; Cerning, 1990; Doco et al., 1990) and *L. delbrueckii* subsp. *bulgaricus* (Cerning et al., 1986; Garcia-Garibay and Marshall, 1991; Gruter et al., 1993; Grobben et al., 2000) to produce "gums" that differ in their chemical nature (see also Cerning et al., 1990; Cerning, 1995; Cerning and Marshall, 1999; Ricciardi and Clementi, 2000).

All the "gums" produced by starter cultures for yogurt are polysaccharides; and, while details of their tertiary structures have not been published, qualitative differences with respect to the constituent monosaccharides have been reported (Charteris et al., 1998); a summary of some typical differences is shown in Table 8.6. In addition, relationship between the "gum" and the bacterial cell can differ, and either (a) the strain of *S. thermophilus* or *L. delbrueckii* subsp. *bulgaricus* secretes a "slimy/ropy" polysaccharide that migrates into the surrounding gel or (b) the polysaccharide forms a capsular envelope around the cell (Charteris et al., 1998; Cerning, 1990; Hassan et al., 1996). To some extent, these differences are reflected in the physical properties of the resultant yogurts; and although the correlation is not precise, it is reasonable to suggest the following:

- Cultures secreting glucan-like polysaccharides give rise to products with low structural stability to applied stresses, so that once the chemical bonds within a chain of monosaccharides are broken during stirring or packaging, they do not reform and any loss of viscosity is permanent (Hassan et al., 1996). To the consumer, the retail yogurts will appear fluid, but have a pleasant and slightly "slimy" mouthfeel.
- Capsular polysaccharides give a thicker, "spoonable" texture to the product, and the gel is less prone to damage during pumping or similar operations. In addition, once the yogurt is in the carton, there is a tendency for part of the viscosity lost during packaging to recover, probably because the encapsulated gells tend to "clump" together and bind the casein micelles as well. Furthermore, if such yogurts are held at ambient temperature prior to chilling, the recovery in viscosity may be enhanced as a result, perhaps, of additional polysaccharide synthesis (Rawson and Marshall, 1997).

TABLE 8.6. Sugars Reported to be Present in Polysaccharides Secreted by the Yogurt Microorganisms

	Sugar[a]								
Starter Culture	Gal	Glu	Fru	Rha	Man	Xyl	Ara	GalA	Neu
S. thermophilus	+	+	–	–	–	–	–	–	–
	+	+	–	+/–	+	+/–	+/–	–	–
	+	+	–	–	–	–	–	+	–
	+	+	–	–	–	+	–	–	+
	+	+	–	+/–	+/–	–	+/–	–	–
L. delbreuckii subsp. *bulgaricus*	+	+	–	–	+	–	+	–	–
	+	+	–	–	–	–	–	–	–
	–	+	+	–	–	–	–	–	–
	+	+	–	+	–	–	–	–	–

[a]Gal, galactose; Glu, glucose; Fru, fructose; Rha, rhamnose; Man, mannose; Xyl, xylose; Ara, arabinose; GalA, galactosamine; Neu, neuramic acid. +, present; –, absent; +/–, trace reported.

Source: Adapted from Tamime and Robinson (1999).

Obviously, any natural fermentation brings about numerous of minor changes in the substrate, with the nature of the substrate (e.g., cow's, sheep's, or goat's milk) and the strains of bacterium giving rise to subtle differences in the end product. Some of the less predictable changes associated with the yogurt fermentation are summarized in Table 8.7.

TABLE 8.7. Some of the Detectable Changes in Yogurt Milk that Result from the Activity of the Starter Organisms

Carbohydrate	Lactose is hydrolyzed inside the bacterial cell by the enzyme β-galactosidase to glucose and galactose. The former is utilized for the production of D(−)- and L(+)-lactic acid, and the galactose accumulates.
	The calcium–caseinate–phosphate complex is destabilized by the lactic acid, and this leads to the formation of the yogurt coagulum.
	Production of flavor components ($\mu g\,g^{-1}$) in yogurt due to fermentation of the milk sugar—for example, acetaldehyde (2.4–41.0), acetone (1.0–4.0), acetoin (2.5–4.0), and diacetyl (0.4–13.0).
Proteolysis	Yogurt starter cultures are slightly proteolytic, and the peptides and amino acids produced act as precursors for the enzymic and chemical reactions that produce the "flavor compounds."
	Protein degradation is associated mainly with *L. delbrueckii* subsp. *bulgaricus*, but peptidase enzymes are produced by both *S. thermophilus* and *L. delbrueckii* subsp. *bulgaricus*; the production of bitter peptides is attributed to the use of incubation temperatures below 30°C and/or enzymic activity during cold storage; strain differences may also be important.
	The free amino acid content of yogurt can vary from 23.6 to 70 mg $100\,ml^{-1}$.
Lipolysis	Some degree of fat degradation by the starter culture does take place, and the end products contribute significantly toward the flavor of yogurt.
	Lipases from the yogurt starter culture are especially active against short-chain triglycerides.
Miscellaneous	The volatile acidity of yogurt can reach 7.5 $meq\,kg^{-1}$, mainly as the result of the metabolism of *L. delbrueckii* subsp. *bulgaricus*.
	On average, there is an increase in niacin and folic acid during fermentation, but the levels of vitamin B_{12}, thiamine, riboflavin, and pantothenic acid decline.

Source: Adapted from Tamime and Robinson (1999).

8.2.2.2.7.4 INHIBITORS OF STARTER ACTIVITY. It is evident that that the growth and metabolism of both *S. thermophilus* and *L. delbrueckii* subsp. *bulgaricus* are essential for the production of a satisfactory yogurt, and hence that any deficiency in starter activity will produce a substandard product. Even with the advent of direct-to-vat cultures, starter failures can occur, and hence it is worth highlighting the principal causes of potential microbiological problems:

- Although strains of both *S. thermophilus* and *L. delbrueckii* subsp. *bulgaricus* are susceptible to infection by host-specific bacteriophages, the presence of phages is not usually as serious as with some mesophilic fermentations. This contrast appears to stem from the facts that (a) the milk for yogurt-making is severely heat-treated compared with that employed for cheese-making, for example, so that the number of virus particles present at the time of inoculation with the starter culture should be low, and (b) the undisturbed yogurt milk coagulates in around 2 h, so that any infected cells of *S. thermophilus* or *L. delbrueckii* subsp. *bulgaricus* quickly become isolated; as cheese milk is stirred continuously during acidification, any bacteriophages released by lysis of infected bacterial cells are distributed throughout the vat. Nevertheless, the risk does exist, and yogurt producers need to maintain high standards of hygiene to avoid the buildup of bacteriophages in pools of "stagnant" whey.
- Changes in the activity of a culture (e.g., rate of acid production or level of aroma/flavor compounds secreted) can arise as a consequence of routine subculturing. Exactly why these changes arise is not clear, but the gradual emergence of a numerical imbalance between *S. thermophilus* and *L. delbrueckii* subsp. *bulgaricus* has been cited as one possible cause.
- The presence of antibiotics or other inhibitory substances in the milk is a major cause of poor fermentations in some countries, and *S. thermophilus* is especially sensitive to antibiotics like penicillin, streptomycin, neomycin, and amphicillin, which are widely used to control mastitis. Levels of contamination as low as 0.004 International Units (IU) of penicillin G ml^{-1} can inhibit cell wall development of *S. thermophilus* (Yamani et al., 1998). Strains of *L. delbrueckii* subsp. *bulgaricus* tend to be more tolerant (0.02 IU of penicillin G ml^{-1}), as are cultures of both species growing together. However, even when the two organisms are present and under optimum conditions, as little as 0.01 IU of penicillin G ml^{-1} can

delay fermentation. Sanitizing agents employed to clean a plant, such as chlorine (100 mg $liter^{-1}$) or iodophors (60 mg $liter^{-1}$), can also cause inhibition of the mixed cultures, and hence the screening of bulk milk for microbiocidal agents is essential.

- Although commercial cultures are selected for their performance under industrial conditions, incompatibility between strains of *S. thermophilus* and *L. delbrueckii* subsp. *bulgaricus* can result in an almost complete absence of protoco-operation between the species. As a consequence, coagulation times may be increased by several hours, and the organoleptic quality of the end product may be extremely poor. In practice, this problem should be a very rare occurrence, but any attempt by a yogurt-maker to create his/her cultures by isolating and blending strains of *S. thermophilus* and *L. delbrueckii* subsp. *bulgaricus* from different sources could lead to disappointment.

8.2.2.2.7.5 MICROBIOLOGICAL QUALITY OF THE END PRODUCT. An examination of yogurt for contaminant organisms is concerned with (a) protection of the consumer from any potentially pathogenic species and (b) assurance that the material will not undergo microbial spoilage during its anticipated shelf life (Stannard, 1997); these issues are of vital importance to any company.

As far as pathogens are concerned, yogurt with an acidity of around 10 kg^{-1} lactic acid is a fairly inhospitable medium, and really troublesome pathogens like *Salmonella* spp. and *Listeria monocytogenes* will be incapable of growth (Hobbs, 1972). A degree of survival of *L. monocytogenes* at pH 4.5 in Labneh has been reported (Gohil et al., 1996), but the counts declined rapidly within 24 h—that is, long before the product would have reached the consumer. Schaack and Marth (1988a–c) observed that *L. monocytogenes* was inhibited during a yogurt fermentation, but Choi et al. (1988) suggested that the final pH of the product was important, as was the precise strain of *L. monocytogenes* (see also Kerr et al., 1992).

Coliforms should also be inactivated by the low pH (Feresu and Nyati, 1990); in addition, some species may be susceptible to antibiotics released by the starter organisms. The acid-sensitivity of *Campylobacter* spp. suggests that this genus will not survive a normal fermentation (Cuk et al., 1987); but whether *Staphylococcus* spp., and in particular coagulase-positive strains (Masud et al., 1993), can survive in yogurt is a matter of some dispute (Arnott et al., 1974; Alkanahl and Gasim, 1993). To date there have been no records of staphylococcal food poisoning being associated with the consumption of yogurt in the United

Kingdom (Gilbert and Wieneke, 1973; Keceli and Robinson, 1997), and Attaie et al. (1987) showed that a virulent strain of *S. aureus* was inhibited during fermentations involving either yogurt cultures or *L. acidophilus*.

However, this general confidence does have to be tempered with caution, because a recent report linked an outbreak of *Escherichia coli* 0157 with the consumption of yogurt (Morgan et al., 1993), so that it should be remembered that low starter activity and/or post-heat-treatment contamination can lead to problems even with this traditionally safe product (Al-Mashhadi et al., 1987; Ibrahim et al., 1987). This latter point has been emphasized by studies with some of the so-called "emerging pathogens" like *Yersinia enterocolitica* and *Aeromonas hydrophila*, in which survival in yogurt has been shown to be closely correlated with pH (Ahmed et al., 1986; Aytaç and Özbas, 1994a,b); the behavior of *Bacillus cereus* in yogurt will follow a similar pattern (Stadhouders and Driessen, 1992). The freak occurrence of *Clostridium botulinum* in hazel-nut yogurt also highlights the need for vigilance.

More significant from the producer's standpoint is the examination for yeasts and molds, because these organisms are capable of spoiling yogurt well within an anticipated sell-by date. Yeasts, whether lactose-utilizers like *K. marxianus* var. *marxianus* and *K. marxianus* var. *lactis*, or more cosmopolitan species, such as *Saccharomyces cerevisiae* or *Torulopsis candida* (Jordano et al., 1991), are a major concern (Fleet, 1990; Onggo and Fleet, 1993). In order to avoid in-carton fermentation—often manifest by a "doming" of the lid of a carton or collapse of the carton (Foschino et al., 1993)—Davis et al. (1971) have suggested that yogurt, at the point of sale, should contain less than 100 viable yeast cells ml^{-1}. Above 10×10^2 cells ml^{-1} would imply a serious risk of deterioration, because a yeast population of $70–100 \times 10^3$ cells ml^{-1} and serious gas/off-flavor development can be achieved quite easily within a 2 to 3-week shelf life (see also Barnes et al., 1979).

Molds tend, on the whole, to develop more slowly than the yeasts; and although some genera like *Aspergillus* can form "button-like" colonies within a coagulum, most fungi require oxygen for growth and sporulation (see Jordano and Salmeron, 1990). Hence, molds are usually visible only in retail cartons of set yogurt, because the surface of stirred yogurt rarely remains undisturbed for any length of time. Nevertheless, occasional problems can arise from such genera as *Mucor*, *Rhizopus*, *Aspergillus*, *Penicillium*, or *Alternaria*, and the unsightly superficial growths of mycelium will lead to consumer complaints (Garcia and Fernandez, 1984). For this reason, a mold count of up to 10 cfu ml^{-1} of retail product has been rated as "doubtful quality" by Davis et al. (1971).

It has been reported by Jordano et al. (1989) that some strains of *Aspergillus flavus* isolated from commercial yogurts were aflatoxigenic; but although the sucrose content of fruit yogurt would be sufficient to support aflatoxin production (Ahmed et al., 1997), it has not been suggested that aflatoxin synthesis does occur in yogurt. Occasionally, aflatoxin M_1 has been identified in the milk for yogurt production, but even this contamination may, depending on the pH of the product, decline during fermentation (Hassanin, 1994).

Because fruit should be heat-treated prior to use, infections from this source should be rare, but airborne spores or yeast cells can prove more difficult to control. The unexpected variety of yeast species isolated from yogurt by Tilbury et al. (1974) and Suriyarachchi and Fleet (1981) can probably be explained by this type of chance contamination, and protection of the filling area is a top priority. Regular monitoring of the air in the processing area may help to identify the route(s) taken by airborne propagules, and the examination of representative samples of the end product employing acidified malt agar or Rose Bengal agar (Brisdon, 1998), yogurt whey agar (Yamani, 1993), or chloramphenicol agar (yeasts) (IDF, 1990a) can provide a warning of impending problems. Some suggested international standards for yogurt and the methods proposed for monitoring compliance are reported by IDF (1990b,c, 1991a,b, 1998) and APHA (1992).

Alternatively, impedance measurements can be employed to determine low levels of yeast in yogurt (Shapton and Cooper, 1984; Pettipher, 1993), and the Direct Epifluorescent Filter Technique (DEFT) has been used successfully by McCann et al. (1991). More recently, the Petrifilm™ method has been recommended by Vlaemynck (1994) for enumerating yeasts and molds in yoghurt, as has the ISO-GRID membrane filtration system in conjunction with YM-11 agar (Entis and Lerner, 1996).

Overall, therefore, it is clear that well-made yogurt should not present a manufacturer with many complaints as far as microbiological quality is concerned, although some small producers have yet to match the standards of the major suppliers (see Tamime et al., 1987a,b).

8.2.2.2.8 Conclusion. The quality of yogurt is influenced by a multitude of factors such as (a) quality and formulation of the milk base, (b) processing conditions and plant design, and (c) the role of starter cultures (i.e., non-EPS and EPS producer) during the incubation period and as cause of post-acidification during storage. Recently, de

Brabandere (1999) and de Brabandere and de Baerdmaeker (1999) evaluated the process conditions that influenced the pH development during the manufacture of yogurt:

- The solids content of the milk base did not affect the rate of pH development.
- Sterilization of the milk base and incubation at optimal temperature of the starter organisms enhanced the pH development.
- The use of EPS starter cultures yielded different pH profiles against time compared with non-EPS cultures.

8.2.2.3 Yogurt-Related Products. Different types of fermented milk products—for example, strained/concentrated, frozen, or dried yogurt—are primarily manufactured from natural stirred yogurt. The physical characteristics of these products, as compared with yogurt, are completely different, but since their popularity is increasing in certain European markets, a brief outline of their production may be relevant. The following are recommended for further reading regarding technological aspects (Robinson and Tamime, 1993; Bylund, 1995; Tamime and Robinson, 1999; Tamime et al., 2001).

8.2.2.3.1 Drinking/Fluid Yogurt. This is really a stirred yogurt with low viscosity (i.e., 11 g $100 g^{-1}$ total solids or lower), and it is usually flavored with fruit juice and/or synthetic flavoring and coloring compounds. Depending on the process employed (Bylund, 1995), three different types of product may be marketed:

- Short shelf life: 3 weeks under refrigeration.
- Medium shelf life: several weeks under refrigeration.
- Long shelf life: several months at room temperature.

These groupings depend on the handling of coagulum after fermentation and, in particular, on the extent of any heat treatment; Figure 8.3 shows the relationship between the process lines for normal yogurt and its fluid variant.

Although in most cases the milk is fermented with a yogurt starter culture, additional cultures can include the following species. *Lactococcus*, *Acetobacter*, *Bifidobacteria*, and probiotic lactobacilli.

8.2.2.3.2 Concentrated/Strained Yogurt and Closely Related Products. This is manufactured from natural, stirred yogurt or any type of

TABLE 8.8. Synonyms for Concentrated Fermented Milk Products in Different Countries

Traditional Names	Countries
Labneh, Labaneh, Lebneh, Labna	Eastern Mediterranean
Tan, Than	Armenia
Laban Zeer	Egypt, Sudan
Stragisto, Sakoulas, Tzatziki	Greece
Torba, Suzme	Turkey
Syuzma	Russia
Mastou, Mast	Iraq, Iran
Basa, Zimne, Kiselo, Mleko-Slano	Yugoslavia, Bulgaria
Ititu	Ethiopia
Greek-style	United Kingdom
Chakka, Shrikhand[a]	India
Ymer[a]	Denmark
Skyr[b]	Iceland

[a]Products are made with mesophilic lactic acid bacteria, while Shrikhand is made from Chakka sweetened with sugar and fortified with cream.
[b]The microflora of Skyr consists of yogurt starter culture, *L. helveticus* and lactose fermenting yeast.
Source: After Tamime and Robinson (1999). Reproduced by courtesy of Woodhead Publishers, Cambridge, England.

fermented milk, and some typical examples, such as Labneh (Middle East), Tan or Than (Armenia), Skyr (Iceland), and Shrikhand (India), are shown in Table 8.8. A further range of products, such as Ymer (Denmark), is derived from milk incubated with mesophilic lactic acid bacteria, while the microflora of Skyr consists of *S. thermophilus, L. delbrueckii* subsp. *bulgaricus*, and *L. helveticus* and with some lactose-fermenting yeasts (Tamime and Robinson, 1988). Traditionally, most of these products were concentrated using the cloth-bag method (Tamime and Crawford, 1984), but more recently the process has been mechanized using one of two available techniques:

- A nozzle separator is capable of concentrating warm, low-fat, natural yogurt to the desired level of solids non-fat, and cream is blended in at a later stage to yield a product of around 24 g $100\,g^{-1}$ total solids and 10 g $100\,g^{-1}$ fat (Tamime and Robinson, 1988, 1999).
- The membrane filtration of full-fat, natural yogurt is also feasible (Tamime et al., 1989a,b), and the process is carried out at 30–50°C and with yogurt of pH 4.6 or thereabouts. Preconcentration of the

milk prior to fermentation provides an alternative route (El-Samragy and Zall, 1988).

- Product formulation involves recombination of dairy ingredients (SMP, AMF, and stabilizer) to the exact chemical composition of strained yogurt (Tamime, 1993). The milk is handled and processed in a way similar to that of set yogurt.

8.2.2.3.3 Pasteurized Yogurt. This is heat-treated after fermentation, either by conventional means through a heat exchanger or by a "heat shock" process (Driessen, 1984). The objective of the procedures is to extend the shelf life of the yogurt by inactivating both organisms from the original starter culture as well as contaminants, such as yeasts and molds. According to Driessen (1984), heat treatment of yogurt in a carton at 58°C for 5 min is sufficient to inactivate both *S. thermophilus* and *L. delbrueckii* subsp *bulgaricus*; with a normal HTST process, a treatment of 75–80°C for 15 s could be anticipated as giving a shelf life of around 3 months (Bylund, 1995; Tamime and Robinson, 1999).

In some countries, these processes are not permitted by law, because yogurt must, by definition, contain "abundant and viable" organisms of starter origin. However, Waes (1987) has indicated that some strains of *L. delbrueckii* subsp *bulgaricus* can survive pasteurization temperatures at pH 4.6, so that at least this species could be present at levels in excess of 20–70 × 10^5 cfu ml^{-1}. Obviously, the use of such strains does give a "live" yogurt free from contaminants, but overacidification could still be a problem with respect to shelf life (see Figure 8.4F,G).

8.2.2.3.4 Frozen Yogurt. This product resembles ice cream in its physical state. Its characteristic is simply described as having the sharp and acidic taste of yogurt combined with the refreshing coolness of ice cream. Depending how the frozen yogurt is made, the product is classified into three main categories: soft, hard, or mousse. According to Tamime and Robinson (1999), the process consists of mixing cold, natural, stirred yogurt with a fruit/syrup base, stabilizers/emulsifiers, and sugar and then freezing the mix in a conventional ice-cream freezer. A suggested chemical composition (g 100 g^{-1}) for hard frozen yogurt consists of fat 2–6, milk SNF 5–14, sugar 8–16, and stabilizer/emulsifier 0.2–1.0. The anticipated percentage of overrun is around 70–80, while the pH varies between 4.5 and 6.0, depending on consumer preferences in different markets.

8.2.2.3.5 Dried Yogurt. Currently, low-fat dried yogurt is produced commercially, and the product is stable and utilized as a food ingredient. Two-stage drying is employed for the manufacture of dried yogurt. The first stage of drying takes place in a spray dryer with an integrated fluid-bed dryer, and the second stage involves an external fluid-bed dryer where the product is dried to 2 g $100\,g^{-1}$ moisture (Tamime, 1993). However, typical survival rates of *S. thermophilus* or *L. delbrueckii* subsp. *bulgaricus* in commercially made low-fat yogurt are low, and they range between 10×10^2 and 10×10^3 cfu g^{-1}.

8.3 YEAST–LACTIC FERMENTATIONS

Although yeast–lactic fermentations of various substrates have been recorded in many countries, it is mainly in eastern Europe that milk provides the raw material. These alcoholic milk beverages have a lactic/sour taste; and depending on the species present in the starter culture, the ethanol content could be as high as 2 g $100\,ml^{-1}$. The traditional products also have a foaming and effervescent characteristic as a result of CO_2 production. The microflora of the starter cultures for these products is not well-defined (see Chapter 7 of this volume; Zourari and Anifantakis, 1988; Mann, 1988, 1989; Thompson et al., 1990; Salof-Coste, 1996; Pintado et al., 1996; Rea et al., 1996; Tamime and Marshall, 1997; Kuo and Lin, 1999; Lin et al., 1999). Typical examples of such beverages are Kefir and related products, Koumiss and Acidophilus–yeast milk. These products originated in central Asia between the Caucasian mountains and Mongolia.

8.3.1 Kefir

The starter cultures is in the form of Kefir grains characterized by irregular, folded, and uneven surfaces. The grains may be white or yellowish in color and have an elastic consistency (Koroleva, 1991). The diameter of the kefir grain may range between 1 mm and 6 mm or more, depending on the extent of agitation during its growth in the milk; however, when the grains are recovered from milk and washed with water, they are of variable sizes ranging from 0.5 cm to 3.5 cm in diameter, and they resemble cauliflower florets in shape and color. The exact origin of the kefir grains is unknown; but according to legend the kefir grains were given to the people living in the Caucasian mountains, possibly by the Prophet Mohammed (Koroleva, 1991).

8.3.1.1 Microflora of Kefir Grains. As mentioned elsewhere, the microflora of the Kefir grains is complex and not always constant, and it consists of an undefined mixture of species of bacteria and yeasts (IDF, 1997). However, Tamime and Marshall (1997) and Takizawa et al. (1998) have reviewed in detail the organisms that have been identified in Kefir grains by many researchers in different countries. Figure 8.7 summarizes the different species of yeast, lactic acid bacteria (primarily *Lactobacillus* spp. plus *Lactococcus* spp., *Leuconostoc* spp. and *S. thermophilus*), acetic acid bacteria (*Acetobacter aceti* and *rasens*), molds (*Geotrichum candidum*), and other bacterial contaminants found in the grains. In some countries, *A. aceti* and/or *A. rasens* and *G. candidum* are regarded as contaminants, while in other countries they are considered desirable (Tamime and Marshall, 1997).

8.3.1.2 Commercially Developed Kefir Starters. This foamy and effervescent drink has always achieved great popularity in eastern Europe; but elsewhere, consumption is limited to a minority market (Kemp, 1984). Nevertheless, attempts are being made in Canada,

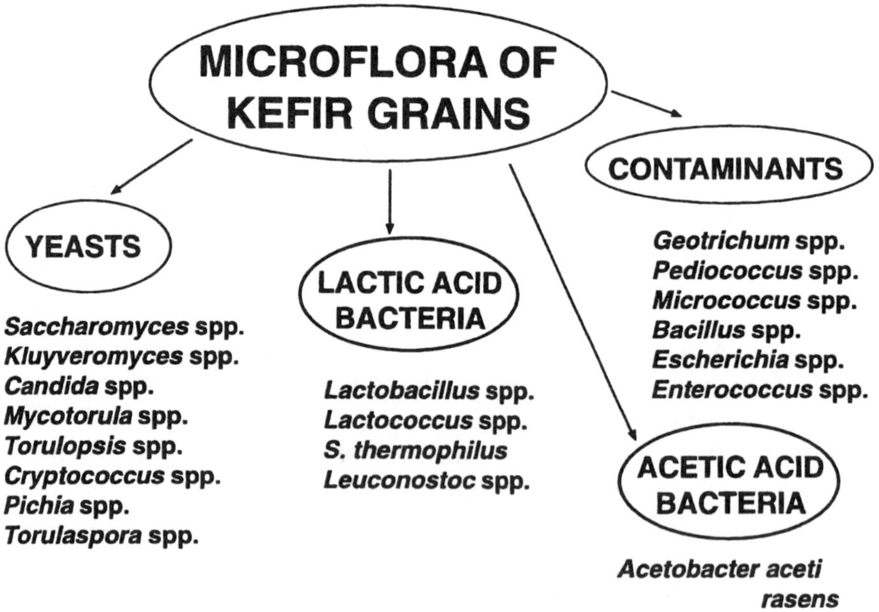

Figure 8.7. Microflora that have been identified in Kefir grains by many researchers. [Adapted from Tamime and Marshall (1997). Reproduced by courtesy of Dairy Industries International (Tamime et al., 1999d).]

Sweden, and Germany to standardize the product and perhaps widen its appeal (Mann, 1985; Duitschaever et al., 1988a,b), and it is notable that neither of the proposed processes is traditional. Thus, the German system employs a mixed culture (2 ml 100 ml^{-1} inoculation rate) of *Lactococcus* spp., *S. thermophilus*, and *Lactobacillus* spp. (*L. acidophilus*, *L. brevis*, *L. delbrueckii* subsp. *lactis*) with a ratio of lactococci: lactobacilli of 20:1; the yeast is *Candida kefyr*. The Swedish process is a three-stage program employing kefir grains, but spread over some 3 days. However, Vayssier (1978a,b) proposed making Kefir with proportions of

L. lactis subsp. *lactis*
subsp. *lactis* biovar *diacetylactis*
L. mesenteroides subsp. *mesenteroides* or
subsp. *dextranicum*
Lactobacillus paracasei subsp. *alactosus*
Lactobacillus brevis/*Lactobacillus cellobiosus*

where the count of each strain is 10^8 cfu g^{-1}, and that of *Saccharomyces florentinus* is 10^6 cfu g^{-1}; acetic acid bacteria was excluded. Koroleva (1991), based on years of experience in the former USSR, suggested different proportions of microorganisms in the Kefir bulk starter, and the counts are

Lactococcus spp. (10^8–10^9 cfu ml^{-1})
Leuconostoc spp. (10^7–10^8 cfu ml^{-1})
Thermophilic lactobacilli (10^5 cfu ml^{-1})
Acetic acid bacteria (10^5–10^6 cfu ml^{-1})
Yeasts (10^5–10^6 cfu ml^{-1})

However, the so-called "buttermilk plant," which is used domestically in Northern Ireland and had its origin in imported Kefir grains, has a microbial flora consisting of predominantly *Lactobacillus* spp., some lactococci, *S. cerevisiae* (lactose$^-$), and *Candida kefyr* (lactose$^+$) (Thompson et al., 1990).

Most kefir is produced by the conventional process using kefir grains as the basic inoculum, and they are composed, as indicated earlier, of a mixture of microorganisms held together in a highly organized pattern. Thus, the peripheral layers of the granules are dominated by various rod-shaped bacteria; but toward the center, yeasts become the major component of the microflora. What is not known, however, is how these units are built up, nor do we know the nature of the relationship between the constituent organisms (Bottazzi and Bianchi, 1980; Duitschaever et al., 1988b), and yet the granules proliferate freely in milk with no detectable changes in character (see Figure 8.8).

The precise, special relationship between the various organisms has been examined by Marshall et al. (1984), and it has been suggested that the grains are, in fact, highly convoluted sheets with yeasts (*Candida kefyr*) on one side and lactobacilli on the other. As mentioned elsewhere, the species composition of the lactic microflora is still in some

1 3

2 4

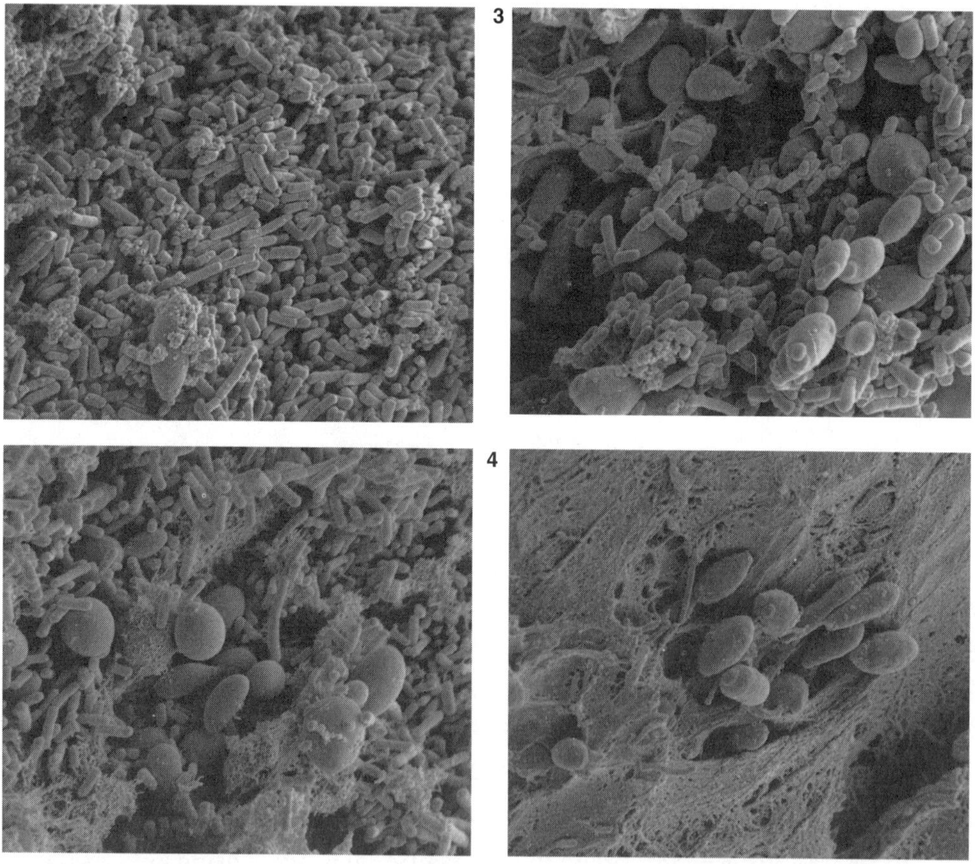

Figure 8.8. Microorganisms observed at various points within a Kefir granule (scanning electron microscope original magnification × 3040). (1) Rod-shaped bacteria forming peripheral layer; (2) and (3) intermediate zones showing increasing abundance of yeasts and decreasing numbers of bacteria; (4) yeasts embedded in matrix of microbial origin dominating the central region of the granule. For further details see Bottazzi and Bianchi (1980). (Reproduced by courtesy of Professor V. Bottazzi.)

doubt, and it is probable that it varies according to the source of the grains. However, it does appear that *Lactobacillus kefir* is frequently present, along with another species that may be responsible for forming the matrix of the grain. Thus, la Rivière et al. (1967) isolated a polysaccharide (kefiran) from kefir grains incorporating *Lactobacillus brevis*, and Marshall et al. (1984) suggest that this material may be composed of branched chains of glucose and galactose. If this view is correct, then it would be reasonable to propose that the *L. brevis* (or a similar species) is responsible for the basic structure of the grain, and that the other species become embedded in the growing mass of polysaccharide. It does not explain, however, the apparent spatial separation of the yeasts and the lactobacilli, and it is relevant that Olsson (1981) has suggested that the supporting matrix may, at least in some grains, be secreted by a constituent yeast. It may be, of course, that there are, in reality, a number of different types of "grain" and that any one of them can give rise to an acceptable end product. Recently, Toba et al. (1990a, 1991b) have isolated a capsular polysaccharide-producing *Lactobacillus kefiranofaciens* K_1 from Kefir, and the fermented milk produced from the isolated strain had a ropy consistency and was resistant to syneresis (see also Neve, 1992).

8.3.1.3 Production of Kefir Bulk Starters. Traditionally, the production of Kefir involves the recovery of the grains, and it also involves their re-use after washing. Although this method of production retains the "typical" characteristics of Kefir, in large-scale operations the method has the following limitations:

- Excessive washing of the grains will ultimately modify the balance between the microfloras and decrease their activity.
- The process is somewhat laborious, and the estimated production time of the starter and product is ~2 days.

Over the years, Koroleva (1991) developed a method for the production of starter cultures for Kefir in the former USSR, and Figure 8.9 illustrates schematically the different manufacturing stages; either starter I or II could be used. According to Koroleva (1991), both starters have been recommended for Kefir production, but starter II is only used in factories if equipment for separating and washing the Kefir grains from the product is not available. It is evident, however, that careful handling of the starter culture is important in maintaining the appropriate ratio between the various microfloras in the Kefir grains.

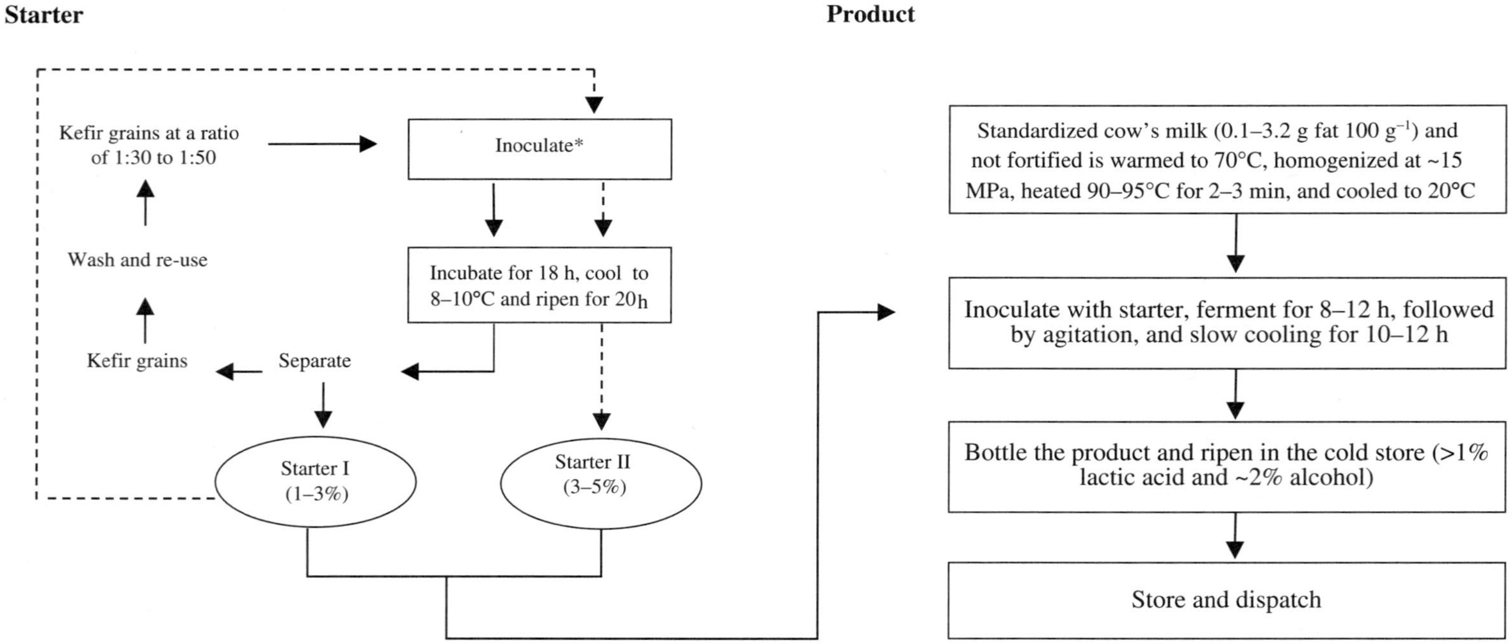

Figure 8.9. Schematic illustration for the production of Kefir starter cultures and the final product. *Skimmed milk heated to 95°C for 10–15 min and cooled to ~20°C. [Adapted from Koroleva (1991).]

These requirements have been detailed by Koroleva (1991) and have recently been reviewed by Tamime and Marshall (1997).

Over the past few decades, scientists in different laboratories have been developing Kefir starter cultures without using the grain (see the review by Tamime and Marshall, 1997). As a consequence, the product has lost its typical characteristics—that is, no CO_2 and with low or zero alcohol content. Although such products appear in some countries packaged in cartons, they do not easily fall within the category justifiably described as "traditional" Kefir. Nevertheless, some starter cultures companies—for example, Rhodia Food UK Ltd. (i.e., through their subsidiary company Biolacta-Texel in Poland)—have marketed freeze-dried Kefir starter cultures for DVI of bulk starter or the product. A typical illustration for the production of a bulk starter for Kefir is shown in Figure 8.10, and it is safe to conclude that this method of processing is used to make a "modern" Kefir.

8.3.1.4 Method of Manufacture of Kefir. The technology of commercially made Kefir is shown in Figure 8.11 (see also Özer and Özer, 2000). The raw material for Kefir is usually whole milk, which is severely heat-treated (95°C for 5 min) to denature the whey proteins. The hydrophilic properties of these denatured proteins improves the viscosity of the end product, as does the frequently employed process of homogenization. A portion of this process milk is also employed to prepare the inoculum; and because of the nature of the "starter culture," strict levels of hygiene are essential.

Thus, the initial stage of culture preparation involves inoculating the pretreated milk with the Kefir grains and then incubating the mixture at around 23°C. After some 20 h, the grains are sieved out of the milk and carefully washed in cold water prior to re-use. The remaining milk provides the bulk starter for the commercial-scale fermentation, and it is added to the process milk at the rate of 3.5 ml 100 ml^{-1}. The final incubation at 23°C will again last around 20 h, and, after cooling, the Kefir is often held for several hours to "ripen." This latter stage allows for maximum stability of the coagulum, and the final packaging stage must be designed in such a way that mechanical damage to the product is kept to a minimum. Thus, while Kefir is envisaged as a refreshing drink, it has a quite distinct viscosity that is regarded by devotees as an essential feature of a good-quality product.

The chemical changes that may take place during fermentation and storage of Kefir are shown in Table 8.9 (see also Guzel-Seydim et al., 2000). Nevertheless, it is safe to conclude that the quality of the product is greatly influenced by (a) the origin and microflora of the Kefir grains

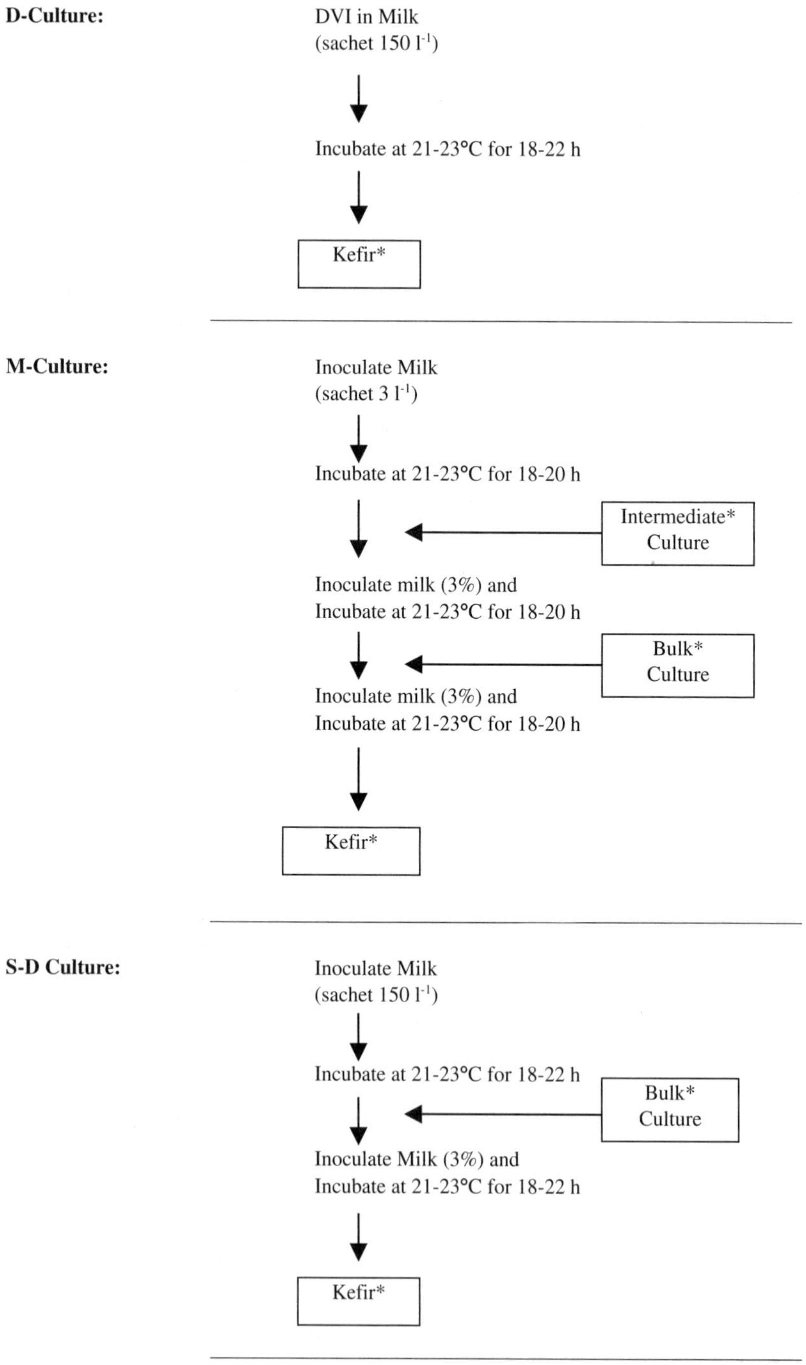

Figure 8.10. Flow diagram for the production of Kefir from Rhodia-Texel freeze-dried cultures. *No Kefir grains produced. (Reproduced by courtesy of Rhodia Food UK Ltd., Stockport.)

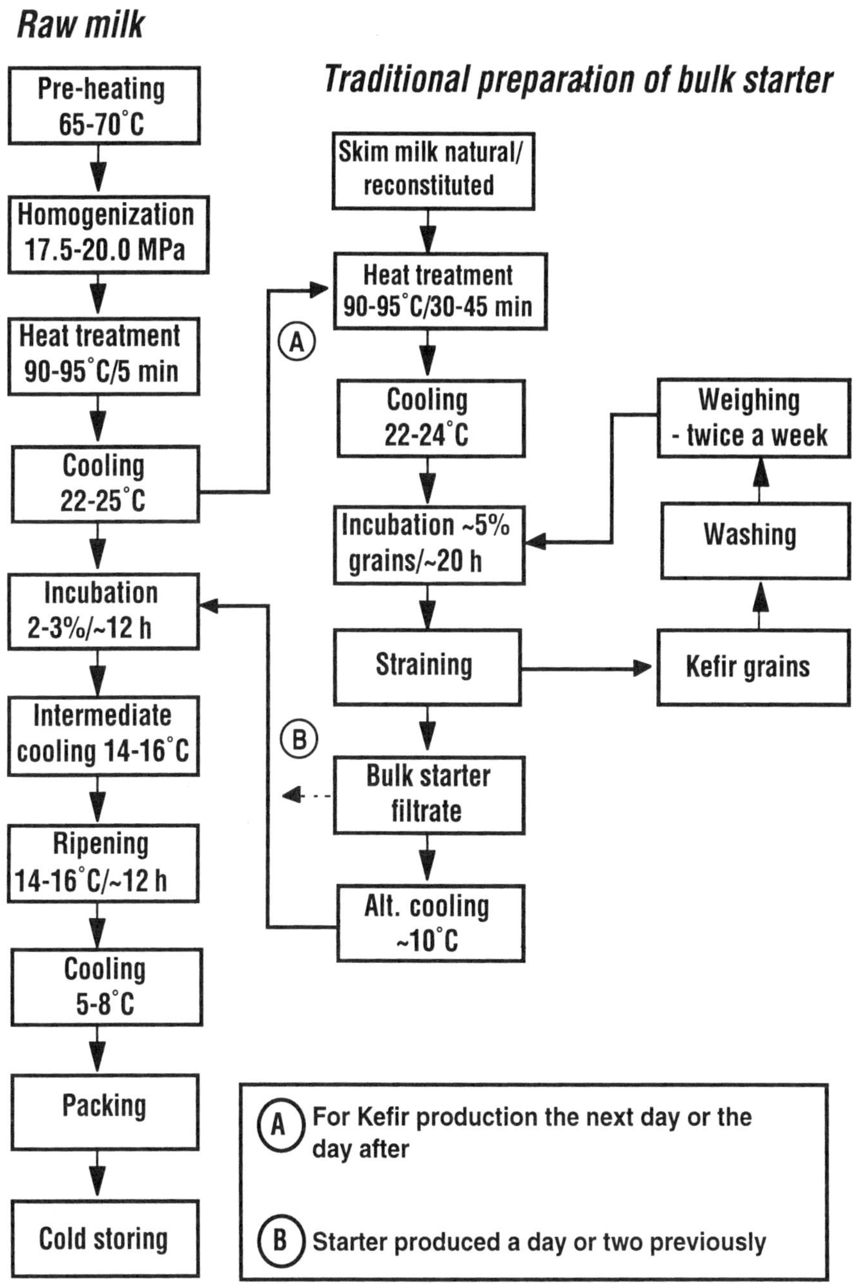

Figure 8.11. Processing stages for Kefir production. [Reproduced by courtesy of Tetra Pak (Processing Systems Division) A/B, Lund, Sweden.]

TABLE 8.9. Gross Chemical Composition (g 100 g^{-21}) of Cow's Milk and Kefir During Storage at ~5°C

		Kefir (d)[a]		
Component	Milk	1	2	3
Moisture	87.4	87.4	87.4	87.4
Fat	3.4	3.4	3.4	3.4
Lactose	4.8	3.9	3.4	3.00
Ash	0.64	0.64	0.64	0.64
Alcohol	—[b]	0.24	0.44	0.72
Lactic acid	0.14	0.73	1.08	1.15
CO_2	—	0.08	0.12	0.20
Casein	3.1	2.9	2.8	2.7
Globulin	0.17	0.14	0.13	0.11
Albumin	0.20	0.13	0.12	0.10
Peptides	ND[c]	0.195	0.386	0.495

[a]Days of storage.
[b]Not detected.
[c]Not determined.
Source: Data complied from Molska (1988).

in the starter culture and (b) the quality and type of raw milk used. Recently, Wszolek et al. (2001) have evaluated the quality of Kefir produced from cow, goat, and sheep milk (i.e., the fat content was standardized to ~3 g 100 g^{-1}), and they used Kefir grains and two DVI Kefir starter cultures. They reported the following observations:

- Lactic acid bacteria and yeasts were abundant in fresh Kefir, but some strains decreased by 1 $\log_{10}$ cycle after 12 days' storage.
- The microbial metabolites (e.g., free fatty acids, diacetyl, and ethanol) were associated with species of milk used, and the former two components were also affected by the storage period.
- The firmness of Kefir was influenced by the type of milk used (sheep > cow > goat); the protein content averaged 6.4, 3.3, and 2.9 g 100 g^{-1}, respectively.
- All the sensory attributes (e.g., flavor, odor, texture/mouth-feel, and acceptability) were influenced by the species of milk, while the storage effect influenced the mouth-feel characters ("serum separation," "chalky," "mouth-coating," and "slimy").

Incidentally, the microflora of the Kefir grains and two DVI starter cultures consisted of the following species: *Lactobacillus*, *Lactococcus*,

S. thermophilus, and yeast, and only one DVI culture contained *Leuconostoc* spp. (Wszolek et al., 2001; see also Tamime et al., 1999d).

8.3.2 Koumiss, Kumiss, Kumys, or Coomys

Koumiss is a traditional drink of Central Asia, and when it is manufactured from mare's milk the chemical composition (g $100\,g^{-1}$) is moisture ~90, protein 2.1 (casein 1.2 and whey proteins 0.9), lactose 6.4, fat 1.8, and ash (Doreau and Boulot, 1989; Doreau et al., 1990; Pagliarini et al., 1993; Lozovich, 1995; Ochirkhuyag et al., 2000). Koumiss is also known as Airag, Arrag, Chige, or Chigo in Mongolia and China. The name of this fermented milk is probably derived from a tribe, the Kumanes, who inhibited the area along the Kumane or Kuma River in the Asiatic Steppes, but it may be of Tartar origin (Tamime and Marshall, 1997).

The overall characteristics of the product prepared from mare's milk are as follows:

- It is a beverage where the milk does not coagulate; however, the thermal sensitivity of mare's milk proteins has been studied by Bonomi et al. (1994).
- It is milky gray in color, light, and fizzy, and it has a sharp alcoholic and acidic taste.
- The main metabolites after fermentation are lactic acid, ethanol, and up to 0.9 ml $100\,ml^{-1}$ carbon dioxide (see Table 8.10).

8.3.2.1 *Microflora of Koumiss.* The microflora of Koumiss is not well-defined, but it consists mainly of lactobacilli (*L. delbrueckii* subsp.

TABLE 8.10. Classification of Koumiss[a] Based on the Extent of Fermentation

Flavor	Acidity (%)	Alcohol (%)
Weak	0.54–0.72	0.7–1.0 (1.0)[b]
Medium	0.73–0.90	1.1–1.8 (1.5)
Strong	0.91–1.08	1.8–2.5 (3.0)

[a]A typical Koumiss may have viable cell counts of 5.0×10^7 cfu ml^{-1} (bacteria) and 1.43×10^7 cfu ml^{-1} (yeast).
[b]Data in parentheses are from Lozovich (1995).
Source: After Berlin (1962).

bulgaricus and *L. acidophilus*), lactose-fermenting yeasts (*Saccharomyces* spp. and *Torula koumiss*), non-lactose-fermenting yeast (*Saccharomyces cartilaginosus*), and a non-carbohydrate-fermenting yeast (*Mycoderma* spp.) (Koroleva, 1991; Oberman and Libudzisz, 1998). However, *Lactococcus* spp. have been found in Mongolian Koumiss, but their presence in the starter culture in some countries may not be desirable because the fast acid development inhibits the growth of yeasts (Koroleva, 1991).

Developments in starter cultures for Koumiss, which include different blends of lactic acid bacteria and yeasts, have been reported by many researchers (Koroleva, 1991; Aguirre and Collins, 1993; Oberman and Libudzisz, 1998), and it is evident that these developments have been carried out to meet the demands of industrial-scale production of Koumiss from cow's milk. In another study, Montanari et al. (1996) and Montanari and Grazia (1997) screened 94 samples of traditional Koumiss in Kazakhstan, and they identified *Saccharomyces unisporus* (galactose-fermenting yeast) as the principal microorganism responsible for the alcoholic fermentation of the product. Incidentally, the yeast is a non-lactose-fermenting organism, and it gives rise to a slower and less clean alcoholic fermentation than *S. cerevisiae* because it produces a wider range of metabolites, such as glycerol, succinic acid, and acetic acid (Montanari and Grazia, 1997).

Recently, Ishii et al. (1997) have identified the lactic acid bacteria and yeasts from Chigo (i.e., Koumiss made from mare's milk in inner Mongolia and China) which were obtained from local markets. A total of 43 strains of lactic acid bacteria were identified in the samples (e.g., *Lactobacillus rhamnosus*, *Lactobacillus paracasei* subsp. *paracasei* and subsp. *tolerans*, and *Lactobacillus curvatus*), and the overall counts ranged between 1.5×10^7 and 1.7×10^7 cfu ml^{-1}. In addition, 20 strains of yeasts (i.e., lactose-fermenting) were identified, and these included *Kluyveromyces marxianus* subsp. *lactis* and *Candida kefyr*; the counts ranged between 3.9×10^6 and 8.0×10^6 cfu ml^{-1}. The alcohol concentrations in the products were 0.5–2.17 g 100 ml^{-1}.

8.3.2.2 Production Systems. The traditional method of production takes place in smoked horse's hides known as tursuks or burduks, which contain the microflora from the previous season (Tamime and Marshall, 1997). These containers are filled with unheated mare's milk, and as the Koumiss is consumed, more milk is added to provide an ongoing fermentation.

In the 1960s, the commercial production of Koumiss developed using *L. delbrueckii* subsp. *bulgaricus* and *Torula* spp. to produce a bulk starter culture in skimmed cow's milk. Then, the starter culture was used to inoculate mare's milk at a rate up to 30 ml 100 ml^{-1}, followed by bottling, incubation at 18–20°C for 2 h, and finally storage at 5°C (Berlin, 1962).

A typical commercial method for the production of Koumiss is based on skimmed cow's milk with added sucrose (2.5 g 100 g^{-1}). The milk base is then heated to 90°C for 2–3 min, cooled to 28°C, inoculated with starter culture (~10 ml 100 ml^{-1}), stirred for 15–20 min, and incubated at 26°C for 5–6 h or until the acidity reaches ~0.9 g 100 g^{-1} lactic acid. Tamime and Marshall (1997) have reviewed other production systems using different milk bases, including the following:

- Recombined milk using whole, skimmed, and whey powders.
- Blends of five parts cow's milk with eight parts of UF rennet whey (i.e., twofold concentration of the protein).
- Cow's milk blended with clarified whey at a ratio 1:1.

8.3.3 Miscellaneous Products

Products such as Acidophilus-yeast milk, Acidophiline, and/or Acidophilin have been developed in the former USSR for the treatment of certain intestinal disorders. Little is known regarding the technology of these fermented milk beverages; for further details refer to Eller (1971), Koroleva (1991), and Tamime and Marshall (1997).

Although the Icelandic product Skyr may be considered within the yeast–lactic group of fermentation products because the starter culture contains yeast, the product has been discussed in Section 8.2.2.3.2 because the technology of production is similar to strained yoghurt or Quarg.

8.4 MOLD–LACTIC FERMENTATIONS

Apart from the distinctive, mold-ripened cheeses, there are few dairy products that deliberately include a mold in the microflora. An exception to this rule, however, is the Finnish product, Viili, for which the starter culture includes *L. lactis* subsp. *lactis* biovar *diacytelactis* and *L. mesenteroides* subsp. *cremoris*, together with the mold *G. candidum*. This product has a distinctive taste, aroma, and appearance. According to Laukkanen et al. (1988) the different types of Viili are as follows:

- Low-fat (~2.5 g $100\,g^{-1}$) Viili, known as Kevytviili
- Whole-fat (3.9 g $100\,g^{-1}$) Viili
- Cream Viili, which contains 12 g $100\,g^{-1}$ fat

The original product is natural, and the fruit-flavored and sweetened fermented milk is known as Marjaviili (Tamime and Robinson, 1988; Tamime and Marshall, 1997).

The fat standardized milk base is heated to 83°C for 20–25 min (or possibly 90–95°C for a few minutes as in yogurt-making) without homogenization, cooled to 20°C, and inoculated with 3–4 ml $100\,ml^{-1}$ starter culture consisting of lactococci, leuconostocs, and the mould (see above). The processed milk is then packaged into retail containers, incubated for ~24 h until the acidity reaches 0.9 ml $100\,ml^{-1}$ lactic acid and finally cooled. Because the milk base was not homogenized, the fat and mold spores rise to the surface during the incubation period. As a consequence, the mold grows on the surface of the milk to give rise to the velvet-like appearance of the product. Thus, the natural Viili consists of two layers (i.e., the coagulated milk and a cream layer plus the mold growth); the Marjaviili consists of three layers in that the added fruit is at the bottom of the cup.

According to Marshall (1986) and Tamime and Marshall (1997) the overall characteristics of Viili are as follows:

- The product is mildly sour, aromatic in taste, and stretchy, and it can be cut easily with a spoon.
- The flavor of Viili is similar to other types of buttermilk product, but with a slightly musty aroma that is attributed to *G. candidum*.
- The mold layer on the surface of the product may be advantageous in preventing the growth of spoilage microorganisms.

The *Lactococcus* strains used in Viili production produce EPS, and the ropy material forms a network attaching the bacterial cells to the protein matrix (see Nakajima and Toyoda, 1995). However, the slime filaments in Viili are more compact or thicker when compared with the microstructure of yogurt (see Figure 8.6), which could be due to differences in the chemical composition of the EPS (Toba et al., 1990b, 1991a). The yogurt starter cultures tend to produce EPS with phosphopolysaccharide components, while the *Lactococcus* spp. of Viili produce EPS made up of polysaccharide and protein (Toba et al., 1990b; see also Nakajima et al., 1990; ZhenNai et al., 1999; Oba et al.,

1999a,b). However, plasmid profiles of lactococcus strains from Viili and other Nordic fermented milks have been reported by Forsén et al. (1985, 1989), Kontusaari and Forsén (1988), and Neve et al. (1988).

Although this brief review has covered the major types of fermented milk, numerous variations can be found around the world. Many of these products are manufactured on a basis of limited microbiological knowledge; and it is fortuitous that although no two products can ever be handled/monitored in exactly the same way, the general principles remain immutable. The intelligent application of basic techniques can, therefore, go a long way toward avoiding serious problems in relation to starter production and/or microbial contamination and spoilage of the end products.

REFERENCES

Accolas, J.-P., Hemme, D., Desmazeaud, M. J., Vassal, L., Bouillane, C., and Veaux, M. (1980) *Lait*, **60**, 487–524.

Aguirre, M., and Collins, M. D. (1993) *J. Appl. Bacteriol.*, **75**, 95–107.

Ahmed, A. A.-H., Moustafa, M. K., and El-Bassiony, T. A. (1986) *J. Food Prot.*, **49**, 983–985.

Ahmed, I. A., Ahmed, A. W. K., and Robinson, R. K. (1997) *J. Sci. Food Agric.*, **74**, 64–68.

Alkanahl, H. A., and Gasim, Z. (1993) *J. Food Prot.*, **55**, 84–87.

Al-Mashhadi, A. S., Saadi, S. R., Ismail, A., and Salji, J. P. (1987) *Cult. Dairy Prod. J.*, **22**(1), 24–33.

APHA (1992) *Standard Methods for the Examination of Dairy Products*, 16th ed., R. T. Marshall, ed., American Public Health Association, Washington, DC.

Arnott, D. R., Duitschaever, C. L., and Bullock, D. H. (1974) *J. Milk Food Technol.*, **37**, 11–13.

Attaie, R., Whalen, P. J., Shahani, K. M., and (VIC) Amer, M. A. (1987) *J. Food Prot.*, **50**, 224–228.

Aytaç, S. A., and Özbas, Z. Y. (1994a) *Milchwissenschaft*, **49**, 322–325.

Aytaç, S. A., and Özbas, Z. Y. (1994b) *Aust. J. Dairy Technol.*, **49**, 90–92.

Barnes, G., Beaton, S., and Goldenberg, N. (1979) *R. Soc. Health J.*, **99**, 107–113.

Berlin, P. J. (1962) Kumiss. In *Annual Bulletin*, Part IV, Section A, International Dairy Federation, Brussels, pp. 4–16.

Bonomi, F., Iametti, S., Pagliarini, E., and Solaroli, G. (1994) *J. Dairy Res.*, **61**, 419–422.

Bottazzi, V., and Bianchi, F. (1980) *J. Appl. Bacteriol.*, **48**, 265–268.

Bottazzi, V., and Bianchi, F. (1986) *Sci. Tech. Lattiero-Casearia*, **37**, 297–315.

Bottazzi, V., and Vescovo, M. (1969) *Netherlands Milk Dairy J.*, **23**, 71–78.

Bottazzi, V., Battistotti, B., and Vescovo, M. (1971) *Milchwissenschaft*, **26**, 214–219.

Bridge, P. D., and Sneath, P. H. A. (1983) *J. Gen. Microbiol.*, **129**, 565–597.

Brisdon, E. Y. (1998) L-S differential medium—CM 495. In *The Oxoid Manual*, 8th ed., Oxoid Ltd., Basingstoke.

Bylund, G. (1995) *Dairy Processing Handbook*, Tetra Pak (Processing System Division) A/B, Lund.

Cerning, J. (1990) *FEMS Microbiol. Rev.*, **87**, 113–130.

Cerning, J. (1995) *Lait*, **75**, 463–472.

Cerning, J., and Marshall, V. M. E. (1999) *Recent Res. Dev. Microbiol.*, **3**, 195–209.

Cerning, J., Bouillanne, C., Desmazeaud, M. J., and Landon, M. (1986) *Biotechnol. Lett.*, **8**, 625–628.

Cerning, J., Bouillanne, C., Desmazeaud, M. J., and Landon, M. (1988) *Biotechnol. Lett.*, **10**, 255–260.

Cerning, J., Bouillanne, C., Landon, M., and Desmazeaud, M. J. (1990) *Sci. Ailments*, **10**, 443–451.

Charteris, W. P., Kelly, P. M., Morelli, L., and Collins, J. K. (1998) *Int. J. Dairy Technol.*, **51**, 123–136.

Cheryan, M. (1998) *Ultrafiltration and Microfiltration Handbook*, Technomic Publishing, Basel.

Choi, H. K., Schaack, M. M., and Marth, E. H. (1988) *Milchwissenschaft*, **42**, 790–792.

Cooper, P. J., and Broomfield, A. R. (1974) *XIX Int. Dairy Congr.*, **IE**, 447–448.

Cuk, Z., Annan-Prah, A., Janc, M., and Zajc-Satler, J. (1987) *J. Appl. Bacteriol.*, **63**, 201–205.

Davis, J. G., Ashton, T. R., and McCaskill, M. (1971) *Dairy Ind.*, **36**, 569–573.

de Brabandere, A. G. (1999) *Measuring Rheological Changes in Yoghurt Fermentation Processes as a Foundation for In-Line Process Monitor*, PhD thesis, Katholieke Universiteit Leuven, Belgium.

de Brabandere, A. G., and De Baerdemaeker, J. G. (1999) *J. Food Eng.*, **41**, 221–227.

Dirar, H. (1993) *The Indigenous Fermented Foods of the Sudan*, CAB International, Wallingford.

Doco, T., Wieruszeski, J.-M., and Fournet, B. (1990) *Carbohydr. Res.*, **198**, 313–321.

Doreau, M.-Y., and Boulot, S. (1989) *Livestock Prod. Sci.*, **22**, 213–235.

Doreau, M.-Y., Boulot, S., Barlet, J.-P., and Patureau-Mirand, P. (1990) *J. Dairy Res.*, **57**, 449–454.

Driessen, F. M. (1984) Modern trends in the manufacture of yoghurt. In *Fermented Milks*, Bulletin No. 179, International Dairy Federation, Brussels, pp. 107–115.

Driessen, F. M., Kingma, F., and Stadhouders, J. (1982) *Netherlands Milk Dairy J.*, **36**, 135–144.

Duitschaever, C. L., Kemp, N., and Emmons, D. (1988a) *Milchwissenschaft*, **43**, 343–345.

Duitschaever, C. L., Kemp, N., and Smith, A. K. (1988b) *Milchwissenschaft*, **43**, 479–481.

Eller, H. (1971) *The Technology of Sour Milk Products*, Ministry of Meat and Milk Industry of Estonia, Tallin, pp. 3–39.

El-Samragy, Y. A., and Zall, R. R. (1988) *Dairy Ind. Int.*, **53**(3), 27–28.

Entis, P., and Lerner, I. (1996) *J. Food Prot.*, **59**, 416–419.

Farah, Z., Streiff, T., and Bachmann, M. R. (1990) *J. Dairy Res.*, **57**, 281–283.

Farrow, J. A. E., and Collins, M. D. (1984) *J. Gen. Microbiol.*, **130**, 357–362.

Feresu, S., and Nyati, H. (1990) *J. Appl. Bacteriol.*, **69**, 814–821.

Fiszman, S. M., Lluch, M. A., and Salvador, A. (1999) *Int. Dairy J.*, **9**, 895–901.

Fleet, G. H. (1990) *J. Appl. Bacteriol.*, **68**, 199–211.

Forsén, R., Niskasaari, K., and Niemitola, S. (1985) *FEMS Microbiol. Lett.*, **26**, 249–253.

Forsén, R., Niskasaari, K., Tasanen, L., and Numiaho-Lassila, E.-L. (1989) *Netherlands Milk Dairy J.*, **43**, 383–393.

Foschino, R., Garzaroli, C., and Ottogalli, G. (1993) *Lait*, **73**, 395–400.

Galesloot, T. E., Hassing, F., and Veringa, H. A. (1968) *Netherlands Milk Dairy J.*, **22**, 50–63.

Garcia, A. M., and Fernandez, G. S. (1984) *J. Food Prot.*, **47**, 629–636.

Garcia-Garibay, M., and Marshall, V. M. E. (1991) *J. Appl. Bacteriol.*, **70**, 325–328.

Ghoddusi, H. B., and Robinson, R. K. (1996) *J. Dairy Res.*, **63**, 151–158.

Gilbert, R. J., and Wieneke, A. A. (1973) *Staphylococcus* food poisoning with special reference to detection of enterotoxin in food. In *The Microbiological Safety of Food*, B. C. Hobbs and J. H. B. Christian, eds., Academic Press, London, pp. 273–285.

Gohil, V. S., Ahmed, M. A., Davies, R., and Robinson, R. K. (1996) *Food Microbiol.*, **13**, 159–164.

Grandison, A. S., and Glover, F. A. (1994) Membrane processing of milk. In *Modern Dairy Technology*, Vol. 1, 2nd ed., R. K. Robinson, ed., Chapman & Hall, London, pp. 273–311.

Grobben, G. J., Boels, I. C., Sikkema, J., Smith, M. R., and de Bont, J. A. M. (2000) *J. Dairy Res.*, **67**, 131–135.

Gruter, M., Leeflang, B. R., Kuiper, J., Kamerling, J. P., and Vliegenthart, J. F. G. (1993) *Carbohydr. Res.*, **239**, 209–226.

Guzel-Seydim, Z., Seydim, A. C., and Greene, A. K. (2000) *J. Dairy Sci.*, **83**, 275–277.

Gyosheva, B. H. (1985) *Die Nahrung*, **29**, 185–190.

Hardie, J. M., and Wiley, R. A. (1995) The genus *Streptococcus*. In *The Genera of Lactic Acid Bacteria*, Vol. 2, B. J. B. Wood and W. H. Holzapfel, eds., Blackie Academic & Professional, London, pp. 55–124.

Harrigan, W. F. (1998) *Laboratory Methods in Food and Dairy Microbiology*, 3rd ed., Academic Press, London.

Hassan, A. N., Frank, J. F., Schmidt, K. A., and Shalab, S. I. (1996) *J. Dairy Sci.*, **79**, 2091–2097.

Hassanin, N. I. (1994) *J. Sci. Food Agric.*, **65**, 31–34.

Hobbs, B. C. (1972) *J. Soc. Dairy Technol.*, **25**, 47–50.

Ibrahim, M. K. E., El-Batawy, M. A., and Girgis, E. S. (1987) *Egypt. J. Dairy Sci.*, **17**, 125–136.

IDF (1969) *Compositional Standard for Fermented Milks*, Standard No. 47, International Dairy Federation, Brussels.

IDF (1988) *Fermented Milks—Science and Technology*, Bulletin No. 227, International Dairy Federation, Brussels.

IDF (1990a) *Enumeration of Yeasts and Moulds*, Standard No. 94B, International Dairy Federation, Brussels.

IDF (1990b) *Detection of* Listeria monocytogenes, Standard No. 143, International Dairy Federation, Brussels.

IDF (1990c) *Enumeration of* Staphylococcus aureus, Standard No. 145, International Dairy Federation, Brussels.

IDF (1991a) *Enumeration of Microorganisms*, Standard No. 100B, International Dairy Federation, Brussels.

IDF (1991b) *Enumeration of Contaminating Microorganisms*, Standard No. 153, International Dairy Federation, Brussels.

IDF (1992a) *General Standard of Identity for Fermented Milks*, Standard No. 163, International Dairy Federation, Brussels.

IDF (1992b) *General Standard of Identity for Milk Products obtained from Fermented Milks Heat-Treated after Fermentation*, Standard No. 164, International Dairy Federation, Brussels.

IDF (1997) *Lactic Acid Starters—Standard of Identity*, Standard No. 149A, International Dairy Federation, Brussels.

IDF (1998) *Enumeration of Coliforms*, Standard No. 73B, International Dairy Federation, Brussels.

Ishii, S., Kikuchi, M., and Takao, S. (1997) *Anim. Sci. Technol.*, **68**, 325–329.

Jordano, R., Jodral, M., Martinez, P., Salmeron, J., and Pozo, R. (1989) *J. Food Prot.*, **52**, 823–824.

Jordano, R., and Salmeron, J. (1990) *Microbiol., Aliments, Nutr.*, **8**, 81–83.

Jordano, R., Medina, L. M., and Salmeron, J. (1991) *J. Food Prot.*, **54**, 131–132.

Jordano, R., Cristina, E., Serrano, M. T., and Salmeron, J. (1992) *J. Food Prot.*, **55**, 999–1000.

Kalab, M. (1993) *Food Struct.*, **12**, 95–114.

Kalab, M., Allen-Wojtas, P., and Phipps-Todd, B. E. (1983) *Food Microstruct.*, **3**, 51–66.

Kandler, O., and Weiss, N. (1986) Regular, non-sporing Gram-positive rods. In *Bergey's Manual of Systematic Bacteriology*, Vol. 2, P. H. A. Sneath, N. A. Mair, M. E. Sharpe, and J. G. Holt, eds., Williams and Wilkins, Baltimore, pp. 1208–1234.

Keceli, T., and Robinson, R. K. (1997) *Dairy Ind. Int.*, **62**(4), 29–33.

Kemp, N. (1984) *Cult. Dairy Prod. J.*, **19**(3), 29–30.

Kerr, K. G., Rotowan, N. A., and Hawkey, P. M. (1992) *J. Nutr. Med.*, **3**, 27–29.

Kilpper-Bälz, R., Fischer, G., and Schleifer, K. (1982) *Curr. Microbiol.*, **7**, 245–250.

Kimonye, J. M., and Robinson, R. K. (1991) *Dairy Ind. Int.*, **56**(2), 34–35.

Kon, S. K. (1972) *Milk and Milk Products in Human Nutrition*, Nutritional Studies No. 27, Food and Agriculture Organization of the United Nations, Rome, pp. 44–45.

Kontusaari, S., and Forsén, R. (1988) *J. Dairy Sci.*, **71**, 3197–3202.

Koroleva, N. S. (1991) Products prepared with lactic acid bacteria and yeasts, in *Therapeutic Properties of Fermented Milks*, R. K. Robinson, ed., Elsevier Applied Science, London, pp. 159–179.

Kuo, C.-Y., and Lin, C.-W. (1999) *Aust. J. Dairy Technol.*, **54**, 19–23.

Kurmann, J. A., Rasic, J. Lj., and Kroger, M. (1992) *Encyclopedia of Fermented Fresh Milk Products*, Van Nostrand Reinhold, New York.

la Rivière, J. W. M., Kooiman, P., and Schmidt, K. (1967) *Arch. Microbiol.*, **59**, 269–278.

Laukkanen, M., Antila, P., Antila, V., and Salminen, K. (1988) *Meijeritiet. Aikak.*, **XLVI**, 7–24.

Lee, S. Y., Vedamuthu, E. R., Washam, C. J., and Reinbold, G. W. (1974) *J. Milk Food Technol.*, **37**, 272–276.

Lin, C.-W., Chen, H.-L., and Liu, J.-P. (1999) *Aust. J. Dairy Technol.*, **54**, 14–18.

Lozovich, S. (1995) *Cult. Dairy Prod. J.*, **30**(1), 18–21.

Macura, D., and Townsley, P. M. (1984) *J. Dairy Sci.*, **67**, 735–744.

Mann, E. J. (1985) *Dairy Ind. Int.*, **50**(12), 11–12.

Mann, E. J. (1988) *Dairy Ind. Int.*, **53**(9), 9–10.

Mann, E. J. (1989) *Dairy Ind. Int.*, **54**(9) 9–10.

Marshall, V. M. E. (1986) The microflora and production of fermented milks. In *Progress in Industrial Microbiology—Microorganisms in the Production of Food*, Vol. 23, M. R. Adams, ed., Elsevier Science, Amsterdam, pp. 1–44.

Marshall, V. M. E. (1987) *FEMS Microbiol. Rev.*, **46**, 327–336.

Marshall, V. M. E., and Mabbitt, L. A. (1980) *J. Soc. Dairy Technol.*, **33**, 129–130.

Marshall, V. M. E., and Tamime A. Y. (1997) Physiology and biochemistry of fermented milks. In *Microbiology and Biochemistry of Cheese and Fermented Milk*, 2nd ed., B. A. Law, ed., Blackie Academic & Professional, London, pp. 153–192.

Marshall, V. M. E., Cole, V. M., and Brooker, B. E. (1984) *J. Appl. Bacteriol.*, **57**, 491–497.

Masud, T., Ali, A. M., and Shah, M. A. (1993) *Aust. J. Dairy Technol.*, **48**, 30–32.

Matalon, M. E., and Sandine, W. E. (1986) *J. Dairy Sci.*, **69**, 2569–2576.

McCann, G., Moran, L., and Rowe, M. (1991) *Milk Ind.*, **93**(2), 15–16.

Metchnikoff, E. (1910) *The Prolongation of Life: Optimistic Studies*, revised edition of 1907, translated by C. Mitchell, Heinemann, London.

Millard, G. E., McKellar, R. C., and Holley, R. A. (1990) *J. Food Prot.*, **53**, 64–66.

Molska, I. (1988) *Zarys Mikrobiologii Mleczarskiej*, PWRIL, Warszawa, pp. 253–262.

Montanari, G., and Grazia, L. (1997) *Food Technol. Biotechnol.*, **35**, 305–308.

Montanari, G., Zambonelli, C., Grazia, L., Kamesheva, G. K., and Shigaeva, M. K. (1996) *J. Dairy Res.*, **63**, 327–331.

Morgan, D., Newman, C. P., Hutchison, D. N., Walker, A. M., Rowe, B., and Majid, F. (1993) *Epidemiol. Infect.*, **111**, 181–187.

Nakajima, H., and Toyoda, S. (1995) *Dairy Sci. Abstr.*, **57**, 301.

Nakajima, H., Toyoda, S., Toba, T., Itoh, T., Mukai, T., Kitazawa, H., and Adachi, S. (1990) *J. Dairy Sci.*, **73**, 1472–1477.

Neve, H. (1992) *Milchwissenschaft*, **47**, 275–278.

Neve, H., Geis, A., and Teuber, M. (1988) *Biochimie*, **70**, 437–442.

Oba, T., Doesburg, K. K., Iwasaki, T., and Sikkema, J. (1999a) *Arch. Microbiol.*, **171**, 343–349.

Oba, T., Higashimura, M., Iwasaki, T., Masters, A. M., Steeneken, P. A. M., Robjin, G. W., and Sikkema, J. (1999b) *Carbohydr. Polym.*, **39**, 275–281.

Oberman, H., and Libudzisz, Z. (1998) Fermented milks. In *Microbiology of Fermented Foods*, Vol. 1, 2nd ed., B. J. B. Wood, ed., Blackie Academic & Professional, London, pp. 308–350.

Ochirkhuyag, B., Chobert, J. M., Dalgalarrondo, M., and Haertlé, T. (2000) *Lait*, **80**, 223–235.

Olsson, G. (1981) *Livsmedelsteknik*, **23**, 428–429.

Onggo, I., and Fleet, G. H. (1993) *Aust. J. Dairy Technol.*, **48**, 89–92.

Orla-Jensen, S. (1931) *Dairy Bacteriology*, 2nd ed., translated by P. S. Arup, J. & A. Churchill, London.

Ott, A., Fay, L. B., and Chaintreau, A. (1997) *J. Agric. Food Chem.*, **45**, 850–858.

Ott, A., Germond, J.-E., Baumgartner, M., and Chaintreau, A. (1999) *J. Agric. Food Chem.*, **47**, 2379–2385.

Ott, A., Germond, J.-E., and Chaintreau, A. (2000a) *J. Agric. Food Chem.*, **48**, 724–731.

Ott, A., Germond, J.-E., and Chaintreau, A. (2000b) *J. Agric. Food Chem.*, **48**, 1512–1517.

Ott, A., Hugi, A., Baumgartner, M., and Chaintreau, A. (2000c) *J. Agric. Food Chem.*, **48**, 441–450.

Özer, D., and Özer, B. H. (2000) Products of Eastern Europe and Asia. In *Encyclopedia of Food Microbiology*, Vol. 2, R. K. Robinson, C. A. Batt, and P. D. Patel, eds., Academic Press, London, pp. 798–805.

Pagliarini, E., Solaroli, G., and Peri, C. (1993) *Ital. J. Food Sci.*, **5**, 323–332.

Pette, J. W., and Lolkema, H. (1950a) *Netherlands Milk Dairy J.*, **4**, 197–208.

Pette, J. W., and Lolkema, H. (1950b) *Netherlands Milk Dairy J.*, **4**, 209–224.

Pettipher, G. L. (1993) Modern laboratory practice—2: Microbiological analysis. In *Modern Dairy Technology*, Vol. 2, 2nd ed., R. K. Robinson, ed., Elsevier Applied Science, London, pp. 417–454.

Piersma, H., and Steenbergen, A. E. (1973) *Off. Orgaan FNZ*, **65**, 94–97.

Pintado, M. E., Lopes da Silva, J. A., Fernandes, P. B., Xavier, Malcata, F., and Hogg, T. A. (1996) *Int. J. Food Sci. Technol.*, **31**, 15–26.

Pirovano, F., Piazza, I., Brambilla, F., and Sozzi, T. (1995) *Lait*, **75**, 285–293.

Radke-Mitchell, L., and Sandine, W. E. (1984) *J. Food Prot.*, **47**, 245–248.

Radke-Mitchell, L., and Sandine, W. E. (1986) *J. Dairy Sci.*, **69**, 2559–2568.

Rasic, J. L. J., and Kurmann, J. A. (1978) *Yoghurt Scientific Grounds, Technology and Preparations*, Technical Dairy Publishing House, Copenhagen.

Rawson, H. L., and Marshall, V. M. E. (1997) *Int. J. Food Sci.*, **32**, 213–220.

Rea, M. C., Lennartsson, T., Dillon, P., Driman, F. D., Reville, W. J., Heaps, M., and Cogan, T. M. (1996) *J. Appl. Bacteriol.*, **81**, 83–94.

Reuter, G. (1985) *Int. J. Food Microbiol.*, **2**, 55–68.

Ricciardi, A., and Clementi, F. (2000) *Ital. J. Food Sci.*, **12**, 23–45.

Robinson, R. K. (1981) *Dairy Ind. Int.*, **46**(3), 31–35.

Robinson, R. K. (1989) *Dairy Ind. Int.*, **54**(7), 23–25.

Robinson, R. K. (2000) Fermented milks. In *Encyclopedia of Food Microbiol.*, Vol. 2, R. K. Robinson, C. A. Batt, and P. D. Patel, eds., Academic Press, London, pp. 784–791.

Robinson, R. K., and Tamime, A. Y. (1975) *J. Soc. Dairy Technol.*, **28**, 149–163.

Robinson, R. K., and Tamime, A. Y. (1976) *J. Soc. Dairy Technol.*, **29**, 148–155.

Robinson, R. K., and Tamime, A. Y. (1993) Manufacture of yoghurt and other fermented milks. In *Modern Dairy Technology*, Vol. 2, 2nd ed., R. K. Robinson, ed., Elsevier Applied Science, London, pp. 1–48.

Rybka, S., and Kailasapathy, K. (1996) *Int. Dairy J.*, **6**, 839–850.

Salof-Coste, C. J. (1996) *Danone-World News*, Issue No. 11, Centre International de Recherche Daniel Carasso, Paris.

Sanchez-Banuelos, J. M., Mantano-Ortga, M., Aguilera-Valencia, G., and Garcia-Garibay, M. (1992) *Cult. Dairy Prod. J.*, **27**(11), 14–18.

Schaack, M. M., and Marth, E. H. (1988a) *J. Food Prot.*, **51**, 600–606.

Schaack, M. M., and Marth, E. H. (1988b) *J. Food Prot.*, **51**, 607–614.

Schaack, M. M., and Marth, E. H. (1988c) *J. Food Prot.*, **51**, 848–852.

Schellhaass, S. M., and Morris, H. A. (1985) *Food Microstruct.*, **4**, 279–287.

Shapton, N., and Cooper, P. J. (1984) *J. Soc. Dairy Technol.*, **37**, 60–65.

Skriver, A., Buchheim, W., and Qvist, K. B. (1995) *Milchwissenschaft*, **50**, 683–686.

Stadhouders, J., and Driessen, F. M. (1992) Other milk products. In *Bacillus cereus* in *Milk and Milk Products*, Bulletin No. 275, International Dairy Federation, Brussels, pp. 40–45.

Stannard, C. (1997) *Food Sci. Technol. Today*, **11**, 137–177.

Suriyarachchi, V. R., and Fleet, G. H. (1981) *Appl. Environ. Microbiol.*, **42**, 574–579.

Takizawa, S., Kojima, S., Tamura, S., Fujinaga, S., Benno, Y., and Nakase, T. (1998) *Syst. Appl. Microbiol.*, **21**, 121–127.

Tamime, A. Y. (1977) *Some Aspects of the Production of Yoghurt and Condensed Yoghurt*, PhD thesis, University of Reading, Reading.

Tamime, A. Y. (1993) Yoghurt-based products. In *Encyclopedia of Food Sciences, Food Technology and Nutrition*, Vol. 7, R. Macrae, R. K. Robinson, and M. J. Saddler, eds., Academic Press, London, pp. 4972–4977.

Tamime, A. Y., and Crawford, R. J. M. (1984) *Egypt. J. Dairy Sci.*, **12**, 299–312.

Tamime, A. Y., and Deeth, H. C. (1980) *J. Food Prot.*, **43**, 939–977.

Tamime, A. Y., and Greig, R. I. W. (1979) *Dairy Ind. Int.*, **44**(9), 8–27.

Tamime, A. Y., and Marshall, V. M. E. (1997) Microbiology and technology of fermented milks. In *Microbiology and Biochemistry of Cheese and Fermented Milk*, 2nd ed. B. A. Law, ed., Blackie Academic & Professional, London, pp. 57–152.

Tamime, A. Y., and McNulty, D. (1999) *Lait*, **79**, 449–456.

Tamime, A. Y., and O'Connor, T. P. (1995) *Int. Dairy J.*, **5**, 109–129.

Tamime, A. Y., and Robinson, R. K. (1985) *Yoghurt Science and Technology*, Pergamon Press, Oxford.

Tamime, A. Y., and Robinson, R. K. (1988) *J. Dairy Res.*, **55**, 281–307.

Tamime, A. Y., and Robinson, R. K. (1999) *Yoghurt Science and Technology*, 2nd Edition Woodhead Publishing, Cambridge.

Tamime, A. Y., Kalab, M., and Davies, G. (1984) *Food Microstruct.*, **3**, 83–92.

Tamime, A. Y., Davies, G., and Hamilton, M. P. (1987a) *Dairy Ind. Int.*, **52**(6), 19–21.

Tamime, A. Y., Davies, G., and Hamilton, M. P. (1987b) *Dairy Ind. Int.*, **52**(7), 40–41.

Tamime, A. Y., Kalab, M., and Davies, G. (1989a) *Food Microstruct.*, **8**, 125–135.

Tamime, A. Y., Davies, G., Chehade, A. S., and Mahdi, H. A. (1989b) *J. Soc. Dairy Technol.*, **42**, 35–39.

Tamime, A. Y., Barclay, M. N. I., Amarowicz, R., and McNulty, D. (1999a) *Lait*, **79**, 317–330.

Tamime, A. Y., Barclay, M. N. I., McNulty, D., and O'Connor, T. P. (1999b) *Lait*, **79**, 435–448.

Tamime, A. Y., Barclay, M. N. I., Law, A. J. R., Leaver, J., Anifantakis, E. M., and O'Connor, T. P. (1999c) *Lait*, **79**, 331–339.

Tamime, A. Y., Muir, D. D., and Wszolek, M. (1999d) *Dairy Ind. Int.*, **65**(5), 32–33.

Tamime, A. Y., Robinson, R. K., and Latrille, E. (2001) Yoghurt and other fermented milks. In *Mechanisation and Automation in Dairy Technology*, A. Y. Tamime and B. A. Law, eds., Sheffield Academic Press, Sheffield (in press).

Thompson, J. K., Johnston, D. E., Murphy, R. J., and Collins, M. A. (1990) *Ir. J. Food Sci. Technol.*, **14**, 35–49.

Tilbury, R. H., Davis, J. G., French, S., Imrie, F. K. E., Campbell-Lent, K., and Orbell, C. (1974) Taxonomy of yeasts in yoghurt and other dairy products. In *Proceedings 4th International Symposium on Yeast*, Part I, **F8**, Vienna, pp. 265–266.

Tinson, W., Broom, M. C., Hillier, A. J., and Jago, G. R. (1982) *Aust. J. Dairy Technol.*, **37**, 14–21.

Toba, T., Arihara, K., and Adachi, S. (1990a) *Int. J. Food Microbiol.*, **10**, 219–224.

Toba, T., Nakajima, H., Tobitani, A., and Adachi, S. (1990b) *Int. J. Food Microbiol.*, **11**, 313–320.

Toba, T., Kotani, T., and Adachi, S. (1991a) *Int. J. Food Microbiol.*, **12**, 167–172.

Toba, T., Uemura, H., Mukai, T., Fuji, T., Itoh, T., and Adachi, S. (1991b) *J. Dairy Res.*, **58**, 497–502.

Ulberth, F., and Kneifel, W. (1992) *Milchwissenschaft*, **47**, 432–434.

van den Berg, J. C. T. (1988) *Dairy Technology in the Tropics and Subtropics*, Pudoc, Wageningen, pp. 157–158.

Vayssier, Y. (1978a) *Rev. Lait Fr.*, **361**, 73–75.

Vayssier, Y. (1978b) *Rev. Lait Fr.*, **362**, 131–134.

Veringa, H. A., Galesloot, T. E., and Davelaar, H. (1968) *Netherlands Milk Dairy J.*, **22**, 115–120.

Vlaemynck, G. M. (1994) *J. Food Prot.*, **57**, 913–914.

Waes, G. (1987) *Milchwissenschaft*, **42**, 146–148.

Walstra, P., and Jenness, R. (1984) *Dairy Chemistry and Physics*, John Wiley & Sons, New York.

Weiss, N., Schillinger, U., and Kandler, O. (1983) *Syst. Appl. Microbiol.*, **4**, 552–557.

Winkelmann, F. (1987) Yoghurt–legal aspects. In *Milk—the Vital Force*, Proceedings XXII International Dairy Congress—The Hague 1986, R. Reidel Publishing, Dordrect, pp. 691–702.

Wszolek, M., Tamime, A. Y., Muir, D. D., and Barclay, M. N. I. (2001) *Lebensm. Wiss. Technol.*, **34**, 251–261.

Yamani, M. I. (1993) *Int. J. Food Sci. Technol.*, **28**, 111–116.

Yamani, M. I., and Ibrahim, S. A. (1996) *J. Soc. Dairy Technol.*, **49**, 103–108.

Yamani, M. I., Al-Kurdi, L. M. A., Haddadin, M. S. Y., and Robinson, R. K. (1998) The detection of inhibitory substances in ex-farm milk supplies. In *Recent Research Developments in Agricultural and Food Chemistry*, Vol. 2, F. G. Pandalai, ed., Signpost Publication, Trivandrum, pp. 611–627.

Zhennai, Y., Huttunen, E., Staaf, M., Widmalm, G., and Tenhu, H. (1999) *Int. Dairy J.*, **9**, 631–638.

Zourari, A., and Anifantakis, E. M. (1988) *Lait*, **68**, 373–392.

CHAPTER 9

MICROBIOLOGY OF THERAPEUTIC MILKS

GILLIAN E. GARDINER, R. PAUL ROSS, PHIL M. KELLY, and CATHERINE STANTON
Teagasc, Dairy Products Research Centre, Moorepark, Fermoy, County Cork, Ireland

J. KEVIN COLLINS and GERALD FITZGERALD
Department of Microbiology, University College, Cork, Ireland

9.1 INTRODUCTION

Fermented milks and dairy products containing beneficial or "probiotic" cultures, such as lactobacilli and bifidobacteria, are currently among the best-known examples of functional foods in Europe. Such probiotic-containing dairy foods are associated with a range of health claims, including alleviation of symptoms of lactose intolerance and treatment of diarrhea to cancer suppression and reduction of blood cholesterol. Milk and dairy products provide the ideal food system for the delivery of these beneficial bacteria to the human gut, given the suitable environment that milk and certain dairy products including yogurt and cheese provide to promote growth and/or support viability of these cultures. Although the value of lactic acid bacteria (LAB) in food fermentations has been recognized for centuries, development of the probiotic idea is attributed to Elie Metchnikoff, who observed that the consumption of fermented milks could reverse putrefactive effects of the gut microflora (Metchnikoff, 1907). At this time, Henri Tissier also suggested that bifidobacteria could be administered to children with diarrhea to help restore the gut microflora balance (Tissier, 1906).

Dairy Microbiology Handbook, Third Edition, Edited by Richard K. Robinson
ISBN 0-471-38596-4

From these beginnings, the probiotic concept has progressed considerably and is now the focus of much research attention worldwide. Significant advances have been made in the selection and characterization of specific cultures and substantiation of health claims relating to their consumption. Consequently, the area of probiotics has advanced from anecdotal reports, with scientific evidence now accumulating to back up nutritional and therapeutic properties of certain strains. Today, the majority of research and commercial attention given to probiotic microoganisms focuses on members of the bacterial genera *Lactobacillus* and *Bifidobacterium*, with the result that an expanding portfolio of probiotic fermented dairy products containing some of these strains is available to the consumer.

This review will examine the morphological and biochemical characteristics of microorganisms with alleged therapeutic properties, including lactobacilli, bifidobacteria, and enterococci. The health claims and evidence for beneficial effects of selected strains, safety issues (e.g., risks from employing antibiotic-resistant cultures such as *Enterococcus faecium*), and the possible role of functional nutrients added to "biomilks" (i.e., prebiotics) will be reviewed. Developments in fermented milk products containing probiotic bacteria will also be discussed.

9.2 PROBIOTIC MICROORGANISMS ASSOCIATED WITH THERAPEUTIC PROPERTIES

The term "probiotic," meaning "for life," originated to describe substances produced by one microorganism which stimulate the growth of others (Lilly and Stillwell, 1965). Since that time, various definitions have been proposed to describe these bacteria (Havenaar and Huis in't Veld, 1992; Guarner and Schaafsma, 1998; Ouwehand and Salminen, 1998; Salminen et al., 1999). The most recent one defines probiotics as "microbial cell preparations or components of microbial cells that have a beneficial effect on health and well being of the host' (Salminen et al., 1999). This definition takes into account the fact that probiotics need not necessarily be viable and that their effects are not restricted to effects on the intestinal microflora. Prior to this, at an international meeting of the Lactic Acid Bacteria Industrial Platform (LABIP) probiotics were described by a consensus definition as "living microorganisms, which, upon ingestion in certain numbers, exert health benefits beyond inherent basic nutrition" (Guarner and Schaafsma, 1998). The most recent consensus requires that probiotics are live and are capable of surviving passage through the digestive tract and have the capability to proliferate in the gut (FAO/WHO, 2001), where they have been redefined as "live micro-organisms which when administered in adequate amounts confer a health benefit on the host".

Lactobacillus and *Bifidobacterium* are the principal bacterial genera central to both probiotic and prebiotic (nondigestible food ingredients that stimulate the growth and/or activity of certain colonic bacteria) approaches to dietary modulation of the intestinal microflora (Table 9.1). This is possibly due to (a) the association of these bacterial groups with human health, (b) their presence in fermented foods, and (c) the fact that they possess generally regarded as safe (GRAS) status and are thus often included in food products. However, microorganisms other than the traditional LAB, such as propionibacteria, *Leuconostoc*, pediococci, and enterococci, have also received attention as candidate probiotic cultures (Table 9.1). Probiotic preparations may contain one or several different microbial strains or species. Enterococci, although not GRAS organisms, have been used as probiotics (Giraffa et al., 1997) (Table 9.1), and so this genus will be discussed along with bifidobacteria and lactobacilli as one with probiotic potential, although it has recenly been recommended that enterococci not be used as a probiotic for human use (FAO/WHO, 2001). While bifidobacteria, lactobacilli, and enterococci are all members of the Gram-positive eubacteria (Prescott et al., 1990), lactobacilli and enterococci are part of the true LAB group within the low G + C group, whereas bifidobacteria are members of the high G + C or actinomycete group (Kandler and Weiss, 1989), and, as such, both groups are phylogenetically unrelated. Therefore, bifidobacteria are not strictly LAB, but can be considered part of this group, given the fact that they have similar morphological and biochemical properties and because they are inhabitants of the gastrointestinal tract (GIT)

TABLE 9.1. Microorganisms Used as Probiotics

Lactobacilli	Bifidobacteria	Enterococci	Others
L. acidophilus	*B. bifidum*	*E. faecium*	*Saccharomyces boulardii*
L. plantarum	*B. infantis*	*E. faecalis*	*Lactococcus lactis* spp. *cremoris*
L. casei	*B. adolescentis*		*Lactococcus lactis* spp. *lactis*
L. rhamnosus	*B. longum*		*Leuconostoc mesenteroides*
L. delbrueckii spp. *bulgaricus*	*B. breve*		*Propionibacterium freudenreichii*
L. fermentum	*B. lactis*		*Pediococcus acidilactici*
L. johnsonii			*Streptococcus salivarius* spp. *thermophilus*
L. gasseri			*Escherichia coli*
L. salivarius			
L. reuteri			

Source: Adapted from Goldin (1998) and Holzhapfel et al. (1998).

(Klein et al., 1998). The following section will discuss the physiological and taxonomic characteristics of lactobacilli, bifidobacteria, and enterococci at the genus level.

9.2.1 Lactobacillus

The genus *Lactobacillus* is a large heterogenous group of microorganisms, currently comprising 54 recognized species and some subspecies (Tannock, 1999), that are ubiquitous in the environment, occupying niches varying from plant surfaces to the GIT of many animals. Morphologically, they are Gram-positive nonmotile rods, often found in pairs or chains, ranging from coccobacilli to long slender rods (Kandler and Weiss, 1989). Morphology depends on the strain or species, and also on factors such as age of the culture and growth medium, but is not as varied as that of bifidobacteria. Lactobacilli lack catalase and cytochromes and are usually microaerophilic, with growth improved under anaerobic conditions (Kandler and Weiss, 1989). They have a strictly fermentative metabolism, and they convert glucose solely or partly to lactic acid. In this respect, they can be classified as either homofermentative (producing predominantly lactic acid) or heterofermentative (producing carbon dioxide, ethanol, and/or acetic acid in addition to lactic acid) (Prescott et al., 1990). Lactobacilli are extremely fastidious, having complex nutritional requirements for many organic substrates, which must therefore be provided in order to achieve optimal growth (Hammes and Vogel, 1995). With respect to growth temperature, the optimum is usually within the mesophilic range of 30–40°C. However, some strains are capable of growth below 15°C and some at temperatures up to 55°C. In addition, lactobacilli are aciduric, with an optimum pH for growth of 5.5–6.2 (Kandler and Weiss, 1989; Hammes and Vogel, 1995). Differentiation of species within the *Lactobacillus* genus does not now depend as much on physiological criteria, such as carbohydrate fermentation, as on molecular characterization. Therefore, like many other genera, the *Lactobacillus* genus is currently undergoing change (Hammes and Vogel, 1995).

9.2.2 Bifidobacterium

Since first described in 1899 as a microorganism predominant in the stools of breast-fed infants (Tissier, 1899), the taxonomy of bifidobacteria has been continuously changed. They were assigned to the genera *Bacillus*, *Bacteroides*, *Tissieria*, *Nocardia*, *Lactobacillus*, *Actinomyces*, *Bacterium*, and *Corynebacterium* (Poupard et al., 1973), before being recognized as a separate genus in 1974. Members of the genus *Bifidobacterium* are generally characterized as Gram-positive, non-spore-forming, nonmotile and catalase-negative (Scardovi, 1989). They are predominantly strict anaerobes, although some species and strains may tolerate oxygen in the presence of carbon dioxide (Modler et al., 1990). Morphologically, the variation and pleomorphism that exist within the genus *Bifidobacterium* are demonstrated by their description in *Bergey's Manual of Systematic Bacteriology*. Here they are described as "short, regular, thin cells with pointed ends, coccoidal regular cells, long cells with slight bends or protuberances or with a large variety of branchings; pointed, slightly bifurcated club-shaped or spatulated extremities; single or in chains of many elements; in star-like clusters or disposed in 'V' or 'palisade' arrangements" (Scardovi, 1989). In addition, morphology may depend not only on the strain or species, but also on cultural conditions used.

Bifidobacteria characteristically ferment hexoses by the fructose-6-phosphate or "bifid" shunt, due to the presence of the enzyme fructose-6-phosphate phosphoketolase, which can be used as a distinguishing feature of bifidobacteria (Poupard et al., 1973; Modler et al., 1990). Fermentation of glucose by means of this pathway yields acetic and lactic acids in a theoretical 3:2 molar ratio, although in practice this exact ratio may not be achieved (Scardovi, 1989). The optimum growth temperature for bifidobacteria is 37–41°C, with 25–28°C and 43–45°C reported as minimum and maximum growth temperatures, respectively (Scardovi, 1989). Given that they have a pH optimum near neutrality (6.5–7.0) and that no growth occurs at pH values less than 4.5 or greater than 8.5 (Scardovi, 1989), bifidobacteria are less acid tolerant than lactobacilli. Nutritionally, different types of bifidobacteria have been described, but in general their requirements are less complex than those of lactobacilli (Poupard et al., 1973). However, in some cases, bifidobacteria do require specific factors (bifidogenic factors) for optimal growth (Modler et al., 1990). Within the genus *Bifidobacterium*, 31 species have been described to date (Tannock, 1999), whereas in the most recent edition of *Bergey's Manual* only 24 are outlined (Scardovi, 1989). This reflects the ongoing examination of the genus, along with the consequences of applying more modern molecular methods, as is the case with the *Lactobacillus* group.

9.2.3 Enterococcus

Although bifidobacteria and lactobacilli are most commonly used as probiotics, some species of enterococci have also found applications as health-promoting cultures (Table 9.1). In this respect, it is predominantly *E. faecium* and *E. faecalis* which have been used as probiotics (Klein et al., 1998). However, enterococci in general are not recognized as GRAS organisms and there is some concern over their potential pathogenicity (Giraffa et al., 1997), which will be discussed in greater detail below. Although enterococci were first described in 1899 by Thiercelin, *Enterococcus* as a genus was not introduced until 1984 (Schleifer and Kilpper-Bälz, 1984) and today it contains 19 species (Franz et al., 1999). All members of the genus *Enterococcus* are Gram-positive, non-spore-forming, catalase-negative facultative anaerobes (Mundt, 1989; Murray, 1990). Morphologically, they appear as spherical or ovoid cocci in pairs or short chains (Mundt, 1989) and in this respect, do not display as much morphological variability as lactobacilli or bifidobacteria. Enterococci are robust micro-organisms, possessing phenotypic characteristics, such as the ability to grow at 10°C and 45°C in 6.5% NaCl and at pH 9.6, and they can survive heating to 60°C for 30 min; these properties are used to differentiate enterococci from other Gram-positive catalase-negative cocci (Mundt, 1989; Franz et al., 1999). However, atypical enterococci have been described which do not possess these properties, while members of other genera may satisfy these criteria, making phenotypic identification of members of the genus *Enterococcus* difficult (Franz et al., 1999). This has therefore led to the development of molecular methods for identification of enterococci to the strain and species level (Descheemaeker et al., 1997). Enterococci are considered homofermentative with respect to glucose metabolism, although some amounts of formic and lactic acids may be produced in some media in addition to lactic acid (Mundt, 1989; Garg and Mital, 1991). As is the case with other LAB, enterococci are nutritionally fastidious microorganisms, requiring B vitamins, certain amino acids, and purine and pyrimidine bases for optimal growth (Garg and Mital, 1991).

9.3 CRITERIA ASSOCIATED WITH PROBIOTIC MICROORGANISMS

Many criteria have been proposed by different researchers as being desirable for potential probiotic cultures (Havenaar and Huis in't Veld, 1992; Lee and Salminen, 1995; Collins et al., 1998; Salminen et al., 1998), including the ability to survive intestinal passage and proliferate and/or colonize the GIT (Figure 9.1). However, in practice the properties of probiotic microorganisms are dependent on the host for which probiotic administration is intended, the anatomical site within the host

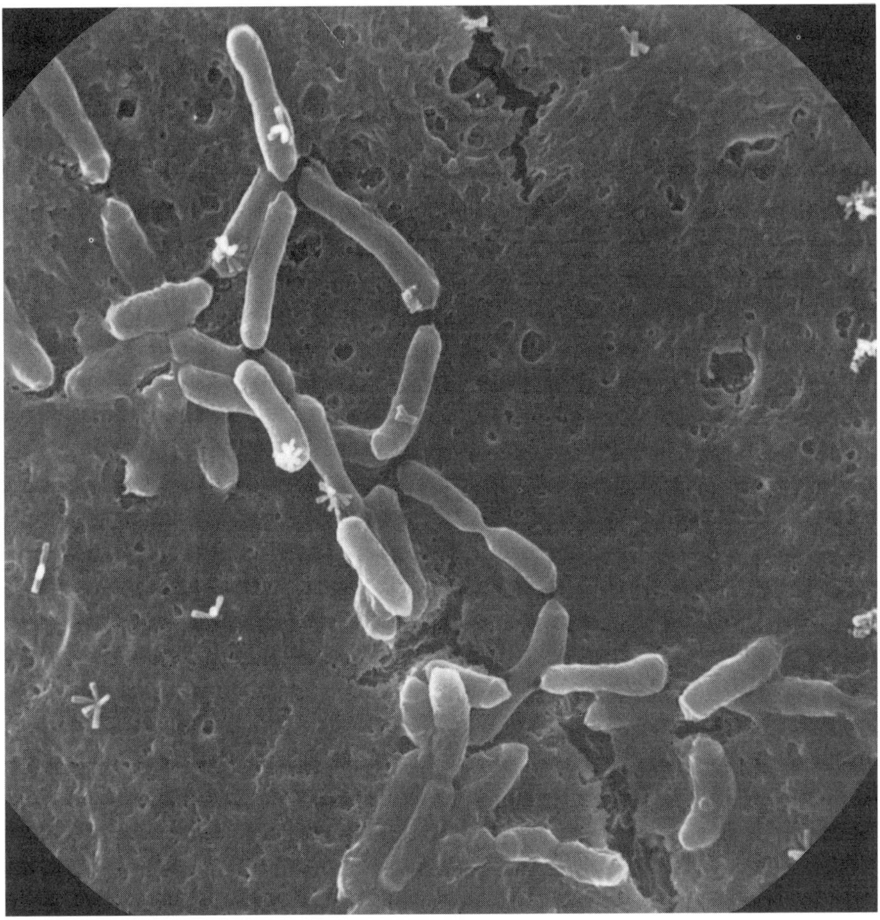

Figure 9.1. Electron micrograph of probiotic lactobacilli adhering to human intestinal epithelial cells. (Electron microscopy courtesy of M. Heapes and W. Reville, E. M. Unit, University College, Cork.)

toward which the probiotic is directed, and the desired effect at that site. The terminology currently in use to define probiotics together with the criteria commonly listed as desirable for probiotic cultures (Salminen et al., 1998) reflect the fact that the GIT is the principal focus of probiotic applications. However, probiotic cultures may find application at sites other than the GIT, such as the respiratory or urogenital tracts, although research on these applications is not as advanced (Havenaar and Huis in't Veld, 1992; Sanders, 1993). Furthermore, although this review focuses on probiotics for human use, health-promoting cultures can also be employed for animal use (Fuller, 1999). In addition, within the human population, probiotic cultures may either

TABLE 9.2. Characteristics Desirable for Probiotic Microorganisms

1. Survival in the environmental conditions in the intended host
2. Proliferation and/or colonization under host environmental conditions
3. Survival in association with the host immune system and noninflammatory
4. Immunostimulatory for the mucosal immune system
5. Production of antimicrobial substances
6. Antagonism against cariogenic and pathogenic bacteria
7. Safety tested; nonpathogenic, nontoxic, nonallergic, nonmutagenic, or anticarcinogenic, even in immunocompromised hosts
8. Genetically stable, no plasmid transfer
9. Technologically suitable for process applications
10. Desirable metabolic activity and antibiotic resistance/sensitivity
11. Desirable health effects
12. Potential for delivery of recombinant proteins and peptides

Source: Adapted from Collins et al. (1998), Salminen et al. (1998), and Havenaar and Huis in't Veld (1992).

be targeted toward specific groups or recommended for general use (Conway, 1996).

A general set of criteria desirable for probiotic microorganisms, regardless of the intended host or site of application, has been compiled (Table 9.2). *In vitro* tests based on these selection criteria, although not a definite means of strain selection, may provide useful initial information. In addition, well-characterized and validated model systems such as the TNO model and the simulator of the human intestinal microbial ecosystem (SHIME), which aim to mimic complex physiological and physicochemical *in vivo* reactions, may also be of value in strain selection, being less expensive than human trials and not having the associated ethical drawbacks (Molly et al., 1994; Huis in't Veld and Shortt, 1996). However, ultimate proof of probiotic effects requires validation in well-designed statistically sound clinical trials, as recommended by the LABIP workshop (Guarner and Schaafsma, 1998). Furthermore, it is important that the probiotic properties are retained during *in vitro* experimentation under laboratory conditions and also during subsequent processing and storage (Conway, 1996). However, suitability for technological applications is only one of the many criteria listed as desirable for probiotic cultures (Table 9.2).

On the other hand, it is neither feasible nor absolutely necessary that a strain intended for probiotic use should fulfill every requirement listed. Furthermore, given that the exact mode of action of probiotic cultures is, for the most part, unknown and, even if elucidated, is unlikely to be identical for all cultures, it is difficult to make generalizations regarding selection of probiotic cultures. It is more likely that a particular culture will have a specific health effect in the host, and, in this respect, each culture should be judged on its own merit (Mattila-Sandholm et al., 1999). Indeed, in a study in which *Bifidobacterium* strains were isolated and characterized for potential probiotic use, no one strain was found to possess all desirable characteristics (Gomez Zavaglia et al., 1998). Perhaps if a broad range of effects is desired, a number of strains may be administered in combination to achieve this (Huis in't Veld and Shortt, 1996). An approach to the selection of strains for probiotic use, suggested by Salminen et al. (1996a), is to first elucidate the properties of successful probiotic strains and from this to draw up a set of criteria under which to assess strains intended for future probiotic use. Furthermore, perhaps strains that are persistent within the gut microflora, such as the *Lactobacillus* and *Bifidobacterium* isolates previously identified by others (McCartney et al., 1996; Kimura et al., 1997), may provide a starting point in the search for new probiotic strains.

9.4 SAFETY ISSUES ASSOCIATED WITH USE OF PROBIOTIC CULTURES FOR HUMANS

It is crucial that probiotic cultures are safe for human use, with recommendations that this criterion be fulfilled even in immunocompromised hosts. A scheme that has been proposed by Donohue and Salminen (1996) for safety assessment of probiotic cultures is outlined in Table 9.3. Tools that may be employed in such assessment include *in vitro* studies of strain properties, pharmacokinetic studies, animal

TABLE 9.3. Scheme for Safety Assessment of Probiotic Cultures

Type of Property Studied	Safety Factor to Be Assessed
Intrinsic strain properties	Adhesion factors, antibiotic resistance, plasmid transfer, enzyme profile
Metabolic products	Concentrations, safety, and other effects
Toxicity	Acute and subacute effects of ingestion of large amounts of culture
Infective properties	*In vitro* with cell lines; *in vivo* with animal models
Dose–response effects	Oral administration in volunteers
Clinical assessment	Potential for side effects and disease-specific effects; nutritional studies
Epidemiological studies	Surveillance of large populations following introduction of new strains and products

Source: Adapted from Donohue and Salminen (1996) and Salminen et al. (1996b).

studies, use of intestinal models, human studies, and epidemiological surveillance (Donohue and Salminen, 1996; Salminen and Marteau, 1998). Each strain needs to be tested separately, with special attention given to genetically modified microorganisms (Mattila-Sandholm et al., 1999). Using the approaches outlined above, many probiotic cultures have undergone safety testing; and with the exception of the examples outlined below, there is no evidence to show that infections have been caused by probiotic bacteria (Donohue and Salminen, 1996; Adams, 1999).

The majority of microorganisms employed as probiotics are LAB (Table 9.1) which, with the exception of enterococci, have GRAS status. LAB are ubiquitous in the environment, part of the indigenous commensal microflora and have a long history of safe use in the production of fermented foods (Donohue and Salminen, 1996; Adams, 1997). Evidence has recently emerged to suggest a role for LAB as opportunistic pathogens associated with human clinical infections, such as endocarditis and urinary tract infections (Aguirre and Collins, 1993; Adams and Marteau, 1995). However, clinical conditions implicating LAB are relatively rare and to date have only been observed in immunocompromised individuals or patients with underlying disease conditions (Aguirre and Collins, 1993; Adams and Marteau, 1995; Adams, 1999). Furthermore, no case of infection has been observed in people exposed to high LAB concentrations in the workplace, nor has any case been linked to consumption of fermented foods, probiotics, or drugs containing LAB (van der Kamp, 1996; Adams, 1997).

A LABIP workshop organized to discuss the safety of LAB recognized the facts outlined above, although enterococci were considered an exception, due to their more frequent involvement in nosocomial infections, ability to acquire antibiotic resistance genes, and possession of potential virulence factors (Jett et al., 1994; Adams and Marteau, 1995; Franz et al., 1999). Of particular concern is the emergence of enterococci with resistance to vancomycin, because this is one of the last effective antibiotics for the treatment of multi-drug-resistant pathogens (Giraffa et al., 1997; Franz et al., 1999). These potential hazards are worthy of consideration, given that *E. faecium* and *E. faecalis* strains are often added to probiotic foods or pharmaceutical preparations. However, the source of enterococci involved in human clinical infection is usually attributed to the patient's own microflora, and it is not clear whether vancomycin-resistant enterococci are transferred through the food chain (Adams, 1997; Franz et al., 1999). Nevertheless, the LABIP safety workshop concluded that if enterococci are to be used intentionally, this must be undertaken in the knowledge of the observations outlined above and that there should be "demonstrable positive effects" (van der Kamp, 1996). Perhaps the benefits of enterococcal probiotic strains such as *E. faecium* PR88 or SF68, which have been successfully used in the treatment of irritable bowel syndrome and diarrhea, respectively (Allen et al., 1996; Franz et al., 1996), may outweigh any associated risks.

While this discussion has focused on the safety of LAB as probiotics, other microorganisms may also be used as probiotics and are perhaps a greater cause for concern. For example, caution should be exercised in the case of *Saccharomyces* spp., considering that infections have been reported in some individuals taking *Saccharomyces boulardii* supplements, although patients often had underlying conditions (Pletincx et al., 1995). Furthermore, a recent case of recurrent septicaemia in an immunocompromised patient was shown to be due to *Bacillus subtilis* strains identical to those found in a probiotic preparation taken by the patient prior to illness (Oggioni et al., 1998). Indeed, Oggioni et al. (1998) conclude that "high numbers of viable microorganisms should not be given to any patient with severe immunodeficiency." On the other hand, these patients may benefit most from probiotic therapy. Although some studies have shown probiotic cultures to be safe and well-tolerated in groups such as HIV patients (Wolf et al., 1998), the unlimited use of probiotics may have undesirable side effects in these "at-risk" subgroups (Guarner and Schaasfsma, 1998).

Given that 143 clinical trials, involving 7526 subjects have been conducted with probiotic LAB in the last 40 years, without adverse effects (Naidu et al., 1999), many probiotic cultures can be considered to have a proven record of safe use. Nevertheless, continued monitoring is essential, and possible risks that have been identified include microbial invasion, deleterious metabolic activities, adjuvant side effects, and gene transfer (Huis in't Veld et al., 1994). Furthermore, although expensive and time-consuming, novel probiotic strains do require safety evaluation.

9.5 BENEFICIAL HEALTH EFFECTS OF PROBIOTIC CULTURES

The potential clinical applications of probiotic bacteria are many and varied; and while some are based only on anecdotal reports and poorly controlled studies, others have been well-substantiated with scientific evidence (Table 9.4). Between 1961 and 1998, 143 probiotic trials involving 7526 human subjects were performed, with many different

TABLE 9.4. Health Effects Attributed to Consumption of Probiotic Cultures

Well-substantiated effects
Alleviation of lactose maldigestion
Prevention/treatment of infections
Reduction of serum cholesterol
Chemopreventative effects
Modulation of the immune system
Potential effects
Treatment/prevention of inflammatory bowel disease
Alleviation of constipation
Improvement of dermatitis
Liver disease therapy

microbial strains and species used either therapeutically or prophylactically in the treatment of various illnesses and physiological disorders (Naidu et al., 1999). This review will focus on the specific health targets for which most scientific evidence exists, including alleviation of lactose maldigestion, treatment of cancer, reduction of infection, reduced serum cholesterol, and enhanced immune system (Table 9.4). In discussing these, emphasis will be placed on evidence that has accumulated from recent human clinical studies.

9.5.1 Alleviation of Lactose Maldigestion

Lactose maldigestion, due to insufficient amounts or activity of lactase in the human gut, affects up to 70% of the world's population, causing varying degrees of abdominal discomfort (Scrimshaw and Murray, 1988). The role of probiotics in alleviation of these symptoms is considered an established and well-substantiated beneficial effect (Table 9.4) (Huis in't Veld et al., 1994; Guarner and Schaasfsma, 1998). However, although certain LAB have yielded positive results in the treatment of lactose maldigestion, there is more evidence of a beneficial effect for yogurt bacteria than for cultures chosen for their probiotic properties (Sanders, 1993), even though yogurt bacteria may not survive well during gastric transit. Undoubtedly, humans can utilize the lactose in yogurt more efficiently than lactose in milk (Kolars et al., 1984). This may be due to the existence of preformed β-galactosidase in the yogurt which reaches the GIT in an active form or due to the presence of viable bacteria in the yogurt which produce the enzyme *in vivo* (Marteau et al., 1997). The latter mechanism is validated by the fact that heat-inactivated bacteria exert a lesser effect than viable cultures (Gilliland and Kim, 1984; Savaiano et al., 1984). However, microbial lactase may not be the only reason why lactose is better digested in yogurt than in milk. The oro-cecal transit time of yogurt is longer than that of milk, thus allowing more effective breakdown of lactose; furthermore, lactose-deficient individuals can adapt to lactose on regular consumption of lactose-containing foods (Jiang and Savaiano, 1997).

Conflicting results have been obtained as to whether or not specific *Lactobacillus* and *Bifidobacterium* strains can ameliorate the symptoms of lactose maldigestion. The end point most commonly measured is breath hydrogen, which is an indication of sugar fermentation in the colon. In lactose maldigestors, lactose reaches the colon undigested where hydrolysis by bacteria takes place, leading to an increase in the level of breath hydrogen (Marteau et al., 1997). While some studies in humans have shown that ingested *L. acidophilus* cultures result in a reduction in breath hydrogen following lactose consumption (Kim and Gilliland, 1983; Mustapha et al., 1997), others have demonstrated little or no effect with these or *B. bifidum* cultures (Lin et al., 1991; Hove et al., 1994; Saltzman et al., 1999). There is, however, strain variability with respect to the demonstration of positive effects, but conflicting opinions exist regarding strain properties important for the exertion of effects in lactose maldigesters. Some highlight the importance of β-galactosidase activity and bile sensitivity in strain selection (Lin et al., 1991), while others, having found no correlation between β-galactosidase activity and lactose digestion *in vivo*, recommend bile and acid tolerance as important criteria (Mustapha et al., 1997). A study by Lin et al. (1998) found nonfermented milk containing *L. bulgaricus* to be more effective in the alleviation of lactose maldigestion than the same milk containing *L. acidophilus*, although both cultures were similar in their β-galactosidase activity, lactose-transporting capability, and bile tolerance. Following ingestion, a *B. longum* strain offered potential in the reduction of symptoms due to lactose maldi-gestion (Jiang et al., 1996), warranting further research on the use of *Bifidobacterium* cultures. Overall, although certain LAB, especially yogurt starter cultures, undoubtedly show efficacy in the alleviation of lactose malabsorption symptoms, it is difficult to draw conclusions regarding the effectiveness of probiotic cultures in the treatment of this disorder.

9.5.2 Prevention/Treatment of Infections

Despite dramatic improvements in medical care and the development of new chemotherapeutic agents for pathogen inhibition, infectious diseases remain a significant health problem. While antibiotics have saved countless lives worldwide, there is considerable concern over their use, given the increasing incidence of microbial antibiotic resistance (Neu, 1992; Bengmark, 1998). Because administration of probiotic microorganisms offers potential in the prevention and/or treatment of certain intestinal and urogenital infections (Table 9.4), these cultures may be useful as alternatives to antibiotic therapy. Many authors have recently advocated the use of oral bacteriotherapy for the treatment and/or prevention of such infections (Reid, 1996; Zoppi, 1997; Bengmark, 1998; Reid et al., 1998); in fact, the World Health Organization (WHO) has recommended the reconsideration of microbial interference therapy for infection control (Bengmark, 1998). Certain probiotic strains have been shown to prevent pathogen attachment and invasion in cell culture, to inhibit the growth of enteropathogens *in vitro* (Gibson and Wang, 1994; Drago et al., 1997), and to enhance the immune response. Considering this, there is at least the possibility that the use of probiotics may decrease reliance on antimicrobials. Proposed mechanisms by which probiotic cultures may act in infection control have been suggested by Fooks et al. (1999) and include competition for nutrients, secretion of antimicrobial substances, reduction of pH, blocking of adhesion sites, attenuation of virulence, blocking of toxin receptor sites, immune stimulation, and suppression of toxin production. Some human studies that demonstrate the potential use of probiotics as therapeutic or prophylactic agents for intestinal infections are outlined below. However, as with other health effects of probiotics, further research is needed to determine efficacy in studies with sufficient subjects, proper controls, and statistical analysis of results.

In humans, diarrhea can result from infection by a number of bacterial or viral agents and in some cases the etiology is unknown (Gibson et al., 1997). Worldwide, many adults are incapacitated by diarrhea and many children die as a result of diarrhea-related illnesses, especially in developing countries (Goldin, 1998). In the United States alone, 21–37 million diarrheal episodes occur in 16.5 million children annually (Glass et al., 1991), with rotavirus the most common agent of infantile gastroenteritis worldwide. One of the most extensively researched probiotics, *Lactobacillus* GG, has been shown to be effective in the treatment of acute viral diarrhea in children, most cases of which were caused by rotavirus, with an associated increase in immunity (Isolauri et al., 1991). This reduction in the duration of rotavirus diarrhea has been observed repeatedly on treatment with *Lactobacillus* GG (Kaila et al., 1992; 1995; Guarino et al., 1997). Other probiotic strains such as *L. casei* Shirota, *B. bifidum* and *S. thermophilus* have also been shown to be effective in both the treatment and prevention of rotavirus diarrhea in children (Saavedra et al., 1994; Sugita and Togawa, 1994). Furthermore, *Lactobacillus* GG was efficacious in promoting recovery from watery diarrhea in children in developing countries, which was either of unidentified aetiology (Raza et al., 1995) or caused by *Salmonella*, *Shigella*, *E. coli*, and rotavirus (Pant et al., 1996). Interestingly, in both of these studies no effect was seen on bloody diarrhea. In a recent study by Oberhelman et al. (1999), *Lactobacillus* GG had an effective prophylactic effect in diarrhea prevention in children in developing countries, but the effect was limited to non-breast-fed children. An uncontrolled study by Hotta et al. (1987) indicated that a strain of *B. breve* administered orally might have potential in the treatment of diarrhea in children and thus warrants further research.

Antibiotic-associated diarrhea (AAD) is a major clinical problem that occurs following antibiotic use, the most serious form of which is pseudomembranous colitis. Diarrhea is caused by pathogen overgrowth due to a microflora imbalance; in 20% of cases, the etiological agent is *Clostridium difficile*, an opportunistic pathogen that is especially persistent and difficult to treat (Lewis and Freedman, 1998). Surprisingly, antibiotics are the treatment of choice for pseudomembranous colitis or other AAD (Corthier, 1997), but relapse often occurs and it is perhaps in these cases that probiotic therapy may be especially useful. For instance, *Lactobacillus* GG has been used successfully in the treatment of colitis, with a concomitant reduction in fecal endotoxin titres and where relapse occurred further probiotic treatment was effective (Gorbach et al., 1987; Biller et al., 1995). Oral therapy with *Lactobacillus* GG was also effective in the prevention of AAD (Siitonen et al., 1990), as were a number of other probiotic microorganisms including *S. boulardii* (McFarland et al., 1995), *B. longum* (Colombel et al., 1987), and *E. faecium* SF68 (Wunderlich et al., 1989). However, studies by Aronsson et al. (1987) and Lewis et al. (1998) which showed the absence of any effect of administration of either *L. acidophilus* or *S. boulardii*, respectively, on the treatment of AAD highlight the importance of proper culture evaluation in clinical trials.

Traveler's diarrhea is a form of diarrhea of unknown etiology which is estimated to occur in 20–50% of travelers to foreign, often developing, countries (Saxelin, 1997). Although usually self-limiting, an effective method of prevention and/or treatment is desirable. Again

Lactobacillus GG has been shown to have a positive effect in traveler's diarrhea (Hilton et al., 1997). However, a study conducted by Oksanen et al. (1990) showed no significant difference in the incidence of diarrhea among travelers to Turkey when the entire study group was assessed as a whole. Nevertheless, in one holiday destination within the country, *Lactobacillus* GG was found to be an effective prophylactic treatment, probably due to the occurrence of different diarrheagenic pathogens at that location. Black et al. (1989) showed a combination of *B. bifidum*, *S. thermophilus*, *L. delbrueckii* spp. *bulgaricus*, and *L. acidophilus* to have efficacy in the prevention of traveler's diarrhea. On the other hand, many studies have shown no effect of probiotic treatment (Clements et al., 1981; Lewis and Freedman, 1998).

Probiotic therapy also shows potential in the treatment of infections of the upper GIT, such as those caused by *Helicobacter pylori*. *H. pylori* is now known to be an important etiological agent in peptic ulcers and has an involvement in gastric cancer (Lambert and Hull, 1996). Studies have demonstrated the *in vitro* inhibition of this pathogen by *L. acidophilus* (Bhatia et al., 1989) and also by other LAB (Midolo et al., 1995). The latter study showed that *H. pylori* inhibition was strain-dependent and correlated with lactic acid concentrations produced by the LAB. Lactic acid production was also shown to be an important means of suppression of *H. pylori* in a gnotobiotic murine model of disease (Aiba et al., 1998). However, conflicting results from human studies make it difficult to conclude whether or not probiotic cultures are capable of inhibiting *H. pylori* colonization of the gastric mucosa *in vivo* (Michetti et al., 1995; Bazzoli et al., 1992).

9.5.3 Reduction of Serum Cholesterol

Hypercholesterolaemia has been identified as a risk factor for cardiovascular disease, and ingestion of probiotic bacteria has been proposed as one means of attaining a reduction in blood cholesterol levels (Table 9.4). However, the role played by LAB in reducing blood cholesterol remains a controversial point, with no clear evidence that such an effect exists. The mechanism by which LAB may reduce cholesterol is currently unclear. Because Gilliland et al. (1985) observed that certain *L. acidophilus* strains can decrease cholesterol concentrations in growth medium, many cultures have been tested *in vitro* for their cholesterol assimilating ability (Walker and Gilliland, 1993; Buck and Gilliland, 1994; Taranto et al., 1998). However, Klaver and van der Meer (1993) later proposed that bacterial uptake of cholesterol did not occur, but rather that a co-precipitation of cholesterol with deconjugated bile salts was the reason for the observed cholesterol reduction. Another possible mechanism that has been suggested is that deconjugation of bile salts in the upper small intestine by ingested probiotics with bile salt hydrolase (BSH) activity lowers cholesterol levels by decreasing the digestibility of lipids and increasing bile salt elimination in the feces (De Smet et al., 1998; Vesa et al., 1998). Thus the presence of BSH has been suggested as a desirable property in bacteria intended for cholesterol-lowering uses, and many studies have involved *in vitro* assay for the presence of this enzyme in bacteria (De Smet et al., 1994; Grill et al., 1995; Corzo and Gilliland, 1999). However, the products of BSH activity may have possible detrimental effects in the host, and consideration should be given as to whether or not this enzyme activity is a desirable property in probiotic bacteria (Vesa et al., 1998). In general, although strategies have been proposed for the selection of bacteria intended to have a hypocholesterolaemic effect (Gilliland and Walker, 1990), the value of *in vitro* measurement of cholesterol assimilation and BSH activity is questionable.

Initial human studies focused on the possible hypocholesterolaemic effect of yogurt and other fermented milks. Early studies that demonstrated a reduction in serum cholesterol have been criticized due to the administration of very high doses of fermented dairy products and the lack of control groups (Mann and Spoerry, 1974; Mann, 1977). Some human studies have indicated no significant reduction in blood cholesterol levels as a result of yogurt consumption (Rossouw et al., 1981; Thompson et al., 1982). While positive effects have been obtained in some cases (Hepner et al., 1979; Howard and Marks, 1979), this is often transient (Jaspers et al., 1984). Initial reductions in blood lipids observed at the beginning of intervention studies have been suggested to be due to changes in dietary and other lifestyle habits that are difficult to control (Taylor and Williams, 1998). Therefore, a baseline or "run-in" period is recommended but not often included in studies.

These initial studies, while indicating the potential for fermented milks in cholesterol reduction, did not identify the particular cultures used. More recent studies have employed defined cultures administered either in dairy products or as pharmaceutical preparations. Positive results have been obtained with an *E. faecium* strain in one human study (Agerbaek et al., 1995). However, other trials conducted using the same culture failed to show a cholesterol-lowering effect (Sessions et al., 1997); or where there was an effect, it was not sustained and was also observed in the control group (Richelsen et al., 1996). Similarly, the hypocholesterolemic effect observed by Anderson and Gilliland (1999) due to *L. acidophilus* consumption was not a persistent one, demonstrating the need for continuous culture consumption in order to appreciate such an effect. Due to intra-individual variation in blood cholesterol observed over time and the variation which occurs in cholesterol analysis (Taylor and Williams, 1998), a large sample size is needed if small changes in blood cholesterol are to be detected in human studies. The usual requirements for human trials (double-blind, placebo-controlled, statistical analyses) also apply. Interestingly, one of the only studies that has been conducted with a large number of subjects (334), which was double-blind and controlled, showed no effect of administration of a commercial preparation of *L. acidophilus* and *L. bulgaricus* on blood cholesterol (Lin et al., 1989). Overall, studies in humans have yielded mixed results with no clear cholesterol-lowering benefit observed due to probiotic consumption; furthermore, the exact mechanism by which probiotic cultures may exert such an effect remains unclear.

9.5.4 Chemopreventative Effects

Epidemiological studies, such as those conducted by van't Veer et al. (1989) which have indicated that fermented milk intake may have a protective effect against carcinogenesis, suggest a potential anti-cancer effect for LAB. Given that the colonic microflora—in particular, bacteroides, eubacteria, and clostridia—are thought to be involved in carcinogenic processes, it is considered that increasing the proportion of LAB in the gut may have beneficial effects in the prevention of cancer (Rowland, 1996). Indeed, there is evidence to suggest that individuals with a high risk of colon cancer harbor lower levels of lactobacilli (Saxelin, 1997). On the contrary, an analysis of the intestinal microflora of individuals considered to have a high risk of colon cancer suggested that *Bifidobacterium* was associated with an increased risk of this disease (Moore and Moore, 1995). Nevertheless, observations exist concerning the protective effects of certain LAB against carcinogenesis. However, although much research concerning the use of LAB in cancer therapy has been conducted *in vitro* and with animal models, limited studies have been conducted in humans due to the associated economic and practical difficulties.

Because DNA damage is considered to be an early event in the process of carcinogenesis, *in vitro* studies demonstrating the antimutagenic effects of certain cultures may be of significance (Pool-Zobel et al., 1993; Nadathur et al., 1994). *In vivo* decreases in DNA damage and genotoxic injury have also been observed in animal models (Pool-Zobel et al., 1996). In addition, selected LAB have the ability to bind or degrade mutagens (Zhang and Ohato, 1991; Orrhage et al., 1994), and *Lactobacillus* consumption has been associated with a decrease in the mutagenicity of feces and urine caused by fried meat consumption (Lidbeck et al., 1992; Hayatsu and Hayatsu, 1993). A further possible mechanism by which probiotics may exert anti-carcinogenic activity involves suppression of members of the microflora with enzyme activities capable of converting pro-carcinogens to carcinogens. Such enzymes include β-glucuronidase, nitroreductase, and azoreductase, and there is evidence from human studies that LAB can decrease fecal levels of these compounds (Goldin and Gorbach, 1984; Ling et al., 1994). Some studies, however, have shown little or no effect of culture administration, perhaps due to strain differences (Marteau et al., 1990). Moreover, the exact involvement of these enzymes in the etiology of cancer is unclear, and the biological significance of such findings is unknown (Sanders, 1993). Beneficial effects of LAB in cancer therapy may be associated with their immunostimulatory properties, which will be discussed below. It has been proposed that probiotics may influence tumor development through their ability to modulate immune parameters (Ouwehand et al., 1999b).

Apart from the indirect evidence outlined above, there are numerous reports of the anti-carcinogenic effects of LAB from *in vivo* experiments conducted in animal models. Although the method of testing cultures in such models is often by intraperitoneal or intravenous injection, this cannot be related to oral consumption, and this discussion will outline only those studies where dietary supplementation with LAB was investigated. In this respect, LAB administration has been shown to reduce the incidence of chemically induced aberrant crypt foci (preneoplastic lesions) in the colon (Kulkarni and Reddy, 1994; Rao et al., 1999). Prebiotic substances, including fructooligosaccharides (FOS) such as neosugar, have also been shown to have potential in this area (Koo and Rao, 1991), and in some cases the combined influence of pro- and prebiotics was more effective than that of either component administered alone (Rowland et al., 1998). Furthermore, *B. longum* has been shown to prevent the induction of colon, liver, and mammary tumors by the cooked food carcinogen IQ (Reddy and Rivenson, 1993). Many other studies that have demonstrated the anti-tumor activity of oral probiotic administration in animal models have been summarized recently (Naidu et al., 1999; Reddy, 1999) and suggest a role for certain LAB in the prevention of cancer.

The value of animal experimentation for assessment of the anti-cancer effects of probiotics is questionable, especially because it has not been correlated with the human system. However, few direct clinical studies have been undertaken in humans. Biasco et al. (1991) demonstrated that consumption of LAB for 3 months by colon cancer patients significantly reduced cell division in colonic crypts, which is considered a pre-neoplastic phenomenon. Studies by Okawa et al. (1989, 1993) have shown that consumption of a *L. casei* strain may have an adjuvant effect in cancer therapy. In this respect, enhanced tumor regression and prolonged survival during radiation therapy was observed. In addition, studies concerning the effect of probiotics in bladder cancer prevention in humans have shown promising results (Aso and Akaza, 1992; Aso et al., 1995). At the present time, there is much indirect evidence for the anti-cancer effects of probiotic consumption and human studies to assess the potential of probiotic cultures for use in cancer therapy are required.

9.5.5 Modulation of the Immune System

The immune system consists of a complex series of interlinked mechanisms, which function in protection against infections and uncontrollably growing tumor cells. Possible stimulation of an immune response by probiotic bacteria may therefore explain potential therapeutic and prophylactic applications of such cultures in the treatment of infections and carcinogenesis. While the immune system appears to ignore or otherwise tolerate most intestinal microbes, the normal commensal gut microflora plays an important role in modulation of host mucosal defenses, as demonstrated by comparison of immune function in germ-free and conventional animals (Collins et al., 1996; Blum et al., 1999; McCracken and Gaskins, 1999). Consequently, there is much interest in modulation of this microflora through either probiotic or prebiotic administration in order to strengthen the gut mucosal barrier and augment the immune response, two protective mechanisms which are related (Brassart and Schiffrin, 1997).

Although the mechanism by which probiotic bacteria stimulate the immune system is not fully understood, it has been suggested that effects may be mediated through interaction with specialized lymphoid aggregates of the gut-associated lymphoid tissue (GALT) (Marteau and Rambaud, 1993; Famularo et al., 1997). Peyer's patches, part of the GALT system present at intervals in the GI wall, are covered by specialized epithelial cells (M cells) that facilitate antigen uptake. The immunostimulatory activity of LAB has been attributed to bacterial cell wall constituents, and perhaps the most probable explanation as to how these substances come into contact with the host immune system is via these antigen-sampling areas of the GIT. Indeed, it has been shown in mice that the barrier effect demonstrated by some probiotic strains can be induced by activated immunocompetent cells of the GALT (Perdigon et al., 1990). It has also been suggested that microorganisms capable of adhering to gut mucosal surfaces are more effective in immune stimulation (Gill, 1998; Ouwehand et al., 1999a), and this may account for strain variation in this respect (Paubert-Braquet et al., 1995; Nagafuchi et al., 1999). On the other hand, it has been suggested that such variation may perhaps be due to structural differences in the cell wall composition of strains (Gill, 1998). Furthermore, there is controversy concerning whether or not culture viability is necessary for immune stimulation. In some cases, only live bacteria have enhanced host defenses (De Simone et al., 1987, 1988; Muscettola et al., 1994; Kaila et al., 1995), whereas in other cases an effect was seen only with dead bacteria (Namba et al., 1981). Regardless of the mechanisms involved or the criteria necessary for exertion of an effect, probiotic cultures have been shown to stimulate both nonspecific (innate) and specific (adaptive) immunity. Evidence of these effects has accumulated from *in vitro*, animal, and human studies, some of which will be outlined below.

The first line of defense of the mammalian immune system is the nonspecific immune response, involving cellular effectors that include mononuclear phagocytes, polymorphonuclear leucocytes, and natural killer (NK) cells. Animal studies have shown an enhancement of nonspecific immunity, including increased macrophage activity and NK activity in mice injected intraperitoneally with lactobacilli (Sato et al., 1988; Kato et al., 1994). Oral administration of LAB has also been shown to increase macrophage phagocytosis in mice and to enhance NK cell activity of mouse spleen cells against tumor cells (Perdigon et al., 1986, 1988; Gill, 1998). Furthermore, the magnitude of the response has also been shown to be dose-dependent, with mice receiving 10^{11} cfu of LAB per day showing significantly greater phagocytic activity than mice receiving 10^{9} or 10^{7} LAB per day (Gill, 1998). Similar dose effects have been observed in humans, with 10^{9} but not 10^{8} cfu/day of *Lactobacillus johnsonii* La1 enhancing both phagocytic and respiratory burst activities (Donnet-Hughes et al., 1999). Furthermore, studies have shown that, at least in very high doses, ingested yogurt bacteria can stimulate the level of NK cells and interferon (IFN)-γ in human volunteers (De Simone et al., 1986). Similarly, further human trials have demonstrated increased phagocytic activity of peripheral blood leucocytes on LAB consumption (Schiffrin et al., 1995, 1997). However, while administration of *Lactobacillus salivarius* UCC 118 has been shown not to affect macrophage phagocytic activity, granulocyte phagocytic activity was significantly increased in subjects consuming the probiotic strain in yogurt as opposed to unfermented milk (Dunne et al., 1999). Such an effect has been seen previously—that is, that lactic cultures delivered in fermented milk products induce a superior immune response compared to cultures delivered in unfermented products (Gill, 1998).

Several studies employing experimental animals and humans have demonstrated the immunostimulatory effects of LAB on both humoral and cell-mediated immune responses of the specific immune system. Perdigon et al. (1990) showed that oral administration of certain LAB increased immunoglobulins in the intestinal fluid and that only pretreated mice had increased secretion of *Salmonella*-specific antibodies when subsequently challenged with *Salmonella*. Other studies have also suggested an oral adjuvant effect for some LAB strains. For example, T-cell proliferative responses to *Candida albicans* antigens in mice co-infected with *C. albicans* and *Lactobacillus* GG or *B. animalis* were stronger than in mice infected with *C. albicans* alone (Wagner et al., 1997). In addition, the administration of yogurt or yogurt supplemented with LAB enhanced mucosal and systemic responses to both cholera toxin and salmonellae (De Simone et al., 1987, 1988; Tejada-Simon et al., 1999). LAB administration in animal studies has also been shown to induce the production of IFN-α, -β, and -γ (Kitazawa et al., 1992; Muscettola et al., 1994). Interestingly, a recent study has shown that *B. lactis* Bb12 administered to lactating mice enhances IgA production not only in the intestine but also in milk (Fukushima et al., 1999). While animal studies are useful in the evaluation of immunostimulatory properties of LAB, ultimately studies in humans are required. While it is not possible to induce infection in humans from an ethical point of view, studies have shown an immune adjuvant effect of probiotic consumption in both healthy and vaccinated individuals and subjects with preexisting infections. For instance, a human trial conducted by Link-Amster et al. (1994) involved consumption of probiotic strains in combination with attenuated *Salmonella typhi* to mimic an enteropathogenic infection. It was found that the titer of specific serum IgA to *S. typhi* in the test group was significantly higher than that in the control group. Furthermore, *Lactobacillus* GG promotes recovery of children with rotavirus diarrhea, possibly through increases in serum IgA titers to rotavirus (Kaila et al., 1992, 1995); in fact, this strain has been used effectively as an adjuvant to rotavirus vaccine (Isolauri et al., 1995).

Interestingly, a study by Fukushima et al. (1998) has shown *Bifidobacterium* consumption to increase fecal levels of total IgA and anti-poliovirus IgA in healthy subjects. Other studies in humans have demonstrated that probiotic consumption can influence immune parameters such as IgA specific to the probiotic strain administered (Dunne et al., 1999) and IFN-γ production (Solis Pereyra and Lemonnier, 1993).

When considering the use of probiotic cultures for immunostimulatory purposes, possible adverse effects of such immune modulation should be considered. This is of particular relevance in individuals with inflammatory diseases (such as inflammatory bowel disease (IBD), rheumatoid arthritis, and autoimmune diseases) as well as in individuals with food allergy, where the microflora itself has been implicated in pathogenesis (Collins et al., 1996). However, it has been suggested that, depending on the disease state and immune status of an individual, up-regulation of the immune response occurs when required, with down-regulation alleviating exaggerated immune responses during infection and hypersensitivity (Pelto et al., 1998; Salminen et al., 1998).

9.5.6 Other Health Benefits of Probiotics

Other potential or less-substantiated health effects of probiotic cultures are listed in Table 9.4 and include treatment/prevention of disorders such as IBD (Dunne et al., 1999; Madsen et al., 1999; Shanahan, 2000), constipation (Motta et al., 1991), liver disease (Adawi et al., 1997), and dermatitis (Majamaa and Isolauri, 1997). Given the range of health benefits attributed to probiotic consumption, it is unlikely that each strain will act in the same way. In general, health effects are related to microflora modification and strengthening of the gut mucosal barrier (Salminen et al., 1996b). In addition, the metabolic activities of probiotic cultures, either in the preparation of fermented foods or in the digestive tract, may lead to nutritional benefits such as an increase in the production or bioavailability of certain vitamins and minerals or an improvement in the digestibility of protein (Friend and Shahani, 1984; Fernandes et al., 1992). Some of the health benefits attributed to probiotic cultures are strain-dependent, stressing the importance of probiotic strain selection and highlighting the fact that claims made for one probiotic culture cannot necessarily be applied to another. There is a need to select probiotic strains on the basis of their functional attributes; based on this, a particular culture or mix of cultures should be chosen for certain health effects.

9.6 EFFECTIVE DAILY INTAKE OF PROBIOTICS

A daily intake of 10^6–10^9 cfu has been recommended as the minimum effective dose for probiotics (Lee and Salminen, 1995), but this has not been validated by scientific studies. In general, high levels of daily consumption are preferred, and there is some evidence to support this. For example, dose–response studies conducted with *Lactobacillus* GG demonstrated that when administered in either freeze-dried powder or gelatine capsules, the minimum dose required to yield fecal recovery was 10^{10} cfu/day (Saxelin et al., 1991; 1995), with lower doses (10^6–10^8 cfu/day) found not to be effective. On the other hand, the intake required to yield fecal recovery was 10-fold lower (10^9 cfu/day) when the strain was administered in either fermented milk or enterocoated tablets, highlighting the importance of the delivery system (Saxelin et al., 1993). However, the use of fecal recovery as a measurable end point in such dose–response studies has been questioned. A recent study by Donnet-Hughes et al. (1999) has addressed this by determining the minimum effective dose of *L. johnsonii* La1 required for immune modulation. Findings showed that although fecal recovery was found in all subjects consuming the culture, 10^9 cfu/day elicited immune effects, whereas a lower dose of 10^8 cfu did not. Taken together, these studies illustrate the difficulties involved in defining a general minimum effective dose for all probiotic cultures, given that variations occur depending on the particular strain or delivery system used. Furthermore, there is the question of whether or not probiotic cultures must be viable in order to exert health benefits (Salminen et al., 1999).

9.7 PROBIOTIC DAIRY PRODUCTS

Because of their associated health benefits, food and pharmaceutical companies have an interest in exploiting probiotic cultures as an opportunity for product development. The addition of probiotic cultures or prebiotic substances to food products can be seen as fortification with biologically active components and, as such, leads to the development of "functional foods," which are described as "foods claimed to have a positive effect on health." Indeed, one of the most active areas within the functional foods sector, from the point of view of both research and commercial development, is that of probiotics (Stanton et al., 2001). Consequently, a variety of food products and supplements containing viable microorganisms with probiotic properties are commercially available (Table 9.5), and many more are in the process of evaluation

TABLE 9.5. Some Examples of Commercially Available Therapeutic Milk Products and Their Health Claims

Product	Health Claim
Actimel	"Reinforces your natural resistance . . . your daily dose of natural protection"
Yakult	"A healthy start to every day"
Actimel Cholesterol Control	"Helps, if taken regularly as part of a healthy and varied diet, to reduce you cholesterol values"
BIO Aloe Vera	"Feeds and hydrates in a very self-evident way: from inside out"
Biotic Plus Oligofructose	"Promotes the natural balance of the gut flora and thus your health . . . oligofructose stimulates the body's own positive bacteria and increases (as dietary fiber) the activity and purification of the gut"
Daily FIT	"A valuable contribution to fitness and health . . . positive action on the gut flora . . . stimulates natural resistance"
Fyos	"Promotes a healthy gut balance . . . take care of your whole health"
Fysiq	"Contributes to a healthy cholesterol level . . . this effect is strengthened by the presence of a dietary fiber"
Jour après Jour	"To help maintain your vitality . . . the bifidogenic fibers promote the development of the bifidobacteria ('good' bacteria in our body) and contribute to the balance and good functioning of the organism"
PROAC	"A valuable contribution to your health . . . cleans the gut in a natural way and stimulates the required natural activity of the organic cells"
Silhouette Plus	"The soluble bifidogenic fibers help to preserve and reestablish the balance of the digestive flora"
ProCult 3	"A positive influence on the gut flora . . . additionally supported by the nutritious substance inulin"

Source: Adapted from Coussement (1997).

(Tamime et al., 1995; Holzhapfel et al., 1998). There is, however, some skepticism in scientific circles regarding the quality and efficacy of probiotic products. This is probably due, in part, to the findings of studies which have shown that some probiotic products contain neither the number nor the type of cultures stated on the label (Iwana et al., 1993; Hamilton-Miller et al., 1996; Micanel et al., 1997; Hamilton-Miller et al., 1999). Nevertheless, the probiotic food industry is flourishing, with the European probiotic yogurt market alone currently estimated to be worth around £520 million (Shortt, 1998). In many European countries, most notably France and Germany, the market is expanding with the result that probiotic yogurts now account for over 10% of all yogurts sold in Europe (Stanton et al., 2001),

Current recommendations are that probiotic microorganisms should be viable in food products, although there is some debate regarding the necessity for this (Ouwehand and Salminen, 1998). Indeed, the use of nonviable microorganisms would have many advantages with regard to product development, and it could allow the expanded use of probiotics in developing countries where it may not be possible to meet strict handling conditions (Ouwehand and Salminen, 1998). Despite this, the emphasis is still on production of foods with high numbers of viable probiotics, although the minimum requirement remains unclear (Hamilton-Miller and Fuller, 1996). There is a need for clear guidelines in this respect (Hamilton-Miller and Gibson, 1999). The minimum therapeutic probiotic level required in a food product is dependent on the recommended daily intake. There are some indications that a probiotic intake of approximately 10^9 cfu per day is necessary to elicit some health effects. Based on consumption of 100 g or ml of a probiotic food per day, a product should therefore contain at least 10^7 cells per gram or milliliter, which is in agreement with current Japanese recommendations (Ishibashi and Shimamura, 1993), but considerably higher numbers have also been suggested (Lee and Salminen, 1995). However, the minimum probiotic level recommended for foods must be founded on the demonstration of health benefits and should not be based on a concentration that is simply easily attainable and cost-effective industrially (Sanders, 1993).

The most popular food delivery systems for probiotic cultures have been fermented milk products, probably because of their traditional association with health and because they already harbor viable microorganisms. Within the dairy foods sector, products with a relatively short shelf life, such as yogurts and fermented milks in addition to unfermented milks with added cultures, have been the most popular choice for probiotic incorporation (Stanton et al., 1998). However, the portfolio of dairy products containing probiotic cultures is expanding to include foods such as cheese (Gardiner et al., 1998, 1999a), ice cream (Christiansen et al., 1996), and frozen yogurts (Laroia and Martin, 1991). In addition, many nondairy foods provide alternative systems for the delivery of viable probiotics to the GIT (Lee and Salminen, 1995). The delivery system may in fact have an important role in determining probiotic viability in the GIT following consumption. For example, 15-month-old Cheddar cheese is at least as effective as, if not superior to, fresh yogurt for delivery of viable probiotic microorganisms to the porcine GIT (Gardiner et al., 1999b), while both fermented milks and enterocoated tablets are more efficacious for delivery of *Lactobacillus* GG than a freeze-dried powder (Saxelin, 1997).

With respect to incorporation of probiotic cultures, singly or in combination into fermented milk products, two approaches can be taken: either the application of a probiotic as a starter culture or as an adjunct to the starter culture. The former is often limited by the inability of probiotic cultures to produce sufficient lactic acid in milk, thereby requiring the addition of growth-promoting supplements such as cysteine, yeast extract, and casein hydrolysates (Poch and Bezkorovainy, 1988; Klaver et al., 1993; Gomes et al., 1998). Consequently, addition of the probiotic culture as an adjunct to the starter culture may be a more favorable option. This could take advantage of any possible symbiotic relationship that may exist between the strains, resulting in increased microbial growth rates and improved flavor of the finished product (Hughes and Hoover, 1991). Furthermore, one of the challenges in probiotic food development is that the probiotic microorganisms should not adversely affect product quality—for example, inferior sensory scores due to acetic acid production by bifidobacteria (Gomes et al., 1995). If there is an effect, it should be a positive one—for example, improvement of flavor, texture, or other organoleptic qualities. In this respect, the use of exopolysaccharide-overproducing strains of bifidobacteria has been suggested for the improvement of texture and mouth-feel in fermented dairy products (Roberts et al., 1995), and an *E. faecium* strain with proven probiotic properties had a positive influence on Cheddar cheese flavor (Gardiner et al., 1999a).

9.7.1 Probiotic Yogurts and Fermented Milks

In 1996, the United Kingdom yogurt market was worth £523 million, with £47 million of this accounted for by probiotic yogurts (Russell, 1996). Indeed, the majority of probiotic-containing dairy products on the European market are yogurts or yogurt-type fermented milks (Young, 1998) (Table 9.5). However, many commercial yogurts surveyed in both Europe and Australia exhibited poor probiotic viability, particularly with respect to *Bifidobacterium* strains (Iwana et al., 1993; Micanel et al., 1997), indicating that these may not be the ideal carriers for some probiotic microorganisms. Nevertheless, these products remain a popular choice for probiotic incorporation (Table 9.5), where many are marketed on the basis of their mild taste rather than for their health-promoting attributes (Fuller, 1993). Of all the dairy markets, that for yogurt, with its existing health image, is well-positioned to capitalize on the growth in healthy foods, benefiting additionally from the fact that it is a food that tastes good and is enjoyable. A number of approaches can be taken toward the production of probiotic yogurts.

For example, depending on the regulations in individual countries, the *L. bulgaricus* component of a yogurt starter culture may be replaced by a probiotic *Lactobacillus* strain such as *L. acidophilus*; or, alternatively, a probiotic *Lactobacillus*, *Enterococcus*, or *Bifidobacterium* culture can be added as an adjunct to the starter. It is recommended that one or both of the traditional starters is used in order to ensure a product with desirable flavor and texture chracteristics (Marshall and Tamime, 1997). Often, changes to the manufacturing process are required; for example, the fermentation temperature is lowered, at the expense of increased manufacturing time, to one that favors probiotic growth—that is, from 45°C to 37°C (Kosikowski and Mistry, 1997). In addition, starter culture selection is of utmost importance in avoiding post-acidification—a principal cause of declining probiotic cell numbers in yogurt during storage. Other forms of yogurt which are available include frozen yogurt, drinking yogurt, dried/instant yogurt (see below), and carbonated yogurt (Tamime and Deeth, 1980), some of which have been investigated as carriers for probiotic cultures (Stanton et al., 1998).

9.7.2 Commercial Developments in Probiotic Yogurts and Fermented Milks

As consumer contact with the probiotic concept increases, the demand for yogurts and fermented milks with therapeutic properties is growing and manufacturers are responding by introducing new products, which will add value to their existing portfolios. The differences in the approach to functional foods in various countries have resulted in a number of different but related developments. Many dairy products containing pro- and prebiotics with associated health claims have been launched onto the market (Table 9.5), and in some countries these are an established market segment. The trend is toward exploitation of the synergistic effect of combining probiotics with prebiotics, while some of the earlier products contained probiotic cultures alone (Young, 1998). Of the yogurts and fermented milk products to which probiotic cultures have been applied, "LC1" (Nestlé), "Vifit" (Campina Melkunie), "Actimel" (Danone), and "Yakult" (Yakult) have emerged as market leaders (Stanton et al., 2001). Many European countries are experiencing considerable growth in demand for existing probiotic products, and there is a surge in the numbers of new products being launched. Some product developments in the area of probiotic milk products are outlined below.

Developed in Japan in the 1930s, Yakult, the fermented milk drink containing *L. casei* Shirota, is now viewed as the world's leading mass-marketed functional food product and is sold worldwide in 23 countries at a volume of 16 million bottles per day (Hasler, 1998; Heasman and Mellentin, 1998). Marketed in 65-ml bottles containing 6.5 billion bacteria, Yakult is considered neither a food nor a pharmaceutical. Since establishing a European production base in The Netherlands in 1994, the Japanese company Yakult Honsha has extended distribution into Belgium, Luxembourg, the United Kingdom, and Germany during 1995 and 1996 and plans to serve all European countries by 2005 (Wright, 1999). Sales in Europe are now estimated to total 388,000 bottles a day; and even in the challenging UK market, since its launch in 1996, Yakult has more than doubled its sales, securing a £7.2 million niche in the yogurt and pot-dessert market. Based on this performance, all the major multiples have moved to national distribution of the product. The advertising expenditure is expected to continue with further advertising backed by sampling of over 1 million bottles in-store and in the community and workplace.

Nestlé's LC1, available either as a set cultured milk or as a drinking product, contains the *L. acidophilus* strain La 1, recently renamed as *L. johnsonii* LJ 1. This *Lactobacillus* strain, chosen for its probiotic characteristics, was the outcome of an extensive research effort conducted by Nestlé. Based on human studies and supported by a strong scientific dossier, this culture is claimed to stimulate the immune system, leading to the statement "helps the body protect itself" (Young, 1996). In 1994, the launch of the LC1 product onto the French market took place, costing $8 million; and by the end of its first year it had seized an 11% share of the French "bio" yogurt market (Heasman and Mellentin, 1998). The market has continued to grow, and in 1997 it had gained a 25% share. LC1 now accounts for 20% of the company's European trade in yogurts and fermented milks and is the leading brand in the German yogurt market, with a 60% share (Hilliam, 1998). However, despite its success on the French and German markets and the fact that it is currently available in most European countries, a slower start for the LC1 product has been evident in a number of other European countries, most notably the United Kingdom (Young, 1998).

While the potential and success of the functional food market are widely acclaimed, the risks and failures are not, and success has not been enjoyed by all pro- or prebiotic products launched. The rate of product failure is high, and an example of a product that has not been successful in some countries is Gaio, a cultured dairy product, for which cholesterol reducing properties were claimed. Introduced in Denmark in 1993, the product containing the "Causido" culture enjoyed great success and seized 15% of the Danish yogurt market in its first year (Young, 1998). It had grown to account for 65% of the Danish probiotic market in 1997, despite a 70% price premium over other yogurts (Heasman and Mellentin, 1998). However, the product has not been as successful in other countries such as the United Kingdom, where it was withdrawn in 1997 due to unsatisfactory sales and negative public relations. The principal difficulties encountered were in relation to the cholesterol-lowering claims that were judged by the Advertising Standards Authority to be exaggerated and misleading (Young, 1998).

9.7.3 Spray-Dried Probiotic Dairy Powders

Spray-drying, described as the transformation of liquid products into dried powder forms by spraying liquid into a controlled flow of hot air within a drying chamber (Masters, 1985), is the predominant method used for drying of milk and milk products in the dairy industry. It causes less scorching of powders and has a higher capacity than roller drying, producing minimal undesirable changes on the nutritive value (Masters, 1985; Caric, 1994). This is reflected in the manner in which temperatures are controlled during the spray drying process: drier inlet and outlet temperatures via modulation of the air heater system and product feed to the atomizing device, respectively (Masters, 1985). Spray-drying is used widely in the dairy industry for the preparation of various products including whole and skim milk powder, whey powder, baby food, caseinate, coffee whitener, and dried yogurt. These dried products have applications in human nutrition either as nutritive additives, in a reconstituted form, or in a wide range of food products, including dairy and meat products, various toppings, coatings, mayonnaises, soups, puddings, and instant breakfasts in addition to having uses in the bakery industry (Caric, 1994). The major advantage of dried dairy products is their long shelf life due to low moisture content (<4%), which inhibits development of microorganisms during storage (Caric, 1994).

Spray-drying can also be applied to the preparation of culture-containing powders for starter or adjunct applications (Knorr, 1998). Preparing bulk cultures is usually a troublesome and time-consuming process, especially where probiotic cultures are concerned due to poor growth rates in milk. From an industrial point of view, a low-cost method of production of reliable cultures for direct addition to process milk is therefore desirable. In general, in the food industry the trend is toward direct vat inoculation with concentrated starter cultures most commonly supplied in frozen concentrated or freeze-dried forms.

Freeze-dried powders and frozen concentrates of probiotic *Bifidobacterium* and *Lactobacillus* spp. have been developed (Gilliland and Lara, 1988; Misra and Kuila, 1991; Gagne et al., 1993); but even though spray-dried probiotic cultures are available commercially, little research attention has focused on the use of spray drying as a means of probiotic culture preparation. Spray-drying has, however, been investigated as a means of preservation of yogurt containing viable microorganisms (Kim and Bhowmik, 1990), dairy starter cultures (Metwally et al., 1989; Teixeira et al., 1995a,b), and bacteriocin-producing cultures (Mauriello et al., 1999), as well as for adjunct culture attenuation (Johnson and Etzel, 1993, 1995). In order to prevent extensive bacterial mortality, lenient drying conditions are employed, whereby low inlet and outlet temperatures are used. Temperatures should, however, be high enough to ensure a moisture content suitable for good powder-keeping quality (~4%) (Masters, 1985). The advantages of spray-drying over freeze-drying as a method of culture preparation are that it is economical, is suitable for large-scale production, and results in high-quality shelf-stable powders that are transportable at low cost. However, a potential disadvantage is that spray-drying may result in thermal and/or dehydration inactivation of cultures (Knorr, 1998). Some studies have investigated spray-drying of *L. acidophilus* cultures chosen on the basis of their health-promoting properties (Espina and Packard, 1979; Prajapati et al., 1986, 1987), and a patent application has been filed which describes a process for spray-drying probiotic LAB, including *Lactobacillus*, *Leuconostoc*, and *Bifidobacterium* spp. (Meister et al., 1998). Furthermore, Gardiner et al. (2000) investigated the use of spray-drying as a method of preservation of human-derived *L. paracasei* and L. *salivarius* strains, and they concluded that this method was applicable to probiotic lactobacilli, although cell damage was encountered, and subsequent viability during storage affected.

9.8 FACTORS AFFECTING PROBIOTIC SURVIVAL IN FOOD SYSTEMS

There are many technological challenges associated with the development of fermented milks that harboring high levels of viable microorganisms with desirable therapeutic activities. In many cases, probiotic cultures are not dairy microorganisms but are of gastrointestinal origin, with the result that milk and dairy products may not provide a stable environment for culture maintenance. This is illustrated by the fact that many probiotic cultures, especially bifidobacteria, do not grow well in milk (Klaver et al., 1993; Svensson, 1999). A further consideration is that a typical cultured milk product may have a pH of 3.5–4.5 and contain 0.5–1.5% (w/v) lactic acid (Lee and Salminen, 1995). Although the pH of yogurt is typically ~4.5 immediately post-fermentation, values may drop to 3.6 during storage, due to post-acidification, which is caused by growth of the starter culture at refrigeration temperatures (Kailasapathy and Rybka, 1997). This is one of the principal factors contributing to poor viability of probiotic cultures—in particular, bifidobacteria—in yogurts and fermented milks (Iwana et al., 1993; Kailasapathy and Rybka, 1997). There are, however, some approaches that may be taken to improve the viability of acid sensitive cultures in food systems. Microencapsulation has proved useful for the protection of cultures from the lethal effects of acid (Rao et al., 1989), and the *L. delbrueckii* spp. *bulgaricus* component of the yogurt starter culture may be replaced to overcome the post-acidification problem. It may also be possible to exploit an acid stress response such as that observed in *B. breve*, whereby survival in acidic and other physiologically stressful conditions was improved upon acid adaptation (Park et al., 1995). Some cultures, when added to fermented milk products early in the fermentation, may acquire this ability to tolerate acidic conditions, thereby improving their chances of survival. Indeed, acid adaptation has been shown to promote the survival of *Salmonella typhimurium* in fermented dairy products (Leyer and Johnson, 1993). Using such an approach, acid-tolerant strains could be either selected or developed for fermented milk applications.

Oxygen sensitivity poses certain problems for the industrial-scale cultivation of probiotic cultures and limits their applications in fermented milk products, where dissolved oxygen concentrations may be high. Oxygen toxicity is of particular relevance to bifidobacteria, as they are strict anaerobes (Scardovi, 1989). Certain measures have been proposed to limit the toxic effects of oxygen during the manufacture and storage of dairy products. Milk may be deaerated prior to fermentation, although this requires expensive specialized equipment (Tamime et al., 1995; Poirier et al., 1998). Alternatively, in yogurt manufacture a *S. thermophilus* starter culture with a high oxygen-utilizing capacity may be employed. This lowers the dissolved oxygen content, thereby creating a more favorable environment for growth and survival of oxygen-sensitive cultures such as bifidobacteria (Ishibashi and Shimamura, 1993; Shah, 1997). Superior survival of *L. acidophilus* and *Bifidobacterium* cultures was reported in yogurt made and stored in glass bottles (Ishibashi and Shimamura, 1993; Dave and Shah, 1997a), indicating that the use of oxygen-impermeable packaging may eliminate the toxic effects of oxygen during product storage. Further approaches that may be taken include the addition of reducing agents such as cysteine (Dave and Shah, 1997b) or oxygen scavengers such as ascorbic acid (Klaver et al., 1993; Dave and Shah, 1997c) and the selection of oxygen-tolerant strains. In addition, the exploitation of oxidative stress responses such as that described in *E. faecalis* (Ross and Claiborne, 1991, 1997) may improve the tolerance of probiotic cultures to oxygen toxicity.

It is also important that cultures chosen for their therapeutic properties are capable of withstanding processing parameters, usually designed to reduce or inactivate microbial populations, and that probiotic properties are maintained following processing. For example, thermotolerance is an important parameter when considering microbial survival in food processes such as spray-drying, which involve the use of relatively high temperatures. In general, probiotic cultures are of intestinal origin with an optimum growth temperature of 37°C and so they are particularly heat-labile. However, within the genera most often employed as probiotics, certain strains and species are more heat-resistant than others—namely, "thermophilic" lactobacilli (Kandler and Weiss, 1989), *E. faecalis and E. faecium* (Mundt, 1989) and the *Bifidobacterium* species *thermophilum* (Scardovi, 1989)—and these may therefore be technologically superior for certain dairy applications. Indeed, recent studies have shown that a probiotic *L. paracasei* strain was considerably more heat-resistant than a *L. salivarius* strain with subsequent superior survival demonstrated for the former during the preparation of dairy-based spray-dried powders (Gardiner et al., 2000). In addition to appropriate culture selection, induction of a heat shock response may improve technological suitability of some probiotic cultures. Indeed, an inducible thermotolerance in *L. bulgaricus* has been exploited to improve microbial survival during spray-drying (Teixeira et al., 1997).

It is evident that many technological hurdles must first be overcome in order to successfully develop therapeutic milk products. Therefore, as well as defining strain criteria necessary for exertion of clinical benefits, there are additional technological properties that deserve consideration in selecting probiotic strains that are sufficiently robust for the intended product application. Some cultures may be more suitable than others, with the result that each strain should be evaluated individually prior to use. However, as outlined above, the development of foods harboring high levels of viable probiotic organisms may require modification of process parameters or the use of techniques to protect cultures. Alternatively, it may be possible to generate stress-induced tolerant variants for food applications or to exploit molecular biology techniques to genetically engineer "superior" strains with desirable technological properties (Knorr, 1998).

9.9 PREBIOTICS

Prebiotics are nondigestible food ingredients that beneficially affect the host by stimulating the growth and/or activity of one or a limited number of bacteria in the colon, thus improving host health (Gibson and Roberfroid, 1995). In addition, when used in combination with probiotics (as synbiotics), prebiotics can provide an accessory to the probiotic concept. As with probiotics, a number of criteria must be satisfied in order for a food ingredient to be considered prebiotic, and these are listed as follows: they should be neither hydrolyzed nor absorbed in the upper part of the GIT; they should represent a selective substrate for one or a limited number of beneficial bacteria commensal to the colon, which are stimulated to grow and/or are metabolically activated; they should be able to alter the colonic microflora in favor of a healthier composition; and should be capable of inducing luminal or systemic effects that are beneficial to host health (Gibson and Roberfroid, 1995). As with probiotics, lactobacilli and bifidobacteria are usually the "beneficial microorganisms" targeted by prebiotics. However, prebiotics overcome the challenges of viability and colonization associated with probiotics (Gibson and McCartney, 1998; Gibson, 1999). Although any undigested food component that is selectively fermented by the beneficial members of the gut microflora may be considered prebiotic, most attention has focused on carbohydrates, and these have recently been reviewed by Crittenden (1999). Prebiotics can be disaccharides (such as lactulose and lactitol), oligosaccharides [such as fructooligosaccharides (FOS)], transgalactosylated oligosaccharides (TOS), and soybean oligosaccharides or polysaccharides (such as inulin and resistant starch) (Crittenden, 1999). While many well-designed trials have assessed the effect of various prebiotics on the human gut microflora, most current attention has focused on nondigestible oligosaccharides (Roberfroid, 1998b). In this respect, FOS have been shown to increase *Bifidobacterium* counts *in vitro* (Wang and Gibson, 1993) and *in vivo* (Buddington et al., 1996; Kleessen et al., 1997), with concomitant decreases in potential pathogens, such as fusobacteria, clostridia, and bacteroides observed in some cases (Gibson et al., 1995). Other oligosaccharides such as TOS and soybean oligosaccharides have also resulted in the numerical predominance of bifidobacteria in feces

(Benno et al., 1987; Bouhnik et al., 1997), while disaccharides, such as lactulose and lactitol, also show potential as selective fermentable substrates for desirable colonic bacteria (Ballongue et al., 1997).

As with probiotics, there is a need to assess the effect of prebiotics on microbial species and strains within total populations in the gut microflora, which should be made possible with the use of new molecular techniques. With regard to prebiotics, these foodstuffs manipulate the existing ecosystem, and therefore may not be of great value in situations where the gut microflora does not already contain beneficial microorganisms. Such cases may be more responsive to the use of synbiotics (mixtures of probiotics and prebiotics), which may improve intestinal survival and implantation of the live microbial supplement (Gibson and Roberfroid, 1995). Furthermore, prebiotics have potential in areas other than modification of the intestinal microflora, such as improvement of calcium bioavailability, reduction of the risk of development of precancerous lesions, and modification of lipid metabolism (for review, see Roberfroid, 1998a). Marketing in the area of prebiotics has concentrated principally on oligosaccharides. Although initially confined to the Japanese market, a number of products have recently appeared in Europe, containing prebiotic substances, in particular, inulin and the application of prebiotics to commercially available fermented milks and yogurts is evident (Table 9.5).

9.10 CONCLUSIONS

Probiotics are the focus of much research attention worldwide; consequently, scientific evidence is accumulating concerning the nutritional and therapeutic benefits of regular consumption of certain cultures. As a result, although it varies depending on the country in question, there are already a number of probiotic products on the market which are supported by considerable scientific research, and there is substantial evidence to demonstrate that there is a successful and expanding market for probiotic functional foods. However, the long-term exploitation of probiotics as health promoters is dependent upon a number of factors, including sound, scientifically proven clinical evidence of health-promoting activity, accurate consumer information, effective marketing strategies and, above all, a quality product that fulfills consumer expectations. Further research is needed in areas such as strain selection and mechanism of action, and consensus is needed regarding a definition of probiotic and functional foods, the minimum therapeutic dose necessary, and legislation relating to products containing probiotic cultures.

ACKNOWLEDGMENTS

Gillian E. Gardiner was supported by a Teagasc Walsh Fellowship. This work was supported by the European Research and Development Fund and by the European Union (SMT-CT98-2235).

REFERENCES

Adams, M. R. (1997) *BNF Nutr. Bull.*, **22**, 91–98.

Adams, M. R. (1999) *J. Biotechnol.*, **68**, 171–178.

Adams, M. R., and Marteau, P. (1995) *Int. J. Food Microbiol.*, **27**, 263–264.

Adawi, D., Kasravi, F. B., Molin, G., and Jeppsson, B. (1997) *Hepatology*, **25**, 642–647.

Agerbaek, M., Gerdes, L. U., and Richelsen, B. (1995) *Eur. J. Clin. Nutr.*, **49**, 346–352.

Aguirre, M., and Collins, M. D. (1993) *J. Appl. Bacteriol.*, **75**, 95–107.

Aiba, Y., Suzuki, N., Kabir, A. M., Takagi, A., and Koga, Y. (1998) *Am. J. Gastroenterol.*, **93**, 2097–2101.

Allen, W. D., Linggood, M. A., and Porter, P. (1996) *European patent* 0508701B1.

Anderson, J. W., and Gilliland, S. E. (1999) *J. Am. Coll. Nutr.*, **18**, 43–50.

Aronsson, B., Barany, P., and Nord, C. E. (1987) *Eur. J. Clin. Microbiol.*, **6**, 352–356.

Aso, Y., Akaza, H., Kotake, T., Tsukamoto, T., Imai, K., and Naito, S. (1995) *Eur. Urol.*, **27**, 104–109.

Aso, Y., and Akaza, H. (1992) *Urol. Int.*, **49**, 125–129.

Ballongue, J., Schumann, C., and Quignon, P. (1997) *Scand. J. Gastroenterol.*, **32**(Suppl.), 41–44.

Bazzoli, E., Zagari, R. M., Fossi, S., Morelli, M. C., Pozzato, P., Novelli, V., Ventrucci, M., Mazzella, G., Festi, D., and Roda, E. (1992) *Gastroenterology*, **102**, A38.

Bengmark, S. (1998) *Gut*, **42**, 2–7.

Benno, Y., Endo, K., Shiragami, N., Sayama, K., and Mitsouka, T. (1987) *Bifidobacteria Microflora*, **6**, 59–63.

Bhatia, S. J., Kochar, N., Abraham, P., Narr, N. G., and Mehta, A. P. (1989) *J. Clin. Microbiol.*, **27**, 2328–2330.

Biasco, G., Paganelli, G. M., Brandi, G., Brillanti, S., Lami, F., Callegari, C., and Gizzi, G. (1991) *Ital. J. Gastroenterol.*, **23**, 142.

Biller, J. A., Katz, A. J., Flores, A. F., Buie, T. M., and Gorbach, S. L. (1995) *J. Pediatr. Gastroenterol. Nutr.*, **21**, 224–226.

Black, F. T., Anderson, P. L., Oeskov, J., Gaarslev, K., and Laulund, S. (1989) *Travel Med.*, **7**, 333–335.

Blum, S., Delneste, Y., Alvarez, S., Haller, D., Perez, P. F., Bode, C. H., Hammes, W. P., Pfeifer, A. M. A., and Schiffrin, E. J. (1999) *Int. Dairy J.*, **9**, 63–68.

Bouhnik, Y. B., Flourie, B., D'Agay-Abensour, L., Pochart, P., Gramet, G., Durand, M., and Rambaud, J. C. (1997) *J. Nutr.* **127**, 444–448.

Brassart D., and Schiffrin, E. (1997) *Trends Food Sci. Technol.*, **8**, 321–326.

Buck, L. M., and Gilliland, S. E. (1994) *J. Dairy Sci.*, **77**, 2925–2933.

Buddington, R. K., Williams, C. H., Chen, S. C., and Witherly, S. A. (1996) *Am. J. Clin. Nutr.*, **63**, 709–716.

Caric, M. (1994) *Concentrated and Dried Dairy Products*, VCH Publishers, New York.

Christiansen, P. S., Edelsten, D., Kristiansen, J. R., and Nielsen, E. W. (1996) *Milchwissenschaft*, **51**, 502–504.

Clements, M. L., Levine, M. M., and Black, R. E. (1981) *Antimicrob. Agents Chemother*, **20**, 104–108.

Collins, J. K., O'Sullivan, G., and Shanahan, F. (1996) In *Gut Flora and Health-Past, Present and Future*, R. A. Leeds and I. R. Rowland, eds., International Congress and Symposium Series No. 219. Royal Society of Medicine Press, London, pp. 13–18.

Collins, J. K., Thornton, G., and O'Sullivan, G. (1998) *Int. Dairy J.*, **8**, 487–490.

Colombel, J. F., Cortot, A., Neut, C., and Romond, C. (1987) *Lancet*, **II**, 43.

Conway, P. L. (1996) *Asia Pac. J. Clin. Nutr.*, **5**, 10–14.

Corthier, G. (1997) In *Probiotics 2: Applications and Practical Aspects*, R. Fuller, ed., Chapman and Hall, London, pp. 41–63.

Corzo, G., and Gilliland, S. E. (1999) *J. Dairy Sci.*, **82**, 466–471.

Coussement, P. (1997) *World of Ingredients*, **7**, 12–17.

Crittenden, R. G. (1999) In *Probiotics: A Critical Review*, G. W. Tannock, ed., Horizon Scientific Press, Wymondham, United Kingdom, pp. 141–156.

Dave, R. I., and Shah, N. P. (1997a) *Int. Dairy J.*, **7**, 31–41.

Dave, R. I., and Shah, N. P. (1997b) *Int. Dairy J.*, **7**, 537–545.

Dave, R. I., and Shah, N. P. (1997c) *Int. Dairy J.*, **7**, 435–443.

De Simone, C., Salvadori, B. B., Negri R., Ferrazzi, M., Baldinelli, L., and Vesely, R. (1986) *Nutr. Rep. Int.*, **33**, 419–433.

De Simone, C., Vesely, R., Negri, R., Bianchi Salvadori, B., Zanzoglu, S., Cilli, A., and Lucci, L. (1987) *Immunopharmacol. Immunotoxicol.*, **9**, 87–100.

De Simone, C., Tzantzoglou, S., Baldinelli, L., Di Fabio, S., Bianchi-Salvadori, B., Jirillo, E., and Vesely, R. (1988) *Immunopharmacol. Immunotoxicol.*, **10**, 399–415.

De Smet, J., Van Hoorde, L., De Saeyer, N., Vande Woestyne, M., and Verstraete, W. (1994) *Microb. Ecol. Health Dis.*, **7**, 315–329.

De Smet, I., De Boever, P., and Verstraete, W. (1998) *Br. J. Nutr.*, **79**, 185–194.

Descheemaeker, P., Lammens, C., Pot, B., Vandamme, P., and Goossens, H. (1997) *Int. J. Syst. Bacteriol.*, **47**, 555–561.

Donnet-Hughes, Rochat, F., Errant, P. S., Aeschlimann, J. M., and Schiffrin, E. (1999) *J. Dairy Sci.*, **82**, 863–869.

Donohue, D. C., and Salminen, S. (1996) *Asia Pac. J. Clin. Nutr.*, **5**, 25–28.

Drago, L., Gismondo, M. R., Lombardi, A., De Haen, C., and Gozzini, L. (1997) *FEMS Microbiol. Lett.*, **153**, 455–463.

Dunne, C., Murphy, L., Flynn, S., O'Mahony, L., O'Halloran, S., Feeney, M., Morrissey, D., Thornton, G., Fitzgerald, G., Daly, C., Kiely, B., Quigley, E. M. M., O'Sullivan, G. C., Shanahan, F., and Collins, J. K. (1999) *Antonie van Leeuwenhoek* **76**, 279–292.

Espina, F., and Packard, V. S. (1979) *J. Food Prot.*, **42**, 149–152.

Famularo, G., Moretti, S., Marcellini, S., and De Simone, C. (1997) In *Probiotics 2: Applications and Practical Aspects*, R. Fuller, ed., Chapman and Hall, London, pp. 133–161.

FAO/WHO (2001) Report on Joint FAO/WHO Expert Consultation on Evaluation of Health and Nutritional Properties of Probiotics in Food including Powder Milk with Live Lactic Acid Bacteria. http://www.fao.org/es/ESN/Probio/Probio.htm.

Fernandes, C. F., Chandan, R. C., and Shahani, K. M. (1992) In *Lactic Acid Bacteria in Health and Disease*, B. J. B. Wood, ed., Elsevier Applied Science Publishers, London, pp. 297–339.

Fooks, L. J., Fuller, R., and Gibson, G. R. (1999) *Int. Dairy J.*, **9**, 53–61.

Franz, C. M. A. P., Holzapfel, W. H., and Stiles, M. E. (1999) *Int. J. Food Microbiol.*, **47**, 1–24.

Friend, B. A., and Shahani, K. M. (1984) *J. Appl. Nutr.*, **36**, 126–153.

Fukushima, Y., Kawata, Y., Mizumachi, K., Kurisaki, J., and Mitsuoka, T. (1999) *Int. J. Food Microbiol.*, **46**, 193–197.

Fukushima, Y., Kawata, Y., Hara, H., Terada, A., and Mitsuoka, T. (1998) *Int. J. Food Microbiol.*, **42**, 39–44.

Fuller, R. (1993) *Int. Food Ingredients*, **3**, 23–26.

Fuller, R. (1999) In *Probiotics: A Critical Review*, G. W. Tannock, ed., Horizon Scientific Press, Wymondham, United Kingdom, pp. 15–22.

Gagne, J., Roy, D., and Gauthier, S. F. (1993) *Milchwissenschaft*, **48**, 501–505.

Gardiner, G., Ross, R. P., Collins, J. K., Fitzgerald, G., and Stanton, C. (1998) *Appl. Environ. Microbiol.*, **64**, 2192–2199.

Gardiner, G., Ross, R. P., Wallace, J. M., Scanlan, F. P., Jagers, P. P. J. M., Fitzgerald, G. F., Collins, J. K., and Stanton, C. (1999a) *J. Agric. Food Chem.*, **47**, 4907–4916.

Gardiner, G., Stanton, C., Lynch, P. B., Collins, J. K., Fitzgerald, G., and Ross, R. P. (1999b) *J. Dairy Sci.*, **82**, 1379–1387.

Gardiner, G. E., O'Sullivan, E., Kelly, J., Auty, M. A. E., Fitzgerald, G. F., Collins, J. K., Ross, R. P., and Stanton, C. (2000) *Appl. Environ. Microbiol.*, **66**, 2605–2612.

Garg, S. K., and Mital, B. K. (1991) *Crit. Rev. Microbiol.*, **18**, 15–45.

Gibson, G. R. (1999) *J. Nutr.*, **129**(Suppl.), 1438S–1441S.

Gibson, G. R., and McCartney, A. L. (1998) Modification of the gut flora by dietary means. *Biochem. Soc. Trans.*, **26**, 222–228.

Gibson, G. R., and Roberfroid, M. B. (1995) *J. Nutr.*, **125**, 1401–1412.

Gibson, G. R., and Wang, X. (1994) *J. Appl. Bacteriol.*, **77**, 412–420.

Gibson, G. R., Beatty, E. R., Wang, X., and Cumings, J. H. (1995) *Gastroenterology*, **108**, 975–982.

Gibson, G. R., Saavedra, J. M., MacFarlane, S., and MacFarlane, G. T. (1997) In *Probiotics 2: Applications and Practical Aspects*, R. Fuller, ed., Chapman and Hall, pp. 10–39.

Gill, H. (1998). *Int. Dairy J.* **8**, 535–544.

Gilliland, S. E., and Kim, H. S. (1984) *J. Dairy Sci.*, **67**, 1–6.

Gilliland, S. E., and Lara, R. C. (1988) *Appl. Environ. Microbiol.*, **54**, 898–902.

Gilliland, S. E., and Walker, D. K. (1990) *J. Dairy Sci.*, **73**, 905–911.

Gilliland, S. E., Nelson, C. R., and Maxwell, C. (1985) *Appl. Environ. Microbiol.*, **49**, 377–381.

Giraffa, G., Carminati, D., and Neviani, E. (1997) *J. Food Protection*, **6**, 732–738.

Glass, R. I., Lew, J. F., Gangarosa, R. E., LeBaron, C. W., and Ho, M. S. (1991) *J. Pediatr.*, **118**(Suppl.), S27–S33.

Goldin, B. R. (1998) *Br. J. Nutr.*, **80**(Suppl.), S203–S207.

Goldin, B. R., and Gorbach, S. L. (1984) *Am. J. Clin. Nutr.*, **33**, 15–18.

Gomes, A. M. P., Malcata, F. X., Klaver, F. A. M., and Grande, H. G. (1995) *Netherlands Milk Dairy J.*, **49**, 71–95.

Gomes, A. M. P., Malcata, F. X., and Klaver, F. A. M. (1998) *J. Dairy Sci.*, **81**, 2817–2825.

Gomez Zavaglia, A., Kociubinski, G., Perez, P., and De Antoni, G. (1998) *J. Food Prot.*, **61**, 865–873.

Gorbach, S. L., Chang, T., and Goldin, B. (1987) *Lancet*, **II**, 1519.

Grill, J. P., Manginot-Durr, C., Schneider, F., and Ballongue, J. (1995) *Curr. Microbiol.*, **31**, 23–27.

Guarino, A., Canani, R. B., Spagnuolo, M. I., Albano, F., and Di Benedetto, L. (1997) *J. Pediatr. Gastroenterol. Nutr.*, **25**, 516–519.

Guarner, F., and Schaafsma, G. J. (1998) *Int. J. Food Microbiol.*, **39**, 237–238.

Hamilton-Miller, J. M. T., and Fuller, R. (1996) *Br. Nutr. Fed. Nutr. Bull.*, **21**, 199–208.

Hamilton-Miller, J. M. T., and Gibson, G. R. (1999) *Br. J. Nutr.*, **82**, 73–75.

Hamilton-Miller, J. M. T., Shah, S., and Smith, C. T. (1996) *Br. Med. J.*, **312**, 55–56.

Hamilton-Miller, J. M., Shah, S., and Winkler, J. T. (1999) *Public Health Nutr.*, **2**(Suppl.), 223–229.

Hammes, W. P., and Vogel, R. F. (1995) In *Genera of Lactic Acid Bacteria*, W. H. N. Holzhapfel, ed., Chapman and Hall, London, pp. 20–54.

Hasler, C. M. (1998) *Chem. Ind.*, **3**, 84–89.

Havenaar, R., and Huis in't Veld, J. H. J. (1992) In *Lactic Acid Bacteria in Health and Disease*, B. J. B. Wood, ed., Elsevier Applied Science Publishers, London, pp. 151–170.

Hayatsu, H., and Hayatsu, T. (1993) *Cancer Lett.*, **73**, 173–179.

Heasman, M., and Mellentin, J. (1998) *Int. Food Ingredients*, **4**, 40–41.

Hepner, G., Fried, R., Joer, S. S., Fusetti, L., and Morin, R. (1979) *Am. J. Clin. Nutr.*, **32**, 19–24.

Hilliam, M. (1998) *World of Ingredients*, **2**, 45–47.

Hilton, E., Kolakowski, P., Smith, M., and Singer, C. (1997) *J. Travel Med.*, **4**, 3–7.

Holzhapfel, W. H., Haberer, P., Snel, J., Schillinger, U., and Huis in't Veld, J. H. J. (1998) *Int. J. Food Microbiol.*, **41**, 85–101.

Hotta, M., Sato, Y., Iwata, S., Yamashita, N., Sunakawa, K., Oikawa, T., Tanaka, R., Watanabe, K., Takayama, H., Yajima, M., Sekiguchi, S., Arai, S., Sakurai, T., and Mutai, M. (1987) *Keio J. Med.*, **36**, 298–314.

Hove, H., Nordgard-Andersen, M., and Mortensen, P. B. (1994) *Am. J. Clin. Nutr.*, **59**, 74–79.

Howard, A. N., and Marks, J. (1979) *Lancet* **II**, 957.

Hughes, D. B., and Hoover, D. G. (1991) *Food Technol.*, **45**, 74–83.

Huis in't Veld, J. H. J., and Shortt, C. (1996) In *Gut Flora and Health-Past, Present and Future*, R. A. Leeds and I. R. Rowland, eds., International Congress and Symposium Series No. 219. Royal Society of Medicine Press, London, pp. 27–36.

Huis in't Veld, J. H. J., Havenaar, R., and Marteau, P. (1994) *Trends Biotechnol.*, **12**, 6–8.

Ishibashi, N., and Shimamura, S. (1993) *Food Technol.*, **46**, 126–135.

Isolauri, E., Juntunen, M., Rautanen, T., Sillanaukee, P., and Koivula, T. (1991) *Pediatrics*, **88**, 90–97.

Isolauri, E., Joensuu, J., Suomalainen, H., Luomala, M., and Vesikari, T. (1995) *Vaccine*, **13**, 310–312.

Iwana, H., Masuda, H., Fujisawa, T., Suzuki, H., and Mitsouka, T. (1993) *Bifidobacteria Microflora*, **12**, 39–45.

Jaspers, D. A., Massey, L. K., and Luedecke, L. O. (1984) *J. Food Sci.*, **49**, 1178–1181.

Jett, B. D., Huycke, M. M., and Gilmore, M. S. (1994) *Clin. Microbiol. Rev.*, **7**, 462–478.

Jiang, T., and Savaiano, D. A. (1997) *J. Nutr.*, **127**, 1489–1495.

Jiang, T., Mustapha, A., and Savaiano, D. A. (1996) *J. Dairy Sci.*, **79**, 750–757.

Johnson, J. A. C., and Etzel, M. R. (1993) In *Food Dehydration*, G. V. Barbosa-Canovas and M. R. Okos, eds., Institute of Chemical Engineering, New York, pp. 98–107.

Johnson, J. A. C., and Etzel, M. R. (1995) *J. Dairy Sci.*, **78**, 761–768.

Kaila, M., Isolauri, E., Soppi, E., Virtanen, E., Laine, S., and Arvilommi, H. (1992) *Pediatr. Res.*, **32**, 141–144.

Kaila, M., Isolauri, E., Saxelin, M., Arvilommi, H., and Vesikari, T. (1995) Viable versus inactivated *Lactobacillus* strain GG in acute rotavirus diarrhea. *Arch. Dis. Child.*, **72**, 51–53.

Kailasapathy, K., and Rybka, S. (1997) *Aust. J. Dairy Technol.*, **52**, 28–35.

Kandler, O., and Weiss, N. (1989) In *Bergey's Manual of Determinative Bacteriology*, Vol. 2, P. H. A. Sneath, ed., Williams & Wilkins, Baltimore, pp. 1208–1234.

Kato, I., Endo, K., and Yokokura, T. (1994) *Int. J. Immunopharmacol.*, **16**, 29–36.

Kim, H. S., and Gilliland, S. E. (1983) *J. Dairy Sci.*, **66**, 959–966.

Kim, S. S., and Bhowmik, S. R. (1990) *J. Food Sci.*, **55**, 1008–1010.

Kimura, K., McCartney, A. L., McConnell, M. A., and Tannock, G. W. (1997) *Appl. Environ. Microbiol.*, **63**, 3394–3398.

Kitazawa, H., Matsumura, K., Itoh, T., and Yamaguchi, T. (1992) *Microbiol. Immunol.*, **36**, 31–315.

Klaver, F. A., and Van der Meer, R. (1993) *Appl. Environ. Microbiol.*, **59**, 1120–1124.

Klaver, F. A. M., Kingma, F., and Weerkamp, A. H. (1993) *Netherlands Milk Dairy J.*, **47**, 151–164.

Kleessen, B., Stoof, G., Proll, J., Schmiedl, D., Noack, J., and Blaut, M. (1997) *J. Animal Sci.*, **75**, 2453–2462.

Klein, G., Pack, A., Bonaparte, C., and Reuter, G. (1998) *Int. J. Food Microbiol.*, **41**, 103–125.

Knorr, D. (1998) *Trends Food Sci. Technol.*, **9**, 295–306.

Kolars, J., Levitt, M., Aouji, M., and Savaiano, D. (1984) *N. Engl. J. Med.*, **310**, 1–3.

Koo, M., and Rao, A. V. (1991) *Nutr. Cancer*, **16**, 249–257.

Kosikowski, F., and Mistry, V. V. (1997) *Cheese and Fermented Milk Foods*, Vol. I: *Origins and Principles*. Edwards Brothers, Ann Arbor, MI.

Kulkarni, N., and Reddy, B. S. (1994) *Proc. Soc. Exp. Biol. Med.*, **207**, 278–283.

Lambert, J., and Hull, R. (1996) *Asia Pac. J. Clin. Nutr.*, **5**, 31–35.

Laroia, S., and Martin, J. H. (1991) *Cult. Dairy Prod. J.*, **26**, 13–21.

Lee, Y. K., and Salminen, S. (1995) *Trends Food Sci. Technol.*, **6**, 241–245.

Lemonnier and Solis Pereya (1993)

Lewis, S. J., and Freedman, A. R. (1998) *Aliment. Pharmacol. Ther.*, **12**, 807–822.

Lewis, S. J., Potts, L. F., and Barry, R. E. (1998) *J. Infect.*, **36**, 171–174.

Leyer, G. J., and Johnson, E. A. (1993) *Appl. Environ. Microbiol.*, **59**, 1842–1847.

Lidbeck, A., Overvick, E., Rafter, J., Nord, C. E., and Gustafsson, J. A. (1992) *Microb. Ecol. Health Dis.*, **5**, 59–67.

Lilly, D. M., and Stillwell, R. H. (1965) *Science*, **147**, 747–748.

Lin, M. Y., Savaiano, D., and Harlander, S. (1991) *J. Dairy Sci.*, **74**, 87–95.

Lin, M. Y., Yen, C. L., and Chen, S. S. (1998) *Dig. Dis. Sci.*, **43**, 133–137.

Lin, S. Y., Ayres, J. W., Winkler, W., Jr., and Sandine, W. E. (1989) *J. Dairy Sci.*, **72**, 2885–2899.

Ling, W. H., Korpela, R., Mykkanen, H., Salminen, S., and Hanninen, O. (1994) *J. Nutr.*, **124**, 18–23.

Link-Amster, H., Rochat, F., Saudan, K. Y., Mignot, O., and Aeschlimann, J. M. (1994) *FEMS Immunol. Med. Microbiol.*, **10**, 55–64.

Madsen, K. L., Doyle, J. S., Jewell, L. D., Tavernini, M. M., and Fedorak, R. N. (1999) *Gastroenterology*, **116**, 1107–1114.

Majamaa, H., and Isolauri, E. (1997) *J. Allergy Clin. Immunol.*, **99**, 179–185.

Mann, G. V. (1977) *Atherosclerosis*, **26**, 335–340.

Mann, G. V., and Spoerry, A. (1974) *Am. J. Clin. Nutr.*, **27**, 464–469.

Marshall, V. M., and Tamime, A. Y. (1997) *Int. J. Dairy Technol.*, **50**, 35–41.

Marteau, P., and Rambaud, J. C. (1993) *FEMS Microbiol. Rev.*, **12**, 207–220.

Marteau, P., Pochart, P., Flourie, B., Pellier, P., Santos, L., Desjeux, J. F., and Rambaud, J. C. (1990) *Am. J. Clin. Nutr.*, **52**, 685–688.

Marteau, P., Minekus, M., Havenaar, R., and Huis in't Veld, J. H. J. (1997) *J. Dairy Sci.*, **80**, 1031–1037.

Masters, K. (1985) In *Evaporation, Membrane Filtration and Spray Drying in Milk Powder and Cheese Production*, R. Hansen, ed., North European Dairy Journal, Vanlose, Denmark, pp. 393–403.

Mattila-Sandholm, T., Matto, J., and Saarela, M. (1999) *Int. Dairy J.*, **9**, 25–35.

Mauriello, G., Aponte, M., Andolfi, R., Moschetti, G., and Villani, F. (1999) *J. Food Prot.*, **62**, 773–777.

McCartney, A. L., Wenzhi, W., and Tannock, G. W. (1996) *Appl. Environ. Microbiol.*, **62**, 4608–4613.

McCracken, V. J., and Gaskins, H. R. (1999) In *Probiotics: A Critical Review*, G. W. Tannock, ed., Horizon Scientific Press, Wymondham, United Kingdom.

McFarland, L. V., Surawicz, C. M., Greenberg, R. N., Elmer, G. W., Moyer, K. A., Melcher, S. A., Bowen, K. E., and Cox, J. L. (1995) *Am. J. Gastroenterol.*, **90**, 439–448.

Meister, N., Sutter, A., and Vikas, M. (1998) Patent No. WO 98/10666.

Metchnikoff, E. (1907) *The Prolongation of Life: Optimistic Studies*, William Heinemann, London.

Metwally, M. M., Abd El Gawad, I. A., el Nockrashy, S. A., and Ahmed, K. E. (1989) *Egypt. J. Dairy Sci.*, **17**, 35–43.

Micanel, N., Haynes, I. N., and Playne, M. J. (1997) *Aust. J. Dairy Technol.*, **52**, 24–27.

Michetti, P., Dorta, G., Brassard, D., and Vouillamoz, D. (1995) *Gastroenterology*, **108**, 166.

Midolo, P. D., Lambert, J. R., Hull, R., Luo, F., and Grayson, M. L. (1995) *J. Appl. Bacteriol.*, **79**, 475–479.

Misra, A. K., and Kuila, R. K. (1991) *Cult. Dairy Prod. J.*, **26**, 4–6.

Modler, H. W., McKellar, R. C., and Yaguchi, M. (1990) *Can. Inst. Food Sci. Technol. J.*, **23**, 29–41.

Molly, K., Vande Woestyne, M., De Smet, I., and Verstraete, W. (1994) *Microb. Ecol. Health Dis.*, **7**, 191–200.

Moore, W. E. C., and Moore, L. H. (1995) *Appl. Environ. Microbiol.*, **61**, 3202–3207.

Motta, L., Blancato, G., Scornavacca, G., De Luca, M., Vasquez, E., Gismondo, M. R., Lo Bue, A., and Chisari, G. (1991) *Clin. Ther.*, **138**, 27–35.

Mundt, O. (1989) In *Bergey's Manual of Determinative Bacteriology*, Vol. 2, P. H. A. Sneath, ed., Williams & Wilkins, Baltimore, pp. 1063–1065.

Murray, B. E. (1990) *Clin. Microbiol. Rev.*, **3**, 46–65.

Muscettola, M., Massai, L., Tanganelli, C., and Grasso, G. (1994) *Ann. N. Y. Acad. Sci.*, **717**, 226–232.

Mustapha, A., Jiang, T., and Savaiano, D. A. (1997) *J. Dairy Sci.*, **80**, 1537–1545.

Nadathur, S. R., Gould, S. J., and Bakalinsky, A. T. (1994) *J. Dairy Sci.*, **77**, 3287–3295.

Nagafuchi, S., Takahashi, T., Yajima, T., Kuwata, T., Hirayama, K., and Itoh, K. (1999) *Biosci. Biotechnol. Biochem.*, **63**, 474–479.

Naidu, A. S., Bidlack, W. R., and Clemens, R. A. (1999) *Crit. Rev. Food Sci. Nutr.*, **38**, 13–126.

Namba, Y., Hidaka, Y., Taki, K., and Morimoto, T. (1981) *Infect. Immun.*, **31**, 580–583.

Neu, H. C. (1992) *Science*, **257**, 1064–1073.

Oberhelman, R. A., Gilman, R. H., Sheen, P., Taylor, D. N., Black, R. E., Cabrera, L., Lescano, A. G., Meza, A. G. R., and Madico, G. (1999) *J. Pediatr.*, **134**, 1–2.

Oggioni, M. R., Pozzi, G., Valensin, P. E., Galieni, P., and Bigazzi, C. (1998) *J. Clin. Microbiol.*, **36**, 325–326.

Okawa, T., Kita, M., Arai, T., Iida, K., Dokiya, T., Takegawa, Y., Hirokawa, Y., Yamazaki, K., and Hashimoto, S. (1989) *Cancer*, **64**, 1769–1776.

Okawa, T., Niibe, H., Arai, K., Sekiba, K. Noda, S., Hashimoto, S., and Ogawa, N. (1993) *Cancer*, **72**, 1949–1954.

Oksanen, P., Salminen, S., Saxelin, M., Hamalainen, P., Ihantola-Vormisto, A., Muurasniemi-Isoviita, L., Nikkari, S., Oksanen, T., Porsti, I., Salminen, E., Siitonen, S., Stuckey, H., Toppila, A., and Vapaatalo, H. (1990) *Ann. Med.*, **22**, 53–56.

Orrhage, K., Sillerstrom, E., Gustafsson, J. A., Nord, C. E., and Rafter, J. (1994) *Mutat. Res.*, **311**, 239–248.

Ouwehand, A. C., and Salminen, S. J. (1998) *Int. Dairy J.* **8**, 749–758.

Ouwehand, A. C., Niemi, P., and Salminen, S. J. (1999a) *FEMS Microbiol. Lett.*, **177**, 35–38.

Ouwehand, A. C., Kirjavainen, P. V., Shortt, C., and Salminen, S. (1999b) *Int. Dairy J.*, **9**, 43–52.

Pant, A. R., Graham, S. M., Allen, S. J., Harikul, S., Sabchaeron, A., Cuevas, L., and Hart, C. A. (1996) *J. Trop. Pediatr.*, **42**, 162–165.

Park, H. K., So, J. S., and Heo, T. R. (1995) *Food Biotechnol.*, **4**, 226–230.

Paubert-Braquet, M., Gan, X. N., Gaudichon, C., Hedef, N., Serikoff, A., Bouley, C., Bonavida, B., and Braquet, P. (1995) *Int. J. Immunother.*, **11**, 153–161.

Pelto, L., Salminen, S., Lilius, E.-M., Nuutila, J., and Isolauri, E. (1998) *Allergy*, **53**, 307–310.

Perdigon, G., Nader de Macias, M. E., Alvarez, S., Medici, M., Oliver, G., and De Ruiz Holgado, A. P. (1986) *J. Food Prot.*, **49**, 986–989.

Perdigon, G., De Macias, M. E. N., Alvarez, S., Oliver, G., and De Ruiz Holgado, A. A. (1988) *Immunology*, **63**, 17–23.

Perdigon, G., Alvarez, S., Nader de Macias, M. E., Roux, M. E., and De Ruiz Holgado, A. P. (1990) *J. Food Prot.*, **53**, 404–410.

Pletincx, M., Legein, J., and Vandenplas, Y. (1995) *J. Pediatr. Gastroenterol. Nutr.*, **21**, 113–115.

Poch, M., and Bezkorovainy, A. (1988) *J. Dairy Sci.*, **71**, 3214–3221.

Poirier, I., Marechal, P. A., Evrard, C., and Gervais, P. (1998) *Appl. Microbiol. Biotechnol.*, **50**, 704–709.

Pool-Zobel, B. L., Bertram, B., Knoll, M., Lambertz, R., Neudecker, C., Schillinger, U., Schmezer, P., and Holzapfel, W. H. (1993) *Nutr. Cancer*, **20**, 271–282.

Pool-Zobel, B. L., Neudecker, C., Domizlaff, I., Ji, S., Schillinger, U., Rumney, C., Moretti, M., Vilarini, I., Scasselati-Sforzolini, R., and Rowland, I. (1996) *Nutr. Cancer*, **26**, 365–380.

Poupard, J. A., Husain, I., and Norris, R. F. (1973) *Bacteriol. Rev.*, **37**, 136–165.

Prajapati, J. B., Shah, R. K., and Dave, J. M. (1986) *Cult. Dairy Prod. J.*, **21**, 16–21.

Prajapati, J. B., Shah, R. K., and Dave, J. M. (1987) *Aust. J. Dairy Technol.*, **42**, 17–21.

Prescott, L. M., Harley, J. P., and Klein, D. A. (1990) *Microbiology*, Wm. C. Browne Publishers, Dubuque, IA.

Rao, A. V., Shiwnarain, N., and Maharaj, I. (1989) *Can. Inst. Food Sci. Technol. J.*, **22**, 345–349.

Rao, C. V., Sanders, M. E., Indranie, C., Simi, B., and Reddy, B. S. (1999) *Int. J. Oncol.*, **14**, 939–944.

Raza, S., Graham, S. M., Allen, S. J., Sultana, S., Cuevas, L., and Hart, C. A. (1995) *Pediatr. Infect. Dis. J.*, **14**, 107–111.

Reddy. B. S. (1999) *J. Nutr.*, **129**(Suppl.), 1478S–1482S.

Reddy, B. S., and Rivenson, A. (1993) *Cancer Res.*, **53**, 3914–318.

Reid, G. (1996) *J. Am. Med. Assoc.*, **276**, 29–30.

Reid, G., Bruce, A. W., and Smeianov, V. (1998) *Int. Dairy J.*, **8**, 555–562.

Richelsen, B., Kristensen, K., and Pedersen, S. B. (1996) *Eur. J. Clin. Nutr.*, **50**, 811–815.

Roberfroid, M. B. (1998a) *Br. J. Nutr.*, **80**(Suppl.), S197–S202.

Roberfroid, M. B. (1998b) *J. Nutr.*, **129**(Suppl.), 1398S–1401S.

Roberts, C. M., Fett, W. F., Osman, S. F., Wijey, C., O'Connor, J. V., and Hoover, D. G. (1995) *J. Appl. Bacteriol.*, **78**, 463–468.

Ross, R. P., and Claiborne, A. (1991) *J. Mol. Biol.*, **221**, 857–871.

Ross, R. P., and Claiborne, A. (1997) *FEMS Microbiol. Lett.*, **151**, 177–183.

Rossouw, J. E., Burger, E. M., van der Vyver, P., and Ferreira, J. J. (1981) *Am. J. Clin. Nutr.*, **34**, 351–356.

Rowland, I. R. (1996) In *Gut Flora and Health-Past, Present and Future*, R. A. Leeds and I. R. Rowland, eds., International Congress and Symposium Series No. 219, Royal Society of Medicine Press, London, pp. 19–25.

Rowland, I. R., Rumney, C. J., Coutts, J. T., and Lievense, L. C. (1998) *Carcinogenesis*, **19**, 281–285.

Russell, P. (1996) *Eur. Dairy Mag.*, **6**, 25–26.

Saavedra, J. M., Bauman, N., Oung, I., Perman, J., and Yolken, R. (1994) *Lancet*, **344**, 1046–1049.

Salminen, S., and Marteau, P. (1998) In *Proceedings Lactic 97*, September 10–12, 1997, Caen, France, pp. 329–340.

Salminen, S., Laine, M., von Wright, A., Vuopio-Varkila, J., Korhonen, T., and Mattila-Sandholm, T. (1996a) *Biosci. Microflora*, **15**, 61–67.

Salminen, S., Isolauri, E., and Salminen, E. (1996b) *Antonie van Leeuwenhoek*, **70**, 347–358.

Salminen, S., Ouwehand, A. G., and Isolauri, E. (1998) *Int. Dairy J.*, **8**, 563–572.

Salminen, S., Ouwehand, A. G., Benno, Y., and Lee, Y. K. (1999) *Trends Food Sci. Technol.*, **10**, 107–110.

Saltzman, J. R., Russell, R. M., Golner, B., Barakat, S., Dallal, G. E., and Goldin, B. R. (1999) *Am. J. Clin. Nutr.*, **69**, 140–146.

Sanders, M. E. (1993) In *Functional Foods: Designer Foods, Pharmafoods and Neutraceuticals*, I. Goldberg, ed., Chapman and Hall, New York, pp. 294–322.

Sato, K., Saito, H., and Tomioka, H. (1988) *Microbiol. Immunol.*, **32**, 689–698.

Savaiano, D. A., El Anouar, A. A., Smith, D. E., and Levitt, M. D. (1984) *Am. J. Clin. Nutr.*, **40**, 1219–1223.

Saxelin, M. (1997) *Food Rev. Int.*, **13**, 293–313.

Saxelin, M., Elo, S., Salminen, S., and Vapaatalo, H. (1991) *Microbiol. Ecol. Health Dis.*, **4**, 209–214.

Saxelin, M., Ahokas, M., and Salminen, S. (1993) *Microb. Ecol. Health Dis.*, **6**, 119–122.

Saxelin, M., Pessi, T., and Salminen, S. (1995) *Int. J. Food Microbiol.*, **25**, 199–203.

Scardovi, V. (1989) In *Bergey's Manual of Determinative Bacteriology*, Vol. 2, P. H. A. Sneath, ed., Williams & Wilkins, Baltimore, pp. 1418–1434.

Schiffrin, E. J., Rochat, F., Link-Amster, H. L., Aeschlimann, J. M., and Donnet-Hughes, A. (1995) *J. Dairy Sci.*, **78**, 491–497.

Schiffrin, E. J., Brassart, D., Servin, A. L., Rochat, F., and Donnet-Hughes, A. (1997) *Am. J. Clin. Nutr.*, **66**(Suppl.), 515S–520S.

Schleifer, K. H., and Kilpper-Balz, R. (1984) *Int. J. Syst. Bacteriol.*, **34**, 31–34.

Scrimshaw, N., and Murray, E. (1988) *Am. J. Clin. Nutr.*, **48**, 1086–1098.

Sessions, V. A., Lovegrove, J. A., Taylor, G. R. J., Dean, T. S., Williams, C. M., Sanders, T. A. B., MacDonald, I., and Salter, A. (1997) *Proc. Nutr. Soc.*, **56**, 120A.

Shah, N. P. (1997) *Milchwissenschaft*, **52**, 16–20.

Shanahan, F. (2000) *Inflammatory Bowel Dis.*, **6**, 107–115.

Shortt, C. (1998) *Chemistry Ind.*, **8**, 300–303.

Siitonen, Stuckey, H., Toppila, A., and Vapaatalo, H. (1990) *Ann. Med.*, **22**, 53–56.

Solis Pereyra, B., and Lemmonier, D. (1993) *Nutr. Res.*, **13**, 1127–1140.

Stanton, C., Gardiner, G., Lynch, P. B., Collins, J. K., Fitzgerald, G., and Ross, R. P. (1998) *Int. Dairy J.*, **8**, 491–496.

Stanton, C., Gardiner, G., Meehan, H., Collins, J. K., Fitzgerald, G., Lynch, P. B., and Ross, R. P. (2001) *Am. J. Clin. Nutr.*, **73**(suppl.), 4765–4835.

Sugita, T., and Togawa, M. (1994) *Jpn. J. Pediatr.*, **47**, 2755–2762.

Svensson, U. (1999) In *Probiotics: A Critical Review*, G. W. Tannock, ed., Horizon Scientific Press, Wymondham, United Kingdom, pp. 57–64.

Tamime, A. Y., and Deeth, H. C. (1980) *J. Food Prot.*, **43**, 939–977.

Tannock, G. W. (1999) In *Probiotics: A Critical Review*, G. W. Tannock, ed., Horizon Scientific Press, Wymondham, United Kingdom, pp. 45–56.

Tamime A. Y., Marshall, V. M., and Robinson, R. K. (1995) *J. Dairy Research*, **62**, 151–187.

Taranto, M. P., Medici, M., Perdigon, G., Ruiz Holgado, A. P., and Valdez, G. F. (1998) *J. Dairy Sci.*, **81**, 2336–2340.

Taylor, G. R. J., and Williams, C. M. (1998) *Br. J. Nutr.*, **80**(Suppl.), S225–S230.

Teixeira, P., Castro, M. H., and Kirby, R. (1995a) *J. Appl. Bacteriol.*, **78**, 456–462.

Teixeira, P. C., Castro, M. H., Malcata, F. X., and Kirby, R. M. (1995b) *J. Dairy Sci.*, **78**, 1025–1031.

Teixeira, P., Castro, H., Mohacsi-Farkas, C., and Kirby, R. (1997) *J. Appl. Microbiol.*, **83**, 219–226.

Tejada-Simon, M. V., Lee, J. H., Ustunol, Z., and Pestka, J. J. (1999) *J. Dairy Sci.*, **82**, 649–660.

Thompson, L. U., Jenkins, D. J. A., Amer, V., Reichert, Jenkins, R., and Kamulsky, J. (1982) *Am. J. Clin. Nutr.*, **36**, 1106–1111.

Tissier, H. (1906) *C. R. Soc. Biol.*, **60**, 359–361.

Tissier, M. H. (1899) *C. R. Acad. Sci.*, **51**, 943–945.

van der Kamp, J. W. (1996) *Int. Dairy Fed. Nutr. Newsl.*, **5**, 27–28.

Van't Veer, P., Dekker, J. M., Lamers, J. W., Kok, F. J., Schouten, E. G., Brants, H. A., Sturmans, F., and Hermus, R. J. (1989) *Cancer Res.*, **49**, 4020–4023.

Vesa, T. H., Hatakka, K., Saxelin, M., Korpeala, R., and Marteau, P. (1998) In *Proceedings Lactic 97*, September 10–12, 1997, Caen, France, pp. 39–50.

Wagner, R. D., Pierson, C., Warner, T., Dohnalek, M., Farmer, J., Roberts, L., Hilty, M., and Balish, E. (1997) *Infect. Immun.*, **65**, 4165–4172.

Walker, D. K., and Gilliland, S. E. (1993) *J. Dairy Sci.*, **76**, 956–961.

Wang, X., and Gibson, G. R. (1993) *J. Appl. Bacteriol.*, **75**, 373–380.

Wolf, B. W., Wheeler, K. B., Ataya, D. G., and Garleb, K. A. (1998) *Food Chem. Toxicol.*, **36**, 1085–1094.

Wright, R. (1999) *Dairy Ind. Int.*, **3**, 27–29.

Wunderlich, P. F., Braun, L., Fumagalli, I., D'Apuzzo, V., Heim, F., Karly, M., Lodi, R., Politta, G., Vonbank, F., and Zeltner, L. (1989) *J. Int. Med. Res.*, **17**, 333–338.

Young, J. (1996) *Functional Foods: Strategies for Successful Product Development*, FT management report, Pearson Professional Publishers, London.

Young, J. (1998) *Br. J. Nutr.*, **80**(Suppl.), S231–S233.

Zhang, X. B., and Ohata, Y. (1991) *J. Dairy Sci.*, **74**, 752–757.

Zoppi, G. (1997) *Acta Paediatr.*, **86**, 1148–1151.

CHAPTER 10

MICROBIOLOGY OF SOFT CHEESES

NANA Y. FARKYE
Dairy Products Technology Center, California Polytechnic State University, San Luis Obispo, CA 93407

EBENEZER R. VEDAMUTHU
994 NW Hayes, Corvallis, OR 97330

10.1 INTRODUCTION

Classification of cheeses as hard, soft, semi-soft, etc. is purely arbitrary and utilitarian. Classification helps to systematically group cheeses that are alike in certain basic features or characteristics. The most widely accepted and used basis for the classification of cheeses is the moisture content, as moisture determines the body, consistency or compactness of cheese. Thus, the term soft cheese denotes that the consistency of the cheese is soft to touch or to pressure applied between fingers. This attribute of "softness" is directly related to the moisture content of the cheese—high moisture cheeses being softer than low moisture cheeses.

The U.S. Code of Federal Regulations does not specify maximum moisture content for soft cheeses. However, a legal minimum of 50% milkfat in the solids phase (fat in dry matter, FDM) of the cheese is specified for soft ripened cheeses. Scott et al. (1998) define soft cheeses as containing water in the fat-free cheese matter (Wff) greater than 61% and fat in the dry matter, FDM of 10–50%. The FAO/WHO classifies soft cheese as cheese containing over 67% Wff (Teuber, 1998). Data on the composition of various soft cheeses in the literature (Kosikowski and Mistry, 1997) reveal moisture contents in the range of 50–80%.

Dairy Microbiology Handbook, Third Edition, Edited by Richard K. Robinson
ISBN 0-471-38596-4

Soft cheeses may be manufactured from whole milk, skimmilk, cream, whey or combinations thereof. Because most soft cheeses are consumed fresh and also, to eliminate the risk of food poisoning, it is important that the milk or other dairy ingredients used for soft cheese manufacture be adequately pasteurized (72°C × 15 s). In several countries, the use of raw milk for cheesemaking is still prevalent. In the U.S., cheese manufactured from raw milk must be stored at a minimum of 1.7°C for at least 60 days before consumption. This regulation limits the manufacture and sale of unripened soft cheeses from raw milk.

10.2 CATEGORIES OF SOFT CHEESES

Soft cheeses can be grouped into four categories—unripened, surface mold-ripened, surface bacterial smear-ripened, and picked (Table 10.1). Most unripened soft cheeses are acid-coagulated cheeses (e.g., Cottage, Cream, Quarg, Fromage Blanc) for which curd is formed by acidification of milk to or near the isoelectric pH of casein (i.e., pH 4.6). Very small amounts of rennet may be added to aid coagulation and increase curd firmness. Some soft cheeses, e.g., Mozzarella and the Hispanic varieties such as Queso Fresco and Panela are rennet-coagulated, i.e., the curd is formed by the proteolytic cleavage of κ-casein by the action of chymosin or other milk-clotting enzyme on milk at pH > 6.2. In others, such as Mascarpone, Ricotta, Queso Blanco Fresco, and Paneer, coagulation is achieved by a combination of heat and acidification at pH ≥5.5.

10.3 UNRIPENED SOFT CHEESES

10.3.1 Acid-Coagulated Varieties

10.3.1.1 Cottage Cheese. A detailed description of the procedure for the manufacture of cottage cheese is given by Emmons and Tuckey (1967). Cottage cheese is manufactured by acid coagulation of skim milk fortified with nonfat dry milk to a total solids content of at least 10%. Coagulation is achieved by a food-grade organic acid (e.g., acetic, lactic or gluconic acid), inorganic acids (e.g., hydrochloric or phosphoric), by acidogen (glucono-δ-lactone, GDL), or by bacterial acidification with mesophilic lactic acid bacteria. Cottage cheese can be made by the short-set (i.e., coagulation occurs in 4–6 h) or long-set (i.e., coagulation in 12–16 h) methods. In the short-set method, milk at 30–32°C is inoculated with 4–5% starter while in the long-set method the milk

TABLE 10.1. Examples of Varieties of Soft Cheeses

Cheese category	Example	Moisture Content	Starter Type/Method of Acidification	Secondary Flora
Unripened soft cheese (acid coagulated)	Cottage	≤80	*Lactococcus lactis* ssp. *cremoris*	*Leuconostoc* ssp.
	Quark		*Lactococcus lactis* ssp. *lactis*	*Lactococcus lactis* ssp. *lactis* cit
	Cream	55% (max.)	*Lactococcus lactis* ssp. *cremoris*	*Lactococcus lactis* ssp. *lactis* cit
	Neufchâtel	65% (max.)	*Lactococcus lactis* ssp. *lactis*	*Leuconostoc* sp.
Unripened soft cheese (rennet-coagulated)	Mozzarella	<60	*Streptococcus thermophilus*	
	Oaxaca		*Lactobacillus delbrueckii* ssp. *bulgaricus*	*Leuconostoc* sp.
			Lactobacillus helveticus	
	Queso Fresco		Little or no starter. Mininal acidification	
	Panela			
Unripened soft cheese (heat + acid coagulated)	Ricotta	<80	No starter. Acidification by food-grade organic acid.	
	Mascarpone			
	Queso Blanco			
	Paneer			
Surface mold-ripened cheese	Camembert	54	*Lactobacillus lactis* ssp. *cremoris*	*Penicillium camemberti*
	Coulommiers	53		Yeast
	Brie			
Surface bacterial smear-ripened cheese	Limburger	48	*Lactobacillus lactis* ssp. *cremoris*	*Brevibacterium linens*
	St. Nectaire		*Lactobacillus lactis* ssp. *lactis*	*Micrococcus* spp.
	Romadour			
Pickled cheese	Domiati		*Streptococcus thermophilus*	
	Feta	55	*Lactobacillus delbrueckii* ssp. *bulgaricus*	
	Teleme			

is set at 20–22°C with 1–2% starter. The starter culture used for cottage cheese manufacture contains primarily, the mesophilic lactic acid bacteria, *Lactococcus lactis* ssp. *cremoris* or *Lactococcus lactis* ssp. *lactis*. Starter that contains citrate$^+$ *Lactococcus lactis* ssp. *lactis* (formerly, *Lactococcus lactis* ssp. *lactis* biovar. *diacetylactis*) is not satisfactory for cottage cheese manufacture because of the production of considerable amounts CO_2 which causes the curd to float. Floating curd is difficult to cook and shatters easily. To obtain a coagulum that is ready to cut at pH of 4.7–4.8, very small amounts of rennet (0.1–0.2 ml/100 L milk for the long-set method or 0.2–0.3 ml/100 L for the short-set method) may be used. The coagulum is cut with small 6.35 mm (0.25-in) wire curd knives to give small-curd cottage cheese or with large 12.7 mm (>0.5-in) knives to give large-curd cottage cheese. The curd is cooked over an hour to 53–57°C until desired firmness is reached. Heating is generally slow and continuous (~1°C every 5 min). The curd is held in the whey for about 30 min at the cooking temperature to attain desired firmness. Then, the whey is drained off and the water in the jacket is drained. Hence, the curd is washed and drained three times with cold water (successive temperatures of 22–26, 10–18 and 3–7°C) to give dry-curd cottage cheese. The amount of water used for washing equals approximately 50–67% of the initial milk volume. The microbiological quality of wash water is important because it influences the shelf life of cottage cheese. Thus, addition of hypochlorite to wash water to give 5 ppm available chlorine with a contact time of at least one min prior to addition to the curd is satisfactory. Also, adjustment of wash water pH to 6.5 with food-grade phosphoric acid before or at the time of chlorination is desirable (Sellars and Babel, 1985).

To manufacture creamed cottage cheese, the dry curd is mixed with freshly made cream dressing containing approximately 11% fat, 3% salt and 0.3% stabilizer. Typically, the ratio of dressing to dry curd is 1:1.5—giving a minimum fat content of 4% in the finished product. The average composition of creamed Cottage cheese is 78.3% moisture, 13.6% protein, 4.2% fat and 1.0% ash while dry-curd cottage cheese contains 79% moisture, 17.0% protein, 0.3% fat and 1.0% ash (Olson, 1974).

Cottage cheese flavor is from the cream dressing. The major flavor compound is diacetyl, which is added in the form of starter distillate or generated by aroma-producing cultures added to the cream dressing during manufacture. Also, concentrated cells—frozen or lyophilized—of flavor bacteria may be added to the cold cheese dressing immediately before it is added to the dry curd. The flavor-producing bacteria metabolize citrate to give diacetyl. To obtain a higher level of flavor, the dressing mix is often fortified with citric acid or citrate salt within the legally allowable limit.

10.3.1.2 Cream, Neufchâtel and Baker's Cheese. Both cream cheese and Neufchâtel cheese are similar soft unripened cheeses. U.S. standards of identity for cream cheese require a minimum of 33% fat and maximum of 55% moisture whereas the standards of identity for Neufchâtel cheese requires a fat content of not less than 20% but less than 33% and a maximum moisture content of 65%. Hence, Neufchatel cheese is the same type of cheese as cream cheese but contains less fat and more moisture than cream cheese.

Cream cheese is manufactured by one of two methods *viz*, hot-pack or cold-pack method. The starting material for cream cheese manufacture is a mixture of milk and cream standardized to a minimum of 11% fat and 8% nonfat milk solids. The mix is pasteurized (63–85°C, preferably 68°C × 30 min, cooled to 49°C and homogenized at 124.14–137.93 MPa (1800–2000 psi) single stage. Single stage homogenization enhances slow whey drainage and uniform distribution of fat globules. The homogenized mix is cooled to 22 or 32°C, respectively, for long or short-set methods. Mesophilic starter is added at the rate of 0.5 and 5%, respectively—resulting in setting times of 15 h (long-set method) or 5 h (short-set method) and cutting pH of 4.7 or titratable acidity (TA) of 0.6% lactic acid. To facilitate faster whey drainage, a small amount of rennet (0.2 ml/100 L milk) may be added at the time of starter addition. At pH 4.7 and TA of 0.6% lactic acid, the coagulum is broken by stirring. The resulting homogeneous mass is heated to 52–55°C, held there until clear visible whey separation occurs. Then, the product is cooled to 32°C and drained. In the traditional process the product is placed in muslin bags and allowed to drain overnight at 4°C. Alternately, in continuous industrial processes, whey is removed from the curd using a continuous centrifugal separator. After complete whey drainage, the moisture of the curd is adjusted with liquid or dried cheese whey, salt (0.8–1.0%) is added and the cheese is packaged (cold-pack method). Typical curd yields are 2.96, 2.59, 2.42 and 2.18 kg cheese per kg fat when cream cheese is made from mixes containing 8, 12, 16 and 18% fat, respectively (Van Slyke and Price, 1952).

In the hot-pack method, after the coagulum is stirred and heated to about 55°C, it is further heated to ~81°C then centrifuged to separate curd and whey. To the hot curd, salt and a hydrocolloid stabilizer (e.g., locust bean gum) are added after moisture adjustment. U.S. standards of identity allow the use of up to 0.5% stabilizer. The resultant product is homogenized at 137.93 MPa (2000 psi) first stage and 34.47 MPa

(500 psi) second stage at 63°C to give cream cheese a smooth consistency and uniform composition. Condiments such as chives, vegetables, onions, pimentos, nuts, pineapple, cherries, etc. may also be added at this stage (Kosikowski and Mistry, 1997). The curd is hot packed into appropriate size packages and chilled to 4°C. Shelf life of hot-packed cream cheese is at least 60 days (Strauss, 1997).

Bakers' cheese is a smooth, soft unripened cheese made by coagulation of skimmilk similar to the traditional Neufchâtel process. It contains 65–78% moisture and is practically fat-free. To manufacture Baker's cheese, good quality skimmilk is pasteurized and cooled to 21–22°C for overnight long-set or 30–31°C for the short-set method. Active mesophilic lactic starter is added at the rate of 0.4% (w/w) for the long-set method or 5% (w/w) for the short-set method. Rennet is added at the rate of 4 ml/454 kg milk. At pH 4.4–4.5 or a TA of 0.6% lactic acid, the coagulum is broken and curd transferred to muslin bags to drain, preferably under refrigeration. The curd may be salted at the rate of 0.5–1%. Typical yields are 13–14% (Emmons and Tuckey, 1967; Van Slyke and Price, 1952).

10.3.1.3 Quarg. Quarg (or Quark) is a soft homogeneous cheese with clean and mildly acid flavor. Traditionally, Quarg is manufactured from good quality skimmilk although whole milk may be used. The procedure for manufacture of Quarg involves pasteurization of skimmilk at 72–85°C × 15 s. The milk is cooled to 20–23°C (for long-set method of manufacture) or ~30°C (for the short-set method). Next, it is inoculated with 1–2% starter culture consisting of *Lc. lactis* ssp. *lactis* or *cremoris* and held at that temperature for 14–18 h (long-set method) or 3–6 h (short-set method) to reach pH of 4.6–4.7. A small quantity of rennet (<1 ml/100 L milk) is added after culture addition to aid in coagulation and produce a firm curd at a higher pH, thereby avoiding over acidification of the cheese. In the traditional method, the coagulum is cut at the desired firmness and curd scooped into muslin bags for whey drainage. In industrial methods, the curd and whey are pumped into special Quarg separators for whey separation. Finished curd is packaged and stored cold. Also, the curd may be further processed (e.g., heating, homogenization, aeration) or condiments, spices, fruits, etc. may be blended into the cheese.

Whey obtained by the traditional Quarg manufacturing process contains ~0.65% protein and ~0.2% non-protein N (Guinee et al., 1993). Therefore, newer industrial technologies aim at reducing the whey protein loss thereby increasing cheese yields. The technologies include the Westfalia Thermoprocess, the Centriwhey process, the Lactal process and ultrafiltration (UF). The Westfalia Thermoprocess, the Centriwhey and Lactal processes all involve heating of whey to 95°C. This causes thermal denaturation of whey proteins and their subsequent coagulation (co-precipitation) with caseins on acidification. For example, in the Westfalia Thermoprocess, 50% of the whey proteins are recovered, resulting in yield increases of 10% (Siggelkow, 1984). In the UF method, fermented skim milk (pH < 4.6) is heated to 38–42°C then concentrated by UF to 17.5% solids. The concentrate is cooled, homogenized and packaged (Siggelkow, 1984). Typical composition of skimmilk (low-fat) Quarg is 17–24% dry matter, 70–83% moisture, 12–18% protein, trace–0.5% fat, 2–4% lactose, 0.94% ash and 0.125% calcium (Puhan and Gallman, 1980; Koth and Richter, 1989). High-fat Quarg contains about 74% water, 10% protein and 12% fat (Walstra et al., 1999). The shelf life of Quarg is limited to ~3 wk when stored at < 8°C (Kroger, 1979).

10.3.1.4 Fromage Blanc. Fromage blanc is a popular French soft cheese that is made from skimmilk (Ramet, 1990). The skimmilk is pasteurized (75–80°C × 10–30 s), cooled to 25–28°C, inoculated with mesophilic starter (0.001–0.002%) and rennet (0.5–1 ml/100 L milk) is added. Clotting occurs in 6–8 h and the curd is left undisturbed for 10–15 h for acidification and hardening. When the pH is 4.5–4.6, the curd is ladled into molds to drain for 20–24 h at 20–25°C before packaging. Alternately, after clotting, the curd is gently stirred for 5–10 min then fed into muslin bags for drainage for 24–36 h at room temperature (20–25°C). Gentle turning and pressing is applied to the curd during drainage. In modern industrial processes, the milk is pasteurized (72°C × 40 s or 86°C × 8–10 s), cooled to 28–30°C and inoculated with 0.5–1% mesophilic starter bacteria. After 1.5 h and when the milk pH reaches about 6.3, rennet (0.5–1.0 ml/100 L) is added. The milk is left quiescent for 16 h to coagulate. At pH 4.5–4.55, the coagulum is stirred and the curd/whey mixture is fed into a separator for separation of curd and whey. The whey is removed and the resultant cheese is cooled to 6–8°C, packaged and stored at 0–4°C. To increase yields by incorporating whey proteins in cheese, skimmilk is heated to 82–92°C for 5–6 min to denature the whey proteins that co-precipitate with casein. Also, UF techniques may be used to trap whey proteins in the cheese.

10.3.2 Rennet-Coagulated Unripened Soft Cheeses

10.3.2.1 Mozzarella Cheese. Mozzarella cheese is a pasta filata (stretched cheese) variety in which the drained curd is heated in hot

TABLE 10.2. US Specifications for Different Types of Mozzarella Cheese

Type of Mozzarella	Moisture	Fat in Dry Matter
Mozzarella	Greater than 52% Less than 60%	Not less than 45%
Low-moisture mozzarella	Greater than 45% Less than 52%	Not less than 45%
Part-skim mozzarella	Greater than 52% Less than 60%	Not less than 30% Not greater than 45%
Low-moisture part-skim mozzarella	Greater than 45% Less than 52%	Not less than 30% Not greater than 45%

whey or water at 75–80°C and subsequently kneaded at low pH to remove calcium from the curd, thereby making the cheese stretchable. The U.S. standards of identity classify Mozzarella based on the composition of the finished cheese (Table 10.2).

The different types of Mozzarella cheese have different functional behavior as their intended uses differ. The general manufacturing process for the different Mozzarella cheeses is similar. However, the initial composition (i.e., casein to fat ratio, C/F) of milk used for manufacture is different, hence differences in composition of the various Mozzarella types. The C/F ratio is adjusted by removal of fat as cream or addition of casein in the form of skimmilk, skimmilk powder, and condensed or concentrated skimmilk. Generally, the typical fat contents of milk for manufacture of the low-moisture part skim varieties range from 1.5–2.5% (C/F of 1.1–1.25) whereas Mozzarella is made from whole milk with a C/F of 0.67–0.72. Mozzarella cheese may be manufactured by direct chemical acidification using a food-grade acid or by bacterial acidification using thermophilic lactic acid bacteria. Pasteurized milk is cooled to 30–32°C. The milk is inoculated with a starter comprising a typical ratio of 1:1 of *Streptococcus thermophilus* and *Lactobacillus delbrueckii* ssp. *bulgaricus* or *Lactobacillus helveticus*. Rennet is added at the rate of 90 ml single strength/454 kg milk. Following coagulation, the curd is cut, cooked to 38°C, and held there for 1 h. The whey is drained and when the curd pH reaches 4.9–5.2, it is plasticized by stretching and kneading in hot (75–85°C) water. In industrial processes, stretching and kneading of curd are done in a mixer/molder. Typical curd temperatures during stretching are in the range, 55–60°C. After kneading and stretching, the curd is molded into desired shape and quickly cooled and salted by placing in cold brine 22–23% Nacl. The duration of brining depends on the size of cheese loaf and brine temperature. Recently, Barbano et al. (1994)

described a no-brine method for Mozzarella cheese in which the curd is dry-salted before it is plasticized. The desired salt content for Mozzarella is 1% and that for the low-moisture part-skim Mozzarella is in the range 1.5–1.7%.

10.3.2.2 Hispanic and Other Soft Cheeses. Hispanic cheeses have become increasingly popular in the U.S. The most popular Hispanic soft cheeses are Queso Fresco (fresh cheese), Panela (meaning, "basket") and Queso Crema (cream cheese but different from traditional cream cheese), Queso Blanco (white cheese), and Oaxaca (which is similar to Mozzarella). Queso Fresco is a soft cheese with crumbly texture and a mild, slightly salty flavor. Queso Fresco is manufactured from pasteurized whole or part-skim milk whereas whole milk is used for Panela. Queso Crema is a higher fat version of Panela. Because Queso Fresco has a characteristic white appearance, the milk may be homogenized to disperse the fat globules. Alternately, a chemical whitening-agent, such as titanium dioxide may be added to the milk. The pasteurized milk is cooled to 28–30°C. Generally, Queso Fresco is made without starter bacteria. When culture is added, typically flavor-producing BD cultures (i.e., cultures containing *Leuconostoc* ssp. and citrate$^+$ *Lactococcus lactis* ssp. *lactis* (formerly, *Lactococcus lactis* ssp. *lactis* biovar. *diacetylactis*) are used. Rennet is added at the rate of 2 ml/100 L milk. Coagulation takes 30 min. The curd is cut and cooked to 40–46°C. Following whey drainage, the curd is allowed to mat then milled by grinding. The finely ground curd is salted, packaged, and cooled immediately.

The procedure for manufacture of Panela is similar to that for Queso Fresco except that after cooking, a portion of the whey is drawn and salt is added to the remaining curd-whey mixture in the vat. Then, the salted curd is dipped and molded in baskets and drained on a drain table. The finished cheese has a unique basket shape and a mild pleasant flavor. The manufacture of Queso Crema is similar to that of Panela except that cream is added to standardize the milk to a higher percentage fat. Typical composition of Queso Fresco from part-skim milk is 51% moisture, 19% fat, 22% protein, 1.7% salt and pH > 6.1. Panela contains 53% moisture, 25% fat, 18% protein, 1.5% salt and pH >6.2. Other popular Hispanic cheeses are Cotija which is a hard cheese and Requeson, a soft cheese similar to Ricotta.

10.3.3 Unripened Soft Cheeses Manufactured by Acid-Heat Coagulation

10.3.3.1 Mascarpone. Mascarpone is a soft cream–style cheese produced by heat-acid coagulation of cream. The procedure for manufacture involves adjusting the composition of cream to ~50% fat and 2.8–6% protein, and then heating to 95°C × 40 min. Then, dilute acidifying agent such as acetic, citric, tartaric, lactic acids or lemon juice is added. The curd is separated from the whey and the finished cheese packaged and cooled to 4°C in 6 h or less. The final pH of the finished cheese is in the range 5.7–6.6 (Franciosa et al., 1999). Mascarpone contains about 50% moisture, 44.5% fat, 3.3% protein and 0.2% salt.

10.3.3.2 Ricotta. Ricotta is a soft unripened soft cheese that originated from Italy. In Latin American and the Hispanic communities, Ricotta is known as Requeson. Traditionally, the starting material for Ricotta cheese is whey from Mozzarella cheese manufacture. At present, Ricotta can be made from almost any type of sweet whey provided that the initial tritatable acidity of the whey is ≤0.16% lactic acid and a pH ≥6.0. The principle for the manufacture of Ricotta is to heat whey to 85–88°C, followed by coagulation of the proteins by addition of acid to hot whey. The curd formed floats on top of the whey and is scooped out to drain. In industrial methods, the whey is neutralized to pH 6.5 with a 25% (w/v) solution of NaOH. The neutralized whey is heated to 65–70°C, then whole milk equal to 25% of the whey volume is added and heating of the whey/milk mixture is continued until a temperature of 75–80°C is reached. Cream may be added at this time if a higher fat product is desired. Next, NaCl (0.5%, w/v) is added and heating continued till 85–95°C is reached. Alternately, calcium may be added. NaCl dehydrates the whey proteins and has a destabilizing effect on bovine serum albumin. Similarly, calcium destabilizes all the whey proteins. Next, dilute food-grade acetic or citric acid is added for coagulation and curd formation. Typically, ~1.5% (v/v) of dilute (~3.85%) acetic acid is needed to clot the whey/milk mixture. The curd is left in the hot whey for about an hour to increase firmness and enhance whey drainage. The curd, which floats on the surface of the whey, is ladled off. Alternately, the whey may be drained from the bottom leaving the curd behind in the vat or kettle. Optimal coagulation pH occurs between 5.6–5.8 to give maximum yields (Weatherup, 1986). Approximately 5 kg of fresh ricotta is obtained from 100 kg whey by the addition of 5 kg whole milk. Typical composition of Ricotta is 2.5% fat, 16.0% protein, 3.5% lactose, 1.0% ash, 20– 23% total solids and pH of 5.6–6.0.

10.3.3.3 Queso Blanco and Paneer. Another Hispanic cheese that may be classified as soft is Queso Blanco (also called Queso Blanco Fresco). This cheese, similar to Paneer, which is produced in India, is made by direct acidification of highly heated milk. The procedure for manufacture of Queso Blanco involves heating milk to 82–85°C, and adding a diluted food grade acid, vinegar or lemon juice to the milk to achieve a pH of 4.6–4.7 in the curd/whey mixture. The acidified milk is gently stirred and the curd is left undisturbed in the whey for about 15 min to firm up. Then the whey is drained and curd salted at 2.5% (Chandan et al., 1979; Kosikowski and Mistry, 1997; Farkye et al., 1995).

10.4 RIPENED SOFT CHEESES

10.4.1 Surface Mold-Ripened Soft Cheeses

The most notable surface mold ripened soft cheeses are Camembert and Brie—both of which originate from France. These cheeses are characterized by the presence of a white mycelial covering (due to the growth of the white mold, *Penicillium camemberti*) that covers the surface of the cheese.

Camembert is made by a traditional Normandy-style process or by an industrial process. Generally, raw or thermized (63°C × 15 s) milk is used in the traditional process while pasteurized (72°C × 15 s) milk is used in the industrial process (see Table 10.3). The heated milk is cooled to 10–12°C, inoculated with a 0.1–0.2% mesophilic lactic starter (*Lc. lactis* ssp. *lactis* or *cremoris* or *Leuconostoc mesenteroides* ssp. *cremoris*) and held for 15–20 h before warming to 30–34°C. In other methods, the milk is cultured at 30–34°C. Rennet (15–20 ml/100 L for the traditional process and 20–25 ml/100 L for the industrial process) is added to the warm milk. The rennet is diluted in cold water before adding to the milk. Coagulation takes about 1.5 h for the traditional process or 30–45 min for the industrial process. In the traditional process, the curd is ladled carefully (to avoid breaking) into perforated molds that are placed on a drain table to allow whey drainage. The mold may be turned a few times to facilitate whey drainage. After 15–20 h, curd is removed from the mold and dry-salted by dusting with fine salt. Simultaneously, a suspension of *Penicillium camemberti* spores is sprayed or rubbed on the curd surface.

Next, the cheese is cured at 12–13°C and R.H. of 85–90% for 3 wk, then at 8–10°C for a week. During ripening of Camembert, a white mycelial covering appears on the cheese surface at about the fifth to sixth day of ripening. The covering becomes 2–3 mm thick after 10–12 days of ripening. The predominant microflora that grow on the cheese surface and that contribute to the ripening consists of lactic acid

TABLE 10.3. Procedure for Manufacture of Camembert/Brie Cheese

Step in Manufacture	Temperature and Duration	Comments
Heat treatment	63°C × 15 s 72°C × 15 s	Thermization Pasteurization
Addition of starter (1–2.5%)	30–34°C × 1 h	Ripen milk
Add $CaCl_2$ (optional) 0.02%		
Addition of rennet (15–20 ml single-strength 100 liter^{-1})		Coagulation takes 30–45 min
Cutting	3 min	Curd is cut into fairly large pieces
Dipping/molding		Ladle into perforated cylindrical forms (8- to 12-cm diameter and 11–15 cm high) on drainning table.
Pressing	Overnight, 22°C	Turn cheese and stack molds on top of one another. Drain overnight. Curd shrinks to half its original weight.
Whey acidity is 0.60–0.70% lactic acid at end of drainage.		
Salting		Sprinkle salt; add mold spores.

bacteria, molds (*P. camemberti*, *Geotrichum candidum*), and yeast (e.g., *Saccharomyces lactis, S. fragilis, Torulopsis sphaerica, Candida pseudotropicalis, Debaryomyces hansenii, Torulopsis candida*). Other yeasts and molds have been isolated Brie and Camembert cheeses (Nooitgedagt and Hartog, 1988).

With the surface development of Penicillium and yeasts, the pH of the cheese rises rapidly—reaching pH 7.0 on the surface and pH 6.0 at the center after about 30 days of ripening. The rise in pH, which is due to the metabolism of lactic acid by the yeasts, results in a decrease in the population of fungi and an increase in the population of non-acid tolerant aerobic bacteria, micrococci and corynebacteria after the fifteenth day of ripening. The growth of large numbers of pigmented corynebacteria gives cheese undesirable appearance and flavor. For a detailed review on the biochemistry of Camembert cheese ripening, see Gripon (1993). Camembert cheese comes in wheels of approximately 10.2 cm diameter, 2.54–3.8 cm thick and weighs 3.6–4.54 kg. The cheese must contain at least 45% FDM according to its standard of identity. Typically Camembert cheese contains ~21% protein and ~24% fat.

Brie is another surface-ripened variety that originates from France. It is made from whole milk (although skimmilk or partly skimmed milk may be used). Brie is made in three wheel sizes, i.e., large (21.6–40.6 cm diameter, 3.8–4.24 cm thick, and weighs 2.72 kg); medium (20.3–30.48 cm diameter, and weighs 1.6 kg), small (14 to <20.32 cm diameter, 3.18 cm thick and weighs 0.454 kg). The principles of manufacture of Brie are similar to those of Camembert; and the standards of identity require that Brie must contain at least 40% FDM and 44% dry matter. Typical composition of Brie is ~19% protein and ~27% fat.

10.4.2 Bacterial Surface "Smear"-Ripened Soft Cheeses

Typical surface-ripened soft cheeses include Limburger and Munster. Munster contains at least 44% total solids and a minimum of 45% FDM. Limburger contains at least 50% total solids and 50% FDM. Generalized cheese making procedure for Limburger is given in Table 10.4. Pasteurized milk at 32–35°C is cultured with mesophilic lactic acid bacteria until the pH reaches 6.2–6.5. Next, rennet is added at a rate of 30 ml/100 L milk. The rennet is diluted in cold water before adding to the milk. Then, the curd is cut, cooked to 36°C, dipped into perforated molds and allowed to drain.

After removal from the mold, the cheese is salted and sprayed with cultures of *Brevibacterium linens*. Limburger is ripened at 15°C and 90–95% R.H. during which the development of surface microflora occurs.

10.5 PICKLED SOFT CHEESES

10.5.1 Feta Cheese

Feta is a white cheese that originates from Greece. It is traditionally produced from sheep or goat milk or their mixtures (Anifantakis, 1991). Industrially, it is manufactured equally from cow or sheep milk, and has been produced from milk concentrated by UF techniques (Tamime and Kirkegaard, 1991). Because Feta is traditionally made from sheep or goat milk which lack carotene, the cheese has a whitish appearance. Therefore, when cow's milk is used for manufacture, the milk is often bleached or decolorized. Feta has a crumbly compact texture with mechanical openings. The cheese can be manufactured from whole or part-skim milk. Feta cheese manufactured from sheep and goat milk is normally standardized to 5.8–6% fat because of the

TABLE 10.4. Typical Procedure for the Manufacture of Limburger Cheese

Step in Manufacture	Temperature and Duration	Comments
Heat treatment	63°C/15 s 72°C/15 s	Thermization Pasteurization
Addition of starter (0.25–0.5%)	30°C for 1 h	Ripen milk
Add $CaCl_2$ (optional) 0.02%		
Addition of rennet (15–20 ml single-strength 100 liter^{-1})		Coagulation takes 30–45 min.
Cutting	3 min	Curd is cut into small pieces.
Cooking	35.6°C	Raise temperature in 30 min.
Draining		Remove whey to the level of curd.
Dipping and hooping		Dip curd into appropriate hoops.
Pressing	Overnight, 22°C	Turn cheese every 20 min for first hour; every 30 min for 2 h, then hourly until properly matted.
	Acidity of 0.40% Lantic acid pH of 5.1	
Salting		Dry salt or brine.
Ripening and smearing	15°C; 90–95% R.H.	Development of surface microflora (mainly non-lactose-fermenting yeast, e.g., *Geotricum candidum*). Spray surface of cheese with *Brevibacterium linens* 24–48 h after yeast development.

high fat content of sheep and goat milk (Anifantakis, 1991). However, an optimum C/F of 0.69–0.73 is ideal (Tamime and Kirkgaard, 1991) and the mean composition of Feta cheese is 52–55% moisture, 18–26% fat, ~16% protein, ~0.17% lactose, ~3% salt and pH of 4.4–4.6. The steps involved in Feta cheese manufacture are given in Table 10.5. The milk is pasteurized (72°C × 15 s) or heat-treated (68°C × 10 min) and may be homogenized. Traditionally, starter used for manufacture is a combination of *Lb. delbrueckii* ssp. *bulgaricus* and *S. thermophilus*. However, in industrial processes, *Lc. lactis* ssp *lactis* or *cremoris* and *Leuconostoc mesenteroides* ssp. *cremoris* may be included. Feta is dry salted by rubbing cheese with granular salt and left overnight to absorb

TABLE 10.5. Typical Procedure for the Manufacture of Feta Cheese

Step in Manufacture	Temp. & Duration	Comments
Heat treatment	72°C/15 s	Pasteurization
Addition of starter (0.25–0.5%)	30–32°C/1 h	Ripen milk
Add $CaCl_2$ (0.02%)		
Addition of rennet (15–20 ml single-strength 100 liter^{-1})		Coagulation takes 50–60 min.
Cutting	3 min	Curd is cut into small cubes (2–3 cm^3)
Heal	5–10 min	
Dipping/hooping/pressing		Dip curd into appropriate hoops. Drain for 2–3 h without pressure. Turn cheese frequently until pH reaches 4.8.
Salting		Dry salt; then store in brine (60–80 g liter^{-1}) for 10 days at 16–18°C.
Ripening	4–5°C	Ripen for 2 months before marketing.

the salt. Granular or coarse salt dissolves slowly on cheese surface and contributes to normal draining of whey. Use of fine grain salt leads to over-salting and leads to the surface of cheese becoming hard (Anifantakis, 1991). After salting, the cheese is cut into 1 kg blocks or smaller and stored in 6–10% brine, and stored for 2 mo at 4–5°C before it is sold commercially.

10.5.2 Domiati Cheese

Domiati cheese is soft pickled cheese variety that originates from Egypt. The manufacture of Domiati involves standardization of milk to a desired fat content. The milk is pasteurized or heated to 65°C × 15 min and cooled to 35–40°C. Next, 5–15% NaCl (depending on the season of the year) is added to inhibit microbial growth. When pasteurized milk is used for manufacture, the amount of NaCl added does not exceed 6%. Alternately, raw milk is divided into two lots. NaCl (5–14%) is added to one lot (which represents two-thirds of the raw milk). The remaining one-third is heated to 77–80°C and both lots mixed together thereby reducing the temperature of the mixed milk to the setting temperature of 35–40°C. Next, the milk is inoculated with salt-tolerant lactobacilli followed by rennet (22–30 ml single strength/100 L milk) addition. Coagulation takes 2–3 h. The curd is carefully ladled into molds lined with cheesecloth and placed on a drain table for whey drainage. After overnight drainage, the cheese is removed from the mold and packaged in salted whey (10–15% NaCl) or brine. Typical composition of Domiati cheese from bovine milk is 55% moisture, 20% fat, 12.9% protein, 44.4% FDM and 4.9% salt (Abou-Donia, 1991).

10.6 STARTER MICROORGANISMS FOR SOFT CHEESE

Starter bacteria used in soft cheeses play important roles in the acid development during curd formation and syneresis. Activities of starter bacteria in addition to nonstarter bacteria, molds, yeast, smear flora, etc in specific varieties contribute to flavor development and unique characteristics of a specific variety. Also, starter and nonstarter lactic acid bacteria contribute to the preservation of some varieties through the activities of metabolites produced. The various starter types used in soft cheese manufacture, their characteristics and specific functions are discussed below.

10.6.1 Genus Lactococcus

With the exception of Mozzarella and other Pasta Filata varieties, the acid-producing microorganisms involved in acid curd formation or in aiding rennet coagulation and syneresis of the curd, or in imparting the pleasant, clean lactic acid flavor belong to the mesophilic lactic acid bacteria grouped under the genus *Lactococcus*. These bacteria were formerly classified under the genus *Streptococcus*. Sherman placed them under the group *lactic*; serologically they fell under Lancefield group N. The dairy lactococci comprise one species, with two sub-species designations. The species name is *Lc. lactis*. The two subspecies are *Lc. lactis* ssp. *lactis* and *Lc. lactis* ssp. *cremoris*. Within the sub-species *lactis* are two types—citrate-negative (cit^-) and citrate-positive (cit^+) *Lc. lactis* ssp. *lactis*. The citrate-positive sub-species are designated cit^+ *Lc. lactis* ssp. *lactis* (formerly, *Lc. lactis* ssp. *lactis* biovar. *diacetylactis*).

Lactoccoci are ellipsoidal cocci found in pairs and in short chains. The original habitat of lactococci is plant or vegetable matter, but they are widely found in dairy environs, milk and dairy products. Lactococci thrive in the mesophilic temperature range and best growth of these bacteria is obtained under microaerophilic conditions. However, their oxygen relationship is quite complex (Condon, 1987). Lactococci do not possess the cytochrome or terminal oxidase systems but they possess direct nicotinamide cofactor oxidase enzyme systems to directly transfer hydrogen from reduced cofactors to molecular oxygen. In certain cofactor regeneration reactions involving oxidases, direct transfer of H_2 to molecular O_2 results in the formation of H_2O_2. Because lactococci lack catalase, the accumulation of H_2O_2 becomes toxic. Alternately, there is another cofactor oxidase system that results in the formation of non-toxic end products, namely, H_2O and O_2.

The lactococci are fermentative, and derive their bond energy by substrate level phosphorylation. Under normal conditions lactococci that are incapable of citrate metabolism, use hexose diphosphate (HDP) pathway for the fermentation of carbohydrates (lactose in milk). The citrate-metabolizing lactococci use the phospho-ketolase (PK) pathway to ferment carbohydrates.

The mechanism for lactose utilization is significant in dairy technology. Lactose uptake is accomplished by phosphoenolpyruvte (PEP)-phosphotransferase system. In this system, lactose is phosphorylated and transported into the cell where another enzyme called phospho-β-galactoside galactohydrolase hydrolyzes the phosphorylated lactose into galactose and glucose-6-phosphate. The galactose is fed through other pathways for conversion to glucose derivatives, which are then fed into HDP or PK pathways. Details of the mechanisms of lactose fermentation in lactococci are given in recent reviews (Axelsson, 1998).

The efficiency of lactose utilization in milk by lactococci is closely linked to their proteolytic system(s). Milk contains only trace amounts of amino acids, and simple organic nitrogenous compounds such as urea. To synthesize the enzymes involved in the fermentation of lactose in milk, the lactococci need proteolytic systems to produce amino acids from complex milk proteins (primarily, caseins). The proteolytic system of lactococci consists of cell wall and cell membrane-associated proteases, oligopeptide transport vehicles and intra-cellular peptidases. Thus, efficiency of metabolism of lactose depends on the efficiency of the proteolytic systems. For a more comprehensive review on the proteolytic systems in lactococci, see Axelsson (1998) and Law and Haandrikman (1997).

The genetic and regulatory mechanisms for lactose utilization and related proteolytic function have been topics of intensive research. The lactose-utilization and proteolytic functions of lactococci are coded on relatively unstable genetic elements, namely, plasmids. In certain strains, both functions are coded on one and the same plasmid, and in others, on separate plasmids. Lactococci often display deterioration of their lactose fermentative capacity, which is attributable to the instability of plasmids. Hence, proper selection, propagation and careful handling of lactococcus starters are very important.

The metabolic functions of lactococci are important in flavor generation in soft cheeses. In very high moisture fresh cheeses, e.g., Cottage cheese and related varieties, diacetyl—a compound that imparts "fresh, nut—meat-like dairy" flavor—is highly desired. Diacetyl is produced from the fermentation of citrate in milk by cit^+ *Lc. lactis* ssp. *lactis*. To enhance diacetyl flavor, milk or cream (used in the manufacture of Cottage cheese dressing) is fortified with citrate. The pathway for the conversion of citrate to diacetyl is very well understood. Understanding of a) the significance of pyruvate in diacetyl synthesis; b) the role of diacetyl reductase in the conversion of diacetyl to acetoin, together with the availability of methods to select mutants that favor the synthesis of the key intermediate, α-acetolactate, have offered immense possibilities in "engineering" high flavor-producing lactococci (Hugenholtz, 1993; Hugenholtz et al., 1994). Metabolism of citrate by citrate-fermenting lactic acid bacteria is dependent on transport of citrate into the cell via a permease system. The permease system optimally functions below pH 6.0. Genetic coding for the permease enzyme is also found on a plasmid.

Fermentation of citrate by citrate-fermenting lactococci starters also results in the generation of CO_2 which plays an indirect role in enhancing the quality of blue-veined mold ripened cheeses such as Blue cheese because internally generated CO_2 promote mechanical openings between curd pieces that aid in the growth of *P. roqueforti*. Excessive gas production within Cottage cheese curds leads to floating curd cubes during the cooking of the cheese curd and causes improper, inadequate and non-uniform syneresis. This leads to inferior texture and, at times, matting of individual curd particles, which is undesirable. The citrate-fermenting lactococci produce significant amounts of CO_2 (some strains producing excessive amounts) and relatively high concentrations of acetaldehyde. This results in a "yogurt-like green apple flavor."

10.6.2 Genus Leuconostoc

Other closely associated microbial species that are used in starter mixtures for soft cheeses belong to the genus *Leuconostoc*. The species associated with dairy starter cultures are *Leuconostoc mesenteroides* ssp. *cremoris* and *Leuconostoc lactis*. The dairy leuconostoc are morphologically relatively small, spherical cells in long chains. They are relatively inert in milk, and prefer mesophilic temperature range. They are more acid tolerant than the lactococci, and possess enzyme systems to ferment citrate. Their citrate permease systems also have pH optima of <6.0. The permease is inducible; and unlike the citrate-fermenting lactococci (where citratase, the first enzyme in the breakdown of citrate, is constitutive), the citratase in leuconostocs is an inducible enzyme (Hugenholtz, 1995; Vedamuthu and Washam, 1983). The dairy leuconostoc impart a clean, diacetyl flavor and produce relatively less CO_2 than cit^+ *Lc. lactis* ssp. *lactis*. The citrate-fermenting lactococci, on the other hand, produce significant amounts of CO_2 (some strains producing excessive amounts) and relatively high concentrations of acetaldehyde, giving a "yogurt-like green apple flavor".

Leuconostocs exhibit high alcohol dehydrogenase activity. Alcohol dehydrogenase catalyzes the reduction of acetaldehyde to ethyl alcohol, thereby reducing "green" off-flavor resulting from excessive levels of acetaldehyde. Thus, leuconostocs are preferred in starter mixtures for soft cheeses because their activities result in less gas production and clean, diacetyl flavor that is free from "greenness". For a more comprehensive discussion of the role of leuconostocs in fermented dairy foods, see reviews by Vedamuthu (1994) and Dessart and Steenson (1995).

Both $citrate^+$ *Lc. lactis* ssp. *lactis* and dairy leuconostocs inhibit psychrotrophic bacteria such as *Pseudomonas fluroscens*, *Pseudomonas fragi,* other dairy spoilage pseudomonads, *Alcaligenes* and *Achromobacter* spp. Their inhibitory action is due to the combined effects of metabolic byproducts such as lactic and acetic acids, diacetyl, CO_2 and other compounds, some of which may be "bacteriocin-like" peptides.

Due to their high moisture contents, soft cheeses like Cottage cheese, Quarg, etc. are highly perishable. Psychrotrophic bacteria that enter the milk or cheese as post-pasteurization contaminants cause most of the spoilage. In creamed Cottage cheese, the dressing mixture may be cultured with selected strains of cit^+ *Lc. lactis* ssp. *lactis* or *Leuconostoc mesenteroides* ssp. *cremoris* or a combination of both to serve a dual purpose of providing good diacetyl flavor and increased shelf life by inhibiting psychrotrophic bacteria. Patented procedures have been developed (Sing, 1976) for adding concentrated cells of flavor-producing bacteria (cit^+ *Lc. lactis* ssp. *lactis*) to non-cultured chilled cream dressing just before mixing with the dry Cottage cheese curd for the purpose of flavor enhancement and to control the growth of psychrotrophic bacteria during refrigerated storage, and distribution through marketing channels. For special markets requiring low acid development or low flavor development without significant loss of psychrotroph-inhibitory properties, specific plasmid-cured derivatives of cit^+ *Lc. lactis* ssp. *lactis* have been developed and patented (Gonzales, 1984; 1986).

10.6.3 High Temperature Starter Genera—Streptococcus and Lactobacillus

The manufacture of soft cheeses such as Mozzarella that are cooked to high temperatures (42–43°C) require thermo-tolerant lactic acid bacteria because good acid production at the high temperatures is necessary for sufficient moisture (whey) expulsion. Starter for Mozzarella consists of *S. thermophilus,* and any one of following: *Lb. delbrueckii* ssp. *bulgaricus, Lb. delbrueckii* ssp. *lactis,* or *Lb. helveticus.* Nath (1993) reported that *S. thermophilus* is sometimes used in small amounts along with the lactococci, which form the major portion of the starter, in the production of other semi-soft and soft cheeses such as Brick and Limburger.

Streptococcus thermophilus is included in Sherman's *varidans* group, but does not fall under any of the serological grouping of Lancefield (Mundt, 1986). In milk, pure cultures of *S. thermophilus* produce just enough acid to form a relatively weak coagulum. Optimal temperature for growth of *S. thermophilus* is between 35 and 42°C. In milk cultures, the cells are spherical and found in long chains. On solid laboratory media, and under stressful growth conditions, the cells are elongated and can be mistaken for rods. *Streptococcus thermophilus* possess a β-galactoside permease that transports lactose into the cell. In the cell, lactose is hydrolyzed by β-galactosidase into glucose and galactose. The glucose moiety is metabolized through the HDP pathway. The galactose moiety is not metabolized by *S. thermophilus* and is excreted out of the cells. *Streptococcus thermophilus* strains are weakly proteolytic but possess urease activity. Thus, they can actively hydrolyse urea in milk to yield NH_3 and CO_2. The ability of *S. thermophilus* to furnish CO_2 and amino acids plays an important role in its symbiotic growth in milk with lactobacilli—especially, *Lb. delbrueckii* ssp. *bulgaricus* (Vedamuthu, 1991; vonWright and Sibakov, 1988). For a comprehensive review on *S. thermophilus,* see Mercenier (1990) and von Wright and Sibakow (1998).

The lactobacilli are rod-shaped bacteria. Their physiology and genetics have been reviewed recently (Arihara and Luchansky, 1995). *Lactobacillus* species used as starters are quite heat tolerant, and grow in milk at optimal temperatures of 43–46°C. Additionally, lactobacilli are very acid tolerant and produce high titratable acidity in milk. For rapid rate of acid production in milk, symbiotic growth with *S. thermophilus* is necessary. In the U.S. cheese industry, mixed cultures containing symbiotic *Lb. delbrueckii* ssp. *bulgaricus* and *S. thermophilus* are referred to as "*rod-coccus*" combinations. *Lb. delbrueckii* ssp. *bulgaricus* generally use β-galotosidase to breakdown lactose, although some species also express phospho-β-galactoside galactohydrolase enzyme in addition to β-galactosidase. Lactobacilli prefer strictly microaerophilic conditions. *Lb. delbrueckii* ssp. *bulgaricus* also cannot use galactose, which is expelled out of the cells. *Lb. helveticus*, on the other hand, can metabolize galactose.

The pizza industry in the U.S. prefers that shredded Mozzarella or pizza cheese used on pizza topping show minimal browning when baked. The presence of residual galactose in cheese enhances browning of cheese on baked pizza. As a result, some cheese manufacturers prefer starters containing *Lb. helveticus* instead of *Lb. delbrueckii* ssp. *bulgaricus*. For more details on rod-coccus cultures, the reader should refer to recent reviews (Zourari et al., 1992; Oberg and Broadbent, 1993).

10.7 BACTERIOPHAGES OF STARTER BACTERIA

One the major obstacles to successful cheese manufacture is the infection of starter bacteria by bacterial viruses or bacteriophages. Bacteriophages (or phages) can disrupt cheesemaking in several ways. In extreme cases, these infective viruses destroy (or lyse) starter bacteria causing cessation of acid production, resulting in what is called a "dead vat". Such a drastic effect occurs when the infection is massive, and the phage numbers or titer is quite high, usually greater than 100,000 plaque-forming units per ml (pfu/ml) of whey. A reduction in the rate of acid production disrupts production schedules. Disruption of production schedules result in overtime wages for production personnel. Also, slow acid production caused by phage usually leads to quality problems in the finished cheese, and may also allow pathogens to grow or permit enterotoxin formation in the cheese. The overall effect is severe economic loss to the cheese industry.

Phages for lactococci have been extensively studied and reviewed (Allison and Klaenhammer, 1998). Their morphology, structural units (i.e., their protein coat and tail organization), and their DNA have been characterized. A classification system based on their serology has been described. The most commonly encountered phages in cheese industry in the Western Hemisphere belong to groups known as 936 (small isometric heads), C2 (prolate headed) and 335 (small isometric heads). The last mentioned group includes newly emerging, virulent phages that show unique evolutionary mechanisms involving exchange of DNA sequences from host cell genome.

The mechanism of infection, factors associated with infective process (e.g., cofactor requirements), the growth cycle and host range of lactococcal phages are well documented. The inherent phage resistance mechanisms found among lactococci have been studied and exploited to derive phage resistant derivatives of starter strains. Resistance mechanisms involve alteration of phage attachment sites (i.e., prevention of the entry of phage DNA), restriction—modification (R/M) systems (i.e., destruction of injected phage DNA), and abortive infection mechanisms (disruption of phage assembly). All of these systems are coded on plasmids, although there could be instances where the information may be located on the genome. These systems have been cloned, transferred, and functionally expressed within the dairy lactococci, as well as in species belonging to a closely related genus, *Streptococcus.*

Efficient control of phages in a manufacturing plant can be achieved by using various commercially available phage inhibitory media (PIM) for starter propagation in addition to maintaining proper sanitation and positive air pressure. The PIM are based on the chelation of divalent cations, especially Ca^{2+} and Mg^{2+}, which are vital for phage attachment to host cells. Various plant, air handling and equipment design features, and sanitation practices have also been developed to control phage-related failures in cheese plants. For details of methods for controlling phage in a dairy processing plant, see Sandine (1979) and Nath (1993).

Phages for thermo-tolerant starter lactic acid bacteria have not been as extensively studied as those for the dairy lactococci. The effects of phage infection of rod-coccus mixtures during cheesemaking are not as dramatic or disruptive as what happens with the lactococci. Reddy (1974) reported that in many cases, when the coccus component is lysed, the lytic products stimulate acid production by the rod cultures and vice versa. However, the cheese industry is plagued with disruptions caused by phages infecting rod-coccus cultures. Many of the *S. thermophilus* phages are capable of infecting their hosts even in the absence of Ca^{2+} ions, and in some instances, Mg^{2+} ions could substitute for calcium ions in phage adsorption. Specially formulated PIM have been developed and marketed for the thermo-tolerant starters. Phage inhibitory media designed for lactococci are unsuitable for rod-coccus cultures. High phosphate concentrations used in PIM for lactococci inhibit the normal and vigorous growth of rod-coccus cells. PIM for rod-coccus cultures have lower phosphate content, substitute chelators like citrates, and stimulants to ensure unimpeded growth of these bacteria.

The detection and enumeration of rod-coccus phages are more difficult, and special techniques and media have been described to study these phages (Reddy, 1974). Recently a classification system for *S. thermophilus* phages has been proposed. Host range studies using a fairly large collection of host coccus strains and phages received from widely separated geographical areas have been reported (La Marrec et al., 1997). Phage resistance systems among *S. thermophilus* strains have also been reported. A report documenting the expression of a lactococcal restriction-modification (R/M) system in a *S. thermophilus* strain has been published. More information on phages infecting rod strains is now available. For a review on phages infecting starter lactic acid bacteria, see Allison and Klaenhammer (1998).

10.8 ASSOCIATED MICROBIAL FLORA OR SUPPLEMENTARY MICROBIAL STARTER FLORA

Microorganisms added for flavor generation include molds, yeast and bacteria. Typically, they do not play a significant role in acid production leading to curd formation or syneresis. Pure cultures of molds used in soft cheeses, contribute to the unique flavor of the cheese variety.

In soft cheese varieties like Camembert, mold (*P. camemberti* and (or) *P. caseicolum*) growth is only on the surface of the cheese. Spore suspensions or broth cultures of the molds are inoculated on the surface of the cheese using a soft brush or cheesecloth soaked in the spore suspension or broth culture. After inoculation the cheeses are placed on wire racks, and periodically turned to get an even, fuzzy white mycelial growth over the entire surfaces of cheeses. Some cheesemakers do not inoculate the cheese with spores or broth culture but rely on natural acquisition of the mold from the wire racks on which cheese is stored. The latter approach is, however, not a good practice because the inoculation may be uneven and spotty, and unwanted contaminant molds may also develop on the cheese. During curing, the brittle, white grainy raw curd is changed to soft, smooth translucent cheese. Too long a ripening would result in a runny consistency. The cheese develops a slightly fruity, mushroom-like flavor at the end of curing. For a more comprehensive review on the role of molds in the ripening and flavor development, see Shaw (1981), Law (1982), Johnson (1998), and Nath (1993).

Brick, Limburger and Port du Salut represent a class of soft cheeses that are ripened exclusively by smear flora. Smear flora also play a minor role in the curing of other soft cheese varieties like Camembert,

Brie, Munster, Bel Paesa, etc. The smear flora includes bacteria, yeasts and molds. However, the predominant bacteria in the smear are *B. linens* and *Micrococcus* spp.

Brevibacterium linens cells are very short, Gram-positive rods that occur singly. On solid surfaces like agar plates or on cheese, these bacteria produce orange-red growth. *B. linens* is aerobic, and grows well within a wide temperature range (7–32°C). A unique feature that favors the dominance of *B. linens* in the smear is its salt-tolerance. *B. linens* can grow well in medium containing 15% salt. Surface ripened cheeses are sometimes first placed in a cooler held at 8–10°C, where they are rubbed with dry dairy salt over a two-day period. During this time the salt concentration is quite high on the surface of the cheese, setting a steep concentration gradient from the outside to the interior of the cheese loaf. The salt gradually migrates to the interior. After the salting is complete, the smear is applied, and the cheeses are moved into the curing room. The high salt concentration provides a selective environment for the "desired and necessary" smear flora. Another widely practiced procedure is to hold the cheese immersed in chilled saturated salt brine for a day, and after draining off excess brine, the smear is applied and the cheese moved into the curing room. Symbiotic relationship between *B. linens* and smear yeasts has been observed. *B. linens* has very strong proteolytic systems, and contribute to strong odors and flavors associated with surface smear ripened cheeses. Lately, there has been considerable interest in using these organisms as a supplementary starter to improve the flavor of low-fat hard cheeses, where greater intensity of flavor is needed to be comparable in flavor to full-fat cheeses. The renewed interest has resulted in intensive research on their proteolytic enzymes, and related functions. Recent reviews of the subject can be found in Rattray and Fox (1999).

The other predominant bacteria in the smear belong to the genus *Micrococcus*. The micrococci are Gram-positive, spherical bacteria occurring as single cells or pairs or clumps. They are aerobic and oxidative. Micrococci are also quite salt tolerant (can grow in media containing 10%) and thermoduric. They grow well at mesophilic temperature range. Many micrococci produce pigmented colonies with colors ranging from yellow to deep red. Micrococci produce both proteolytic and lipolytic enzymes, but their lipolytic activities are generally weak. Three micrococci, namely *M. varians, M. caselyticus, and M. freudenreichii* (some of the species are not recognized in current bacterial taxonomy) may be found in smear microflora (Johnson, 1998).

Yeasts found in the smear also contribute to the enzymatic pool involved in flavor development. *Geotrichum candidum,* one of the species in the mixed flora, produces a lipase that is very active against milk fat. Other yeasts include *Candida* spp. and the pinkish-pigmented *Debramyces* spp. The yeasts metabolize lactic acid in the cheese, and raise the pH of the microenvironment in the area adjacent to the surface, and allow good growth and metabolic activities of the bacteria in the smear. Additionally, the lysis of yeast cells liberates vitamins and amino acids, which stimulate the bacteria, and provide flavor precursors. Yeast metabolism yields alcohols, esters etc. which may contribute to the overall flavor in surface ripened soft cheeses. In addition to yeasts, in certain systems, some molds may also be associated with the smear flora. Although the individual contribution, and the exact role of the different microbial species to the flavor is unknown, the importance of this microbial community occupying this unique ecological niche, in the collective generation of the flavor(s) unique to surface ripened soft cheeses is well established. The role of smear flora in semi-soft and soft varieties is discussed in Olson (1969).

10.9 MICROBIAL SPOILAGE OF SOFT CHEESE

The relatively high moisture content of soft cheeses makes them easily susceptible to microbial spoilage. To lengthen shelf life, dry Cottage cheese curd, Feta and Domiati are held in cold brine until ready to sell. The high salt content of brine and its dehydrating effect provide a few more days of shelf life. The best way to ensure good shelf life of soft cheeses is to observe strict sanitary practices throughout the manufacturing steps, and post-manufacture handling. For example, during Cottage cheese manufacture, it is important that all the equipment used be properly cleaned and sanitized. This is especially important with dial thermometers, knives, agitators, paddles, scoops or any equipment or utensil that come into contact with the milk or cheese. After the whey is removed, the curd is washed with chilled water to remove excess lactose and to harden the curd thereby preventing shattering during later operations. The quality and the pH of the water used for washing the curd are very important. The water should be of potable quality, and pH preferably slightly acidic or neutral. Hard water or water with an alkaline pH is unacceptable. When well water is used, the microbiological quality of the water should be monitored closely. Coliforms and psychrotrophic bacteria are of major concern in Cottage cheese. To ensure destruction of any contaminants that may have inadvertently entered the cheese vat, or carried through the wash water, the wash water is chlorinated to give 5 ppm available chlorine. If the water is alkaline, food-grade acids such as phosphoric or citric are used to adjust the pH to slightly acidic values (≤pH 6.5), so that the chlorine will be effective. The use of citric acid to adjust the pH not only helps the effectiveness of chlorine, but also provides some precursor for diacetyl synthesis after the dressing is added. Further chances for contamination occur during the mixing of the dressing and filling. Because of intricate working parts, fillers are difficult to clean, and they are often the source of psychrotrophic flora. The individual parts of fillers should be disassembled after each use and hand brushed to remove residual curd particles, milk components, etc. Equipment should be periodically examined with a black light (UV light) to detect milk stones and should be treated with acid cleaner to remove such films. The filler lines, bowls and filler ports, valves, or pistons should be treated with scalding water flush followed by a light chlorine rinse just before filling the cheese. Another source of psychrotrophic contamination is dripping condensate from cold overhead pipes. Condensate drippings could fall into open vats where dressing is mixed with the curd, or into open filler bowls. Sometimes aerosolization of water sprayed from high-pressure water lines used to clean spills on the floor during filling operation also introduces contaminants into the product.

Psychrotrophic bacteria of major concern are *Pseudomonas* spp., *Alcaligenes* spp., *Achromobacter* spp. and *Flavobacterium* spp. These are Gram-negative rods, that are aerobic and are easily destroyed by pasteurization and cooking temperatures used for Cottage cheese manufacture. Their presence indicates post-manufacturing contamination. Among the pseudomonads, *P. fluorescens*, *P. fragi* and *P. putida* are important. All these species grow rapidly at normal refrigeration temperature (~7°C) used in the dairy industry. These bacteria produce very active proteolytic and lipolytic enzymes. They cause bitterness, putrefactive and rancid odors, liquefaction of curd particles, gelatinization of curd (white, opaque curd turning translucent), slimy, mucous appearance of the curd surface, and rancidity. *Pseudomonas fluorescens* also causes discoloration because of the formation of water-soluble florescent pigments (which glow under UV light). Other pseudomonads also discolor cheese surface as the casein is hydrolyzed, causing darkening or yellowing of the curd. Pseudomonads and other psychrotrophic Gram-negative bacteria have strong diacetyl reductase enzyme, which reduces diacetyl to acetoin and other reduction products, thereby causing loss of desirable flavor. *Alcaligenes viscolactis* produces ropiness and sliminess in Cottage cheese. *Alcaligenes metacaligenes* reduces diacetyl flavor resulting in "flat, flavorless" Cottage cheese. The flavobacter produce yellow flavin pigments that discolor cheese.

The occurrence of psychrotrophic *Bacillus* spp. is of concern. These bacilli are Gram-positive spore-forming rods that survive pasteurization. Their presence in Cottage cheese results in bitterness and other proteolytic defects. Some species can produce dark pigments, and *B. cereus* strains can cause food poisoning.

Other psychrotrophic bacteria of importance in Cottage cheese spoilage are yeasts and molds. Yeasts often form yellow, orange or pink colonies on the surface of cottage cheese. They also induce yeasty fruity off-flavors and gassiness. Because molds are very aerobic, they are generally found on the surface of cheese or on the underside of package lids. Molds form fuzzy, pigmented colonies (black, green or dark brown) on the cheese surface. They impart a musty, moldy odor and produce discoloration, and liquefaction of the curd due to their powerful proteolytic enzymes. Yeasts and mold generally enter Cottage cheese as contaminants from the air, or improperly stored (dusty) containers used for packaging the product. In many dairy plants the packaging room is designed to have positive air pressure with laminated airflow through HEPA (high efficiency particulate air) filters. To prevent fungal contaminants from packaging materials, it is necessary to store containers and lids in closed boxes (especially, partially used boxes) in a cool, dry area.

Preservatives—"natural" and chemical—are also used to control psychrotrophic spoilage. The commonly used preservatives are propionates and sorbates to control yeasts and molds. The levels added are regulated in the U.S. and many other countries. A "natural" dried, fermentation product using *Propionibacterium freudenreichii* marketed under the trade name MicroGARD™, is widely used in the U.S. to control Gram-negative psychrotrophs, yeasts and molds in Cottage cheese. The propionate and acetate in the fermentate, as well as an undefined "bacteriocin-like" compound in the fermentate are considered to be the active components. There are other similar "natural" commercial preparations. Also, live cultures of *Leuconostoc* spp. and cit$^+$ *Lc. lactis* ssp. *lactis* are used to control psychrotrophic bacteria. A direct CO_2 injection of the dressing is also promoted for controlling psychrotrophs (Kosikowski and Mistry, 1997). Psychrotrophic Gram-negative bacteria and molds are quite aerobic, and saturation of the cream dressing with carbon dioxide exerts a suppressive effect. To control psychrotrophic *Bacillus* spp. the use of low acid producing, nisin (a bacteriocin) ex-creting *Lc. lactis* ssp. *lactis* strain in Cottage cheese dressing is suggested. The use of tamper-proof sealed packaging for Cottage cheese precludes the use of gas-generating cultures or CO_2 saturation of cream dressing because of bloating or bulging of containers caused by interior gas pressure especially with some temperature abuse during distribution.

In soft cheeses (e.g., Camembert) that are cured by promoting external mold growth external contaminant molds sometimes appear. *Penicillium roqueforti* is a common contaminant. Spots of green to blue mycelia in the mass of white mycelial growth of *Penicillium camemberti* indicate such contamination. Kosikowski and Mistry (1997) suggest the use of a high dosage of *P. camemberti* and (or) *P. caseicolum* spores, and periodic cleaning and fumigation of cheese curing rooms, manufacturing premises, and equipment to control unwanted molds. Sometimes, a black mold, *Mucor rasmusen*, also causes chronic discoloration problems in Camembert cheese. The addition of more salt, strict sanitation and fumigation will control *M. rasmusen*, but the mold tends to reappear off and on. The surface of Camembert cheese also supports limited growth of some of the smear bacteria (especially, *B. linens*) that helps to develop full flavor. Excessive growth of contaminating smear bacteria may cause over ripening, and ammonia off-flavors.

Defects caused by microorganisms in smear surface-ripened cheeses (e.g., Brick cheese) have been reviewed (Olson, 1969). When the pH of Brick cheese gets below 5.0, good smear development is retarded as growth of *B. linens* is inhibited at pH values below 5.0. In cheeses that are too acidic, flavor development is poor because of little or no smear flora. The use of lactococci solely as starter in Brick cheese manufacture may lead to poor smear development as lactococci are active at low temperatures during later stages of whey draining and curing, thus lowering the pH below 5.0. To prevent excessive acid development, the curd is washed after whey drainage to remove excess lactose. Inclusion of *S. thermophilus* to partially replace lactococci helps reduce over-acidification because the "cocci" do not produce as much acid, and as rapidly as lactococci. Also, the "cocci" do not grow or metabolize lactose as well as lactococci at low temperatures. The presence of too many enterococci also results in excessive acidity (enterococci grow well at low temperatures) and bitterness. Fruity, fermented and metallic flavors are attributed to anaerobic spore-formers, which also induce "late gas openings" in the cheese. *Bacillus polymyxa* is reported to produce "early gas defects". Mold contamination in the smear induces body and flavor defects. Proper pasteurization of cheese milk, care in the selection and propagation of starters, good sanitation in the cheese plant and proper manufacturing procedures are important in preventing microbial-induced defects.

Other high-moisture whey cheeses like Ricotta are easily perishable. The high moisture can allow any contaminants found in the cheese to multiply rapidly and cause spoilage. Therefore, proper sanitation, good packaging and refrigeration are mandatory for such cheeses. For Mozzarella and other pasta filata varieties, selection of proper starter strains ("rod-coccus" and lactococci) is important to obtain good rheological and melting properties in the cheese. The stretching and flowing properties of the cheese used in pizza are not only dependent on the manufacturing processes, but also on the proteolytic and acid forming capabilities of the starter bacteria. Browning or darkening of cheese on pizza during baking is directly related to the removal of sugar by starter bacteria. Some of these aspects were discussed earlier in this chapter.

Other defects result from mold contamination. Also, surface defects may be caused by bacterial contamination from brine tanks that are not properly maintained (i.e., lack of circulation, failure to maintain proper salt level in the brine, improper temperature control, lack of filtration through screens during circulation to remove cheese particles and scum that may collect on the brine surface). Defects of Mozzarella and other Italian cheese varieties have been discussed by Reinbold (1963).

10.10 PATHOGENIC MICROFLORA IN SOFT CHEESE

The microbiological safety of cheeses has been recently reviewed (Johnson et al., 1990). The prevalence of pathogenic microorganisms in soft cheeses depend upon the quality of milk used, heat treatment of milk, general sanitation in the cheese plant, quality of starters, occurrence of phages, cheese handling procedures, temperature of holding at the plant, during transport and through marketing channels. Because of the relatively higher moisture levels in soft cheeses, they provide a favorable environment for the growth of pathogens. Generally, in soft cheeses made from pasteurized milk, unless the cheese is handled carelessly, there is not much danger of heavy pathogenic contamination. However, recently there have been more than a few reports of cheese borne food infections, and food poisoning cases reported. Many such outbreaks have been associated with *L. monocytogenes.* In a few cases there have been fatalities associated with cheese-borne listeriosis. Mexican-style white soft cheese (Queso Fresco) has been implicated in some of these outbreaks (Johnson et al., 1995). In France, listeriosis was associated with Brie cheese. Another report from Switzerland implicated another ripened soft cheese. *Listeria monocytogenes* is Gram-positive rod shaped bacterium that is psychrotrophic, fairly heat-tolerant, widely distributed in dairy farms, and frequently found in raw milk. In dairy plants they are found in drains, floor conveyer belts, crates etc. Some strains survive pasteurization, and adapt to acidic environments. Because of their psychrotrophic nature, listeria grows well during refrigerated storage. The occurrence of *L. monocytogenes* in soft cheeses could be controlled by regular environmental monitoring of the cheese plant coupled with assaying for listeria regularly in incoming raw milk and materials going into the cheese vat, proper pasteurization, and strict sanitation in every step of the process. The numbers of listeria can also be controlled in cheeses by using "natural" preservatives such as bacteriocins that are inhibitory to these bacteria. Examples are pediocins (bacteriocin produced by *Pediococcus* spp. and other lactic acid bacteria), and nisin. Pediocins are very effective against listeria, more so than nisin. These bacteriocins are more effective in acidic pH range than at neutral or alkaline pH range, are easily bound up by protein and fat, and can be hydrolyzed and inactivated by any proteolytic enzymes that may be present. Glass et al. (1995) reported differences in the efficacies of different organic acids and a bacteriocin type product in the control of *L. monocytogenes* in Queso Blanco–type cheese.

Escherichia coli serotype O157: H7, which is widely found in raw milk is a pathogen that has recently caused a lot of concern. These bacteria have been found to exhibit unusual heat and acid tolerances, and have a very threatening potential to cause serious cheese-related food infections and fatalities. *E. coli* has been implicated in several cases of food poisoning resulting from the consumption of French Brie and Camembert in the U.S. and Scandinavia (D'Aoust, 1989).

Raw milk soft cheeses may potentially carry different *Salmonella* strains. Salmonella have been implicated in food-borne illness resulting from the consumption of Vacherin in Switzerland, French Brie and Camembert in Scandinavia (D'Aoust, 1989). Because salmonella are enteropathogens that cause serious food infections and raw milk often contains salmonellae, careful monitoring of raw milk supply and plant personnel for "healthy carriers", strict sanitation, active starters, and good refrigeration are necessary to combat these pathogens. Cit^+ *Lc. lactis* ssp. *lactis* have been reported to inhibit salmonellae in broth, agar and cheese systems (Daly et al., 1972; Vedamuthu et al., 1966). Addition of high numbers of selected, low acid producing strains (for example, cured of lactose-plasmid) of these bacteria to cheese milk may be beneficial.

Another pathogen of concern is enterotoxigenic *Staphylococcus aureus*. *S. aureus* is a Gram-positive coccus that occurs singly or in clumps. It is salt tolerant and can grow in systems containing 10% NaCl. It can also grow at relatively low water activity (A_w ~ 0.86). *S. aureus* can produce heat-stable enterotoxin in neutral and low acid systems (pH > 5.0). Growth of *S. aureus* in food does not cause any perceptible changes in appearance, smell or taste. The preformed toxin in the food causes intoxication with symptoms appearing within 2–4 h. Symptoms of *S. aureus* food poisoning consist of abdominal cramps, vomiting, diarrhea, sometimes sweating and exhaustion that last for a very short time. Because of their ability to grow at relatively low A_w and at high NaCl concentrations, the potential for toxin production in low acid soft cheeses exists. Therefore, good sanitation among plant workers, general sanitation in cheese plant, effective refrigeration, the use of active starters and good packaging are important in preventing *S. aureus*-related food poisoning.

The growth of *Clostridium botulinum* and its toxin production in some cheese varieties can lead to the outbreak of botulism. In France and Switzerland in 1974, the outbreak of botulism was attributed to the consumption of Brie cheese containing botulinal toxins (Johnson et al., 1990). Recently, Franciosa et al. (1999) analyzed 1017 Mascarpone cheese samples in Italy and found that 32.5% were positive for botulinal spores. Growth requirements for *C. botulinum* include a low-oxygen atmosphere and minimum temperature, pH and A_w of 10°C, 4.6 and 0.93, respectively, for proteolytic strains and 3.3°C, 5.0, and 0.97 for non-proteolytic strains.

There is a potential for mycotoxin formation in some of the mold-ripened cheeses, but so far the level of mycotoxins found in various kinds of cheeses have not been of much concern. Generally, there have been very few disease outbreaks caused by the ingestion of cheese. If the use of high quality raw materials, proper sanitation, proper adherence to established cheesemaking procedures, use of active starter cultures, proper concentration of salt application, immediate and effective refrigeration at various stages of manufacture and distribution, and sound packaging are practiced, pathogens could be effectively controlled and kept out of soft cheeses. For a comprehensive discussion of public health concerns pertaining to dairy products, the reader should consult the review by Ryser (1988).

ACKNOWLEDGMENT

The preparation of this review was made possible by a grant from the California Dairy Research Foundation.

REFERENCES

Abou-Donia, S. A. (1991) Manufacture of Egyptian, soft and pickled cheese. In *Feta and related cheeses*, R. K. Robinson and A. Y. Tamime, eds., Ellis Horwood Ltd., West Sussex, England. pp 161–208.

Allison, G. E., and Klaenhammer, T. R. (1998) Phage resistance mechanisms in lactic acid bacteria. *Int. Dairy J.* **8**, 207–226.

Anifantakis, E. M. (1991) Traditional Feta cheese. In *Feta and related cheeses*, R. K. Robinson and A. Y. Tamime, eds., Ellis Horwood Ltd. England. pp 49–69.

Arihara, K., and Luchansky, J. B. (1995) Dairy lactobacilli. In *Food biotechnology: Microorganisms*, Y. H. Hui and G. G. Khachtourians, eds., VCH Publishers Inc., New York, NY.

Axelsson, L. (1998) Lactic acid bacteria: Classification and physiology. In *Lactic acid bacteria: Microbiology and functional aspects*, 2nd ed., S. Salminan and A. von Wright, eds., Marcel Dekker Inc., New York, NY.

Barbano, D. M., Yun, J. J., and Kindstedt, P. S. (1994) Mozzarella cheese making by a stirred curd no brine method. *J. Dairy Sci.* **77**, 2687–2694.

Chandan, R. C., Marin, H., Nakrani, K. R., and Zehner, M. D. (1979) Production and consumer acceptance of Latin American white cheese. *J. Dairy Sci.* **62**, 691–696.

Chapman, H. R., and Sharpe, M. E. (1990) Microbiology of cheese. In *Dairy Microbiology*, Vol 2: *Microbiology of milk products*, 2nd ed., R. K. Robinson, ed., Elsevier Applied Science, New York. pp 203–289.

Condon, S. (1987) Responses of lactic acid bacteria to oxygen. *FEMS Microbiol. Rev.* **46**, 268–280.

D'Aoust, J.-Y. (1989) Manufacture of dairy products from unpasteurized milk: A safety assessment. *J. Food Prot.* **52**, 906–912.

Daly, C., Sandine, W. E., and Elliker, P. R. (1972) Interactions of food starter cultures and food-borne pathogens: *Streptococcus diacetilactis* versus food pathogens. *J. Milk Food Technol.* **35**, 349–357.

Dessart, S. R., and Steenson, L. R. (1995) Biotechnology of dairy *Leuconostoc*. In *Food biotechnology: Microorganisms*, Y. H. Hui and G. G. Khachatourians, eds., VCH Publishers Inc., New York, NY.

Emmons, D. B., and Tuckey, S. L. (1967) Cottage cheese and other cultured milk products. Pfizer Cheese Monographs, Vol. III. Chas Pfizer & Co. Inc., New York.

Farkye, N. Y., Prasad, B. B., Rossi, R., and Noyes, O. R. (1995) Sensory and textural properties of Queso Blanco-type cheese influenced by acid type. *J. Dairy Sci.* **78**, 1649–1656.

Franciosa, G., Pourshaban, M., Gianfranceschi, M., Gattuso, A., Fenicia, L., Ferrini, A. M., Mannoni, V., De Luca, G., and Aureli, P. (1999) *Clostridium botulinum* spores and toxin in Mascarpone cheese and other milk products. *J. Food Prot.* **62**, 867–871.

Glass, K. A., Bhanu Prasad, B., Schlyter, J. M., Uljas, H. E., Farkye, N. Y., and Luchansky, J. B. (1995) Effects of acid type and Alta™2341 on *Listeria monocytogenes* in Queso Blanco type of cheese. *J. Food Prot.* **58**, 737–741.

Gonzalez, Carlos. F. (1984) Preservation of foods with non-lactose fermenting *Streptococcus lactis* subspecies *diacetilactis.* U.S. Patent 4,477,471.

Gonzalez, Carlos. F. (1986) Preservation of foods with non-lactose fermenting *Streptococcus lactis* subspecies *diacetilactis*. U.S. Patent 4,599,313.

Gripon, J. C. (1993) Mould-ripened cheeses. In *Cheese: Chemistry, Physics and Microbiology*, Vol. 2. *Major Cheese Groups*. P. F. Fox, ed., Chapman and Hall, London.

Guinee, T. P., Pudja, P. D., and Farkye, N. Y. (1993) Fresh acid-curd cheese varieties. In *Cheese: Chemitry, Physics and Microbiology*, Vol. 2. *Major Cheese Groups*. P. F. Fox, ed., Chapman and Hall, London.

Hugenholtz, J. (1993) Citrate metabolism in lactic acid bacteria. *FEMS Microbiol. Rev.* **12**, 165–178.

Hugenholtz, J. M., Starrenburg, J. C., and Weerkamp, A. H. (1994) Diacetyl production by *Lactococcus lacis*: Optimilisation and metabolic engineering. *Proc. 6th European Congr. Biotechnol.* 225–228.

Johnson, E. A., Nelson, J. H., and Johnson, M. (1990) Microbiological safety of cheese made from heat-treated milk, Part II. Microbiology. *J. Food Prot.* **53**, 519–540.

Johnson, M. E. (1998) Cheese products. In *Applied dairy microbiology*, E. H. Marth and J. L. Steele, eds., Marcel Dekker Inc., New York, NY.

Kosikowski, F. V., and Mistry, V. V. (1997) Cheese and fermented milk foods. Vol 1: *Origins and principles*, 2nd ed., F. V. Kosikowski LLC, Westport, CT.

Koth A. B., and Richter, R. L. (1989) A limited test market evaluation for Quarg sales potential. *Cult. Dairy Prod. J.* **24**(3), 4–8.

Kroger, M. (1979) The manufacture of Quarg cheese. In *1st Biennial Marschall Int. Cheese Conference*, Dane County Exposition Ctr., Madison, WI. Sept. 10–14.

La Marrec, C., van Sinderen, D., Walsh, L., Stanley, E., Vlegels, E., Moineau, S., Heinze, P., Fitzgerald, G., and Fayard, B. (1997) Two groups of bacteriophages infecting *Streptococcus thermophilus* can be distinguished on the basis of packaging and genetic determinants for major structural proteins. *Appl. Environ. Microbiol.* **63**, 3246–3253.

Law, B. A. (1982) Cheeses. In *Fermented Foods. Economic microbiology*, Vol. 7. A. H. Rose, ed., Academic Press, New York, NY.

Law, J., and Haandrikman (1997) Proteolytic enzymes of lactic acid bac-teria. *Int. Dairy J.* **7**, 1–11.

Mercenier, A. (1990) Molecular genetics of Streptococcus thermophilus. *FEMS Microbiol. Rev.* **87**, 61.

Mundt, J. O. (1986) Lactic acid streptococci. In *Bergey's Manual of Determinative Microbiology*, Vol 2. P. H. Sneath, N. S. Mair, M. E. Sharpe, and J. G. Holt, eds., William and Wilkins, Baltimore, MD.

Nath, K. R. (1993) Cheese. In *Dairy Science and Technology Handbook*, Vol 2: *Product manufacturing*, Y. H. Hui, ed., VCH Publishers Inc., New York, NY.

Nooitgedagt, A. J., and Hartog, B. J. (1988) A survey of the microbiological quality of Brie and camembert cheese. *Neth. Milk Dairy J.* **42**, 57–72.

Oberg, C. J., and Broadbent, J. R. (1993) Thermophilic starter cultures: Another set of problems. *J. Dairy Sci.* **76**, 2392–2406.

Olson, H. C. (1974) Composition of cottage cheese. *Cult. D. Prod. J.* **9**(3), 8–10.

Olson, N. F. (1969) Ripened Semi-soft Cheeses. Pfizer Cheese Monographs. Vol 4. Chas. Pfizer & Co. Inc., New York, NY.

Puhan, Z., and Gallmann, P. (1980) Ultrafiltration in the manufacture of Kumys and Quark. *Cult. Dairy Prod. J.* **15**(1), 12–16.

Ramet, J. P. (1990) The production of fresh cheese cheeses in France. *Dairy Industries Int.* **55**(6), 49–52.

Rattray, F. P., and Fox, P. F. (1999) Aspects of enzymology and biochemical properties of *Brevibacterium linens* relavant to cheese ripening: a review. *J. Dairy Sci.* **82**, 891–909.

Reddy, M. S. (1974) Development of cultural techniques for the study of *Streptococcus thermophilus* and *Lactobacillus* bacteriphages. Ph.D. thesis, Iowa State University, Ames, IA.

Reinbold, G. W. (1963) Italian cheese varieties. Pfizer Cheese Monographs. Vol 1. Chas. Pfizer & Co. Inc., New York, NY.

Ryser, E. T. (1998) Public health concern. In *Applied Dairy Microbiology*, E. H. Marth and J. L. Steele, eds., Marcel Dekker Inc., New York, NY.

Sandine, W. E. (1979) Lactic starter culture technology. Pfizer Cheese Monographs, vol. 4, Pfizer Inc., New York, NY.

Scott, R., Robinson, R. K., and Wilbey, R. A. (1988) Cheesemaking practice. 3rd ed. Aspen Publishers, Inc. Gaithersburg, MD.

Sellars, R. L., and Babel, F. J. (1985) Cultures for the manufacture of dairy products. Chr. Hansen's Laboratory, Inc. Milwaukee, WI. pp 37.

Shaw, M. B. (1981) The manufacture of soft, surface mould, ripened in France with particular reference to Camembert. *J. Soc. Dairy Technol.* **34**, 131–138.

Siggelkow, M. A. (1984) Quarg production for consumer sale. *Dairy Ind. Int.* **49**(6), 17–20.

Sing, E. L. (1976) Preparation of cottage cheese. U. S. Patent 3,968,256.

Strauss, K. (1997) Cream cheese. In *Cultures for the Manufacture of Dairy Products*. Chr. Hansen's Laboratory Inc. Milwaukee, WI. pp 116.

Tamime, A. Y. and Kirkegaard, J. (1991) Manufacture of Feta cheese—industrial. In *Feta and related cheeses*, R. K. Robinson and A. Y. Tamime, eds., Ellis Horwood, Ltd. England.

Teubner, C. (1998) The Cheese Bible. Penguin Putnam Inc., New York, NY.

Van Slyke, L. L., and Price, W.V. (1952) Cheese. Orange Judd Publishing Co., Inc. New York.

Vedamuthu, E. R., Hauser, B. A., Henning, D. R., Sandine, W. E., and Elliker, P. R. (1966) Competitive growth of *Streptococcus diacetilactis* in mixed strain lactic cultures and cheese. *Proc. Intern. Dairy Sci. Congr. Munich.*, **D**(2), 611–618.

Vedamuthu, E. R., and Washam, C. J. (1983) Cheese. In *Biotechnology*, vol 5, *Food and Feed Production with Microorganisms*, G. Reed, ed., Verlag Chemie, Deerfield Beach, FL.

Vedamuthu, E. R. (1991) The yogurt story. Past, present and future. *Dairy Food Envoiron. Sanitation* **11**, 265–267.

Vedamuthu, E. R. (1994) The dairy *Leuconostoc*: Use in dairy products. *J. Dairy Sci.* **77**, 2725–2737.

von Wright, A., and Sibakov, M. (1998) Genetic modification of lactic acid bacteria. In *Lactic acid bacteria—Microbiology and functional aspects*, 2nd ed., S. Salminen and A. von Wright, eds., Marcel Dekker Inc., New York, NY.

Weatherup, W. (1986) The effect of processing variables on the yield and quality of Ricotta. *Dairy Ind. Int.* **5**(8), 41–45.

Zourari, A., Accolas, J. P., and Desmazeaud, M. J. (1992) Metabolism and biochemical characteristics of yogurt bacteria. A review. *Lait*, **72**, 1.

CHAPTER 11

MICROBIOLOGY OF HARD CHEESE

TIMOTHY M. COGAN and THOMAS P. BERESFORD
Dairy Products Research Centre, Teagasc, Fermoy, County Cork, Ireland

11.1 INTRODUCTION

It is generally believed that cheese-making developed around the Tigris and Euphrates rivers in Iraq around 8000 years ago when the first animals (probably sheep and/or goats) were domesticated. Until the end of the eighteenth century, cheese-making was a farmhouse or artisanal skill. The first industrial cheese-making plant in the United States was opened in Rome, New York in 1851, and the first one in the United Kingdom was opened in Langford, Derbyshire in 1870. Today, cheese is made all over the world but especially in Europe, North America, Australia, and New Zealand. Annual worldwide production of cheese is ~15×10^6 tonnes in which ~35% of the total amount of milk produced is used. Cheese is mostly made from cow's milk, but sheep's, goat's, and water buffalo's milk is also used. Production of sheep's milk cheese is particularly high in Mediterranean countries and in Portugal.

Cheese is essentially a microbial fermentation of milk by selected lactic acid bacteria whose major function is to produce lactic acid from lactose, which, in turn, causes the pH of the curd to decrease. The final pH after manufacture ranges from 4.6 to 5.3, depending on the buffering capacity of the curd. A reduction in the moisture content of the milk (dehydration) also occurs during cheese-making from an initial value of ~88%, in the case of cows' milk, to ≤50%. In all hard cheeses and, indeed, in most other cheeses, an enzyme called rennet or chymosin, originally from the abomasum of the calf but now produced mostly by

Dairy Microbiology Handbook, Third Edition, Edited by Richard K. Robinson
ISBN 0-471-38596-4

fermentation, is used to coagulate casein, the major milk protein. Other coagulants—for example, pepsin from various sources and extracts of cardoon flowers, especially in Iberian cheeses—are also used.

Essentially there are seven key steps in cheese-making:

- Addition of lactic acid starter bacteria and coagulant
- Coagulation followed by cutting the curd
- Cooking to temperatures from 32°C to 54°C, which, together with acid production, assist expulsion of whey from the curd
- Separation of the curds and whey
- Molding and pressing the curd at low (soft cheese) or relatively high (hard cheese) pressure
- Salting or brining
- Ripening at temperatures of 6°C to 24°C to allow the characteristic flavor and texture to develop

Most cheeses are brined after pressing, but Cheddar is an exception; it is dry-salted and milled before being pressed.

Originally the primary objective of cheese manufacture was to extend the shelf life of milk and conserve its nutritive value. This was achieved by either acid production and/or dehydration. While all acid-coagulated cheeses are consumed fresh, most rennet-coagulated cheese undergo a period of ripening which can range from about 3 weeks for Mozzarella to 2 years or more for Parmesan and extra-mature Cheddar. Cheese ripening is a complex process involving a range of microbiological and biochemical reactions. High densities of microorganisms are present in cheese throughout ripening, and they play a significant role in the manufacturing and ripening process. In this chapter the microbiology of hard cheeses will be considered emphasizing principles, which can be applied to all cheeses; what happens in individual cheeses will be used as examples to highlight important points. In the interests of saving space, reviews rather than the original literature will be cited wherever possible; the reviews should contain the original references.

11.2 STARTER BACTERIA

The microflora of cheese may be divided into two groups: starter lactic acid bacteria and secondary microorganisms. Starter lactic acid bacteria are involved in acid production during manufacture and contribute to the ripening process. Secondary microorganisms do not contribute to acid production during manufacture but play a significant role during ripening. The secondary microflora consist of (a) nonstarter lactic acid bacteria (NSLAB), which grow internally in most cheese varieties, and (b) other bacteria, yeasts, and/or molds, which grow internally or externally and are usually unique to the specific cheese variety. A list of the most important bacteria that are found in cheese, along with their major functions, is shown in Table 11.1.

Basically, two types of starter bacteria are used in cheese-making: thermophilic with optimum temperatures of ~42°C and mesophilic, with optimum temperatures of ~30°C (see Chapter 1 in this volume). They are invariably lactic acid bacteria (LAB) and are usually deliberately added to the milk at the beginning of cheese-making. However, in some cheese—for example, Pecorino Sardo (from Sardinia) and Majorero (from Spain)—starters are not deliberately added to the milk. Instead, the cheese-maker relies on adventitious lactic acid bacteria, which are naturally present in the cheese milk, to grow and produce the necessary acid during cheese manufacture. Mesophilic cultures are used in the production of Cheddar, Gouda, Edam, Blue, and Camembert, while thermophilic cultures are used for hard cheeses such as Emmental, Gruyère, Comté, Parmigiano Reggiano, and Grana Padano, which are cooked to high temperatures (50–55°C). The starter bacteria are members of the genera *Lactococcus*, *Lactobacillus*, *Streptococcus*, *Leuconostoc*, and *Enterococcus*.

The primary function of the starter bacteria is to produce acid during the fermentation process. They also provide a suitable environment, with respect to redox potential, pH, moisture, and salt contents, to allow enzyme activity from the chymosin and starter (along with growth of the secondary flora) to proceed favorably in the cheese. Their enzymes are involved in the conversion of protein and fat into amino acids and fatty acids, repectively, from which the flavor compounds are produced (Urbach, 1995; Fox and Wallace, 1997). The starters are either (a) pregrown in milk, whey, or phage inhibitory media or (b) added directly to the milk as frozen or freeze-dried concentrates. The initial number of starter cells in the milk depends on the inoculation rate appropriate to the cheese being made and ranges from 10^5 to 10^7 cfu/ml of milk. The starter bacteria grow during manufacture and typically attain numbers of 10^9 cfu/g 24 h from the addition of the starter to the milk.

Defined cultures consisting of two or more strains of *Lactococcus lactis* are generally used to make Cheddar cheese. *Streptococcus thermophilus* may be used as an adjunct culture. Undefined mixed-strain mesophilic cultures are also used to some extent. These contain several

TABLE 11.1. Microorganisms Commonly Found in Cheese and Their Major Functions

Genus or Species	Major Function	Other Important Information
Starter bacteria		
Lactococcus lactis	Acid production; flavor formation during ripening	Citrate positive strains metabolize citrate
Leuconostoc spp.	Metabolism of citrate	
Lactobacillus helveticus	Acid production; flavor formation during ripening	
Lactobacillus delbrueckii	Acid production; flavor formation during ripening	
Streptococcus thermophilus	Acid production; flavor formation during ripening	Unable to ferment galactose
Enterococcus spp.	Flavor formation during ripening	
Nonstarter lactic acid bacteria		
Lactobacillus casei	Probably involved in flavor formation during ripening	Conversion of L- to D-lactate
Lactobacillus paracasei	Probably involved in flavor formation during ripening	Conversion of L- to D-lactate
Pediococcus spp.	Probably involved in flavor formation during ripening	Conversion of L- to D-lactate
Secondary flora		
Prop. shermanii	Eye formation in Swiss-type cheese	
Corynebacterium spp.	Production of methanthiol	Surface of smear cheese
Microbacterium spp.	Production of methanthiol probably	Surface of smear cheese
Brevibacterium linens	Production of methanthiol	Surface of smear cheese
Brachybacterium spp.		Surface of smear cheese
Micrococcus spp.		Surface of smear cheese
Kokuria spp.		Surface of smear cheese
Staphyloccocus spp.		Surface of smear cheese
Penicillium camemberti	Mainly involved in proteolysis	White growth on the surface of Camembert and Brie cheeses
Penicillium roqueforti	Mainly involved in lipolysis; some proteolysis	Seams in blue cheeses, e.g., Stilton, Roquefort
Yeasts	Deacidification (i.e., conversion of lactate to CO_2 and H_2O)	

strains each of *Lc. lactis* and *Leuconostoc* spp., but the exact number is not known (see Chapter 1 in this volume). Undefined or mixed-strain thermophilic cultures are used in the production of Comté, Emmental, and Grana cheeses of France, Switzerland, and Italy. These cultures are generally produced by incubating whey from the previous day's production overnight at 40–45°C. The flora is composed primarily of *S. thermophilus* (though enterococci, lactococci, and leuconostocs may also be present) and several species of lactobacilli, including *Lactobacillus helveticus*, *Lactobacillus delbrueckii*, *Lactobacillus acidophilus*, and *Lactobacillus fermentum*. A similar natural whey starter containing thermophilic LAB is used to make Parmigiano Reggiano, a hard, cooked cheese from Northern Italy. This starter consists primarily of undefined strains of *Lc. helveticus* (75%) and *Lc. delbrueckii* ssp. *bulgaricus* (25%); however, their precise composition is uncontrolled and significant variations at strain level are likely to occur (Coppola et al., 1997; Nanni et al., 1997).

11.3 GROWTH OF STARTERS DURING MANUFACTURE

Early in the manufacture of most cheeses a coagulum, called curd, is formed as a result of chymosin activity on casein. The coagulum is subsequently cut into different size particles, depending on the type of cheese being made. The curds then begin to expel whey and contract (this is called syneresis) at a rate that is dependent on the size of the particle, the rate of stirring, the rate of acid production by the starter, and the temperature. Whey expulsion results in a reduction in the water content within the curd particle. Small curd particles, rapid rates of stirring, and high temperatures promote syneresis, whereas lower rates of acid production, slow rates of stirring, and low temperatures retard it. Acid production or, more precisely, the decrease in pH also promotes syneresis of the curd and expulsion of the whey. Most of the starter cells are concentrated in the curd. Consequently, there is a higher concentration of lactate and a lower concentration of lactose within the curd particle than in the surrounding whey. In fact, a common way of following acid production in cheese-making is to squeeze the curd and measure the pH in the whey expressed from it.

Acid production during cheese-making is the direct result of growth of the starter. During the first few hours, acid production will depend primarily on the rate of inoculation, the time of renneting, and the subsequent rate of cooking the curd. Low inocula will result in lower total amounts of lactate and higher pH values at a particular time. In Stilton cheese, acid production is very slow and generally the curd is kept in contact with the whey overnight to promote it. In contrast, rates of acid production in Cheddar, Cheshire, and other British Territorial cheeses are relatively rapid from the beginning; for example, in Cheddar cheese the pH decreases from 6.6, the initial pH of the milk, to pH 6.1, the pH at which the whey is drained, about 3h after the addition of the starter. The more rapid acid production results in greater rates of expulsion of whey, and consequently of lactose, from the curd. During this time, lactate levels in the curd particles are continuously increasing.

Cooking temperatures are also important in controlling the growth of, and acid production by, starters. Mesophilic cultures have optimum temperatures of ~30°C and are used in the manufacture of Cheshire, Cheddar, and Gouda cheese. Low cooking temperatures (e.g., ~32°C) as occur in the manufacture of Cheshire cheese will have no effect on acid production. In contrast, Cheddar cheese is cooked to ~40°C, and this will significantly reduce acid production by the mesophilic starter. This cooking temperature will have no effect on any thermophilic organisms, which may be used as adjunct cultures. In Gouda and Edam cheese the cooking temperature is ~35°C, which has little inhibitory effect on the cultures. In the latter two cheeses, part of the whey is removed during manufacture and replaced with warm water. This reduces the lactose and lactate levels in the curd and hence controls acid production. Washing of the curd occurs at the stage during manufacture when significant syneresis of the curd is occurring. Hard cooked cheeses like Emmental and Parmigiano Reggiano, which are made with thermophilic cultures, are cooked to ~54°C. In these cheeses, most of the acid is produced after pressing. Thermophilic culture have optimum temperatures of ~40°C, and little or no acid production occurs at ~54°C. However, the cultures are not inactivated and begin to produce acid again as soon as the temperature falls. It is no accident that these cheese are also very large; this slows down the rates of cooling, and plenty of acid is produced as soon as the temperature falls to ≤45°C.

Most strains of *Lc. lactis* are unable to metabolize citrate. However, many mixed-strain, mesophilic starters contain small numbers of *Lactococcus lactis* and *Leuconostoc* spp, which metabolize citrate to acetate, acetoin, and CO_2 during cheese manufacture. Small amounts of diacetyl are also produced. This is an important flavor compound in fresh (unripened) cheese, and it is also involved in determining the flavor of hard and semihard cheeses. Citrate metabolism is also important in many cheeses (e.g., Edam and Gouda) because the CO_2

produced is responsible for eye formation. In these cheeses, virtually all the citrate is metabolized in the first 24 h of cheese manufacture.

11.4 GROWTH OF STARTERS DURING RIPENING

Within 24 h from the addition of the starter, the number of starter cells will have increased to $\sim 10^9$ cfu/g and most of the lactose will have been transformed to lactate. Starter cells are the dominant flora in the young curd, and these high numbers represent considerable biocatalytic potential during cheese ripening. Most of the lactate is the L-isomer, but in cheeses made with thermophilic starters a small amount of D-lactate will also be produced by the thermophilic lactobacillus.

Much (~90%) of the lactose is lost in the whey, but there is still sufficiently large amounts of it retained in the curd to result in production of at least 1.0 g lactic acid/100 g cheese within a day of the beginning of manufacture. The amount depends on the particular cheese being made. Generally, most cheeses are brined 1 or 2 days after manufacture. As a general rule, the salt diffuses relatively slowly into the cheese, and thus brining has no immediate effect on the growth of the starter LAB. All of the lactose disappears very quickly from the curd in brined cheeses due to the ability of the high numbers of starter bacteria present to metabolize lactose. Cheddar is an exception; this cheese is dry-salted at milling, which occurs about 5.5 h after the starter is added, and this reduces the ability of the starter culture to produce lactate. At milling, Cheddar cheese may contain ~0.8% lactose and the subsequent metabolism of the lactose depends directly on the salt or, more specifically, on the SM ratio. Salt is a major factor controlling microbial growth in cheese. It is all dissolved in the moisture of the cheese. Consequently the salt is generally calculated as a percentage of the level of moisture as

$$\frac{\text{Salt, g/100 g} \times 100}{\text{Moisture, g/100 g}}$$

This is called the percentage salt-in-moisture (SM) and has no units.

An example of the effect of the SM ratio on lactose metabolism in Cheddar cheese is shown in Figure 11.1 (Turner and Thomas, 1980). At low SM levels (4.1%) all the lactose was utilized within 8 days, while at high SM levels (6.3%) the lactose remained high in the cheese for several weeks post-manufacture. SM levels of 4.9% and 5.6% lay

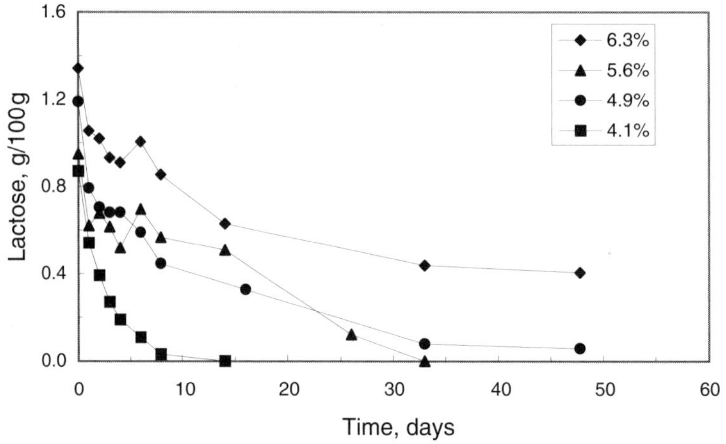

Figure 11.1. Effect of percent salt-in-moisture on the rate of lactose fermentation in Cheddar cheese during ripening. [Redrawn from Turner and Thomas (1980).]

between these extremes, but there was little difference in the rates of lactose metabolism observed between them. These data were obtained with one particular pair of cultures, and the cheese was ripened at 12°C. Other combinations of starters will probably show greater or lesser sensitivities to salt. In addition, ripening at lower temperatures will increase the time at which the lactose disappears from the cheese. This residual lactose can then be utilized by nonstarter lactic acid bacteria (NSLAB) (see later), but there is no strong correlation between residual lactose and growth of NSLAB, indicating that they use other energy sources besides lactose and/or other factors for growth.

Strains of *Lc. lactis* ssp. *lactis* survive much better than *Lc. lactis* ssp. *cremoris* strains in the presence of 4% to 5% NaCl (Martley and Lawrence, 1972). The levels of lactate in cheese early in ripening when all the lactose has been metabolized vary with the type of cheese. In Camembert, Swiss, and Cheddar, they are 1.0, 1.4, and 1.5 g/100 g, respectively (Karahadian and Lindsay, 1987; Turner et al., 1983; Turner and Thomas, 1980).

The final pH of the cheese generally lies between 4.6 and 5.3 and is reached within 1 day of manufacture. Starter lactococci have an optimum pH of 6.3, and growth rates will be much lower at these low pH values. However, in smear cheeses, the pH can increase significantly during subsequent ripening (see below). The final pH in the curd depends to a large extent on its buffering capacity (BC), which, in turn, depends indirectly on the moisture level in the cheese. A high-moisture cheese will have a low BC and therefore a low pH, whereas a low-moisture cheese will have a high BC and, consequently, a relatively high pH. Between pH 6.6 and 5.5, the BC of fresh curd is fairly low, but it increases rapidly below 5.5. BC is probably mainly due to the amount of soluble salts, especially phosphate and citrate, present in the cheese. The protein is probably also important. Even though the protein is virtually insoluble in fresh cheese, it is present in high concentrations and will have a large number of carboxyl groups exposed on its surface, each of which will buffer. Lactate is also present in large amounts, but it will only have a small effect on BC at pH 5.3 because the pK_a of lactic acid is 3.85. In highly buffered cheeses like hard cheese, extensive acid production in the early days of ripening may only have a small effect on the pH of cheese (Lucey and Fox, 1993).

11.5 AUTOLYSIS OF STARTERS

Proteolysis is probably the most important biochemical reaction in developing the flavor of most hard cheeses. This is mainly due to chymosin and starter proteinase activity, although in cheeses cooked to high temperatures (>50°C), milk plasmin also plays a role as chymosin is inactivated in these cheeses. The proteinases hydrolyze the casein into small and large peptides. Autolysis of starters is also important in this regard because intracellular enzymes, particularly peptidases, are released that will hydrolyze further any peptides, which are present (Crow et al., 1995). The major autolytic activity in lactococci is due to a muraminidase (Niskasaari, 1989). Autolysis of starter cells is also likely to be influenced by the starter itself, the NaCl concentration, and the cooking temperatures. The latter is an indirect effect because high cooking temperatures induce temperate phage in at least some lactococci (Feirtag and McKay, 1987). Variation in the autolysis of 5 different starters in Cheddar cheese is shown in Figure 11.2 (Martley and Lawrence, 1972). Cell numbers of strains AM1 and AM2 were much lower in the cheese at the beginning of ripening. These strains are heat-sensitive and are prevented from growing at the maximum cooking temperatures for Cheddar cheese (~40°C).

Should the starter reach too high a population or survive too long, flavor defects (such as bitterness) that mask or detract from cheese flavor are produced. Bitter peptides are mainly produced from the hydrophobic regions of the different caseins by chymosin and starter proteinases. It is generally believed that intracellular peptidases released through autolysis of the starter cells will hydrolyze the bitter

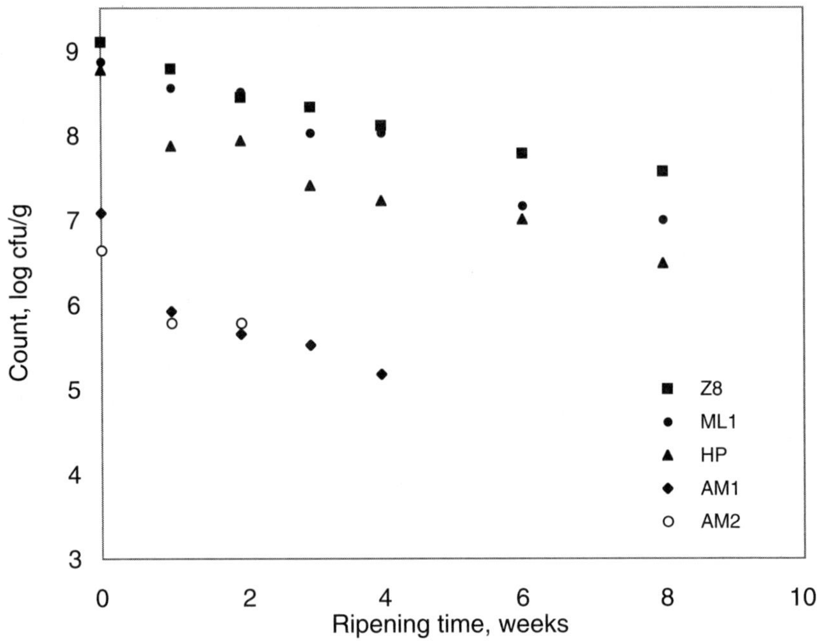

Figure 11.2. Autolysis of different strains of *Lactococcus lactis* in Cheddar cheese during ripening. [Redrawn from Martley and Lawrence (1972).]

peptides to smaller nonbitter peptides and amino acids. In a study by Wilkinson et al. (1994) in which starter viability and release of intracellular starter enzymes into cheese was monitored, it was concluded that *Lc. lactis* ssp. *cremoris* strains had different autolytic patterns. The most autolytic strain produced cheese with the highest levels of free amino acids. Crow et al. (1995) reported that intact starter cells fermented lactose, removed oxygen, and probably initiated a number of flavor reactions, while autolyzed cells accelerated peptidase activities. Amino acids are the products of peptidolysis. They accumulate faster following starter autolysis and are the major precursors of the compounds required for flavor production (Urbach, 1995; Fox and Wallace, 1997). Some amino acids (e.g., glycine, proline, and alanine) are quite sweet, whereas others (e.g., glutamate) may potentiate flavor. The amino acids are also the precursors for further flavor-generating reactions in cheese. Autolysis of thermophilic cultures also occurs, but it has not been studied to the same extent as mesophilic cultures. However, one would expect that it would act in the same way as in mesophilic cultures. Autolysis of *Lc. helveticus* has also been observed in Grana and Swiss-type cheese (Valence et al., 1998). Cheddar cheese is made with mesophilic cultures, but the use of strains of *Sc. thermophilus* and/or *Lc. helveticus* as starter adjuncts in the manufacture of this cheese is now quiet common. Kiernan et al. (2000) demonstrated that *Lc. helveticus* autolyzed very rapidly in Cheddar cheese and resulted in significantly higher levels of free amino acids and improved the flavor of the cheese.

11.6 SECONDARY FLORA

In all hard cheeses, a secondary flora also develops during ripening. These include nonstarter lactic acid bacteria (NSLAB), propionic acid bacteria (PAB), staphylococci, micrococci, and coryneform bacteria, yeasts, and molds. They are either (a) adventitious contaminants (e.g., lactobacilli, staphylococci, some coryneforms and some yeasts and molds) or (b) deliberately added (e.g., other coryneforms, *Brevibacterium linens*, PAB, and other yeasts and molds).

11.6.1 Nonstarter Lactic Acid Bacteria

Mesophilic lactobacilli form a significant portion of the microbial flora of most cheese varieties during ripening and, after the starter, are probably the next most important group of organisms in cheese. They are not part of the normal starter flora and generally do not grow well in milk; they are usually referred to as the NSLAB. In cheeses made from pasteurized milk, they are normally present in relatively low numbers, probably $<10^3$ cfu/g at the beginning of ripening, but they grow relatively rapidly during ripening to levels of $\sim10^8$ cfu/g within 2 to 4 months, depending on the species, the cheese, and the ripening temperature. Growth of starter and NSLAB in a typical Cheddar cheese during ripening at 6°C is shown in Figure 11.3. The starters were only enumerated until the 60th day of ripening because the NSLAB also grow on the LM17 medium used for counting the starter and make the starter counts inaccurate. A selective medium, Rogosa agar, was used to determine the NSLAB. NSLAB generally remain at $\sim10^8$ cfu/g in cheese, but in Parmigiano-Reggiano they decrease from $\sim10^8$ cfu/g at 5 months to $\sim10^4$ cfu/g at 24 months of ripening (Coppola et al., 1997). In cheese made from raw milk, the initial levels of lactobacilli usually vary and depend on the amount of care taken in producing the milk.

Pediococci are occasionally encountered in cheese, and *Pediococcus acidilactici* and *Pediococcus pentosaceus* are the most frequent species.

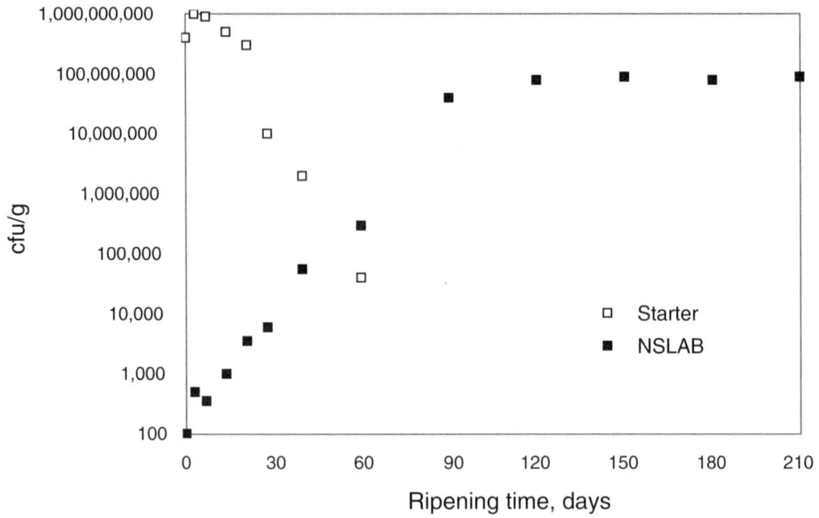

Figure 11.3. Numbers of starter and NSLAB in Cheddar cheese, ripened at 6°C.

Unlike lactobacilli, most pediococci do not ferment lactose; however, mesophilic lactobacilli and pediococci grow well in the presence of 6.5% salt.

Lactobacilli are catalase-negative, Gram-positive rods that taxonomically are divided into three groups: obligatory homofermentative, facultatively heterofermentative, or obligatory heterofermentative. The nonstarter lactobacilli in cheese are members of the facultatively heterofermentative group and are sometimes referred to as facultatively heterofermentative lactobacilli (FHL). They are also called the mesophilic lactobacilli to distinguish them from the starter lactobacilli, which are thermophilic. They ferment lactose homofermentatively and gluconic acid heterofermentatively. The most frequently encountered species are *Lactobacillus casei*/*Lactobacillus paracasei*, *Lactobacillus plantarum*, *Lactobacillus rhamnosus*, and *Lactobacillus curvatus* (Coppola et al., 1997; Fitzsimons et al., 1999). There is a problem with the taxonomy of *Lc. casei* because the type species *Lc. casei* ATCC 393 is actually a strain of *Lc. zeae* and is not related to the other common strains of *Lc. casei*. Therefore, it has been proposed that a new species, *Lc. paracasei*, should be formed to include these other *Lc. casei* strains. This has not been resolved, but for convenience we will use the name *Lc. paracasei* here.

The dominant species of NSLAB change during ripening. *Lc. paracasei*, *L. plantarum*, and *Lc. brevis* dominate young cheese, but only *Lc. paracasei* dominate the ripened chees. Evidence for the appearance, disappearance, and recurrence of different strains of *Lc. paracasei* during ripening has also been observed (Fitzsimons et al., 2001).

Swiss-type cheeses (e.g., Emmental, Gruyère, and Comté), are generally manufactured from raw milk. During ripening, these cheeses are held at 15–24°C for several weeks in a so-called "warm room" to promote the propionic acid fermentation (see below). Such temperatures will also promote rapid growth of lactobacilli (Demarigny et al., 1996; Beuvier et al., 1997). NSLAB levels at the end of the warm-room ripening were higher in cheese made from raw milk cheese (10^8 cfu/g) than in cheese made from pasteurized milk (10^6 cfu/g). The reason for this is not clear, but it has been suggested that pasteurization retards the growth of NSLAB during ripening in some way.

NSLAB require sugar for growth and energy production. As all the lactose is fermented to lactic acid by the starter bacteria within the first few days of ripening in most cheeses, the energy source used by the NSLAB in cheese is not clear. NSLAB can transform the L-isomer of lactate to the D-isomer, but this will not result in energy production (Thomas and Crow, 1983). Several substrates have been suggested, including citrate, ribose, and amino acids. Citrate is present in young Cheddar cheese at ~8 mmol/kg, and some investigators have suggested that it may be a potential energy source for NSLAB (Jimeno et al., 1995). However, more recent data indicated that citrate is not used as an energy source by NSLAB (Palles et al., 1998). Ribose released from RNA after starter autolysis could be used as a energy source. However, many strains of NSLAB isolated from mature Cheddar cheese are unable to ferment ribose. Mesophilic lactobacilli possess glycoside hydrolase activity and, in model systems, can utilize the sugars of the glycoproteins of the milk fat globule membrane as an energy source (Diggin et al., 1999).

The source of NSLAB in cheese has been the focus of considerable debate because they are found in cheeses made from both raw and pasteurized milk. In the case of cheeses made from raw milk, the main source is likely to have been the cheese milk. This is also the most likely source in cheeses manufactured from pasteurized milk, because some of them withstand pasteurization to some extent (Jordan and Cogan, 1999); postpasteurization contamination may also occur. NSLAB have a mean generation time of 8.5 days in Cheddar cheese ripened at 6°C; thus, low levels of contamination will result in NSLAB rapidly becoming a significant proportion of the total cheese flora.

The microbial flora of cheese, particularly those made from raw milk, is very complex. It is likely that interactions occur between strains of NSLAB and other strains in the cheese. The study of these interactions is difficult because of the complex ecosystem in the cheese (Martley and Crow, 1993). However, some progress has been made. *Lc. casei*, *Lc. rhamnosus*, and *Lc. plantarum* inhibit propionic acid bacteria and enterococci in cheese. The mechanism of the inhibition, however, remains obscure. Inhibition is normally observed only in cheese or cheese juice; indeed, in some culture media stimulation is observed (Jimeno et al., 1995). This implies that the inhibition is not mediated by an inhibitor, such as a bacteriocin, but may be through competition for limiting metabolites.

NSLAB are responsible for transforming L-lactate to D-lactate during ripening (Figure 11.4), but this has no effect on flavor development but can effect the appearance of the cheese. Usually a racemic mixture of the L- and D-isomers of lactate is produced. Calcium D-lactate is much more insoluble than calcium L-lactate and is responsible for the white specks that can occur in many old cheeses during ripening. In the cheese shown in Figure 11.4, the level of L-lactate increased during the first 10 days of ripening. This is due mainly to continued acid production, by the starter bacteria, from the small amounts of lactose remaining in the cheese; the high numbers of NSLAB may also be involved toward the end of the period.

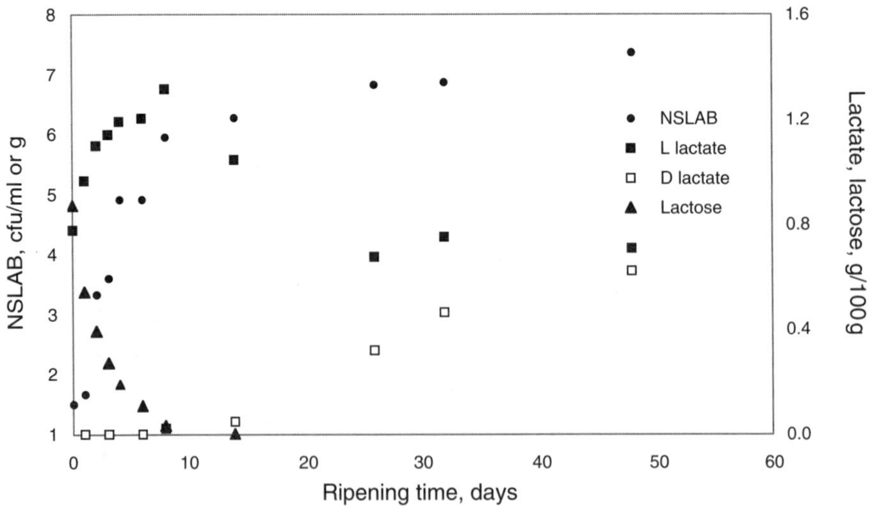

Figure 11.4. Growth of NSLAB and the transformation of L-lactate to D-lactate in Cheddar cheese during ripening. [Redrawn from Turner and Thomas (1980).]

The role of NSLAB in the development of Cheddar cheese flavor has been a contentious issue for many years. Numerous studies have shown that lactobacilli, used as adjuncts, can affect flavor development, particularly in Cheddar cheese. Most researchers report enhanced levels of proteolysis and enhanced flavor intensity [see Lynch et al. (1996) for references]. It would appear that selection of the adjunct strain is crucial, because certain strains of *Lc. casei* produced high-quality Cheddar, while other strains of these species resulted in cheese with acid and bitter flavor defects (Lawrence and Gilles, 1987). They can also produce open texture in cheese by producing CO_2 from citrate (Martley and Crow, 1996). NSLAB, particularly pediococci, can also oxidise lactate to acetate. This requires O_2 and would only occur on or close to the surface of cheese, which has been removed from its packet.

11.6.2 Propionic Acid Bacteria

Propionic acid bacteria (PAB) grow in many cheese varieties during ripening and are part of the characteristic microflora of Swiss-type cheeses such as Emmental, Gruyère, Appenzell, and Comté, where they metabolize lactate:

$$3 \text{ Lactate} \rightarrow 2 \text{ Propionate} + 1 \text{ Acetate} + 1 \text{ CO}_2$$

The CO_2 is responsible for the holes or eyes in these cheeses, and the propionic acid is thought to contribute to the development of the characteristic sweet, nutty flavor of these cheeses. Emmental is ripened at higher temperatures (~22°C) than Comté (~18°C), and hence the eyes are much larger in the former than in the latter. However, in Italian-type cheeses (e.g., Grana and Parmigiano-Reggiano), gas production by PAB is considered a defect. During the "warm-room" ripening of Swiss-type cheese, the levels of PAB increase from 10^4 cfu/g to 10^8 to 10^9 cfu/g cheese (Steffen et al., 1993).

PAB are Gram-positive, short, rod-shaped bacteria, and two major groups are recognized; the "cutaneous" and the "classic" or "dairy" PAB. The classic PAB are the most important with respect to cheese microbiology; and of the five species currently recognized, *Propiouibacterium freudenreichii* is the most important. PAB are slow-growing bacteria and are difficult to separate into species by classic methods. However, a number of molecular methods for detection and identification of PAB have been reported (Riedel et al., 1998; Rossi et al., 1999).

Emmental cheese is traditionally made from raw milk and the PAB are adventitious contaminants, which withstand the high cooking temperature (54°C) that this cheese is given during manufacture. Nowadays, they are normally added deliberately to the milk in Emmental manufacture. After the eyes have developed, the cheese is stored at low temperatures to retard further growth and metabolism of the PAB. The relationship between the development of PAB and the conversion of lactate to propionate and acetate in Emmental cheese is shown in Figure 11.5. Unusually for a starter bacterium, *Sc. thermophilus* is unable to metabolize galactose and excretes it in direct proportion to the amount of lactose taken up. The galactose is then metabolized by the thermophilic lactobacilli during molding and early ripening.

Growth of PAB does not occur in milk or whey at low cell densities (~10^5 cfu/ml). However, growth will occur if the initial cell density is >10^6 cfu/ml. Inhibition of growth is due to a heat-stable inhibitor(s) present in the milk (Piveteau et al., 2000). Pre-growth of some thermophilic starter LAB (used as starter cultures in Swiss-type manufacture) in the milk medium removed the inhibition, which explains why

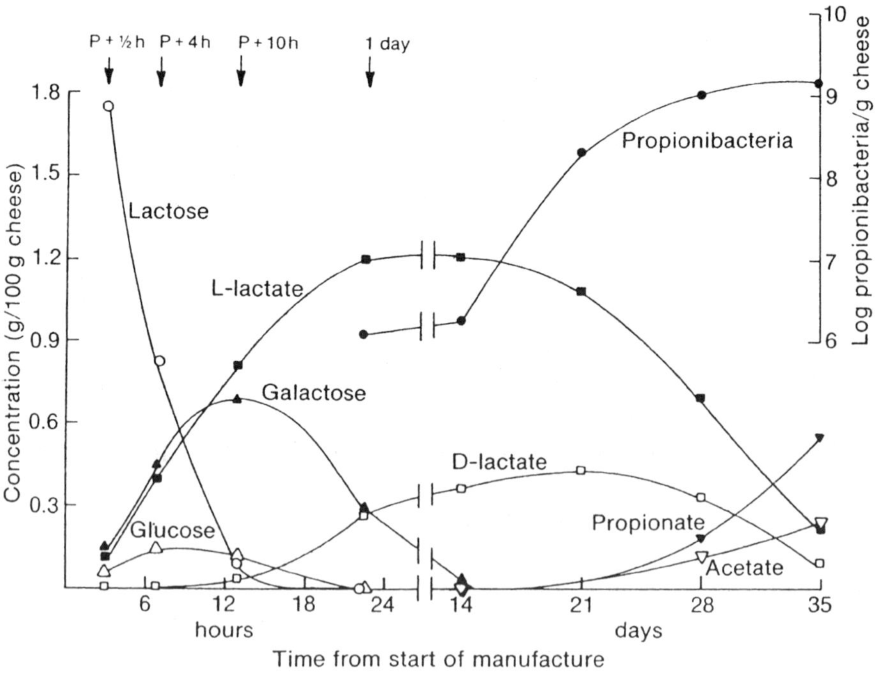

Figure 11.5. Relationship between lactose and lactate metabolism, growth of propionic acid bacteria, and production of propionate and acetate in Swiss-type cheese. [From Turner et al. (1983).]

PAB develop in Swiss-type cheese from low densities even though they are inhibited in milk.

Although spontaneous autolysis of *P. freudenreichii* has been demonstrated in synthetic media, no evidence of autolysis of *P. freudenreichii* during ripening as measured by release of intracellular enzymes was detected by Valence et al. (1998). Bacteriophage infection of *P. freudenreichii* during growth in Swiss-type cheese also can occur, but its relevance to growth of PAB in cheese is not clear.

Interactions between PAB and other bacteria (e.g., starter bacteria and NSLAB) probably play a significant role in the cheese during cheese ripening, although the evidence is not strong. Jimeno et al. (1995) reported that *Lc. rhamnosus* and *Lc. casei* inhibited the growth of *Pc. freudenreichii* in Swiss-type cheese. In contrast, Piveteau et al. (1995) showed stimulation of PAB by several strains of LAB, including strains of *Lc. helveticus*, *Lc. acidophilus*, *Lc. lactis*, *Sc. thermophilus*, and *Lc. lactis* in whey; stimulation of PAB by *Lc. helveticus* and *Sc. thermophilus* was particularly strong. The stimulants appeared to be due to peptides, but they were recalcitrant to purification.

11.6.3 Enterococci

Cheese may also contain enterococci. These are also lactic acid bacteria and are often present in natural starters and in artisanal cheeses. They can also be considered as NSLAB. Currently, 20 species of enterococci are recognized, but the commonest ones in cheese are *Enterococcus faecalis* and *Enterococcus faecium*. These bacteria withstand pasteurization, can grow at 10°C and in the presence of 6.5% salt, and are found in high numbers (up to 10^7 cfu/g) in Comté and in cheeses made in Mediterranean countries (Litopoulou-Tzanetaki et al., 1993; Demarigny et al., 1996; Freitas, 1996). They are also found in the traditional natural whey cultures used in Mozzarella cheese manufacture (Coppola et al., 1990), and it has been suggested that they should be used more generally as starter cultures. Their main source is the gastrointestinal tract of humans and animals, and contamination from this source is likely in raw milk. Despite this, they are felt to have a positive role in flavor formation in those cheeses in which they are found. Enterococci are the most proteolytic of the LAB, which goes a long way to explain why their growth results in good-flavored cheese. One species, *Enterococcus malodoratus*, has been incriminated in off-flavor production in Gouda cheese.

Their use as starters in foods has been questioned because they are considered to be opportunistic pathogens. They have been occasionally implicated as the cause of bacteremia, urinary tract infections, and endocarditis. *Ee. faecalis* and *Ee. faecium* are also promiscuous and take up plasmids encoding antibiotic resistance relatively easily. The antibiotic of choice in treating diseases caused by antibiotic-resistant enterococci is vancomycin, and vancomycin-resistant enterococci (VRE) are also becoming prevalent in hospitals. This a serious situation because there are very few alternative therapies available for people suffering from VRE infection (Franz et al., 1999).

11.6.4 Coryneform Bacteria

Surface-ripened cheeses can be divided into those in which molds and yeast colonize the surface during ripening (e.g., Camembert and Brie) and those in which bacteria and yeast colonise the surface (e.g., Tilsit and Limburger). The former are called mold-ripened cheeses, and the latter are also called smear-ripened cheeses because the surface of the cheese has a glistening, viscous appearance, reminiscent of a smear. Smear-ripened cheeses can be hard (e.g., Beaufort, Comté, and Gruyère), semihard (e.g., Tilsit), or soft (e.g., Limburger and Reblochon). Beaufort and Comté are French cheeses and Gruyère is a Swiss cheese, and all are made with thermophilic cultures. Beaufort cheese is brine-salted, whereas Comté and Gruyère cheeses are dry-salted.

In the past it was generally believed that *Brevibacterium linens* formed the major proportion of the bacteria growing on the surface of smear-ripened cheeses. However, recent studies suggest that it represents no more than 15–20% of isolates. This organism occurs in Gruyère cheese (Grand et al., 1992) and probably also in Beaufort and Comté, but this has not been reported. The organism also produces orange- to red-colored colonies. These are carotenoids and are also responsible for the red or orange or red color of the surface of these cheeses. *B. linens* also produces methanethiol from methionine, which is considered to be responsible for the "smelly sock" odor of many of these cheeses. DNA/DNA hybridization studies have shown that *B. linens* is a mixture of two different homology groups. These are represented by *B. linens* ATCC 9172, which is the type species, and *B. linens* ATCC 9175, which has not been renamed (Fiedler et al., 1981). The microbiology, biochemistry, and enzymology of this organism have been reviewed (Boyoval and Desmazeaud, 1983; Rattray and Fox, 1999).

Recent studies have indicated that several other species of bacteria including *Arthrobacter citreus*, *Arthrobacter globiformis*,

Arthrobacter nicotianae, *Brevibacterium imperiale*, *Brevibacterium fuscum*, *Brevibacterium oxydans*, *Brevibacterium helvolum*, *Corynebacterium ammoniagenes*, *Corynebacterium betae*, *Corynebacterium insidiosum*, *Corynebacterium variabilis*, *Curtobacterium poinsettiae*, *Microbacterium imperiale*, and *Rhodoccoccus fascians* are also found on the surface of semihard and soft cheeses (Eliskases-Lechner and Ginzinger, 1995a; Valdes-Stauber et al., 1997).

Arthrobacter, *Brachybacterium*, *Brevibacterium*, *Corynebacterium*, *Microbacterium*, and *Rhodococcus* spp. are all considered to be coryneform bacteria. They are all Gram-positive, irregularly shaped rods, but their identification is difficult and usually the genus to which an isolate belongs is determined before the species. Microscopic examination of the isolates during growth is necessary because some of these genera go through a rod–coccus transformation during growth with rod-shaped organisms dominating exponential growth and coccoid-shaped organisms dominating stationary phase growth. In addition, chemical analysis of several components of the cell is undertaken, including the type of peptidoglycan, the GC content of the DNA, the presence or absence of mycolic acids, and the types of menaquinones, fatty acids, and polar lipids present. This analysis is called *chemotaxonomy*. The differences between the genera of coryneforms are summarized in Table 11.2. Once the genus has been identified, phenotypic analysis can usually be used to determine the species.

Molecular approaches area also useful in identification of this group of bacteria—for example, sequencing the 16S rRNA and the 16S rDNA gene. This is becoming much more common with the availability of fast and automated sequencing equipment. In addition, DNA probes for *Brevibacterium*, *Microbacterium/Aureobacterium*, and *Micrococcus/Arthrobacter* have been developed (Kolloffel et al., 1997) and proved to be very reliable in a recent study of the microflora of a surface-ripened cheese (Brennan, unpublished observations).

Some of the coryneform bacteria isolated from cheese have been misclassified. For example, *Caseobacter polymorphus*, which was originally isolated from the surface of Limburger and Meshanger cheese, is now considered to be *C. variabilis* (Collins et al., 1989); and *Microbacterium flavum*, which was isolated from cheese (the type was not specified) by Orla-Jensen, is now considered to be *C. flavescens* (Barksdale et al., 1979) while *B. ammoniagenes*, which is commonly isolated from cheese, has been reclassified as *C. ammoniagenes* (Collins, 1987). In addition, methanethiol-producing coryneforms, isolated from Cheddar cheese, have been identified as *B. casei* (Collins et al., 1983).

TABLE 11.2. Differential Characteristics of Some Coryneform and Micrococcal Taxa[a] Involved in Cheese Ripening

Taxon	Shape	Peptidoglycan Type	Peptidoglycan Variation	Peptidoglycan *N*-Glycolyl	Mol % GC	Mycolic Acids	Fatty Acid Types[b]	Major Menaquinones	Polar Lipids
Corynebacteria									
Arthrobacter	Rod/coccus cycle	L-Lys-MCA	A3α	–	59–66	—	S, A, I	MK-9(H_2)	DPG, PG, PI, DMDG, (DGDG), (MGDG)
Brevibacterium	Rod/coccus cycle	*meso*-DAP	A1γ	–	60–64	—	S, A, I	MK-8(H_2)	DPG, PG, DMDG
Brachybacterium	Rod	*meso*-DAP-DCA	A4γ		73	—	S, A, I	MK-7; MK-8	DPG, PG, GL
Coryebacterium	Rod	*meso*-DAP	A1γ	–	51–60	22–36 Carbon atoms	S, U (T)	MK-9(H_2); MK-8(H_2)	DPG, PI, (PG), PIM, (G)
Curtobacterium	Rod	D-Ornithine	B2β	–	67–75	—	S, A, I	MK-9	DPG, PG, G
Microbacterium	Rod	L-Lysine	B1α or B1β	+	69–75		S, A, I	MK-11; MK-12	DPG, PG, DMDG, (MMDG), (PGL)
Micrococci									
Micrococcus luteus	Cocci	Lys-peptide subunit	A2		70–76		S, A, I	MK-8, MK-8(H_2)	DPG, PI, PG, PL, GL
Micrococcus lylae	Cocci	Lys-DCA	A4α		69		A, I	MK-8	DPG, PI, PL, GL
Kocuria	Cocci	Lys-MCA	A3α		66–75		S, A, I		DPG, PG, (PI, PL, GL)
Nesterenkonia	Cocci	Lys-MCA, DCA	A4α		70–72		(S), A, I	MK-8; MK-9	DPG, PI, PG, PL, GL
Kytococcus	Cocci	Lys-DCA	A4α		68–69		(S), A, I	MK-8; MK-9; MK-10	DPG, PI, PG,
Dermacoccus	Cocci	Lys-MCA, DCA	A4α		66–71		S, A, I	MK-8(H_2)	DPG, PI, PG,

[a]S, straight-chain saturated; A, anteiso-methyl-branched; I, iso-methyl-branched; U, monounsaturated; T, 10-methyl-branched acids; ( ) may be present; DPG, diphosphotidylglycerol; PG, phosphotidylglycerol; G, unknown; GL, glycolipid; PE, phosphotidylethanolimine; PGL, phosphoglycolipid; PI, phosphotidylinositol; PIM, phosphotidylinositol mannoside; DMDG, dimannosyldiacylglycerol; DGDG, digalactosyldiacylglycerol; MGDG, monogalactosyldiacylglycerol; DGD, diglucosyldiglyceride; PL, phospholipid; MCA, monocarboxylic amino acid; DCA, dicarboxylic amino acid; Lys, lysine; Orn, ornithine; DAP, diamionopimelic acid.

Source: *Bergey's Manual of Systematic Bacteriology* and Stackebrandt et al. (1995).

11.6.5 Micrococci and Staphylococci

Staphylococcus and *Micrococcus* spp. are also present on the surface of smear-ripened cheeses. These are catalase-positive, Gram-positive cocci. Both genera are found in the family, *Micrococcaceae*, but they are totally unrelated to each other. The former have a low-mol% GC (30–39%) and are found in the clostridial branch of the Gram-positive bacteria, while the latter have a high-mol% GC (65–75%) and are found in the actinomycete branch of Gram-positive baacteria. They are readily distinguished from each other: staphylococci produce acid from glucose anaerobically, are sensitive to lysostaphin, and contain glycine in their peptidoglycans; micrococci do not have any of these features, but, unlike staphylococci, they can grow in the presence of furazolidone. Recently, the genus *Micrococcus* was divided into five genera: *Kocuria*, *Nesterenkonia*, *Kythococcus*, *Dermacoccus*, and *Micrococcus* (Stackebrandt et al., 1995); chemotaxonomic analyses are required to distinguish them from each other (Table 11.2). *Micrococcus* contains two species: *Micrococcus luteus*, the type species, and *Micrococcus lylae*. Although both species have different cell wall compositions, they are genetically closely related to each other. *Kocuria* also contains two species: *Kocuria varians* and *Kocuria roseus* and the other genera only one species.

11.7 SMEAR-RIPENED CHEESES

Smear-ripened cheeses are produced in relatively large amounts in many European countries and are characterized by the development of an ill-defined complex of bacteria and yeasts on the surface of the cheese during ripening. The principles of making smear cheeses is the same as other cheeses except that several times during the early days of ripening, these cheeses are rubbed with slighly salted water or with dry salt and smear. This, together with relatively high incubation temperatures during ripening (8–19°C for several months, depending on the cheese), leads to the development of the smear on the cheese surface. Between the washings, microcolonies of yeasts and bacteria develop, and washing the cheese spreads the microcolonies more uniformly on the cheese surface.

Smear-ripened cheeses have one characteristic that differentiates them from other cheeses, namely, the pH on the surface increases during ripening. This is termed *deacidification* and is traditionally associated with the growth of yeasts, which oxidize the lactate on the cheese surface to CO_2 and H_2O. Production of NH_3 from deamination of amino acids helps to increase the pH also. The pH can increase from ~5.0 to more than 7.5, and the increase is much more rapid on the outside surface than in the internal part of the cheese. It is generally believed that the yeast grow first and when the pH has increased sufficiently, growth of the bacteria begins. Recent studies (Brennan, unpublished observations) suggest that many of the bacteria on the surface can also metabolize lactate at pH values of ~5.0. The rise in pH can be relatively fast; for example, the pH of the surface of Tilsit increases from an initial level of ~5 to >7.5 during the first 10 days of ripening (Eliskases-Lechner and Ginzinger, 1995b).

In Beaufort, Comté, and Gruyère, the yeasts on the surface attain densities of 10^8–10^9 cfu/g of smear within the first 3 weeks of ripening, after which they decrease to approximately 10^5 cfu/g of smear for the remainder of the ripening period. The bacteria on the surface of these cheeses grow to densities of 10^{11} cfu/g of smear and remain at that level throughout ripening (Grand et al., 1992; Piton-Malleret and Gorrieri, 1992). Similar results have been found in Tilsit cheese. In this cheese, the number of yeasts increased from 10^3 to 5×10^7 cfu/cm^2 within 2 weeks, after which they either remained constant or decreased gradually to 10^5 cfu/cm^2, while the number of bacteria increased from 10^3 to 10^4 cfu/cm^2 to $>10^9$ cfu/cm^2 after 3 weeks of ripening, after which they remained constant (Eliskases-Lechner and Ginzinger, 1995a,b). To transform counts per cm^2 to counts per g of smear, one should multiply the counts per g by 100. The relationships between pH, yeast, and bacterial counts on the surface of a semihard cheese are shown in Figure 11.6.

The yeasts on Beaufort, Comté, and Gruyère cheeses have not been identified; but in other surface-ripened cheeses, a large number of yeasts including *Candida*, *Debaryomyces*, *Kluyveromyces* and *Rhodotorula*, *Pichia*, *Rhodotorula*, *Saccharomyces*, *Torulaspora*, *Trichosporon* and *Zygosaccharomyces* spp, and *Geotrichum candidum* have been found (Prillinger et al., 1999; Wyder and Puhan, 1999). *Geotrichum candidum* is commonly known as the dairy mold and has properties of both yeasts and moulds; nowadays it is considered to be a yeast. It is generally believed that *G. candidum* is found in all smear and mold-ripened cheeses, although in some studies its presence was not reported. The reason for this may be that some workers probably consider it to be a mold and did not report its presence in the various cheeses in which yeast isolates were identified.

The bacteria occurring on the surface of Beaufort, Comté, and Gruyère cheese have been poorly described and are mainly distin-

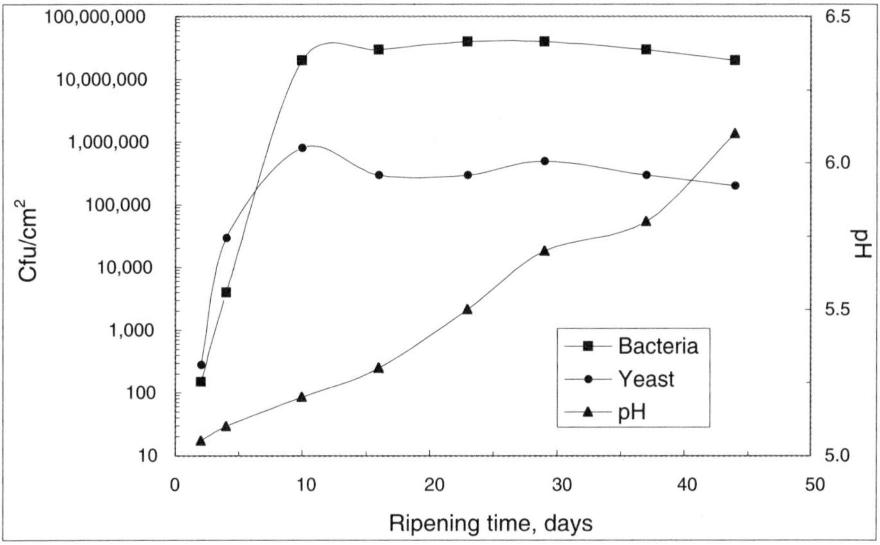

Figure 11.6. Growth of salt-tolerant bacteria and yeast and change in pH on the surface of a smear-ripened cheese during ripening.

guished on the basis of the color of the colonies—that is, lemon, orange, or nonpigmented (presumably white or cream) (Grand et al., 1992; Piton-Malleret and Gorrieri, 1992). All three colored colony types were dominated (>75% of strains) by *Micrococcease* (whether they were *Staphylococcus* or *Micrococcus* spp. was not reported); the remainder were mainly coryneforms, with 11% of the yellow colonies being neither. Several isolates from Beaufort and Comté cheese have been identified as *Brachybacterium tyrofermentans* and *Brachybacterium alimentarium* (Schubert et al., 1996). *B. linens* forms a major proportion of the coryneform bacteria, at least on Gruyère cheese, while Tilsit cheese has been shown to contain numerous *Arthrobacter*, *Corynebacterium*, *Brevibacterium*, *Microbacterium*, and *Staphylococcus* spp. (Eliskases-Lechner and Ginzinger, 1995a). In several smear-ripened chesees, the micrococci have been identified as *Micrococcus luteus*, *Micrococcus lylae*, *Kocuria kristinae* and *Kocuria roseus*, and the staphylococci have been identified as *Staphylococcus equorum*, *Staphylococcus vitulus*, *Staphylococcus xylosus*, *Staphylococcus saptrophyticus*, *Staphylococcus lentus*, and *Staphylococcus sciuri* (Irlinger et al., 1997; Irlinger and Bergere, 1999).

Whether a progression of bacteria occurs in the smear has not been determined, but a recent study of a semihard cheese showed that cocci are the dominant organism found early in ripening (within 4 days) and coryneform bacteria dominate the later stages of ripening (Brennan, unpublished observations). The stage of ripening at which the bacteria listed above were isolated is not clear.

The role of the smear microorganisms in the ripening of the cheese has not been studied to any great extent. The most intensively studied has been *B. linens*, which produces proteinases, peptidases, and lipases, which, in turn, produce amino acids and fatty acids, which are the precursors of many of the flavor compounds in the cheese (Boyoval and Desmazeaud, 1983; Rattray and Fox, 1999). *B. linens* produces methanethiol from methionine. Yeast, particularly *G. candidum*, also produces sulfur compounds, including methanethiol, in model cheese systems (Berger et al., 1999). Methanethiol inhibits the growth of molds, and this may be the reason that molds are not found on the surface of smear cheese.

Many cheese-makers deliberately inoculate the surface of smear cheese with various combinations of *B. linens*, *G. candidum*, and/or *D. hansenii* at the beginning of ripening. Except where deliberate inoculation of *B. linens* is practiced, the other bacteria, which grow on the cheese, are more than likely adventitious contaminants, which grow well in the high salt concentrations and relatively high pH of the cheese surface. Their source has not been identified, but likely ones include the brine, the wooden shelves on which the cheese rests during ripening, and skin because many smear-ripened cheeses receive a lot of manual handling during ripening and coryneform bacteria, staphylococci, and micrococci are major components of the skin microflora. Interestingly, most *Brevibacterium* isolates from clinical sources have been identified as *B. casei* which was first isolated form cheese (Funke and Carlotti, 1994).

11.7.1 Mold-Ripened Cheeses

Mold-ripened cheeses are divided into two groups: (a) those that are ripened with *Pencillium roqueforti*, which grows at the interfaces between the curd particles and forms blue veins within the cheese, such as Stilton, Roquefort, Gorgonzola, and Danish Blue, and (b) those that are ripened with *Penicillium camemberti*, which grows as a fuzzy mass on the surface of the cheese, such as Camembert and Brie. Blue cheeses are often called internally mold-ripened cheese, while Camembert and Brie are called externally (surface) mold-ripened cheeses to distinguish them from the bacterial surface (smear)-ripened cheeses. Molds are associated with a range of other cheese varieties also; however, the molds and their impact on ripening in these cheeses are less well understood. The surface of the French cheeses St. Nectaire and Tome de Savoie is covered by a complex fungal flora containing *Penicillium*,

Mucor, *Cladosporium*, *Geotrichum*, *Epicoccum*, and *Sporotrichum*, while *Penicillium* and *Mucor* have been reported on the surface of the Italian cheese Taleggio. *Penicillium roqueforti* are generally added to the cheese milk with the lactic starter. Interior- or surface-mold-ripened cheeses have different appearances, and the high biochemical activities of these molds produce the typical aroma and taste (Gripon, 1993). Recently, *P. roqueforti* has been split into three species—*P. roqueforti*, *P. carneum*, and *P. paneum*—based on the sequences of the internal transcribed spacers between the 18s and 28S rRNA (Boysen et al., 1996).

In the production of Stilton cheese a low level of starter is used—for example, 10ml/100 liters of milk (equivalent to an inoculation rate of 0.01%). Consequently, acid production in the cheese is slow. Part of the whey is run off during manufacture, but the curd is left in the remaining whey overnight, after which the whey is run off and the curd is milled, weighed, salted, and placed in the molds or hoops. Blue cheeses contain relatively high levels of salt; for example, Stilton cheese may contain 9% SM. The hoops are turned daily for 3 or 4 days. This is known as the "hastening" period, during which the curd continues to loose moisture and the acidity increases (Anonymous, 1994). The cheeses are then ripened at ~13°C for 4 weeks at an RH of 85%. Then the cheese is pierced with a series of stainless steel needles to allow air to enter the cheese and promote growth of *P. roqueforti*. Then the cheese is further ripened at 16°C for 8 weeks at a RH of 95%. An aqueous suspension of *P. roqueforti* spores is added to the milk just prior to setting, or spores are dusted onto the curd. Gas production by yeasts and heterofermentative LAB, particularly *Leuconostoc* spp., results in curd-openness, which is deemed to be very important for the subsequent development of *P. roqueforti* and hence good flavor. It has been suggested that the methyl ketones produced by *P. roqueforti* are inhibitory to further mold growth, and they may be a factor in preventing excessive mold development in blue veined cheese.

Like smear cheeses, mold-ripened cheeses also undergo deacidification by both yeasts and molds during ripening. The interrelationships between lactate, NH_3, and pH in a Camembert cheese are shown in Figure 11.7 (Karahadian and Lindsay, 1987). Utilization of the lactate occurs relatively rapidly and is much faster on the surface than in the center of the cheese. NH_3 production does not begin until most of the lactate has been used, and the mold has stopped growing and is higher on the surface than in the internal part of the cheese. The fact that the pH and NH_3 are higher on the surface than in the center of the cheese means that gradients develop in the ripening cheese. In Camembert

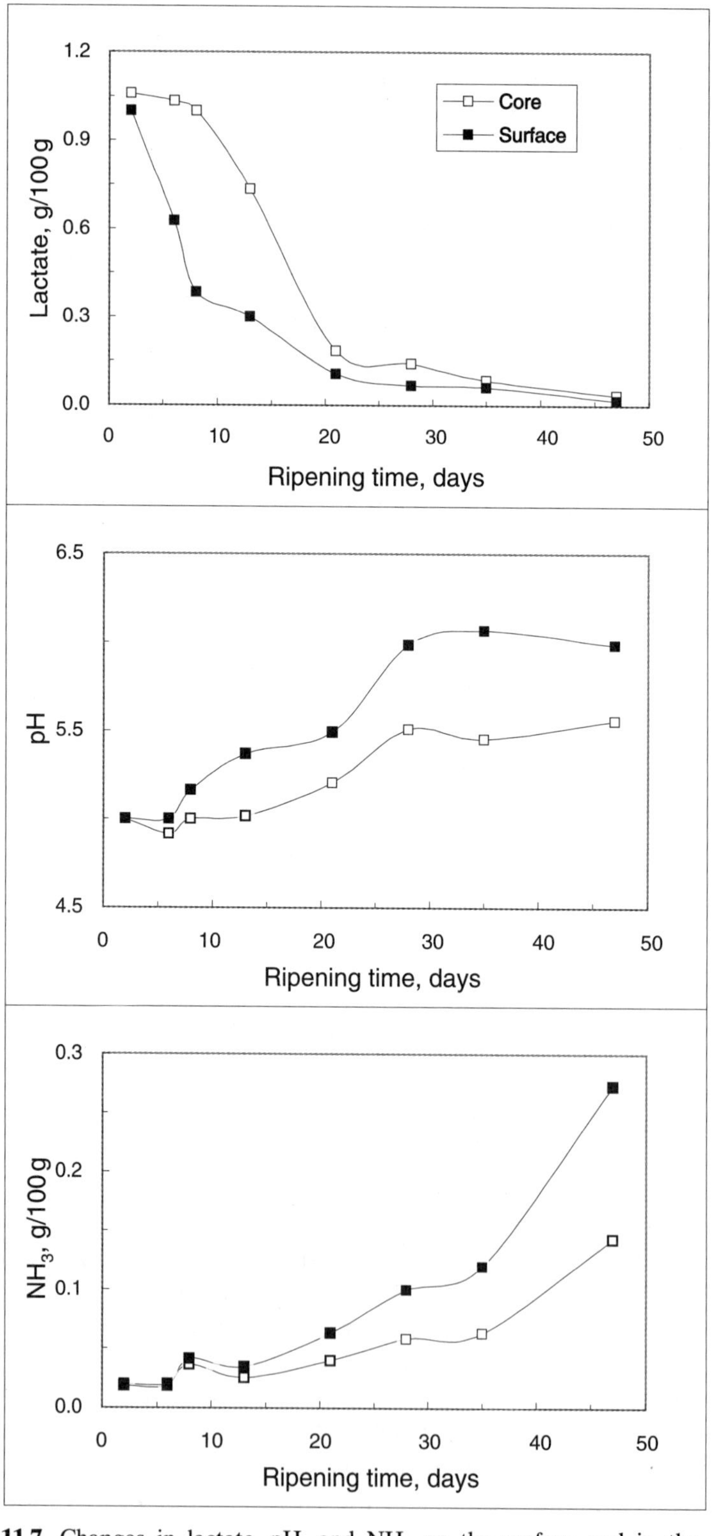

Figure 11.7. Changes in lactate, pH, and NH_3 on the surface and in the core of Camembert cheese during ripening. [Redrawn from Karahadian and Lindsay (1987).]

and Brie cheeses, Ca and phosphate also migrate from the center to the surface as the cheese is ripened. In these cheeses, the curd softens from the outside to the inside, and at one time it was thought the proteinase produced by *P. camemberti* was responsible for this. Subsequently it was shown that the proteinases do not diffuse to any great extent. The results of Karahadian and Linday (1987) give a much better explanation for the softening of these cheeses, because the NH_3 and pH gradients cause the curd to solublize, giving the impression that the cheese is ripening from the outside to the inside.

P. roqueforti produces a range of secondary metabolites during growth including *P. roqueforti* toxin (PRT), roquefortine C, marcfontines, fumigaclivine A, festuglavine, mycophenolic acid, patulin, cyclopaldic acid, penicillic acid, and roquecin. Several of these, particularly PRT and mycophenolic acid, are toxic to humans, but the amounts produced by cheese strains are so small that they do not represent a public health hazard, even when the cheese is consumed in large amounts. In fact, smaller amounts of these compounds are produced in cheese than in conventional media. In addition, PRT-aldehyde, which is the toxic form of PRT, is transformed into PRT-imine and PR-amide, which are nontoxic forms, during growth of the mold (Engel et al., 1982; Chang et al., 1993).

Curiously, there is little detailed information on the growth and development of the mold in mold-ripened cheese. In both internally and externally ripened mold cheeses, the activity of the proteinases and, in the case of Blue cheeses, that of the lipases are the dominant biochemical activities that occur during ripening. Free fatty acid development in two batches of Stilton cheese was reported by Madkor et al. (1987). The concentration of the individual fatty acids increased between 4- and 20-fold during ripening, and the levels of the long-chain saturated and unsaturated acids were higher than those of the short-chain ones (C4 to C10) in the cheeses. In addition to the short-chain acids, methyl ketones formed from β-oxidation of the fatty acids are also important in flavor development in Blue cheese. Unlike other molds, *P. roqueforti* can grow at low O_2 levels and sufficient O_2 enters the cheese when it is pierced to allow visible mycelium to develop.

SEM studies of Camembert cheese show the presence of several different layers containing different organisms in the cheese (Rousseau, 1984). The LAB are associated with the curd, whereas the yeasts were concentrated in the deep zone of the rind, while the micrococci and coryneforms were attached to the surface of the *Penicillium*, after which they migrated to the inner zone of the rind.

The micrococci on the surface of Roquefort cheese have been identified as *Micrococcus luteus*, *Micrococcus lylae*, *Dermacoccus sedentarius*, *Kocuria varians*, and *Kocuria rosea* (Vivier et al., 1994). Yeasts are also found in Blue cheeses; however, in most cases, their role in cheese ripening is unclear (Fleet, 1990). Presumably they play some role in deacidification of the cheese surface, because the pH of mold cheeses also increases during maturation. Whether a progression in the species of yeast occurs during ripening is also not clear because in many of these studies the stage of ripening at which the yeasts were isolated was not defined. This point was addressed by van den Tempel and Jakobsen (1998), who found that *D. hansenii*, *Zygosaccharomyces* spp., *Y. lipolytica*, and *Candida rugosa* were the dominant species in Danish Blue ripened for 1 and 14 days, while only *D. hansenii* and *C. rugosa* were found after 28 days of ripening. These yeasts showed significant lipolytic activity on tributyrin agar but no proteolytic activity in casein agar, implying that their major role is in lipolysis and deacidification.

Yeasts contribute positively to flavor and texture development (Roostita and Fleet, 1996). In Roquefort cheese, some surface ripening has been attributed to proteolytic activity of a surface slime composed in part of yeast, which is scrubbed off prior to packaging (Kinsella and Hwang, 1976). Gas production by heterofermentative LAB, particularly leuconostocs, is deemed to be important for the subsequent development of *P. roqueforti* and hence good flavor. This results in more curd-openness; the leuconostocs are also thought to be stimulated by the yeast. The degree of piercing of Roquefort cheese is also important because it increases the internal oxygen content and allows the multiplication of yeast capable of oxidizing lactic acid, which, in turn, leads to deacidification (Galzin et al., 1970).

Brines are a potent source of yeasts, and populations of yeasts in Danish Blue brines ranged from 1.9×10^4 to 2.3×10^6 cfu/g, depending on the dairy. The brines had a fairly typical composition (~22% NaCl, pH 4.5) and were held at 19°C. *Debarymyces hansenii* was the dominant yeast in three of the brines and *Corynebacterium glutosa* in the fourth (van den Tempel and Jakobsen, 1998).

Generally, commercial strains of *B. linens*, *G. candidum*, *D. hansenii*, *P. roqueforti*, and/or *P. camemberti* are added deliberately to either the milk or the cheese after brining. Several methods of inoculation are used. The milk for the mold-ripened varieties—Blue, Camembert, and Brie cheeses—is inoculated with pure cultures of the relevant species of *Penicillium*, at the same time as the starters. The curd of Blue cheese is subsequently pierced to allow limited entry of O_2 to promote growth of *P. roqueforti*. Surface- or smear-ripened cheese (e.g., Tilsit, Muenster,

Limburger, etc.) is dipped, sprayed, or brushed with aqueous suspensions of *B. linens*, *D. hansenii*, and *G. candidum*, soon after the cheeses are removed from the brine. Different combinations of organisms may also be used. Both mold- and bacterial-ripened cheese are then ripened at 10–15°C to promote microbial growth and activity and at a high RH to prevent loss of moisture from the cheese surface.

The involvement of yeast in the maturation of Cheddar cheese is unclear. Most studies on the microflora of this cheese do not report on the presence of yeast; this may be due to a lack of specific examination for yeast. Fleet and Mian (1987) reported that almost 50% of Australian Cheddar cheeses examined contained 10^4–10^6 cfu/g. A study of South African Cheddar by Welthagen and Vijoen (1999) indicated that all 42 cheeses examined contained yeast at levels varying from $<10^2$ to $>10^7$ cfu/g; however, 88% of the cheeses had $<10^5$ cfu/g, a level deemed necessary to influence flavor development. Monitoring of yeast growth during ripening indicated that their density increased from 10^2 to 10^3 cfu/g over the first 30 days of ripening, from 30 to 40 days of ripening yeast counts increased to 10^6 cfu/g, before decreasing again. The reason for this growth pattern is not clear.

11.8 SALT AND ACID TOLERANCE

All brined cheeses contain high levels of salt in the surface layers, and therefore the microorganisms growing on the cheese surface must be salt-tolerant. This is true for all of them, except *Propionibacterium shermanii*, which, in any case, only grows inside the cheese. *Brachybacterium tyrofermentans*, *Brachybacterium alimentarius*, *Kocuria rosea*, *Kocuria varians*, *Dermacoccus sedantarius*, *Micrococcus lylae*, *Micrococcus luteus*, and most strains of coryneforms found in cheese can grow in the presence of 15% NaCl, and most strains of staphylococci grow in the presence of 10% salt. NSLAB can grow in the presence of 6.5% salt. The growth of *P. camemberti* is largely unaffected by 10% NaCl, and some strains of *P. roqueforti* can tolerate 20% NaCl. Several yeasts including *Debaryomyces hansenii* can also grow in 10% salt, and most of them can also grow in 15% salt (A_w = 0.88). *Geotrichium candidum* is an exception and is quite sensitive to salt. A slight reduction in its growth occurs in the presence of 1% NaCl, and it is completely inhibited at ~6%. Therefore, too much brining will prevent its growth on the cheese surface. Perhaps its intolerance to salt explains why *G. candidum* is generally deliberately added in the manufacture of surface-ripened cheeses, the hope being that some cells will grow.

Yeast and molds grow much better than bacteria at the pH of cheese, and this is the reason they grow first on the cheese surface. They grow quite well at pH values of 2 to 4, where bacteria either do not grow or only grow very poorly. The low pH of freshly made cheese is therefore partially selective for their growth. *Brevibacterium linens* is generally considered not to grow at pH values below 6.0, and this is also probably true of the other bacteria found on the surface of cheese.

Generally, yeast and molds are not nutritionally demanding, but they grow more slowly than bacteria. Therefore, they do not compete with bacteria in environments in which bacteria grow (e.g., at pH values around 7).

Generally, yeasts are facultative anaerobes, whereas molds are considered to be obligate aerobes. However, *P. roqueforti* can grow in the presence of limited levels of O_2, which is exemplified by its growth throughout the mass of blue cheese. Yeast and molds are generally heat sensitive and are killed by pasteurization.

11.9 FACTORS INFLUENCING GROWTH OF MICROORGANISMS IN CHEESE

A number of physical parameters are involved in controlling the growth of microorganisms in cheese during ripening. These include moisture, salt, ripening temperature, redox potential, NO_3^- and pH. The extent of variation in these parameters is influenced by the cheesemaking process.

11.9.1 Moisture and Salt

Salt and moisture are very much interrelated and are considered together. All microorganisms require water for growth, and one of the most effective ways of controlling their growth is to reduce the available water through dehydration and/or addition of some water-soluble compound such as sugar or salt. In cheese, the major compound is salt. It is the availability of the water rather than the absolute amount present that is important. The availability is mainly determined by the concept of water activity, A_w. This was developed by Scott (1957) and has provided a basis for understanding the relationships between microorganisms and water. A_w is a thermodynamic concept defined as the ratio between the vapor pressure of the water present in a system (p) and that of pure water (p_0) at the same temperature:

$$a_w = p/p_0$$

A_w ranges from 0 to 1. In general, bacteria have higher minimum A_w requirements than do yeast, which, in turn, have higher requirements than do molds. During the first stages of cheese manufacture, the A_w is ~0.99, which supports the growth and activity of the starter culture. However, during ripening the A_w levels are much lower, ranging from 0.917 to 0.988 (Rüegg and Blanc, 1981). The A_w of cheese is mainly a reflection of its salt content. The higher the salt, the lower the A_w.

The concentration of salt required to inhibit bacteria depends on the nature of the food, its pH and its moisture content, but, generally, 10–100 g/kg is sufficient. The relationship between salt concentration and A_w is almost linear:

$$A_w = -0.0007x + 1.0042$$

where x is the g of salt per kg of cheese. The correlation coefficient (r^2) was 0.997. The salt concentration in cheese ranges from 0.7 to 7 g/100 g, corresponding to A_w values of 0.99 to 0.95, respectively. Many spoilage microorganisms can grow under such conditions, but they normally do not because of the other factors involved in controlling growth. It is the concentration of salt dissolved in the moisture of the cheese rather than the actual concentration of salt added that is important. This is usually measured as percent salt-in-moisture (SM), which is equivalent to 100 times the ratio of salt to moisture, both in g/100 g. Cheeses of 38% moisture with salt contents of 0.7 and 7.0 g salt/100 g will have a percent salt in moisture value of 1.84% and 18.4%, respectively.

A_w contributes significantly to the control of the metabolic activity and multiplication of microorganisms in cheese. The minimum A_w for *Lc. lactis*, *Sc. thermophilus*, *Lc. helveticus*, and *P. freudenreichii* ssp. *shermanii* are 0.93, >0.98, >0.96, and 0.96, respectively (Weber and Ramet, 1987). Other factors are also involved. A reduction in A_w during cheese ripening occurs due to evaporation of water from the surface. This does not occur in commercially produced Cheddar cheese, because it is packed in plastic bags, which prevent evaporation of the moisture. Other factors include hydrolysis of proteins to peptides and amino acids and triglycerides to glycerol and fatty acids and involve the take-up of 1 molecule of water for each peptide or ester bond hydrolyzed. Because significant proteolysis occurs in cheese, the A_w will also decrease due to this during maturation. Moisture loss is controlled by increasing the relative humidity in the ripening room (e.g., in the case of soft cheeses, like Limburger, Tilsit, and Camembert) or by packing the cheese in wax or plastic. This improves the surface growth on these cheeses. There may be variation in the A_w values in different zones in cheese. This occurs particularly in large brine-salted hard and semihard cheeses like Emmental, where the values are usually higher toward the center. Cheddar cheese is dry-salted and vacuum-packed; therefore, no loss of moisture will occur and there will be no change in A_w because the salt is uniformly distributed in the cheese.

11.9.2 pH and Organic Acids

The optimum pH for the growth of most common bacteria is around neutral, and growth is often poor at pH values <5.0. Lactobacilli are an exception to this rule because many of them will grow quite well at pH values of <4.5. In addition, yeasts and molds will also grow quite well at pH 4.5. In fact, the main function of the yeasts in the surface of cheese is to metabolize the lactic acid to CO_2 and H_2O. This increases the pH to a point where the surface bacteria can grow and is called *deacidification*. Due to the accumulation of organic acids, cheese curd post-manufacture has a pH ranging between 4.5 and 5.3; such low pH values will not allow the survival of acid-sensitive species. The real inhibitor is thought to be the undissociated form of the organic acid. The principal organic acids found in cheese are lactic, acetic and propionic acids, which have pK_a values of 3.08, 4.75, and 4.87, respectively. Therefore, at the pH of cheese, lactic acid is the least and propionic acid the most effective inhibitor at the same concentration. However, lactate is invariably present at much greater concentrations than either of the other two acids—except in the case of Swiss cheese, where the concentration of propionic acid may be higher than that of lactic acid in the ripened cheese (Steffen et al., 1993).

11.9.3 Ripening Temperature

The microorganisms involved in cheese manufacture and ripening are either mesophilic or thermophilic, having temperature optima of ~30°C or 42°C, respectively; at temperatures below these values, growth will be retarded. Many Swiss-type and Italian cheeses are cooked at temperatures of ~54°C for 30–60 min, and thus the thermophilic starters used in the manufacture of these cheeses must be capable of withstanding these temperatures. They do not grow at the cooking temperature, and little acid production occurs in these cheeses until the cheese is put in the molds and the temperature begins to decrease. The temperature at which cheese is ripened is a compromise between the need to promote ripening reactions and growth of the desirable secondary flora and the need to prevent the propagation of potential spoilage and pathogenic bacteria. Generally, Cheddar cheese is ripened at 6–8°C, while mold and smear-ripened cheeses are ripened at 10–15°C. Emmental cheese goes through several different time– temperatures required during ripening. Initially, this cheese is ripened at ~12°C for 2–3 weeks, after which the temperature is increased to ~22°C for 2–4 weeks to promote the growth of the propionic acid bacteria, after which the temperature is reduced to ~4°C. Higher temperatures promote accelerated ripening (Folkertsma et al., 1996), but the changes to body and flavor are often detrimental.

11.9.4 Redox Potential

The redox or oxidation–reduction potential (E_h) is a measure of the ability of chemical/biochemical systems to oxidize (lose electrons) or reduce (gain electrons). E_h is usually measured in mV, and the oxidized and reduced states are indicated by positive and negative signs, respectively. The E_h of milk is about +150 mV, while that of cheese is about –250 mV. While the exact mechanism of the reduction in E_h in cheese is not fully established, it is most probably related to (a) fermentation of lactic acid by the starter during growth and (b) the reduction of small amounts of O_2 in the milk to water (Crow et al., 1995). As a consequence of these reactions, the interior of cheese is essentially an anaerobic system, which can only support the growth of obligate or facultative anaerobic microbes. The E_h of cheese is one of the major factors in determining the types of microorganisms that will grow in cheese; therefore, obligate aerobes, such as *Pseudomonas*, *Brevibacterium*, *Bacillus*, and *Micrococcus* spp. are excluded from growth in the interior of the cheese. Bacteria that develop on the cheese surface are predominantly obligate aerobes.

11.9.5 Nitrate

In the production of some cheeses—especially the Dutch varieties, Edam and Gouda—saltpeter (KNO_3) or $NaNO_3$ is added to the cheese milk to prevent growth of *Clostridium tyrobutyricum*, which ferments lactate to butyrate, H_2 and CO_2:

$$2 \text{ Lactate} \rightarrow 1 \text{ Butyrate} + 2\, CO_2 + 2\, H_2$$

The H_2 (mainly) and CO_2 are responsible for the large holes present in the cheese, while the butyrate is responsible for the development of a nasty rancid off-flavor. In these cheeses, the rate of NaCl migration is relatively slow and equalization of the salt concentration throughout the cheese can take several weeks. In addition, the moisture content is high. Therefore, growth inhibition of spoilage microorganisms such as *Cl. tyrobutyricum* is necessary prior to achieving salt equilibrium, and nitrate fulfills this function. The concentration added is usually 20 g/100 liters of milk, and most of it is lost in the whey. During ripening, nitrate is reduced to nitrite (the actual growth inhibitor) by the indigenous xanthine oxidase present in the milk/curd. The maximum amount of nitrite permitted in the cheese is 50 mg/kg. By the time the cheese is ready for consumption, the level of nitrite is usually well below this level.

Nitrite does not affect the growth of LAB, but does inhibit the growth of propionibacteria, which are essential for eye formation in Emmental cheese. Thus NO_3^- is not suitable for control of *C. tyrobutyricum* in cheese where growth of propionibacteria is required. Nitrite can react with aromatic amino acids in cheese to produce nitrosamines, many of which are carcinogenic. The reaction is pH-dependent, and is optimal in the pH range of 2 to 4.5. Nitrate also prevents the formation of early gas by coliform. This is due to H_2 formation from formate by formic hydrogenlyase. However, it does not prevent their growth. Instead the NO_3^- acts as an alternative electron acceptor to O_2 and allows complete oxidation of the lactose to CO_2 and H_2O rather than incomplete oxidation of lactose to lactate acid formate.

11.10 SPOILAGE OF CHEESE

The major spoilage organisms in cheese are coliforms and *Cl. tyrobutyricum*, which cause the development of early and late gas, respectively. As its name implies, early gas occurs within 1 or 2 days after manufacture, while late gas may require several months to develop. Early gas may occasionally also be due to growth of lactose fermenting yeasts, and gas may also be produced from citrate by citrate utilizing lactobacilli. Today, spoilage of cheese by any of these microorganisms is unlikely because of improvements in raw milk quality, better hygiene and control of microbial growth in cheese factories, and rapid acid producing starters. Growth of molds may occasionally also cause discoloration of the cheese surfaces; waxing, as occurs in some cheeses, will prevent molds from growing.

11.11 PATHOGENS IN CHEESE

In the past 30 years, there have been only 32 confirmed outbreaks of food poisoning due to cheese in Western Europe, the United States, and Canada, during which an estimated 235,000,000 tonnes of cheese was produced; 28% of the outbreaks involved cheese made from raw milk (Fox et al., 2000). These data indicate that cheese is a very safe food product. Several organisms were involved, but *Salmonella* spp., *Staph. aureus*, and *Listeria monocytogenes* were the most common; various enterotoxigenic *E. coli*, including O157, were also involved. Staphylococcal food poisoning is due to a heat-stable toxin produced by the cells during growth, and a general rule is that 10^6 staphylococci are required to produce sufficient toxin(s) to cause problems. In contrast, it is thought that 10 cells of *E. coli* O157 is sufficient. Growth of some pathogens, particularly *E. coli* and *S. aureus*, can occur during manufacture, but this will depend on how rapid acid production is and the cooking temperatures used; for example, the high cooking temperature (54°C) and the length of time the curd is held at this temperature (~60 min) in Swiss and Italian cheeses will kill most, if not all, pathogens that might be present. Generally, it is only semihard and soft cheeses—particularly those in which the pH of the surface increases during ripening—that cause problems. The effect of pH is clearly seen in the study of Hargrove et al. (1969) on the survival of salmonella in Cheddar cheese (Figure 11.8). At pH 5.03 and 5.23, which is close to that of a normal

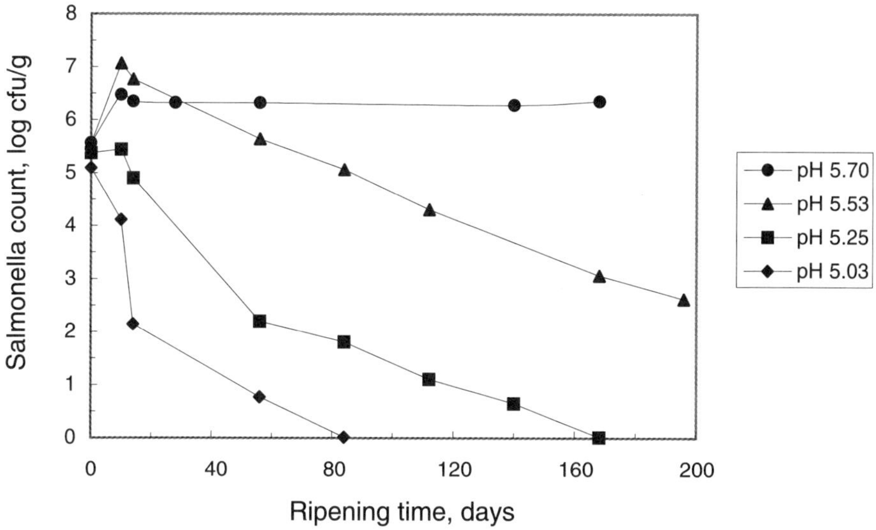

Figure 11.8. Effect of pH on the survival of Salmonella in Cheddar cheese during ripening. [Redrawn from Hargrove et al. (1969).]

Cheddar cheese, they died off quickly, whereas at pH 5.7, which could occur as a result of attack by phage, they did not die off at all. In soft cheeses there is an increase in pH during ripening and growth of some pathogens; for example, *L. monocytogenes*, and *Hafnia alvei*, but not *Enterobacter aerogenes* nor *E. coli*, can occur (Fox et al., 2000).

11.12 RAW MILK CHEESES

Cheese made from raw milk is generally considered to have a much more intense flavor than the same cheese made from pasteurized milk, and this is an important marketing advantage for raw milk cheeses. This has been ascribed mainly to more proteolysis and lipolysis by the raw milk microflora in the raw milk cheese. Recent evidence showed that NSLAB, PAB, enterococci, and yeasts grow to higher cell numbers in hard cheeses (Comté and Cheddar) made from raw milk than in the same cheeses made from pasteurized milk. The reasons for this are unclear, but the results are consistent (McSweeney et al., 1993; Demarigny et al., 1996, Beuvier et al., 1997). One suggestion was that pasteurization retards the growth of NSLAB in some unspecified way. The results imply that the indigenous microflora have a significant effect on the development of flavor of the cheese. The flavor of the raw milk Swiss-type cheese was more intense than the pasteurized milk cheese, while that of the Cheddar was atypical and was downgraded by the official graders. However, the authors considered that this intensely flavored cheese could be an attractive cheese for connoisseurs of Cheddar cheese.

Staph. aureus is the commonest cause of mastitis in dairy cows. Both it and other pathogens (e.g., *E. coli*, *Listeria*, etc.) can be present in the raw milk and can grow to high numbers during cheese manufacture and ripening. Despite this, significant amounts of raw milk cheeses are produced, particularly on the European mainland, and relatively few outbreaks of food poisoning have involved cheese, made from raw milk. To prevent growth of pathogens in cheese, several precautions can be taken. The milk should be produced under very good conditions and be held at 4°C until cheese manufacture begins. A total count of <20,000 cfu/ml of raw milk at the beginning of manufacture (i.e., before the starter is added) should be attainable. An active starter, good hygiene, attention to detail, and a HAACP system will all help to produce a cheese free of pathogens. Care should also be taken in handling the cheese during ripening, especially in the washing of smear cheeses, where the pH increases during ripening. Many of these factors will also help to control the growth of pathogens in pasteurized milk cheese.

11.13 MICROBIOLOGICAL ANALYSIS OF CHEESE

Cheese is a complex microbial ecosystem containing starter bacteria, NSLAB, non-lactic-acid bacteria, yeasts, and molds. Most cheese varieties contain these groups of organisms, and only a limited number of effective selective media are available for enumerating them. Some of the more common ones used are listed in Table 11.3.

The total number of bacteria has been counted in cheese, but the results are not very meaningful in terms of the care taken in manufacturing the cheese because most of the bacteria in the cheese will be either starter or NSLAB. Not all starter and NSLAB grow well on plate count agar (PCA). The reason for this is that the levels of nutrients and buffering capacity of PCA are too low to sustain good growth of them. LAB are quite fastidious and require several amino acids and vitamins for growth. Therefore, media for LAB must contain rich sources of amino acids and peptides (peptones and yeast extract) and vitamins (yeast extract) and high levels of buffer to neutralize the large amounts of lactate produced from sugar metabolism during growth.

Many media have been developed to enumerate starter bacteria. Today, the medium of choice for enumerating lactococci is Medium 17, containing lactose (LM-17), which was originally developed to estimate lactococcal phage. It contains sufficient amounts of all the nutrients necessary to support the growth of lactococci and a high concentration of β-glycerophosphate (19 g/liter) as a buffer, which, unlike phosphate, does not chelate the Ca^{2+} required for adsorption of phage to its host. This medium is non-selective; nevertheless, it is used to count lactococci in cheese because these bacteria outnumber all other microorganisms in the cheese, especially during the early stages of ripening. As the cheese ripens, LM-17 becomes less selective because enterococci, NSLAB, and leuconostocs can grow on it.

Incubation of LM-17 plates at 45°C makes it relatively selective for *Sc. thermophilus*, because lactococci do not grow at this temperature and thermophilic lactobacilli grow poorly if at all in this medium. However, enterococci will grow on the medium, and colonies should be examined to determine if they are enterococci or *Sc.. thermophilus*.

Thermophilic cultures are often used today as adjuncts in commercial cheeses made with mesophilic cultures. If high counts of *Sc.*

TABLE 11.3. Media and Incubation Conditions Commonly Used to Enumerate Different Types of Bacteria in Cheese

Group	Medium	Temperature (°C)	Incubation Time and Conditions
Lactococcus spp.	LM-17 agar	30	3 days; spread or pour plate, aerobic
Leuconostoc spp.	MRS agar +20 μg vancomycin/ml	30	3 days; spread or pour plate, aerobic
Streptococcus thermophilus	LM-17 agar	45	2 days; spread or pour plate, aerobic
Lactobacillus helveticus	MRS agar, pH 5.4 agar	45	3 days; spread plate, anaerobic
Lactobacillus lactis	MRS agar, pH 5.4 agar	45	3 days; spread plate, anaerobic
Nonstarter lactic acid bacteria	Rogosa agar (RA)	30	5 days; pour plate, aerobic with overlay
Citrate utilizers	Calcium-citrate agar (KCA)	30	2–3 days; spread plate, anaerobic
Staphylococcus aureus	Baird Parker agar	37	72 h; spread plate, aerobic
Coliforms	Violet Red bile agar	30	18 h; pour plate, aerobic
Enterococcus spp.	Kanamycin aesculin azide agar	37	24 h; pour plate, aerobic
Smear bacteria	Plate count agar containing 70 g/liter NaCl	25	7–10 days; spread plate, aerobic; plates should be spread with natamycin to inhibit growth of yeasts
Yeasts/molds	Yeast extract glucose Chloramphenicol agar	25	3–5 days; pour plate, aerobic

thermophilus (on LM-17 at 45°C) are found in a cheese made with a mesophilic culture, the count of *Lactococcus* (on LM-17 at 30°C) must be adjusted, since *Sc. thermophilus* will also grow on LM-17 at 30°C. Sometimes, the colonies of *Sc. thermophilus* are much smaller than those of *Lactococcus* spp., and the difference in size could be used to differentiate between them, but a smaller size is by no means an absolute feature of *Sc. iteanophilus*

MRS agar is a general-purpose medium for the enumeration of lactobacilli. Most other LAB can grow in it, but reducing the pH to 5.4 and increasing the temperature of incubation to 45°C make the medium more or less selective for the thermophilic lactobacilli found in starters. The mesophilic lactobacilli found in cheese are usually counted on Rogosa agar (RA) (Rogosa et al., 1951). This medium contains a high concentration of acetate (0.225 M) and has a low pH (5.4), which make it quite selective for mesophilic lactobacilli. Some leuconostocs and pediococci may grow on it, but the thermophilic lactobacilli present in starters generally do not grow on this medium. It is currently widely used in cheese microbiology and appears to be quite selective for cheese containing different LAB.

In the past, the medium of Mayeaux, Sandine, and Elliker (1962) (MSE) was used to enumerate leuconostocs. This medium contains sucrose as the energy source, and many authors consider it to be selective for *Leuconostoc*. This is not so. The medium is nutritionally rich, and many, if not at all, LAB will grow on it. It may be selective if only dextran producers (very large colonies) are counted, because dectran formation is generally confined to *Leuconostoc* spp. However, not all leuconostocs produce dextrans. Because most leuconostocs are resistant to the antibiotic vancomycin, the addition of vancomycin (20 μg/ml) to an otherwise nutritionally adequate medium (e.g., MRS) makes it selective for these microorganisms. This method is acceptable when applied to starters, but mesophilic lactobacilli and pediococci are also naturally resistant to vancomycin. Both mesophilic lactobacilli and pediococci are often found in large numbers in cheese, which limits the usefulness of the medium, unless colonies are also examined microscopically and any doubtful ones examined by additional tests.

Calcium citrate agar (Anonymous, 1997) is very useful for enumerating citrate utilizers present in mesophilic cultures and dairy products. This is a differential, non-selective medium that is opaque owing to the presence of insoluble calcium citrate. As the citrate is metabolized, a clear halo develops around the colony. This medium can also be used to estimate total starter numbers if triphenyltetrazolium chloride (TTC) is added to it (1 ml of a filter-sterilized 1% [w/v] TTC solution per 100 ml of medium). TTC is reduced by the starter bacteria from a colorless soluble form to a red or pink insoluble form that precipitates around the starter colonies, making them more readily apparent. If this medium is used to enumerate citrate-utilizing bacteria in cheese, colonies surrounded by halos should be examined microscopically, because many mesophilic lactobacilli, which are found in high numbers in ripened cheese, metabolize citrate and will produce halos around the colonies.

There are no specific selective media for the bacteria found in the smear on cheese. All these bacteria are salt-tolerant, so the addition of NaCl (e.g., 70 g/liter) to an otherwise suitable medium (e.g., plate count agar) will inhibit the starter bacteria without affecting the smear bacteria. These bacteria grow slowly; therefore, plates must be incubated for at least 5 days.

Numerous selective media have been proposed for enumerating enterococci, but none is completely reliable. The most commonly used ones are m-*Enterococcus*, KF, and kanamycin aesculin azide (KAA) agars. Good selective media are available for *Staph. aureus* (Baird Parker agar), coliforms (Violet Red Bile agar), and yeasts and molds (Potato Dextrose Agar acidified to pH 3.5 with lactic acid or Yeast Glucose Chlorotetracycline agar). The antibiotic (chlorotetracycline) inhibits the bacteria without affecting yeast and molds. Yeasts and molds are easily distinguished, because the latter produce large fluffy colonies rather than the small, opaque, sometimes glistening colonies of yeast.

11.14 FLAVOR DEVELOPMENT DURING RIPENING

Cheese flavor cannot develop without the involvement of LAB. During the ripening of cheese, three major biochemical events—glycolysis, lipolysis, and proteolysis—occur, each of which is involved in flavor formation. The latter is probably the most important and also the most complex.

Glycolysis is the conversion of lactose to lactic acid and is almost exclusively due to the growth of the starter bacteria (see Chapter 1 in this volume) and the lactate produced gives the freshly made cheese its overall acidic taste. They can also produce other compounds—for example, diacetyl, acetate, and acetaldehyde, which are important compounds in flavor formation in fresh cheeses; diacetyl is also an important flavor compound in hard cheeses.

Lipolysis results in hydrolysis of the milk fat and the production of glycerol and free fatty acids, many of which, particularly the short-chain ones, have strong characteristic flavors. There are several potential sources of lipase in cheese, including milk, pregastric esterase in rennet pastes for Italian cheeses, starter, NSLAB, and the secondary flora. Native milk lipase is inactivated by pasteurization and is of no significance in lipolysis in cheeses made from pasteurized milk. In hard cheeses made from pasteurized milk the major source of lipases are the starter and NSLAB. Even though these bacteria only produce small amounts of these enzymes, they result in significant activity in the cheese during the long ripening time. *P. roqueforti* possesses considerable lipolytic activity and is responsible for the strong rancid taste of Blue cheeses, which can contain ~30g of fatty acids per kilogram. The fatty acids can be further metabolized to methyl ketones by the molds present in blue mold cheeses. The ketones contain one less carbon than the fatty acid from which they are produced. Fat also acts as a solvent for many of the flavor compounds produced in cheese.

The sources of proteinases in cheese are milk itself, chymosin, starter, NSLAB, and the secondary microflora (coryneforms, micrococci, staphylococci, yeasts, and molds). Native milk proteinase is called plasmin and is only significant in cheeses, which are cooked to high temperatures because the cooking process inactivates chymosin. A generalized and therefore oversimplified view of proteolysis is as follows. Chymosin is responsible for the initial hydrolysis of the casein into large and small peptides, which the proteinases and peptidases, which are released from the starter by autolysis, act on to produce smaller peptides and free amino acids. Many of the small peptides and amino acids contribute directly to flavor, but the starter bacteria also have a plethora of enzymes that degrade the amino acids to amines, acids, alcohols, carbonyls, sulfur-containing compounds, and so on, which are also involved in flavor formation. Proteolysis can be limited (e.g., in Mozzarella cheese, due to the high heat treatment it gets during the kneading process) or extensive (e.g., in mold-ripened cheeses). At one time it was considered that the major role of the starter organisms was to produce the right environment for purely chemical reactions to occur, but it is now felt that chemical reactions play no role in flavor development. Instead, biochemical reactions are involved, for which enzymes are necessary. For a more exhaustive treatment of the biochemistry of cheese ripening and flavor formation, see Urbach (1995), Fox and Wallace (1997), and Fox et al. (2000).

11.15 ACCELERATION OF RIPENING

Ripening of hard cheeses is a slow process and can take as long as 3 years in the case of Parmigiano–Reggiano cheese. Long ripening times make ripening an expensive process; consequently, numerous methods have been studied to accelerate it and reduce the cost. The methods include raising the temperature of ripening, addition of various enzymes, addition of chemically, physically, or genetically modified cells, addition of adjunct cultures, or increasing the moisture. The ripening temperatures for cheese varies from 7°C for Cheddar to ~22°C for Swiss-type cheeses; Dutch, mold, and smear cheeses are ripened at ~15°C. Increasing the temperature is the easiest and cheapest way of accelerating ripening. There is a risk of increased spoilage of the cheese, but this is unlikely if the cheese has been made to the correct composition.

Proteolysis is thought to be the limiting step in cheese ripening, and various proteinases (from molds and bacteria) and peptidases (from starter cultures) have been suggested. One of the disadvantages of these is that much of the enzyme is removed with the whey, and relatively little remains in the cheese.

This increases the cost, and so most investigators add the enzyme to the curd at salting. This method is only suitable for Cheddar cheese where the curd particles are relatively small because diffusion of enzymes is very low. Encapsulation of enzymes has also been studied. The microcapsules are incorporated into the cheese efficiently, but the efficiency of encapsulation itself is low and increases the cost. Rapid lysing cultures have also been tried especially as a source of peptidases. Novel methods involves the use of cultures containing prophages that are induced by the cooking temperatures or cultures that are sensitive to a particular bacteriocin, produced by one of the starter cultures.

Because the starter plays a key role in cheese ripening, it might be expected that increasing cell numbers would accelerate ripening, but high cell numbers have been associated with the development of bitterness, at least in Cheddar cheese. An alternative is to add attenuated starter cells, which have lost their acid-producing ability (excessively rapid acid production is undesirable in cheese) but retain their proteolytic and peptidolytic activities. Physically heated cells (e.g., by heat or freeze-shocking) treated with lysozyme (to hydrolyze the cell walls), lactose, and proteinase negative mutants have also been shown to be effective, at least at the pilot scale level. Adjunct cultures, particularly mesophilic lactobacilli, have also been shown to improve the flavor, particularly of Cheddar cheese. Elevated ripening temperatures are the simplest and most effective method of accelerating the ripening of Cheddar cheeses (Fox et al., 1996). The use of exogenous enzymes has not been commercially successful. Fox et al. (1996) concluded that the key to accelerating the ripening of cheese ultimately depends on identifying the key sapid flavour compounds. This has been a rather intractable problem: Work on the subject commenced nearly 100 years ago and has been quite intense since about 1960—that is, since the development of gas chromatography. Although as many as 400 compounds, which might be expected to influence cheese taste and aroma, have been identified, it is not possible to describe cheese flavor precisely. Until such information is available, attempts to accelerate ripening will be speculative and empirical.

REFERENCES

Anonymous (1994) *Dairy Ind. Int.*, **59**(11), 34–35.

Anonymous (1997) *International Dairy Federation*, Standard 149A.

Barksdale, L., Lanéelle, M. A., Pollice, M. C., Pollice, M. C., Asselineau, J., Welby, M., and Norgard, M. V. (1979) *Int. J. Syst. Bacteriol.*, **29**, 222–233.

Berger, C., Khan, J. A., Molimard, P., Martin, N., and Spinnler, H. E. (1999) *Appl. Environ. Microbiol.*, **65**, 5510–5514.

Beuvier, E., Berthaud, K., Cegarra, S., Dasen, A., Pochet, S., Buchin, S., and Duboz, G. (1997) *Int. Dairy J.*, **7**, 311–323.

Boyaval, P., and Desmazeaud, M. (1983) *Lait*, **63**, 187–216.

Boysen, M., Skouboe, P., Frisvad, J., and Rossen, L. (1996) *Microbiology*, **142**, 542–549.

Centano, J. A., Menendez, S., and Rodriguez-Otero, J. L. (1996) *Int. J. Food Microbiol.*, **33**, 307–317.

Chang, S.-C., Lu, K.-L., and Yeh, S.-F. (1993) *Appl. Environ. Microbiol.*, **59**, 981–986.

Collins, M. D. (1987) *Int. J. Syst. Bacteriol.*, **37**, 442–443.

Collins, M. D., Farrow, J. A. E., Goodfellow, M., and Minnikin, D. E. (1983) *Syst. Appl. Microbiol.*, **4**, 388–395.

Collins, M. D., Smida, J., and Stackebrandt, E. (1989) *Int. J. Syst. Bacteriol.*, **39**, 7–9.

Coppola, S., Villani, F., Coppola, R., and Parente, E. (1990) *Lait*, **70**, 411–423.

Coppola, R., Nanni, M., Iorizzo, M., Sorrentino, A., Sorrentino, E., and Grazia, L. (1997) *J. Dairy Res.*, **64**, 305–310.

Crow, V. L., Coolbear, T., Gopal, P. K., Martley, F. G., McKay, L. L., and Riepe, H. (1995) *Int. Dairy J.*, **5**, 855–875.

Demarigny, Y., Beuvier, E., Dasen, A., and Duboz, G. (1996) *Lait*, **76**, 371–387.

Diggin, M. B., Waldron, D. S., McGoldrick, M. A., Cogan, T. M., and Fox, P. F. (1999) *Ir. J. Agric. Food Res.*, **38**, 183.

Eliskases-Lechner, F., and Ginzinger, W. (1995a) *Lait*, **75**, 571–584.

Eliskases-Lechner, F., and Ginzinger, W. (1995b) *Milchwissenschaft*, **50**, 458–462.

Engel, G., von Milczewski, K. E., Prokopek, D., and Teuber, M. (1982) *Appl. Environ. Microbiol.*, **43**, 1034–1040.

Feirtag, J. M., and McKay, L. L. (1987) *J. Dairy Sci.* **70**, 1779–1784.

Fiedler, F., Schaffler, M. J., and Stackebrandt, E. (1981) *Arch. Microbiol.*, **129**, 85–93.

Fitzsimons, N. A., Cogan, T. M., Condon, S., and Beresford, T. (1999) *Appl. Environ. Microbiol.*, **65**, 3418–3426.

Fitzsimons, N. A., Cogan, T. M., Condon, S., and Beresford, T. (2001) *J. Appl. Microbiol.*, **90**, 600–608.

Fleet, G. H. (1990) *J. Appl. Bacteriol.*, **68**, 199–211.

Fleet, G. H., and Mian, M. A. (1987) *Int. J. Food Microbiol.*, **4**, 145–155.

Folkertsma B., Fox, P. F., and McSweeney, P. L. H. (1996) *Int. Dairy J.*, 6.

Fox, P. F., and Wallace, J. M. (1997) *Adv. Food Microbiol.*, **45**, 17–85.

Fox, P. F., Wallace, J. M., Morgan, S., Lynch, C. M., Niland, E. J., and Tobin, J. (1996). *Antonie von Leeuwenhoek*, **70**, 271–297.

Fox, P. F., Guinee, T. P., Cogan, T. M., and McSweeney, P. L. H. (2000) *Fundamentals of Cheese Science*, Aspen Publishers, Gaithersburg, MD.

Franz, C. M. A. P., Holzapfel, W. H., and Stiles, M. E. (1999) *Int. J. Food Microbiol.*, **47**, 1–24.

Freitas, A. C., Pais, C., Malcata, F. X., and Hogg, T. A. (1996) *J. Appl. Microbiol.*, **59**, 155–160.

Funke, G., and Carlotti, A. (1994) *J. Clin. Microbiol.*, **32**, 1729–1732.

Galzin, M., Galzy, P., and Bret, G. (1970) *Lait*, **50**, 1–37.

Grand, M., Weber, A., Perret, J., Zenther, U., and Glattli, H. (1992) *Schweiz. Milchwirtsch. Forsch.*, **21**, 3–5.

Gripon, J. C. (1993) In *Cheese: Chemistry, Physics and Microbiology*, Vol. 2, P. F. Fox, ed., Chapman and Hall, London. pp. 111–136.

Hargrove, R. E., McDonagh, F. E., and Mattingly, W. A. (1969) *J. Milk Food Technol.*, **32**, 580–584.

Irlinger, F., and Bergère, J. L. (1999) *J. Dairy Res.*, **66**, 91–103.

Irlinger, F., Morvan, A., El Solh, N., and Bergere, J. L. (1997) *Syst. Appl. Microbiol.*, **20**, 319–328.

Jimeno, J., Làzaro, M. J., and Sollberger, H. (1995) *Lait*, **75**, 401–413.

Jordan, K. N., and Cogan, T. M. (1999) *Lett. Appl. Microbiol.*, **29**, 136–140.

Karahadian, C., and Lindsay, R. C. (1987) *J. Dairy Sci.*, **70**, 909–918.

Kiernan, R. C., Beresford, T. P., O'Cuinn, G., and Jordan, K. N. (2000) *Ir. J. Agric. Food Res.*, **39**, 95.

Kinsella, J. E., and Hwang, D. H. (1976) *Crit. Rev. Food Sci. Nutr.*, **8**, 191–228.

Kolloffel, B., Burri, S., Meile, L., and Teuber, M. (1997) *Syst. Appl. Microbiol.*, **20**, 409–417.

Lawrence, R. C., and Gilles, J. (1987) In *Cheese: Chemistry, Physics and Microbiology*, Vol. 2, P. F. Fox, ed. Elsevier Applied Science, London, p. 1–44.

Litopoulou-Tzanetaki, E., Tzanetakis, N., and Valopoulou-Mastrojiannaki, A. (1993) *Food Microbiol.*, **10**, 31–41.

Lucey, J. A., and Fox, P. F. (1993) *J. Dairy Sci.*, **76**, 1714–1724.

Lynch, C. M., McSweeney, P. L. H., Fox, P. F., Cogan, T. M., and Drinan, F. D. (1996) *Int. Dairy J.*, **6**, 851–867.

Madkor, S., Fox, P. F., Shalabi, S. I., and Metwalli, N. H. (1987) *Food Chem.*, **25**, 93–109.

Martley, F. G., and Lawrence, R. C. (1972) *N. Z. J. Dairy Sci. Technol.*, **7**, 38–44.

Martley, F. G., and Crow, V. L. (1993) *Int. Dairy J.*, **3**, 461–483.

Mayeaux, J. V., Sandine, W. E., and Elliker, P. R. (1962) *J. Dairy Sci.*, **45**, 655–656.

McSweeney, P. L. H., Fox, P. F., Lucey, J. A., Jordan, K. N., and Cogan, T. M. (1993) *Int. Dairy J.*, **3**, 613–634.

Nanni, M., Coppola, R., Iorizzo, M., Sorrentino, A., Sorrentino, E., and Grazia, L. (1997) *Sci. Tec. Lattiero-Casearia*, **48**, 211–216.

Niskasaari, K. (1989) *J. Dairy Res.*, **56**, 639–649.

Palles, T., Beresford, T., Condon, S., and Cogan, T. M. (1993) *J. Appl. Microbiol.*, **85**, 147–154.

Piton-Malleret, C., and Gorrieri, M. (1992) *Lait*, **72**, 143–164.

Piveteau, P. G., Condon, S., and Cogan, T. M. (1995) *Lait*, **75**, 331–343.

Piveteau, P. G., Condon, S., and Cogan, T. M. (2000) *J. Dairy Res.*, **67**, 65–71.

Prillinger, H., Molnar, O., Eliskases-Lechner, F., and Lopandic, K. (1999) *Antonie van Leeuwenhoek* **75**, 267–283.

Rattray, F. P., and Fox, P. F. (1999) *J. Dairy Sci.*, **82**, 891–909.

Riedel, K. H. J., Wingfield B. D., and Britz, T. J. (1998) *Syst. Appl. Microbiol.*, **21**, 419–428.

Rogosa, M., Mitchell, J. A., and Wiseman, R. F. (1951) *J. Dent. Res.*, **30**, 682–689.

Roostita, R., and Fleet, G. H. (1996) *Int. J. Food Microbiol.*, **28**, 393–404.

Rossi, F., Torriani, S., and Dellaglio, F. (1999) *Appl. Environ. Microbiol.*, **65**, 4241–4244.

Rousseau, M. (1984) *Milchwissenschaft*, **39**, 129–135.

Rüegg, M., and Blanc, B. (1981) In *Water Activity: Influences on Food Quality*, L. B. Rockland and G. F. Stewart, eds. Academic Press, New York, pp. 791–811.

Schubert, K., Ludwig, W., Springer, N., Kroppenstedt, R. M., Accolas, J. P., and Fiedler, F. (1996) *Int. J. Syst. Bacteriol.*, **46**, 81–87.

Scott, W. J. (1957) *Adv. Food Res.*, **7**, 83–101.

Stackebrandt, E., Koch, C., Gvozdiak, O., and Schumann, P. (1995) *Int. J. Syst. Bacteriol.*, **45**, 682–692.

Steffen, C., Eberhard, P., Bosset, J. O., and Rüegg, M. (1993) In *Cheese: Chemistry, Physics and Microbiology*, Vol. 2, P. F. Fox, ed., Elsevier Applied Science, London, pp. 83–110.

Thomas, T. D., and Crow, V. L. (1983) *N. Z. J. Dairy Sci. Technol.*, **18**, 131–141.

Turner, K. W., and Thomas, T. D. (1980) *N. Z. J. Dairy Sci. Technol.*, **15**, 265–276.

Turner, K. W., Morris, H., and Martley, F. G. (1983) *N. Z. J. Dairy Sci. Technol.*, **18**, 117–124.

Urbach, G. (1995) *Int. Dairy J.*, **5**, 877–903.

Valdes-Stauber, N., Scherer, S., and Seiler, H. (1997) *Int. J. Food Microbiol.*, **34**, 115–129.

Valence, F., Richoux, R., Thierry, A., Palva, A., and Lortal, S. (1998) *J. Dairy Res.*, **65**, 609–620.

van den Tempel, T., and Jakobsen, M. (1998) *Int. Dairy J.*, **8**, 25–31.

Vivier, D., Ratomahenina, R., and Galzy, P. (1994) *J. Appl. Bacteriol.*, **76**, 646–552.

Weber, F., and Ramet, J. P. (1987) In *Cheesemaking, Science and Technology*, A. Eck, ed., Lavoiser Publishing, New York, pp. 293–309.

Welthagen, J. J., and Vijoen, B. C. (1999) *Food Microbiol.*, **16**, 63–73.

Wilkinson, M. G., Guinee, T. P., O'Callaghan, D. M., and Fox, P. F. (1994) *J. Dairy Res.*, **61**, 249–262.

Wyder, M. T., and Puhan, Z. (1999) *Milchwissenschaft*, **54**, 330–333.

CHAPTER 12

MAINTAINING A CLEAN WORKING ENVIRONMENT

RICHARD K. ROBINSON
School of Food Biosciences, The University of Reading, Reading, England

ADNAN Y. TAMIME
Scottish Agricultural College, Ayr, Scotland

12.1 INTRODUCTION

While all plants used for the production of pasteurized liquid milk are basically similar, the layout of the factory may be surprisingly idiosyncratic. There are, of course, general guidelines for maintaining a clean working environment that apply to all dairy operations; but, even so, there may be numerous solutions to the same problem. For example, the "ideal" detergent for one type of food residue may be quite inappropriate for another; similarly, one sterilant formulation may be highly effective against Gram-positive bacteria but leave populations of Gram-negative bacteria largely unscathed.

It is vitally important, therefore, that factory managers seek expert advice with respect to the hygienic design and operation of any given process line and, in addition, are fully aware of the microbiological hazards that may be associated with every facet of the manufacturing facility.

12.2 LIKELY SOURCES OF CONTAMINATION

Just occasionally, food ingredients can be contaminated at the source; the presence of *Camplylobacter* spp. in milk as it leaves the udder

Dairy Microbiology Handbook, Third Edition, Edited by Richard K. Robinson
ISBN 0-471-38596-4

(Keceli and Robinson, 1997) or *Salmonella enteritidis* in eggs (Delves-Broughton and Board, 2000) are cases in point. Alternatively, raw materials can attract an undesirable microflora during collection and/or storage, and the presence of spores of *Bacillus* spp. in milk powders is usually the result of contamination prior to processing. Consequently, an end-user should always purchase ingredients against an agreed specification that is relevant to the application. The same restrictions should be extended to packaging materials as well, because there is little point in maintaining a positive pressure of sterile air over a filling line if the cartons are already contaminated. However, while manufacturers need to be aware of this essential requirement, it is likely that most incidents of product contamination arise within the premises of manufacture and from one or more of the following sources:

- The environment
- Improperly cleaned plant and equipment
- Food handlers
- Spoiled or waste products

Effective control over each of these facets of a food-related operation will be essential to ensure that the microbiological or chemical integrity of the product(s) is not compromised.

12.3 THE ENVIRONMENT

On many occasions, a new development may have to be located in an existing factory; but if the choice is green-field site, it is worth surveying the site in advance for potential microbiological hazards. If the prevailing wind passes over a rubbish dump prior to reaching the factory, airborne pollution will be a major hazard, while the location of the facility on a flood plain could lead to equally serious problems of a different kind. Access by unpaved roads should be avoided as well, because the accumulation of water in winter or swirling dust in summer can make the attainment of hygienic working conditions extremely difficult (IDF, 1994). Annual temperature and rainfall profiles are also worth noting because, although such figures are unlikely to determine the location of a factory, the hygienic design of the interior must take into account the predicted range of relative humidities and temperatures.

12.3.1 Airborne Problems

The air entering any building will contain nonviable dust along with endospores of *Bacillus* spp., various species of non-spore-forming bacteria and yeasts, and a range of mold spores. The balance between the different groups will depend upon the location of the factory, the time of year, and climatic conditions. Yeasts, for example, tend to dominate the air flora in the spring following their release from opening leaf or flower buds, while mold spores reach their highest level after showers of rain. The end result is that bacterial counts in the air circulating within a factory may range from 5.8×10^2 colony-forming units (cfu) m^{-3} to more than 4.0×10^3 cfu m^{-3} with yeasts and molds in the range of 2.0–10.0 $\times 10^2$ cfu m^{-3} (Lück and Gavron, 1990). The main sources of this microflora are air movement into the plant, open drains in the floor, and the passage of personnel.

Little can be done about movement of air into a factory because, apart from the natural opening of doors for ingredient, product, or personnel movements, ventilation is essential to remove heat and moisture released during processing. However, with air filtration, the risk of airborne contamination declines dramatically, and the ducting of incoming air through a primary filter to remove gross contamination (5.0- to 10.0-μm diameter) followed by a filter capable of removing 90–99% of particles above 1.0 μm is essential for areas containing open vats of food. In selected parts of a factory, the use of filters capable of removing 99.9% of all particles in the 0.1- to 0.2-μm range or 99.9% of units down to 0.01 μm may be essential to achieve a desired shelf life. However, the location of air intakes and outlets, the method and frequency of cleaning filter-holders, and the routine for replacing filters are aspects that merit inclusion within the hazard analysis critical control points (HACCP) plan (see Chapter 14 in this volume), and some useful guidelines with respect to recommended standards for air cleanliness in the food industry, contamination control in a processing environment, and maintaining efficiency of air filtration systems have been compiled by Brunderer and Schicht (1987), Anonymous (1988), Schicht (1989, 1991), Ligugnana and Fung (1990), Fitzpatrick (1990), Blümke (1993), Hampson and Kaiser (1995), and Audidier (1996). In addition, it is important to remember that excessive movements by personnel or too high a density of staff can negate attempts to attain clean air, as can the storage of inappropriate packing materials (e.g., cardboard outer sleeves), in a sensitive area (refer to Chapter 14 in this volume for further details).

In general, it has been proposed that good-quality air should have total bacterial count of $<2.0 \times 10^2$ cfu m^{-3} and a yeast and mold count of $<1.0 \times 10^2$ cfu m^{-3} (Lück and Gavron, 1990), and these standards should be attainable over a section of a production line for wrapping consumer portions of butter, for example. If the area in question can be physically isolated, then the use of filtered air to achieve a positive pressure in the room can further reduce the risk of casual microbial contaminants entering from outside.

Additional security can be gained through the installation of ozone generators. If these are allowed to operate in the confines of a production for 5–6 h overnight, then most bacteria (including *Escherichia coli* O157) or mold spores exposed to the ozone (O_3) will be killed (Anonymous, 1999a). Average concentrations of ozone as low as 1.17 mg liter^{-1} of air can achieve 100% mortality in 5–6 h of exposed microorganisms—that is, those not on soiled surfaces and protected by food residues. Enough time must be allowed for the ozone level to decline to below the level recommended for a working environment (0.2 mg liter^{-1} of air) before the next shift starts, and it is advisable to install ozone monitors to ensure that workers are not placed at risk. Two further attractions of this system are that (a) ozone will oxidize many of the organic compounds associated with unpleasant odours, thereby making for a much more pleasant atmosphere, and (b) during its antimicrobial action, each ozone molecule reverts to oxygen, so that no toxic residues are left on any contact surfaces.

The alternative of "fogging" has been explored in the cheese industry—"mists" of hypochlorite in starter culture rooms, for example—but, aside from the fact that chlorine is corrosive to some metal surfaces, lingering residues may make the atmosphere unpleasant for the work force. Occasional "fogging" with more dangerous reagents like formaldehyde or peracetic acid is practiced, but health and safety issues tend to limit their use (Anonymous, 1998). Some practical guidelines to the use of disinfectant "mists" are available from agencies, such as Codex Alimentarius or the Department of Environment, Food and Rural Affairs (DEFRA) in the United Kingdom, but the effectiveness of "fogging" is controversial. Thus, while Hedrick (1975) reported that "chlorine fogs" could reduce the counts of airborne microorganisms in dairy plants, Holah et al. (1995) described "fogging" as an ineffective and uncontrollable system of air disinfection. However, "uncontrollable" may be the key word, because Burfoot et al. (1999) suggest that "fogging" is most effective when the median diameter of the fog droplets ranges from 10 μm to 20 μm; water droplets in this range disperse well and settle within 45 min.

Consequently, it may be more appropriate to consider "fogging" as an additional safeguard rather than as a replacement for other routines for reducing the microbial load in the atmosphere.

12.3.2 Water Supplies

Water is required in the dairy industry for incorporation into a product—for example, for the recombination of powder, for process operations like heating and cooling, and for the cleaning of plant, equipment, and fabric of the factory. Consequently, the overall water requirements of a dairy plant are large and, while not all supplies need to be of potable quality or better, water conservation and management have become major issues. Obviously, the volume of water required for direct incorporation or process applications will vary from product to product, but it is of note that factories in different countries have different demands, even when manufacturing the same product. According to Hiddink (1995), the water-to-milk ratios for liquid milk and/or dessert processing ranged between 0.5 and 6.71 kg^{-1} of milk in Belgium, Germany, Hungary, Ireland, and United Kingdom.

In most industrialized countries, domestic water supplies are safe to drink, and only occasionally do blooms of toxic algae or the development of protozoa (e.g., *Cryptosporidium* spp.) in reservoirs cause problems. One of the reasons for this general level of confidence is that pathogens like *E. coli*, *Listeria monocytogenes*, or *Salmonella* spp. cannot grow in water, and, in any event, such organisms are sensitive to the levels of chlorine found in drinking water. Nevertheless, inability to grow in water does *not* imply an inability to survive, and hence it is important that water used for rinsing items of equipment or the curd and/or flavoring components of cottage cheese, (e.g., chives or cucumber) should be of a higher quality than mains water; additionally, any vats of water used for washing purposes should be changed frequently. The importance of this latter point was highlighted by Gohil et al. (1995), who noted that, while fresh salad vegetables purchased direct from a market were free from *L. monocytogenes*, the proposed standard for *L. monocytogenes* in washed, prepacked salad vegetables is $<1.0 \times 10^2$ cfu g^{-1} (Lund, 1993). This latter figure suggests that there is a real risk of contamination during the washing/chopping/mixing of fresh vegetables, and the role of polluted water could be important. If the vegetables are to be blended with a dressing of pH 4.3 or below, then spoilage problems from airborne yeasts will the principal concern, but otherwise the quality of any rinsing/washing water must be carefully monitored.

Even in the absence of any risk from pathogens, the level of chlorination in potable water is usually ineffective against spoilage groups like *Pseudomonas* spp., which, under aerobic conditions, can cause taints or other defects in any dairy products of neutral pH; this risk even applies to foods stored at 3–4°C. Given this latter dimension, WHO (1993) and Hiddink (1995) suggested that water likely to come into direct contact with components of a meal should have a psychrotrophic count of $<1.0 \times 10^2$ cfu ml^{-1} and that a count of $>1.0 \times 10^3$ cfu ml^{-1} should be rated as "unsatisfactory"; the proposed standards for other bacterial groups are give in Table 12.1. Consequently, the introduction of a plant to chlorinate water supplies to >20 mg liter^{-1} and monitor that the level is maintained throughout a normal shift can prove a worthwhile investment.

Treatment of process or washing water with short-wave (254 nm) ultraviolet light could be considered as an alternative to chlorination (WHO, 1994; Sharma, 2000; Bintsis et al., 2000). The advantage of UV disinfection is that it leaves no taints or taint-generating residues in the water, but the disadvantage is that the system has to be carefully engineered to be successful. In particular, the depth of water to be treated has to be restricted, the flow through the region under the UV source has to be turbulent, and the water must be adequately filtered before exposure to the germicidal waves. This filtration stage is essential to remove not only particulates that might protect bacteria from the UV light, but also "clumps" of bacteria; in the latter case, bacteria on the outside of the "clump" will absorb the UV and leave the centrally located cells intact.

12.3.3 Animal Vectors

Infestation implies the presence of storage pests other than microorganisms, and hence that problems may be associated with rodents, birds, or insects. An outline of their potential activities is shown in Table 12.2, but some specific features of each group are worthy of mention.

12.3.3.1 Rodents. In some locations, squirrels can be a problem, but contamination of foodstuffs is more usually linked with the activities of rats and mice—principally the Norway rat, the roof rat, and the house mouse. These three animals enjoy worldwide distribution, and their omnivorous diets render most food stores attractive targets. Any unguarded gap in masonry (including drains) or woodwork can provide access and, while the roof rat enjoys the best reputation for

TABLE 12.1. Detailed Specifications of Drinking and Process Waters

Parameter	European Union GL[a]	European Union MAC[a]	WHO
Physicochemical			
pH	6.5–8.5	—[b]	<8.0
Conductivity (μSm^{-1})	400	—	—
Chloride (mg liter^{-1})	25	—	<250
Sulfate (mg liter^{-1})	25	250	<250
Calcium (mg liter^{-1})	100	—	—
Magnesium (mg liter^{-1})	30	50	—
Sodium (mg liter^{-1})	20	150	<200
Potassium (mg liter^{-1})	10	20	—
Aluminum (mg liter^{-1})	0.05	0.2	<0.2
Undesirable substances			
Nitrate (mg liter^{-1})	25	50	<50
Nitrite (mg liter^{-1})	—	0.1	<3
Ammonium (mg liter^{-1})	0.05	0.5	<1.5
Nitrogen (mg liter^{-1})	1	—	—
Oxygen (mg liter^{-1})	2	5	—
Mineral oil (μg liter^{-1})	—	10	—
Iron (μg liter^{-1})	50	200	<300
Manganese (μg liter^{-1})	20	50	<100
Copper (μg liter^{-1})	100	3000	<1000
Zinc (μg liter^{-1})	100	5000	<3000
Phosphorus (μg liter^{-1})	400	5000	—
Toxic substances (μg liter^{-1})			
Arsenic		50	<10
Cadmium		5	<3
Cyanides		50	<70
Chromium		50	<50
Mercury		1	<1
Nickel		50	<20
Lead		50	<50
Antimony		10	<5
Selenium		10	<10
Pesticides		0.5[c]	0.03[d]
PACS		0.2	<0.1
Microorganism			
Total viable count (22°C) (cfu ml^{-1})		<100	—
Total viable count (37°C) (cfu ml^{-1})		<10	—
Coliform (cfu 100 ml^{-1})		<1	0
Fecal coliforms (cfu 100 ml^{-1})		<1	0
Fecal streptococci (cfu 100 ml^{-1})		<1	
Sulfite-reducing *Clostridia* spp. (cfu 20 ml^{-1})		<1	

[a]GL, guide level; MAC, maximum admissible concentration.
[b]Data not reported.
[c]Total level of all pesticides.
[d]Per individual substance.
Source: After Hiddink (1995).

TABLE 12.2. Some Examples of Problems Associated with Animal Vectors

Group	Result of Attack
Rodents	Loss by consumption and partial eating of products.
	Contamination with droppings, hair; health hazard from bacteria.
	Indirect losses from damaged containers.
	Possible blockage of pipework, and considerable contamination if carcass accidently macerated during product handling.
Birds	Consumption and contamination by droppings, feathers.
Insects	Limited losses, but high risk of bacterial contamination.

climbing, the agility of most rodents makes entry at roof level a feasible proposition.

While the food consumption per animal at any one point in time/location is not high, large volumes of food can become contaminated with saliva, hairs, urine, or feces. As all such contaminated food has to be discarded, the prevention of infestation is essential, and this approach will involve the following:

- Careful observation of possible entry points. Rats, in particular, are creatures of habit, and they soon leave black greasemarks on favored runways. In addition, rodent urine and hairs fluoresce brightly in UVA, so that occasional inspections with "black" lamps will help to confirm that rats have not found an entry point (Anonymous, 1999b);
- Rodent-proofing of vulnerable points with, depending upon the location, sheet metal or wire mesh.

Control through the use of rodenticides is always possible, but prevention is far better than cure!

12.3.3.2 Birds. Birds are a potentially serious nuisance because their droppings may contain salmonellae or other pathogens. However, it is usually easy to deny access to any birds that stray into the vicinity, and even potential perches on the outside of buildings can be coated with materials to deter habitual use.

12.3.3.3 Insects. Insects must be deterred from entering food premises at all times because some (e.g., the ubiquitous house fly) will feed on animal feces and human food alike, and the potential for disease transfer is self-evident. The use of aerosol insecticides on dairy premises tends to be viewed with some concern, partly because of the potential toxicity to humans if a foodstuff becomes contaminated and partly because a dying insect could easily drop into an open vat of milk without being noticed—except perhaps by the eventual consumer. For this reason, it is better to employ fine gauze screens over store-room windows, along with electrified grids backed by ultraviolet fluorescent tubes in food processing areas; the latter are excellent for trapping all flying insects (WHO, 1994). The control of cockroaches, on the other hand, does require the skilled application of insecticides. Thus while, in most cases, cockroaches do not pose a serious health risk, they have been reported to carry pathogenic bacteria, and hence population numbers must be kept under strict control.

12.3.4 Structural Features

One of the most obvious means of avoiding aerial contamination involves the physical separation of specific processes (e.g., dry mixing from wet mixing operations) or the separation of storage facilities for milk powder—for example, away from stores holding product cartons or wrapping foils. However, achieving the optimum degree of compartmentalization may not be easy, because old buildings are often not amenable to conversion, and the allocation of large floor areas for storage may not prove popular with the accounts department. Nevertheless, it is an aspect of any operation that should be evaluated on a regular basis, and with particular diligence if a new product line is being introduced that is potentially susceptible to spoilage. For example, a draught originating in a packaging store may have to be deflected away from a newly sensitive area by the introduction of heavy-duty, polythene strip curtains, and such foresight can prove well worth the effort (Brolchain, 1993).

The same diligent appraisal should be given to monitoring the flow of material(s) through a factory, and the general principles are as follows:

- Ingredients should enter at one end of a building and be held in a designated stores or silos.
- The processing and cooling stages should follow.
- The flow of end product should then meet with a flow of packaging materials coming from a separate store.
- The packaged food should then enter a finished product store for bulk packaging and dispatch.

This logical approach, together with the isolation of plant services and office facilities, can help to prevent casual incidents of airborne contamination (Brolchain, 1993), the more so if none of the sensitive areas are allowed to function as passageways for personnel. In addition, all internal doors should be self-closing and, if in frequent use, the separation should be enhanced with strip curtains or, if appropriate, an efficient air curtain; only emergency exits should lead from a processing area to the outside.

The flow of materials should also take account of the concept of "clean" and "unclean" areas. For example, the reception area for raw milk will, by the very nature of the operation, be classed as microbiologically "unclean," and it is vital that adventitious bacteria and fungal spores arriving on milk tankers should be denied access to a "clean" room holding exposed cartons of retail product—that is, food that will not be subject to a further process that will destroy microorganisms. Similarly, the presence of bacteria in raw milk is not important because it will be subject to pasteurization or sterilization, but, ahead of the heat-treatment stage, raw milk must not be allowed to contaminate any surfaces that will come into contact with the retail product(s).

12.3.4.1 Floors. The other consistent source of contamination will be the floor, which should be constructed of high-quality concrete with a "topping" that offers a range of properties. For example, hygiene will be a major consideration, but, because many food production areas are wet, a high degree of slip resistance is important to ensure operator safety. In addition, the surface must be hard-wearing: "Scuffing" can be important where pieces of equipment have to be moved on a regular basis, be resistant to cleaning chemicals, and, if steam cleaning is employed, be able to withstand elevated temperatures. This tolerance of thermal shock must apply equally to the bond with the underlying concrete, because lifted finishes can soon crack and leave cavities for detritus/microbial growth to buildup.

Resin flooring provides a seamless covering that is produced by a chemical reaction between liquid components. Because one or more of the components may be volatile, it is essential to check that the floor, once hardened, does not give off any volatile compounds that might give rise to taints. Epoxy resin floors exhibit excellent durability and resistance to chemicals, but they are intolerant of high temperatures and the floor may retain a residue of volatile materials long after application. Consequently, polyurethanes have become the flooring material of choice for the food industry, and they can be laid as thin seals intended merely to prevent through to heavy duty finishes up to 9 mm thick (Cattel, 1988; Weatherburn, 1997; Jackson, 1997;

Anonymous, 1999c). Further advantages of polyurethane finishes are that the floor will be resistant to chemicals, such as the acidic or alkaline detergents and sanitizing agents that may be used to clean plant or equipment (Jolly, 1993), and they are tolerant of the extremes of temperature associated with hot water/steam cleaning. In this context, "sanitizing" implies a reduction in microbial loading on a surface—as opposed to sterilization, which should make surfaces free from all contaminants.

Given this inherent compatibility of resin floors with cleaning regimes, finishes containing bacteriocidal agents may be best avoided because (a) their use may engender a false sense of security (i.e., traditional cleaning regimes may be scaled-down), (b) it will only be a matter of time before resistant strains of bacteria arise within a factory, and (c) if one of the first species gaining resistance happens to be a pathogen, then the absence of competition could enable the pathogen to pose a serious risk (Anonymous, 1999c). Obviously this latter risk may be exaggerated, but it remains an aspect of floor construction that merits serious consideration.

Quarry tiles are impervious to water and chemicals; and if embedded in, and grouted with, a similarly resistant mixture, they can provide a long-lasting alternative to resin floors.

Both forms of flooring have advantages and disadvantages. For example, tiles are extremely durable, but, if the grouting between the tiles becomes loose, the resulting cavities will provide a serious reservoir of potential contamination. Synthetic finishes can crack or peel, particularly if laid by inexperienced contractors; and if carelessly drilled after laying, water can slowly migrate along the concrete/resin interface until the entire floor begins to lift. However, as knowledge of synthetic finishes improves, so should their durability, and there is little doubt that a joint-free surface has much to recommend it.

By their very nature, certain areas of floor will be designated as walkways, while other areas will be covered with tables for handling ingredients, or with mixers or other items of equipment. These divisions mean that a definite strategy for cleaning will have to laid down, and some considerations may include the following:

- Mechanical cleaning equipment of the correct capacity for the floor area should be available (Anonymous, 1999d).
- Detergents and sanitizing agents must be selected that are appropriate for the type of soil that may be dropped onto the floor and, perhaps, the types of microorganisms that may colonize any food residues (Anonymous, 1999e).

- Tables on castors/clamps are useful to allow easy floor cleaning, but high-pressure jets may be necessary to scour corners or crevices that are inaccessible for cleaning machines.
- A timetable for cleaning must be agreed upon, along with procedures to cope with unexpected spillages. In other words, while spilled food must not be left on a floor to "ferment" until the end of a shift, the use of a filthy mop to clean the area may create more problems than it solves.

Because all cleaning/rinsing procedures will involve copious supplies of water, a fall of least 1:60 toward suitable floor drains is essential (Brolchain, 1993). The design of the actual drain merits careful thought as well, because not only must it be accessible for manual cleaning and disinfection, but also access to the outside must not provide an entry point for insects or rodents. Although they can handle large quantities of water, trench drains can be difficult to clean and, in general, hub drains can prove much easier to sanitize.

12.3.4.2 Walls and Ceilings. As with floors, walls must free from crevices that could harbor pockets of moist food and live colonies of bacteria/fungi, and the surfaces must be resistant to hot water and chemical detergents/sanitizing agents; in some situations, insulation may be desirable to prevent surface condensation. The junctions between floors and walls must have smooth concave surfaces to aid cleaning, and the same consideration must be given to the links between ceilings and walls; any contaminants lodged in these latter joints could fall directly onto food below. In areas where high-risk foods are being prepared (i.e., foods of neutral pH which may be eaten cold or after mild warming), such overhead sources of contamination would constitute a severe hazard. The same risk could arise from poorly grouted ceramic wall tiles, and the option of a seamless surface employing a polyurethane-based coating could prove a safer choice. As with floors, conventional cleaning is straightforward; and if the basic resin does not distort at high temperature, steam cleaning can be employed on a occasional basis to ensure high standards of hygiene. Furthermore, the introduction of color no longer presents any obstacle to contractors, so that walls can both be hygienic and provide an attractive working environment.

Horizontal ledges like shelves or window sills should be avoided; and if a ledge is unavoidable, a downward sloping surface of at least 45° should avoid the accumulation of dust and other debris (Brolchain,

1993; Timperley, 1993, 1994; IDF, 1997a). To avoid the problem of droplets of condensation carrying microorganisms into food, windows should be double-glazed, easily accessible for cleaning, and, in processing and packaging areas, made of clear polyvinylchloride; broken glass is difficult to detect in a retail food item, and it provides a common source of consumer complaints.

The general attitude to ceilings is often "out-of-sight/out-of-mind"; and yet every aspect of ceiling design from light fittings to ventilator grills must be given a high priority with respect to hygiene, especially because the ceiling is often the least accessible area for cleaning. Internal girder work should be avoided; and if panels are used to provide thermal insulation, the joints must be sealed with a bonding material that does not crack or loosen during routine cleaning.

12.4 PLANT AND EQUIPMENT

The types of factory that handle milk and milk products range from simple labor-intensive units through to highly automated plants relying on sophisticated unit operations under computer control. Consequently, a wide range of processing equipment including heat exchangers, vats, mixers, and containers may be needed; but whatever the configuration, every item must be accessible for cleaning both inside and outside. For example, product contact surfaces should be stainless steel, and the cleaning/sanitizing regime must allow thorough disinfection of any surface irregularities. Equally important is the area behind or beneath an item of equipment, because if dust or food is allowed to build up in such crevices, then occasional incidents of spoilage are sure to follow and, being sporadic in nature, the source(s) of such complaints will be a nightmare to identify (Hayes, 1985).

For small-scale operations, two interrelated pressures can cause problems. Firstly, the investment necessary to "gut" a building and install new plant and equipment may not be available; and secondly, a particular process may be little more than a industrial version of a traditional or village-scale procedure. If the end result is a hybrid system—part industrial and part traditional—then the operators may not realize the risks involved in scaling-up a process in order to supply a more extensive market. The use of a wooden chopping board to dice a low-volume, high-cost ingredient for a speciality cheese could be case in point, because expansion to an industrial-scale operation necessitates a shift to an impervious, heat-resistant, synthetic surface that will not allow the penetration of food or bacteria.

In large plants, pipework must be installed with neither dead ends from which food is never removed during cleaning, nor unsupported runs of pipe that "sag" in the center and provide "pools" of dilute food for bacteria or yeasts (Holm, 1980). If the latter situation cannot be improved, then filling the cleaned and disinfected plant with water may be the only option to prevent the buildup of microbial contamination overnight. This latter approach can prove effective if followed by sanitization/rinsing immediately prior to start-up next morning, and the routine also denies insects or rodents entrance to the pipework.

However, whatever the scale of the operation, it is important to remember that similar plants will be in use elsewhere in the world, and that organizations in many countries publish documents covering equipment standards with respect to hygiene. For example, the International Association of Milk, Food and Environmental Sanitarians (3-A Sanitary Standards, USA), the European Committee for Standardisation (CEN/TC—Belgium), the European Hygienic Design Group (Belgium), and the International Standards Organisation (ISO/TC 199/WG2) can all provide advice on the hygienic design of process plant. In addition, the International Dairy Federation issues regular monographs on the same topic, and some relevant publications are cited in Table 12.3 (see also Clark, 1993; Tuthill et al., 1997; Holah, 1998).

12.4.1 Cleaning of Process Plant

Irrespective of the capacity of the plant, cleaning to remove food residues on both the inside and outside of each component must have

TABLE 12.3. A List of International Dairy Federation Publications Dealing with Equipment Specification, Cleaning, and Hygiene

Topic	Reference
Equipment	
Corrosion of metals	IDF (1980a, 1981, 1983, 1988)
Milk tankers	IDF (1980b)
Surface finish of stainless steel	IDF (1985)
Hygienic design	IDF (1987, 1996)
Hygienic code of practice	
Dairy powders	IDF (1984, 1991)
Factories	IDF (1992)
Milk and dairy products	IDF (1994, 1995a, 1997b)
Equipment	IDF (1995b, 1997c)

a high priority. Final decisions relating to the overall cleaning regime will have to consider the following:

- Identification of any points in the layout of the plant that may be especially prone to the buildup of food residues.
- Whether these component(s) need to be dismantled for manual washing or whether they can be cleaned *in situ*; hand washing means using lower water temperatures, but the scrubbing and visual inspections involved may leave the surfaces really clean.
- The nature of any residues. Each major ingredient of a food reacts differently to cleaning agents. For example, sugars are easily soluble in warm water, but they caramelize in hot surfaces and become difficult to remove, whereas fats, although dispersible in warm water, are liberated more readily in the presence of surface-active agents.

The selection of the best detergent for the task(s) should be completed with expert advice because, in addition to the variables mentioned above, factors like available contact-time and hardness of local water can alter the expected efficiency of a selected detergent (Anonymous, 1999e).

In general, cleaning of static items of equipment will involve three stages:

- Removal of gross soil through mechanical action in some form—for example, hand washing with brushes or the use of high-velocity jets of water/circulating water, and the action of an appropriate detergent.
- Removal of fine residues of food that may still be adhering to a contact surface or may have been deposited during drainage of the first wash.
- Rinsing to remove detergent residues that might hinder the action of any chemical sanitizing agents.

The options for cleaning are summarized in Table 12.4, and cleaning-in-place (CIP) is the preferred option for large items of plant with the pipework necessary to link into a CIP circuit. The advantages of CIP include (a) the use of the highest temperature compatible with the chemical composition of the soil and (b) the scouring action of cleaning fluid which will pass through pipework at around 1–2 $m\,s^{-1}$. The use of rotating jets or spray-balls enables enclosed vessels to be thoroughly cleaned as well (Romney, 1990; Tamime and Robinson, 1999).

TABLE 12.4. Some of the Options for Cleaning Stainless Steel or Other Hard Surfaces in a Food Factory[a]

	Hoses			
	High-Pressure	Low-Pressure	CIP[b]	Manual
Type of soil				
Tenacious (coagulated protein, carmelized sugar)	++	+	++	++
Water-soluble	++	++	++	++
Amount of soil				
High	++	+	++	++
Low	++	++	++	++
Open vessels				
Easy access	++	++	–	++
Difficult access	+	–	–	–
Tables and benches				
Horizontal surfaces	++	+	–	++
Vertical surfaces	++	–	–	++
Enclosed processing equipment				
Pipework and tanks	++	–	++	–

[a]The application of detergent foams can be useful for cleaning inaccessible surfaces.
[b]CIP, cleaning-in-place; (++/+), suitable if managed effectively; (–), inappropriate.
Source: After Harrigan and Park (1991).

The slight "downside" to the installation of a CIP system is the cost of (a) the centrifugal pumps needed to provide the appropriate fluid velocity, (b) the tank for holding and heating the detergent solution, (c) the valves that isolate the product and CIP circuits, and (d) some level of automatic control. The extent of automation can, of course, be adjusted to suit the complexity of the plant and the level of investment available, but it is essential that the operatives have a visual display showing which tanks/pipework are full of product and which are being cleaned. This degree of control also makes it possible to incorporate flexible cleaning schedules into a normal shift pattern, because the familiar Gantt charts enable production and cleaning schedules to be manipulated to meet the varying pressures that may be placed on the manufacturing facilities. Fluctuations in demand for a particular product line, for example, or the need to meet a "just in time" delivery date can be accommodated much more easily if cleaning of critical sections of the plant is under CIP/computer control (Tamime and Robinson, 1999).

It is evident also from Table 12.4 that not all vessels and surfaces are suitable for CIP treatment; in these situations, manual cleaning with a suitable detergent followed by rinsing with a high-pressure hose can prove just as effective, a point that highlights the necessity of ensuring that drains are correctly sited to coincide with the slope of the floor.

12.4.2 Sanitization of Process Plant

The same care must be exercised with respect to the selection and use of chemical sterilants, and the factors that merit consideration are discussed in the following subsections.

12.4.2.1 Concentration of Detergent/Sterilant Solution. Ensuring that the concentration of the selected chemical compound in the sterilising solution is correct at the start of a cleaning operation, and that the strength does not decline with time. Three types of chemical agent tend to be used most widely, and Table 12.5 highlights some of their advantages and disadvantages as contrasted with steam. The fact that there is no "perfect" sterilant for all operations makes it imperative that the choice of compound and its optimum working concentration be discussed with a reputable supplier during the design of the sanitizing system.

If a CIP configuration is present, then probes will be fitted to measure the concentration of the sterilant automatically, and any decline in concentration will "trigger" a dosing mechanism to rectify the shortfall. For manual applications, the sterilant will have to be measured out and added by an appointed operative, and samples of the active solution will have to be checked in the laboratory at prescribed intervals. Experience in cleaning a specific section of plant or piece of equipment will be necessary to determine the optimum frequency for checking the strength of a sanitizing solution, and the decision will have to be included in the appropriate Process Manual along with details of the test procedure(s).

In general, halogen-based sanitizing agents are the most effective; and while iodophores have the advantages of causing little corrosion and leaving visible stains in a poorly cleaned plant, chlorine-based agents are usually more microbiocidal (Romney, 1990). Sodium or calcium hypochlorites are the most popular sources of chlorine, and a concentration of 200–250 mg liter^{-1} of solution is a feasible working target.

TABLE 12.5. A Comparison of the Activities of Some Commonly Used Sanitizing Agents

Properties	Steam	Chlorine	Iodophors	QAC[a]
Effect against:				
Gram-positive bacteria (e.g., starter cultures, clostridia, *Bacillus* spp., *Staphylococcus* spp.)	Best	Good	Good	Good
Gram-negative bacteria (e.g., *E. coli*, *Salmonella* spp., psychrotrophs)	Best	Good	Good	Poor
Spores	Good	Good	Fair	Poor
Yeasts	Best	Good	Fair	Poor
Effective at neutral pH	Yes	Yes	No	Yes
Corrosive to metals	No	Yes	Slightly	No
Effect of hard water	None	None	Some	Variable
Incompatible	Heat-sensitive materials	Phenols, amines, soft metals	Starch	Anionic wetting agents, soaps, wood, cloth, cellulose, nylon
Stability of working solution	—	Dissipates rapidly	Dissipates slowly	Stable
Stability in hot water (>65°C)[b]	—	Unstable	Unstable	Stable
Leaves active residue	No	No	Yes	Yes
Tests for residual chemical	None	Simple	Simple	Difficult
Cost	Expensive	Cheapest	Cheap	Expensive

[a]QAC, quaternary ammonium compounds.
[b]Some halogen compounds are stable.
Source: After Harrigan and Park (1991).

12.4.2.2 Contact Time. The contact time between the solution and the surface of the equipment is also critical, as is the temperature of the solution. In practice, a contact time of 10 min at 40°C or 15–20 min at ambient temperature should be regarded as the minimum conditions to achieve total counts of 2.0×10^{-2} to $<1.0 \times 10^{3}$ cfu 100 cm^{-2} using hypochlorites. For surfaces in contact with pasteurized retail items that will not be reheated, the appropriate counts should be <50 cfu 100 cm^{-2}.

12.4.2.3 Cleaning Cycle. If the cleaning/rinsing stage is not completed efficiently, presence of residual food on the surface may protect bacteria from the action of the biocidal agent, or there may be an interaction with residual detergent (Tamime and Robinson, 1999).

This latter point may be especially relevant to manual operations because, on some production lines, a "clean as you go" policy means that ingredient containers and associated utensils may be removed at regular intervals for washing and sanitizing. This policy is clearly advantageous in removing soiled items to an area away from the processing operation, particularly because the adhering food could allow the growth of bacteria or fungi. However, it is important that the washing facilities for such items are physically separated from the main process room, and that the sanitizing regime produces surfaces with microbial counts at the levels suggested earlier.

For a persistent problem of contamination from plant or equipment, heat is often the best option. One traditional option is the circulation of water at 80–85°C for 30 min; and for complex pieces of plant-like plate coolers, this approach can be most effective in eliminating microorganisms from crevices where chlorine has failed to penetrate. Similarly, oven-drying can be used to sterilize small components with difficult surfaces, and steam under pressure can achieve the same end-point—that is, <1 cfu 100 cm^{-2} with large items of plant. However, high-pressure steam is a difficult and dangerous medium to handle; and unless the food itself is commercally sterile (e.g., UHT treated), surfaces that have been sanitized should be quite adequate.

12.4.3 Hygiene Monitoring

Targets for hygiene are of little value unless they can be monitored, and while there are microbiological methods available to monitor the total counts on items of plant, most methods require the establishment of a laboratory designed to handle live cultures. For large factories, the investment in laboratory facilities and personnel can be justified, particularly because the workload can be readily expanded to check end-product specifications as well. Small- or medium-sized operations may, by contrast, find it difficult to justify the expense of an on-site laboratory, and yet routine (i.e., at least once a week) checks on cleaning efficiency are essential.

Thus, every sanitized item of plant or equipment will retain a "background" count that should vary only within well-defined boundaries and where the actual mean count (cfu 100 cm^{-2}) will depend on (a) the use to which the item is put, (b) the nature of the surface, and (c) the method of cleaning/sanitizing. As mentioned above, the mean background count designated as *acceptable* can be adjusted in light of the food in contact with the surface (e.g., raw material or pasteurized end-product), and it should be anticipated that this figure will be broadly stable. Hence it is "wild" fluctuations outside the preset boundaries or a general upward trend in counts that are causes for concern, and such changes can only be identified by routine observations over an extended period of time (Jarvis, 2000).

12.4.4 Procedures for Hygiene Assessment

The value of visual appearance, smell, and even touch in determining cleanliness should not be ignored; but in recent years a cheap, convenient, and objective method of monitoring the hygienic state of food contact surfaces has become available, namely, the ATP Bioluminescence System. It is based upon the fact that all biological materials (i.e., food residues or microorganisms) contain a chemical compound—adenosine triphosphate (ATP)—that is an essential component of all metabolic reactions where the transfer of energy is involved. Consequently, on any food contact surface, there will be a "background" level of ATP deposited by food residues and/or microorganisms (Anonymous, 1999f). If this level of ATP rises above a predetermined "norm," then it can be assumed that the surface has been inadequately cleaned and that microbial counts may have increased as well (see Chapter 14 in this volume).

While the ATP Bioluminescence System is an excellent guide to general hygiene, it reveals nothing about the presence of pathogens, and some manufacturers may feel that this "absence of information" should be remedied. Obviously, many supplier/buyer contracts stipulate that the supplier must routinely send random samples of end-product for microbiological analysis by an accredited laboratory; and while this procedure is a useful safeguard for consumers, routine monitoring does increase producer/buyer confidence. However, as mentioned earlier, the size of many companies does not justify the establishment of extensive laboratory facilities, but this limitation does not mean that the employment of "low-technology" procedures does not merit consideration.

The examination of plant or equipment for specific pathogens is rarely justified, but there is no reason why routine checks for so-called *indicator organisms* should not be introduced. One group of bacteria that is useful in this context are the "coliforms," which are easily detected (see Chapter 14 in this volume); if detected on a piece of plant or equipment, their presence suggests that (a) standards of plant sanitation need to be raised and (b) there is the possibility that serious human pathogens could be present as well.

The most common procedure is to rub a food-contact surface with a sterile swab and then disperse any bacteria adhering to the swab into 10ml of a physiologiclly inert diluent (0.1% peptone water or maximum recovery diluent) (Brisdon, 1998). One-milliliter aliquots of the contaminated diluent are then dispensed into three tubes (with inverted Durham tubes inside) of MacConkey broth (Brisdon, 1998) which are then incubated at 37°C for 72h. If an estimate of numbers of coliforms is required, then two or more 10-fold dilutions of the original diluent can be made, and three tubes of MacConkey broth can be inoculated from each dilution.

If the indicator dye in the MacConkey broth changes color to confirm that acid has been produced and gas is present in the Durham tube, then the tube(s) is positive for coliforms. An estimate of the number of coliforms ml^{-1} of the original diluent can be obtained by counting the number of positive tubes at each of three dilutions, and calculating the most probable number (MPN) of coliforms from a published "table." If the presence of this group, which includes inhabitants of the human intestinal tract like the genus *Escherichia*, is confirmed on a plant surface, it could, *perhaps*, indicate contamination with *Salmonella* or another intestinal pathogen (Hobbs and Roberts, 1995). A wide variety of other techniques can be used for the identification of coliforms (Patel and Williams, 1994) or *E. coli*, in particular, but the sheer simplicity of inoculating tubes of MacConkey broth has much to recommend it for routine testing.

The same technique can be used also to check for the presence of *Listeria*, because *L. monocytogenes* is one pathogen that can survive in soft cheeses like Brie and can grow at refrigeration temperatures, and it is relevant also that the infective dose for *L. monocytogenes* may be as low as 1000 cells (Prentice and Neaves, 1993; Martin and Fisher, 2000). To test for *Listeria*, one approach is to replace the tubes of MacConkey broth with UVM I broth (Brisdon, 1998) and then incubate them at 35°C for 24h. Aliquots from each tube (0.1ml) are then transferred to tubes of Fraser broth (Brisdon, 1998), and any tubes of Fraser broth that turn black after 24h at 35°C are presumptive positive for *Listeria*. There is a small risk of "false" negatives with this technique (<2.0%), but the procedure is quicker and less technically demanding than most classic methods (Gohil et al., 1996).

Identification of suspected *Listeria* to species level will involve help from an accredited laboratory because, while *Listeria* spp. are widespread in occurrence, it is important to establish that *L. monocytogenes* is not among those being isolated.

12.5 THE HUMAN ELEMENT

If contamination from the environment is under control and the food contact surfaces are clean, then plant operatives remain the final avenue for the possible transfer of spoilage or pathogenic organizms to food.

Medical advice should be sought concerning such matters as screening for possible symptomless carriers of *Salmonella* or the time that should be taken off work following an intestinal infection, but Harrigan and Park (1991) concluded that routine medical examinations should have low priority. Nevertheless, as with so many aspects of hygiene, a sympathetic and constructive collaboration between management and employees will bring considerable benefits.

12.5.1 Training

Assuming that the medical aspects can be agreed, the next line of defense against the spread of a food-borne disease and/or an incident of spoilage must involve training. Initially, this training should cover general food hygiene, and certificate courses specifically designed for the dairy industry are provided by most environmental health agencies. Because these one-day courses are intended to provide only a basic knowledge of food microbiology, the training element should be extended in-house through the use of hygiene videos or special lectures. Thus, every food handler should appreciate the following:

- The factors that control microbial growth and metabolic activity of specific groups of microorganism (see Table 12.6).
- The essential difference between growth and survival. For example, the water activity (A_w) of Grade A milk powder (moisture content 3–4 g $100\,g^{-1}$) will be below 0.6, and hence *Salmonella* spp. will not grow in milk powder, but cells may well survive and be capable of growth once the powder is reconstituted.
- In particular, how the component(s) for which he/she is responsible *should* be handled, and the risk(s) associated with noncompliance; the HACCP schedule will detail the extent to which any deviations from "best practice" can be tolerated.

12.5.2 Protection of the Food

If food is to be protected from contamination by factory personnel, it is essential that protective clothing must, in the processing or

TABLE 12.6. The Major Factors that Control the Growth of Microorganisms in Foodstuffs or on Process Plant or Other Surfaces

Factors	Comments
Substrates	Lactose is the dominant sugar in dairy products, and in liquid milk the microflora is dominated by lactose-positive genera like *Lactobacillus*, *Lactococcus*, and *Bacillus* at ambient temperatures or *Pseudomonas* at <5°C. Vitamins, mineral salts, and amino acids are rarely limiting, and proteolytic species will often allow nonproteolytic species to grow.
Moisture	An available water/water activity (A_w) value below 0.60–0.7 (e.g., skimmed milk powder) prevents the growth of all microorganisms; most mycelial fungi need an $A_w > 0.80$ to grow (e.g., sweetened condensed milk), yeasts > 0.88 (e.g., fruit yogurt) and bacteria > 0.91 (e.g., liquid milk, soft cheeses); the A_w of pure water is 1.0. The alarm water content (AWC) of a product—defined as the total moisture content equivalent to an A_w of 0.7—should be noted as well, because the total water content is easier to measure. For skimmed milk powder, the AWC is ~15 g 100 g^{-1} moisture and ~8.0 g 100 g^{-1} for whole milk powder.
Temperature	Thermophiles grow best at 55–65°C (e.g., *Bacillus* spp.); mesophiles grow best at 30–37°C (e.g., most spoilage bacteria and yeasts, as well as pathogens like *E. coli* O157 or *Salmonella*) and cease growing at 5–7°C—most mycelial fungi are mesophiles, but have optimum growth temperatures of 22–25°C; psychrotropic organisms are mesophilic species that are also able to grow at 0–5°C (e.g., *Pseudomonas* spp., *L. monocytogenes*, and some strains of *B. cereus*).
Acidity	Most spoilage and pathogenic bacteria are best adapted to pH 5.5–6.5 and cannot grow below pH 4.3—*Lactobacillus* spp. are important exceptions; the endospores of *Bacillus* spp. or *Clostridium* spp. cannot germinate either. Many yeasts and molds can tolerate ~pH 3.0.
Atmosphere	Many mycelial fungi will only grow in the presence of oxygen, but, while yeasts often grow best aerobically, most species will actively metabolize in conditions of oxygen deprivation. A few bacteria are obligate anaerobes like *Clostridium* spp., but most are aerophiles (*B. cereus*) or microaerophilic—that is, grow best with reduced levels of oxygen (dairy starter cultures).

packaging area, be designed to protect the food rather than the operative.

12.5.2.1 Hair. Hair is a potential problem not only as a source of loose strands falling into food, but also as a haven for pathogenic bacteria—for example, coagulase-positive strains of *Staphylococcus* *aureus*, or the spores of molds associated with spoilage. Mob or snood caps are the best solution for operatives handling the components of a meal (Anonymous, 1999g), because total enclosure of the hair is the only sensible option; hair nets may prevent stray hairs from falling into a vat of milk or carton of yourt, but they cannot retain bacteria or mould spores. Face masks are essential for men with beards or moustaches, and they may be desirable also for anyone with a tendency to sneeze or cough (e.g., sufferers from hay fever).

12.5.2.2 Hands and Washing Facilities. Hands are the most obvious source of pathogenic bacteria, and lightweight disposable gloves have become the popular option for people handling any ingredient that will not be heated thoroughly before consumption. There are disadvantages in that (a) gloves may become uncomfortable during prolonged wear, and (b) wearers may forget that gloves can become contaminated just as easily as hands.

It is important also that this trend toward the use of gloves does not mean that hand-washing facilities can be neglected, because the outside surfaces of gloves may well be touched during retrieval from a dispenser. For this reason, toilet facilities should be maintained to a very high standard of cleanliness, and washing procedures should ensure that the residual bacterial flora on the skin is minimal; that is, taps should be knee/foot operated, the soap should have bacteriocidal properties, the use of a final sanitising rinse with a germicidal solution should be encouraged, and the system for hand-drying should prevent recontamination. To further enhance this regime, fingernails should be short enough to allow easy cleaning without the use of a nailbrush; unless disposable, nailbrushes can deposit more bacteria on the fingers than they remove.

12.5.2.3 General Protection. Protective body and footwear will be needed to protect both the day-wear of the operative from stains and the food from contamination. Washable polyester/cotton coats or overalls are widely used (Anonymous, 1999g); and if a particular task is likely to cause gross soiling of a coat within a shift, disposable aprons or oversleeves may provide additional protection. However, the value of these precautions will be undermined unless:

- Changing facilities are designed so that outer garments (coats, scarves, gloves, shoes) can be left in a room separate from factory wear.

- Clean coats or overalls are in a separate changing room, and they are wrapped so that they cannot become contaminated prior to use.
- Dirty coats are bagged-up at the end of each shift and then sent to a specialist laundry—that is, a laundry with (a) the facility to handle all food industry clothing separately from other garments, (b) washing machines that load at one end and empty at the other, so that there is a physical barrier between the dirty and clean clothes, (c) full traceability for each garment, (d) a failsafe system covering aqueous and thermal disinfection, and (e) isolated pressing and packing rooms with high-quality filtered air.

It is important also that protective garments should *never* leave their designated location, and the color coding of coats/overalls/footwear can provide an easily visible safeguard.

Clearly specific rules will come into operation for changes of clothing during meal or other breaks, but observation of these rules becomes vital if an operative has more than one function. For example, it would be quite acceptable for a person to spend the morning receiving raw materials like milk into a reception area, and then switch later in the day to help package an urgent order. However, unless the system acknowledges that the garments worn during the morning may be contaminated with bacteria like *Campylobacter* spp. or *E. coli* O157, there will be a serious risk of postprocessing contamination. Wellington boots can pick up such contaminants as well, so that unless germicidal footwells are sited between high/low-risk areas, movement of personnel must be restricted. Obviously the risk of such transfers will be low, but, given the severity of infections caused by *E. coli* O157, for example, any risk becomes unacceptable.

12.5.2.4 Nonmicrobiological Hazards. The major concern with dairy products tends to revolve around the risk of food-borne disease because, humanitarian considerations aside, the economic losses that can follow from a well-publicized and traceable incident of food poisoning should be sufficient to encourage caution. However, product recalls resulting from the presence of glass or metal fragments can be expensive as well, so that a firm policy over the wearing of watches, jewelery, and hair clips is essential.

Watches must, of course, never be worn in a process area, and jewelery must be banned as well if there is a possibility of loss into the product. Even pens that may be essential for keeping records must be metal and of a design that is readily detected under a scanner. Fortunately, metal detection systems can examine retail items before they leave the factory and automatically reject those that are suspect (Wallin and Haycock, 1998; Lock, 1999), but, with glass, plastics, or fragments of stone, prevention of entry is the only option.

Contamination of foods with heavy metals like lead or mercury appears to be infrequent (Hobbs and Roberts, 1995); thus, as long as the ingredient specifications detail the levels of such elements, serious complaints of chemical origin should be rare. However, it is worth remembering that stores used by an engineering/maintenance section may house all types of toxic materials for treating wood or metal, and hence their use within the food processing area must be rigorously monitored.

12.6 WASTE DISPOSAL

If the packaging of a retail item is damaged during manufacture, two interrelated consequences may follow. First, the faulty item will be discarded and, perhaps, be left undisturbed for several days; second, the food inside the package will be exposed to contamination from the atmosphere. Once the latter has occurred, the growth of yeasts, bacteria, or molds will surely follow. This inevitable pattern may seem unexceptional, but what makes it exceptional is that the microorganisms concerned must, *a priori*, be well adapted to the foodstuff in question. In other words, the spoiled product is, in effect, selecting from the atmosphere those bacteria or fungi that can utilize the nutrients available and that are able to grow under the prevailing conditions of pH or available water. Bearing in mind that a bacterial colony of 1000 cells can climb to over 500,000 in 3 h under good conditions, or that a small colony of *Penicillium* spp. can liberate several million spores, it is evident that a minor incident can fast become a major source of potential contamination.

Waste disposal units with polythene liners or holders (with lids) to take paper sacks must, therefore, be located at strategic points for receipt of any damaged food items, and a system must be in place for the units to be routinely emptied. Sealing of the liners or bags prior to dumping in a "skip" is another preventative measure that will cost nothing, but containing the microflora could save the company a great deal of money.

ACKNOWLEDGMENT

The author wishes to thank all the companies listed in the references for their help in providing information on current commercial practices within Europe, and Hui (1992) can be consulted for similar information applicable to North America.

REFERENCES

Anonymous (1988) *J. Environ. Sci.*, **31**(5), 53–76.

Anonymous (1998) *A Practical Guide to the Disinfection of Food Processing Factories and Equipment Using Fogging*, Silsoe Research Institute, West Park, Silsoe, MK45 4HS, United Kingdom.

Anonymous (1999a) *Hygiene Control in the Food Industry*, Ozone Industries Ltd., Farnborough GU14 0NR, United Kingdom.

Anonymous (1999b) *Application Bulletin*, Ultra-Violet Products Ltd., Science Park, Milton Road, Cambridge CB4 4FH, United Kingdom.

Anonymous (1999c) *Industrial Resin Flooring*, Resdev Ltd., Ainleys Industrial Estate, Elland HX5 9JP, United Kingdom.

Anonymous (1999d) *Hard Floor Cleaning Manual*, Dowding and Plummer Ltd., Stockfield Road, Acocks Green, Birmingham B27 6AP, United Kingdom.

Anonmyous (1999e) *Food Safe Sanitisers*, Darenas Ltd., 8 Gravelly Park, Birmingham B24 8TB, United Kingdom.

Anonymous (1999f) *Hygiene Monitoring*, Merck Ltd., Poole, Dorset BH15 1TD, United Kingdom.

Anonymous (1999g) *Clothing for the Food Industry*, Shermond Disposable Hygiene Products, Meridian Industrial Estate, Newton Road, Peacehaven BN10 8Q, United Kingdom.

Audidier, Y. (1996) Ultra clean workshop concept. In *Minimal Processing and Ready Made Foods*, T. Ohlsson, R. Ahvenainen, and T. Mattila-Sandholm, eds., SIK, Göteborg, pp. 91–111.

Bintsis, T., Litopoulou-Tzanataki, E., and Robinson, R. K. (2000) *J. Food Agric. Sci.*, **80**, 637–645.

Blümke, H. (1993) Sterile room techniques in the food industry. In *Aseptic Processing of Foods*, H. Reuter, ed., Behr's Verlag, Hamburg, pp. 265–270.

Brolchain, M. O. (1993) Factory contruction—environmental considerations. In *Encyclopaedia of Food Science, Food Technology and Nutrition*, Vol. 3, R. Macrae, R. K. Robinson, and M. Sadler, eds., Academic Press, London, pp. 1710–1713.

Brisdon, E. Y. (1998) *The Oxoid Manual*, 8th ed., Oxoid Ltd., Basingstoke.

Brunderer, J., and Schicht, H. H. (1987) *Swiss Food*, **9**(12), 14–17.

Burfoot, D., Hall, K., Brown, K., and Xu, Y. (1999) *Trends Food Sci. Technol.*, **10**, 205–210.

Cattel, D. (1988) *Specialist Floor Finishes: Design and Installation*, Blackie and Sons Ltd., Glasgow.

Clark, P. J. (1993) Plant design. In *Encyclopaedia of Food Science, Food Technology and Nutrition*, Vol. 6, R. Macrae, R. K. Robinson, and M. Sadler, eds., Academic Press, London, pp. 3605–3617.

Delves-Broughton, J., and Board, R. G. (2000) The microbiology of eggs. In *Encyclopedia of Food Microbiology*, Vol. 1, R. K. Robinson, C. Batt, and P. Patel, eds., Academic Press, London, pp. 569–573.

Fitzpatrick, B. W. F. (1990) *Swiss Contam. Control*, **3**(4A), 348-349.

Gohil, V. S., Ahmed, M. A., Davies, R., and Robinson, R. K. (1995) *J. Food Prot.*, **58**, 102–104.

Gohil, V. S., Ahmed, M. A., Davies, R., and Robinson, R. K. (1996) *Food Control*, **6**, 365–369.

Hampson, B. C., and Kaiser, D. (1995) *Dairy Food Environ. Sanit.*, **15**, 371–374.

Harrigan, W. F., and Park, R. W. A. (1991) *Making Safe Food*, Academic Press, London.

Hayes, P. A. (1985) *Food Microbiology and Hygiene*, Elsevier Applied Science, London.

Hedrick, T. I. (1975) *Chem. Ind.*, **20**, 868–872.

Hiddink, J. (1995) *Water Supply Sources, Quality and Water Treatment in the Dairy Industry*, Document No. 308, International Dairy Federation, Brussels, Belgium, pp. 16–32.

Hobbs, B. C., and Roberts, D. (1995) *Food Poisoning and Food Hygiene*, Edward Arnold, London.

Holah, J. T. (1998) *Dairy Food Environ. Sanit.*, **18**, 212–220.

Holah, J. T., Rogers, S. J., Holder, J., Hall, K. E., Taylor, J., and Brown, K. L. (1995) *The Evaluation of Air Disinfection Systems*, Research and Development Report No. 13, Campden & Chorleywood Food Research Association, Chippen Campden.

Holm, S. (1980) Hygienic design of food plants and equipment. In *Hygienic Design and Operation of Food Plant*, R. Jowitt and Ellis Horwood, eds., Chichester, pp. 17–44.

Hui, Y. H. (ed.) (1992) *Dairy Science and Technology Handbook*, Vol. 3, VCH, New York.

IDF (1980a) *Corrosion in The Dairy Industry*, Document No. 127, International Dairy Federation, Brussels.

IDF (1980b) *Code of Practice for the Design & Construction of Milk Collection Tankers*, Document No. 128, International Dairy Federation, Brussels.

IDF (1981) *Corrosion in the Dairy Industry—Selected Cases*, Document No. 139, International Dairy Federation, Brussels.

IDF (1983) *Corrosion in the Dairy Industry—Selected Cases*, Document No. 161, International Dairy Federation, Brussels.

IDF (1984) *General Code of Hygienic Practice for the Dairy Industry & Advisory Microbiological Criteria for Dried Milk, Edible Rennet Casein & Food Grade Whey Powders*, Document No. 178, International Dairy Federation, Brussels.

IDF (1985) *Surface Finishes of Stainless Steel—New Stainless Steel*, Document No. 189, International Dairy Federation, Brussels.

IDF (1987) *Hygienic Design of Dairy Processing Equipment*, Document No. 218, International Dairy Federation, Brussels.

IDF (1988) *Corrosion: Prozacetic Acid Solutions, Electrochemical Principles, Glossary of Terms*, Document No. 236, International Dairy Federation, Brussels.

IDF (1991) *IDF Recommendations for the Hygienic Manufacture of Spray-Dried Milk Powders*, Document No. 267, International Dairy Federation, Brussels.

IDF (1992) *Hygienic Management in Dairy Plants*, Document No. 276, International Dairy Federation, Brussels.

IDF (1994) *Recommendations for the Hygienic Manufacture of Milk and Milk Based Products*, Document No. 292, International Dairy Federation, Brussels.

IDF (1995a) *IDF Recommendations for the Hygienic Storage, Transport and Distribution of Milk and Milk Based Products*, Document No. 302, International Dairy Federation, Brussels.

IDF (1995b) *Fouling and Cleaning in Pressure Driven Membrane Processes*, Special Issue No. 9504, International Dairy Federation, Brussels.

IDF (1996) *IDF General Recommendations for the Hygienic Design of Dairy Equipment*, Document No. 310, International Dairy Federation, Brussels.

IDF (1997a) *Hygienic Design and Maintenance of Dairy Buildings and Services*, Document No. 324, International Dairy Federation, Brussels.

IDF (1997b) *Implication of Microfiltration on Hygiene and Identity of Dairy Products*, Document No. 320, International Dairy Federation, Brussels.

IDF (1997c) *Fouling and Cleaning of Heat Treatment Equipment*, Document No. 328, International Dairy Federation, Brussels.

Jackson, R. (1997) *Food Process.*, **66**(9), 40–41.

Jarvis, B. (2000) Good manufacturing practice. In *Encyclopedia of Food Microbiology*, Vol. 2, R. K. Robinson, C. Batt, and P. Patel, eds., Academic Press, London, pp. 961–972.

Jolly, A. C. (1993) Factory construction—materials for internal surfaces. In *Encyclopaedia of Food Science, Food Technology and Nutrition*, Vol. 3, R. Macrae, R. K. Robinson, and M. Sadler, eds., Academic Press, London, pp. 1713–1719.

Keceli, T., and Robinson, R. K. (1998) *Dairy Ind. Int.*, **62**(4), 29–33.

Lock, A. (1999) *The Guide to Reducing Metal Contamination in the Food Processing Industry*, Safeline Limited, Montford Street, Salford, Lancashire.

Lück, H., and Gavron, H. (1990) Quality control in the dairy industry. In *Dairy Microbiology*, Vol. 2, 2nd ed., R. K. Robinson, ed., Elsevier Applied Sciences, London, pp. 345–392.

Lund, B. M. (1993) *Food Technol. Int. Eur.*, A. Turner, ed., pp. 196–200.

Martin, S. E., and Fisher, C. W. (2000) *Listeria monocytogenes.* In *Encyclopedia of Food Microbiology*, Vol. 2, R. K. Robinson, C. Batt, and P. Patel, eds., Academic Press, London, pp. 1228–1238.

Patel, P. D., and Williams, D. W. (1994) Evaluation of commercial kits and instruments for the detection of foodborne bacterial pathogens and toxins. In *Rapid Analysis Techniques in Food Microbiology*, P. D. Patel, ed., Blackie Academic & Professional, London, pp. 61–103.

Prentice, G., and Neaves, P. (1993) Chilled storage—microbiological considerations. In *Encyclopaedia of Food Science, Food Technology and Nutrition*, Vol. 2, R. Macrae, R. K. Robinson, and M. Sadler, eds., Academic Press, London, pp. 887–891.

Romney, A. J. D. (ed.) (1990) *CIP: Cleaning in Place*, Society of Dairy Technology, Long Hanborough United Kingdom.

Schicht, H. H. (1989) *Swiss Pharma*, **11**(11), 15–23.

Schicht, H. H. (1991) *Eur. Food Drink Rev.*, **Summer**, 5–10.

Sharma, G. (2000) Ultraviolet light. In *Encyclopedia of Food Microbiology*, Vol. 3, R. K. Robinson, C. Batt, and P. Patel, eds., Academic Press, London, pp. 2208–2214.

Tamime, A. Y., and Robinson, R. K. (1999) *Yoghurt—Science and Technology*, Woodhead Publishers, Cambridge, England.

Timperley, D. A. (1993) *Guidelines for the Design and Construction of Floors for Food Production Areas*, Technical Manual No. 40, Campden & Chorleywood Food Research Association, Chippen Campden.

Timperley, D. A. (1994) *Guidelines for the Design and Construction of Walls, Ceilings and Services for Food Production Areas*, Technical Manual No. 44, Campden & Chorleywood Food Research Association, Chippen Campden.

Tuthill, A. H., Avery, R. E., and Covert, R. A. (1997) *Dairy Food Environ. Sanit.*, **17**, 718–725.

Wallin, P., and Haycock, P. (1998) *Foreign Body Prevention, Detection and Control*, Blackie Academic & Professional, London.

Weatherburn, D. (1997) *Food Process.*, **66**(9), 33–34.

WHO (1993) *Guidelines for Drinking-Water Quality*, World Health Organization, Geneva, Switzerland.

WHO (1994) *Environmental Health Criteria 160*, World Health Organization, Vammala, Finland.

CHAPTER 13

APPLICATION OF PROCESS CONTROL

DAVID JERVIS (deceased)
Unigate Group Technical Centre, Wootton Bassett, Wiltshire, England

13.1 INTRODUCTION

Process control can be defined as the management of all elements of a process that control the legality, safety, contractual, and commercial requirements of the product. The scope is, therefore, from farm to consumer and embraces raw materials, formulation, bacteriocidal or bacteriostatic treatments, plant and equipment hygiene, personnel practices and hygiene, packaging, distribution conditions, and consumer use.

Historically, process requirements have evolved on a basis of need to respond to incidents of product failure and changing marketing criteria. Pasteurization of drinking milk was introduced in the 1930s to address public health risks associated with changing patterns of distribution in cities. Global incidents of contamination of milk powders in the 1960s resulted in stricter control of plant and environmental hygiene in spray-drying units. More recently, *Listeria monocytogenes* has emerged as a pathogen of concern following a soft cheese outbreak in Los Angeles, with 86 cases and 29 deaths. Following this outbreak, surveillance showed that this microrgenism was a global problem, particularly with mold-ripened soft cheeses, and enhancement of process hygiene served to minimize risks. Currently, enterpathogenic *Escherichia coli* is a pathogen of concern in milk and milk products by virtue of an apparently increasing incidence at farm level, a relatively high tolerance of reduced pH, and low infective dose. This should

Dairy Microbiology Handbook, Third Edition, Edited by Richard K. Robinson
ISBN 0-471-38596-4

trigger a review of current process control parameters, particularly for mold-ripened and other soft cheeses to minimize risk from this pathogen.

In parallel with the emergence of public health failures has been the trend toward novel and efficient processes, changes in formulations to reduce manufacturing costs per unit of product by increased throughput on high capital plant, and the use of cheaper ingredients more likely to be obtained on a global basis. There has also been an ongoing trend toward healthier foods—for example, lower fat, less salt, and the elimination of preservatives. These factors, together with the commercial demand for longer shelf life to accommodate consumer shopping patterns and reduce distribution costs, can have a significant effect on the microbiological stability of products. Clearly, process control requirements need to be constantly reviewed and amended to accommodate change, and this should be done using a disciplined and documented approach that is amenable to constant review.

It is the contention of this chapter that the required disciplined approach is best provided by the hazard analysis critical control point (HACCP) procedure applied as an integral element of total quality management (TQM) principles, which include good manufacturing practice (GMP), good hygiene practice (GHP), and document control (e.g., ISO 9000 Quality Systems). HACCP is an internationally accepted hazard management tool that can, and should, be applied to all stages of food manufacture from farm to consumer, irrespective of the level of development or size of business. The approach is sufficiently flexible to accommodate the varying levels of sophistication and complexity implied.

13.2 MANAGEMENT TOOLS

13.2.1 Total Quality Management (TQM)

TQM schemes address the approach that a manufacturing organization needs to take to ensure product quality. They aim to involve every member of the organization in the achievement of management objectives to produce safe, wholesome food, enhance customer satisfaction and confidence, and identify means of ongoing improvement. The fundamental requirements of the TQM approach is communication at all levels, so that process and product requirements can be translated from the corporate quality statement to the operatives running the process. TQM schemes embracing HACCP and document control form an

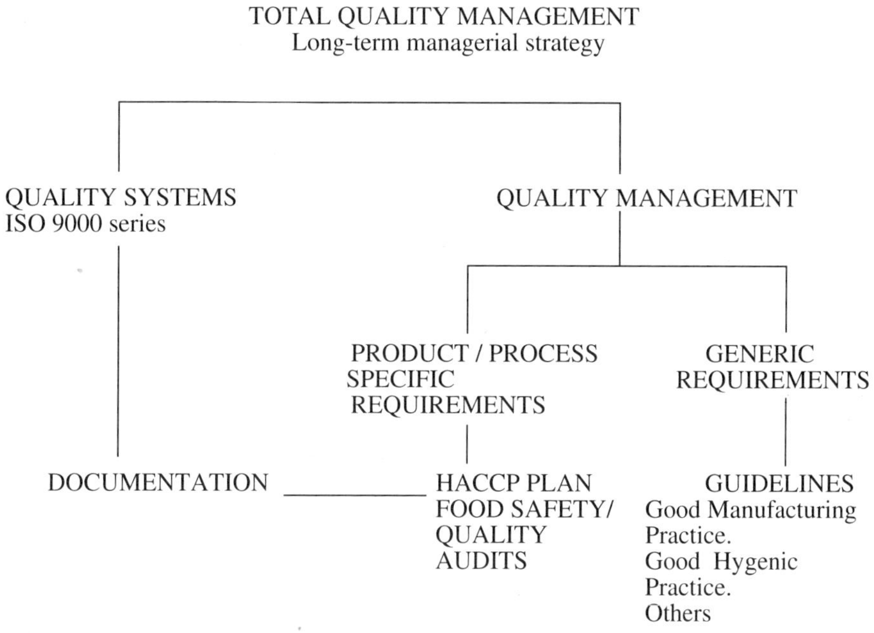

Figure 13.1. The elements that are included under the heading of Total Quality Management (TQM).

important framework within which quality requirements can be communicated effectively and in a way that can be demonstrated and audited. The overall approach is summarized in Figure 13.1.

13.2.2 Quality Systems

ISO 9000:1994 Quality Systems and associated standards in the ISO 9000 series (9001, 9002) are widely recognized as documentation systems applicable to food manufacture. The scope of these standards calls for the development of controlled issue documents that cover the business quality policy, management structure and responsibility, training program, and corrective and remedial actions, as well as the establishment of process control parameters to meet product quality requirements and the recording and review of related monitoring programs. Such quality systems are, therefore, an ideal vehicle for the control of all quality-related documentation, including HACCP plans and the related monitoring and verification systems, together with amendments introduced as a result of HACCP reviews. A quality system also permits a transparent translation of HACCP control measures and critical limits into specific process instructions and record sheets, respectively. This interrelationship between ISO-style quality systems and HACCP plans forms a basis for an HACCP-based audit of the process. This will be amplified in the following sections.

13.2.3 Quality Management

13.2.3.1 Generic Requirements. The microbiological quality and safety of foods depends on (a) manufacture and handling in an environment that meets basic standards for hygiene and (b) the management of hygiene. This requirement has a broad scope and is best addressed through guidelines for good hygiene practice that are focused on the particular product or commodity under consideration. The scope of these Guidelines should include the following headings and address the objectives summarized.

- *Primary Production.* Control of hazards relevant to the process and product under consideration.
- *Site Location.* Avoidance of external sources of contamination that are a hazard to the process and product.
- *Design and Layout.* Provision of adequate space for all process stages with a layout that prevents cross-contamination between process stages.
- *Building Fabric.* Design and maintenance to prevent ingress of external contamination.

  Internal surfaces that minimize soiling and contamination buildup and are easily cleaned.
- *Equipment.* Design and construction to minimize accumulation of food material and easy cleaning.
- *Waste Disposal.* Provision of segregated containers/systems for collection and continual removal from production areas in order to prevent contamination buildup/cross-contamination.
- *Water Supply.* (a) Provision of potable water for all food contact uses including product formulation, food washing, and cleaning of food contact surfaces to prevent recontamination. (b) Provision of microbiologically potable water for cooling purposes (plate heat exchangers, tank jacket) to prevent recontamination. (c) Complete segregation and identification of nonpotable water.
- *Drainage.* (a) Designed and constructed to flow from clean to dirty areas and have sufficient capacity to avoid backup under the heaviest loading conditions.

- *Cleaning.* (a) Provision of facilities for the cleaning of utensils and equipment and factory environment that avoids product contamination or cross-contamination between process stages. (b) For clean-in-place (CIP) systems, design and maintenance should ensure specified detergent strength, temperature, and flow rates throughout the system. (c) Design and operation of CIP systems to avoid cross-contamination between raw food and processed food duties.
- *Personnel.* (a) Provision of changing rooms, toilets, and suitably located hand-washing facilities. (b) Application of food hygiene training appropriate to the process requirements.
- *Air Quality.* (a) Design and operation of ventilation systems to (1) prevent condensation on surfaces that might be product contamination sources either directly or via aerosol droplets, (2) minimize airborne contamination, (3) direct air flow from clean to dirty areas, (4) control ambient temperature, and (5) control odor transfer. (b) Filtration of air that impinges directly onto food (e.g., drying air, compressed air) to a standard that prevents product contamination.
- *Storage Areas.* Provision of adequate facilities for the storage of (1) food, food ingredients, and packaging, (2) nonfood materials (cleaning materials, lubricants, engineering spares).
- *Pest Control.* Application of control programs and monitoring to prevent the access and/or harborage of pests.
- *Foreign-Body Control.* Application of control programs and monitoring to minimize the risk of foreign-body contamination of food (glass, glass-like plastics, metal, or other debris).
- *Personal Hygiene.* Application of procedures and guidelines for health screening, sickness and injury reporting, personal behavior, and use of designated protective clothing to prevent contamination of food and food contact surfaces.
- *Training.* Provision of training of all food handlers covering food hygiene requirements and relevant (to job) process requirements to prevent food contamination.
- *Transportation.* Application of packaging and warehouse/transport conditions that prevent recontamination of food and maintain the required temperature regime.
- *Recall Procedures.* Maintenance of records and procedures that enable the identification and recovery of any unacceptable product for investigation and disposal.

- *Control of Process.* The objective is to reduce the risk of unsafe and unacceptable food reaching the consumer by taking preventative measures to control hazards at an appropriate stage in the process. This is, in effect, the objective of an HACCP program (see Section 13.4).

The headings outlined above should be considered as a template on which to base codes of good manufactoring practice (GMP) and good hygiene practice (GHP). Such codes may be developed internationally (e.g., Codex, 1999) as commodity codes (e.g., IDF, 1991), as National Trade Association codes of practice, or as manufacturers site-specific documents produced as an element of local process control procedures.

It is always advisable to develop site-specific codes of GMP/GHP so that the detailed conditions and circumstances of the establishment and process can be fully addressed and carried forward to process instructions.

13.2.3.1.1 Barrier Hygiene. An area of GMP/GHP not fully covered in generic templates is the requirements for handling designated high-risk foods. General Principles of Food Hygiene documents do outline the need for enhanced precautions where risks are high, but it is important that the relevant control measures are understood and considered in the development of GMP/GHP codes of practice.

High-risk foods can be defined as those foods that (1) are handled and/or exposed to the environment after any bacteriocidal process and before wrapping; (2) can support the growth or survival of relevant pathogens at a level hazardous to the consumer and/or support the growth of spoilage microorganisms to undesirable levels within the designed shelf life of the product, and (3) are designed to be eaten without further cooking.

High-risk foods (e.g., some dairy desserts) are at risk from cross-contamination from other process areas, environmental contamination, and transfer of relevant microorganisms from personnel. They should be handled in segregated production areas that, in addition to the general hygiene requirements, have the following enhanced provisions.

- The designated high-risk production area should be separated from all other process areas by permanent barriers.
- Drainage should be suitably trapped and direct to a main drain with no possibility of back flooding. Drainage from other areas should not transit the high-risk area.

- Floors walls, ceilings, and all fittings should be designed and maintained to prevent accumulation of soiling and be easily cleaned and sanitised.
- The air supply to the high-risk room(s) should be filtered to remove relevant microorganisms, should maintain the room at a positive pressure, and should be supplied at a sufficient rate and at appropriate humidity to prevent condensation on all surfaces at all times.
- There should be dedicated equipment, utensils, and tools that are not used or cleaned in other areas.
- Cleaning equipment should be dedicated to the high-risk area with color or other coding to assist supervision.
- Material brought into the high-risk area should be handled in a way that prevents the transfer of contamination. Control measures to consider include the removal of outer packaging in an annex, with materials moved into the designated area through a restricted access using sanitizer dips if appropriate (e.g., canned or sealed plastic packaging). Trucks and trolleys should be dedicated to the designated area, and particular attention should be paid to the buildup of contamination on wheels and under frames.
- Personnel should change from general factory clothing into specific "high-risk" protective clothing and footwear at the access to the high-risk area. There should be hand-washing facilities available before high-risk clothing is handled. The high-risk protective clothing and footwear should be color-coded for easy identification and should not be used elsewhere on the site; a change back to general clothing and footwear should occur whenever personnel leave the designated high-risk area. These provisions should apply to all personnel entering the designated high-risk area, including engineering staff, management, and visitors.

  Note: As an alternative to footwear change, the use of walk-through baths, mechanized footwear cleaners, and sanitized foot mats have been considered, but with little success. The principal disadvantages are that (a) footwear is wetted and trails water onto high-risk floors, (b) soiling and contamination is not always effectively eliminated, and (c) the units can become a focus of contamination if not thoroughly cleaned and sanitized daily. Footwear cleaners can also be a source of aerosol contamination.
- Personnel hygiene should be maintained at a high standard with close supervision of hand washing and protective clothing disciplines and adherence to safe food handling procedures. Sickness

and contact with enteric disease cases and outbreaks should be reported and investigated by an occupational health department or medical practitioner, with appropriate exclusion from high-risk operations until clearance is confirmed.

These barrier hygiene principles refer particularly to designated high-risk rooms. Another operation where barrier hygiene is an important consideration is the cooling of unpacked product in static or blast chillers. Examples are the manufacture of clotted cream and some scalded dairy dessert derivatives that can be cooled unwrapped in blast chillers. The principles of barrier hygiene should be applied noting that, in addition to the handling of product from cooker to blast chill and blast chill to static chill and packing, the principal hazard is likely to be air in the blast chiller impinging on the product in high volume, along with the tendency for condensation to form on the unit walls and ceilings—a known focus of *L. monocytogenes* contamination.

The barrier hygiene principles given in this section are of necessity generic. In any manufacturing situation a specific hazard analysis should be applied to identify which of the barrier hygiene requirements are relevant. This is, in fact, the case with all GMP/GHP requirements outlined in generic codes and identifies the important principle that hazard analysis using HACCP should always be applied in the development of product-specific documents as an element of process control.

13.2.3.2 Product/Process Specific Requirements. Against a background structure afforded by GMP/GHP documents, there is a need to assess individual products/processes in their manufacturing environment to identify potential hazards to food safety and quality and derive appropriate control measures, monitoring, procedures, protocols for corrective action, and documentation. HACCP, correctly integrated into a total quality management scheme, is the preferred risk management tool (see Figure 13.2).

13.3 RISK ANALYSIS

Risk analysis is a structured and formalized approach to quantifying risk and setting levels to which casual agents should be controlled to assure safety. Risk analysis has three components: risk assessment, risk management, and risk communication.

Microbiological risk analysis protocols are being addressed internationally and at national levels, and they are becoming a key element in determining the level of consumer protection necessary.

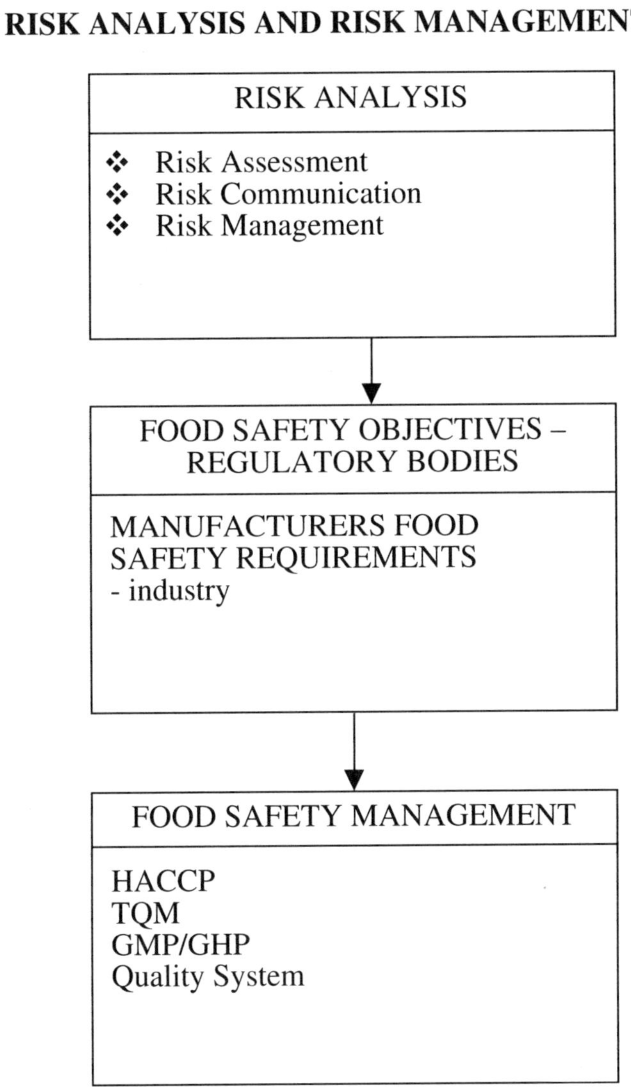

Figure 13.2. Food safety objectives as the link between risk analysis and food management.

13.3.1 Codex Alimentarius Commission (CAC)

The Codex Committee on Food Hygiene (CCFH) is developing a document covering the principles of microbiological risk assessment (Codex Alimentarius, 1999a); this document addresses the risk assessment element of risk analysis. The CCFH is also developing a similar document on risk management (Codex Alimentarius, 1999b), which addresses the risk management element of risk assessment.

It should be noted that Codex "standards" and related texts are intended to be applied by governments—either directly through regulations, or indirectly as guides in national food control programs. The GATT Sanitary and Phytosanitary (SPS) agreement in 1994 has resulted in increased implementation of Codex standards in national and European Union rules, and the Codex standards are the only reference for the World Trade Organization in trade conflicts. It should also be noted that the term "standards" in this context includes horizontal issues (e.g., general principle documents on food hygiene) as well as commodity specific standards (e.g., milk and milk products).

It is to be expected, therefore, that standards and codes of practice generated by CAC will increasingly influence national legislation with respect to food hygiene and food safety. This will include (a) the General Principles of Food Hygiene document (Codex, 1997), (b) appended supplements covering HACCP Guidelines and Principles for the Establishment and Application of Microbiological Criteria for Foods, and (c) the documents on microbiological risk assessment and microbiological risk management.

While the risk analysis guidelines elaborated by the CCFH are targeted for application by national regulatory bodies, the principles enshrined are equally valid in the development of food industry quality management strategies.

13.3.2 Microbiological Risk Assessment

The outline notes above are derived from the CCFH document—Proposed Draft Principles and Guidelines for the Conduct of Microbiological Risk Assessment (Codex Alimentarius, 1999a)—which has as its scope the risk assessment of microbiological hazards in foods. In this document the importance of a functional separation between risk assessment and risk management is emphasized as is the importance that a risk assessment should be based on sound science, and it should include (a) recognition of uncertainty in the data used and interpreted and (b) the need for reassessment in the light of new data, including human health data.

The elements of risk assessment can be summarized as follows:

Statement of Purpose: Definition of the food or group of foods under consideration and the form of assessment output required (e.g., annual rate of illness per 100,000 consumers).

Hazard Identification: The identification of microorganisms and toxins that could be present in the food or food group and be the cause of adverse health effects. Data derived from epidemiological studies, surveillance, and scientific literature and databases, including industry QC records, are relevant in this element.

Exposure Assessment: An estimate of the likely intake of pathogenic agents based on the frequency of contamination at consumption, as well as consumption patterns and habits. An exposure assessment can be either qualitative or quantitative.

Hazard Characteristics: An evaluation of the nature of an adverse health effect associated with the microorganisim(s) and toxin(s) under consideration. A dose-response assessment for relevant target populations is important to this evaluation.

Risk Characterization: A statement of the probability of occurrence and severity of adverse health effects in a target population based on the outputs of hazard identification, exposure assessment, and hazard characterisation. Risk characterization can be either qualitative or quantitative, depending on the quality of data used in the stages of risk assessment. The degree of confidence in the stated estimate should be recorded, together with working notes of the risk assessment deliberations for future reference.

It is important to accept that the "formal" microbiological risk assessment procedure developed by the CCFH is intended for official governmental application in the context of international trade under the SPS requirements. For this reason it is necessarily a complex procedure requiring extensive information input and significant input of time and other resources to achieve the ideal of a quantitative assessment with minimal variability and uncertainty. From the food industry perspective, such resources are not usually available, and a qualitative risk assessment based on in-house experience and available literature and Trade Association data is a more realistic prospect. It must be emphasized that such qualitative risk assessments are by definition imprecise with significant variabilities and uncertainties, so application must always make due allowance and err on the side of safety.

13.3.3 Risk Management

Risk management is the element of risk analysis that determines policy options in the light of a microbiological risk assessment output. The options that might be available range from (1) withdrawal of the food from use to be destroyed or reprocessed, (2) a recommendation that is not suitable for sensitive groups such as infants and immunocompromised consumers, or (3) modification of the process to meet the limits indicated in the output of the microbiological risk assessment. This last option could involve regulatory measures, but could be equally well achieved through process review and application of appropriate remedial measures by the manufacturer.

While risk management planning of necessity will involve both risk assessors and risk managers, it is important that a functional separation is maintained between the risk assessment process and the risk management to protect the integrity of the former and the transparency of the latter.

13.3.2 Risk Communication

This third element of risk analysis is the exchange of information and opinions between risk assessors, risk managers, consumers, and other interested parties in order to reach a consensus on the level of consumer protection that is acceptable. Clearly, the processes of risk management and risk communication interact to arrive at a consensus on the balance between risk to consumer and the cost and other consequences of modifying a process or withdrawing a food from the market place.

13.3.5 Food Safety Objectives

Figure 13.2 shows Food Safety Objectives as the bridge between risk analysis and risk management. Food safety objectives are proposed as the output of a risk analysis in Codex procedures and, as such, can be seen as a statement by regulatory bodies of the maximum level of a microbiological hazard in a food considered to be acceptable for human consumption. This statement would be relevant to the consumer groupings defined in the risk assessment scope (e.g., whole population, excluding infants or excluding immune-compromised). From a regulatory point of view, the food safety objective sets the limit against which risk management options are judged.

13.3.6 Manufacturer's Food Safety Requirements

A manufacturer has a fundamental requirement to provide safe, wholesome product if the business is to thrive. While a level of food spoilage might be tolerable in the context of abnormal distribution and home storage conditions, food poisoning incidents (or worse) damage the reputation and economy of a business. Against this background, manufacturers should have a clearly defined statement of food safety requirements within their quality policy; and while the risk of the presence of relevant pathogens cannot be reduced to zero, the statement would usually require that there to be no food poisoning incidents or outbreaks associated with the product. Therefore, while the manufacturer will always need to adopt regulatory food safety objectives as a minimum requirement in food safety, his food safety requirement would usually need to be more demanding.

As shown in Figure 13.2, food safety objectives and/or manufacturer's food safety requirements advise food safety management and HACCP program of risk analysis outputs. The availability of such information to a HACCP team enables a more focused assessment in hazard analysis and therefore facilitates the process of determining critical control points (Section 13.5.8).

13.4 HAZARD ANALYSIS CRITICAL CONTROL POINTS (HACCP)

13.4.1 Background

The origins of HACCP are traced to the 1960s and the United States of America when the Pillsbury Company, the United States Army Laboratories at Natick, and the National Aeronautics and Space Administration collaborated to develop the system as a means of managing safe food production for manned space flight. The outcome was the HACCP concept, which has been adopted and developed to its current status as the food safety management tool recommended by the Codex Alimutarius Commission to advise on consumer protection under the Sanitary and Phytosanitary Measures (1994) agreed at the Uraguay round of Gatt negotiations. As such, HACCP is a reference point in international trade disputes, and it is increasingly enshrined in national legislation. Examples are the European Community Directive on the Hygiene of Foodstuffs (EEC, 1993) and current European

Parliament and Council Regulation proposals on the Hygiene of Foodstuffs (VI/1181/98 rev 2–111/5227/98 rev 4). Article 5 in this proposal requires food business operators to "put in place and maintain a permanent procedure developed and implemented in accordance with the principles of HACCP."

The HACCP system offers a structured approach to the control of hazards in food processing and, properly applied, identifies areas of concern and appropriate control measures *before* product failure is experienced. It represents a shift from retrospective quality control through end-product testing to a preventative quality assurance approach. As will be illustrated, end-product testing against microbiological criteria is shifted to the role of verification in a HACCP program.

The HACCP procedure is generally targeted at food safety management (pathogenic microorganisms and their toxins), but, as an approach in the context of broader quality management, it can be effectively applied to microbiological spoilage, foreign-body contaminations or pesticide contamination. It is preferable to conduct an HACCP program with a narrow scope (a single pathogen or possibly pathogens) rather than attempt to cover an extended list of hazard areas when documentation will become complex. However, an experienced team might choose to cover the whole spectrum of hazard areas, depending on (a) the resources available to produce and maintain a composite HACCP plan and (b) the way in which it is to be incorporated into the local quality plan and quality system.

13.4.2 Limitations in Microbiological Sampling Plans

Sampling plans frequently incorporated into legislative and commercial contract specifications are based on the two-class and three-class attribute plans described by the International Commission for Microbiological Specifications for Foods (ICMSF, 1986).

The two-class plan is usually applied in criteria for pathogens, and the operating characteristics curves for this plan will show if there is a significant risk of accepting a batch or lot of food with some level of contamination with the target pathogen. An example can be found in the case of European criteria for *Salmonella* in milk powder (EEC, 1992), which is in the format of an ICMSF 2-class plan and requires absence in 25 g, $n = 10$, $c = 0$; that is, ten 25-g samples taken from the lot of powder under test, *Salmonella* should not be detected using an approved methodology. With this sample plan there is a 35% probability of accepting a lot with 10% defectives, and there is a 3% probability of accepting a lot with 30% defectives. While a microbiological risk analysis might show that compliance with the sample plan is an appropriate food safety objective in international trade because it maintains contamination below a level that is a significant risk to consumers, it can produce conflict in commercial transactions. The key issue is that a vendor might perform end-product testing and show a compliance with the criteria, while the buyer might take a different ten samples and show noncompliance. In this respect, end-product testing is not a reliable means of assuring quality; and process control, established on HACCP principles, is to be preferred as a means of eliminating the source and cause of a low-level contamination. The manufacturer's Food Safety Requirement is tighter than the regulatory Food Safety Objective.

The reader is referred to the Codex Alimentarus Commission publication (Codex, 1997) and the Institute of Food Science and Technology (UK) publication (1999) for further information on Microbiological Criteria and protocols for developing Microbiological Criteria.

13.4.3 Benefits of HACCP

The benefits that can be derived from the application of HACCP in the Food Industry are many, and the key benefits can be listed as follows:

- HACCP has the potential to identify all hazards in a food process so that controls can be established to assure food safety/quality.
- HACCP is a systematic approach relevant to all stages of food processing covering agricultural and horticultural practices, harvesting, processing, product distribution, and customer practices.
- HACCP is the preferred risk management tool in total quality management.
- HACCP focuses technical resources on critical parts of the process and provides a cost-effective control of food-borne hazards.
- HACCP facilitates the move from retrospective end-product testing to a preventative quality assurance approach enabling the manufacturer to get it right the first time and reduce reject waste.
- HACCP is recognized and promoted by international bodies (such as the Codex Alimentarius Commission) as the system of choice

for ensuring food safety and is becoming enshrined in national legislation. Proactive application in the food industry will facilitate compliance with developing legislation and demonstrates a diligent approach to food safety.

13.4.4 Principles of the HACCP System

The HACCP system consists of seven principles (Codex, 1997) that can be addressed in 14 sequential stages (Campden and Chorleywood Food Research Association, 1997). The principles and stages are as follows:

principle 1: conduct a hazard analysis

Stage 1 Define the terms of reference.
Stage 2 Select a HACCP team.
Stage 3 Describe the product.
Stage 4 Identify intended use.
Stage 5 Construct a flow diagram.
Stage 6 On-site confirmation of flow diagram.
Stage 7 List all potential hazards associated with each process step, conduct a hazard analysis, and consider any measures to control the identified hazards.
A hazard is a biological, chemical, or physical agent in a food with the potential to cause an adverse health effect.

principle 2: determine critical control points (ccps)

Stage 8 Determine CCPs.
A CCP is a step at which control can be applied and is essential to prevent or eliminate a hazard or reduce it to an acceptable level.

principle 3: establish critical limits

Stage 9 Establish critical limits for each CCP.
A critical limit is a criterion that separates acceptability from unacceptability.

principle 4: establish a system to monitor control of ccps

Stage 10 Establish a monitoring system for each CCP.

principle 5: establish the corrective action to be taken when monitoring indicates that a particular ccp is not under control

Stage 11 Establish corrective action plans.

principle 6: establish procedures of verification to confirm that the haccp system is working effectively

Stage 12 Verification

principle 7: establish documentation concerning all procedures and records appropriate to these principles and their application

Stage 13 Establish documentation and record keeping.
Stage 14 Review the HACCP plan.

In any HACCP application it is vital that the stages be considered in order and that the required information and conclusion be completed for each stage before moving on to the next; HACCP is designed as a structured approach, and the proper sequencing of activities is crucial to obtaining an effective output. The HACCP procedure can be applied in product/process development, but it should be noted that the resulting HACCP plan will be based on anticipated production plant conditions and, therefore, will need to be kept under constant review until the new product/process is established and the HACCP plan can be finalized.

13.5 APPLICATION OF HACCP

This section addresses in detail what needs to be done at each of the HACCP stages, and it refers to generic flow diagrams and HACCP plan records that have been produced in order to illustrate the points made. **It is essential that each HACCP study be based on the specific process and product details, and generic plans should never be adopted as a shortcut to save time and resources.**

13.5.1 Stage 1: Define Terms of Reference

Terms of reference should clearly define the scope of the intended HACCP study and address the following points:

- The product to be considered.
- The process site and, if relevant, the process line within that site. It is not advisable to group together apparently similar products and processes where what might be minor variations in formulation and/or process conditions could significantly change the preservation characteristics of the product (see Section 13.5.3).
- What the study will cover—biological, chemical, or physical hazards (or combinations of these)—and whether the study will be limited to food safety considerations or cover broader quality issues (i.e., spoilage). The study will proceed more quickly if the terms of reference are limited to biological food safety issues, or even the consideration of one pathogen relevant to the food. (The single-pathogen study would be appropriate for food safety management to meet Food Safety Objectives/Requirements.)
- The point in the process at which safety or other quality attributes are to met: at point of manufacture or at point of consumption?

13.5.2 Stage 2: Select a HACCP Team

It is important that senior management in the company be made aware of the resources necessary to carry out an effective HACCP study (personal time, appropriate meeting room, secretarial support, and the need to consult outside resources for information) and are committed to providing these resources. The time required to complete the study will depend on the complexity of the process and the terms of reference agreed at Stage 1. If resources cannot be assured to meet the study defined in Stage 1, then the study should not be progressed. HACCP requires a multidisciplinary approach, and the HACCP team should include the following skills:

- **A quality assurance/quality control specialist** who understands the hazards and risks for the product and process under study. Depending on the study terms of reference, this might involve a microbiologist or chemist; and, if this resource is not available in-company, consultation with an external resource might be necessary to obtain information relating to microbiological risks and hazards.
- **A production specialist** to contribute details of what actually happens on the production line throughout all shift patterns.

- **An engineer** to provide information on (a) the operating characteristics of the process equipment under study and (b) the hygienic design of equipment and buildings.
- **Others** co-opted onto the team as necessary. These might include specialist equipment operators, hygiene manager, ingredient and packaging buyers, and distribution managers. It might also be appropriate to consider co-opting specialist technicians from companies to which various scheduled maintenance and calibration functions are contracted (e.g., temperature measurement equipment, pasteurizer plate and jacketed silo integrity, spray-drier chamber integrity, clean-in-place systems).

An individual experienced in HACCP should be nominated as chairman to be responsible for managing the study. The chairman should have received training in the principles of HACCP and be experienced in HACCP team work. In the United Kingdom, The Royal Institute for Public Health and Hygiene (RIPHH) in consultation with industry and government representatives has published an advanced HACCP Training Standard (RIPHH, 1999), and training courses conforming to this standard with certification are available. This advanced standard is suitable for use in the training of individuals who will lead HACCP teams.

While HACCP team members will be selected for their specialist knowledge, it is important that they will also have a working knowledge of the HACCP procedure so that they can contribute effectively to the study. Team members may need some training before commencement of the study, and this can be provided either internally by the HACCP team or externally.

It is important that an HACCP team member or co-opted person is identified to keep notes as the work progresses and from which both the HACCP plan and the HACCP study notes can be derived. HACCP study notes should record background information and the basis for conclusions reached in sufficient detail to be helpful when the HACCP plan is reviewed (see Section 13.5.14) The HACCP study notes might also be used as background information in trouble-shooting (see Section 13.6) in the event of product failure or inadequate outcome from the verification program.

13.5.3 Stage 3: Describe the Product

The product under study should be fully described. This stage often tends to be inadequately covered, but diligent attention to detail here is crucial to the identification of hazards at Stage 7. The product description should be considered against the following headings and recorded as HACCP study notes.

Composition. All factors that might influence the preservative characteristics of the food should be recorded. Basic compositional data should be noted including that on solids/moisture level, fat level, salt and/or other solute level, pH, and level/type of preservative, if used. The water activity (A_w) of the product might be useful as an indicator of the preservative effect derived from dissolved solutes in moisture, but, if this analytical data is not available, a simple calculation of dissolved solids (specified) in moisture can be used.

Where parameters will change as part of the product maturation process, data should be gathered to cover all points within the terms of reference of the study, which is usually up to and including the point of consumption (end of shelf life). This consideration is particularly relevant to mold-ripened soft cheese where the pH will increase from <5.0 to >6.0 before consumption, significantly affecting the ability of contaminating pathogens to multiply.

Where the final product is made up from different components (e.g., layered dairy-based desserts), data should be collected for each layer. Also, compositional data should be recorded for any additives used, particularly where these are supplied as fresh, hydrated materials.

Structure. The structure of a product can be considered under two headings:

1. To note whether the final product unit is of one type throughout (e.g., butter, hard cheese, drinking milk) or whether it is made up of two or more "layers" of differing composition (e.g., some dairy-based desserts). Also the decorative dressing of some products (e.g., high-fat, soft cheese units rolled in oat flakes or nuts) should be recorded.
2. The microstructure of products can have a significant effect on the ability of contaminating microorganisms to multiply (Brocklehurst and Wilson, 2000). This is relevant in dairy-based products, particularly those consisting of oil-in-water emulsions (creams, dairy dessert custard bases) and water-in-oil emulsions (butters, low fat spreads) where a significant preservative mechanism has been shown to be the immobilization of contaminating microorganisms in water droplets (typically <10 µm diameter) which inhibits multiplication by nutrient depletion and accumulation of metabolic by-products.

The significance of microstructure in various foods, including some dairy-based products, was the subject of a LINK Research project carried out at the Institute of Food Research, Norwich. This work showed that what might have been expected to be minor adjustments in product composition can have a significant effect on the growth rate of contaminating pathogens (e.g., a reduction in fat level in dessert custard allowing an increased growth rate for psychrotrophic *Bacillus cereus*). The work illustrates the significance of controlling formulation and process to maintain microstructure as one element in the *hurdle* approach to food preservation.

Processing. All relevant processing parameters should be recorded. This list will include the time and temperature of all heating stages and the temperatures (and duration) of intermediate stages in manufacture. For example, in Cheddar cheese making, it would be appropriate to record pasteurizing conditions (and, if relevant, the conditions for preparation of the bulk starter medium), the milk temperature in the vat, the curd cooking time and temperature, and the maturation temperature. In the case of dairy dessert manufacture, it would be necessary to record for each component the pasteurizing conditions, time, and temperature that components might be held in bulk before final assembly and filling/despatch temperatures. For spray-dried powder manufacture, it has been suggested that pasteurizing conditions for skim milk, evaporator input and output temperatures, and the time/temperature conditions for preheat of the concentrated feed to the drier, together with drier temperature conditions, might be a minimum requirement. Other bacteriocidal thermal processes (e.g., UHT, sterilization) should be similarly recorded.

Alternatives to thermal processing are becoming increasingly available. These include:

- Electric Pulse Technology
- Ultrasonics (combined with reduced heating temperature)
- High-pressure treatment (300 MPa or greater)
- Microfiltration
- Irradiation

In all cases where these technologies are employed, they should be validated as giving the required effect with respect to microorganisms of concern and the appropriate operating conditions recorded at this stage in a HACCP study. A likely area of development for alternative technologies to conventional HTST pasteurization is their adoption to meet food safety objectives in cheeses manufactured from raw milk, or as possible strategies to eliminate *Mycobacterium avium* subsp. *paratuberculosis* from drinking milk as an alternative to increases in HTST holding times that could be detrimental to the organoleptic quality of the milk. A combination of ultrasonic treatment and HTST pasteurization might be an avenue for development.

Packaging System. The type of packaging should be noted. This note will include differentiation between shrink wrapping, vacuum packing, modified atmosphere/gas flush wrapping (specifying gas mix used) and sealed plastic tub packing. Aseptic or ultraclean packaging regimes should also be noted where appropriate. In the context of dairy products, it is useful to record the conditions of storage of intermediate stages of production (e.g., bulk powder; pasteurized milk held in bulk tanks prior to filling; dessert components; cheeses in maturing rooms). The degree of exposure to the process plant environment during filling should also be recorded, and this will range from zero in aseptic fill systems to 30s or longer for the filling of some layered desserts.

Storage and Distribution Conditions. The storage temperature regimes (ambient, chilled, frozen) throughout the product shelf life should be recorded where possible, and this should include anticipated variations (e.g., retail display, customers shopping bag, home storage conditions).

Required Shelf Life. The total shelf-life requirement together with "life after opening," where appropriate, should be recorded.

Instructions of Use. Dairy products are usually consumed without further processing (heating), so that this section should record instructions given with regard to refrigerated storage (where appropriate) and "use within" times, after opening, together with overall "use by or best before" dates. Suitability for home freezing is also significant and may be relevant to product stability after thawing if the microstructure of the product is changed.

13.5.4 Stage 4: Identify Intended Use

The consumer target group for the product should be noted. This will range from suitable for all consumer groups through to not suitable for infants or immunocompromised consumers (e.g., where there may be a level of *L. monocytogenes* at a point of consumption in soft cheeses).

13.5.5 Stage 5: Construct a Flow Diagram

The purpose of a flow diagram in a HACCP study is to elicit a thorough examination of the process, which is recorded in a way that assists and directs subsequent stages. There is no specified format to be used in HACCP flow diagrams, but they should sequentially set out all steps in the process together with relevant technical data. Consideration should be given to the following:

- The sequence of all process steps within the scope of the study including rework/recycle loops.
- Interaction of services (e.g., cooling water, air, compressed air, clean-in-place systems).
- Temperature/time history for all raw materials, intermediate products, and final products within the scope of the study, together with microbiological and analytical data where appropriate.
- Floor plans and equipment layout with particular reference to segregation of process steps, which might represent a source of cross-contamination (e.g., raw milk areas to heat-treated milk handling).
- Environmental hygiene to include potential for external contamination (e.g., roof soiling, air intake) to contaminate vulnerable internal environments and process equipment.
- Equipment design with particular attention to ease of cleaning and presence of void spaces that might accumulate contamination.
- Personnel routes and hygiene disciplines.

One format that has proved to be effective is that illustrated in Figure 13.4 (skim milk powder manufacture) and Figure 13.5 (Cheddar cheese manufacture). It should be noted that these are much simplified presentations generated to illustrate principles. For example, process step 8 in Figure 13.4 (Skim milk powder) would be subdivided into several steps in an actual HACCP study. Also, because this is a generated example, the diagrams do not include all of the points listed above (e.g., floor plans, equipment design, personnel routes). Flow diagrams can be annotated with details of composition and process conditions, but an alternative is to consider notes generated at HACCP Stage 3 (Describe the Product; see Section 13.5.3) and Stage 4 (Identify Intended Use; see Section 13.5.4) together with the flow diagram as the required basis for Stage 7 (Listing of Potential Hazards). In any event, Stages 3 to 5 (as confirmed in Stage 6) are crucial to an effective HACCP study, and all information gathered should be retained as part of the HACCP Study notes.

13.5.6 Stage 6: On-Site Confirmation of Flow Diagrams

The flow diagrams produced at Stage 5 should be confirmed, on site, by the HACCP team. Points to be confirmed are that any effect of shift patterns and weekend working are included on the flow diagram, together with circumstances of any reclaim or rework activity that might be introduced from time to time. If the HACCP Study is being applied to a proposed new process line/product, flow diagram confirmation will not be possible. In this case the HACCP plan can be completed, but it must be subject to review as the line/product is finalised.

13.5.7 Stage 7: List All Potential Hazards within Each Process Step, Conduct a Hazard Analysis, and Consider Any Measure(s) to Control Identified Hazards

This is the final stage in HACCP Principle 1 (Conduct a Hazard Analysis), **and it should be emphasized that no attempt should be made here to preempt HACCP principle 2 by considering Critical Control Points.** Stage 7 is best considered in three parts: listing hazards, conducting a hazard analysis, and identifying control measures.

13.5.7.1 List All Potential Hazards. Using the flow diagram (Stage 5) and the product description (Stages 3 and 4), the HACCP team should list all potential hazards relevant to the Terms of Reference of the Study (Stage 1). This activity should involve all disciplines in the HACCP team (QA/QC, production, engineering) in a "brainstorming" session that identifies all actual and potential hazards. The chairman should ensure that the following areas are considered:

- Hazards in raw materials.
- Hazards introduced during the process (cross-contamination, factory environment, equipment design, equipment cleaning, introductions by process air or personnel).
- Hazards that survive the process steps.
- The microbiological stability of the product during distribution and in the home.

In these considerations, the intrinsic factors of the product (e.g., pH, A_w, E_h, structure, preservatives, temperature) will be important from the point of view of both (a) the lethal effect of a heating or other process and (b) the way in which the potential for pathogen multiplication might occur before consumption (e.g., the increase in levels of *L. monocytogenes* on mold-ripened soft cheeses as pH increases to ~6.0 during cheese ripening, or the growth of *Salmonella* in "cold reconstituted agglomerated skim milk powder").

It is emphasized here that all potential hazards should be listed. This requirement should not be undermined by the concept of **Prerequisite Programs** that is being developed by Codex Alimentarius and actively applied in some national programs.

13.5.7.1.1 Prerequisite Programs. Prerequisite programs can be defined as "universal steps that control the operational conditions within a food establishment allowing for environmental and other conditions that are favorable for the production of safe/acceptable food."

Corlett (1999) discusses in detail the concept of prerequisite programs in the context of HACCP application and gives as examples of prerequisites the design, maintenance, and layout of production facilities, supplier control, specifications, equipment design, cleaning and sanitation, personal hygiene, training, and suggested quality assurance procedures, process operating procedures, and product formulation.

The concern here is that if the proposed prerequisite programs are in place for what are, in effect, good manufacturing and good hygiene practice matters, there would be a tendency not to consider these areas at Stage 7 of an HACCP study; that is, a generic view would be taken that precluded consideration of points specific to the target product manufactured in the unique circumstances of the process plant under consideration.

There is a place for generic codes of hygiene practice covering food manufacture in general, irrespective of commodity and food type (Codex, 1997), or as sector-specific codes or as site-specific documents, and these should be seen as guidance on the general levels of hygiene required. In all cases, a HACCP study should consider, independently, the potential hazards relevant to the study and progress them through the procedure. If it is found that the control measures recommended in existing codes and guidelines are adequate for specific hazards, the code or guideline can be referred to in the HACCP plan as identified control measures with no need to consider subsequent HACCP stages.

If process/product hazards are identified but are not covered, these need to be progressed either in the HACCP plan or by amendment of the existing code or guideline (if these are site-specific).

13.5.7.2 Hazard Analysis. The process of collecting and evaluating information on hazards and conditions leading to their presence and to decide which are significant within the scope of the exercise should be addressed in the HACCP plan.

The objective of this section of Stage 7 is to consider all of the potential hazards identified (see Section 13.5.7.1) and identify those that need to be eliminated or reduced to an acceptable level if food meeting the established Food Safety Requirements (or any other objective set out in the terms of reference) is to be produced. To a large extent, expert judgement and opinion will be involved and, if the necessary expertise is not available in house, external experts may need to be consulted or co-opted to the HACCP team. Hazard analysis should consider the following points:

- The consequence of the target microorganism(s) or toxins being present at harmful levels in the final product at the point of consumption.
- The likelihood of the target microorganism(s) or toxins being present at harmful levels in the final product at the point of consumption. Conclusions for this and the previous point might be based on previous company or industry experience, on epidemiological data, or on a microbiological risk assessment output.
- The survival and/or multiplication of target microorganism(s) in the product or the potential for production of toxin that will persist to the point of consumption at significant (toxic) levels.
- The hurdle effect (the synergistic preservative effect of two or more inhibitory factors) is relevant to these assessments, and information can be found in reference texts [e.g., ICMSF (1980) or Lund et al. (2000)]. It should be noted, however, that unless the conclusions with respect to the ability of a formulation to inhibit the growth of, or eliminate, the target microorganism is definitive, it might be necessary to carry out "spiking trials" to validate the formulation.
- The numbers of consumers potentially exposed and their vulnerability.
- Any relevant food safety objectives (see Section 13.3.5) or manufacturer's food safety requirements (see Section 13.3.6).

The data from microbiological risk assessments in the context of risk analysis (see Section 13.3) will be useful in the hazard analysis stage of HACCP. In the absence of a formal risk analysis output, hazard analysis in HACCP will be made on quantitative data with appropriate expert input and/or reference to external data sources.

13.5.7.3 Identification of Control Measures. For each of the hazards concluded to be significant in the hazard analysis, the HACCP team should identify control measures that will eliminate the hazard or reduce it to an acceptable level. There may be more than one control measure required to control a hazard (e.g., pasteurization, formulation, and refrigeration to reduce the growth of *B. cereus* in some dairy desserts to an acceptable level). In other cases, one control measure at a single control point can control more than one hazard (e.g., pasteurization eliminates all vegetative pathogens and spoilage microorganisms). One control measure can be relevant to several process steps where a hazard is repeated (e.g., application of CIP cleaning or environmental cleaning to control recontamination). Where no control measure can be identified to control a hazard, redesign or modification of the process or product formulation may need to be considered. A final point to note is that in identifying control measures in a HACCP study on an established product and process, the team should not restrict consideration to the measures already in place but should be prepared to propose other control measures that might be appropriate.

The completion of Stage 13.5.7 completes Principle 1 of HACCP. **It is emphasized that, up to this point, no consideration has been given to the identification of critical control points.**

13.5.8 Stage 8: Determine CCPs (Principle 2)

The objective of Stage 8 (Principle 2) is to systematically assess the hazards and related control measures identified in Step 7 by considering each process step (as recorded in the flow diagram) in turn and reaching a conclusion on its "CCP" status before moving on to the next process step—that is, to identify process steps at which control can be applied and which are essential to prevent or eliminate a hazard or reduce it to an acceptable level.

It is useful to be guided by a CCP decision tree, and the version recommended by the Campden and Chorleywood Food Research Association (1997) is reproduced as Figure 13.3. The application of such decision trees should be flexible and requires professional judgment and common sense.

Answer each question in sequence at each process step for each identified hazard.

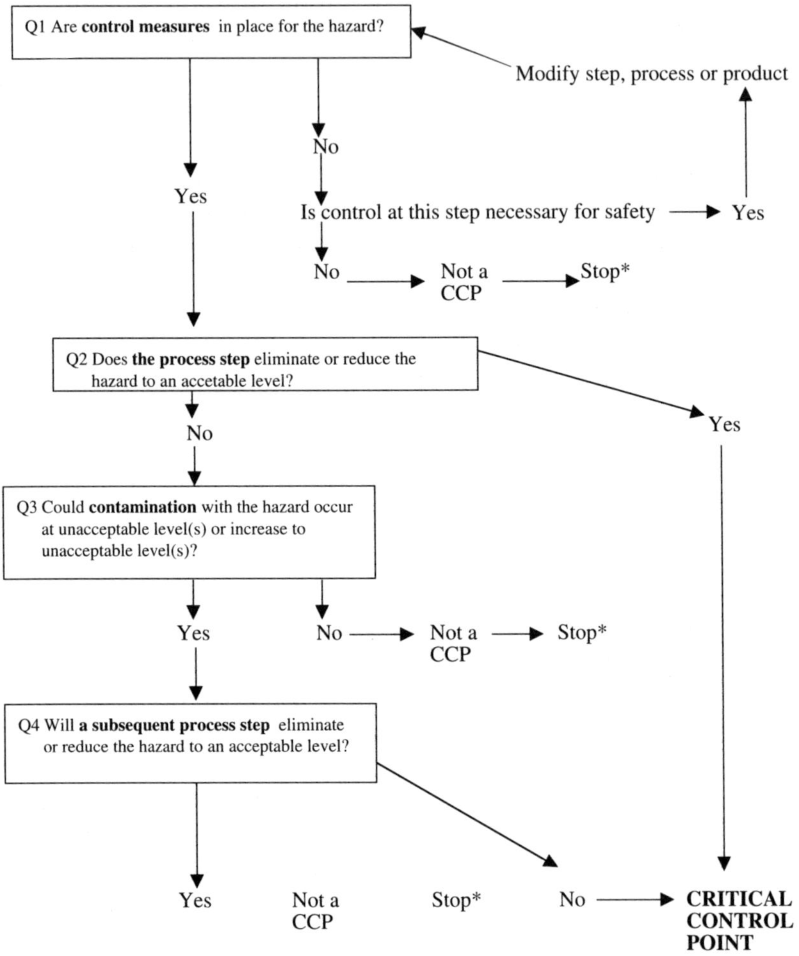

• Proceed to next step in the described process
Leaper (1997)

Figure 13.3. A CCP decision tree for the determination of critical points in HACCP plans.

The determination of CCPs is discussed below using two generic HACCP plans. These are Figure 13.5 (skim milk powder) to be considered with the related flow diagram (Figure 13.4) and Figure 13.7 (Cheddar cheese) to be considered with the related flow diagram (Figure 13.6). A flow diagram for a dairy dessert (Figure 13.8) is also presented for more general comment.

It is emphasized that these figures are not intended to represent the outcome of a full HACCP study. They have been prepared to illustrate points and are based on generic experience.

13.5.8.1 Skim Milk Powder—Salmonellae. Reference is made to Figure 13.4 (generic flow diagram) and Figure 13.5 (generic HACCP plan) that follow this section.

The terms of reference limit this illustrative example to the control of contamination with salmonellae. The food safety requirement, as recorded in regulatory specifications, is typically absent in 25 g, $n = 10$, $c = 0$ (EEC, 1992).

Figure 13.5 illustrates a typical output of Principle 1 (to Stage 7) with agreed hazards and identified control measures recorded against process step numbering that is taken from the flow diagram (Figure 13.4). The outcomes of the application of the CCP questions following the decision tree (Figure 3) are recorded in the fourth column. Bearing in mind that application of the decision tree should be flexible and based on informed judgment, the following points can be made.

1. The hazard of contamination of the environment (by raw milk) is recorded for process steps 1 to 4, but control is not considered to be a critical control point because subsequent process steps are designed to control this hazard. This is recorded as Step 9 (after Step 1).
2. Effective cleaning (of equipment) is recorded as a control measure at Process Steps 2, 4, 5, 6, 7, 8.1, and 8.4 and as an element of an identified CCP from Step 6 onwards. The control measure, effective cleaning, would need to be defined by detailed specification and procedures relevant to each process step in support of the HACCP plan—if this detail is not to be repeated in the plan. The HACCP team would need to validate cleaning procedures as being suitable to remove relevant contamination at each step, as would be the case for environmental cleaning.
3. The pasteurization step (Process Step 4) is recorded as *not* being a critical control point in the process under study. If the procedure is followed rigidly and each process step is considered and the CCP status decided before moving to the next process step, the CCP question response would be: Q1 yes, Q2 yes—a CCP. However, a focused HACCP team should be aware that a subsequent step (Step 6, evaporator preheat) affords a time/tempera-

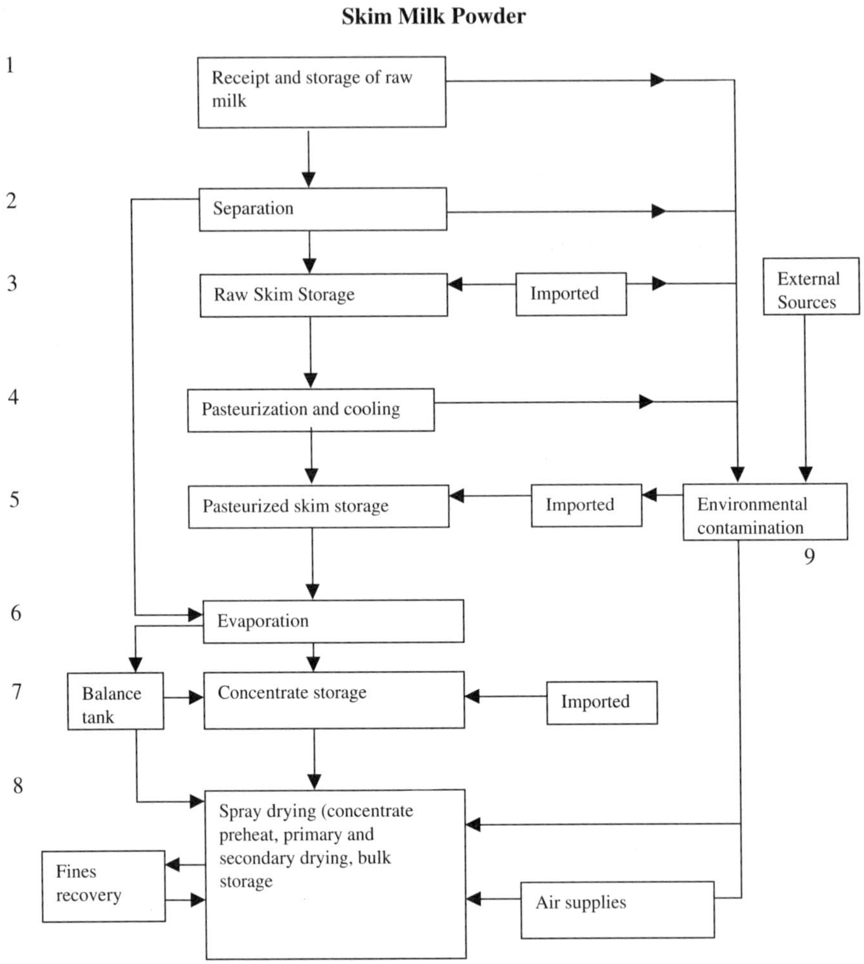

Figure 13.4. Generic flow diagram for the manufacture of skim milk powder. *Notes*: (a) Single cooling water supply: stages 1, 2, 3, 4, 5, and 7. (b) CIP: unit 1 to stages 1, 2, and 3; unit 2 to stage 4; unit 3 to stages 5, 6, 7, and 8. (c) Compressed air supplies are filtered at delivery. (d) All storage silos are chill water jacketed: stages 1, 3, 5, and 7. (e) Main drier chamber has insulation material between inner and outer skin; cyclones are all single-skin. (f) Ventilation intake from roof area above cyclone matrix.

ture condition that will eliminate salmonellae. This potential for confusion can be overcome in one of two ways: (a) by assigning Step 4 as a CCP and then reviewing when Step 6 is also identified as an effective control measure or (b) by interpreting Question 2 on the decision "Free" as the pasteurization at Step 4 *not* being

Process Conditions - Low-Heat Skim Milk Powder

Raw milk held at below 5°C, maximum 48 h

Separation at 55°C

Raw skim held at below 5°C, maximum 24 h

Pasteurization not less than 71.7°C, 15 s

Pasteurized skim storage at below 5°C, maximum 48 h

Feed to evaporator preheated to 72°C, 40 s (regeneration)

Final evaporator effect temperature 45°C at 50% solids

Concentrate feed to drier 65°C, time estimated < 30 s

Primary air temperature 215°C in drier, cold-filtered to Eurovent EU10

- There is no effective segregation of process areas from raw milk through to evaporator feed.
- Silos are vented to an environment which is potentially contaminated with salmonellae from raw process stages.
- Concentrate silos are in the yard area, next to raw milk intake bay and silos.

Figure 13.4. *Continued.*

included (designed*) to eliminate salmonellae and then continuing through to the "Not a CCP" decision. It must be emphasized that this interpretation is only valid within the terms of reference of the study and the specific process equipment and layout.

4. The second hazard at Process Step 6.1, leakage of skim milk from regenerator coils to final effect product at a temperature that would not be expected to eliminate salmonellae, can be judged as a CCP as indicated. An HACCP team would need to be assured that leakage was unlikely to occur, provided that the unit was maintained and validated to schedule. If this assurance could *not* be given, then the control measure at this point would need to be the delivery of pasteurized skim milk that had not been recontaminated and Step 4 becomes a CCP (i.e., consideration of imported skim, tanks and lines, and cleaning).

**Note.* Codex (1997) words Question 2 as follows: Is the step specifically *designed* to eliminate or reduce the likely occurrence of a hazard to an acceptable level. (refer to Q2 in Figure 13.3).

Generic HACCP plan for management of *Salmonella* in the manufacture of Skim Milk Powder.

Skim Milk Powder – *Salmonella*

PROCESS STEP	HAZARD	CONTROL MEASURES	CCP QUESTIONS					CRITICAL LIMITS	MONITORING	CORRECTIVE ACTION	RESPONS-IBILITY
			1	2	3	4	CCP				
1. Receipt and storage of raw milk	• Salmonellae present in raw milk supply.	• None.	N	N			No				
9. Contamination of environment (also at steps 2,3 and 4)	• Contamination of environment	• Environmental cleaning	Y	N	Y	Y	No				
2. Separation	• Increase in level of salmonellae • Contamination of environment	• Immediate cooling • Effective cleaning • Environmental cleaning	Y	N	N		No				
3. Raw skim storage	• Contamination of environment • Imported raw skim	• Environmental cleaning • None	N	N			No				

Figure 13.5. Generic HACCP plan for management of *Salmonella* in the manufacture of skim milk powder.

PROCESS STEP	HAZARD	CONTROL MEASURES	CCP QUESTIONS 1	2	3	4	CCP	CRITICAL LIMITS	MONITORING	CORRECTIVE ACTION	RESPONSIBILITY
4. Pasteurization and cooling of skim milk	• Survival of salmonellae	• Heating temperature • Heating time	Y	N	Y	Y	No				
	• Contamination of environment	• Environmental cleaning									
	• Recontamination from raw skim or cooling water	• Flow diversion system to prevent forward flow of underheated skim • Process side at positive pressure • Integrety of plate packs • Status of cooling water • Effective cleaning and sanitation	Y	N	Y	Y	No				

Figure 13.5. *Continued.*

PROCESS STEP	HAZARD	CONTROL MEASURES	CCP QUESTIONS 1	2	3	4	CCP	CRITICAL LIMITS	MONITORING	CORRECTIVE ACTION	RESPONS-IBILITY
5. Pasteurized skim storage	• Recontamination from equipment and/or environment	• Integrity of silos • Status of cooling water • Siting of storage silos and status of air drawn in during emptying • Effective cleaning	Y	N	Y	Y	No				
	• Contaminated imported pasteurized skim	• (Specification for product)	Y	N	Y	Y	No				

Figure 13.5. *Continued.*

PROCESS STEP	HAZARD	CONTROL MEASURES	CCP QUESTIONS 1	2	3	4	CCP	CRITICAL LIMITS	MONITORING	CORRECTIVE ACTION	RESPONS-IBILITY
6. Evaporation **6.1.** Skim feed	• Contaminated skim delivered to preheat • Leakage of regeneration coils to final effect product	• Step 4 o 5 controls • Integrity of regeneration coils	Y	Y			CCP	• No leakage	Scheduled pressure checks	• Repair • Review product quality	E TM
6.2. Preheat	• Survival of salmonellae in preheat process	• Heating temperature • Heating time	Y	Y			CCP	• Minimum 71.7°C • Minimum 15 s	Continuous temperature log Flow rate monitored	• Correct fault • Review product quality • Adjust • Review product quality	E TM E TM
6.3. Evaporator	• Evaporator and/or lines to/from contaminated	• Start up and restart procedures • Effective cleaning	Y	Y			CCP	• To process instruction • To C/P schedule	Process log sheets CIP logs	• Shut down clean and restart • Reclean • Review product quality	P P TM

Figure 13.5. *Continued.*

PROCESS STEP	HAZARD	CONTROL MEASURES	CCP QUESTIONS 1	2	3	4	CCP	CRITICAL LIMITS	MONITORING	CORRECTIVE ACTION	RESPONS-IBILITY
7. Concentrate storage (silos in yard next to raw milk intake)	• Recontamination from equipment or environment	• Integrity of silos	Y	Y			CCP	• No leaks	Annual crack Detection check	• Repair • Increase monitor frequency	E
		• Status of cooling water						• Microbiologically potable • < 1 ppm free Cl_2	• Weekly checks TBC and coiliform • Daily check	• Review treatment • Adjust dose rate	E E
		• Filtered air to silo head space						• No contamination	• Scheduled inspection of filters	• Replace	E
		• Effective cleaning						• To CIP schedule	• CIP logs	• Reclean • Review product quality	P TM
		• Lidded / enclosed balance tank						• Lid in place	• Supervision	• Replace lid	P
	• Contaminated imported concentrate	• Specification for product and delivery hygume	Y	N	Y	N	CCP	• *Salmonella* absent in 25 g, $n = 10$, $C = 0$	• Suppliers QA and QC records	• Review product quality	TM

Figure 13.5. *Continued.*

PROCESS STEP	HAZARD	CONTROL MEASURES	CCP QUESTIONS 1	2	3	4	CCP	CRITICAL LIMITS	MONITORING	CORRECTIVE ACTION	RESPONSIBILITY
8. Spray drying (primary, secondary, transport to bulk silos)											
8.1. Hygiene	• System contaminated	Hygiene procedures for drier invasion	Y	Y			CCP	• *Salmonella* not introduced	• Environmental sample plan	• Focused clean of affected areas	P/TM
		• Effective wet cleaning						• To cleaning instruction standards	• Cleaning logs	• Reclean	P
		• Effective dry out						• No residual water in system	• Temperature and humidity at remote air exit point	• Continue dry out	P
		• Barriers to control spread of water in wet cleans						• In place and no wetting of dry zones	• Supervision	• Wet clean whole system	P
		• Clean and dry after fire water activation						• As above			

Figure 13.5. *Continued.*

PROCESS STEP	HAZARD	CONTROL MEASURES	CCP QUESTIONS 1	2	3	4	CCP	CRITICAL LIMITS	MONITORING	CORRECTIVE ACTION	RESPONS-IBILITY
8.2. Start-up	Contamination of system before primary air reaches operating temperature	• None identified • *Modify process to introduce new control measure	N	Y			*				
		• Filter primary air supply	* Y	Y			CCP	• Air filtered to Eurovent EU 10	• Scheduled inspection of filters	• Replace	E
8.3. Air sweep door	• Contamination of system via air sweep door	• Environmental hygiene • Environmental air quality	Y	N	Y	N	CCP	• Salmonellae not introduced • Salmonellae absent	• Environmental Sampling plan	• Focused clean of affected areas • Review source	P/TM TM
8.4. Concentrate preheating	• Delivery of contaminated concentrate to atomizer • Survival of salmonellae in concentrate droplets in primary drying	• Preview steps • Effective cleaning and sanitising	Y	N	Y	N	CCP	• To cleaning instruction standards	• Cleaning logs	• Reclean	P

Figure 13.5. *Continued.*

PROCESS STEP	HAZARD	CONTROL MEASURES	CCP QUESTIONS 1	2	3	4	CCP	CRITICAL LIMITS	MONITORING	CORRECTIVE ACTION	RESPONS-IBILITY
8.5. Drying and powder handling systems	Recontamination of powder by:		Y	Y			CCP			• Repair	E
	Contamination from insulation on drier chamber (and other double skin units in the system)	• Maintain product contact surfaces crack free						• No cracks/pin holes	• Annual inspection and test	• Increase monitoring frequency • Review product quality	E TM
		• Keep insulation material dry						• Insulation dry	• Monthly check at base of drier and check for salmonellae	• Investigate source, repair and replace lagging • Review product quality	TM/E E
		• Drain points on double skin vessels						• No evidence of leakage	• Daily inspection	• Repairs	E
	Accumulation of under dried product in system	• Shutdown and clean/sanitize						• No accumulation	• Daily inspection	• Shutdown and clean	P/TM
	Airborne contamination via secondary drying air and transport air	• Air filtration						• Air filtered to Eurovent EU1O	• Scheduled inspection of filters	• Replace	E

Figure 13.5. *Continued.*

PROCESS STEP	HAZARD	CONTROL MEASURES	CCP QUESTIONS 1	2	3	4	CCP	CRITICAL LIMITS	MONITORING	CORRECTIVE ACTION	RESPONS-IBILITY
	Environmental air drawn into system via trunking joints, hatch seals, etc.	• Maintenance						• No air ingress	• Daily inspection	• Repair	E
		• Environmental hygiene						• Salmonellae not detected in environmental dust samples	• Weekly samples to plan	• Focused clean of affected areas	TM
	Contaminated trunking gaskets	• Maintenance/ replacement						• Not contaminated	• Annual inspection and check for salmonellae	• Replace	E/TM
8.9. Environmental contamination	• Contamination of building from external sources	• Building maintenance to prevent ingress of soiling	Y	Y			CCP	• No ingress of water or other external contamination	• Ongoing inspection of building	• Repair • Clean effect areas including structure voids	E P
		• Specified and scheduled environmental cleaning						• To cleaning instruction standards	• Cleaning log	• Reclean	P
								• Salmonellae not detected	• Weekly samples to plan	• Focused clean of affected areas • Review product quality	P TM

Figure 13.5. *Continued.*

5. At Process Step 8.2, (drier start-up), a hazard is identified in the initial hazard analysis that has no control measure, but control is considered necessary for food safety. Under these circumstances, the HACCP team would consider how to modify the process, and in this case the filtration of primary air to the drier is proposed and identified as a CCP.
6. At Process Step 8.4 (concentrate preheating), the hazard of salmonellae surviving in drying droplets during primary drying was considered. Concentrate preheat conditions (60°C for ≤1 min) might be considered as a control measure; but it would be concluded that at the solids levels involved (50%), this would not be an effective control (Mackey and Derrick, 1987), nor could this process step be modified (higher temperature/longer hold time) without altering the physical quality of the product. The delivery of salmonellae-free concentrate is, therefore, identified as the critical control point relying on control measures at previous process steps and the control measure of effective cleaning and sanitizing at Step 8.
7. Process Step 8.5 records hazards representing routes of contamination of dried powder in the process (secondary drying, powder/air separation in cyclones, and powder transport). It is generally accepted that the most frequent route of recontamination of spray-dried powder is from the environment into the drier system; a particular focus of contamination is damp or damaged insulation on drier systems. In many incidents of salmonellae-positive powders, the original source has been ingress of roof soiling (powder deposits, bird activity). The view should be taken at Question 2 that this process step is "specifically designed to eliminate or reduce the likely occurrence of the hazard of powder recontamination." In a detailed, targeted HACCP study, however, this process step would be subdivided into several steps representing the various process elements from primary drying through to delivery of the powder to bulk silos. In this detail, individual elements (steps) might be judged differently; for example, spray-drier chambers are not designed to eliminate the hazard of insulation material becoming a focus of contamination, although this has proved to be a route of contamination in many incidents.

13.5.8.2 Cheddar Cheese—Control of Salmonellae. Figure 13.6 presents a generic flow diagram, and Figure 13.7 presents a generic HACCP plan.

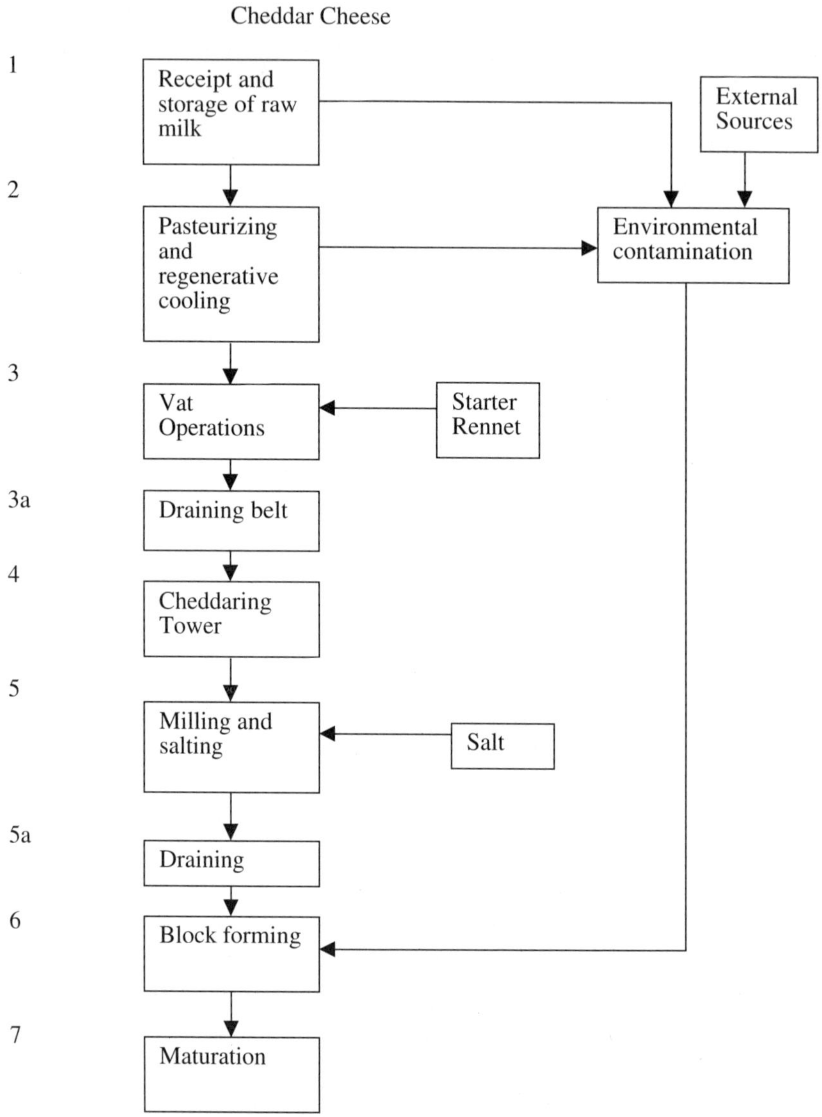

Figure 13.6. Generic flow diagram for the manufacture of Cheddar cheese. *Notes*: (a) Cooling water supply to stage 1 only. (b) CIP: Unit 1 to stages 1 to 3; Unit 2 to 3a, 5, and 5a; others have dedicated units. (c) Air for salt and curd transport-filtered to Eurovent EU110.

The terms of reference limit this illustrative example to the control of contamination with salmonellae. The food safety requirement, as recorded in regulatory specifications, is typically absent in 25 g, $n = 10$, $c = 0$ (EEC, 1992). Figure 13.7 illustrates a typical output of Principle

PROCESS CONDITIONS – CHEDDAR CHEESE

Raw milk held:	At below 5°C, maximum 48 h
Pasteurization:	At not less than 71.7°C, 15 s
Milk to Vat:	30 ± 1°C
Starter DVI:	Added to supplier's instruction within 30 min of vat fill
Rennet:	Added to supplier's instruction
Scalding:	30–39°C over 45–60 min, hold at 39°C for 45–60 min
Mill:	At not less than 0.4% lactic acid
Salt addition:	2–3%
Maturing temperature:	11°–12°C

Figure 13.6. *Continued.*

1 (to Stage 7) with agreed hazards and identified control measures recorded against process step numbering that is taken from the flow diagram (Figure 13.6). The outcomes of the CCP questions following the decision tree (Figure 13.3) are recorded in the fourth column. Bearing in mind that application of the decision tree should be flexible and based on informed judgment, the following points can be made.

1. Process Step 1 (receipt and storage of raw milk) is identified as a CCP with respect to the hazard of inhibitory substances (antibiotics) that might inhibit starter activity in the cheese vat. This process step would not be considered to be a CCP with respect to the elimination or control of salmonellae.

2. The view taken with respect to environmental contamination (Process Step 9) and effective cleaning (of equipment) is the same as that for the skim milk powder example and would apply throughout a focused, specific HACCP Study.

3. Process Step 3 (vat operations) is identified as a CCP. The key issue here is the rate and level of acid development, which directly influences the microbiological status of subsequent process steps and indirectly influences the physical condition of the cheese. With slow acid

Cheddar Cheese – *Salmonella*

PROCESS STEP	HAZARD	CONTROL MEASURES	CCP QUESTIONS 1	2	3	4	CCP	CRITICAL LIMITS	MONITORING	CORRECTIVE ACTION	RESPONSIBILITY
1. Receipt and strorage of raw milk	• Salmonellae present in raw milk supply	• None									
	• Presence of inhibitory substances	• Specification and supply audit	Y	N	Y	N	CCP	Not greater than 0.6 IU inhibitory substances	• Each intake	• Review effect on silo mixture do not use milk if limits are exceeded	P/TM
9. Contamination of environment (also at step 2)	• Contamination	• Environmental cleaning	Y	N	Y	Y	CCP	• To cleaning instruction standards	• Cleaning logs	• Reclean	P
2. Pasteurization and cooling to vat temperature	• Survival of salmonellae	• Heating temperature	Y	Y			CCP	• Minimum 71.7°C	• Continuous record	• Embargo product for quality and disposal review	TM
										• Clean affected equipment before restart	P
		• Heating time						• Minimum 15 s	• Annual holding time check	• Adjust	
	• Contamination of environment	• Environmental cleaning • Lidded pasteurizer balance control						• As at step 9 • In place	• Supervisor	• Put in place	P

Figure 13.7. Generic HACCP plan for the management of salmonellae in the manufacture of Cheddar cheese.

PROCESS STEP	HAZARD	CONTROL MEASURES	CCP QUESTIONS 1	2	3	4	CCP	CRITICAL LIMITS	MONITORING	CORRECTIVE ACTION	RESPONSI-BILITY
	• Recontamination from raw milk (regeneration)	• Flow diversion system to prevent forward flow of under heated milk	Y	Y			CCP	• All underheated milk diverted before cooling stage	• Continuous record • Daily start up FDV operation check	• Embargo product for quality and disposal review • Clean affected equipment before restart	TM/P
		• Process side at positive pressure						• < +4 psi		• Adjust	
		• Integrity of plate packs						• No leaks from raw to pasteurized side	• Continuous record	• Replace affected plates	E
								• to CIP standard	• 6 monthly leak detection checks	• Reclean	E
		• Effective cleaning and sanitation							• CIP logs		P
	• Milk not subject to full pasteurization	• All of the above plus lines	Y	Y			CCP	• Pass on phosphatose test at cooler exit	• Start-up and 2 hourly	• Embargo product for quality and disposal review	TM

Figure 13.7. *Continued.*

PROCESS STEP	HAZARD	CONTROL MEASURES	CCP QUESTIONS 1	2	3	4	CCP	CRITICAL LIMITS	MONITORING	CORRECTIVE ACTION	RESPONSIBILITY
3. Vat operations 3.1. Acidity development	• Abnormal fermentation-failure to reach required acidity/ptt	• Pasteurized milk temperature • Correct starter addition rate/time • Starter activity • Inhibitory substances minimized • Bacteriophage control protocols	Y	Y			CCP	• 30 ± 1°C • To recipe • See step 1 • To site limits	• Pasteurize cooler exit continuous record • Process log • Process log • Acid development • Daily phage detection test	• Adjust • Embargo for quality and disposal review • Change strates rotation	P TM P/TM
3.2. Scalding cutting striring drainning	• Failure to shrink curd to required level (effect on whey expulsion and cheese moisture levels) • Recontamination	• Scald time and temperature profile • Stir time and temperature • Effective equipment cleaning and sanitizing (vats, lines and valving	Y	Y			CCP	• To 39°C over 45–60 min and hold at 39 ± 1°C 45 to 60 min • To CIP standard	• Process records • Curd quality • CIP log	• adjust • Reclean	P
4. Cheddaring (draining belt 3a and tower)	• Abnormal transit time and/or loading to affect cheese moisture • Recontamination	• Manufactory profile • System adjustment • Effective equipment cleaning and sanitising	Y	Y			CCP	• To product recipe conditions • To CIP standard	• Process logs • Curd quality • CIP log	• Adjust timings • Reclean	P P

Figure 13.7. *Continued.*

PROCESS STEP	HAZARD	CONTROL MEASURES	CCP QUESTIONS 1	2	3	4	CCP	CRITICAL LIMITS	MONITORING	CORRECTIVE ACTION	RESPONSIBILITY
5. Milling, salting and draining (5a)	• Failure to meet minimum acidity level in curd • Wrong level of salt or uneven distribution	• Control at step 3 • Maintenance and calibration of salting mechanism • Inadequate milling and mixing	Y	Y			CCP	• Acidity develop to < 0.4% LA within 6 h of vat filling • Salt levels within 20% of target value throughout blocks	• Acidity check each batch • Random checks on salt distribution daily	• Cheese embargoed for quality checks and disposal decision • Adjust system • Review cheese disposal	TM P TM
	• Recontamination	• Effective cleaning and sanitizing • Salt and curd transported using filtered air	Y	Y			CCP	• To CIP standard • Air filtered to environment EU 10 standard	• CIP log • Scheduled inspection of filters	• Reclean • Replace	P E
6. Block forming	• Failure to develop to acidity (ptt requirement)	• Controlled at steps 3, 4, 5									
7. Maturation	• None relevant to scope										

TM = Technical Manager, P = Production, E = Engineering

Figure 13.7. *Continued.*

production, salmonellae (and other pathogens—i.e., *Staphylococcus aureus* and enteropathogenic *Escherichia coli*) can grow rapidly during cheese-making (6–7 generations after overnight pressing) and salmonellae can survive during cheese maturation (IDF, 1980). Cheddar cheese has been involved in salmonellae outbreaks, and one incident involving 2700 cases was attributed to incomplete pasteurization of vat milk, poor acid development, and poor equipment cleaning (Ratnam and March, 1986). The strain linked with this outbreak was reported to have a low infective dose (<10 cells per portion as consumed). Instances of contamination of Cheddar cheeses with salmonellae are known in the industry; and it is generally accepted that a minimum acid development in the cheese vat should be 0.4% lactic acid, and reasonable time limit for this is within 6h of vat filling. Application of these critical limits has proved to be an effective control; that is, there should be a minimum of 0.4% lactic acid at milling and salting.

However, some processes seek to maximize throughput by accelerating passage through the vat, and salting and milling takes place at acidities as low as 0.3% lactic acid with obvious consequences in terms of the control of salmonellae should the curd be contaminated. Should this situation be presented to an HACCP team, it should elicit a "no" to Question 1 and a "yes" to the supplementary question calling for a modification to the process at this step. One solution might be the selection of starter cultures that are able to continue to develop acidity at an adequate rate after salting, so that the desired minimum acidity is achieved within the required time. Before acceptance, such a change in process would need to be fully validated and an effective means of monitoring identified.

4. Process Step 5 (milling and salting) is identified as a CCP, with the key issue being the level and distribution of the salt. In poorly run systems, salt variations with a 10-kg block of cheese can be as wide as 1% to 3% against a recipe level of 2%. Apart from adversely affecting the organoleptic quality of the cheese, such extremes could have a significant effect on further acidity development (see 3 above).

13.5.8.3 Dairy Dessert—Bacillus cereus. Figure 13.8 is a schematic flow diagram for the manufacture of a single-layer, mousse-type dairy dessert. An illustrative HACCP plan has not been produced, but the following notes illustrate some key points that an HACCP team would need to consider against the terms of reference for control of psychrotrophic *B. cereus* in the product.

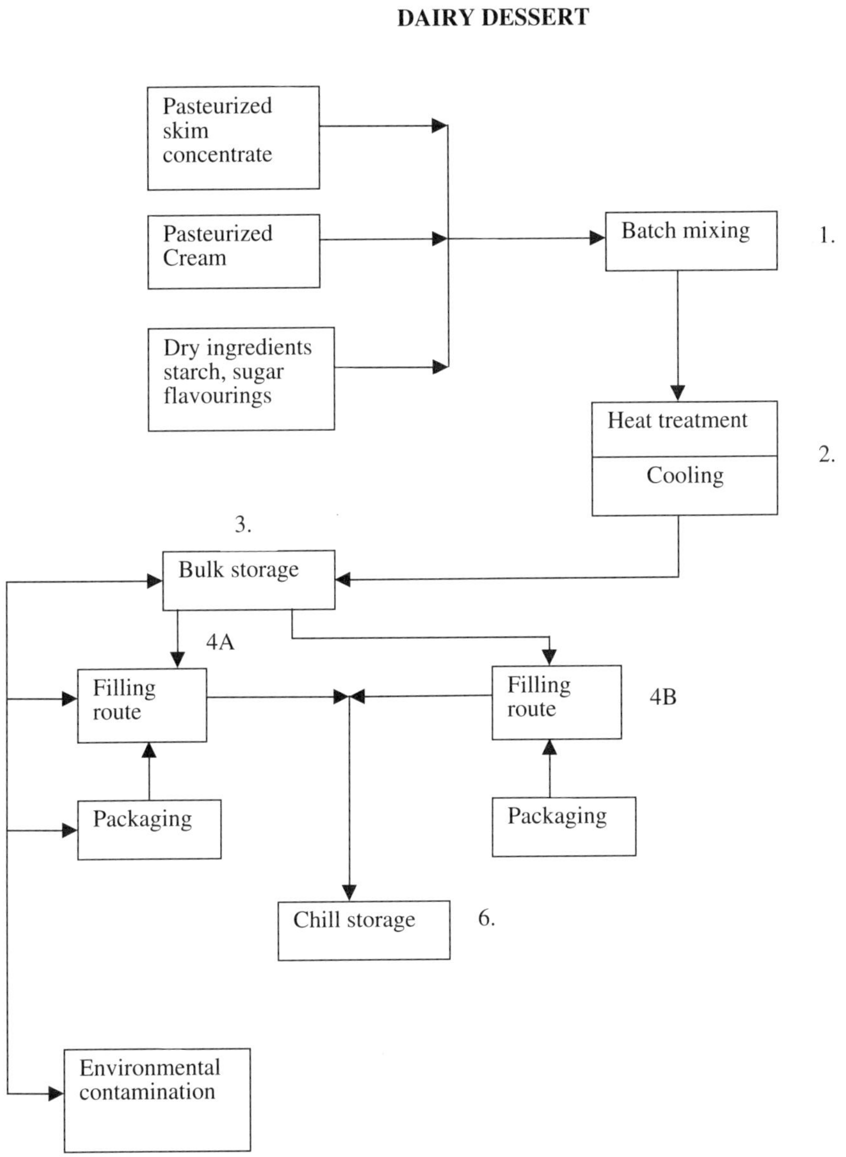

Note: (A) Cooling water supply to Stage 2 and 3
(B) CIP-Unit 1 to Stage 1
Unit 2 to Stage 2 and 3 and 4A
Stage 2 and 4B have dedicated Units
(C) Compressed air supplies

Figure 13.8. Generic flow diagram for the manufacture of a dairy dessert. *Note*: (a) Cooling water supply to stages 2 and 3. (b) CIP: Unit 1 to stage 1; Unit 2 to stages 2, 3, and 4A; stages 2 and 4B have dedicated units.

1. While there is no epidemiological evidence that this type of product is implicated in food poisoning cases, there are guidelines indicating that 10^4/cfu g^{-1} should be the maximum at point of sale. This figure should be adopted as the manufacturer's food safety requirement.

2. The level of psychrotrophic *B. cereus* in ingredients is significant with respect to the base level in the batch presented for heat treatment.

3. The heat treatment may or may not be considered to be sufficient to eliminate the spores of psychrotrophic *B. cereus*; for example, an estimate based on some reference data suggests that a thermal process of 110°C for 36 s will effect a 6 log reduction. If the thermal process is to be accepted as an effective control measure, a detailed review of databases for thermal death data and validation of the chosen process would be necessary.

4. The thermal process also has the function of cooking the starch component; this has a direct bearing on the structure of final product, and possibly on the growth rate of *B. cereus* in the product.

5. Changes in formulation can influence the rate of growth of *B. cereus* in the product—for example, a change in sugar levels (A_W effect) or a change in product structure (starch matrix or fat level).

6. Bulk storage time and temperature can significantly influence the shelf life of the packed product because psychrotrophic strains of *B. cereus* will multiply relatively quickly at slightly elevated temperatures.

7. Filling route 4A represents a conventional pot filler with all the attendant opportunities for recontamination with Gram-negative organisms (e.g., *Pseudomonas* spp.), so psychrotrophic *B. cereus* does not limit shelf life.

8. Filling route 4B represents an "aseptic" filling system aimed at extended shelf life (20 days or more), and in the absence of other contaminants the multiplication of psychrotrophic *B. cereus* (or others psychrotrophic spore-formers) is likely to limit shelf life. Points 1 to 7 are the more significant in this situation.

Again, it is emphasized that the notes above relate to specific areas for consideration in an HACCP study with the stated terms of reference. In a "live" HACCP study, other relevant points would need to be considered.

The decision tree approach can be applied to process/product development projects when the HACCP team would either (a) consider control measures available at Question 1 or (b) use the procedure to specify controls that would need to be established.

The output of Principle 2/Stage 8 is identification of critical control points (CCPs) at process steps at which control is essential to prevent or eliminate a hazard or reduce it to an acceptable levels. This ensures that control and management at CCPs is given priority to assure food safety (or other quality requirements specified in the terms of reference for the study).

13.5.9 Stage 9: Establish Critical Limits for Each CCP (Principle 3)

A critical limit is a criterion that separates acceptability from unacceptability at each CCP. It should be measurable in real time (while the process is running) and might include measurements of temperature/time, pH or acidity, moisture, the phosphatase test for pasteurized milk, ATP methodology to assess cleaning efficiency, or other observations. A critical limit might be mandatory (e.g., pasteurization temperature and time) or based on data collected under good manufacturing practice where a specific target level and tolerances are set.

Typical critical limits are illustrated against CCPs in Figures 13.5 and 13.7.

13.5.10 Stage 10: Establish a Monitoring System for a CCP (Principle 4)

Monitoring involves a planned sequence of observations or measurements against Critical Limits to assess whether a CCP is under control. Ideally, monitoring should identify a trend toward a critical limit maximum or minimum so that corrective action can be taken before the process is out of control and, in any event, should aim to identify violation of critical limits as soon as possible to minimize the amount of embargoed/rejected product. Monitoring can be on-line with automated corrective action (e.g., flow diversion systems on pasteurizers), or they can be off-line when corrective action might involve the embargo/rejection of any product implicated. Physical and chemical measurements are preferred to microbiological testing because they can be completed rapidly and often be indicative of conditions that control the microbiology of the product (e.g., phosphatase test on pasteurized milk, acidity of cheese curd).

13.5.11 Stage 11: Establish a Corrective Action Plan (Principle 5)

This specifies the action(s) necessary when monitoring shows a potential or actual loss of control at a CCP. The action(s) will aim to bring the process back into control before critical limits are reached (e.g., a temperature drift from a target of 5°C to near the tolerance value of 7°C would call for an engineer to adjust the refrigerator plant), or it will specify the disposal of product that has breached a critical limit (e.g., handling a vat of cheese milk failing a phosphatase test could involve affected product embargo, a decision on disposal, an investigation of the reasons for failure and correction, and cleaning and sanitizing all equipment affected; this would be the responsibility of the technical manager).

Monitoring requirements and corrective action plans should be considered together by the HACCP team, and a clear decision should be reached and recorded on responsibilities for corrective actions. Outline monitoring and corrective actions are illustrated in Figures 13.5 and 13.7, but see also Stage 13 (Principle 7).

13.5.12 Stage 12: Verification (Principle 6)

Verification applies methods, procedures, product tests, and other evaluations, other than monitoring, to determine compliance with the HACCP plan; that is, it demonstrates that the HACCP plan and its application is consistently controlling the process so that product meets the food safety requirements (or quality requirement). The HACCP team should specify methods and frequency of verification procedures, which might include the following:

1. Microbiological examination of intermediate and final product samples. In the examples used in Section 13.5.8, these steps would include the examination of skim milk powder for salmonellae, the examination of the dessert (at the end of shelf life) for *B. cereus*, and the examination of cheese blocks (at 24 h, for example) for acidity/pH, salt level and distribution, and moisture.
2. Review of complaints from consumers or regulatory bodies and outcomes of investigations into these complaints, if they were substantiated, indicating that the HACCP plan did not completely control the process.

3. Auditing all monitoring and corrective actions records to establish whether the HACCP plan is being fully implemented and demonstrates control.
4. A review of validation records and, if appropriate, the application of more searching tests at selected CCPs to confirm the efficacy of the control measure (e.g., does the thermal process specified for the dessert reduce spores of psychrotropic *B. cereus* by the specified log reductions?).

Note: Codex Alimentarius defines validation as "obtaining evidence that the elements of the HACCP plan are effective." There is some confusion on the differentiation between verification and validation in HACCP procedures, with some suggesting that validation extends to demonstrating the correct choice of critical limits criteria. Unless and until this is clarified, confusion can be avoided by (a) interpreting validation as confirming the efficacy of control measures *within* the HACCP plan and (b) interpreting verification as tests and observations *outside* of the HACCP plan to confirm that control is achieved by the combined effect of the control measures and monitoring of the critical limits chosen.

13.5.13 Stage 13: Establish Documentation and Record Keeping (Principle 7)

The complexity and quality of documentation necessary will depend on the size and type of operation. A large unit manufacturing only cheese would tend to be easier to document than a smaller unit manufacturing a range of dairy desserts with varying formulations, structures, and packaging regimes.

The key point is that the manufacturer must be able to demonstrate that the seven principles of HACCP have been correctly applied.

To be effective, HACCP must be fully integrated into the unit quality systems as an element of total quality management. Figure 13.1 shows the relationship between HACCP and quality systems through documentation.

13.5.13.1 Documentation Control Under ISO 9000. The following documentation should be issued as controlled documents:

1. The finalized HACCP plan, for example, uses the format illustrated in Figures 13.5 and 13.7 together with flow diagrams (Figures 13.4 and 13.6). Process steps assessed as *not* being CCP should also have

critical limits, monitoring procedures, and corrective actions identified on the HACCP plan, and they can be designated as control points that contribute to good manufacturing practice.

2. **Guidelines, Procedures, and Work Instructions/Record Sheets.** *Guidelines* on good hygienic practice (GHP) are an essential element of the documentation required. The output of HACCP plans should be used to produce site-specific guidelines that address all of the points listed in the template given in Section 13.2.3.1. It is important that HACCP drives the content of such guidelines so that all the control measures are fully covered. It is good practice to refer to the current site Hygiene Guidelines at the flow diagram stage of the HACCP study (Stage 5, Section 13.5.5), so that their contents are "placed on the table" for review in subsequent stages. Any issues specific to the HACCP study that are missing can be covered either by amendment of the Guidelines or by inclusion in the HACCP plan.

Procedures cover the following:

- Training for hygiene and operation.
- Personnel hygiene and sickness reporting.
- On-site food services.
- Use of protective clothing.
- Inspection and maintenance of equipment, manufacturing services (water, compressed air, drainage), and the building/site.
- Raw materials/ingredients—specification/audit/sourcing.
- Waste disposal.
- Cleaning—equipment/environment; CIP/manual.

In all cases the procedures should state clearly what should be done, how equipment or materials should be used, and by whom and how defects should be recorded, remedial action initiated, and action signed-off when completed.

As with guidelines, current procedures should be reviewed in an HACCP study and modified, if necessary, on the basis of the hazard analysis.

Work instructions give detailed instruction to operatives as to what has to be done at each process step. This will include, as appropriate, equipment manufacture's instructions, product recipe (ingredient quantities, process times and temperatures, routing of intermediate product and final product through the factory), and action to be taken in abnormal circumstances. Monitoring record sheets should be prepared to support, as necessary, work instructions, preferably with critical limits shown, and instructions on how to complete them and action to be taken if critical limits are challenged (process adjustment and/or notify management). The work instructions should be generated directly from the HACCP plan, and the monitoring record sheet gives the detail that would otherwise complicate HACCP documents. Furthermore, there should be a clearly defined mechanism by which abnormal (critical limits breached) results are notified on an exception reporting system that calls for a traceable record of corrective or remedial action taken and the outcome of these actions, signed-off at a designated management level.

Note: Where guidelines, procedures, and work instructions serve to carry forward HACCP plan requirements, cross-referencing by document number is recommended.

13.5.13.2 HACCP Study Notes. While the HACCP plan should be issued as a controlled document as part of site quality systems, it is important that the background notes made during the HACCP study be kept as a file for reference in HACCP review or trouble-shooting exercises. As a minimum, these notes should include the following:

- Product description notes (Stage 3).
- Basis for decisions taken in Stage 7 (Hazard Analysis).
- A note of any "judgment" decision taken at Stage 8 (Determination of CCPs), together with data refered to and/or external expert advice source.
- Recommended verification schedule (Stage 12).
- Notes on any validation exercise undertaken (Stage 12).
- A schedule of other quality system documents that are derived from/support the HACCP plan.
- Data derived from HACCP reviews.

The detail required in the HACCP study notes will depend on the complexity and novelty of the process. A long-established process (powder drying, cheese-making) should be relatively straightforward with few judgmental decisions. However, novel processes or products, such as desserts where formulations may be changed and structure/intrinsic factors altered, tend to move into areas where information is incomplete and informed judgments have to be made. It is important that the basis for these judgments be carefully recorded, with all data available to the HACCP team attached, so that HACCP reviews initiated by product/process change or adverse verification results have a basis for analysis.

13.5.14 Stage 14: Review of HACCP Plans

The review of a HACCP plan evaluates any changes in process, product, or manufacturing site against the current HACCP plan to determine whether new hazards have been introduced that are not covered by existing control measures at critical control points or control points. HACCP study notes will afford a valuable background to the review process. If new hazards that are not adequately controlled are identified, the HACCP plan should be amended accordingly and notes of the review should be added to HACCP study notes.

HACCP plan reviews should be triggered under the following circumstances:

- By routine schedule at a frequency determined by the HACCP team based on risk.
- Change in product formulation (e.g., change of sugar or fat level in a dairy dessert).
- Change in process (e.g., salting and milling of Cheddar curd at 0.3% lactic acid instead of 0.4%).
- Change in raw materials (e.g., a new source of starch for dairy dessert manufacture—possibility of higher *B. cereus* levels).
- Change in consumer use/longer shelf life assigned.
- Evidence of health or spoilage risk in the market place.
- Emergence of "new" food-borne pathogens (e.g., verotoxigenic *E. coli*).
- Change to factory layout and environment (e.g., additional process installed alongside the one in the HACCP plan).
- Modification to process equipment (e.g., different filler for desserts).
- Changes in packaging, storage, and distribution.
- Change in cleaning and sanitation program (e.g., new equipment, change in chemicals used, change in time/temperature regime in CIP or extension of CIP circuits).
- Change in staff levels and responsibilities (e.g., initiation of a night shift).

- Verification findings. **Of particular importance are the findings of food safety/quality audits based on HACCP plans.** Properly constructed, these audits will review monitoring records derived from the HACCP plan, including records of corrective actions, product disposals, and consumer complaints. In addition, the audit would cover all areas under the site Good Hygienic Practice Guideline.

Food safety/quality audits should be formally recorded and issued within the site Quality System and list nonconformances (including those in application of an HACCP plan) and set timetables for remedial action. **This is an effective route for the integration of the HACCP procedure into the application of process control, and it ensures that process control is focused on issues that are relevant to the safety/quality of the product.**

13.6 TROUBLE-SHOOTING

Product failure, whether identified by in-house quality control data or by customer complaint, should be investigated using a disciplined and structured approach. It is not helpful, and often counterproductive, to collect together all available samples and test them in the hope of demonstrating that the problem was a "one-off" (e.g., a laboratory somewhere got it wrong!). This wastes valuable time while microbiological results are awaited, during which time the problem might be ongoing, or the process closed down in the event of an implied public health risk. A preferred way to investigate is to call the HACCP team together, who should then follow a prepared procedure involving the following steps:

1. Confirm the identification of the microorganism implicated; it is not unknown for investigations to run for several days along the wrong path on the basis of misunderstandings about the causal microorganism.
2. Review the HACCP plan together with HACCP study notes in order to list what could have gone wrong. For example, if the complaint is of high *B. cereus* levels in a dessert, the list might include thermal process, time, temperature before filling and in distribution, change of ingredient supply, and change of cleaning regime.
3. Based on the outcome of step 2, a detailed review of relevant monitoring and verification records, including equipment main-

tenance logs, cleaning logs, and personnel issues, should be undertaken.

4. If any suspicions are raised in step 3, the HACCP team might need to ask for some specific tests to confirm suspicions.
5. The outcome of step 4, assuming a cause is discovered, should lead to an amendment of the HACCP plan and/or the implicated procedure/work instruction.
6. In the absence of a conclusion in step 4, the HACCP team should initiate a time-limited program of verification testing to confirm that adequate control is in place and that the complaint was in fact a "one-off" or spurious.

In all such trouble-shooting exercises, it is important that *all* members of the HACCP team be involved so that all disciplines can make an input.

The discussion points and outcome of the trouble-shooting exercise should be recorded with the HACCP study notes for future reference.

13.7 CONCLUSION

As was stated in the Introduction to this chapter, Process Control embraces all elements of a process that influence the legality, safety, contractual, and commercial requirements of a product. The scope is, therefore, from farm to consumer and includes raw materials, formulation, bacteriocidal or bacteriostatic treatments, plant and equipment hygiene, personnel practices and hygiene, packaging, distribution conditions, and consumer use.

Proper attention to such a broad scope requires a disciplined and documented approach, and Figure 13.9 summarizes the outline of such an approach based on HACCP. Details have been addressed in the chapter from a procedural point of view rather than the listing of process control requirements for specific products, although some generic examples are given in Figures 13.5 and 13.7. This approach was because other chapters in these volumes will give more specific information and, more importantly, each product in a particular manufacturing unit is unique and cannot be covered generically.

As has been illustrated, the fundamental requirement in the establishment of technically sound and cost-effective process control is, by addressing the food safety (or quality) requirements through HACCP

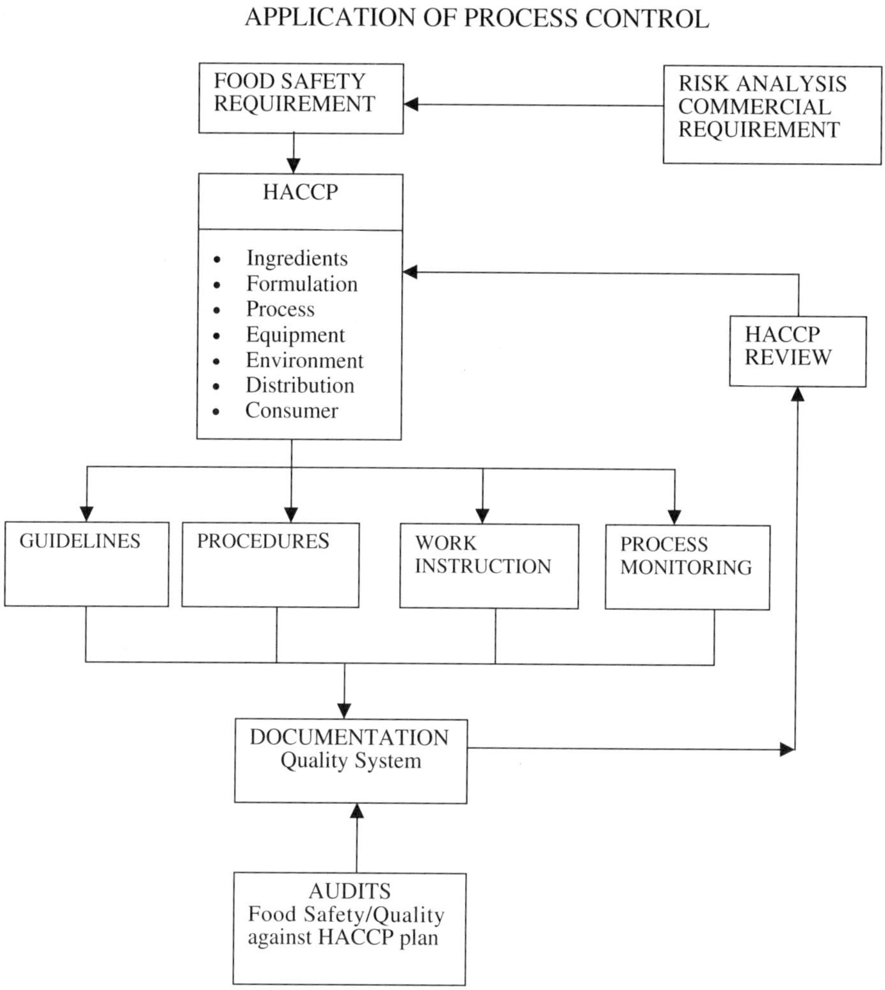

Figure 13.9. Schematic diagram showing an integrated approach to the application of process control.

studies, the derivation of the necessary control measures and monitoring methods to demonstrate compliance. HACCP should drive on-site guidelines (hygiene), procedures, work instructions, and process monitoring, and it should be incorporated into the documentation of a site quality system if it is to advise on process control. The importance of documented food safety/quality audits against the HACCP plan as one element of HACCP verification, and one of the triggers for a HACCP plan review is apparent. Figure 13.9 summarizes the elements of an integrated scheme.

The benefits of such an integrated system of process control can be summarized as follows:

- Hazards (to food safety/quality) are identified and control measures are confirmed.
- Monitoring requirements derived from a HACCP plan are focused on real-time measurements and observations wherever possible to allow rapid corrective action and avoidance of undue wastage of out-of-specification product—a move from retrospective QC to preventive QA.
- Available technical resources are focused on critical control points.
- It presents documented and relevant information to facilitate trouble-shooting exercises.
- It is a means of demonstrating diligence in process control to regulatory bodies and customers.

The HACCP principles discussed are relevant to all sizes of food manufacturing businesses. A World Health Organization consultation (WHO, 1999), while recognizing the difficulties is implementing HACCP in small and/or less developed businesses, stated that "the seven principles of HACCP can be applied to all businesses processing or preparing food, irrespective of size or nature of their work."

The value of HACCP will, in all cases, be enhanced by its inclusion as an element of any quality system documentation, and through this it will become a direct influence on the correct choice of process control parameters.

REFERENCES

Brocklehurst and Wilson (2000)

Campden and Chorleywood Food Research Association (1997) *HACCP—A Practical Guide*, 2nd ed. Technical Manual 38.

Codex (1997) *General Requirements (Food Hygiene)*, Supplement to Vol. 1B, FAO/WHO Rome.

Codex (1999)

Codex Alimentarius (1999a) *Proposed Draft Principles and Guidelines for the Conduct of Microbiological Risk Assessment*, Alinorm 99/13, Appendix 1V.

Codex Alimentarius (1999b) *Proposed Draft Principles and Guidelines for the Conduct of Microbiological Risk Management*, Alinorm 01/13, Appendix IV.

Corlett, D. A., Jr. (1999) *HACCP Users Manual*, Aspen Publishers, Gaithersburg, MD.

EEC (1992) Council Directive 92/46/EEC laying down the health rules for the production and placing on the market of raw milk, heat treated milk and milk based products. *Official Journal of the European Communities*, 14 9 92 No. L268/1.

EEC (1993) Council Directive 93/43/EEC on the Hygiene of Foodstuffs. *Official Journal of the European Communities* No. L175/1-11 (and proposed revisions 111/5227/98 rev 4 and V1/1181/98 rev 2).

IDF (1980)

IDF (1991) IDF Recommendations for the Hygienic Manufacture of Spray Dried Milk Powders. International Dairy Federation, Bulletin No. 267, Brussels.

Institute of Food Science and Technology (UK) (1999) *Development and Use of Microbiological Criteria for Foods*, IFST, London.

International Commission on Microbiological Specification for Foods (ICMSF) (1980) *Microbiological Ecology of Foods 1, Factors Affecting Life and Death of Micro-organisms*, Academic Press, London.

International Commission on Microbiological Specification for Foods (ICMSF) (1986) *Microorganisms in Foods 2. Sampling for Microbiological Analysis: Principles and Specific Applications*, 2nd ed., University of Toronto Press, Toronto, Canada.

Lund, B. M., Baird-Parker, A., and Gould, G. W. (2000) *The Microbiological Safety and Quality of Food*, Vols. 1 and 2, Aspen Publishers, Gaithersburg, MD.

Mortimer, S., and Wallace, C. (1994) *HACCP: A Practical Approach*, Chapman and Hall, London.

Mackey, B. M., and Derrick, C. M. (1987) *Lett. Appl. Microbiol.*, **5**, 115–118.

Public Health Laboratory Service (UK) (1998) *PHLS Microbiol. Dig.*, **13**(1), 41–43.

Ratnam, S., and March, S. B. (1986) *J. Appl. Bacteriol.*, **61**, 51–56.

Royal Institute for Public Health and Hygiene (RIPHH) (1999) *HACCP Principles and Their Application in Food Safety (Advanced Level) Training Standards*, RIPHH, London.

World Health Organization (WHO) (1999) *Strategies for HACCP in Small and/or Less Well Developed Businesses*, WHO/SDE/PHE/FOS 99.7, Geneva.

References Included as Background Information

Institute of Food Science & Technology (UK) (1998) *Food and Drink—Good Manufacturing Practice—A Guide to its Responsible Management*, 4th ed., IFST, London.

World Health Organization (1998) *Guidance in Regulatory Assessment of HACCP, WHO/FSF/FOS* 98.5, Geneva.

World Health Organization (1997) *HACCP—Introducing the Hazard Analysis and Critical Control Point System, WHO/FSF/FOS*/97.2, Geneva.

World Health Organization (1998) *Guidance in Regulatory Assessment of HACCP, WHO/FSF/FOS* 98.5, Geneva.

CHAPTER 14

QUALITY CONTROL IN THE DAIRY INDUSTRY

J. FERDIE MOSTERT and PETER J. JOOSTE
ARC—Animal Nutrition and Animal Products Institute, Irene, South Africa

14.1 INTRODUCTION

The quality of food, such as milk and dairy products, may be defined as that sum of characteristics which enables the food to satisfy definite requirements and which determines its fitness for consumption (Molnar, 1993). In this sense, quality can be judged by means of sensory evaluation, its nutritive value, and according to its chemical, physical, and microbiological characteristics.

When referring more specifically to the microbiological characteristics of a food, it implies the measurement of the hygienic quality of that food. The concept of hygienic quality, in turn, requires that undesirable microorganisms or residues do not gain access to milk or dairy products because these may prove harmful to human health, cause spoilage or deterioration of the food product, or simply be frowned upon from an aesthetic point of view.

Cases of food-borne disease and food poisoning are becoming more and more common throughout the world. Both of these public health problems and the microbiological spoilage of foods can be minimized by the careful choice of raw materials and correct manufacturing and storage procedures. Achievement of such objectives requires, in many cases, monitoring at various stages to assess microbiological load or to look for particular microbial types.

In this chapter the accent will be on monitoring procedures and microbiological analytical methods, and no attempt will be made to

Dairy Microbiology Handbook, Third Edition, Edited by Richard K. Robinson
ISBN 0-471-38596-4

discuss concepts such as good manufacturing practice (GMP), ISO 9000 certification, or hazard analysis critical control points (HACCP) (Harrigan and Park, 1991).

Monitoring procedures to be dealt with will include the air and water supplies in a factory environment, the hygiene of packaging material, and the sampling and testing of raw materials and end products.

The analytical procedures will include standard and rapid methods for assessing microbiological load and for enumerating and detecting specific microbial genera or groups. The principles of these methods will be outlined, and their advantages and disadvantages will be discussed.

14.2 CONTROL OF AIRBORNE MICROORGANISMS IN DAIRY PLANTS

The demand for extended shelf life and safety of dairy products has put increased emphasis on the microbial quality of air in dairy environments. Air quality in the processing and packaging areas is a critical control point in the processing of dairy products because airborne contamination reduces shelf life and may serve as a vehicle for transmitting spoilage organisms and, if pathogens are present, transmission of diseases (Anonymous, 1988; Kang and Frank, 1989a). Every precaution should therefore be taken to prevent airborne contamination of the product during and after processing (Kang and Frank, 1989b; Lück and Gavron, 1990; Hickey et al., 1993). Air quality in processing areas, the factory environment (e.g., walls, floors, drains), and air used in the manufacturing of dairy products should be monitored on a regular basis. Airborne microorganisms in dairy plants include bacteria, molds, yeasts, and viruses. Data on air counts and types of different microorganisms in processing areas, reported by various authors, are comprehensively reviewed by Kang and Frank (1989b). The generic composition and the levels of microorganisms can vary widely within and among plants, and on a day-to-day basis within the same plant. The variations can consequently be attributed to differences in plant design, airflow, personnel activities, and status of factory hygiene. Installation of air filters, application of UV-irradiation, and regular chemical disinfection (bactericidal, fungicidal, and viricidal agents) of air can be applied to critical areas to control airborne microorganisms (Singh et al., 1986; Homleid, 1997; Rockmann, 1998; Arnould and Guichard, 1999).

14.2.1 Sources and Routes of Airborne Microorganisms

Airborne microorganisms can be attached to solid particles like dust, are present in aerosol droplets, or occur as individual organisms due to the evaporation of water droplets or growth of certain mold species. The main sources of airborne microbes may include the activity of factory personnel, ventilation and air-conditioning systems, inflow of outdoor air, and packaging materials (Heldman et al., 1965; Hedrick and Heldman, 1969; Hedrick, 1975; Kang and Frank, 1989b; Lück and Gavron, 1990). Frontini (2000) found that factory personnel, dairy equipment, building materials, and ventilation systems are responsible for 50–60%, 25–35%, 10–20%, and 1–5%, respectively, of airborne contaminants. An increase in viable aerosols have been detected during flooding of floor drains (Heldman and Hedrick, 1971) and after rinsing the floor with a pressure water hose, thereby illustrating the ability of microorganisms to be disseminated from drains and wet surfaces as a result of physical activities (Kang and Frank, 1990; Mettler and Carpentier, 1998). Whenever possible, wet cleaning should not be used during the processing of milk products in areas in which the product is exposed and can be contaminated by aerosols. Once a high concentration of viable aerosol is generated, it can take more than 40 min to return to the normal background level. It is important to minimize the generation of aerosol droplets from bubbles bursting at a water surface—for example, during rinsing activities or during raw milk handling. It has been shown that drains, floors, and standing or condensed water can be a source of pathogens in dairy plants (El-Shenawy, 1998). The potential contamination from bacterial biofilms is of major concern because microbial cells may attach, grow, and colonize on open exposed wet surfaces—for example, floors, floor drains, walls, and conveyor belts (Wong and Cerf, 1995; Carpentier et al., 1998; Mettler and Carpentier, 1998). Floors in dairy plants are one of the main reservoirs of *Listeria monocytogenes* (Davis et al., 1996; Fenlon et al., 1996).

Water used in open circulation systems is another significant source of airborne microbial populations (Lighthart and Frisch, 1976). These authors found that as many as 10^{10} viable bacteria per second can be released into the air from a 15-m-high cooling tower. It should be recognized that in a cooling tower, not only does evaporation of water occur, but also the formation of small water droplets. This spray from cooling towers, if contaminated, may be a possible source of certain pathogens (Hiddink, 1995) and consequently airborne contamination. Microorganisms and small particles are commonly found in the immediate vicinity of water surfaces (Al-Dagal and Fung, 1990).

14.2.2 Outdoor Environment

The control of airborne microorganisms in the immediate surroundings of dairy premises is more difficult than in closed, indoor environments where more controlled measures can be taken. According to Al-Dagal and Fung (1990), one aspect, that could be helpful in reducing the microbial load outdoors is the control of organic materials. Natural agents such as UV light, humidity, temperature, wind direction, and speed have a significant influence on the total number of airborne microorganisms in the outdoor atmosphere.

14.2.3 Processing Rooms

The most positive approach to controlling airborne contaminants indoors is to remove all contamination sources from the area where the product might be exposed to air. Good ventilation is necessary to remove moisture released during the processing of dairy products. It will also prevent condensation and subsequent mold growth on surfaces. More attention is given nowadays to air cleaning in food plants, among other things, by establishing air flow barriers against cross-contamination from the environment (Jervis, 1992; Kosikowski and Mistry, 1997a,b). In modern dairy plants, the air entering processing rooms is chilled and filtered to remove practically all bacteria, yeasts, and molds. It is essential that filtered sterilized air be supplied to areas where sterile operations are to be carried out. Rigid frame filters or closely packed glass fibers are available to achieve contamination-free air for (a) culture transfer and (b) manufacturing and packaging of sterilized milk and milk products (Shah et al., 1996, 1997). The use of high-efficiency particulate (HEPA) filters will remove 99.99% of airborne particles 0.3 μm and larger (Everson, 1991), while new ultra HEPA (ULPA) filters remove 99.999% of particles as small as 0.12 μm (Shah et al., 1997). Passage of air through a combined HEPA/ULPA filter is usually considered suitable for use where contamination-free work is to be carried out. Standard high-efficiency air filter systems allow more air into the room than normal, thereby establishing a positive air pressure. Upon opening a door, filtered air flows out, thus blocking the entry of untreated air and minimizing microbial contamination. For optimal ventilation, sufficient air changes have to be made to prevent the buildup of condensation on surfaces. This is usually achieved by not less than 10 changes per hour and often up to 20 changes per hour in rooms where moisture is generated. The ventilation unit must also be able to accommodate extreme conditions—for example, during cleaning periods (Jervis, 1992).

Rooms in which direct exposure to outside air is inevitable can have air flow barriers installed, mounted over open doorways to secure a significant downward velocity of air flow, preventing contamination from outside (Kosikowski and Mistry, 1997a). Outside air should be filtered and free of condensate.

Compressed air is commonly used in various processing operations and can contribute to contamination of products by dust and microorganisms and, in the case of lubricated compression systems, by oil fumes (Guyader, 1995; Wainess, 1995a). Whenever air under pressure comes into direct contact with the product (pneumatic filling, agitation, or emptying of tanks) or is directed at milk contact surfaces, it should be of the highest quality. Sterile compressed air can be obtained by drying the air after compression in adsorption filters (e.g., chemically pure cotton, polyester, or polypropylene) and by installing a series of filters with 0.2-μm pore size downstream, immediately preceding the equipment where the air is needed (Bylund, 1995; Guyader, 1995; Anonymous, 1997).

Walls and ceilings must also be of the highest standard with no opportunity for accumulation of dust and other deposits. To achieve the required standards, suitable sealants and sterilants, as well as coatings with antimicrobial properties, are available with effectivity against a wide range of bacteria, molds, and yeasts (Vedani, 1996; Russell, 1997a; Botta, 1998). The use of clean-room clothing, head covering, masks, and gloves largely eliminates the release of microorganisms into the processing environment. A good hygiene training program for factory personnel will contribute to reducing contamination by workers (Al-Dagal and Fung, 1990).

Standards for airborne counts in various processing areas have been proposed by various authors (Kang and Frank, 1989b). Proposed standards in this regard are presented in Table 14.1. Although these standards are relatively strict, experience has shown that they are achievable (Lück and Gavron, 1990).

14.2.4 Methods of Air Sampling

The main reason for sampling air in the dairy plant is to evaluate its quality and to obtain information about the hygienic condition in certain critical areas where microorganisms may contaminate the product directly, or indirectly. Samples may be taken from the following locations:

TABLE 14.1. Suggested Standards for Air Counts in Various Processing Areas

Processing Area	Plate Count ($cfu\ m^{-3}$)		Yeasts and Molds ($cfu\ m^{-3}$)	
	Satisfactory	Unsatisfactory	Satisfactory	Unsatisfactory
Cultured milk and cream, cottage cheese	<150	>1500	<50	>1000
Milk and cream	<150	>1500	<50	>1000
Butter	<100	>1000	<50	>1000
Powdered milk	<200	>2000	<100	>1000
Ripened cheese	<200	>2000	<100	>1000

Source: Adapted from Lück and Gavron (1990).

1. At openings of processing equipment that may be subjected to potential contamination by air currents.
2. At selected points in a room e.g. where products are filled and packed.
3. In areas where employees are concentrated (IDF, 1987).

Although various air samplers have been designed for sampling airborne organisms (Kang and Frank, 1989b; Al-Dagal and Fung, 1990), none of these recover viable particles without some inactivation or losses during or after sampling. The effectiveness of air quality monitoring depends on the type of sampler used, as well as on the nature of air in the specific environment to be monitored (Kang and Frank, 1989a). There are two main principles by which airborne microbes can be sampled:

(a) Collection onto solid and semiliquid media or filters.
(b) Collection into a liquid solution or medium.

The objective in each case is to determine the number of organisms on the plates, in the filter, or in the liquid media. Sampling time for all collection methods is usually standardized at 15, 30, and 60 min (IDF, 1987). The basic methods include techniques such as sedimentation (gravitation settling), impaction on solid surfaces, and impingement in liquids, as well as centrifugation and filtration (Kang and Frank, 1989b; Al-Dagal and Fung, 1990; Hickey et al., 1993; Neve et al., 1995). Comparative studies of air sampling devices have indicated that there is often no obvious choice of the correct sampler to use. Results in this regard are discussed by Kang and Frank (1989a–c, 1990).

14.3 MICROBIAL CONTROL OF WATER SUPPLIES

Water has many applications in the dairy industry and the quality requirements vary with different applications. Because water is an important commodity for dairy product manufacture, special attention should be paid to the supply and quality of water. The dairy industry consumes large quantities of water for various purposes, such as direct preparation of products, cleaning and disinfection, cooling, and steam generation. Without sufficient good-quality water, it is impossible to produce high-quality dairy products. Water systems can present a hazard if the microbiological quality is not monitored and appropriate water treatment applied (Hiddink, 1995). Water used in dairy operations must be safe and must be practically free from organisms that could contaminate the product and initiate spoilage. Spoilage of refrigerated milk and milk products by water-borne organisms—for example, psychrotrophs (Witter, 1961)—can occur either directly, through product contact with the water itself, or indirectly, by microbes metabolizing nutrient residues on improperly cleaned equipment surfaces (Hickey et al., 1993). It is necessary to check the quality of water regularly and to incorporate such practices into the quality management system, specifying the frequency and parameters to be monitored.

14.3.1 Water Used for Processing

Process water is water that can come into contact with the product, either directly or indirectly. Therefore, it must be of the highest quality, meeting the requirements for drinking water quality (Table 14.2) or, preferably, exceeding these standards. It should consequently be clear, free from odor, color, and taste, soft, and virtually sterile (Bylund, 1995). In the dairy industry process water is, for example, used for the direct preparation of products, for starting-up pasteurizers and evapo-

TABLE 14.2. Proposed Standards for Drinking Water

Parameters	Count (cfu)
Total bacterial count (22°C)	<100 ml^{-1}
Total bacterial count (37°C)	<10 ml^{-1}
Total coliforms	<1 100 ml^{-1}
Fecal coliforms	<1 100 ml^{-1}
Fecal streptococci	<1 100 ml^{-1}
Sulfite-reducing clostridia	<1 20 ml^{-1}

Source: Adapted from Hiddink (1995).

rators, for flushing-out product from process equipment at the end of production, for rinsing equipment after cleaning, for washing cheese and butter, for CIP-cleaning for post-finishing of process equipment, for regeneration of water treatment equipment, and for air-conditioning humidity control in stores (Hiddink, 1995).

Water with suitable microbiological quality (very low counts) should be used for the final rinsing of equipment after cleaning; otherwise recontamination of the cleaned surface can take place. In this case, disinfection of the water is necessary. Suitable standards for process water are usually <100 and a maximum of 1000 cfu ml^{-1} for total viable count at 22°C and absence of coliform organisms in 100 ml of water. Any deviation from this standard should be investigated to identify the source of contamination. Whatever the origin of water, it should be routinely examined microbiologically at point of entry and especially at the most critical place—that is, at the point of use (Hiddink, 1995; Jervis, 1992).

The sources of water supply to the dairy industry are surface water, ground water, condensate from evaporators, and public mains or tap water. Public mains or tap water can mostly be used without further treatment. However, in some countries, additional disinfection at the factory may be required. Considerable attention should be given to meet quality requirements if process water is produced at the factory itself from sources such as surface water and ground water. Recovered water (e.g., evaporator condensate) is usually not recommended for use in food contact applications, including the final rinse in cleaning (Jervis, 1992). Since most of the micro-organisms in water are destroyed by chlorination or heat, many plants treat all water to keep contamination at a minimum (Hickey et al., 1993). Stored water is usually chlorinated in order to maintain suitable microbiological quality, levels of up to 1 ppm available chlorine usually being sufficient although up to 2 ppm available chlorine is advised if water is softened (Jervis, 1992). Present-day techniques for the treatment of water include disinfection with chlorine, chlorine dioxide, ultraviolet-light, ozone, microfiltration, and other processes (Hiddink, 1995).

14.3.2 Water Used in Cooling Systems

Cooling water is used for the removal of heat from process streams and products. The quality requirements for cooling water used in plate heat exchangers to cool milk products is critical, because with this type of equipment there is a risk of failure and leakage of cooling water to the product. In such situations, cooling water should be of drinking water quality. Mains or tap water, ground water, surface water, and condensate can be used. To obtain drinking water quality, the same treatments as for process water are applicable. To prevent problems like corrosion and fouling in cooling systems by microbial biofilms, chemical conditioning is usually applied (Mattila-Sandholm and Wirtanen, 1992; Assink and Van Deventer, 1995; Hiddink, 1995).

14.3.3 Microbiological Tests

The type of sample taken depends on the purpose of sampling. General bacteriological sampling (for *E. coli* and coliforms) involves the collection of relatively small volumes of water (500 ml to 3 liters), whereas samples for detection of specific pathogens involve larger volumes (10 to 1000 liters) (IDF, 1987; Clesceri et al., 1989; Fricker, 1993). It must be stressed that only sterile sampling containers should be used and that these should be completely filled. The samples should be examined as soon as possible after collection, preferably within 6 h (Fricker, 1993).

The methods for the microbiological examination of water are intended to give an indication of the degree of contamination and to ensure the safety of supply. In general, the tests are based on indicator organisms, the presence or absence of which provides a measure of the microbiological quality of the water. Detailed procedures for the sampling and testing of the microbiological quality of water is outlined in the 17th edition of *Standard Methods for the Examination of Water and Waste Water* (Clesceri et al., 1989).

14.4 ASSESSMENT OF DAIRY EQUIPMENT HYGIENE

Hygiene monitoring of dairy equipment is a routine exercise that must be carried out to verify that the cleaning and sanitation/sterilization operations have been properly conducted. Plant hygiene/sanitation is dependent on the efficiency of cleaning (removal of residual soil from surfaces) as well as on the effective destruction of most (sanitation) or all (sterilisation) of the remaining microorganisms. Verification of cleaning and sterilization of dairy equipment surfaces usually comprises sensory (sight, feel, smell) and bacteriological examination. Although modern dairy plants are highly automated and processing lines seldom assessable for visible inspection, it is often very useful in detecting inadequately cleansed equipment (IDF, 1987; Zall, 1990; Tamime and Robinson, 1999a).

14.4.1 Biofilm Formation on Dairy Equipment Surfaces

Failure by cleaning procedures to adequately remove residual soil from surfaces (especially from milk/product contact surfaces), or the ineffective destruction of the residual microorganisms, may have serious implications. Microorganisms remaining on equipment surfaces may survive for prolonged periods, depending on the amount and nature of residual soil, temperature, and relative humidity. Milk is a highly nutritious medium, hence any residue not removed can promote bacterial growth, bacterial adhesion to the surface, and consequently biofilm development (Wong and Cerf, 1995). Biofilm is a convenient term to designate microorganisms adhering to and growing on wet surfaces and acquiring, within a matter of hours, resistance to adverse environmental conditions (Carpentier et al., 1998).

Biofilm formation is not a new phenomenon and has been and is being studied extensively (Mattila-Sandholm and Wirtanen, 1992; Carpentier and Cerf, 1993; Austin and Bergeron, 1995; Lindsay et al., 2000). It has been established that biofilm accumulation in the dairy environment and especially on milk/product contact surfaces has, for example, the following potential implications:

- Postpasteurization contamination, decreased shelf life, or potential spoilage of products (Koutzayiotis, 1992; Koutzayiotis et al., 1992; Austin and Bergeron, 1995) and, if pathogens are present, transmission of diseases (Dunsmore et al., 1981; Ronner and Wong, 1993; Blackman and Frank, 1996; Miettinen et al., 1999).
- Adhered cells in a biofilm are more resistant to adverse conditions than planktonic (free-living) cells and have, for example, increased resistance to antibacterial agents (antibiotics, disinfectants), chemical shock, desiccation, starvation, inconsistent nutrient supply, and extreme heat or cold (Frank and Koffi, 1990; Costerton et al., 1995; Wong and Cerf, 1995). The existence of viable but nonculturable cells within a biofilm, which survive these stressful conditions, are also possible (Leriche and Carpentier, 1995; Mettler and Carpentier, 1997). The occurrence of bacteria in such a state would not be easily detected under normal microbiological culture conditions (Wong and Cerf, 1995) (see Section 14.11.1).
- Attached cells become irreversibly adsorbed to the surface, which enables the organisms to resist mechanical and chemical cleaning procedures (Lundén et al., 2000).
- Biofilms can be found in apparently extreme environments, such as crevices between gaskets and pasteurizer plates where they

survive repeated cycles of pasteurization, cleaning, and sanitation (Austin and Bergeron, 1995; Mettler and Carpentier, 1997; Lindsay et al., 2000).

Areas in which biofilms most often develop are those that are the most difficult to rinse, clean, and sanitize, and are also more difficult to sample, regardless of the method used. Dead ends, joints, grooves, surface roughness, bypass valves, sampling cocks, overflow siphons in filters, and corrosion patches, and so on, are hard-to-reach areas (Wong and Cerf, 1995). Chemical cleaning and sanitation/sterilization are indispensable tools for dairy plant hygiene operations; however, other means of ensuring the hygiene of contact surfaces, at least at critical points, are needed. New ideas to improve surface hygiene are, for example, the modification of surfaces by incorporation of biocides, antimicrobial agents, or catalysts; improved/new processes and methods for sanitation/sterilization; and biocontrol (Carpentier et al., 1998).

14.4.2 Methods for Assessment of Dairy Equipment Hygiene

Different methods and/or techniques have been devised to monitor the hygiene of dairy equipment surfaces (IDF, 1987; BSI, 1991; Hickey et al., 1993; Wong and Cerf, 1995; Tamime and Robinson, 1999a), thus contributing to maintaining production of high-quality products and at the same time ensuring compliance with legal requirements. Whatever tests are employed, it is essential that they be applied routinely, because individual observations are in themselves meaningless; only when values for a typical, high standard of hygiene have been established for a given plant, along with acceptable tolerances, do the results of any microbiological/hygiene test become valuable (Tamime and Robinson, 1999b).

Enumeration of total counts of bacteria, coliforms, yeasts, and molds are the most common microbiological examinations carried out to assess the bacteriological contamination of surfaces. The types of microorganisms present reflect to some extent the standard of plant hygiene (Tamime and Robinson, 1999a). Selective and differential culture media may also be used to test specifically for given groups of organisms. Although a given assessment method may not remove all the organisms from the surface being tested, its consistent use in specific areas can still provide valuable information as long as it is realized that not all organisms are being removed (Jay, 1992). The most commonly methods for surface assessment are presented below:

Swab/Swab-Rinse Method. The swab method is applicable to any surface (flat or curved, horizontal, vertical, or sloped) that can be reached with hand-held sticks containing either cotton or alginate gauze swabs (or other approved alternatives). The swab technique can be used for hard-to-reach areas such as surfaces with cracks, corners, or crevices (Hickey et al., 1993). A sterile swab, moistened in an appropriate solution, is rubbed over a designated area of the contact surface. Sterile templates, with openings corresponding to the size of the area to be swabbed, are often very useful. The swab is transferred to its holder (test tube) with a known volume of a physiological neutral solution and vigorously agitated (preferably in an automated shaker to ensure reproducibility). When calcium alginate swabs are used, the organisms are released into the diluent after dissolving the alginate, for example, in 1% sodium hexametaphosphate solution. Samples of the solution, or decimal dilutions if necessary, are examined by, for example, the plate count method (Jay, 1992; Wong and Cerf, 1995).

The cellulose sponge swab method (Hickey et al., 1993) is another technique that could be used to assess dairy equipment hygiene. Little pieces of sponge (free from bacterial inhibitors) held by tweezers or by hand (using sterile gloves) are used to sample surfaces. This technique is particularly useful to examine large surface areas. Numbers of organisms recovered by alginate swabs are reportedly higher than those obtained by cotton swabs (Jay, 1992). The reproducibility of the swab/swab-rinse techniques is variable due to the unreliable efficiency of swabbing, and the proportion of bacteria removed from the surface is unknown. Furthermore, it is highly operator-, day-, and time-dependent (Wong and Cerf, 1995). The swab method is, despite its limitations, very useful and almost universally applied in the dairy industry (Tamime and Robinson, 1999b). The swab and rinse methods may also be supplemented by a bioluminescence test for total adenosine-5-triphosphate (ATP) (Pettipher, 1993; Anonymous, 1995; Werlein and Wucherpfennig, 1999) (see Section 14.8.3.1) whereby an indication of the state of hygiene of the plant surface is acquired. Obviously, the readings are not intended to correlate with the microbial count, but there is an excellent correlation between clean surfaces and low levels of ATP (Tamime and Robinson, 1999b).

Surface Rinse Method. The effectivity of cleaning and sanitation of containers and equipment can be assessed by rinsing the container or equipment with a measured volume of sterile water or Ringer's solution and analyzing the sample for total bacterial numbers or the presence of different types of organisms. The rinse (solution) method is more appropriate for assessing internal surface contamination of containers (Lück and Gavron, 1990). In cases where the volume of the rinse is large, or the microbial load is low, it is advisable to use the membrane filter technique (see Section 14.3.3) whereby a known volume of the rinse sample is filtered through an appropriate membrane (generally 0.45 μm), retaining any microorganisms that may be present. The membrane is placed onto the surface of a pre-poured agar plate and inoculated, and visible colony growth is observed between 48 and 72 h. Rinse water could also be examined by the direct epifluorescent filter technique (DEFT), using fluorescent dyes and fluorescence microscopy (Holah et al., 1988; Jay, 1992; Tamime and Robinson, 1999a) (see Section 14.7.1.2). An advantage of the DEFT is that results can be obtained within 25–30 min.

Agar Flooding Method. The agar flooding method is used for assessing the hygiene of internal surfaces of pieces of equipment (tubing, valves, pumps, etc.), cans, and bottles. A molten nutritive agar medium is poured into the item, which is immediately closed and rolled by hand, or by an automated system to form a thin and continuous layer, until the agar sets. After incubation, the colony-forming units are counted visually through the wall if it is transparent, or with an endoscope (Wong and Cerf, 1995).

Agar Contact Plate Methods. Flat or slightly bent surfaces that are smooth and nonporous can be sampled by applying a solidified piece of appropriate nutritive agar medium. Microcolonies detached from the sampled surfaces and sticking to the agar can grow and form visible colonies when the agar is incubated. A number of commercial products are also available in this regard:

RODAC Plate Count. The replicate organism direct agar contact (RODAC) method employs special commercially available plastic plates in which the agar medium protrudes slightly above the rim. The agar surface is pressed onto the test area, removed, the lid replaced and incubated (Lück and Gavron, 1990; Jay, 1992; Hickey et al., 1993).

Agar Slice Methods. A sterile 100-ml syringe (modified by removing the needle end to create a hollow cylinder) is filled with agar medium. A portion of the agar is pushed out to make contact with the test surface, cut off, and placed into a petri dish and incubated (Jay, 1992). Similarly, an artificial (plastic) sausage casing can also be used in this way (Ten Cate, 1965). Drawbacks to these methods are, for example, the covering of the agar surface by spreading colonies and its ineffectiveness for heavy surface contamination (Jay, 1992). Unless caution is taken to apply agar to the sample surface with constant pressure and time, reproducibility of sampling can be questionable (Wong and Cerf, 1995).

Dry Rehydratable Film Method. The dry rehydratable film (Petrifilm aerobic count) method provides a simple direct-count technique for detecting bacterial contamination on both flat and curved surfaces (Jay, 1992; Hickey et al., 1993). Petrifilm methods exist for the detection and enumeration of specific groups, such as coliforms (see Section 14.7.3.1.6). This procedure is less applicable for surfaces with cracks or crevices (Hickey et al., 1993) or when surfaces are heavily contaminated (Wong and Cerf, 1995).

Other Methods. Various other methods are described in the literature—for example, the adhesive (sticky) tape method (Tamminga and Kampelmacher, 1977) and rapid methods for monitoring the hygiene of dairy equipment surfaces (Russell, 1997b).

14.4.3 Suggested Standards

Some suggested standards for dairy equipment in contact with products prior to pasteurization/heat treatment are shown in Table 14.3. With improved cleaning and sanitation regimes, a total colony count of 200 cfu 100 cm^{-2} would be expected nowadays, and a value of <50 cfu 100 cm^{-2} would be expected for any equipment containing pasteurized product (Lück and Gavron, 1990).

Reliable methods for sampling and enumeration of microorganisms remaining on dairy equipment surfaces, especially techniques to

TABLE 14.3. Suggested Standards for Dairy Equipment Surfaces Prior to Pasteurization/Heat Treatment

cfu 100 cm^{-2}	Conclusion
500 (coliforms < 10)	Satisfactory
500–2500	Dubious
>2500 (coliforms > 100)	Unsatisfactory

Source: Adapted from Harrigan and McCance (1976); Tamime and Robinson (1999b).

detect and enumerate adhering bacteria, will contribute toward the effective monitoring of dairy equipment hygiene (Wong and Cerf, 1995). At the same time, research into the microbial ecology of surfaces in the dairy industry to minimize biofilm formation (e.g. inhibiting/preventing colonization, adhesion, and/or growth of unwanted bacteria), as well as the removal of biofilms, is also needed (Carpentier et al., 1998).

14.5 HYGIENE OF PACKAGING MATERIAL

The primary purpose of packaging is to ensure that milk and dairy products reach the ultimate consumer in a safe, sound, and convenient condition. Packaging is in fact an integral part of modern production processing and is usually considered to be the key to successful plant operation. During the last 25 years, the packaging of dairy products has made tremendous strides in improving the hygienic quality and shelf life of the product. Packaging equipment requires special attention because the product reaching this equipment will no longer be treated to reduce its microbial content. Ideally, equipment design should not allow any contamination to either the product or the package (Wainess, 1995a).

14.5.1 Manufacturing of Packaging Materials

Basically, the same general hygiene requirements that apply to dairy plants should apply to plants manufacturing packaging materials (also those in which containers are formed and filled). Requirements for the hygienic manufacture of packaging materials, containers, and closures, especially for single-service containers, have been suggested by Wainess (1995b). Uncoated paper stock, prior to lamination, should meet a microbiological standard of not more than 250 cfu per gram as determined by a disintegration test (Hickey et al., 1993). Where a rinse test can be used, the residual microbial count should not exceed 50 cfu per package except that in packages of less than 100 ml the count should not exceed 10. Where the swab test technique is used (e.g., laminated board, sheet, wrapping, etc.) the microbial count should not exceed 1 cfu cm^{-2} of product contact surface. Product contact surfaces should be free from coliform organisms. It is evident that packages or packaging material must arrive at the dairy plant with an "acceptable" low microbial count and be formed, filled, and sealed employing the proper hygienic measures to preclude additional contamination.

14.5.2 Retail Packaging

Although the use of returnable containers is largely restricted to liquid milk, it may extend to other products such as cream and fermented milks. For retail volumes, the container can be made of glass, polycarbonate, or polyethylene and sealed by single-service aluminum, paper, or plastic caps. For wholesale quantities, stainless steel or aluminum cans are used. With modern mechanical bottle washing operations, the combination of jetting and the bactericidal power of the cleaning solution normally gives a very high standard of cleanliness. The residual colony count should not exceed 50 per container. In containers of less than 100 ml, the colony count should not exceed 10 cfu per container. The number of residual microorganisms on the inner surface of returnable metal cans, normally used for distribution of pasteurized milk or cream in bulk (5–45 liters), should not exceed 50 cfu per container (Wainess, 1995b).

With the advent of plastic-coated packages and closures and the development of vacuum-formed and blow-molded plastic packages, plastic bags, and extruded and fabricated sheets of plastic for packaging, hygienic problems that could not be solved by treatment after forming the package have become evident. It is obvious that physical impurities such as dust or particles released from the material should not gain access to the product. The influence of packaging on the contamination of dairy products may be direct, due to the presence of microorganisms on the material, or indirect due to the permeability of the material to bacteria. The packaging material used for heat-treated milk should first of all be free from pathogenic bacteria, but also from other microorganisms that are able to multiply in the milk or product under the prevailing conditions (Ronkilde Poulsen et al., 1995). The trend toward extended shelf-life products demands that special attention be devoted to the microbial content of air in packaging areas and filling areas. Radmore (1986) found that a correlation ($r = 0.93$) existed between the number of airborne organisms present in a packaging environment and the number of organisms contaminating the final product. He calculated that during a 60s exposure period, 2.2% of the organisms in 1 m^3 air would be able to contaminate 1 liter of a product that is being packed in a container with an opening of 100 cm^2. The microbial count of plastics and plastic-coated cartons is about 0.1 cfu cm^{-2}, provided that no recontamination has taken place after manufacture (Kelsey, 1974).

In aseptic packaging, only one spore originating from the packaging material is admissible per 10,000 containers. The surface of 10,000 containers (e.g., 1-liter Tetra Brik) equals 800 m^2. Assuming a level of 0.1 organism cm^{-2}, this surface comprises a total of 800,000 organisms before it is sterilized, of which 24,000 are spores (Cerny, 1976). Consequently, sterilization of the packaging material must reduce the spore count by at least four decimals (4D reduction). The most widespread technique to obtain sterile milk contact surfaces using thermolabile packaging material is by in-line sterilization using 15–35% hydrogen peroxide (H_2O_2). In practice, removal of H_2O_2 residues from the packaging material surface is usually achieved at temperatures exceeding 100°C, with the result that the 4D reduction could take place within a few seconds. Wetting agents improve the sporicidal effect of H_2O_2 (Kelsey, 1974). Ultraviolet and high-energy irradiation are other alternatives for the sterilization of packaging materials during aseptic filling and packaging (Flückiger, 1995; Van den Berg, 1995).

New developments in the use of high-intensity pulsed light technology to sterilize packaging materials without chemicals provide new possibilities in terms of quality, monitoring, and controlling the destruction of microorganisms (Harrysson, 1998). The innovative development of bioactive packaging material to inhibit pathogens, mycotoxin-producing molds, and spoilage organisms is also very promising, although further work is necessary to evaluate the performance of these materials in food systems (Scannell et al., 1999; Floros et al., 2000; Han, 2000). The wrapping of retail quantities of butter and cheese in coated paper, aluminum, plastic, and many combinations has changed very little in recent years. Nevertheless, the development of new materials, laminates, cups, and pots has widened the choice, improved hygiene, and provided better protection for various products.

14.5.3 Methods for the Assessment of Hygiene

Detailed information on the sampling of packaging material, containers and closures is outlined by Grace et al. (1993). Methods for the assessment of microorganisms on packaging material must reliably detect bacteria, molds, and yeasts. Various methods—that is, the disintegration test, rinse test, coating technique, membrane filter, and direct inoculation techniques—are used for this purpose (Hickey et al., 1993; Wainess, 1995b; Tacker and Hametner, 1999). The following methods that can be used are described by Hickey et al. (1993):

Disintegration Method. The disintegration method comprises the blending of paper, paperboard, or molded pulp samples in sterile phosphate dilution water using a disintegrator blender. Ten milliliters of the disintegrated suspension (representing 0.1 g of the sample when, e.g., 3 g of packaging material is blended in 300 ml of dilution water) is equally divided among three Petri dishes and pour-plated with standard methods agar or appropriate differential media to determine specific microorganisms or groups of organisms. After incubation, the sum of the colonies developed on the three plates from 0.1 g of sample is multiplied by 10, and the result is reported as the number of colonies per gram of packaging material. A total count of not more than 250 cfu g^{-1} is usually regarded as acceptable (Wainess, 1995b).

Rinsing Methods. A suitable method for containers is the rinse solution test in which a measured volume of a sterile buffer solution or nutrient broth is repeatedly flushed over the interior surfaces and the bacterial population is determined by plating or membrane filter techniques (Clesceri et al., 1989). Various amounts of rinsing solution (20, 50, or 100 ml) are used, depending on the size of the container. Containers smaller than 1 liter are, for example, rinsed with 20 ml solution, whereafter 5 ml is divided between two Petri dishes and pour-plated with standard methods agar. After incubation at 32°C for 48 h, the residual bacterial count (RBC) per specified container size is calculated by multiplying the number of colonies by the volume of the rinse solution divided by the volume of the sample plated. For example, if the volume of the rinse solution is 20 ml, the volume of sample plated is 5 ml, and the number of colonies is 15, the RBC is

$$15 \times (20 \div 5) = 60$$

For coliforms, 10 ml of rinse solution is divided among three plates. After incubation the coliform count per specified container capacity is calculated by multiplying the number of coliform colonies by the volume of rinse solution divided by the volume of the sample plated. The direct epifluorescent filter technique (DEFT) system could also be used to examine rinse water (Pettipher, 1993).

Screening Method for Retail Milk Containers. Retail milk containers can also be evaluated by rinsing the interior surfaces of, for example, 50 containers thoroughly with 20-ml portions each of nutrient broth. The containers with nutrient broth are then incubated at 32°C for 48 h, and the percentage of containers showing growth is calculated (Hickey et al., 1993).

Membrane Filter Technique. Coliforms, yeasts, molds, proteolytic bacteria or other specific microorganisms can also be determined by the membrane filter technique using appropriate differential media, prescribed temperatures, and incubation temperatures (Clesceri et al., 1989; Hickey et al., 1993).

Direct Plating Method. According to Lück and Gavron (1990) the surface count of nonabsorbent packing materials based on paper, cardboard, plastics, aluminum foil, etc. can also easily be determined by the direct surface agar plating method. A specified area of the packing material is aseptically placed on the solid agar medium of a Petri dish and then overlaid with the same medium. After incubation, the colonies on both sides of the material can be counted. The use of selective media also allows the counting of coliforms, yeasts, molds, or other organisms.

14.6 SAMPLING OF PRODUCTS FOR MICROBIOLOGICAL EVALUATION

Correct sampling procedures for microbiological analysis require careful attention during sampling, storage, and transport of samples before analysis. Special precautions to prevent direct contamination by microorganisms and subsequent growth of such contaminants have to be taken. Sampling should therefore only be undertaken by experienced persons trained in the appropriate techniques. Emphasis cannot be too strongly placed on the necessity of obtaining a representative sample, using appropriate aseptic techniques. It is imperative that the sample drawn gives a true reflection of the compositional and microbiological quality of the product from which it has been selected. Detailed information on the general requirements and technical instructions for sampling, sampling equipment, and sampling techniques, as well as guidance on the storage and transport of milk and milk product samples, is described in various national and international standards [e.g., IDF (1995a)] and in *Standard Methods for the Examination of Dairy Products* (Grace et al., 1993). Reference methods for sampling of milk and milk products are summarized in Table 14.4. Sampling equipment for milk collection tankers (IDF, 1990a) and the sampling of milk for quality payment schemes (IDF, 2000) are described by the International Dairy Federation.

TABLE 14.4. Reference Methods for Sampling of Milk and Milk Products

Product	Description	IDF	ISO	AOAC
Butter	General instructions	50C: 1995	707: 1997	970.29/930.31
Cheese	Collection of sample			920.122
	General instructions	50C: 1995	707: 1997	970:30
Condensed milk (sweetened)	General instructions	50C: 1995	707: 1997	970:27
Cream	Collection of sample			925.25
Dried milk and dried milk products	General instructions	50C: 1995	707: 1997	935.41A/970.28
Edible ices, ice cream, frozen desserts	General instructions	50C: 1995	707: 1997	
Evaporated milk	General instructions	50C: 1995	707: 1997	968.12
Infant formula (milk-based)				985.30
Milk	Collection of sample			925.20
	General instructions	50C: 1995	707: 1997	968.12
	Automated method (bulk tanks)			970.26
Milk and milk products	Attributes sampling schemes	113A: 1990	5538: 1987	
	General instructions	50C: 1995	707: 1997	968.12
	Variables sampling schemes	136A: 1992	8197: 1988	
Milk fat products	General instructions	50C: 1995	707: 1997	
Sterilized milk	Sampling technique	48: 1969		

Source: Adapted from Webber et al. (2000).

14.6.1 Sampling Equipment and Containers

All sampling equipment and instruments must be clean, sterile, and dry prior to use. Sterilization by hot air (170–175°C for at least 2h) or steaming (121 ± 1°C for at least 20min in an autoclave) is normally recommended, although the following alternative methods can also be used if these methods are unpractical (IDF, 1995a):

- Direct exposure of sampler surfaces to a suitable flame
- Immersion in at least 70% (v/v) ethanol solution
- Ignition after immersion in 96% (v/v) ethanol
- Exposure to sufficient gamma-radiation

After thermal sterilization, the equipment should be allowed to cool down before using for sampling. It is essential that the sample containers and closures should be clean, sterile, and dry and that they should be securely closed to prevent contamination from external sources. Any deviations from the prescribed sampling instructions, abnormal sampling conditions, or additional information concerning the samples to be tested should be noted in the sampling report to ensure scientifically sound interpretation of the test results.

14.6.2 Sampling Techniques

Samples for microbiological examination are always taken first and, whenever possible, from the same product containers as those for chemical, physical, and sensory evaluation. Specific sampling techniques for milk and milk products are described in various standard methods (Grace et al., 1993; IDF, 1995a). The mixing of milk and liquid milk products, for example, can be achieved by pouring the milk from one container to another, by using a stirrer (plunger) of a suitable design, by mechanical agitation, or, sometimes, by clean, filtered compressed air. When air is used, care should be taken to avoid foaming of milk because it may cause oxidative lipolysis. The milk is usually agitated for 5min when the tank capacity is 500–4000 liters, for at least 10–15min when the volume is more than 4000 liters, and for 30–60min in large factory storage tanks (Lück and Gavron, 1990). If tanks are equipped with time-programmed agitation systems, samples may be taken after agitation for shorter periods.

The collection of a representative sample from large vessels, storage tanks, and tankers may present problems. In a large vessel with a bottom discharge outlet, samples should preferably be taken through the manhole. If taken from the discharge outlet valve or the sampling cock, sufficient milk must be discharged to ensure that the sample is representative of the whole. Proportionate sampling is done by taking representative quantities from each container and mixing the portions in amounts that are proportional to the quantity in the container from which they were taken. With raw milk, bulk portions must be split with care, and the homogeneity of the samples must regularly be validated. A useful method is to determine the butterfat levels because the distribution of microorganisms in raw milk closely follows that of the fat (Reuter and Quente, 1977).

Special procedures must be followed for sampling other dairy products (Grace et al., 1993; IDF, 1995a). The recommended sample sizes for various products are shown in Table 14.5.

In-Line Sampling. Flow lines of modern dairy plants are complex, and improper designs may cause recontamination of heat-treated products (Dickerson, 1987). A quality control program may include sampling of milk at different sites after pasteurization to assess the microbiological quality or possible postpasteurization contamination. Samples can be withdrawn from different critical points in the processing line using modern sampling devices. Various commercially available devices (e.g., membrane/rubber septums for syringe sampling, valves, and cock-types) of different designs may be obtained for in-line sampling. In general, septa are used, for microbiological sampling and valves for chemical sampling. Sampling devices should be hygienically designed without dead spaces or difficult-to-clean areas. The seal design should not harbor bacteria and should be drainable. Devices that can be cleaned or sterilized independently of flow-line (or tank) cleaning are ideal for sampling (Anonymous, 1998). Multiple samples from the same batch, or taken at specific time intervals, can consequently be obtained without the risk of cross-contamination of samples. Great care should be taken when sampling unmixed milk, because significant carry-over from one sample to the next can invalidate sensitive tests. Stepaniak and Abrahamsen (1995) found no effect of the type of seven different sampling valves on the total plate count of freshly pasteurized milk or on the count of cold-stored samples. However, the sample volume and sample storage container influenced the plate count of cold-stored samples.

Automatic Sampling Systems. With automatic sampling it is also essential that the sample taken is sterile and representative of the milk

TABLE 14.5. General Guidelines that Might Be Used for Storage Temperatures, Storage Times, and Minimum Sample Sizes for Microbiological Analysis

Product	Storage temperature (°C)	Maximum Time Before Examination	Minimum Sample Size
Nonsterilized milk and liquid milk products	0–4	24 h	100 ml or g
Sterilized milk, UHT milk and sterilized liquid milk products in unopened containers	Ambient, max. 30	7 days	100 ml or g
Sterilized milk, UHT milk and sterilized liquid milk products after sampling from the production line	0–4	24 h	100 ml or g
Evaporated milk, sweetened condensed milk, and milk concentrates in unopened containers	Ambient, max. 30	7 days	100 g
Semisolid and solid milk products (except butter and cheese)	0–4	24 h	100 g
Edible ices and semifinished ice products	–18 (or lower)	7 days	100 g
Dried milk and dried milk products	Ambient, max. 30	7 days	100 g
Butter, butter products, butter fat (butteroil), and similar products	0–4 (in the dark)	48 h	50 g
Fresh cheese	0–8	48 h	100 g
Processed cheese	Ambient, max. 30	7 days	100 g
Other cheeses	4–8	48 h	100 g

Source: Adapted from Prentice and Langridge (1992); IDF (1995a).

or milk product from which the sample is taken. Automatic sampling systems should be constructed in such a way that it is guaranteed that there are no milk residues left from the previous sample, which might lead to carry-over from one sample to the next, leading to erroneous results (IDF, 1990a; Lück and Gavron, 1990). With modern automatic sterile sampling systems it is, however, possible to automatically clean and sterilize the system between samplings. Several samples could also be taken simultaneously at various sampling points from pipelines, production tanks, and storage tanks. One of the advantages of these systems is that sampling valves can be fitted, for example, at normally inaccessible sampling points.

14.6.3 Numerical Selection of Samples

It is normal practice that the producer and buyer should come to an agreement as to what the quality of the product should be. The critical major and minor defects (IDF, 1992) should be clearly defined before selection of a sampling plan, so that they are unambiguously understood by all users of the contract or specification, when referring to the sampling plan. A single test on any individual sample may suggest that the product is better or worse than it actually is. Therefore, an element of risk is introduced, because only 100% sampling will give 100% certainty. In practice, this is not possible, and for normal quality control procedures the producer and the buyer accept a certain range of error—for example, 5% and 10%, respectively (Lück and Gavron, 1990). When the quality of dairy products is tested, the number of units to be sampled from a bulk consignment or batch depends on the size of the unit (large containers or small retail units) and the purpose of the test (determination of qualitative or quantitative characteristics). The batch is accepted or rejected according to the sampling plan that is based on the batch or lot size and acceptable quality levels (e.g., 1–10%). An acceptable quality level is considered to be the average of quality which, if maintained by a provider, could result in the acceptance of most of his production (IDF, 1990b). Tables are available which show how sample-taking should be carried out in order to obtain a statistically reasonable basis for the assessment of quality—that is, the number of consumer units that should be taken (randomly) from a consignment (batch) of a certain size. Sampling plans for milk and milk products are usually based on inspection by attributes (IDF, 1990b) or inspection by variables (IDF, 1992).

Attribute Sampling. Attribute sampling is used to qualitatively classify whether a unit is "good" or "defective." A "good" unit is one that meets the requirements of a specification, while a "defective" unit is one that does not. There are therefore only two answers when checking for an attribute (characteristic). For example, a UHT product can either be sterile or not.

Inspection by Variables. A variable may be described as a characteristic that can be measured quantitatively and that may have any value within certain limits—for example, total bacterial count, coliform count, percentage butter fat, and so on. Inspection by variables should not be used for critical defects—for example, one that would make the product unacceptable. The large variation in microbiological properties necessitates that more units have to be sampled to determine the microbiological quality than to determine the chemical quality of a batch (Lück and Gavron, 1990). The latter authors suggested the following number of samples (per batch) to be taken for routine microbiological control purposes.

- Pasteurized milk and pasteurized milk products: at least 10.
- Condensed milk, evaporated milk: at least 20.
- Dried milk, dried milk products: at least 10.
- UHT products: at least 75 packages per product per machine.

It is also recommended that the number of samples be increased should any drop in quality be noted. Samples could also be drawn over a period of time (for example, at the start, midway through and at the end of production) to identify time-related problems during production.

14.6.4 Storage and Transport of Samples

The most effective way to stabilize the microbial content of milk and perishable milk product samples prior to analysis is by storage and transport of the samples in crushed ice (0–4°C). At these temperatures, especially between 0°C and 2°C, the microbial numbers will remain virtually unchanged for up to 24–36 h (Harding, 1995). This ideal is sometimes difficult to achieve, especially in less temperate countries. To prevent microbiological, physical, or chemical deterioration of the milk, consideration can be given to the use of chemical preservatives for samples to be tested for compositional quality (Heeschen et al., 1969; Lück et al., 1982; Grace et al., 1993; De Wet, 1998). In many countries, however, the use of preservatives are not permitted.

Samples must reach the laboratory as quickly as possible, preferably within 24h. If cooling is necessary, the minimum requirements to be met are the temperature ranges that are either legally required or prescribed by the manufacturer. General guidelines that can be used in this regard are also presented in Table 14.5. Thermally insulated containers are used for the storage of cooled, frozen and quick-frozen samples to the laboratory. Crushed ice, pre-frozen icepacks, or dry ice (solid CO_2) may be used as refrigerant agents (IDF, 1995a). For most analyses, however, freezing should be avoided because it can cause disruption of bacterial cells.

14.6.5 Preparation of Samples for Microbiological Testing

The preparation of samples, prior to microbiological examination, is just as important as taking representative samples. The correct standard procedures—for example, those prescribed by IDF (1996a)—should consequently be followed carefully. The precise procedure for the preparation of the test portion varies with the nature of the product. All samples should be thoroughly mixed by shaking and inverting, using a rotary blender, a peristaltic blender (stomacher), or glass beads, depending on the type of product. Only specified diluents for general or special purposes are used for primary and further decimal dilutions. Damaging of microorganisms by sudden changes in temperature should be avoided, for example, when transferring a portion of the test sample to a diluent. The normal aseptic precautions during weighing and mixing of test portions, or transferring suspensions, should always be taken.

A myriad of microbiological tests are described in the literature that can be used in the dairy industry to assess the quality of milk and milk products. The selection of a method for a specific test should be carefully considered, and aspects such as the purpose of analysis and the required sensitivity of the method will, for example, determine which method is to be used. It is, however, recommended that officially prescribed or generally recognized procedures such as those of the IDF/ISO/AOAC or other standard procedures—for example, those recommended by the APHA (Marshall, 1993)—be used. Reference methods for determining the microbiological quality of various dairy products are given in Table 14.6. These and other methods will be dealt with in more detail in the ensuing sections.

14.7 PROCEDURES FOR THE DIRECT ASSESSMENT OF THE MICROBIAL CONTENT OF MILK AND MILK PRODUCTS

The procedures described in this section are referred to as "direct" methods on the basis that they are able to give an estimate of the microbial numbers in the food product by counting the cells directly or the colonies developing from viable cells on a nutrient medium.

14.7.1 Cell Counting Procedures

14.7.1.1 The Breed Microscopic Count. The Breed smear or direct microscopic count (DMC) was developed as a rapid method for counting bacterial cells in milk, and the procedure is outlined in *Standard Methods for the Examination of Dairy Products* (Packard et al., 1993). This method suffers from major disadvantages:

- The staining method does not distinguish between dead and viable cells.
- The small sample volume renders the technique insensitive and subject to considerable error.
- For most microscopes a single organism per field represents 3–6 × 10^5 organisms ml^{-1} of milk.

Because of this detection limit the DMC is unsuitable as a quality test for dairying nations that have milk quality standards of less than 100,000 per milliliter (Hill, 1991a). The DMC is therefore mainly of historical interest. Nevertheless, the technique may be of some use as a diagnostic tool where rapid screening of milk supplies with high bacterial counts is required. It has a further advantage that a skilled operator can recognize the morphological characteristics of the bacterial cells and infer whether the contamination has arisen from improperly cleansed utensils, dirty cows, aged or stale milk, or an udder infection.

14.7.1.2 The Direct Epifluorescent Filter Technique (DEFT). The direct epifluorescent filter technique is a modern approach to the direct microscope count (Hill, 1991b). With this technique the milk is first pretreated with a proteolytic enzyme and a surfactant which lyses the

TABLE 14.6. Reference Methods for Determining the Microbiological Quality of Different Dairy Products

Product	Test	Principle	IDF	ISO	AOAC
Butter	Lipolytic microorganisms	Colony count (30°C)	41: 1966		
	Contaminating microorganisms	Colony count (30°C)	153: 1991	CD 13559	
	Molds	Colony count (25°C, 5d)	94B: 1990	6611: 1992	*not* 984.29A
Cheese	*Escherichia coli*	Colony count (44°C, using membranes)	170A: 1999	11 866–3: 1997	
	Staphylococcus aureus	Colony count (37°C)	145A: 1997	CD 11 867	
Cheese (fresh)	Contaminating microorganisms	Colony count (30°C)	153: 1991	CD 13 559	
Cheese (powders)	*Salmonella*	Membrane filter			987.42A
Dried milk	*Salmonella*	MSRV medium			995.07
	Staphylococcus aureus	Colony count (37°C)	138: 1986	CD 8869	
Dried milk products	*Bacillus cereus*	MPN		DIS 17086	
	Coliforms	Colony count (30°C)	73B: 1998	5541–1: 1986	
		MPN (30°C)	73B: 1998	5541–2: 1986	
	Microorganisms (enumeration)	Colony count (30°C)	100B: 1991	6610: 1992	
Dried milk (nonfat)	*Salmonella*	Membrane filter			985.48
Dried milk (whole)	*Salmonella*	Selective broth			967.25
Fermented milks	Contaminating microorganisms	Colony count (30°C)	153: 1991	CD 13 559	
Milk	Bacteriological quality	Guidance on evaluation of routine methods	161A: 1995	CD 12 079	
	Coliforms	Dry rehydratable film method (high sensitivity)			996.02
	Coliform counts	Dry rehydratable film			986.33
	Lipolytic microorganisms	Colony count (30°C)	41: 1966		
	Microorganisms (enumeration)	Dry rehydratable film colony count			986.33
		Plate loop colony count (30°C)	131: 1985	CD 8553	
		Colony count (30°C)	100B: 1991	6610: 1992	
	Psychrotrophic microorganisms	Colony count (6.5°C, 10 days)	101A: 1991	6730: 1992	
		Colony count (21°C, 25 h)	132A: 1991	DIS 8552	
	Staphylococcus aureus	Colony count (37°C)	145A: 1997		
Milk and milk products	Aflatoxin M_1	Thin-layer chromatography			980.21
		Immunoaffinity column & HPLC	171: 1995	14 501: 1998	
	Aflatoxin M_1 & M_2	Liquid chromatography			986.16

	Bacterial and coliform counts	Dry rehydratable film			989.10
	Coliforms	Colony count (30°C)	73B: 1998	5541–1: 1986	
	Coliforms	MPN (30°C)		5541–2: 1986	
	Coliforms	Pectin gel			989.11
	E. coli	MPN	170A: 1999	13 366–1: 1997	
	E. coli	MPN with MUG	170A: 1999	13 366–2: 1997	
	E. coli	Colony count (44°C, using membranes)	170A: 1999	13 366–3: 1997	
	E. coli 0157 : H7	Visual immunoprecipitate assay			996.09
	Listeria	DNA hybridization (Gene-Trak)			993.09
	Listeria	Immunosorbent assay (Listeria-Tek)			994.03
	Listeria monocytogenes	Selective enrichment, isolation, presumptive identification, confirmation	143A: 1995	10 560: 1993	993.12
	Listeria monocytogenes and related species	Visual immunoprecipitate assay			997.03
	Microorganisms, enumeration	Colony count (30°C)	100B: 1991	6610: 1992	
Milk-based desserts Milk and milk products	*Salmonella*	Pre-enrichment, selective media, identification, confirmation	93B: 1995	6785: 1985	
	Staphylococci, thermonuclease	Color zones on toluidine blue-O-DNA medium	83A: 1998	CD 8870	
Yogurt	Yeasts and molds	Colony count (25°C, 5 days)	94B: 1990	6611: 1992	
	Staphylococcus aureus	Colony count (37°C)	145A: 1997	CD 11867	
	Coliforms	Colony count (30°C)	73B: 1998	5541–1: 1986	
		MPN (30°C)	73B: 1998	5541–2: 1986	
	Lactobacillus delbruekii spp. *bulgaricus*	Colony count (37°C)	117B: 1997	DIS 7889	
	Lactobacillus delbruekii spp. *bulgaricus*	Tests for identification	146A: 1998	CD 9232	
	Streptococcus salivarius spp. *thermophilus*	Colony count (37°C)	117B: 1997	DIS 7889	
	Streptococcus salivarius spp. *thermophilus*	Tests for identification	146A: 1998	CD 9232	

Source: Adapted from Webber et al. (2000).

somatic cells and modifies fat globules sufficiently to filter the sample through a 0.6-μm pore size polycarbonate filter. The filtration concentrates bacteria on the surface, after which they are stained with acridine orange. The mounted filter is examined through an epifluorescence microscope. Metabolically active bacteria fluoresce orange-red, while inactive bacteria fluoresce green.

The clumps of orange-red fluorescing bacteria are counted in the field of view, and a DEFT count ml^{-1} of the milk sample is calculated over several fields. The technique has the advantage over the DMC that viable cells can be distinguished from inactive ones. The detection limit is also enhanced by filter concentration of the bacterial cells to a level that is useful for assessing the quality of milk produced by modern dairy industries. In line with the DMC a "same-day" result makes it possible to give producers rapid feedback of information.

Disadvantages of the method include the number of samples that can be examined because the DEFT is a microscopic count. A semi-automated system for slide examination overcomes this problem, but the sample preparation is also a limitation (Hill, 1991b). Another drawback is that the DEFT cannot be applied to heat-treated products due to some nonviable bacteria in such products fluorescing orange (Kroll, 1989).

14.7.1.3 Automatic Fluorescent Microscopic Count of Bacteria (Bactoscan 8000). Both the DEFT and the Bactoscan 8000 method belong to the group of direct microscopic counting methods (Suhren, 1989). In both methods, samples are pretreated with lysing and proteolytic reagents. Where separation and concentration with the DEFT technique is achieved by membrane filtration, gradient centrifugation is used in the Bactoscan method. In the latter method, cells are also stained with the fluorochrome acridine orange but are counted electronically as light impulses of single bacteria.

The Bactoscan technique has been found (Suhren et al., 1991) to be an acceptable alternative to the standard plate count in that it can, depending on the mode of estimation, reliably analyze milk samples with a colony-forming unit content of 40,000 to 80,000 cfu ml^{-1}.

The main advantages of the Bactoscan 8000 method is its speed (80 samples/hour) and the rapid availability of the results (approximately 15 min). Samples can also be preserved (Suhren et al., 1991) for not more than 7 days by addition of chemicals such as boric acid or sodium azide. A possible disadvantage is the fact that the statistical relationship between the standard plate count and Bactoscan values at lower plate counts (e.g., <10,000 cfu ml^{-1}) are less consistent than at higher cfu levels.

14.7.1.4 Flow Cytometry (Bactoscan FC). Flow cytometry is the science of measuring components (cells) and the properties of individual cells in liquid suspension. In essence, suspended cells are brought to a detector by means of a flow channel (Jay, 1992), Flow cytometric analysis of bacteria usually requires staining of the cells with fluorescent dyes binding to specific cell constituents in order to distinguish cells from other particulate matter (Suhren and Walte, 1998). In the Bactoscan FC technique the DNA/RNA of the bacteria is stained with the fluorescent dye ethidium bromide, and "disturbing" milk constituents are reduced/dispersed by buffers, detergents, and enzymes during sample preparation. Fluorescence is excited by laser [usually an argon ion laser (Suhren and Walte, 1998)]. The light emitted is detected when the stained particles pass in a hydrodynamically focused stream by a fluorescence detector and are indicated as Bactoscan counts (BC-FC). With respect to sample preparation, staining, and measuring, the Bactoscan-FC method differs principally from the Bactoscan 8000 (BC 8000) procedure. The accuracy of the estimation by which the SPC ml^{-1} can be estimated from the results of the routine method was slightly superior for BC-FC over BC-8000 and is improved when not single results but the average of two samples of dairy farm milk taken at different times is considered (Suhren and Walte, 1998).

Advantages of this method are similar to those of the BC-8000. Disadvantages are the cost of the equipment and the fact that somatic cell counts exceeding 1 million ml^{-1} might influence the counts.

14.7.2 Electronic Counting of Microcolonies

In this method (Suhren and Heeschen, 1991a) the milk to be tested is mixed with a liquid nutrient gelatin solution to give a milk dilution of 1:500. The mixture is pipetted into tubes (±3-cm depth) and overlayed with 1 ml of the nutrient gelatin. The mixture is allowed to solidify and incubated at 20 ± 1 h at 21°C. The developed colonies in the medium are fixed by overlaying the medium with a 2 ml formaldehyde—hydrochloric acid mixture or, alternatively, glutaraldehyde/hydrochloric acid or potassium dichromate. This step is followed by liquefaction of the medium in a waterbath at 35°C for 30 min. A volume of 7 ml of electrolyte (sodium chloride and formaldehyde) is added and the mixture is stirred gently. The microcolonies are then counted using a Coulter counter. The counter is adjusted to count all particles >600 μm^3.

Calibration is done using 20.5-μm^3-diameter latex particles or by comparing the electronic and direct microscopic count of the microcolonies (Suhren and Heeschen, 1991a).

Milk samples can be preserved before testing by adding a mixture of orthoboric and sorbic acid (final concentrations 0.6% m/v and 0.009% m/v, respectively) for 24h at 5–20°C and for a further 24h at refrigeration temperature.

The correlation coefficient between the microcolony count and the standard plate count (SPC) was $r = 0.85$, and 100,000 microcolonies ml^{-1} corresponded to a SPC of 170,000 cfu ml^{-1}.

Advantages of this method are that the counts are available within 24h and the electronic counter also speeds up the counting procedure. A disadvantage is the fact that the growth of aerobic bacteria such as *Pseudomonas* is suppressed in the tubed medium compared to facultative anaerobes (Suhren and Heeschen, 1991a). This might lead to the microcolony method underestimating counts of samples in which pseudomonads predominate.

14.7.3 Macrocolony Count Procedures

14.7.3.1 "Total" Counts

14.7.3.1.1 Conventional Standard Plate Count. Most standards and regulations refer to macrocolony counts determined by a reference or official method (Suhren, 1989). The problematic nature of this parameter will continue to be debated. Sharpe as quoted by Suhren (1989) has stated that: "the plate count is a totally unique datum; nothing in the physical, chemical, biochemical or immunological world corresponds to it and no test based on these properties can ever correlate with it reliably." Nevertheless, the macrocolony count remains the internationally accepted standard.

In the standard plate count method, those microorganisms that are able to produce colony-forming units in a specific growth medium are enumerated after decimal dilutions of the sample have been plated and incubated aerobically at 30°C ± 1°C for 72 ± 2h (Webber et al., 2000) or at 30–32°C for 48h (Houghtby et al., 1993). Bacterial colonies that grow on the surface and in the various depths of the solid growth medium are counted, and calculations are done according to the specified procedures to determine the number of colony-forming units (cfu's) per milliliter of the original samples (Brazis, 1991).

The method described in IDF (1991a) is recommended for a wide range of dairy products such as milk, liquid and dried milk products, lactose, caseins, caseinates, processed cheese, butter, frozen milk products, custard, desserts, and cream. Plate count standards have been developed with a view to ensuring satisfactory production hygiene and to ensure that the product is safe (Brazis, 1991). This method has also been used as a valuable adjunct to guide sanitarians in correcting sanitation failures and in improving the bacteriological quality of milk. A few disadvantages of the standard plate count include the following:

1. The long incubation time, often yielding counts after the product has been processed.
2. Inherent limitations brought on by a specific culture medium, the aerobic incubation conditions, and temperature of incubation.
3. The fact that both viable single cells and clumps of cells are counted as single colony-forming units.

14.7.3.1.2 Surface Count Technique. Attempts to speed up the standard plate count or decrease the amount of agar medium or number of plates and dilutions have resulted in a range of modifications. One of these modifications is the surface count technique (Lück and Gavron, 1990). Surface counts result in more rapid development of colonies, and these can be counted after 24h. In producing such colonies, the spread or drop method can be applied. In the spread method, 0.1ml of the 10-fold dilutions are transferred to and spread over the dry surface of a solid agar medium. After incubation, countable plates, selected according to the usual procedures, are counted. Advantages of this method are as follows:

1. Plates of media can be prepared and dried beforehand.
2. All colonies are on the surface and easily visible.
3. Aerobic colonies develop more rapidly.

Disadvantages are as follows:

1. Only 0.1ml of dilution can be plated per conventional Petri dish.
2. Undetected contaminant colonies occurring on the prepared plate are spread over the entire plate during the spreading procedure, resulting in a film of growth that masks the development of colonies to be counted.
3. The spreading technique itself is time-consuming.

14.7.3.1.3 Plate Loop Technique. The plate loop technique (Hill, 1991c) substantially decreases the time it takes to process a milk sample (from the sample bottle to the Petri dish). Because of reduced media requirements and the elimination of the necessity for serial dilutions, a single operator can process a greatly increased number of samples compared with the reference method. Disadvantages of this method are as follows:

1. The test result is still governed by the incubation time (e.g., 72h at 30°C),
2. Several features of the technique itself influence the precision and accuracy of the final count obtained.

These factors relate to the loop itself, the manner in which the loop is used by the operator, and the characteristics of the milk under analysis (Hill, 1991c).

14.7.3.1.4 Roll Tube Method. The roll tube method (Slaghuis, 1991) has been developed to save labor and money and is in fact a mechanization of the plate count method. The principle of the method entails transferring a fixed volume of the milk dilution into a thin-walled glass "roll tube" in which the pre-sterilized, melted medium is contained. The tube is sealed by means of a special rubber stopper, and the tube is placed in a horizontal position on an apparatus that spins the tube on its horizontal axis while cold water is sprayed onto the external surface of the tube. The medium solidifies and adheres to the inside surface of the tube. After incubation for 3 days at 30°C, the colonies are counted and the cfu ml^{-1} count is determined as usual.

A modification of this method in which the loop method and the roll-tube method is combined has been applied in the Netherlands in the quality payment scheme of that country (Slaghuis, 1991). The results of the roll tube method have been found to be virtually identical to those of the standard plate count (Slaghuis, 1991). Limitations of the roll tube method are:

1. Similar to those related to colony count methods in general.
2. Those attributable to the roll tube method itself, namely a higher agar concentration necessary to attach the thin layer of medium to the wall; the fact that water does not evaporate in the sealed tube as compared to a Petri dish, which results in occasional surface spreaders; and no replenishment of oxygen as compared

to the Petri dish method. This may result in strict aerobes not developing as well in the roll tube (Slaghuis, 1991).

14.7.3.1.5 Spiral Plate Count. The spiral plate count method (Harding, 1995) is another version of the SPC in which a spiral plating instrument inoculates the surface of a prepared agar plate in such a way that between 500 and 500,000 cfu ml^{-1} of sample can be counted. The instrument deposits a decreasing amount of milk on the surface of the agar plate by means of an Archimedean spiral, such that the volume of the sample deposited on any portion of the plate is known. Colonies on a portion of the plate are counted using a special grid that associates a calibrated volume with each area. An advantage of this method is that it removes the need for multiple dilutions necessary in the SPC techniques. In collaborative studies (Jay, 1992) on milk sample testing, the spiral count compared favorably with the SPC. Spiral plating is an official Association of Official Analytical Chemists method (AOAC, 1990). Other advantages of the method are that less agar is used as well as fewer plates, dilution blanks, and pipettes. Three to four times more samples per hour can be examined, compared to the conventional method (Jay, 1992).

A disadvantage of the method is the expense of the device, and it is not likely to be available in laboratories that do not analyze large numbers of plates; it is also more suited to liquid food such as milk, because more particulate foods can lead to blockage of the dispensing stylus (Jay, 1992).

14.7.3.1.6 Dry Rehydratable Film Technique. The dry rehydratable film consists of two plastic films attached on one side and coated with culture medium ingredients and a cold water-soluble jelling agent. The film was developed by the 3M Company and designated *Petrifilm* [McAllister et al. (1984) as quoted by Jay (1992)].

In applying the film, 1 ml of diluent is placed in the shallow 6-cm-diameter well and is sandwiched in the nutrient area by pressing the two plastic sheets together. Following incubation the microcolonies appear red on the nonselective film due to the presence of a tetrazolium dye in the nutrient phase (Jay, 1992). Petrifilm test methods are available for the aerobic ("total") plate count, the coliform count, and the *Escherichia coli* count.

Use of this method to date indicates that it is an acceptable alternative to the conventional plate count methods (Ginn et al., 1984) and has been approved by the AOAC, the ISO, and the International Dairy Federation (IDF, 1996b).

14.7.3.1.7 Hydrophobic Grid Membrane Technique. The hydrophobic grid membrane filter (HGMF) combines desirable features of the conventional plate count methods, the principles of membrane filtration, and most probable number (MPN) counts and offers additional benefits (Sharpe, 1989).

"Conventional" membrane filtration became the main tool for analytical water microbiology after World War II. Its success was due not only to its permeability to nutrients, its chemical stability, and, to a moderate extent at least, its ability to keep developing colonies separate, but to the improved limits of detection it allowed by concentrating the bacteria in a liquid (Sharpe, 1989).

The HGMF is a square membrane filter (60 × 60 mm, pore size 0.45 μm) printed on one side with a black hydrophobic ("waxy") grid outlining 1600 (40 × 40) small squares. Its unique properties result from the confining of colony growth to the grid cells in which the individual cell/cell clump was captured originally. The typical appearance after incubation is of a grid bearing "square" colonies distributed among the 1600 available locations (Sharpe, 1989). The additional advantages offered by HGMF include the following:

1. An automated counter and a variety of filtration equipment and items supplied by two Canadian companies.
2. With agar plates and ordinary membrane filters, colony overlaps at high cfu densities limit the numerical range necessitating sequential dilutions. The HGMF counts, however, follow a most probable number mathematical principle (Sharpe, 1989) where each grid cell can be linked to one tube in an MPN count done at a single dilution, thus

   $$\text{MPNGU} = 1600 \log \frac{(1600 - X)}{X}$$

   where X is the number of positive grid cells and MPNGU is the most probable number of growth units filtered onto the HGMF. There is therefore no need to prepare serial dilutions of the sample.
3. Plating of duplicate HGMFs is unnecessary because the precision is better than the plate count,
4. The HGMF regiments colonies into arrays that electronics can deal with more easily.
5. HGMF-based analyses are available for all the common foodborne organisms and many of the HGMF techniques (e.g., aerobic

plate count, coliforms, fecal coliforms, *E. coli*, *E. coli* 0157, and *Salmonella* enjoy AOAC official action (Sharpe, 1989).

HGMFs do have some disadvantages (Sharpe, 1989):

1. Each grid (ISO-GRID HGMF) is relatively expensive.
2. An automated colony counter is recommended if large numbers of HGMFs are to be counted, which also brings in a cost factor.
3. For some food suspensions, filterability needs to be improved by means of enzyme digestion.

14.7.3.2 Contaminating Organisms. The so-called "contaminating organisms" count is a version of the standard plate count. It differs from the conventional SPC technique in that the culture medium is carbohydrate-free. Lactic acid starter organisms requiring this carbon source are consequently not able to develop in the medium or at best are only capable of developing pinpoint colonies. The rationale is that non-lactic acid organisms (e.g., Gram-negative spoilage organisms) can be selectively detected in products such as butter, fermented milks, and fresh cheese in which beneficial (flavor- or acid-producing) organisms such as viable lactic acid starter bacteria may be present.

This method was developed by a joint IDF/ISO/AOAC group of experts and has been published as an international standard (IDF, 1991b). The medium consists of peptone from casein, peptone from gelatin, sodium chloride, agar, and water, and all medium components are carbohydrate-free. The incubation temperature/time is 30°C for 72h. Pinpoint colonies do not represent typical "contaminants" and should not be counted. All other colonies should be counted. The results are reported as "contaminating microorganisms per gram of product."

Advantages of the method include the detection of nonstarter contaminants in fermented milk products. Disadvantages are that pinpoint colonies may erroneously be counted as contaminants.

14.7.3.3 Psychrotrophic Bacteria. Psychrotrophic bacteria are those bacteria able to grow at 7°C or less regardless of their optimal growth temperature (Frank et al., 1993). Psychrotrophic bacteria commonly isolated from dairy products belong to a variety of Gram-negative and Gram-positive genera. The most detrimental of these are the oxidative Gram-negative rods belonging to the genus *Pseudomonas*.

This is also the psychrotroph genus most commonly isolated from milk (Frank et al., 1993). In raw milk, high counts of this group of bacteria are related to unsanitary conditions during production and to temperature abuse during storage before pasteurisation. These organisms are inactivated by pasteurization, and their presence in pasteurized milk indicates either improper pasteurization or postpasteurization contamination. In pasteurized milk these organisms severely limit the shelf life of the milk. They also produce proteases and lipases that, when produced in the raw milk, can survive heat treatment and cause sensory and textural defects in the processed dairy product (Frank et al., 1993).

Reference methods for counting psychrotrophic bacteria in milk include the 7°C 10-day incubation period of the American Public Health Association standard method (Frank et al., 1993) and the IDF international standard (IDF, 1991c) (6.5°C for 10 days). The culture media and plating procedures are as for the standard plate count.

The reference methods are of limited use in practice because of the 10-day incubation period they require. Consequently, more rapid methods have been developed. One set of methods is based on using higher temperatures of incubation such as 21°C for 25h (Griffiths et al., 1980) or 18°C for 45h (Oehlrich and McKellar, 1983). Lück et al. (1984) found that no significant differences existed between the standard psychrotrophic counts and the 21°C/25-h counts when poor-quality raw milk that contained a high percentage of psychrotrophs was tested. Significant differences did, however, exist in good-quality raw milk that contained relatively low numbers of psychrotrophs. They are of the opinion that the 21°C/25-h regime can be recommended as a method by which an estimate of the psychrotrophic population can be obtained. This method has been approved as an International Dairy Federation Standard (IDF, 1985).

Oehlrich and McKellar (1983), who proposed an 18°C/45-h temperature–time combination, found that this method led to a more uniform and visible colony size than was the case at 21°C for 25h. With the latter method the majority of colonies were found to be pinpoint colonies. The 18°C/45-h method was also found by Fischer et al. (1986) to be well-correlated with the reference method (r = 0.911), and the percentage distribution of bacterial types in the 18°C/45-h method corresponded very well with those in the reference method.

Media that are selective for Gram-negative bacteria have also been used for estimating the psychrotrophic population. An SPC medium containing crystal violet and tetrazolium is recommended by the American Public Health Association (Frank et al., 1993) with an incubation time and temperature of 48h and 21°C, respectively. Red colonies are counted on this medium. Fischer et al. (1986) found that a count made on SPC medium containing alkyldimethyl benzyl ammonium chloride ("Merquat") and incubated at 18°C for 45 h was well-correlated with the psychrotrophic count reference method ($r = 0.920$). Lück and Gavron (1990) are of the opinion that differences in counts between the reference method and the methods in which inhibitory substances are added to the media to selectively suppress the growth of Gram-positive organisms are often rather great.

The counting of oxidase positive colonies at elevated temperatures has also been proposed as a rapid test to determine the presence of potential psychrotrophs. This test is based on the ability of certain bacteria, which contain a strong cytochrome C oxidase system, to oxidize chemicals and to form dyes (Lück and Gavron, 1990). For estimating the psychrotrophic count, the oxidase-positive count at 27°C or 32°C (r = 0.59 and 0.57, respectively) showed no advantage over the SPC method (r = 0.61 and 0.61, respectively; Lück et al., 1971).

14.7.3.4 Proteolytic and Lipolytic Bacteria. Many bacteria responsible for spoilage of refrigerated dairy products are highly proteolytic and/or lipolytic and can cause flavor defects. Proteolytic enzymes produced by psychrotrophic bacteria during growth in milk often remain active after HTST and even UHT heat treatment and reduce the quality of stored, heat-treated products. Two methods for the detection or enumeration of proteolytic bacteria are recommended by the American Public Health Association (Frank et al., 1993).

The first method entails using standard plate count medium with 10% added skim milk. Plates are incubated at 32°C for 48–72 h. After incubation, plates are flooded with 1% hydrochloric acid and are left for 1 min before decanting the excess acid and counting the colonies surrounded by clear zones. A disadvantage of this method is that acid-producing bacteria can produce false-positive reactions on this agar. In the author's opinion a better counting procedure is the standard methods caseinate agar method. This method is based on the addition to standard methods agar of sodium caseinate dissolved in a citrate solution (Frank et al., 1993). After sterilization and cooling of the medium to 45°C, a sterile calcium chloride solution is added to the molten agar and mixed immediately. Dilutions of the milk sample are surface-plated and spread on the solidified and pre-dried plates of caseinate agar and incubated at 32°C for 48–72 h. Colonies surrounded by a white or off-white zone of casein precipitate are proteolytic. Highly proteolytic colonies will also produce a clear inner zone with an outer opaque precipitate zone.

Disadvantages of this method are that only uncrowded plates (fewer than 80 colonies per plate: Frank et al., 1993) can be read accurately. Calcium chloride added at too high a concentration or temperature results in turbid plates, rendering the plates useless. The caseinate plates must be completely clear after solidification.

Growth of lipase-producing microorganisms can contribute to flavor defects in milk and high-fat dairy products. Some of the free fatty acids released by the action of lipolytic enzymes have a low flavor threshold and can impart a rancid flavor at low concentrations. The less volatile fatty acids are more susceptible to oxidation following hydrolysis, which leads to oxidative flavor defects (Frank et al., 1993). Heat-stable bacterial lipases are of particular concern because they affect products stored for long periods such as cheese, butter, and UHT products.

Several different methods are available for the enumeration of lipolytic microorganisms. The growth medium recommended by the American Public Health Association is preferred over the Victoria blue butter-fat agar (Lück and Gavron, 1990) because of difficulties in preparing the latter medium and problems in interpreting the reaction (Shelley et al., 1987).

The method described by Frank et al. (1993) employs spirit blue agar. This medium is commercially available; and after cooling the melted sterile agar, 3% v/v of lipase reagent (containing tributyrin as a substrate) is added to the medium and thoroughly mixed to emulsify the reagent uniformly in the agar.

The prepared and dried plates are surface-inoculated with 0.1 ml of the milk sample dilutions and incubated at 32°C for 48 h. For psychrotrophic lipolytic organisms, plates are incubated at 21°C for 72 h. The complete medium is light blue and translucent when prepared as above. Colonies of lipolytic microorganisms develop a clear zone and/or a deep blue color around or underneath each colony.

14.7.3.5 Thermoduric Bacteria and Sporeformers. In the dairy industry the term *thermoduric bacteria* refers to microorganisms that survive pasteurization but do not grow at pasteurization temperatures (Frank et al., 1993). Thermoduric bacteria isolated from milk usually include spore-formers such as *Bacillus* and *Clostridium* and also non-spore-forming cocci (e.g., *Micrococcus* spp. and *Streptococcus* spp.) and rods such as *Microbacterium* and other members of the coryneform group (Thomas et al., 1967).

Primary sources of contamination of milk by thermoduric microorganisms are poorly cleaned and sanitized udders, utensils, and equipment. Because high thermoduric counts are consistently associated with unhygienic production practices, the thermoduric or laboratory pasteurization count is used to indicate the thoroughness of equipment sanitation and to detect milk supplies that can be responsible for high-count pasteurized milk products (Frank et al., 1993). The *laboratory pasteurization count* (LPC) is performed by heating a 5-ml sample of milk to 62.8°C for 30 min. Plating and incubation of plates are the same as for the conventional standard plate count method (Frank et al., 1993). The LPC simulates low-temperature long-time (LTLT) pasteurization. For enumeration of these bacteria according to the HTST (high-temperature short-time) method of pasteurization, a standardized loopful of milk sample is mixed with 10 ml of melted plate count agar at 74°C in a water bath. After exactly 15 s, the medium is poured into a Petri dish. The solidified plates are incubated at 30–32°C for 48–72 h (Lück and Gavron, 1990). The above thermoduric count techniques (LPC and the HTST methods) yield counts that are significantly different (Lück and Gavron, 1990).

The spore-forming component of the thermoduric population can be more selectively counted by applying higher temperature treatments of the milk sample. Various temperature–time intervals have been recommended—for example, 80°C for 10 min, 85°C for 10–15 min, and 100°C for 12 min. According to Lück and Gavron (1990), a combination of 80°C for 10 min is not sufficient to kill all vegetative bacteria in milk, and hence a treatment of 85°C for 10 min is recommended. After this treatment the milk sample is immediately cooled to 10°C and the SPC dilution and plating technique is applied with incubation of the plates at 30°C for 72 h.

When spores of anaerobic bacteria are to be counted, a broth of reinforced clostridial medium, plus a "seal" of 2% sterile water agar containing 0.5 g $liter^{-1}$ sodium thioglycolate, is recommended (Lück and Gavron, 1990). Because of the low numbers of anaerobic spores usually found in dairy products, the most probable number method should be applied. The inoculated and sealed (25-mm-thick seal) tubes are incubated at 30°C for 14 days. The tubes are examined on a daily basis, and those showing evidence of gas production are positive. For enumerating hydrogen sulfide-producing clostridia, differential reinforced clostridial medium is used (Lück and Gavron, 1990). Tubes that show blackening of their contents are positive and can be used to isolate and further characterise the clostridia.

14.7.3.6 Thermophilic Organisms. In the dairy industry, the term *thermophilic bacteria* refers to organisms that grow in milk or milk products held at elevated temperatures (55°C or higher), which include LTLT pasteurization conditions. Thermophilic bacteria may also accumulate in certain areas of HTST pasteurizers that have been in continuous operation for extended periods (Frank et al., 1993).

Thermophilic bacteria are usually species of *Bacillus*, which enter into milk from various sources on the farm or from poorly cleansed equipment in the processing plant. These bacteria rapidly increase in numbers when present in milk or dairy products that are held at high temperatures for long periods.

To enumerate these organisms in milk and milk products, prepare dilutions the same as for the standard plate count, but use 15–18 ml of agar medium per plate. Incubate the inverted plates for 48 h at 55°C and maintain humidity during incubation to prevent drying of agar medium. Report the colony count as "thermophilic bacterial count per milliliter" or "per gram" (TBC ml^{-1} or g^{-1}).

14.7.3.7 Enterococci. The *Enterococcus* count is more reliable than the coliform count as an index of the sanitary quality of cultured butter [Blankenagel et al. (1967) and Saraswat et al. (1965) as quoted by Frank et al. (1993)]. This is because enterococci are better able than coliforms to survive in the unfavorable microenvironment of the butter. In addition, the *Enterococcus* count may be a more reliable indicator of the sanitary quality of yogurt than the coliform count because coliforms are inactivated in the low-pH environment whereas enterococci are not [Jordano Salinas (1984) as quoted by Frank et al. (1993)].

Citrate azide agar has been recommended by the American Public Health Association (Frank et al., 1993) for enumerating enterococci in dairy products. Petri dishes are inoculated and poured using citrate azide agar in the usual way. After the medium has solidified, a thin (3–4 ml) overlay of the same medium is poured onto the surface and tilted to allow the overlay to cover the surface of the solidified agar completely. The plates are incubated at 37°C for 48–72 h and only the blue colonies are counted. A white sheet of paper placed under the Petri dish on the illuminated colony counter enhances the contrast between the colonies and the background (Frank et al., 1993). An *Enterococcus* count of fewer than 10 colonies per gram of butter is not considered too stringent for a well-managed butter manufacturing plant [Saraswat et al. (1965) as quoted by Frank et al. (1993)].

14.8 PROCEDURES FOR THE INDIRECT ASSESSMENT OF THE MICROBIAL CONTENT OF MILK AND MILK PRODUCTS

14.8.1 Most Probable Number Method

The MPN method makes use of a statistical technique to detect low counts of bacteria in dairy products (Lück and Gavron, 1990). Liquid media are usually employed, with the nature of the medium depending on the nature of the microorganisms to be counted. Three sets of three or five tubes, each containing the sterile medium, are prepared and inoculated from each of three consecutive 10-fold dilutions. Tubes showing bacterial growth after incubation are positive. From the number of positive tubes in each set of three or five tubes, the most probable number of bacteria per unit of sample is read off from McCrady's tables (McCrady, 1918). When more than three dilutions (more than three sets of tubes) are made, only the results from any three consecutive dilutions are significant. The highest dilution that gives positive results in all tubes, along with the next two succeeding (higher) dilutions, should be chosen. When the mass or volume of sample in the first dilution is 10 or 100 times less than the mass or volume listed in McCrady's tables, then the count tabled must be multiplied by 10 and 100, respectively (Lück and Gavron, 1990).

This method of analysis has gained popularity. According to Jay (1992) the advantages that it offers are the following:

1. It is relatively simple.
2. Results from one laboratory are more likely than SPC results to agree with those from another laboratory.
3. Specific groups of organisms can be determined by use of appropriate selective and differential media.
4. It is the method of choice for determining fecal coliform densities.

Among the drawbacks to its use is the large volume of glassware required (especially for the five-tube method), the lack of opportunity to observe the colonial morphology of the organisms, and the method's lack of precision (Jay, 1992).

14.8.2 Methods Based on the Metabolic Activity of the Microorganisms

14.8.2.1 Dye Reduction. The principle of these tests is to add oxidation–reduction-sensitive dyes, such as methylene blue, resazurin, or triphenyltetrazolium chloride, to milk or liquid dairy products and to measure the color change after incubation. The color change is based on the dehydrogenase activity of the bacteria present. Dehydrogenases (i.e., mainly flavine enzymes) transfer hydrogen from a substrate to biological acceptors or to the dyes added. The period of time needed to change or to decolorise the dye is an index of the bacterial load of the product (Lück and Gavron, 1990).

Dye reduction tests are, however, of little value as an index of the bacterial count of refrigerated milk, because this relationship is poorly correlated [$r = 0.36$ to -0.62 (Lück et al., 1970a; Lück, 1972)]. The reason is that most of the bacteria in refrigerated milk are in a dormant state. Furthermore, a relatively large proportion of the bacteria present are psychrotrophs. These microorganisms have, compared to lactic acid bacteria, a low dehydrogenase activity, a characteristic that contributes to the low correlation between bacterial count and methylene blue reduction time or resazurin disc reading. In order to achieve the same reliability of, say the methylene blue test for nonrefrigerated milk, approximately twice as many samples of refrigerated milk have to be tested (Lück and Andrew, 1975). Preincubation (13–18°C for 16–24h) has been shown to be unsuccessful in improving the relationship between bacterial count before incubation and the results of a metabolic activity test after preincubation (Lück and Gavron, 1990).

Advantages of the dye reduction tests are that the results are available within a few hours and that the test can be carried out in a small laboratory without expensive equipment (Lück, 1991a). Reduction tests are, therefore, suitable to improve the bacteriological quality of milk in developing countries.

Disadvantages of dye reduction tests are that they are of very little value for the estimation of the plate count of cold-stored milk, especially when the "total" bacterial count is below 100,000 ml^{-1}, no matter whether the samples are or are not preincubated before testing (Lück, 1991a). The tests are also imprecise, because some bacteria have a high rate of reduction whereas others have a low rate. Furthermore, somatic cells are also capable of reducing the dyes (Harding, 1995).

14.8.2.2 Nitrate Reduction. The nitrate reduction test (NRT) makes use of the ability of several bacterial species to reduce nitrate to nitrite, which is a multienzyme reaction catalyzed by different flavine enzymes (Lück, 1991b).

The test entails adding 1 ml KNO_3 solution (0.3% w/v) to 10 ml milk. The milk is then incubated at 30°C for a specified period, and a spot test is done for nitrite formation—that is, 1 drop of milk plus 1 drop of nitrite reagent (5 g sulfanilamide + 1 g α-naphthylamine + 100 ml glacial acetic acid + 100 ml H_2O; the solution is made up weekly and refrigerated in an amber glass bottle). A pink to red color indicates a positive reaction.

The NRT is suitable as a method for detecting raw milk samples with psychrotrophic or other contaminating bacteria in numbers exceeding 200,000 ml^{-1}. For this purpose a 5-h NRT at 30°C without preincubation or a 3-h NRT at 30°C after preincubation at 15°C for 16 h can be used (Lück et al., 1972).

The advantages of the NRT is that it gives a better indication of the psychrotroph and coliform content in cold-stored raw milk than the dye reduction tests and that it can be used in smaller laboratories anywhere.

A disadvantage of the NRT is that it is not suitable for testing milk with a bacterial count of less than 100,000 ml^{-1}.

14.8.2.3 Pyruvate Determination. Pyruvate is an intermediate metabolite in a wide variety of metabolic pathways and is therefore a constituent of all microbial cells. Tolle et al. (1972) suggested the measurement of pyruvate as an indication of the hygienic quality of milk. Immediately after milking, the pyruvate level in normal milk is 0.5–1.5 $\mu g\, ml^{-1}$, with much of this pyruvate being derived from sources other than bacteria e.g. leucocytes, (Easter and Prentice, 1989). There is therefore a background level of pyruvate in milk, which limits usefulness for determining low levels of microorganisms. The level of microbial pyruvate also varies according to the storage conditions of the milk (Easter and Prentice, 1989). For these reasons, pyruvate measurement is not a method of choice for determining the hygienic quality of milk with low microbial load.

If it is only necessary to detect milk with high levels of bacteria (e.g., $>10^6\, ml^{-1}$), the pyruvate test has the advantage of being rapid, easily automated, and able to be carried out on preserved milk. For the latter reason, the pyruvate test has been used in a quality control test program in the Federal Republic of Germany (Easter and Prentice, 1989).

14.8.2.4 Catalase Production. The psychrotrophic bacterial population in milk consists largely of Gram-negative, catalase-producing bacterial genera. The enzyme catalase catalyzes the following reaction:

$$2H_2O_2 \overset{\text{catalase}}{\rightleftharpoons} 2H_2O + O_2$$

It may be argued that the amount of oxygen released by catalase activity in milk is related to the microbial load of the milk (Easter and Prentice, 1989). Lück (1991c), however, came to the conclusion that the catalase test cannot be recommended as an accurate index of the bacterial load of raw milk. The activity of the catalase present was found not to be correlated well enough with the number of microorganisms in the milk.

14.8.2.5 Oxygen (O_2) Tension Measurement. Another metabolic activity test described by Lück and Gavron (1990) is the oxygen (O_2) test carried out by means of an O_2 electrode. The amount of dissolved oxygen in the milk decreases as the initial bacterial count increases.

Different procedures of handling milk and liquid dairy products affect the dissolved oxygen content. Such procedures include time lapses until testing, temperature of the milk, and/or the time and speed of stirring or agitation, and so on. For this reason the direct measurement of the O_2 content of the milk does not give a clear indication of the bacterial count.

The milk samples have to be saturated with air first (e.g., by shaking), followed by a specified period of incubation to obtain a close relationship between O_2 content and bacterial count. The multiple regression relationship between O_2 content and bacterial counts under these conditions was consistent enough ($r = -0.70$) to recommend the 6-h O_2 test as an index of the total mesophilic and psychrotrophic count (Lück et al., 1970b; Lück, 1991d). The necessity to saturate the samples with air followed by a 6-h incubation period at 25°C, together with the necessity for having a oxygen meter, has so far discouraged the routine implementation of the O_2 depletion test (Lück, 1991d).

14.8.2.6 Impedance. It is known that growth of microorganisms in a medium alters the electrical properties of the growth medium (Easter and Prentice, 1989). This occurs in two ways:

1. The catabolic activity of microorganisms breaks down large molecules into small molecules that are frequently charged (e.g., undissociated sugars are broken down into lactic and acetic acids, and proteins are broken down into fatty acids). These metabolic products frequently have a greater charge than the larger molecule from which they originate.
2. The smaller molecules are more mobile than the larger molecules and are able to conduct electricity more readily in a solvent.

Microbial growth in a growth medium tends to increase the ability of a medium to conduct electricity; that is, conductance is an increasing function of microbial growth. Because conductance is inversely dependent on impedance, there is an inverse relationship between microbial growth and impedance (Easter and Prentice, 1989).

The electrical impedance in a culture medium remains fairly consistent until a threshold of 10^6–10^7 cells ml^{-1} is reached, when major changes in impedance start to occur. The time taken to reach the threshold value is indicative of the initial bacterial load in raw milk (Cady et al., 1978).

An advantage of this method is that a result can be obtained in 8.5 h compared to 72 h using a conventional SPC (Easter and Prentice, 1989). In addition, this method can be used to detect specific microorganisms in the food. By designing media that will inhibit the growth of all but the test organism, while still allowing satisfactory growth curves of the latter organisms, it is possible to detect a wide range of indicator organisms and pathogens—for example, enterococci, coliforms, and *Salmonella* (Easter and Prentice, 1989).

The impedance detection time (IDT) can, however, also be used to screen raw milk samples, and in a specific study a 7-h cutoff time (equivalent to 10^5 cfu ml^{-1}) was successfully used to screen the samples (Gnan and Luedecke, 1982). Impedimetry was also used to determine the potential shelf life of pasteurized whole milk (Bishop et al., 1984).

Attention must be paid to culture media for impedimetry because not all media sustain satisfactory growth of a given organism. In brain heart infusion (BHI) broth, *Pseudomonas* spp. exhibited a triphasic type growth curve, making it difficult to determine the true IDT (Firstenberg-Eden and Tricarico, 1983). The incubation temperature is also important (Easter and Prentice, 1989). A method for estimating the bacterial content of a milk sample by impedimetry is described in *Standard Methods for the Examination of Dairy Products* (Marshall, 1993).

14.8.3 Methods Based on Specific Cellular Components

14.8.3.1 Adenosine Triphosphate (ATP) Bioluminescence. All living cells contain ATP, which acts as a substrate in the bioluminescence firefly enzymes system luciferin–luciferase, giving rise to light emission (Harding, 1995). This very sensitive light emission reaction can be used as a measure of low levels of bacteria via their ATP content according to the following general reaction:

$$\text{ATP} + \text{luciferin} + \text{oxygen} \xrightarrow[\text{Mg}^{2+}]{\text{luciferase}} \text{reaction products} + \text{light}$$

The light emitted as a result of the above reaction is measured using a liquid scintillation spectrometer or luminometer. The amount of light produced by firefly luciferase is directly proportional to the amount of ATP added (Jay, 1992).

Most foods, however, contain both free ATP and ATP associated with plant and animal cells from which that food was derived. ATP from these sources is often present in amounts greatly in excess of ATP from any contaminating bacteria that may be present (Van Crombrugge and Waes, 1991). The usefulness of the ATP determination as a means of detecting contamination in food depends on the efficiency with which the bacterial ATP can be separated from the nonbacterial ATP. In milk the principal source of ATP is somatic cells and the calcium phosphate–citrate–caseinate micelles (Van Crombrugge and Waes, 1991).

In a review article on the application of ATP bioluminescence in the food industry, Griffiths (1996) refers to various authors who agree that ATP determination can be used successfully as a rapid assay method for the microbial loads in raw milk. Reybroeck and Schram (1995), for example, outlined a test that took less than 6 min. In this test the milk is incubated in the presence of a somatic cell-lysing agent and then filtered through a bacterial cell-retaining membrane. The concentrated cells on the membrane are then lysed using a second extraction agent (specific for the extraction of bacterial ATP) and assayed by measurement of the resultant bioluminescence. Griffiths (1996) stressed that using this method, microbial populations down to 10^4 cfu ml^{-1} can be detected with greater precision than when using the standard plate count. Other workers have also reported that the use of the ATP-bioluminescence method is a practical and reliable screening test for assessing the hygienic quality of bovine raw milk (Bell et al., 1996; Brovko et al., 1999).

Using the bioluminescence principle, the hygienic status of tankers, plants, and equipment can be assessed in as little as 2 min (Anonymous, 1995). Two methods are available, one to check surfaces using swabs and the other to check rinse waters from CIP systems. Both tests determine not only the microbial contamination but also product residues left behind on surfaces and in closed systems. Apart from the undesirable aesthetic nature of the remaining residues, they also serve as nutrients for the growth of microorganisms. Because of the measurement of both soiling and microbial contamination by this method, direct correlation with more traditional methods is not always possible.

In the swabbing procedure (Anonymous, 1995) an area of $100\,cm^2$ is swabbed, the swab is vortexed in a cuvette containing the ATP extraction reagents, and the luciferin–luciferase reagent is subsequently added to produce the bioluminescence. The measurement by a luminometer is recorded in relative light units (RLU). Less than 500 RLU is regarded as a "clean" value, whereas more than 500 RLU is regarded as a "contaminated" value (Anonymous, 1995).

The rinse water test (Anonymous, 1995) is a rapid miniaturized filtration system, and end results are obtained in approximately 5 min. A volume of 20 ml of the rinse water is filtered through an 8-mm-diameter filter by applying vacuum. The filter is removed aseptically and transferred to a flat-bottom cuvette. The reagents to release the microbial ATP and to catalyze the bioluminescence reaction are added, and the light output is measured in the luminometer. Two cutoff limits based on correlation with traditional plate counts have been established (Anonymous, 1995). A value of above 200 RLU indicates residual contamination, while a result above 1000 RLU should necessitate re-cleansing of the equipment.

The advantages of the bioluminescence methods are in keeping with other rapid methods in that reference or conventional methods usually give a result after the milk has been processed or the equipment to be monitored has been used. The bioluminescence method, on the other hand, gives a rapid and also reliable result that will assist in ensuring good manufacturing practice proactively.

Limitations of the bioluminescence system that should be kept in mind are that the bioluminescence assay of ATP is affected by certain interfering factors causing a reduction in measurable photons (Van Crombrugge and Waes, 1991). First, it is important to adhere to the pH optimum of 7.75. Lower or higher pH affects the reaction rate and decreases the light emission. Temperatures higher than 25°C may inactivate the luciferase, and at lower temperatures the reaction is progressively slower. Certain ionic and other compounds can interfere with the reaction, causing a decreased light output. In practice, however, the interfering factors can be minimized by dilution and be corrected for by internal standardization (Van Crombrugge and Waes, 1991).

14.8.3.2 Gram-Negative Endotoxins (Limulus Test). Gram-negative bacteria are characterized by their production of endotoxins, which consist of lipopolysaccharides (LPS) of the outer membrane of the cell envelope (Jay, 1992). The LPS is pyrogenic and responsible for some of the symptoms that accompany infections caused by Gram-negative bacteria.

The *Limulus* amebocyte lysate (LAL) test employs a lysate protein obtained from the hemolymph or blood cells (amebocytes) of the horseshoe *crab* (*Limulus polyphemus*). The lysate protein is the most sensitive substance known for the detection of endotoxins (Jay, 1992). A freeze-dried extract of the *Limulus* amoebocytes forms a clot or a gel when dissolved and brought into contact with the lipopolysaccharides of Gram-negative bacteria (Easter and Prentice, 1989).

The assay technique involves the use of tubes or microtiter plates (Suhren and Heeschen, 1991b) that are prepared by loading the tubes or wells with freeze-dried LAL. For quantitative determination, the milk sample is diluted with pyrogen-free water and 30 µl of the milk or dilution thereof pipetted into the wells of the microtiter plates. Raw milk samples are prepared by heating to 80°C for 10 min and cooling before dilution. The inoculated microtiter plates are then incubated for 1 h at 37°C. Reading the plates involves detecting those wells that contain a gelled or clotted reaction mixture. The detection can be performed by adding a dye solution (toluidine blue) and applying moderate suction using a capillary tube. In the case of positive wells the colored gel is firm and resists suction. Negative reactions are represented by empty wells (Suhren and Heeschen, 1991b). The concentration of the LPS can be calculated by obtaining the reciprocal of the highest positive dilution.

Because this test provides an estimate of the Gram-negative bacterial content of the milk, one nanogram (ng) of LPS can be said to approximate a Gram-negative colony count of 20,000 cfu ml^{-1} with a standard deviation ranging from 5600 to 80,000 cfu ml^{-1} (Suhren and Heeschen, 1991b). This method has been found suitable for the rapid evaluation of the hygienic quality of milk before or after pasteurization (Terplan et al., 1975; Zaadhof and Terplan, 1981; Jaksch et al., 1982; Jay, 1992).

Commercial substrates are available that contain amino acid sequences similar to *Limulus* coagulogen (Jay, 1992). These substrates are rendered chromogenic by linkage to *p*-nitroaniline. When the endotoxin-activated enzyme attacks the chromogenic substrate, free *p*-nitroaniline is released and can be read spectrophotometrically at 405 nm. The amount of chromogenic compound liberated is proportional to the quantity of endotoxin in the sample. An automated method for endotoxin assay was devised by Tsuji et al. (1984), and the method was shown to be sensitive to as little as 30 pg of endotoxin ml^{-1}.

Advantages of the *Limulus* test are as follows:

1. The speed at which results can be obtained.
2. Bacterial lipopolysaccharide is heat-resistant and will even withstand UHT time and temperature treatment (Easter and Prentice, 1989). The application of the *Limulus* test to heated milk therefore gives an indication of the bacteriological quality of the milk prior to processing.

A "disadvantage" of the test is that the relationship between the lipopolysaccharide concentrations and "total" bacterial numbers is dependent on the composition of the microbial population. However, because this test per se is aimed at the Gram-negative component of the population, which would include important spoilage organisms such as the psychrotrophs or indicator organisms such as coliforms, this attribute can also be regarded as a positive one.

14.9 METHODS FOR DETERMINING THE SHELF LIFE OF MILK

An important factor today is the open dating of perishable dairy products to indicate when the products were packed or when they should be removed from the supermarket shelf (Lück and Gavron, 1990). Many dairy factories consequently need accelerated tests to determine the shelf life or keeping quality of milk. The shelf life of milk is defined as the period between manufacture or processing and that point when the milk becomes unsuitable for use by the consumer. The milk is deemed to be unsuitable due to the presence of flavor defects or change of physical appearance (Manners, 1993).

The key to predicting shelf life is not the "total" bacterial count immediately after processing, because this count does not differentiate between contaminants and these bacteria surviving pasteurization (Lück and Gavron, 1990). The shelf life is, however, dependent on the number of postpasteurization contaminants and especially those that multiply rapidly at refrigeration temperatures namely the Gram-negative psychrotrophs (and especially the pseudomonads). Numbers of such microorganisms as low as one or two bacteria per liter can frustrate any attempt to manufacture a product with an extended shelf life.

The shelf life is also related to the time–temperature history of the product. In doing shelf-life tests it is necessary to simulate the marketing conditions to which the product will actually be subjected. For this reason a storage temperature of 5–7°C is recommended (Lück and

Gavron, 1990). Regardless of the test itself, the key to predicting the shelf life of milk and milk products (White, 1998) is that the method must be rapid. Reliable and meaningful results must be obtained within 72h and ideally within 24h. For this reason the classic Moseley count [e.g., the plate count after 5 days of storage at 7°C minus the plate count directly after processing. (Moseley, 1958)] has not proved a popular shelf-life predictive test for pasteurized milk in practice. This despite the fact that it has been shown to be a good index of postpasteurization contamination (Lück and Gavron, 1990).

Determination of the Actual Shelf Life. It is important to note that tests to predict shelf-life must be evaluated against the actual product shelf life. The actual shelf life can be determined by incubating the milk sample at the relevant temperature (e.g., 7°C) and testing at regular intervals until a definite off-flavor is detected by sensory evaluation (White, 1998). More objective tests for determining the shelf-life end point include the alcohol test and the clot-on-boiling tests (Jooste and Groeneveld, 1971). The alcohol test entails the rapid mixing of 2ml of 68% ethanol (alizarin can be added to serve as indicator) with a 2-ml milk sample previously pipetted into a test tube. Any definite evidence of precipitation is regarded as positive (Jooste and Groeneveld, 1971). With the clot-on-boiling test, 5ml of the milk sample in a test tube is placed into boiling water for 5min. Positive results vary from a loose curd to a solid coagulum. The alcohol test was found to be highly correlated with both the clot-on-boiling ($r = 0.9488$) and the sensory evaluation ($r = 0.9072$) test results (Jooste and Groeneveld, 1971). The latter study also showed that at an incubation temperature of 18°C, the milk often became alcohol-positive simultaneously with the appearance of a taint in the milk. Clot-on-boiling usually occurred 1 or 2h later. At that time most of the samples had become completely unpalatable.

Preincubation of the Milk Sample Followed by Plating. Because of the low initial numbers of bacteria in freshly pasteurized milk, most predictive shelf-life tests consist of preincubating the product (in its original container) at 21°C for 18h followed by some rapid bacterial detection test (White, 1998). Such detection methods include the standard plate count, the accelerated psychrotrophic count (21°C for 25h or 18°C for 48h) described previously, or plating on agar plates containing crystal violet and triphenyl tetrazolium chloride (White, 1998). The American Public Health Association (White et al., 1993) recommends the 21°C/25h count method using plate count agar or 21°C for 48h using the Petrifilm aerobic count method following preincubation (PI) of the sample at 21°C for 18h. The relationship between these PI counts and sensory tests (White et al., 1993) gave rise to the following shelf-life estimates:

PI cfu count ml^{-1}	Estimated Shelf Life (days)
<1,000	>14
1,000–200,000	9–14
>200,000	<10

Preincubation of Milk Sample Followed by DEFT. The DEFT microscopic count referred to earlier can also be used as a rapid test to predict shelf life after PI at 21°C for 18h. Alternatively milk with added benzalkonium chloride and crystal violet (for inhibiting Gram-positive bacteria) can be pre-incubated at 30°C for 18h (White et al., 1993) before subjecting the milk to the DEFT microscopic count.

Impedance Method. The use of impedance detection time (DT) with milk samples has been shown to be superior to a traditional keeping quality method such as the Moseley test (Marshall, 1993). Results correlated better with actual keeping quality (r = 0.94 vs. 0.75 for the Moseley test) and are available within 48h. In this test, 5ml of milk sample is added to 5ml of sterile broth medium, and the mixture is subsequently incubated for 18h at 18°C. Modified plate count agar (White et al., 1993) is added and impedance testing is done at 21°C for up to 48h.

According to White (1998), there is no single ideal test for predicting the shelf life of fluid milk products. They suggest that processors should carefully select one or two tests that best fit their overall quality assurance program. The key points regarding shelf-life prediction (White, 1998) include the following:

1. Know the actual shelf life of the product as measured at 7°C.
2. Select the test to predict shelf-life that best fits the total program.
3. Routinely do the tests and develop a history categorizing the results.
4. Define a course of action in case product failure is projected by these tests.

14.10 STERILITY TESTS

The aim of the quality control of ultra-high-temperature (UHT)-treated milk products that have been aseptically processed and packed is to limit the number of containers spoiled by microbial growth to a level acceptable to the market (Lück and Gavron, 1990). The term "sterility test" in this context implies stability rather than sterility in the microbiological sense. Even though a small number of bacteria may have survived the heat treatment, the product is regarded as commercially sterile as long as they do not grow during a commercially acceptable period. Commercial sterility therefore means the absence of all pathogenic or toxigenic microorganisms and the absence of any microorganisms that are capable of multiplication under the conditions of storage and distribution (Lück and Gavron, 1990).

Spore-forming bacteria are usually the cause of microbiological spoilage in UHT-treated milk products. The spore-forming anaerobes, principally of the genus *Clostridium*, are less thermoresistant and are therefore of less concern than the spores of mesophilic and thermophilic aerobes of the genus *Bacillus* (Lück and Gavron, 1990).

UHT treatment reduces the number of spores by 8–10$\log_{10}$ cycles. This means that after treatment, milk with an original spore content of 100 spores ml^{-1} (10^5 spores $liter^{-1}$) will contain 0.001 to 0.00001 spores per liter pack, and thus the number of nonsterile packs will be approximately 1 in 1000 to 1 in 100,000. This level is acceptable (Lück and Gavron, 1990).

To detect unsterile packages, the plate count method per se will not be practically feasible with survivors at a level of less than one per milliliter or as low as one per liter directly after the heating and aseptic packaging. For this reason, incubation of the packaged product at different incubation times and temperatures have been suggested. A 2-week incubation period as suggested by the International Dairy Federation (IDF, 1969) is commercially not practicable (Lück and Gavron, 1990). Tests carried out on UHT milk revealed that 18% of unsterile packs were detected after 3 days at 30°C, 41% after 5 days, and 68% after 7 days (Lück et al., 1978). The results indicated that incubation times of 3 and 5 days were not sufficiently long, and an incubation period of 7 days was consequently deemed advisable (Lück and Gavron, 1990).

With regard to incubation temperature, a thermophilic incubation temperature of 55°C is not necessary because most of the thermophilic spores causing spoilage at 55°C remain dormant at ambient storage temperatures. Only one incubation temperature is consequently needed (Lück and Gavron, 1990), namely in the region of 35–37°C. Facultative thermophiles may be able to grow at 35°C, but one should bear in mind that damaged (stressed) bacteria recover and grow more readily at 27°C than at 35°C.

On termination of the incubation period the UHT product is adjudged defective or not. Defectiveness or spoilage is recorded when the sensory properties are different from those normally obtained by prolonged incubation, when the titratable acidity differs from what it was before incubation by more than 0.02% (expressed as grams of lactic acid per 100g of milk), or when the colony count exceeds 10 per 0.1ml of milk. Normal inspection should consist of 50–100 samples per machine per production run per day (IDF 1969; IDF 1981; Lück and Gavron, 1990).

Other tests that have been recommended for testing the sterility of UHT treated milks include the impedance method on milk preincubated at 25–35°C (time not specified; White et al., 1993) and the bioluminescence method (Anonymous, 1995). In the latter method a dairy products sterility kit is used. A representative sample (0.1–0.3% of total production run) is taken. The unopened packs are incubated at 28–30°C for 48h. The milk is then tested for bioluminescence as recommended in the sterility kit instructions (Anonymous, 1995). Sterile samples have an RLU reading of $<2 \times$ RLU background, while unsterile samples have a RLU reading of $>2 \times$ RLU background.

The possibilities of ultrasound imaging as a nondestructive quality control method was tested among others on UHT milk packages (Mattila et al., 1989). The results showed that spoilage could be detected in all the milk products tested within 2–3 days. The ultrasound could penetrate all the common plastic packaging materials. Cardboard packaging material, however, suppressed penetration. Lower frequencies were used, but this decreased the sensitivity.

14.11 METHODS FOR DETECTING PATHOGENIC MICROORGANISMS AND THEIR TOXINS

Classic methods of detecting and enumerating microorganisms depend on the recognition of microbial colonies, often after growth on selective agar (Waites, 1997). Such methods are time-consuming and labor-intensive and require well-trained staff able not only to carry out aseptic work but also to recognize different morphological types of microorganisms. Because most food-borne pathogenic microorganisms generally occur in low numbers, selective enrichment is also usually required in order to allow detection and prevent overgrowth by other organisms. In addition, some cells, particularly from food processing environments, may be damaged and a pre-enrichment step becomes necessary to assist cells to recover before being exposed to selective agents (Waites, 1997).

Although standard methods based on culturing the organisms will be referred to in this section, the primary aim will be to review a range of more rapid techniques for detecting pathogenic microorganisms and their toxins. The latter methods will include a range of more recent molecular techniques (Table 14.7), immunological methods (Table 14.8), methods for dealing with stressed or damaged cells, and then currently

TABLE 14.7. Molecular Methods for Detecting Microbial Pathogens

Method	Principle and Application	Reference
DNA/RNA probes	A probe is a nucleic acid sequence typical of the organism of interest, used to detect homologous DNA or RNA sequences in the target organism. RNA as target sequence has an advantage in having 10^4 copies per cell versus DNA which has only one or two copies per cell. Nucleic acid fragments for testing are prepared using restriction endonucleases.	Wallbanks (1989) Jay (1992) Waites (1997)
Polymerase chain reaction (PCR)	An amplification technique to increase low numbers of DNA molecules to detectable levels (10^6 copies of the target sequence). PCR can increase one molecule of DNA to produce 10^7 identical copies. Advantages: speed (4 h), sensitivity (1 cell) and high specificity.	Waites (1997)
Amplified fragment length polymorphism (AFLP)	A method for the genomic typing of microorganisms based on DNA sequence polymorphism. The method is reproducible and highly discriminatory and is used for identification and classification of bacteria, fungi and yeasts. It is an advancement on related techniques such as PFGE, BRENDA, and RAPD.	Aarts (1999)

TABLE 14.8. Immunological Methods for the Detection of Microbial Pathogens and Toxins

Method	Principle and Application	Reference
Fluorescent antibody (FA) method	Fluorescent antibody reagent reacts with test antigen. Resultant antigen–antibody complex emits fluorescence and is detected using a fluorescence microscope. AOAC-approved method for detecting *Salmonella* yields results within 52 h.	Jay (1992) Flowers et al. (1993)
Latex agglutination test	Simplest of immunoassays. Reagent consists of latex particles coated with antibodies. Specific antigens in test sample combine with antibodies. Latex particles remain in suspension compared to negative test in which latex particles form a sediment.	Notermans (1989)
Enzyme immunoassays (EIA)	Example of a so-called noncompetitive EIA is the well-known ELISA (enzyme-linked immunosorbent assay). Immobilized antibodies react with specific antigens in test sample. Adsorbed antigen is measured using enzyme-labeled antibodies added to reaction mix (antigen must have two binding sites). The ELISA technique is used to detect and quantitate microorganisms or their metabolic products.	Notermans (1989) Jay (1992)
Immunodiffusion	Gel diffusion methods are widely used for detecting and quantifying bacterial toxins and enterotoxins, for example, *Staphylococcus aureus* enterotoxin and *Clostridium botulinum* toxins. Most widely used test is the Crowle modification of the Ouchterlony slide.	Casman and Bennett (1965) Duncan and Somers (1972) Bennett and McClure (1976) Jay (1992)

TABLE 14.8. *Continued*

Method	Principle and Application	Reference
Radioimmunoassay (RIA)	A radioactive label is attached to the antigen. The labeled antigen is allowed to react with specific antibody in the test sample. The amount of antigen–antibody complex is quantified using a scintillation counter. The RIA is used to examine foods for biohazards such as mycotoxins, endo- and enterotoxins.	Miller et al. (1978) Bergdoll and Reiser (1980) Jay (1992)
Immunomagnetic separation (IMS)	The Dynal system uses Dynabeads®, which are superparamagnetic polystyrene beads coated with specific antibodies. The antibodies combine with the target organism in the test sample, and the bead–bacterial complexes are separated using a magnetic particle concentrator. Incorporation of the IMS step greatly reduces the isolation time for the target organism. IMS can be used for isolation of *Salmonella*, *Listeria*, and *E. coli 0157*.	Wachsmuth et al. (1990) Safarik et al. (1995) Waites (1997) Kaclikovà et al. (1999)

recognized techniques for detecting specific pathogenic and toxigenic organisms (or their toxins) in milk and milk products.

14.11.1 Dealing with Stressed Food-Borne Pathogens

Legislation requires food manufacturers to demonstrate that certain foods they produce are free from contamination by pathogens. For many pathogens, this is interpreted to mean absence of a single cell in a 25-g sample of food. It is also interpreted to imply a single cell that is possibly in an injured or stressed state. The importance of these specifications is reinforced by the epidemiological data obtained from many well-documented outbreaks of food poisoning linked to very low levels of contamination, in some cases less than 10 organisms per portion consumed (Stephens, 1999).

When microorganisms are subjected to environmental stresses such as sublethal heat and freezing, many of the individual cells undergo metabolic injury resulting in their inability to form colonies on selective media that uninjured cells can tolerate. Whether organisms have suffered metabolic injury can be determined by plating aliqots of the sample separately on a nonselective and a selective medium and enumerating the colonies that develop after suitable incubation. The colonies that develop on the nonselective medium represent both injured and uninjured cells, while only the uninjured cells develop on the selective medium (Jay, 1992). The difference between the number of colonies on the two media is a measure of the number of the injured cells in the original sample.

Injury of food-borne microorganisms has been shown by a large number of investigations to be induced not only by sublethal heat and freezing but also by freeze-drying, drying, irradiation, aerosolization, antibiotics, and sanitizing compounds (Jay, 1992). The protection of cells from heat and freeze injury is favored by complex media or specific components thereof. Milk, for example, provides more protection than saline or mixtures of amino acids (Moats et al., 1971; Jay, 1992) and milk components that are most influential appear to be phosphate, lactose, and casein.

Metabolically injured microorganisms have increased nutritional requirements. Foods inoculated with salmonellae were frozen, and the fate of the organisms during freezer storage was studied (Jay, 1992). More organisms could be recovered on highly nutritive nonselective media than on selective media such as MacConkey, deoxycholate, or violet red bile agars. In another example, dry-injured *Staphylococcus aureus* cells failed to recover on the nonselective recovery medium (tryptone soya agar) but did recover when pyruvate was added to the medium (Jay, 1992). Pyruvate is well established as an injury repair agent, not only for injured *S. aureus* cells but for other organisms such as *E. coli*. Higher counts were obtained on media containing this compound when the organisms were injured by a variety of agents (Jay, 1992). Catalase is another compound that increases recovery of injured aerobic organisms. Various investigations (Jay, 1992) have shown that sublethally injured *S. aureus*, *Pseudomonas fluorescens*, *Salmonella typhimurium*, and *E. coli* effectively recovered in the presence of this enzyme. More information on cell injury and methods of recovery (Jay, 1992) can be obtained in the book of Andrew and Russell (1984).

14.11.2 Specific Pathogenic, Indicator or Toxigenic Microorganisms

14.11.2.1 Enterobacteriaceae, *Coliforms, and* Escherichia coli

Enterobacteriaceae. The presence of any member of the *Enterobacteriaceae* family is undesirable in pasteurized dairy products. Methods for detecting *Enterobacteriaceae* are much less specific than those developed for detecting coliforms. The methods, however, do provide increased sensitivity for detecting postpasteurization contamination, and results are available in 12h using the impedance method or in 24h using the modified MacConkey glucose agar plating method (Christen et al., 1993).

Coliform Bacteria. The coliform group of bacteria comprises all aerobic and facultatively anaerobic, Gram-negative, non-spore-forming rods able to ferment lactose with the production of acid and gas at 32°C or 35°C within 48h (Christen et al., 1993). One source of these organisms is the intestinal tract of warm-blooded animals, although certain bacteria of nonfecal origin are also members of this group. Typically, this group is represented by the genera *Escherichia*, *Enterobacter*, and *Klebsiella*, but a few lactose-fermenting species of other genera are also included in the coliform group. In proportion to the numbers present, the existence of any of these species in dairy products is suggestive of unsanitary conditions or practices during production, processing, or storage (Christen et al., 1993).

Reference methods for enumerating these organisms in milk and milk products are referred to in Table 14.6 and include a colony count at 30°C, a dry rehydratable film method, a MPN method, and a pectin gel method. Similar standard methods are described by Christen et al. (1993).

Escherichia coli. Escherichia coli is currently the best-known fecal indicator, and its recovery from dairy products suggests that other organisms of fecal origin, including pathogens, may be present. Reference methods for detecting and enumerating this organism are summarized in Table 14.6. Other standard methods, including hydrophobic grid membrane filtration, an impedance method, and a fluorogenic assay (MUG test), are described by Christen et al. (1993).

Pathogenic E. coli. Most *E. coli* strains are harmless commensals common to the intestinal tract of humans and animals. Some strains have, however, been found to be pathogenic. Final determination of the enteropathogenicity of any strain of *E. coli* is based on that strain's ability to "elicit a specific, consistently positive response in two or more standardized model pathogenicity systems." These model systems demonstrate invasiveness, production of a heat-labile toxin, and production of a heat-stable toxin. These methods are fully described in the American FDA's *Bacteriological Analytical Manual* and in the AOAC's OMA (Flowers et al., 1993).

Based on distinct virulence properties, different interactions with the intestinal mucosa, distinct clinical symptoms, differences in epidemiology, and variations in O (somatic) and H (flagellar) antigens, more than 60 distinct strains causing different forms of diarrhea in humans have been identified (Ryser, 1998). These strains are grouped into the following five categories: classic enteropathogenic *E. coli* (EPEC), enterotoxigenic *E. coli* (ETEC), enteroinvasive *E. coli* (EIEC), enterohaemorrhagic *E. coli* (EHEC), and, most recently, enteroadherent *E. coli* (EAEC) (Ryser, 1998).

Selective recovery of *E. coli* 0157:H7 (EHEC) from food samples is based on the inability of typical isolates to ferment sorbitol and hydrolyze 4-methyl-umbelliferone glucuronide (MUG) to a fluorogenic product. Several selective enrichment broths containing novobiocin or other antibiotics can also be used to enhance recovery. Presumptive *E. coli* 0157:H7 must be serologically confirmed with antisera or with a serotype specific DNA probe. Verotoxin production can be confirmed using traditional cell culture techniques or the newly developed DNA probe and PCR assays for VT-1 (Ryser, 1998).

14.11.2.2 Salmonella. The gastroenteritic form of nontyphoid salmonellosis was not clearly linked to raw milk consumption until the mid-1940s. Interest in milk-borne salmonellosis has peaked twice since the 1940s, first in 1966 when several large outbreaks were traced to nonfat milkpowder and again in 1985 when one of the largest recorded outbreaks of food-borne salmonellosis involving more than 180,000 cases was traced to consumption of a particular brand of pasteurized milk in the Chicago area (Ryser, 1998). Today *Salmonella* and *Campylobacter* are generally recognized as the two leading causes of dairy-related illness in the United States and Western Europe, with rates of infection being particularly high in regions where raw milk is neither pasteurized or boiled.

Examination of milk and milk products according to the International Dairy Federation standard method (IDF, 1995a) begins with pre-enrichment in a buffered peptone water broth to resuscitate injured cells. Following 16–20 h incubation at 37°C, two different media—Rappaport–Vassiliadis and selenite–cystine broths—are inoculated from the pre-enrichment broth and incubated at 42°C and 37°C, respectively, for a total of 48 h. Immunomagnetic concentration of salmonellae present in the enrichment broth can remove the need for enrichment (Waites, 1997) and decrease detection time by 2 days. Streaking out and recognition can be done on two selective media—for example, Brilliant Green/Phenol Red Agar (IDF, 1995b) and any other suitable solid selective medium (e.g., Hektoen Enteric; Xylose Lysine Desoxycholate or Bismuth Sulfite Agar; Ryser, 1998). The plates are then incubated at 35°C for 24–48 h. Biochemical or serological confirmation of isolated colonies can be done according to procedures described by Flowers et al. (1993) or IDF (1995b).

Alternatively, several rapid methods using fluorescent antibodies, immunodiffusion, enzyme immunoassays, DNA hybridization, hydrophobic grid membrane filtration, or impedance determination may be applied (Flowers et al., 1993). All positive findings must be confirmed by culturing. Commercial test kits available are described by Ryser (1998).

Standard methods for detecting *Salmonella* in cheese powder and dried milk are referred to in Table 14.6.

14.11.2.3 Campylobacter. *Campylobacter jejuni* has been recognized since 1909 as an important cause of abortion in cattle and sheep. Improved isolation strategies have also implicated this organism as a causative agent of human diarrhea. Altogether, 45 food-borne campylobacteriosis outbreaks (1308 cases) were reported in the United States between 1978 and 1986, over half of which involved ingestion of raw milk (Ryser, 1998). Similar reports linking raw or inadequately pasteurized milk to 13 outbreaks in Great Britain from 1978 to 1980 (Ryser, 1998) helped to further substantiate *C. jejuni* and also *C. coli* (Duim et al., 1999) as important milk-borne pathogens that have come to rival or even surpass *Salmonella* as an etiological agent of human gastroenteritis worldwide (Ryser, 1998).

Campylobacter is likely to be greatly outnumbered by the normal bacterial flora of milk. Consequently, selective enrichment under microaerobic conditions is a crucial initial step in procedures for recovering *Campylobacter* from raw milk (Ryser, 1998). The latter author refers to three methods that are currently recommended for detecting *Campylobacter* in raw milk, namely, Flowers et al. (1993), Hunt and Abeyta (1995), and Stern et al. (1992). These methods are complicated and require initial centrifugation of the raw milk sample and selective enrichment of the pellet at 42°C under microaerobic conditions (5% O_2, 10% CO_2, 85% N_2). Subsequent plating of the enrichment medium is then done on two selective media followed by similar incubation. Additional steps are required in the FDA procedure (Hunt and Abeyta, 1995).

Phenotypic methods (Duim et al., 1999) have the disadvantage that a large proportion of *C. jejuni* and *C. coli* strains are nontypable by such methods. The latter authors reported that amplified fragment length polymorphism (AFLP) fingerprinting resulted in satisfactory and highly discriminatory identification of the campylobacters. They concluded that AFLP is suitable for epidemiological studies on *Campylobacter*. Mouwen and Prieto (1999) also reported that Fourier-transformed infrared (FT-IR) spectroscopy was found to be a simple, rapid, and accurate procedure for identifying, typing, and grouping *Campylobacter* species and subspecies. A disadvantage of the latter method is the cost of the equipment and the fact that reproducibility of the method needs to be improved.

14.11.2.4 Listeria. *Listeria monocytogenes* has only recently emerged as a serious food-borne pathogen that can cause abortion in pregnant women and meningitis, encephalitis, and septicaemia in newborn infants and immunocompromised adults. The outcome of listeric infection can be particularly devastating, with a mortality rate of 20–30% (Ryser, 1998). Three major dairy-related outbreaks resulting in more than 100 deaths prompted the United States to institute a policy of "zero tolerance" for *L. monocytogenes* in all cooked and ready-to-eat foods including dairy products (Ryser, 1998). Recalls issued for contaminated dairy products, principally ice cream and cheese, in the United States in 1994 and 1995 suggests that *Listeria* contamination within dairy processing plants has not yet been fully controlled (Ryser, 1998).

Reference methods for determining *L. monocytogenes* in dairy products (IDF, 1995c; Hitchins, 1995) both commence with a selective liquid medium containing acriflavin, nalidixic acid, and cycloheximide as selective agents. The inoculated selective medium is incubated at 30°C for up to 48h. If the immunomagnetic separation (IMS) technique is used, the selective medium is replaced by half-Fraser broth and incubated for only 24h at 30°C. This saves 24h on the selective medium step. Oxford medium as main medium and PALCAM as additional medium is recommended in the above reference procedures as isolation media.

Species identification of presumptive *Listeria* isolates is based on morphological, physiological, and biochemical reactions that can take up to 7 days to complete (IDF, 1995c; Hitchins, 1995).

The time required for phenotypic confirmation can be shortened using commercially available test kits (e.g., API 20 S, API-ZYM, API Listeria, Micro ID; Ryser, 1998). Alternatively, several DNA hybridization (Accuprobe, Gene Trak) and ELISA assays such as VIDAS can be used to screen IMS concentrates or enrichment broths for *Listeria* spp. (Ryser, 1998; see also Table 14.6). Molecular methods studied by Harvey et al. (1999) included multilocus enzyme electrophoresis (MEE), pulsed field gel electrophoresis (PFGE), repetitive element sequence-based PCR (REP.PCR), and plasmid profiling. They found PFGE to be the most discriminatory typing method, followed by MEE and lastly REP.PCR. MEE and PFGE were largely in agreement and tended to group the same strains together. The sensitivity of MEE could be improved (Harvey et al., 1999) by increasing the number of enzymes examined. While plasmid profiling was not useful for typing *L. monocytogenes* from a wide range of sources, the presence and stability of plasmid DNA in certain strains could serve as a useful marker in ecological studies. REP.PCR proved to be a rapid and reliable method, but was less discriminatory than MEE or PFGE (Harvey et al., 1999).

14.11.2.5 Staphylococcus aureus. Staphylococcal food poisoning is a classic food-borne intoxication resulting from the ingestion of a preformed, heat-stable toxin (termed enterotoxin) and produced by the bacterium *Staphylococcus aureus*. Dairy products are well-known vehicles of staphylococcal poisoning, with cheese and raw milk linked to outbreaks before the turn of the century (Ryser, 1998). By the late 1930s, staphylococcal poisoning emerged as a major milk-borne illness accounting for 26%, 50%, and 30% of all milk-borne diseases reported in the United States during the 1940s, 1950s, and 1960s, respectively (Ryser, 1998). These cases of staphylococcal poisoning involved various dairy products including raw milk, pasteurized milk, cheese, ice cream, butter, and nonfat dry milk. Improvements in milk pasteurization and dairy sanitation standards in the United States, England, and most other industrialized countries have made dairy-related outbreaks rare (Ryser, 1998).

The significance of finding *S. aureus* in foods suspected of causing staphylococcal poisoning should be interpreted with caution. Although foods must contain at least 10^6 enterotoxigenic *S. aureus* cfu g^{-1} to induce illness, small numbers of *S. aureus* present in thermally processed foods may represent the survivors of very large populations. Consequently, actual or potential staphylococcal poisoning can only be verified by isolating enterotoxigenic staphylococci from the food or demonstrating the presence of enterotoxin in the food (Ryser, 1998).

The enumeration of coagulase-positive staphylococci in milk and milk-based products using the colony count technique is described in an IDF standard (IDF, 1997a). The principle of this technique is the inoculation of serial dilutions of product using a solidified culture medium, namely, Baird–Parker agar medium or rabbit plasma fibrinogen (RPF) agar medium. The use of the latter medium is recommended when the Baird–Parker medium is not selective enough. After incubation for 24–48 h, coagulase-positive staphylococci on RPF agar form gray or black colonies surrounded by an opaque or cloudy zone indicating coagulase activity. Confirmation of colonies is done by means of a coagulase test (IDF, 1997a). In products where staphylococci are expected to be stressed and in low numbers—as, for example, in dried milk products—the IDF recommends a MPN technique (IDF, 1997b). A 9-tube battery of Giolitti and Cantoni broth is inoculated with three consecutive dilutions of the dried milk sample. After incubation at 37°C, tubes are subcultured to Baird–Parker agar. Presumptively positive colonies are subjected to the coagulase test.

An international standard (IDF, 1998a) specifies a method for the detection of heat-stable DNase (thermonuclease) produced by coagulase-positive staphylocci in samples of milk or milk-based products. The enzyme may be used as an indicator of staphylococcal growth to hazardous levels and the potential presence of staphylococcal enterotoxins.

The enzyme is extracted by means of a procedure involving acidification, centrifugation, treatment of the supernatant with trichloroacetic acid, redissolving the resultant precipitate, heating the solution, and testing for thermonuclease activity in toluidine blue O-DNA agar. A positive thermonuclease test indicates that coagulase-positive staphylococci have grown to levels of 10^6 or more cfu g^{-1}. This may imply toxic levels of enterotoxin, and testing for the presence of enterotoxin should be conducted.

Staphylococcal enterotoxin detection techniques include a microslide gel double diffusion immunoassay and a radioimmunoassay technique described by Flowers et al. (1993). The latter technique is rapid and highly sensitive, whereas the former involves a long and complicated extraction and concentration procedure.

An ELISA method, namely the TECRA staphylococcal enterotoxin visual immunoassay (SETVIA), has been designed for rapid detection of staphylococcal enterotoxins in foods. The SETVIA was awarded AOAC first action approval in 1993. More recently the food extraction protocols have been simplified and sample preparation protocols have been improved. Benson et al. (1999) undertook a study to validate these changes and found the SETVIA to be a highly sensitive and specific test for staphylococcal enterotoxins. It gave results within approximately 4h including 30min for the extraction. The microtiter plate assay may be read visually with the aid of the color comparison card provided. DNA hybridization assays and latex agglutination tests are also available for identifying enterotoxins in culture fluids (Ryser, 1998).

14.11.2.6 Bacillus cereus. *Bacillus cereus* is an aerobic spore-forming bacterium that is involved in food intoxification as a result of the production of two enterotoxins. The first is a diarrheal enterotoxin, and the second is an emetic enterotoxin (Ryser, 1998). *Bacillus cereus* is a common contaminant in the dairy environment and the raw milk supply and has also been involved in spoilage phenomena such as "sweet curdling" and "bitty cream" in milk. The ability of *B. cereus* to persist in powdered milk and to grow in the reconstituted product, as evidenced by a large outbreak in Chile involving newborne infants (Ryser, 1998), has led to the establishment of rigid international standards for *B. cereus* in infant formulae (Becker et al., 1994). Most outbreaks of *B. cereus* poisoning have been traced to foods containing at least 10^6 cfu g^{-1}. Small numbers of surviving spores can, however, germinate and grow when milk or baby formulae are reconstituted and attain dangerous levels during storage (Ryser, 1998).

A standard MPN technique is described for the enumeration of *B. cereus* in dried-milk-based infant food and dried milk products intended for use as ingredients of milk-based baby or infant foods and dietary foods (IDF, 1998b). The three- by three-tube MPN method employs tryptone soya polymyxin broth. After inoculation and incubation the tubes are subcultured onto polymyxin pyruvate egg yolk mannitol bromothymol blue agar (PEMBA) or mannitol egg yolk polymyxin agar (MEPA). Identification of colonies is confirmed by biochemical and morphological tests, and a MPN count g^{-1} is read off from the statistical tables.

To confirm that the suspect isolate is toxigenic, the strain should be tested for the diarrheal and emetic enterotoxin. A serologically based microslide gel double diffusion assay has been developed for the diarrheal enterotoxin, with several fluorescence-based immunoblot and reverse passive latex agglutination assays also available commercially

(Ryser, 1998). Production of the emetic (mitochondrio) toxin can be detected by means of a recently developed sensitive bioassay known as the boar spermatozoa toxicity test (Andersson et al., 1998).

14.11.2.7 Aflatoxin. Although mycotoxin production is not limited to aflatoxigenic molds (Ryser, 1998), only aflatoxin will be dealt with in this section. The aflatoxins are highly potent carcinogens produced by certain strains of *Aspergillus flavus*, *A parasiticus*, and *A. nemius* and is the primary mycotoxin of public health concern (Ryser, 1998). Four major forms of aflatoxin are currently recognized, namely, AFB_1, AFB_2, AFG_1, and AFG_2. AFB_1 is the most potent and is most often found in moldy peanuts and animal feeds containing corn (maize) and other grains. When cattle ingest contaminated feed, AFB_1 is metabolized to AFM_1 (M_1), some of which is shed in the milk (Ryser, 1998). Most countries today have legislation in place specifying the maximum levels of AFM_1 that are acceptable in fluid milk and dairy products.

A provisional standard drafted by the IDF (IDF, 1995d) specifies a method for the determination of at least $0.008\,\mu g\ liter^{-1}$ in milk and $0.08\,\mu g\,kg^{-1}$ in whole milk powder. The method is also applicable to fluid low-fat and skim milk and low-fat and skim milk powder.

The principle of the method involves extraction of AFM_1 by passing the sample through an immunoaffinity column. The column contains specific antibodies immobilized on a solid support material. These antibodies bind any AFM_1 passing through the column. After washing the column to remove other sample components, AFM_1 is eluted from the column and the eluate collected. The amount of AFM_1 present is determined by high-performance liquid chromatography.

14.12 MICROBIOLOGICAL STANDARDS FOR DIFFERENT DAIRY PRODUCTS

Limitations of poor-quality products and health protection form the basis for food standards. Legal and voluntary bacteriological standards vary widely from country to country. Lück and Gavron (1990) proposed certain limits that may be a useful tool to improve the quality of dairy products (Table 14.9). A product complies with the bacteriological specifications when at least four out of five portions or samples contain less than the maximum bacterial count specified. Legal action should only be taken when more than one out of the five samples exceeds the upper limit.

TABLE 14.9. Suggested Microbiological Standards for Different Dairy Products

Product	Test	Count or Results[a]
Raw milk intended for further processing	Total bacterial count	<100,000 (30,000) cfu ml^{-1}
	Total coliform count	<500 (100) cfu ml^{-1}
	Escherichia coli (fecal type)	<1 cfu ml^{-1}
	Thermoduric count	<1000 cfu ml^{-1}
	Spore-formers	<10 cfu ml^{-1}
	Bacillus cereus (spores)	<1 cfu ml^{-1}
	Staphylococcus aureus (coagulase-positive)	100 (10) cfu ml^{-1}
	Methylene blue reduction time (at 36°C)	Not less than 6h
	3-h Resazurin test (at 36°C Lovibond disc No. 4/9)	Not less than 4h
	Somatic cell count	<750,000 (500,000) ml^{-1}
Raw milk or raw cream for direct consumption	Total bacterial count	<30,000 (10,000) cfu ml^{-1}
	Total coliform count	<30 (10) cfu ml^{-1}
	E. coli (fecal type)	1 in 10 cfu ml^{-1}
	Methylene blue reduction time (at 36°C)	Not less than 7h
	3-h Resazurin test (at 36°C Lovibond disc No. 4/9)	Not less than 4h
	S. aureus (coagulase-positive)	<10 (1) cfu ml^{-1}
	Somatic cell count	<500,000 (250,000) ml^{-1}
Pasteurized market milk or pasteurized cream (after processing)	Total bacterial count	<30,000 (5000) cfu ml^{-1}
	Total coliforms	<1 (0.1) cfu ml^{-1}
	E. coli (fecal type)	Absent in 1 ml
Dried milk products (including casein and whey powder, etc.)	Total bacterial count	<10,000 (1000) cfu g^{-1}
	Total coliforms	<10 (1) cfu g^{-1}
	E. coli (fecal type)	Absent in 25 g
	S. aureus (coagulase-positive)	Absent in 25 g
	Yeasts and molds	<10 cfu g^{-1}
	Salmonellae (and other pathogens if required)	Absent in 200 (500) cfu g^{-1}
Ice cream	Total bacterial count	<50,000 (5000) cfu g^{-1}
	Total coliforms	<10 cfu g^{-1}
	E. coli (fecal type)	Absent in 1 g
	S. aureus (coagulase-positive)	Absent in 1 g
	Yeasts and molds	<10 cfu g^{-1}
	Salmonellae (and other pathogens if required)	Absent in 25 (100) g
Cultured milk products (including yogurt, cottage cheese, cultured milk, sour cream, etc.)	Contaminating organisms (non-lactic acid bacteria)	<1000 cfu g^{-1}
	Total coliforms	<10 (1) cfu g^{-1}
	E. coli (fecal type)	Absent in 1 g
	S. aureus (coagulase-positive)	Absent in 1 g
	Yeasts and molds	<10 (1) cfu g^{-1}
	Salmonellae (and other pathogens if required)	Absent in 15 (100) g

TABLE 14.9. *Continued*

Product	Test	Count or Results[a]
Sweetened condensed milk	Total bacterial count	<100 cfu g^{-1}
	Spore-formers	Absent in 1 g
	Total coliforms	<1 cfu g^{-1}
	Yeasts and molds	<1 cfu g^{-1}
Butter	Contaminating organisms (non-lactic acid bacteria)	<10,000 (5000) cfu g^{-1}
	Total bacterial count (noncultured butter only)	<50,000 cfu g^{-1}
	Total coliforms	<10 (1) cfu g^{-1}
	E. coli (fecal type)	Absent in 1 g
	S. aureus (coagulase-positive)	Absent in 1 g
	Yeasts and molds	<10 cfu g^{-1}
	Proteolytic organisms	<100 cfu g^{-1}
	Lipolytic organisms	<50 cfu g^{-1}
Cheese (ripened—standards at 1 month)	Total coliforms	<10 cfu g^{-1}
	E. coli (fecal type)	Absent in 1 g
	Spore-formers	<10 (1) cfu g^{-1}
	Fecal streptococci	<100 cfu g^{-1}
	S. aureus	Absent in 1 g
	Yeasts and molds	<10 (1) cfu g^{-1}
	Other pathogens (if required)	Absent in 25 (100) g

[a]Figures in parentheses mean standards that should be strived for.

Source: Adapted from Lück and Gavron (1990).

During recent years, the variation of counts in different samples has been taken into consideration, and the hygiene requirements are, for instance, often expressed as follows: Examine five (n) samples of dried milk, and allow two (c) samples to exceed 50,000 bacteria g^{-1} (m), but none to exceed 200,000 g^{-1} (M) or abbreviated: $n = 5$, $c = 2$, $m = 50{,}000$, $M = 200{,}000$.

14.13 RELEVANCE OF TECHNIQUES AND INTERPRETATION OF RESULTS

No single laboratory test performed on a food product can produce the full information that is required. No bacteriological test yet evolved is above criticism. There is no "best" test, and yet often much money is wasted on elaborate tests for which fictitious accuracy are claimed. It is now recognized that regular testing is of far greater importance and that any test that is better than mere haphazard classification will contribute to an improvement in quality. The real value of a test is whether it can detect products of unsatisfactory quality, and in this way make a contribution to improving production hygiene. It is, therefore, not necessary that a test should give an absolute measure of the quality, nor need it be in complete agreement with the results of other tests (Lück and Gavron, 1990).

Because none of the different methods employed to determine a specified bacterial content yield exactly the same result, one test cannot be automatically substituted for another. Two different tests can be compared statistically in order to arrive at comparable standards, but not necessarily to determine their value. The only conditions of a suitable test are as follows:

1. There must be a significant correlation between the results of the test and the quality required.
2. The operator must be fully informed on how to obtain this quality.

Due to the error inherent in a single bacterial count—for example, limitations of the particular bacterial count/test, seasonal and local variation of the microflora, day-to-day variation of the bacterial count, and counting only colony-forming units or clumps instead of individual bacteria—only approximately fivefold differences in plate counts can be regarded as significant when grading dairy products. Hence the quality categories have to be established in such a way that the differences between bacterial counts are large enough to be significant. The realization of this fact led to the development of rapid screening procedures to meet the requirements of the dairy industry.

Quality control is planned and introduced by management, but diagrams indicating real or anticipated quality levels should, however, concern not only the management but also the operators. One operator should be responsible for a specific machine or a specific process, and the relevant diagrams can take the form of control charts showing, on the horizontal axis, the sequence of sample or the date of manufacture and, on the vertical axis, the quality characteristics (shelf life, log bacterial count, etc.). A line can represent, for instance, the expected average log bacterial count of a product; above this line the upper limit line is drawn, such that only 1 in 20 of the plotted points should lie outside this line (Lück and Gavron, 1990).

For shelf-life tests, a lower limit line is drawn. The number of points outside the limit lines indicates whether the process has been altered in some way; and they are, therefore, essentially indicators of a need for corrective action.

It is often necessary to summarize results or to calculate mean counts. When there is a small variation in the bacterial counts of different samples, or when the counts are low, the arithmetic mean of the bacterial counts may be calculated. When there are, however, large variations between counts, and the counts are high, the geometric mean (logarithmic average) should be used (geometric mean = arithmetic mean of the log count, which is subsequently transformed into a bacterial count again). When the plate count method supplies positive results, more emphasis should be attached to this method than to the most probable number test, because the repeatability of the plate count method is better than that of the MPN values (Lück and Gavron, 1990).

REFERENCES

Aarts, H. J. M. (1999) In *Proceedings of the 17th International Conference of the International Committee on Food Microbiology and Hygiene (ICMFH)*, A. C. J. Tuijtelaars, R. A. Samson, F. M. Rombouts, and S. Notermans, eds., Veldhoven, The Netherlands, pp. 457–461.

Al-Dagal, M., and Fung, D. Y. C. (1990) *Crit. Rev. Food Sci. Nutr.*, **29**, 333–340.

Andersson, M. A., Mikkola, R., Helin, J., Andersson, M. C., and Salkinoja-Salonen, M. (1998) *Appl. Environ. Microbiol.*, **64**, 1338.

Andrew, M. H. E., and Russell, A. D. (1984) *The Revival of Injured Microbes*, Academic Press, London.

Anonymous (1988) *Dairy Food Sanit.*, **8**, 52–56.

Anonymous (1995) In *HACCP and the Lumac Solution*, Lumac B. V., Landgraaf, Netherlands.

Anonymous (1997) *Dtsch. Milchwirtsch.*, **48**, 601–603.

Anonymous (1998) *Food Technol. Int.*, pp. 64–65.

AOAC (1990) *Association of Official Analytical Chemists Methods of Analysis*, 15th ed., Washington, DC, AOAC.

Arnould, P., and Guichard, L. (1999) *Latte*, **24**(3), 44–46.

Assink, J. W., and van Deventer, H. C. (1995) *Eur. Water Pollut. Control*, **5**, 39–45.

Austin, J. W., and Bergeron, G. (1995) *J. Dairy Res.*, **62**, 509–519.

Becker, H., Schaller, G., von Wiese, W., and Terplan, G. (1994) *Int. J. Food Microbiol.*, **23**, 1.

Bell, C., Bowles, C. D., Toszeghy, M. J. K., and Neaves, P. (1996) *Int. Dairy Fed.*, **6**(7), 709.

Bennett, R. W., and McClure, F. (1976) *J. Assoc. Off. Anal. Chem.*, **59**, 594–600.

Benson, J., Walker, L., Dailianis, A., Hughes, D., and Melhuish, R. (1999) In *Proceedings of the 17th International Conference of the International Committee on Food Microbiology and Hygiene (ICMFH)*, A. C. J. Tuijtelaars, R. A. Samson, F. M. Rombouts, and S. Notermans, eds., Veldhoven, The Netherlands, p. 480.

Bergdoll, M. S., and Reiser, R. (1980) *J. Food Prot.*, **43**, 68–72.

Bishop, J. R., White, C. H., and Firstenberg-Eden, R. A. (1984) *J. Food Prot.*, **47**, 471–475.

Blackman, I. C., and Frank, J. F. (1996) *J. Food Prot.*, **59**, 827–831.

Botta, G. (1998) *Latte*, **23**(10), 138–140.

Brazis, A. R. (1991) In *Methods for Assessing the Bacteriological Quality of Raw Milk from the Farm*, W. Heeschen, ed., Bulletin 256, International Dairy Federation, Brussels, Belgium, pp. 4–8.

Brovko, L. Y., Froundjian, V. G., Babunova, V. S., and Ugarova, N. N. (1999) *J. Dairy Res.*, **66**, 627–631.

BSI (1991) In *Methods of Microbiological Examination for Dairy Purpose*, BS 4285: Part 4, British Standards Institution, London.

Bylund, G. (1995) In *Tetra Pak Dairy Processing Handbook*, Tetra Pak Processing Systyms AB, Lund, Sweden, pp. 175–177.

Cady, P., Hardy, D., Martins, S., Dufour, S., and Kraeger, S. J. (1978) *J. Food Prot.*, **41**, 277–283.

Carpentier, B., and Cerf, O. (1993) *J. Appl. Bacteriol.*, **75**, 499–511.

Carpentier, B., Wong, A. C. L., and Cerf, O. (1998) In *Biofilms on Dairy Plant Surfaces: What's New?*, Bull. 329, International Dairy Federation, Brussels, Belgium, pp. 32–35.

Casman, E. P., and Bennett, R. W. (1965) *Appl. Microbiol*, **13**, 181–189.

Cerny, G. (1976) *Verpackungs-Rundschau*, **27**, Technisch Wissenschaftliche Beilage, pp. 27–32.

Christen, G. L., Davidson, P. M., McAllister, J. S., and Roth, L. A. (1993) In *Standard Methods for the Examination of Dairy Products*, R. J. Marshall, ed., American Public Health Association, Washington, DC, pp. 247–269.

Clesceri, L. S., Greenberg, A. E., and Trussell, R. R. (eds.) (1989) *Standard Methods for the Examination of Water and Waste Water*, 17th ed., APHA, Washington, DC, pp 9.1–9.227.

Costerton, J. W., Lewandowski, K., Caldwell, D. E., Korber, D. R., and Lappin-Scott, H. M. (1995) *Ann. Rev. Microbiol.*, **49**, 711–745.

Davis, M. J., Coote, P. J., and O'Byrne, C. P. (1996) *Microbiology*, **142**, 2975–2982.

De Wet, H. (1998) In *Payment Systems for Ex-Farm Milk—Results of IDF Questionnaire 2296/A*, Bulletin 331, International Dairy Federation, Brussels, Belgium, pp. 6–25.

Dickerson, R. W. (1987) *J. Food Prot.*, **50**, 964–967.

Duim, B., Bolder, N., Rigter, A., Jacobs-Reitsma, W., van Leeuwen, N., and Wagenaar, J. (1999) In *Proceedings of the 17th International Conference of the International Committee on Food Microbiology and Hygiene (ICMFH)*, A. C. J. Tuijtelaars, R. A. Samson, F. M. Rombouts, and S. Notermans, eds., Veldhoven, The Netherlands, pp. 499–502.

Duncan, C. L., and Somers, E. B. (1972) *Appl. Microbiol.*, **24**, 801–804.

Dunsmore, D. G., Twomey, A., Whittlestone, W. G., and Morgan, H. W. (1981) *J. Food Prot.*, **44**, 220–240.

Easter, M. C., and Prentice, G. A. (1989) In *IDF Special Issue No. 8901*, International Dairy Federation, Brussels, Belgium, pp. 202–218.

El-Shenawy, M. A. (1998) *Int. J. Environ. Health Res.*, **8**, 241–251.

Everson, T. C. (1991) In *Practical Phage Control*, Bulletin 263, International Dairy Federation, Brussels, Belgium, pp. 24–28.

Fenlon, D. R., Wilson, J., and Donachie, W. (1996) *J. Appl. Bacteriol.*, **81**, 641–650.

Firstenberg-Eden, R., and Tricarico, M. K. (1983) *J. Food Sci.*, **48**, 1750–1754.

Fischer, P. L., Jooste, P. J., and Novello, J. C. (1986) *S. Afr. J. Dairy Sci.*, 18, 137.

Floros, J. D., Nielsen, P. V., and Farkas, J. K. (2000) In *Packaging of Milk Products*, Bulletin 346, International Dairy Federation, Brussels, Belgium, pp. 22–28.

Flowers, R. S., Andrews, W., Donnelly, C. W., and Koenig, E. (1993) In *Standard Methods for the Examinaiton of Dairy Products*, 16th ed., R. T. Marshall, ed., American Public Health Association, Washington, DC, pp. 103–212.

Flückiger, E. (1995) In *Technical Guide for the Packaging of Milk and Milk Products*, 3rd ed., Bulletin 300, International Dairy Federation, Brussels, Belgium, pp. 52–56.

Frank, J. F., Christen, G. L., and Bullerman, L. B. (1993) In *Standard Methods for the Examination of Dairy Products*, 16th ed., R. J. Marshall, ed., American Public Health Association, Washington, DC, pp. 271–286.

Frank, J. F., and Koffi, R. A. (1990) *J. Food Prot.*, **53**, 550–554.

Fricker, C. (1993) In *Encyclopaedia of Food Science, Food Technology and Nutrition*, Vol. 7, R. Macrae, R. K. Robinson, and M. J. Sadler, Academic Press, London, pp. 4859–4861.

Frontini, S. (2000) *Latte*, **25**(1), 28–34.

Ginn, R. E., Packard, V. S., and Fox, T. L. (1984) *J. Food Prot.*, **47**, 753–755.

Gnan, S., and Luedecke, L. O. (1982) *J. Food Prot.*, **45**, 4–7.

Grace, V., Houghtby, G. A., Rudnick, H., Whaley, K., and Lindamood, J. (1993) In *Standard Methods for the Examination of Dairy Products*, 16th ed., R. T. Marshall, ed., American Public Health Association, Washington, DC, pp. 59–83.

Griffiths, M. W. (1996) *Food Technol.*, **50**(6), 62–72.

Griffiths, M. W., Phillips, J. D., and Muir, D. D. (1980) *J. Soc. Dairy Technol.*, **33**, 8–10.

Guyader, P. (1995) *Dairy Sci. Abstr.*, **58**, 208.

Han, J. H. (2000) *Food Technol.*, **54**(3), 56–65.

Harding, F. (1995) In *Milk Quality*, F. Harding, ed., Blackie Academic and Professional, London, pp. 29, 40–59.

Harrigan, W. F., and McCance, M. E. (1976) In *Laboratory Methods in Food and Dairy Microbiology*, 2nd ed., Academic Press, London.

Harrigan, W. F., and Park, R. W. A. (1991) *Marking Safe Food: A Management Guide for Microbiological Quality*, Academic Press, Harcourt Brace Jovanovich, London, p. 178.

Harrysson, G. (1998) In *Proceedings of the 25th International Dairy Congress*, Danish National Committee of the IDF, Aarhus, Denmark, pp. 249–264.

Harvey, J., Curron, S., and Gilmour, A. (1999) In *Proceedings of the 17th International Conference of the International Committee on Food Microbiology and Hygiene (ICMFH)*, A. C. J. Tuijtelaars, R. A. Samson, F. M. Rombouts, and S. Notermans, eds., Veldhoven, The Netherlands, pp. 521–525.

Hedrick, T. I. (1975) *Chem. Ind.*, **20**, 868–872.

Hedrick, T. I., and Heldman, D. R. (1969) *J. Milk Food Technol.*, **32**, 265–269.

Heeschen, W., Reichmut, J., Tolle, A., and Zeidler, H. (1969) *Milchwissenschaft*, **24**, 729–734.

Heldman, D. R., and Hedrick, T. I. (1971) *Research Bulletin*, **33**, Michigan State University Agricultural Experimental Station, East Lansing, p. 76.

Heldman, D. R., Hedrick, T. I., and Hall, C. W. (1965) *J. Milk Food Technol.*, **28**, 41–45.

Hickey, P. J., Beckelheimer, C. E., and Parrow, T. (1993) In *Standard Methods for the Examination of Dairy Products*, 16th ed., R. T. Marshall, ed., American Public Health Association, Washington, DC, pp. 397–412.

Hiddink, J. (1995) In *Water Supply, Sources, Quality and Water Treatment in the Dairy Industry*, Bulletin 308, International Dairy Federation, Brussels, Belgium, pp. 16–32.

Hill, B. M. (1991a) In *Methods for Assessing the Bacteriological Quality of Raw Milk from the Farm*, W. Heeschen, ed., Bulletin 256, International Dairy Federation, Brussels, Belgium, pp. 17–20.

Hill, B. M. (1991b) In *Methods for Assessing the Bacteriological Quality of Raw Milk from the Farm*, W. Heeschen, ed., Bulletin 256, International Dairy Federation, Brussels, Belgium, pp. 20–24.

Hill, B. M. (1991c) In *Methods for Assessing the Bacteriological Quality of Raw Milk from the Farm*, W. Heeschen, ed., Bulletin 256, International Dairy Federation, Brussels, Belgium, pp. 9–12.

Hitchins, A. D. (1995) In *FDA Bacteriological Analytical Manual*, 8th ed., G. J. Jackson, ed., AOAC International, Gaithersburg, MD, p. 10.01.

Holah, J. T., Betts, R. P., and Thorpe, R. H. (1988) *J. Appl. Bacteriol.*, **65**, 215–221.

Homleid, J. P. (1997) *Meieriposten*, **86**, 111–112.

Houghtby, G. A., Maturin, L. J., and Koenig, E. K. (1993) In *Standard Methods for the Examination of Dairy Products*, 16th ed., R. T. Marshall, ed., American Public Health Association, Washington, DC, pp. 213–246.

Hunt, J., and Abeyta, C. (1995) *Campylobacter* In *FDA Bacteriological Analytical Manual*, 8th ed., M. D. Gaithersburg, ed., AOAC International, p. 7.01.

IDF (1969) *Control Methods for Sterilized Milk*, International IDF Standard 48, International Dairy Federation, Brussels, Belgium.

IDF (1981) *New Monograph on UHT Milk*, Document no. 133, International Dairy Federation, Brussels, Belgium.

IDF (1985) *Milk Estimation of Psychrotrophic Micro-organisms—Rapid Colony Count Technique 25 Hours at 21°C*, International IDF Standard 132. International Dairy Federation, Brussels, Belgium.

IDF (1987) *Hygienic Conditions—General Guide On Sampling and Inspection Procedures*, International IDF Standard 121A: 1987. International Dairy Federation, Brussels, Belgium.

IDF (1990a) *Guidelines for Sampling Equipment and Data Collection on Milk Collection Tankers*, Bulletin 252, International Dairy Federation, Brussels, Belgium, pp. 35–48.

IDF (1990b) *Milk and Milk Products—Sampling Inspection by Attributes*, International IDF Standard 113A: 1990. International Dairy Federation, Brussels, Belgium.

IDF (1991a) *Milk and Milk Products—Enumeration of Micro-organisms, Colony Count at 30°C*, International IDF Standard 100B: 1991. International Dairy Federation, Brussels, Belgium.

IDF (1991b) *Butter, Fermented Milks and Fresh Cheese—Enumeration of Contaminating Micro-organisms, Colony Count Technique at 30°C*, International IDF Standard 153: 1991. International Dairy Federation, Brussels, Belgium.

IDF (1991c) *Liquid Milk—Enumeration of Psychrotrophic Micro-organisms, Colony Count at 6.5°C*, International IDF Standard 101A: 1991. International Dairy Federation, Brussels, Belgium.

IDF (1992) *Milk and Milk Products—Sampling Inspection by Variables*, International IDF Standard 136A: 1992. International Dairy Federation, Brussels, Belgium.

IDF (1995a) *Milk and Milk Products—Guidance on Sampling*, International IDF Standard 50C: 1995. International Dairy Federation, Brussels, Belgium.

IDF (1995b) *Milk and Milk Products—Detection of Salmonella*, International IDF Standard 93B: 1995. International Dairy Federation, Brussels, Belgium.

IDF (1995c) *Milk and Milk Products—Detection of Listeria monocytogens*, International IDF Standard 143A: 1995. International Dairy Federation, Brussels, Belgium.

IDF (1995d) *Milk and Milk Powder—Determination of Aflatoxin M_1 Content—Clean-up by Immunoafinity Chromatography and Determination by HPLC*, International IDF Standard 171: 1995. International Dairy Federation, Brussels, Belgium.

IDF (1996a) *Milk and Milk Products—Preparations of Samples and Dilutions for Microbiological Examination*, International IDF Standard 122C: 1996. International Dairy Federation, Brussels, Belgium.

IDF (1996b) *Inventory of IDF/ISO/AOAC International Adapted Methods of Analysis and Sampling for Milk and Milk Products*, 5th ed. Bulletin 312, International Dairy Federation, Brussels, Belgium.

IDF (1997a) *Milk and Milk-Based Products—Enumeration of Coagulase—Positive Staphylococci—Colony Count Technique*, International Dairy Federation Standard 145A: 1997. International Dairy Federation, Brussels, Belgium.

IDF (1997b) *Milk and Milk-Based Products—Enumeration of Coagulase—Positive Products Most Probable Number Technique*, International Dairy Federation Standard 60C: 1997. International Dairy Federation, Brussels, Belgium.

IDF (1998a) *Milk and Milk-Based Products—Detection of Thermonuclease Produced by Coagulase—Positive Staphylococci in Milk and Milk-Based Products*, International Dairy Federation Standard 83A: 1998. International Dairy Federation, Brussels, Belgium.

IDF (1998b) *Dried Milk Products—Enumeration of Bacillus cereus—Most Probable Number*, International Dairy Federation Standard 181: 1998. International Dairy Federation, Brussels, Belgium.

IDF (2000) *Payment Systems for Ex-Farm Milk—Results of IDF Questionnaire*, Bulletin 348, International Dairy Federation, Brussels, Belgium, pp. 15–42.

Jaksch, V. P., Zaadhof, K.-J., and Terplan, G. (1982) *Molk-Ztg Welt Milch*, **36**, 5–8.

Jay, J. M. (1992) *Modern Food Microbiology*, 4th ed., Van Nostrand Reinhold, New York.

Jervis, D. I. (1992) In *The Technology of Dairy Products*, R. Early, ed., Blackie, London, pp. 272–299.

Jooste, P. J., and Groeneveld, H. T. (1971) *S. Afr. J. Dairy Technol.*, **3**(4) 193.

Kacliková, E., Kolenčiková, B., and Kuchta, T. (1999) In *Proceedings of the 17th International Conference of the International Committee on Food Microbiology and Hygiene (ICMFH)*, A. C. J. Tuijtelaars, R. A. Samson, F. M. Rombouts, and S. Notermans, eds., Veldhoven, The Netherlands, pp. 541–542.

Kang, Y-J., and Frank, J. F. (1989a) *J. Food Prot.*, **52**, 655–659.

Kang, Y-J., and Frank, J. F. (1989b) *J. Food Prot.*, **52**, 512–524.

Kang, Y-J., and Frank, J. F. (1989c) *J. Food Prot.*, **52**, 877–880.

Kang, Y-J., and Frank, J. F. (1990) *J. Dairy Sci.*, **73**, 621–626.

Kelsey, R. J. (1974) *Mod. Packaging*, **47**, 37–40.

Kosikowski, F. V., and Mistry, V. V. (1997a) *Cheese and Fermented Milk Foods*, 3rd ed., Vol. 2, F. V. Kosikowski, L. L. C., Westport, CT, p. 308.

Kosikowski, F. V., and Mistry, V. V. (1997b) *Cheese and Fermented Milk Foods*, 3rd ed., Vol. 2, F. V. Kosikowski, L. L. C., Westport, CT, p. 46.

Koutzayiotis, C. (1992) *S Afr. J. Dairy Sci.*, **24**, 19–22.

Koutzayiotis, C., Mostert, J. F., Jooste, P. J., and McDonald, J. J. (1992) In *Advances in the Taxonomy and Significance of* Flavobacterium, Cytophaga *and Related Bacteria*, P. J. Jooste, ed., University Press, UOFS, Bloemfontein, South Africa, pp. 103–109.

Kroll, R. G. (1989) In *Modern Microbiological Methods for Dairy Products*, International Dairy Federation Special Issue 8901, pp. 173–182.

Leriche, V., and Carpentier, B. (1995) *J. Food Prot.*, **58**, 1186–1191.

Lighthart, B., and Frisch, A. S. (1976) *Appl. Environ. Microbiol.*, **31**, 700–704.

Lindsay, D., Brözel, V. S., Mostert, J. F., and von Holy, A. (2000) *Int. J. Food Microbiol.*, **54**, 49–62.

Lück, H. (1972) *Dairy Sci. Abstr.*, **34**, 101.

Lück, H., and Gavron, H. (1990) In *Dairy Microbiology—The Microbiology of Milk Products*, Vol. 2, 2nd ed., R. K. Robinson, ed., Elsevier Applied Science Publishers, London, pp. 345–392.

Lück, H. (1991a) In *Methods for Assessing the Bacteriological Quality of Raw Milk from the Farm*, W. Heeschen, ed., Bulletin 256, International Dairy Federation, Brussels, Belgium, pp. 31–34.

Lück, H. (1991b) In *Methods for Assessing the Bacteriological Quality of Raw Milk from the Farm*, W. Heeschen, ed., Bulletin 256, International Dairy Federation, Brussels, Belgium, pp. 35–37.

Lück, H. (1991c) In *Methods for Assessing the Bacteriological Quality of Raw Milk from the Farm*, W. Heeschen, ed., Bulletin 256, International Dairy Federation, Brussels, Belgium, pp. 45–47.

Lück, H. (1991d) In *Methods for Assessing the Bacteriological Quality of Raw Milk from the Farm*, W. Heeschen, ed., Bulletin 256, International Dairy Federation, Brussels, Belgium, pp. 47–49.

Lück, H., and Andrew, M. J. A. (1975) *S. Afr. J. Dairy Technol.*, **7**, 39–42.

Lück, H., Clark, P. C., and Groeneveld, H. T. (1970a) *Agro Anim.*, **2**, 69.

Lück, H., Clarke, P. C., and van Tonder, J. L. (1970b) *Milchwissenschaft*, **25**, 155–160.

Lück, H., Holzapfel, W. H., and Becker, P. J. (1971) *Milchwissenschaft*, **26**, 421.

Lück, H., and Dunkeld, M. (1972) *S. Afr. J. Dairy Technol.*, **4**, 93–99.

Lück, H., Dunkeld, M., and Becker, P. J. (1972) *S. Afr. J. Dairy Technol.*, **4**, 179–184.

Lück, H., Walthew, J., and Joubert, B. (1978) *S. Afr. J. Dairy Technol.*, **10**, 3.

Lück, H., Gavron, H., and Lategan, B. (1982) *S. Afr. J. Dairy Technol.*, **14**, 63–66.

Lück, H., McDonald, J. J., and Lategan, B. (1984) *S. Afr. J. Dairy Technol.*, **16**, 125–127.

Lundén, J. M., Miettinen, M. K., Autio, T. J., and Korkeala, H. J. (2000) In *Proceedings of the 17th International Conference of the International Committee on Food Microbiology and Hygiene (ICMFH)*, A. C. J. Tuijtelaars, R. A. Samson, F. M. Rombouts, and S. Notermans, eds., Veldhoven, The Netherlands, pp. 803–804.

Manners, J. G. (1993) In *Encyclopaedia of Food Science, Food Technology and Nutrition*, Vol. 5, R. Macrae, R. K. Robinson, and M. J. Sadler, Academic Press, London, p. 3074.

Marshall, R. T. (ed.) (1993) *Standard Methods for the Examination of Dairy Products*, 16th ed., Washington, DC, American Public Health Association.

Mattila, T., Ahvenainen, R., Wirtanen, G., and Manninen, M. (1989) In *IDF Special Issue No. 8901*, International Dairy Federation, Brussels, pp. 113–115.

Mattila-Sandholm, T., and Wirtanen, G. (1992) *Food Rev. Int.*, **8**, 573–603.

McCrady, M. H. (1918) *Public J.* (Toronto), **9**, 201.

Mettler, E., and Carpentier, B. (1997) *Lait*, **77**, 484–503.

Mettler, E., and Carpentier, B. (1998) *J. Food Prot.*, **61**, 57–65.

Miettinen, M. K., Björkroth, K. J., and Korkeala, H. J. (1999) *Int. J. Food Microbiol.*, **46**, 187–192.

Miller, B. A., Reiser, R. F., and Bergdoll, M. S. (1978) *Appl. Environ. Microbiol.*, **36**, 421–426.

Moats, W. A., Dabbah, R., and Edwards, V. M. (1971) *Appl. Microbiol.*, **21**, 476.

Molnar, P. J. (1993) *Encyclopaedia of Food Science, Food Technology and Nutrition*, Vol. 6, R. Macrae, R. K. Robinson, and M. J. Sadler, eds., Academic Press, Harcourt Brace Jovanovich, London, pp. 3846–3850.

Mosely, W. K. (1958) *Proceedings, 51st Annual Convention Milk Industry Foundation* (Laboratory Section), p. 27.

Mossel, D. A. A., Corry, J. E., Burt, S. A., and van Netten, P. (1989) In *Modern Microbiological Methods for Dairy Products*, International Dairy Federation Special Issue 8901, pp. 46–80.

Mouwen, D. J. M., and Prieto, M. (1999) In *Proceedings of the 17th International Conference of the International Committee on Food Microbiology and Hygiene (ICMFH)*, A. C. J. Tuijtelaars, R. A. Samson, F. M. Rombouts, and S. Notermans, eds., Veldhoven, The Netherlands, pp. 928–930.

Neve, H., Berger, A., and Heller, K. J. (1995) *Kieler Milchwirtsch. Forschungsber.*, **47**(3), 193–207.

Notermans, S. (1989) In *IDF Special Issue no. 8901*, International Dairy Federation, Brussels, Belgium, pp. 323–336.

Oehlrich, H. K., and McKellar, R. C. (1983) *J. Food Prot.*, **46**, 528–529.

Packard, V. S., Tatini, S., Fugua, R., Heady, J., and Gilman, C. (1993) In *Standard Methods for the Examination of Dairy Products*, 16th ed., R. T. Marshall, ed., American Public Health Association, Washington, DC, pp. 309–325.

Pettipher, G. L. (1993) In *Modern Dairy Technology*, Vol. 2, 2nd ed., R. K. Robinson, ed., Chapman and Hall, London, pp. 417–454.

Prentice, G. A., and Langridge, E. W. (1992) In *The Technology of Dairy Products*, R. Early, ed., Blackie, London, pp. 247–271.

Radmore, K. (1986) A microbiological study of air in dairy processing and packaging plants, M.Sc. thesis, University of Pretoria, RSA.

Reuter, H., and Quente, J. (1977) *Milchwissenschaft*, **32**, 395–399.

Reybroeck, W., and Schram, E. (1995) *Netherlands Milk Dairy J.*, **49**, 1–14.

Rockmann, R. (1998) *Dtsch. Milchwirtschaft*, **49**(1), 32–34.

Ronkilde Poulsen, P., Odet, G., and Grancher, M. (1995) In *Technical Guide for the Packaging of Milk and Milk Products*, 3rd ed., Bulletin 300, International Dairy Federation, Brussels, Belgium, pp. 95–97.

Ronner, A. B., and Wong, A. C. L. (1993) *J. Food Prot.*, **56**, 750–758.

Russell, P. (1997a) *Milk Ind. Int.*, **99**(11), Technical and Research Supplement, pp. 4–7.

Russell, P. (1997b) *Milk Ind. Int.*, **99**(1), 25–29.

Ryser, E. T. (1998) In *Applied Dairy Microbiology*, E. H. Marth and J. L. Steele, Marcel Dekker, New York, pp. 263–404.

Safarik, I., Safariková, M., and Forsythe, S. J. (1995) *J Appl. Bacteriol.*, **78**, 575–585.

Scannell, A. G. M., Hill, C., Ross, R. P., Marx, S., Hartmeier, W., and Arendt, E. K. (1999) In *Food Microbiology and Food Safety into the Next Millennium*, A. C. J. Tuijtelaars, R. A. Samson, F. M. Rombouts, and S. Notermans, eds., Foundation Food Micro '99, TNO Nutrition and Food Research Institute, A. J. Zeist, The Netherlands, pp. 303–307.

Shah, B. P., Shah, U. S., and Siripurapa, S. C. B. (1996) *Indian Dairyman*, **48**, 19–21.

Shah, B. P., Shah, U. S., and Siripurapa, S. C. B. (1997) *Indian Dairyman*, **49**, 23–27.

Sharpe, A. N. (1989) In *Modern Microbiological Methods for Dairy Products*, International Dairy Federation Special Issue 8901, pp. 165–172.

Shelley, A. Q., Deeth, H. C., and MacCrae, I. C. (1987) *J. Dairy Res.*, **54**, 413.

Singh, L., Leela, R. K., Mohan, M. S., and Sankaran, R. (1986) *Food Microbiol.*, **3**, 307–313.

Slaghuis, B. A. (1991) In *Methods for Assessing the Bacteriological Quality of Raw Milk from the Farm*, W. Heeschen, ed., Bulletin 256, International Dairy Federation, Brussels, Belgium, pp. 12–13.

Stepaniak, L., and Abrahamsen, R. K. (1995) *Milchwissenschaft*, **50**, 22–26.

Stephens, P. J. (1999) In *Proceedings of the 17th International Conference of the International Committee on Food Microbiology and Hygiene (ICMFH)*, A. C. J. Tuijtelaars, R. A. Samson, F. M. Rombouts, and S. Notermans, eds., Veldhoven, The Netherlands, pp. 599–601.

Stern, N. J., Patton, C. M., Doyle, M. P., Park, C. E., and McCardell, B. A. (1992) *Campylobacter* In *Compendium of Methods for the Microbiological Examination of Foods*, C. Vanderzaint and D. F. Splittstoeser, American Public Health Association, Washington, DC, p. 475.

Suhren, G. (1989) In *Modern Microbiological Methods for Dairy Products*, International Dairy Federation Special Issue 8901, pp. 184–200.

Suhren, G., and Heeschen, W. (1991a) In *Methods for Assessing the Bacteriological Quality of Raw Milk from the Farm*, W. Heeschen, ed., Bulletin 256, International Dairy Federation, Brussels, Belgium, pp. 14–16.

Suhren, G., and Heeschen, W. (1991b) In *Methods for Assessing the Bacteriological Quality of Raw Milk from the Farm*, W. Heeschen, ed., Bulletin 256, International Dairy Federation, Brussels, Belgium, pp. 49–52.

Suhren, G., Reichmuth, J., and Heeschen, W. (1991) In *Methods for Assessing the Bacteriological Quality of Raw Milk from the Farm*, W. Heeschen, ed., Bulletin 256, International Dairy Federation, Brussels, Belgium, pp. 24–30.

Suhren, G., and Walte, H. G. (1998) *Kieler Milchwirtsch. Forschungsber.*, **50**(3), 249–275.

Tacker, M., and Hametner, C. (1999) *Dtsch. Lebensm. Rundschau*, **95**(5), 176–180.

Tamime, A. Y., and Robinson, R. K. (1999a) In *Yoghurt Science and Technology—Plant Cleaning, Hygiene and Effluent Treatment*, 2nd ed., Woodhead Publishing Limited, Cambridge, England, pp. 249–305.

Tamime, A. Y., and Robinson, R. K. (1999b) In *Yoghurt Science and Technology—Quality Control in Yoghurt Manufacture*, 2nd ed., Woodhead Publishing Limited, Cambridge, England, pp. 535–587.

Tamminga, S. K., and Kampelmacher, E. H. (1977) *Zentralbl. Bakteriol. Parasitenkd. Hyge. I. Abeilung Orig.*, **B. 165**, 423.

Ten Cate, L. (1965) *J. Appl. Bacteriol.*, **28**, 221.

Terplan, G., Zaadhof, K.-J., and Buchholtz-Berchtold, S. (1975) *Arch. Lebensm.*, **26**, 217–221.

Thomas, S. B., Druce, R. G., Peters, G. J., and Griffiths, D. G. (1967) *J. Appl. Bacteriol.*, **30**, 265–298.

Tolle, A., Heeschen, W., Wernery, H., Reichmuth, J., and Suhren, G. (1972) *Milchwissenschaft*, **27**, 343–352.

Tsuji, K., Martin, P. A., and Bussey, D. M. (1984) *Appl. Environ. Microbiol.*, **48**, 550–555.

Van Crombrugge, J., and Waes, A. (1991) In *Methods for Assessing the Bacteriological Quality of Raw Milk from the Farm*, W. Heeschen, ed., Bulletin 256, International Dairy Federation, Brussels, Belgium, pp. 53–60.

Van Den Berg, M. G. (1995) In *Technical Guide for the Packaging of Milk and Milk Products*, 3rd ed., Bulletin 300, International Dairy Federation, Brussels, Belgium, pp. 57–64.

Vedani, G. (1996) *Latte*, **21**(10), 64–71.

Wachsmuth, K., Barrel, T. J., Griffin, P. M., Toth, I., Stockbine, N. A., and Wells, J. G. (1990) In *Application of Molecular Biology in Diagnosis of Infectious Diseases*, Ø. Olsvik and G. Buckholm, eds., Norwegian College of Veterinary Medicine, Oslo, pp. 5–15.

Wainess, H. (1995a) In *Technical Guide for the Packaging of Milk and Milk Products*, 3rd ed., Bulletin 300, International Dairy Federation, Brussels, Belgium, pp. 47–51.

Wainess, H. (1995b) In *Technical Guide for the Packaging of Milk and Milk Products*, 3rd ed., Bulletin 300, International Dairy Federation, Brussels, Belgium, pp. 40–46.

Waites, W. M. (1997) *Int. J. Dairy Technol.*, **50**, 57–60.

Wallbanks, S., Collins, M. D., Rodrigues, U. M., and Kroll, R. G. (1989) In *IDF Special Issue no. 8901*, International Dairy Federation, Brussels, Belgium, pp. 342–344.

Webber, G., Lauwaars, M., and van Schaik, M. (2000) In *Inventory of IDF/ISO/AOAC International Adopted Methods of Analysis and Sampling for Milk and Milk Products*, 6th ed., Bulletin 350, International Dairy Federation, Brussels, Belgium, pp. 3–42.

Werlein, H.-D., and Wucherpfennig, H. (1999) In *Proceedings of the 17th International Conference of the International Committee on Food Microbiology and Hygiene (ICMFH)*, A. C. J. Tuijtelaars, R. A. Samson, F. M. Rombouts, and S. Notermans, eds., Veldhoven, The Netherlands, p. 605.

White, C. H. (1998) In *Applied Dairy Microbiology*, E. H. Marth and J. L. Steele, eds., Marcel Dekker, New York, pp. 431–460.

White, C. H., Bishop, J. R., and Morgan, D. M. (1993) In *Standard Methods for the Examination of Dairy Products*, 16th ed., R. T. Marshall, ed., American Public Health Association, Washington, DC, pp. 287–308.

Witter, L. D. (1961) *J. Dairy Sci.*, **44**, 983–1015.

Wong, A. C. L., and Cerf, O. (1995) In *Biofilms: Implications for Hygiene Monitoring of Dairy Plant Surfaces*, Bulletin 302, International Dairy Federation, Brussels, Belgium, pp. 40–44.

Zaadhof, K.-J., and Terplan, G. (1981) *Dtsch. Molk Ztg*, **102**, (34/81), 1094–1098.

Zall, R. R. (1990) In *Dairy Microbiology—Control and Destruction of Microorganisms*, Vol. 1, 2nd ed., R. K. Robinson, ed., Elsevier Applied Science Publishers, London, pp. 115–161.

INDEX